Geophysical Monograph Series

Including
IUGG Volumes
Maurice Ewing Volumes
Mineral Physics Volumes

Geophysical Monograph Series

164 **Archean Geodynamics and Environments** *Keith Benn, Jean-Claude Mareschal, and Kent C. Condie (Eds.)*

165 **Solar Eruptions and Energetic Particles** *Natchimuthukonar Gopalswamy, Richard Mewaldt, and Jarmo Torsti (Eds.)*

166 **Back-Arc Spreading Systems: Geological, Biological, Chemical, and Physical Interactions** *David M. Christie, Charles Fisher, Sang-Mook Lee, and Sharon Givens (Eds.)*

167 **Recurrent Magnetic Storms: Corotating Solar Wind Streams** *Bruce Tsurutani, Robert McPherron, Walter Gonzalez, Gang Lu, José H. A. Sobral, and Natchimuthukonar Gopalswamy (Eds.)*

168 **Earth's Deep Water Cycle** *Steven D. Jacobsen and Suzan van der Lee (Eds.)*

169 **Magnetospheric ULF Waves: Synthesis and New Directions** *Kazue Takahashi, Peter J. Chi, Richard E. Denton, and Robert L. Lysak (Eds.)*

170 **Earthquakes: Radiated Energy and the Physics of Faulting** *Rachel Abercrombie, Art McGarr, Hiroo Kanamori, and Giulio Di Toro (Eds.)*

171 **Subsurface Hydrology: Data Integration for Properties and Processes** *David W. Hyndman, Frederick D. Day-Lewis, and Kamini Singha (Eds.)*

172 **Volcanism and Subduction: The Kamchatka Region** *John Eichelberger, Evgenii Gordeev, Minoru Kasahara, Pavel Izbekov, and Johnathan Lees (Eds.)*

173 **Ocean Circulation: Mechanisms and Impacts—Past and Future Changes of Meridional Overturning** *Andreas Schmittner, John C. H. Chiang, and Sidney R. Hemming (Eds.)*

174 **Post-Perovskite: The Last Mantle Phase Transition** *Kei Hirose, John Brodholt, Thorne Lay, and David Yuen (Eds.)*

175 **A Continental Plate Boundary: Tectonics at South Island, New Zealand** *David Okaya, Tim Stem, and Fred Davey (Eds.)*

176 **Exploring Venus as a Terrestrial Planet** *Larry W. Esposito, Ellen R. Stofan, and Thomas E. Cravens (Eds.)*

177 **Ocean Modeling in an Eddying Regime** *Matthew Hecht and Hiroyasu Hasumi (Eds.)*

178 **Magma to Microbe: Modeling Hydrothermal Processes at Oceanic Spreading Centers** *Robert P. Lowell, Jeffrey S. Seewald, Anna Metaxas, and Michael R. Perfit (Eds.)*

179 **Active Tectonics and Seismic Potential of Alaska** *Jeffrey T. Freymueller, Peter J. Haeussler, Robert L. Wesson, and Göran Ekström (Eds.)*

180 **Arctic Sea Ice Decline: Observations, Projections, Mechanisms, and Implications** *Eric T. DeWeaver, Cecilia M. Bitz, and L.-Bruno Tremblay (Eds.)*

181 **Midlatitude Ionospheric Dynamics and Disturbances** *Paul M. Kintner, Jr., Anthea J. Coster, Tim Fuller-Rowell, Anthony J. Mannucci, Michael Mendillo, and Roderick Heelis (Eds.)*

182 **The Stromboli Volcano: An Integrated Study of the 2002-2003 Eruption** *Sonia Calvari, Salvatore Inguaggiato, Giuseppe Puglisi, Maurizio Ripepe, and Mauro Rosi (Eds.)*

183 **Carbon Sequestration and Its Role in the Global Carbon Cycle** *Brian J. McPherson and Eric T. Sundquist (Eds.)*

184 **Carbon Cycling in Northern Peatlands** *Andrew J. Baird, Lisa R. Belyea, Xavier Comas, A. S. Reeve, and Lee D. Slater (Eds.)*

185 **Indian Ocean Biogeochemical Processes and Ecological Variability** *Jerry D. Wiggert, Raleigh R. Hood, S. Wajih A. Naqvi, Kenneth H. Brink, and Sharon L. Smith (Eds.)*

186 **Amazonia and Global Change** *Michael Keller, Mercedes Bustamante, John Gash, and Pedro Silva Dias (Eds.)*

187 **Surface Ocean–Lower Atmosphere Processes** *Corinne Le Quèré and Eric S. Saltzman (Eds.)*

188 **Diversity of Hydrothermal Systems on Slow Spreading Ocean Ridges** *Peter A. Rona, Colin W. Devey, Jérôme Dyment, and Bramley J. Murton (Eds.)*

189 **Climate Dynamics: Why Does Climate Vary?** *De-Zheng Sun and Frank Bryan (Eds.)*

190 **The Stratosphere: Dynamics, Transport, and Chemistry** *L. M. Polvani, A. H. Sobel, and D. W. Waugh (Eds.)*

191 **Rainfall: State of the Science** *Firat Y. Testik and Mekonnen Gebremichael (Eds.)*

192 **Antarctic Subglacial Aquatic Environments** *Martin J. Siegert, Mahlon C. Kennicut II, and Robert A. Bindschadler*

193 **Abrupt Climate Change: Mechanisms, Patterns, and Impacts** *Harunur Rashid, Leonid Polyak, and Ellen Mosley-Thompson (Eds.)*

194 **Stream Restoration in Dynamic Fluvial Systems: Scientific Approaches, Analyses, and Tools** *Andrew Simon, Sean J. Bennett, and Janine M. Castro (Eds.)*

195 **Monitoring and Modeling the Deepwater Horizon Oil Spill: A Record-Breaking Enterprise** *Yonggang Liu, Amy MacFadyen, Zhen-Gang Ji, and Robert H. Weisberg (Eds.)*

196 **Extreme Events and Natural Hazards: The Complexity Perspective** *A. Surjalal Sharma, Armin Bunde, Vijay P. Dimri, and Daniel N. Baker (Eds.)*

197 **Auroral Phenomenology and Magnetospheric Processes: Earth and Other Planets** *Andreas Keiling, Eric Donovan, Fran Bagenal, and Tomas Karlsson (Eds.)*

198 **Climates, Landscapes, and Civilizations** *Liviu Giosan, Dorian Q. Fuller, Kathleen Nicoll, Rowan K. Flad, and Peter D. Clift (Eds.)*

199 **Dynamics of the Earth's Radiation Belts and Inner Magnetosphere** *Danny Summers, Ian R. Mann, Daniel N. Baker, Michael Schulz (Eds.)*

200 **Lagrangian Modeling of the Atmosphere** *John Lin (Ed.)*

201 **Modeling the Ionosphere-Thermosphere** *Jospeh D. Huba, Robert W. Schunk, and George V Khazanov (Eds.)*

202 **The Mediterranean Sea: Temporal Variability and Spatial Patterns** *Gian Luca Eusebi Borzelli, Miroslav GaCiC, Piero Lionello and Paola Malanotte-Rizzoli (Eds.)*

203 **Future Earth - Advancing Civic Understanding of the Anthropocene** *Diana Dalbotten, Gillian Roehrig, and Patrick Hamilton (Eds.)*

204 **The Galapagos as a Natural Laboratory for the Earth Sciences** *Karen S. Harpp, Eric Mittelstaedt, David W. Graham, Noémi d'Ozouville (Eds.)*

Geophysical Monograph 205

Modeling Atmospheric and Oceanic Flows
Insights from Laboratory Experiments and Numerical Simulations

Thomas von Larcher
Paul D. Williams
Editors

This Work is a co-publication between the American Geophysical Union and John Wiley & Sons, Inc.

WILEY

This Work is a co-publication between the American Geophysical Union and John Wiley & Sons, Inc.

Published under the aegis of the AGU Publications Committee

Brooks Hanson, Director of Publications
Robert van der Hilst, Chair, Publications Committee
Richard Blakely, Vice Chair, Publications Committee

© 2015 by the American Geophysical Union, 2000 Florida Avenue, N.W., Washington, D.C. 20009
For details about the American Geophysical Union, see www.agu.org.

Published by John Wiley & Sons, Inc., Hoboken, New Jersey
Published simultaneously in Canada

No part of this publication may be reproduced, stored in a retrieval system, or transmitted in any form or by any means, electronic, mechanical, photocopying, recording, scanning, or otherwise, except as permitted under Section 107 or 108 of the 1976 United States Copyright Act, without either the prior written permission of the Publisher, or authorization through payment of the appropriate per-copy fee to the Copyright Clearance Center, Inc., 222 Rosewood Drive, Danvers, MA 01923, (978) 750-8400, fax (978) 750-4470, or on the web at www.copyright.com. Requests to the Publisher for permission should be addressed to the Permissions Department, John Wiley & Sons, Inc., 111 River Street, Hoboken, NJ 07030, (201) 748-6011, fax (201) 748-6008, or online at http://www.wiley.com/go/permission.

Limit of Liability/Disclaimer of Warranty: While the publisher and author have used their best efforts in preparing this book, they make no representations or warranties with respect to the accuracy or completeness of the contents of this book and specifically disclaim any implied warranties of merchantability or fitness for a particular purpose. No warranty may be created or extended by sales representatives or written sales materials. The advice and strategies contained herein may not be suitable for your situation. You should consult with a professional where appropriate. Neither the publisher nor author shall be liable for any loss of profit or any other commercial damages, including but not limited to special, incidental, consequential, or other damages.

For general information on our other products and services or for technical support, please contact our Customer Care Department within the United States at (800) 762-2974, outside the United States at (317) 572-3993 or fax (317) 572-4002.

Wiley also publishes its books in a variety of electronic formats. Some content that appears in print may not be available in electronic formats. For more information about Wiley products, visit our web site at www.wiley.com.

Library of Congress Cataloging-in-Publication Data is available.

ISBN: 978-1-118-85593-5

Cover image: The main image shows the sunset over St. Brides Bay viewed from Broad Haven, Pembrokeshire, Wales on 4 August 2012, as photographed by Paul D. Williams. The inset image shows a rotating fluid surface visualized by optical altimetry. Different colors correspond to different values of the two components of the horizontal gradient of the surface elevation. The flow is induced by a heated disk on a polar beta plane. Baroclinic instability produces multiple meanders and eddies, which are advected by a zonal current that is flowing anti-clockwise and is driven by the sink in the top-right sector of the image. This flow is a laboratory model of flows occurring in the atmospheres of rotating planets such as Earth and Saturn. More details on the experiment are given in Chapter 5. The image was produced at Memorial University of Newfoundland, Canada, by Y. D. Afanasyev and Y. Sui.

Printed in the United States of America

10 9 8 7 6 5 4 3 2 1

CONTENTS

Contributors vii

Preface xi

Acknowledgments xiii

Introduction: Simulations of Natural Flows in the Laboratory and on a Computer
Paul F. Linden 1

Section I: Baroclinic-Driven Flows

1 **General Circulation of Planetary Atmospheres: Insights from Rotating Annulus and Related Experiments**
  Peter L. Read, Edgar P. Pérez, Irene M. Moroz, and Roland M. B. Young 9

2 **Primary Flow Transitions in the Baroclinic Annulus: Prandtl Number Effects**
  Gregory M. Lewis, Nicolas Périnet, and Lennaert van Veen 45

3 **Amplitude Vacillation in Baroclinic Flows**
  Wolf-Gerrit Früh 61

Section II: Balanced and Unbalanced Flows

4 **Rotation Effects on Wall-Bounded Flows: Some Laboratory Experiments**
  P. Henrik Alfredsson and Rebecca J. Lingwood 85

5 **Altimetry in a GFD Laboratory and Flows on the Polar β-Plane**
  Yakov D. Afanasyev 101

6 **Instabilities of Shallow-Water Flows with Vertical Shear in the Rotating Annulus**
  Jonathan Gula and Vladimir Zeitlin 119

7 **Laboratory Experiments on Flows Over Bottom Topography**
  Luis Zavala Sansón and Gert-Jan van Heijst 139

8 **Direct Numerical Simulations of Laboratory-Scale Stratified Turbulence**
  Michael L. Waite 159

Section III: Atmospheric Flows

9 **Numerical Simulation (DNS, LES) of Geophysical Laboratory Experiments: Quasi-Biennial Oscillation (QBO) Analogue and Simulations Toward Madden–Julian Oscillation (MJO) Analogue**
  Nils P. Wedi 179

10 **Internal Waves in Laboratory Experiments**
   Bruce Sutherland, Thierry Dauxois, and Thomas Peacock 193

11 **Frontal Instabilities at Density–Shear Interfaces in Rotating Two-Layer Stratified Fluids**
   Hélène Scolan, Roberto Verzicco, and Jan-Bert Flór 213

Section IV: Oceanic Flows

12 Large-Amplitude Coastal Shelf Waves
Andrew L. Stewart, Paul J. Dellar, and Edward R. Johnson 231

13 Laboratory Experiments With Abrupt Thermohaline Transitions and Oscillations
John A. Whitehead 255

14 Oceanic Island Wake Flows in the Laboratory
Alexandre Stegner 265

Section V: Advances in Methodology

15 Lagrangian Methods in Experimental Fluid Mechanics
Mickael Bourgoin, Jean-François Pinton, and Romain Volk 279

16 A High-Resolution Method for Direct Numerical Simulation of Instabilities and Transitions in a Baroclinic Cavity
Anthony Randriamampianina and Emilia Crespo del Arco 297

17 Orthogonal Decomposition Methods to Analyze PIV, LDV, and Thermography Data of Thermally Driven Rotating Annulus Laboratory Experiments
Uwe Harlander, Thomas von Larcher, Grady B. Wright, Michael Hoff, Kiril Alexandrov, and Christoph Egbers 315

Index 337

CONTRIBUTORS

Yakov D. Afanasyev
Professor
Department of Physics and Physical Oceanography
Memorial University of Newfoundland, St. John's
Newfoundland, Canada

Kiril Alexandrov
Formerly: Researcher
Department of Aerodynamics and Fluid Mechanics
Brandenburg University of Technology (BTU)
Cottbus-Senftenberg, Germany

P. Henrik Alfredsson
Professor of Fluid Physics
Royal Institute of Technology
Linné FLOW Centre, KTH Mechanics
Stockholm, Sweden

Mickael Bourgoin
Researcher
Laboratoire des Écoulements Géophysiques et
Industriels
Université de Grenoble & CNRS
Grenoble, France

Emilia Crespo del Arco
Professor of Applied Physics
Departamento de Física Fundamental
Universidad Nacional de Educación a Distancia (UNED)
Madrid, Spain

Thierry Dauxois
Directeur de Recherche
Laboratoire de Physique
École Normale Supérieure
Lyon, France

Paul J. Dellar
University Lecturer
Oxford Centre for Industrial and Applied Mathematics
University of Oxford
Oxford, United Kingdom

Christoph Egbers
Professor
Department of Aerodynamics and Fluid Mechanics
Brandenburg University of Technology (BTU)
Cottbus-Senftenberg, Germany

Jan-Bert Flór
Directeur de Recherche
Laboratoire des Ecoulements Géophysiques et
Industriels (LEGI)
Grenoble, France

Wolf-Gerrit Früh
Senior Lecturer
School of Engineering & Physical Sciences
Heriot-Watt University
Edinburgh, United Kingdom

Jonathan Gula
Assistant Researcher
Institute for Geophysics and Planetary Physics
University of California, Los Angeles
Los Angeles, California, United States of America

Uwe Harlander
Professor
Department of Aerodynamics and Fluid Mechanics
Brandenburg University of Technology (BTU)
Cottbus-Senftenberg, Germany

Gert-Jan van Heijst
Professor
Department of Applied Physics
Eindhoven University of Technology
Eindhoven, The Netherlands

Michael Hoff
Researcher
Leipzig Institute for Meteorology
University of Leipzig
Leipzig, Germany

Edward R. Johnson
Professor
Department of Mathematics
University College London
London, United Kingdom

Thomas von Larcher
Postdoctoral Researcher
Institute for Mathematics
Freie Universität Berlin
Berlin, Germany

CONTRIBUTORS

Gregory M. Lewis
Associate Professor
Institute of Technology
University of Ontario
Oshawa, Ontario, Canada

Paul F. Linden
GI Taylor Professor of Fluid Mechanics
Department of Applied Mathematics and Theoretical Physics
University of Cambridge
Cambridge, United Kingdom

Rebecca J. Lingwood
Guest Professor of Hydrodynamic Stability
Royal Institute of Technology
Linné FLOW Centre, KTH Mechanics
Stockholm, Sweden;
Director of Continuing Education
Institute of Continuing Education
University of Cambridge
Cambridge, United Kingdom

Irene M. Moroz
Lecturer in Mathematics
Mathematical Institute
University of Oxford
Oxford, United Kingdom

Thomas Peacock
Associate Professor of Mechanical Engineering
Department of Mechanical Engineering
Massachusetts Institute of Technology
Cambridge, Massachusetts, United States of America

Edgar P. Pérez
Formerly: Graduate Student
Mathematical Institute
University of Oxford
Oxford, United Kingdom

Nicolas Périnet
Postdoctoral Fellow
Institute of Technology
University of Ontario
Oshawa, Ontario, Canada

Jean-François Pinton
Directeur de Recherche, CNRS
Laboratoire de Physique
École Normale Supérieure de Lyon
Lyon, France

Anthony Randriamampianina
Directeur de Recherche, CNRS
Laboratoire Mécanique, Modélisation & Procédés Propres, UMR 7340 CNRS
Aix Marseille Université
Marseille, France

Peter L. Read
Professor of Physics
Atmospheric, Oceanic & Planetary Physics
University of Oxford
Oxford, United Kingdom

Luis Zavala Sansón
Researcher
Departamento de Oceanografía Física
CICESE, Ensenada
Baja California, México

Hélène Scolan
Postdoctoral Research Assistant
Atmospheric, Oceanic & Planetary Physics
University of Oxford
Oxford, United Kingdom

Alexandre Stegner
Researcher
Laboratoire de Météorologie Dynamique, CNRS
France;
Associate Professor
École Polytechnique
Palaiseau, France

Andrew L. Stewart
Postdoctoral Scholar
Department of Atmospheric and Oceanic Sciences
University of California, Los Angeles
Los Angeles, California, United States of America

Bruce Sutherland
Professor of Fluid Mechanics
Departments of Physics and of Earth & Atmospheric Sciences
University of Alberta
Edmonton, Alberta, Canada

Lennaert van Veen
Associate Professor
Institute of Technology
University of Ontario
Oshawa, Ontario, Canada

Roberto Verzicco
Professor in Fluid Dynamics
Department of Mechanical Engineering
Università di Roma Tor Vergata
Rome, Italy

Romain Volk
Lecturer
Laboratoire de Physique
École Normale Supérieure de Lyon
Lyon, France

Michael L. Waite
Assistant Professor
Department of Applied Mathematics
University of Waterloo
Waterloo, Ontario, Canada

Nils P. Wedi
Principal Scientist
Numerical Aspects Section
European Centre for Medium-Range Weather Forecasts
Reading, United Kingdom

John A. Whitehead
Scientist Emeritus
Department of Physical Oceanography
Woods Hole Oceanographic Institution
Woods Hole, Massachusetts, United States of America

Grady B. Wright
Associate Professor
Department of Mathematics
Boise State University
Boise, Idaho, United States of America

Roland M. B. Young
Postdoctoral Research Assistant
Atmospheric, Oceanic & Planetary Physics
University of Oxford
Oxford, United Kingdom

Vladimir Zeitlin
Professor
Laboratoire de Météorologie Dynamique
École Normale Supérieure
Paris, France;
Université Pierre et Marie Curie
Paris, France

PREFACE

The flow of fluid in Earth's atmosphere and ocean affects life on global and local scales. The general circulation of the atmosphere transports energy, mass, momentum, and chemical tracers across the entire planet, and the giant currents of the thermohaline circulation and wind-driven circulation perform the same function in the ocean. These established flows are reasonably steady on long time scales. In contrast, short-lived instabilities may develop and result in transient features such as waves, oscillations, turbulence, and eddies. For example, in the atmosphere, small-scale instabilities are able to grow into heavy storms if the conditions are right.

It is therefore of interest to understand atmospheric and oceanic fluid motions on all scales and their interactions across different scales. Unfortunately, due to the complex physical mechanisms at play and the wide range of scales in space and time, research in geophysical fluid dynamics remains a challenging and intriguing task. Despite the great progress that has been made, we are still far from achieving a comprehensive understanding.

As tools for making progress with the above challenge, laboratory experiments are well suited to studying flows in the atmosphere and ocean. The crucial ingredients of rotation, stratification, and large-scale forcing can all be included in laboratory settings. Such experiments offer the possibility of investigating, under controlled and reproducible conditions, many flow phenomena that are observed in nature. Furthermore, immense computational resources are becoming available at low economic cost, enabling laboratory experiments to be simulated numerically in more detail than ever before. The interplay between numerical simulations and laboratory experiments is of increasing importance within the scientific community.

The purpose of this book is to provide a comprehensive survey of some of the laboratory experiments and numerical simulations that are being performed to improve our understanding of atmospheric and oceanic fluid motion.

On the experimental side, new designs of experiments on the laboratory scale are discussed together with developments in instrumentation and data acquisition techniques and the computer-based analysis of experimental data. On the numerical side, we address recent developments in simulation techniques, from model formulation to initialization and forcing. The presentation of results from laboratory experiments and the corresponding numerical models brings the two sides together for mutual benefit.

The book contains five sections. Section I covers baroclinic instability, which plays a prominent role in atmosphere and ocean dynamics. The thermally driven, rotating annulus is the corresponding laboratory setup, having been used for experiments in geophysical fluid dynamics since the 1940s by Dave Fultz, Raymond Hide, and others. Section II covers balanced and unbalanced flows. Sections III and IV cover laboratory experiments and numerical studies devoted to specific atmospheric and oceanic phenomena, respectively. Section V adresses some new achievements in the computer-based analysis of experimental data and some recent developments in experimental methodology and numerical methods.

We hope this book will give the reader a clear picture of the experiments that are being performed in today's laboratories to study atmospheric and oceanic flows together with the corresponding numerical simulations. We further hope that the lessons learnt from the comparisons between laboratory and model will act as a source of inspiration for the next generation of experiments and simulations.

Thomas von Larcher
Freie Universität Berlin
Germany

Paul D. Williams
University of Reading
United Kingdom

ACKNOWLEDGMENTS

We gratefully acknowledge support from the staff at the American Geophysical Union (AGU), particularly Colleen Matan for guiding us through the book proposal process and Telicia Collick for assisting with the external peer review of the chapters. We also extend our thanks to Rituparna Bose at John Wiley & Sons, Inc., for smoothly and professionally handling all aspects of the book production process.

Introduction: Simulations of Natural Flows in the Laboratory and on a Computer

Paul F. Linden

Humans have always been associated with natural flows. The first civilizations began near rivers, and humans developed an early pragmatic view of water flow and the effects of wind. Experiments and calculations in fluid mechanics can be traced back to Archimedes in his work "On floating bodies" around 250 B.C., in which he calculates the position of equilibrium of a solid body floating in a fluid. He is, of course, attributed with the law of buoyancy known as Archimedes principle. The ancient Greeks also elucidated the principle of the syphon and the pump. This work is essentially concerned with fluid statics, and the first attempts to investigate the motion of fluids is attributed to Sextus Julius Frontinus, the inspector of public fountains in Rome, who made extensive measurements of flow in aqueducts and, using conservation-of-volume principles, was able to detect when water was being diverted fraudulently.

Possibly the first laboratory experiment designed to examine a natural flow was by *Marsigli* [1681] who devised a demonstration of the buoyancy-driven flow associated with horizontal density differences in an attempt to explain the undercurrent in the Bosphorus that flows toward the Black Sea [*Gill*, 1982]. This is a remarkable experiment in that it provides an unequivocal demonstration that flow, now known as baroclinic flow with no net transport, is possible even when the free surface is level, so that there is no barotropic (depth-averaged) flow. These buoyancy-driven flows occur almost ubiquitously in the oceans and atmosphere and are an active area of current research.

Another example of the early use of a laboratory experiment is the explanation of the "dead water" phenomenon observed by the Norwegian scientist *Fridtjof Nansen* [1897], who experienced an unexpected drag on his boat during his expedition to reach the North Pole in 1892. The responsible mechanism, the drag associated with interfacial waves on the pycnocline, was studied by *Ekman* [1904] in his Ph.D. thesis and a review of his work and some modern extensions using synthetic schlieren can be found elsewhere [*Mercier et al.*, 2011].

This last reference nicely demonstrates one role of modern laboratory experiments. Although the basic mechanics has been known since Ekman's study, by careful observation of the flow and making quantitative measurements of the wave fields made possible with new image processing techniques, it has been shown that the dead water phenomenon is nonlinear. The coupling of the large-amplitude interfacial and internal waves with significant accelerations of the boat are an intrinsic feature of the energy transmission from the boat to the waves. Although the essential features have been known for over a century, these recent data provide new insights into the physics of the flows and show that the drag on the boat depends on the forms of the waves generated. Experiments like this provide insight and inspiration about the underlying dynamical processes, ideally motivating theory which can subsequently refocus the experiments.

Numerical methods were first devised to solve potential flow problems in the 1930s, and as far as I am aware the first numerical solution of the Navier-Stokes equations, i.e., the first computational fluid dynamics (CFD) calculation, applied to two-dimensional swirling flow, was published by *Fromm* [1963]. Since then there has been enormous growth in computational power, and this has led to developments in both CFD and laboratory experimentation. The reasons for the improvement in CFD are clear. In order to calculate a flow accurately, it is necessary that the discrete forms of the governing equations are a faithful representation of the continuous partial

Department of Applied Mathematics and Theoretical Physics, University of Cambridge, Cambridge, United Kingdom

Modeling Atmospheric and Oceanic Flows: Insights from Laboratory Experiments and Numerical Simulations, First Edition. Edited by Thomas von Larcher and Paul D. Williams.
© 2015 American Geophysical Union. Published 2015 by John Wiley & Sons, Inc.

2 INTRODUCTION

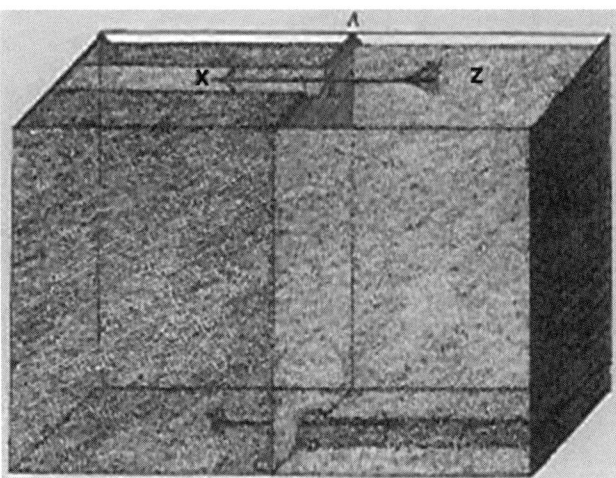

Figure 0.1. A sketch of *Marsigli's* [1681] experiment illustrating the counterflow driven by the density difference between the two fluids in either side of the barrier with flow along the surface toward the denser fluid and a countercurrent along the bottom in the opposite direction.

differential equations. For geophysical flows, which are typically turbulent, this means that in order to avoid approximations it is necessary to compute all the scales of motion, which range from the energy input scales down to the smallest scales, where viscous dissipation occurs. This represents a huge range of scales. Energy is input on global scales (10^6 m) and dissipated at the Kolmogorov scale $(v^3/\epsilon)^{1/4} \sim 10^{-3}$ m. This nine decade range of length (and associated time) scales remains well beyond the capabilities of current (and foreseeable) computing power and represents a huge challenge to the computation of geophysical flows.

Geophysical flows are stably stratified (buoyant fluid naturally lies on top of denser fluid) and occur on a rotating planet. The stratification, characterized by the buoyancy frequency N, is of the order of 10^{-2} s^{-1} and is roughly the same in the atmosphere and the oceans. The rotation of Earth is characterized by the Coriolis parameter f, which, with values of order 10^{-4}s^{-1}, introduces longer time scales than those associated with the stratification. Thus, the atmosphere and the oceans, viewed on global scales, are strongly stratified, weakly rotating fluids.

Stratification provides a restoring force to *vertical* motions through the buoyancy force associated with the density difference between the displaced fluid particle and the background stratification. Rotation provides a restoring force due to *horizontal* motions through the Coriolis force (or by conservation of angular momentum viewed in an inertial frame). For motion with horizontal scale L and vertical scale H, the balance between these forces is given by the Burger number B:

$$B \equiv \frac{NH}{fL}. \qquad (0.1)$$

Stratification dominates when $B \gg 1$, i.e., when horizontal scales are relatively small compared with the Rossby deformation radius $R_D \equiv NH/f$, while rotation dominates when horizontal scales are large compared with R_D and $B \ll 1$. On global scales the oceans and the atmosphere are thin layers of fluids with vertical to horizontal aspect ratios H/L of order 10^{-3}. Consequently, for motion on global scales in the atmosphere or basin scales in the oceans, $B \sim 10^{-1}$ and rotational effects dominate. These flows can be modeled as essentially unstratified flows, with Coriolis forces providing the main constraints. Motion on smaller scales will generally lead to increasing values of B and increasing effects of stratification. Mesoscale motions, in which buoyancy and Coriolis forces balance, are typified by values of $B \sim 1$, in which case the horizontal scale of the motion is comparable to the Rossby deformation radius R_D, which is on the order of 1000 km in the atmosphere and 100 km in the oceans.

In order to examine the effects of rotation, experiments are conducted on rotating platforms. These are generally high-precision turntables capable of carrying significant weight, and they present a significant engineering challenge in their construction. The requirements and performance of these turntables are discussed in Chapter 7, which illustrates these by considering flows of thin fluid layers in rotating containers of different diameters from 0.1 to 10 m. As the size of the turntable is increased, the engineering requirements become more demanding and the cost increases. Furthermore, larger flow domains require more fluid, and if stratified, this requires a more stratifying agent, such as (the commonly used) sodium chloride. For these reasons most laboratory turntables range up to about 1 m in diameter, and there needs to be a compelling reason to work on large-diameter turntables.

One reason for increasing the experimental scale from 1 to 10 m is to reduce frictional effects. Reynolds numbers Re $\equiv UL/v$, where U and L are typical velocity and length scales and v is the kinematic viscosity, are increased by a factor of 10 (equivalently Ekman numbers $E \equiv v/fL^2$ are reduced by a factor of 100), and so the effects of boundary friction are reduced and damping times are increased at large scale. This can be an dominant factor when studying flows where separation or turbulence is dominant or, as in the case discussed in Chapter 7, the motion of vortices driven by vortex interactions or interactions with topography.

On the other hand, there is little to be gained in terms of the overall Rossby number or Burger number, both of which involve the product fL of rotation and length scale. This product is the speed of the rim of the turntable, and it is difficult, for safety and operational reasons, to increase this very much above a few meters per second, which is easily obtainable with a 1 m diameter turntable. However, the size of the flow domain impacts the spatial and temporal resolution that is possible and needed. On larger turntables, structures such as vortices are increased in size, allowing better spatial resolution and lower Rossby numbers.

There has been a major resurgence in interest in laboratory experiments over the past decade due to new developments in optics and in computing power, memory, and the ability to read and write data rapidly using solid-state media. Velocity measurements using particle image velocimetry and particle tracking (see Chapter 15) now have the capabilities to resolve three-dimensional flow fields with subpixel (0.1 pixel) spatial accuracy and millisecond time resolution using high-speed cameras. With currently available megapixel cameras, this implies data rates of gigabytes per second, which requires dedicated data storage and processing, which are now currently available and affordable. There is every reason to expect further technological improvements that will render these techniques increasingly effective in the future and inspire the development of new data analysis methodologies (see Chapter 17) and also increasing study of multiscale phenomena.

Other methods have been developed that also allow nonintrusive measurements of the flow. For example, in Chapter 5 Afanasyev describes optical altimetry which provides measurements of the free surface slope. In a rotating fluid it is then possible to infer the flow provided certain assumptions are made about the dynamics. If the fluid is homogeneous and the flow is in geostrophic and hydrostatic balance, the free surface height is the stream function for the flow, from which the velocity field can be inferred. This method can be extended to two-layer flows if the depth of one layer can be measured independently. This can be done by adding dye to one layer and then measuring the absorption of light through the layer.

Another method for nonintrusive measurements is synthetic schlieren, described in Chapter 10. This method uses the apparent movement of an image (say an array of dots) placed behind an experiment as a result of refractive index changes in the fluid between the camera and the image. This method has found many applications in the study of internal waves, which produce refractive index changes by moving the stratified fluid as they propagate. From the apparent movement of the dots, say, the changes to the density field can be measured and then, assuming these are caused by linear internal waves, the associated motion can be determined. This method has led to successful measurements of momentum and energy fluxes over a wave field and led to new interpretations of internal wave dynamics. Both synthetic schlieren and the optical thickness method produce data integrated over the light path. Thus, in their simplest forms it is assumed that there are no variations in the flow along the light path. Recently attempts have been made to overcome this limitation so that fully three-dimensional flows can be measured, and this has had success in limited circumstances (see Chapter 10).

Computations in the form of numerical solution of model equations or of approximate forms of the full Navier-Stokes equations have also developed significantly. However, perhaps the most interesting development is the increasing capacity to carry our direct numerical simulations (DNSs) of the Navier-Stokes equations without making any approximations. This capability has come from a combination of improved numerical schemes to deal with the discretized forms of the equations and, of course, from the rapid and continuing improvements in computer power. To that extent the development of high-capacity computing has been the key to recent developments in both computations and experiments, which leads to the interesting issue of how well either represent real geophysical flows.

The challenge is to replicate the physical processes that occur on geophysical scales accurately in the laboratory and in computations. In the laboratory it is physically impossible, of course, to work at full scale, and numerically the issue is to deal with the huge range of length and time scales that are dynamically significant. In the case of stratified flows recent work on turbulence discussed in Chapter 8 shows that, in order to avoid viscous effects, the buoyancy Reynolds number

$$\mathcal{R} \equiv \frac{U^3}{\nu N^2 L}, \qquad (0.2)$$

where U is a typical velocity scale and L is a typical *horizontal* scale, needs to be large. Mesoscale motions in the atmosphere have $\mathcal{R} \sim 10^4$ and in the oceans $\mathcal{R} \sim 10^3$, and recent DNS calculations have achieved values up to $\mathcal{R} \sim 10^2$. This is very encouraging and these types of calculations have led to a significant reevaluation of the energetics of stratified turbulence [*Waite*, 2013].

Experiments, on the other hand, have so far been characterized by values of $\mathcal{R} \sim 1$ or less (see Chapter 8), which is mainly a result of the fact that most of the relevant experiments are concerned with decaying turbulence (see *Brethouwer et al.* [2007], Figure 18). It is however possible by directly forcing the flow to achieve high values of $\mathcal{R}$. An example is shown in Figure 0.2: In this experiment a water tank is partitioned into two sections both containing salt water with different densities. A square duct that may be inclined at an angle θ to the horizontal passes throughout the central divide of the tank and connects the two sections of the tank, which act as large reservoirs.

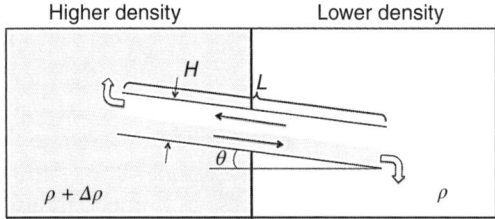

Figure 0.2. Experimental setup.

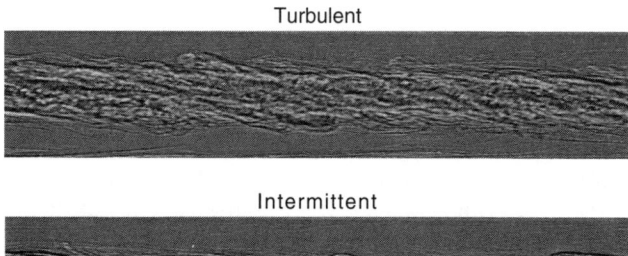

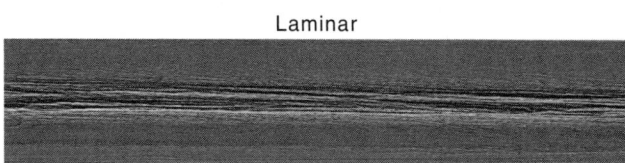

Figure 0.3. Shadowgraphs of the turbulent, intermittent, and laminar flow regimes in buoyancy-driven flow in an inclined duct. Buoyant fluid flows to the left above dense fluid flowing to the right. Images taken by Colin Meyer [*Meyer*, 2014].

The flow is initiated by opening one end of the duct and maintained until the fluid entering each reservoir begins to affect the input at either end of the duct. The flow is characterized by the inclination θ and the density difference $\Delta\rho$ between the two sections of the tank. In these experiments we explored the flows in the range $0° \leq \theta \leq 3°$ ($\theta > 0$ means that the duct is inclined up toward the dense reservoir) and $10 \leq \Delta\rho \leq 210$ kg/m^3.

The buoyancy forces establish a flow with the light fluid flowing uphill in the upper part of the duct and the dense fluid flowing downhill in the lower part. Between these counterflowing layers there is an interfacial region that takes different forms depending on the density difference and the angle of the duct. At high angles and larger density differences, the interfacial region is strongly turbulent with three-dimensional structures of Kelvin-Helmholtz type across its whole width. Mass is transferred from the lower layer directly to the upper layer through eddies that span the entire thickness of the region. As the density difference and/or angle decrease, the turbulence becomes less intense. At a critical density difference and angle combination, a spatiotemporal intermittent regime develops where turbulent bursts and relaminarization events occur. Both Kelvin-Helmholtz and Holmboe-type modes appear in this regime, and the interfacial region has a complicated structure consisting of thinner layers of high-density gradients within it. These layers and the instability modes display significant variability in space and time. At even smaller angles and density differences, the flow is essentially laminar (or at least weakly dissipative), with a relatively sharp interface supporting Holmboe-type wave modes with the occasional breaking event. Shadowgraph images of these flow regimes are shown in Figure 0.3.

Typical values of the flow in this experiment are $U \sim \sqrt{g'H} \sim 0.1$ m/s, $L = 3$ m, $N \sim \sqrt{g'/H} \sim 1$ s^{-1}, $\nu \approx 10^{-6}$ m^2/s gives $\mathcal{R} \sim 10^2$, which is comparable with the best numerical computations. Here we have taken a conservative estimate of the flow speed and the length of the duct as the horizontal scale, although Figure 0.3 suggests that a more likely value is $L \sim 0.1$ m, which will increase the range of values of $\mathcal{R}$ by an order of magnitude and in the range of geophysical flows.

As shown in Chapter 9, other examples where it is possible to match the parameter ranges of DNS and laboratory experiments are the numerical simulations of experiments that model the quasi-biennial oscillation and the Madden-Julian oscillation. These comparisons show that the laboratory experiments by *Plumb and McEwan* [1978] represent the atmosphere rather better than has been previously thought. Once agreement between the experiments and computations has been established, the latter can be used to investigate other effects, such as the dependence on the fluid properties, for example, the Prandtl number (see Chapter 16), which are difficult to vary in experiments.

By revealing flow structures and allowing the physics to be explored by varying parameters under controlled conditions, both experiments and numerical simulations allow fundamental insights to be revealed. Nothing really beats looking at a flow either physically or using the wonderful graphics that are currently available to visualize the outputs of numerical data to get a feel for and develop intuition about the dynamics. There is still a fascination with observing wakes (Chapter 14), the flows associated with boundary layers in rotating systems (Chapter 4), abrupt transitions in flow regimes caused by buoyancy forcing (Chapter 13), the forms of instabilities on fronts (Chapter 11), the form of shelf waves (Chapter 12), and the amazing rich dynamics revealed by observations of the flow in a rotating heated annulus (Chapter 1).

Nevertheless, as illuminating and as necessary as these studies are, extracting the underlying physics and developing an *understanding* of what these observations reveal requires a further component, that is, a model. Without an underlying model, which can be a sophisticated mathematical model of, say, nonlinear processes in baroclinic instability (see Chapters 6 and 3) or nonlinear waves captured in a shallow water context (see Chapter 2) or indeed a more simplistic view based on dimensional analysis, experiments *and* numerical simulations only produce a series of dots in some parameter space. It is the model and the understanding that are inherent in the simplified representation of reality that joins the dots. And when the dots are joined, then one really has something!

REFERENCES

Brethouwer, G., P. Billant, E. Lindborg, and J.-M. Chomaz (2007), Scaling analysis and simulation of strongly stratified turbulent flows, *J. Fluid Mech.*, *585*, 343–368.

Ekman, V. W. (1904), On dead water. Norw. N. Polar Exped. 1893–1896: Sci. results, XV Christiana, Ph.D. thesis.

Fromm, J. E. (1963), A method for computing non-steady, incompressible fluid flows, Tech. Rep. LA-2910, Los Alamos Sci. Lab., Los Alamos, N. Mex.

Gill, A. E. (1982), *Atmosphere-Ocean Dynamics*, Academic Press, New York.

Marsigli, L. M. (1681), Osservazioni intorno al bosforo tracio o vero canale di constantinopli, rappresentate in lettera alla sacra real maesta cristina regina di svezia, roma, *Boll. Pesca, Piscic. Idrobiol*, *11*, 734 – 758.

Mercier, M. J., R. Vasseur, and T Dauxois (2011), Resurrecting dead-water phenomenon, *Nonlin. Processes Geophys.*, *18*, 193–208.

Meyer, C. R., and P. F. Linden (2014), Stratified shear flow: experiments in an inclined duct. *J. Fluid Mech.*, *753*, 242–253, doi:10.1017/jfm.2014.358

Nansen, F. (1897), *Farthest North: The Epic Adventure of a Visionar Explorer*, Library of Congress Cataloging-in-Publication Data.

Plumb, R. A., and A. D. McEwan (1978), The instability of a forced standing wave in a viscous stratified fluid: A laboratory analogue of the quasi-biennial oscillation, *J. Atmos. Sci.*, *35*, 1827–1839.

Waite, M. L. (2013), The vortex instability pathway in stratified turbulence, *J. Fluid Mech.*, *716*, 1–4.

Section I: Baroclinic-Driven Flows

1
General Circulation of Planetary Atmospheres: Insights from Rotating Annulus and Related Experiments

Peter L. Read[1], Edgar P. Pérez[2], Irene M. Moroz[2], and Roland M. B. Young[1]

1.1. LABORATORY EXPERIMENTS AS "MODELS" OF PHYSICAL SYSTEMS

In engineering and the applied sciences, the term "model" is typically used to denote a device or concept that imitates the behavior of a physical system as closely as possible, but on a different (usually smaller) scale, possibly with some simplifications. The aim of such a model is normally to evaluate the performance of such a system for reasons connected with its exploitation for economic, social, military, or other purposes. In the context of the atmosphere or oceans, numerical weather and climate prediction models clearly fall into this category. Such models are extremely complicated entities that seek to represent the topography, composition, radiative transfer, and dynamics of the atmosphere, oceans, and surface in great detail. As a result, it is generally impossible to comprehend fully the complex interactions of physical processes and scales of motion that occur within any given simulation. The success of such models can only be judged by the accuracy of their predictions as directly verified (in the case of numerical weather prediction) against subsequent observations and measurements. Similar models used for climate prediction, however, are often comparable in complexity to those used for weather prediction but are frequently used as tools in attempts to address questions of economic, social, or political importance (e.g., concerning the impact of increasing anthropogenic greenhouse gas emissions) for which little or no verifying data may be available.

In formulating such models and interpreting their results, it is necessary to make use of a different class of model, the "conceptual" or "theoretical" model, which may represent only a small subset of the geographical detail and physical processes active in the much larger, applications-oriented model but whose behavior may be much more completely understood from first principles. To arrive at such a complete level of understanding, however, it is usually necessary to make such models as simple as possible ("but no simpler") and in geometric domains that may be much less complicated than found in typical geophysical contexts. An important prototype of such a model in fluid mechanics is that of dimensional (or "scale") analysis, in which the entire problem reduces to one of determining the leading order balance of terms in the governing equations and the consequent dependence of one or more observable parameters in the form of power law exponents. Following such a scale analysis, it is often possible to arrive at a scheme of mathematical approximations that may even permit analytical solutions to be obtained and analyzed. The well-known quasi-geostrophic approximation is an important example of this approach [e.g., see *Holton*, 1972; *Vallis*, 2006] that has enabled a vast number of essential dynamical processes in large-scale atmospheric and oceanic dynamics to be studied in simplified (but nonetheless representative) forms.

For the fundamental researcher, such simplified "conceptual" models are an essential device to aid and advance understanding. The latter is achievable because simplified, approximated models enable theories and hypotheses to be formulated in ways that can be tested (i.e., falsified, in the best traditions of the scientific method) against observations and/or experiments. The ultimate aim of such studies in the context of atmospheric and oceanic sciences is to develop an overarching framework that sets in perspective *all* planetary atmospheres and oceans, of which Earth represents but one set of examples [*Lorenz*, 1967; *Hide*, 1970; *Hoskins*, 1983].

---

[1]Atmospheric, Oceanic & Planetary Physics, University of Oxford, Oxford, United Kingdom.
[2]Mathematical Institute, University of Oxford, Oxford, United Kingdom.

Modeling Atmospheric and Oceanic Flows: Insights from Laboratory Experiments and Numerical Simulations,
First Edition. Edited by Thomas von Larcher and Paul D. Williams.
© 2015 American Geophysical Union. Published 2015 by John Wiley & Sons, Inc.

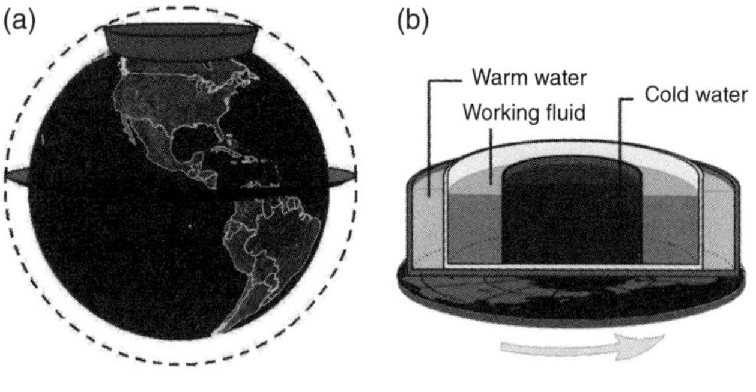

Figure 1.1. (a) Schematic diagram of a rotating annulus; (b) schematic equivalent configuration in a spherical fluid shell (cf. an atmosphere).

The role of laboratory experiments in fluid mechanics in this scheme would seem at first sight to be as models firmly in the second category. Compared with a planetary atmosphere or ocean, they are clearly much simpler in their geometry, boundary conditions, and forcing processes (diabatic and mechanical), e.g., see Figure 1.1. Their behavior is often governed by a system of equations that can be stated exactly (i.e., with no controversial parameterizations being necessary), although even then exact mathematical solutions (e.g., to the Boussinesq Navier-Stokes equations) may still be impossible to obtain. Unlike atmospheres and oceans, however, it is possible to carry out controlled experiments to study dynamical processes in a real fluid without recourse to dubious approximations (necessary to both analytical studies and numerical simulation). Laboratory experiments can therefore complement other studies using complex numerical models, especially since fluids experiments (a) have effectively infinite resolution compared to their numerical counterparts (though can only be measured to finite precision and resolution), (b) are often significantly less diffusive than the equivalent fluid, e.g., in eddy-permitting ocean models, and yet (c) are relatively cheap to run!

In discussing the role of laboratory experiments, however, it is not correct to conclude that they have no direct role in the construction of more complex, applications-oriented models and associated numerical tools (such as in data assimilation). Because the numerical techniques used in such models (e.g., finite-difference schemes, eddy or turbulence parameterizations) are also components of models used to simulate flows in the laboratory under similar scaling assumptions, laboratory experiments can also serve as useful "test beds" for directly evaluating and verifying the accuracy of such techniques in ways that are far more rigorous than may be possible by comparing complex model simulations solely with atmospheric or oceanic observations. Despite many advances in the formulation and development of sophisticated numerical models, there remain many phenomena (especially those involving nonlinear interactions of widely differing scales of motion) that continue to pose serious challenges to even state-of-the-art numerical models yet may be readily realizable in the laboratory. This is especially true of large-scale flow in atmospheres and oceans, for which relatively close dynamical similarity between geophysical and laboratory systems is readily achievable. Laboratory experiments in this vein therefore still have much to offer in the way of quantitative insight and inspiration to experienced researchers and fresh students alike.

1.2. ROTATING, STRATIFIED EXPERIMENTS AND GLOBAL CIRCULATION OF ATMOSPHERES AND OCEANS

At its most fundamental level, the general circulation of the atmosphere is but one example of thermal convection in response to impressed differential heating by heat sources and sinks that are displaced in both the vertical *and/or the horizontal* in a rotating fluid of low viscosity and thermal conductivity. Laboratory experiments investigating such a problem should therefore include at least these attributes and be capable of satisfying at least some of the key scaling requirements for dynamical similarity to the relevant phenomena in the atmospheric or oceanic system in question. Such experimental systems may then be regarded [e.g., *Hide*, 1970; *Read*, 1988] as schematically representing key features of the circulation in the absence of various complexities associated, for example, with radiative transfer, atmospheric chemistry, boundary layer turbulence, water vapor, and clouds in a way that is directly equivalent to many other simplified and approximated mathematical models of dynamical phenomena in atmospheres and oceans.

Experiments of this type are by no means a recent phenomenon, with examples published as long ago as the mid to late nineteenth century [e.g., *Vettin*, 1857, 1884; *Exner*, 1923]; see *Fultz* [1951] for a comprehensive review of this early work. *Vettin* [1857, 1884] had the insight to appreciate that much of the essence of the large-scale atmospheric circulation could be emulated, at least in principle, by the flow between a cold body (representing the cold, polar regions) placed at the center of a rotating, cylindrical container and a heated region (representing the warm tropics) toward the outside of the container (see Figure 1.2). Vettin's experiments used air as the convecting fluid, contained within a bell jar on a rotating platform. As one might expect of a nineteenth century gentleman, he then used cigar smoke to visualize the flow patterns, demonstrating phenomena such as convective vortices and larger scale overturning circulations. However, these experiments only really explored the regime we now know as the axisymmetric or "Hadley"

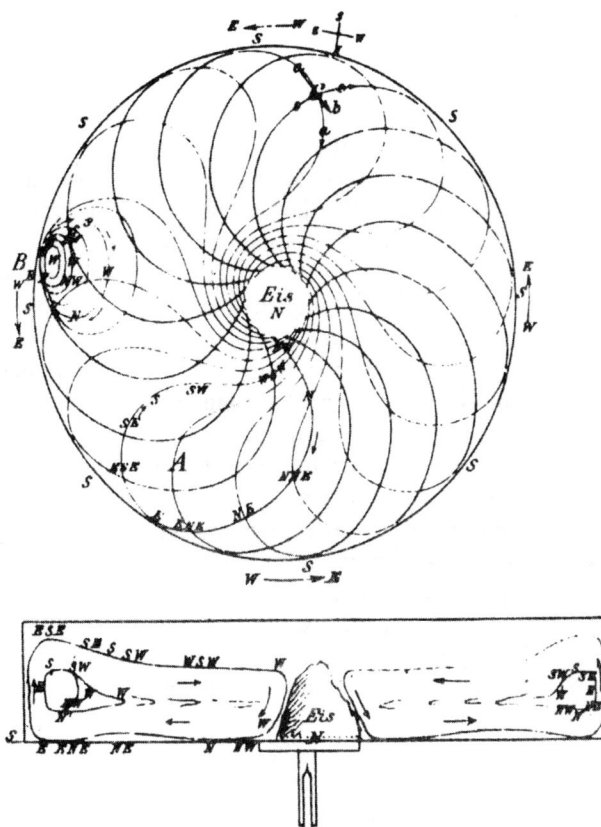

Figure 1.2. Selection of images adapted from *Vettin* [1884] (see http://www.schweizerbart.de). Reproduced with permission from the publishers, showing the layout of his rotating convection experiment and some results.

regime, since the flows Vettin observed showed little evidence for the instabilities we now know as "baroclinic instability" or "sloping convection" [*Hide and Mason*, 1975].

As an historical aside, it is interesting to note that early meteorologists such as *Abbe* [1907] intended for laboratory experiments of this type to serve also as models of the first kind, i.e., as application-oriented, predictive model atmospheres. They realized that, while it might be possible in principle to use the equations of atmospheric dynamics to determine future weather, they were beyond the capacity of mathematical analysis to solve. They hoped to use these so-called mechanical integrators [*Rossby*, 1926] under complicated external forcing corresponding to the observations of the day to reproduce and predict very specific flow phenomena observed in the atmosphere. It was anticipated that many such experiments would be built representing different regions of Earth's surface or different times of year, such as when the cross-equatorial airflow is perturbed by the monsoon [*Abbe*, 1907] (although it is not clear whether such an experiment was ever constructed). However, following the development of the electronic computer during the first half of the twentieth century and *Richardson's* [1922] pioneering work on numerical weather prediction, these more complex laboratory representations of the atmosphere were superseded.

The later experiments of *Exner* [1923] explored a different regime in which baroclinic instability seems to have been present. The flows he demonstrated were evidently quite disordered and irregular, likely due in part to the parameter regime he was working in but also perhaps because of inadequate control of the key parameters. It was not until the late 1940s, however, that Fultz began a systematic series of experiments at the University of Chicago on rotating fluids subject to horizontal differential heating in an open cylinder (hence resulting in the obsolete term "dishpan experiment") and set the subject onto a firm footing. Independently and around the same time, *Hide* [1958] began his first series of experiments at the University of Cambridge on flows in a heated rotating annulus, initially in the context of fluid motions in Earth's liquid core. By carrying out an extensive and detailed exploration of their respective parameter spaces, both of these pioneering studies effectively laid the foundations for a huge amount of subsequent work on elucidating the nature of the various circulation regimes identified by Fultz and Hide, subsequently establishing their bifurcations and routes to chaotic behavior, developing new methods of modeling the flows using numerical techniques, and measuring them using ever more sophisticated methods, especially via multiple arrays of in situ probes and optical techniques that exert minimal perturbations to the flow itself.

An important aspect of the studies by Fultz and Hide was their overall agreement in terms of robustly identifying many of the key classes of circulation regimes and locating them within a dimensionless parameter space. A notable exception to this, at least in early work, was the lack of a regular wave regime in Fultz's open cylinder experiments, in sharp contrast to the clear demonstration of such a regime in Hide's annulus. As further discussed below, this led to some initial suggestions [*Davies*, 1959] that the existence of this regime was somehow dependent on having a rigid inner cylinder bounding the flow near the rotation axis. This was subsequently shown not to be the case in open cylinder experiments by Fultz himself [*Spence and Fultz*, 1977] and by Hide and his co-workers [*Hide and Mason*, 1970; *Bastin and Read*, 1998] and two-layer [*Hart*, 1972, 1985] experiments that clearly showed that persistent, near-monochromatic baroclinic wave flows could be readily sustained in a system without a substantial inner cylinder. It is likely, therefore, that early efforts failed to observe such a regular regime in the thermally driven, open cylinder geometry because of a lack of close experimental control, e.g., of the rotation rate or the static stability in the interior.

Earlier studies in this vein were extensively reviewed by *Hide* [1970] and *Hide and Mason* [1975]. More recently, significant advances have been presented by various groups around the world, including highly detailed experimental studies in the "classical" axisymmetric annulus of synoptic variability, vacillations, and the transitions to geostrophically turbulent motions by groups at the Florida State University [e.g., *Pfeffer et al.*, 1980; *Buzyna et al.*, 1984], the UK Met Office and Oxford University [e.g., *Read et al.*, 1992; *Früh and Read*, 1997; *Bastin and Read*, 1997, 1998; *Wordsworth et al.*, 2008], several Japanese universities [e.g., *Ukaji and Tamaki*, 1989; *Sugata and Yoden*, 1994; *Tajima et al.*, 1995, 1999; *Tamaki and Ukaji*, 2003], and, most recently, the Bremen/Cottbus group in Germany [*Sitte and Egbers*, 2000; *von Larcher and Egbers*, 2005; *Harlander et al.*, 2011] and the Budapest group in Hungary [*Jnosi et al.*, 2010]. These have been complemented by various numerical modeling studies [e.g., *Hignett et al.*, 1985; *Sugata and Yoden*, 1992; *Read et al.*, 2000; *Maubert and Randriamampianina*, 2002; *Lewis and Nagata*, 2004; *Randriamampianina et al.*, 2006; *Young and Read*, 2008; *Jacoby et al.*, 2011]. In addition, the range of phenomena studied in the context of annulus experiments have been extended through modifications to the annulus configuration to emulate the effects of planetary curvature (i.e., a β-effect) [e.g., *Mason*, 1975; *Bastin and Read*, 1997, 1998; *Tamaki and Ukaji*, 2003; *Wordsworth et al.*, 2008; *von Larcher et al.*, 2013] and zonally asymmetric topography [e.g., *Leach*, 1981; *Li et al.*, 1986; *Bernadet et al.*, 1990; *Read and Risch*, 2011]; see also Chapters 2, 3, 7, 16, and 17 in this volume.

The existence of regular, periodic, quasi-periodic, or chaotic regimes in an open cylinder was also a major feature of another related class of rotating, stratified flow experiments using discrete two-layer stratification and mechanically-imposed shears. *Hart* [1972] introduced this experimental configuration in the early 1970s, inspired by the theoretical work of *Phillips* [1954] and *Pedlosky* [1970, 1971] on linear and weakly nonlinear instabilities of such a two-layer, rotating flow system. Because of its simpler mode of forcing and absence of complicated boundary layer circulations, these kinds of two-layer system were more straightforward to analyze theoretically, allowing a more direct verification of theoretical predictions in the laboratory than has typically proved the case with the thermally driven systems. Subsequent studies by Hart [*Hart*, 1979, 1980, 1985, 1986; *Ohlsen and Hart*, 1989a,1989b] and others [e.g., *King*, 1979; *Appleby*, 1982; *Lovegrove et al.*, 2000; *Williams et al.*, 2005, 2008] have extensively explored this system, identifying various forms of vacillation and low-dimensional chaotic behaviors as well as the excitation of small-scale, interfacial inertia-gravity waves through interactions with the quasi-geostrophic baroclinic waves.

In this chapter, we focus on the "classical" thermally driven, rotating annulus system. In Section 1.3 we review the current state of understanding of the rich and diverse range of flow regimes that may be exhibited in thermal annulus experiments from the viewpoint of experimental observation, numerical simulation, and fundamental (mainly quasi-geostrophic) theory. This will include the interpretation of various empirical experimental observations in relation to both linear and weakly nonlinear baroclinic instability theory. One of the key attributes of baroclinic instability and "sloping convection" is its role in the transfer of heat within a baroclinic flow. In Section 1.4 we examine in some detail how heat is transported within the baroclinic annulus across the full range of control parameters, associated with both the boundary layer circulation and baroclinically unstable eddies. This leads naturally to a consideration of how axisymmetric boundary layer transport and baroclinic eddy transports scale with key parameters and hence how to parameterize these transport processes, both diagnostically and prognostically, in a numerical model for direct comparison with recent practice in the ocean modeling community. Finally, in Section 1.5 we consider the overall role of annulus experiments in the laboratory in continuing to advance understanding of the global circulation of planetary atmospheres and oceans, reviewing the current state of research on delineating circulation regimes obtained in large-scale circulation models in direct comparison with the sequences of flow regimes and transitions in the laboratory. The results strongly support many parallels between laboratory systems and planetary atmospheres, at least in simplified models, suggesting a continuing important role for the former in providing insights for the latter.

1.3. FLOW REGIMES AND TRANSITIONS

The typical construction of the annulus is illustrated schematically in Figure 1.1 and consists of a working fluid (usually a viscous liquid, such as water or silicone oil, though this can also include air [e.g., see *Maubert and Randriamampianina*, 2002; *Randriamampianina et al.*, 2006; *Castrejón-Pita and Read*, 2007] or other fluids, including liquid metals such as mercury [*Fein and Pfeffer*, 1976]) contained in the annular gap between two coaxial circular, thermally conducting cylinders, that can be rotated about their common (vertical) axis. The cylindrical sidewalls are maintained at constant but different temperatures, with a (usually horizontal) thermally insulating lower boundary and an upper boundary that is also thermally insulating and either rigid or free (i.e., without a lid).

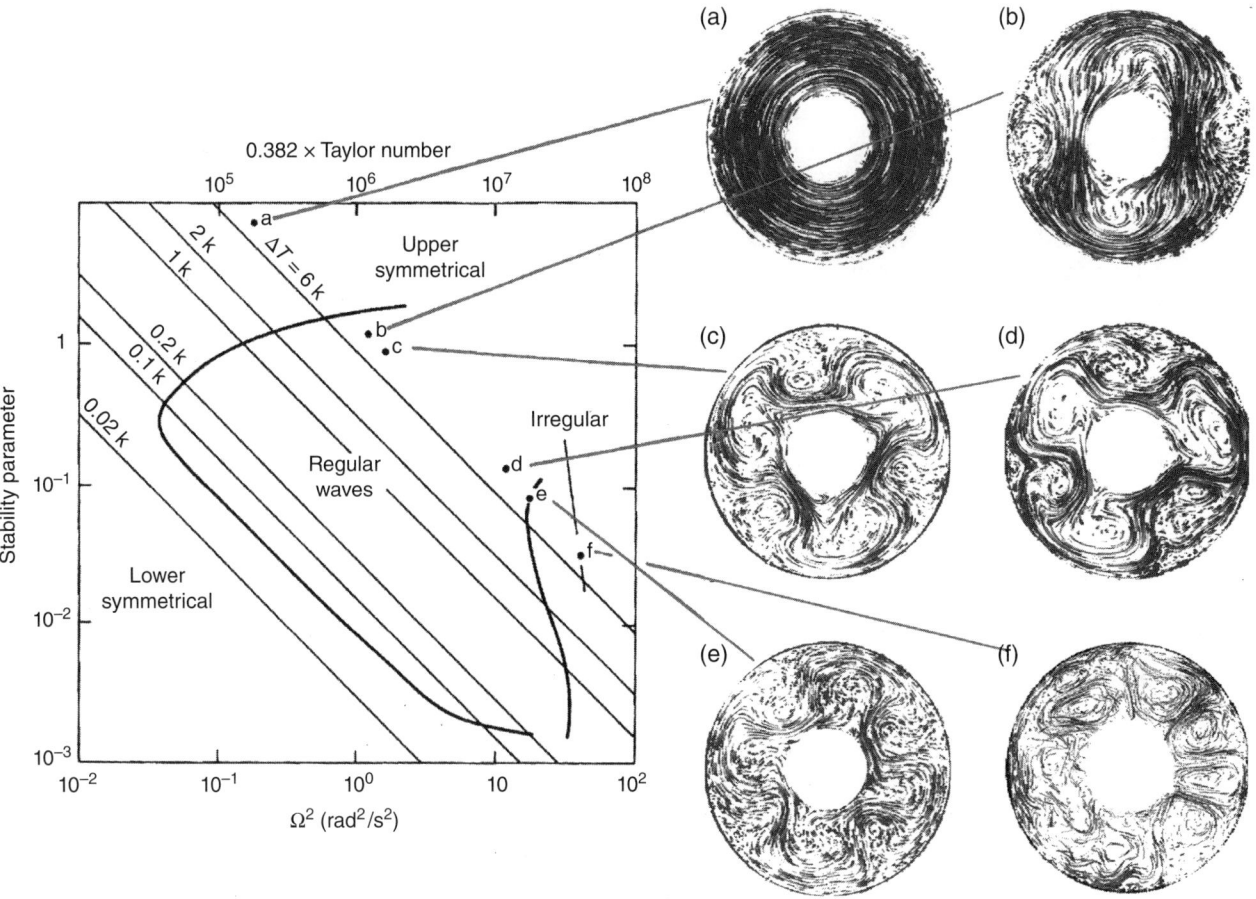

Figure 1.3. Schematic regime diagram for the thermally driven rotating annulus in relation to the thermal Rossby number Θ (or stability parameter, $\propto \Omega^{-2}$) and Taylor number $\mathcal{T} \propto \Omega^2$, showing some typical horizontal flow patterns at the top surface, visualized as streak images at upper levels of the experiment.

1.3.1. Principal Flow Regimes

Although a number of variations in these boundary conditions have been investigated experimentally, almost all such experiments are found to exhibit the same three or four principal flow regimes, as parameters such as the rotation rate Ω or temperature contrast ΔT are varied. These consist of (I) axisymmetric flow (in some respects analogous to Hadley flow in Earth's tropics and frequently referred to as the "upper-symmetric regime"; see below) at very low Ω for a given ΔT (that is not too small); (II) regular waves at moderate Ω; and (III) highly irregular, aperiodic flow at the highest values of Ω attainable. In addition, (IV) axisymmetric flows occur at all values of Ω at a sufficiently low temperature difference ΔT (a diffusively dominated regime termed "lower symmetric" [*Hide and Mason*, 1975, *Ghil and Childress*, 1987] to distinguish it from the physically distinct "upper-symmetric" mentioned above). The location of these regimes are usually plotted on a "regime diagram" with respect to the two (or three) most significant dimensionless parameters. These are typically

(a) a stability parameter or "thermal Rossby number"

$$\Theta = \frac{g\alpha \Delta T d}{[\Omega(b-a)]^2}, \quad (1.1)$$

providing a measure of the strength of buoyancy forces relative to Coriolis accelerations;

(b) a Taylor number

$$\mathcal{T} = \frac{\Omega^2 (b-a)^5}{\nu^2 d}, \quad (1.2)$$

measuring the strength of Coriolis accelerations relative to viscous dissipation; and

(c) the Prandtl number

$$\mathrm{Pr} = \frac{\nu}{\kappa}. \quad (1.3)$$

Here g is the acceleration due to gravity, α the thermal expansion coefficient of the fluid, ν the kinematic viscosity, κ the thermal diffusivity, and a, b, and d the radii of the inner and outer cylinder and the depth of the annulus, respectively. Figure 1.3 shows a schematic form of this diagram with the locations of the main regimes indicated.

From a consideration of the conditions under which waves occur in the annulus (especially the location in the parameter space of the upper-symmetric transition) and a comparison with the results of linear instability theory, it is clear that the waves in the annulus are fully developed manifestations of baroclinic instability (often referred to as "sloping convection" from the geometry of typical fluid trajectories; for example, see *Hide and Mason* [1975]). Since these flows occur in the interior of the annulus (i.e., outside ageostrophic boundary layers) under conditions appropriate to quasi-geostrophic scaling, a dynamical similarity to the large-scale midlatitude cyclones in Earth's atmosphere is readily apparent, though with rather different boundary conditions. A more detailed discussion of the properties of these flows is given below and by *Hide and Mason* [1975] and *Ghil and Childress* [1987]. Associated with this conclusion is the implication that the waves develop in order to assist in the transfer of heat both upward (enhancing the static stability) and horizontally down the impressed thermal gradient (i.e., tending to reduce the impressed horizontal gradient). The action of heat transport by the waves and axisymmetric flows will be considered in the next section.

1.3.2. Axisymmetric/Wave Transition and Linear Instability Theory

The previous section indicated the conditions under which baroclinic waves occur in the annulus and their role as a means of transferring heat upward and against the horizontal temperature gradient. The Eady model of baroclinic instability has been commonly invoked as an idealized, linearized conceptual model to account for the onset of waves from axisymmetric flow [*Hide*, 1970; *Hide and Mason*, 1975, *Ghil and Childress*, 1987]. Although the Eady model is highly idealized, it does seem to predict the location of the onset of large-amplitude waves remarkably close to the conditions actually observed, at least at high Taylor number (note that the Eady problem in its "classical form" is inviscid). Apparent agreement can be made even closer if the Eady problem is modified to include Ekman boundary layers by replacing the $w = 0$ boundary condition with the Ekman compatibility condition

$$w = \frac{\mathcal{E}^{1/2}}{\sqrt{2}\text{Ro}}\nabla^2\psi, \qquad z = 0, 1, \qquad (1.4)$$

where ψ is the stream function for the horizontal flow. This naturally brings in the Taylor number familiar to experimentalists (via Ekman number $\mathcal{E}$) and leads to a plausibly realistic envelope of instability at low

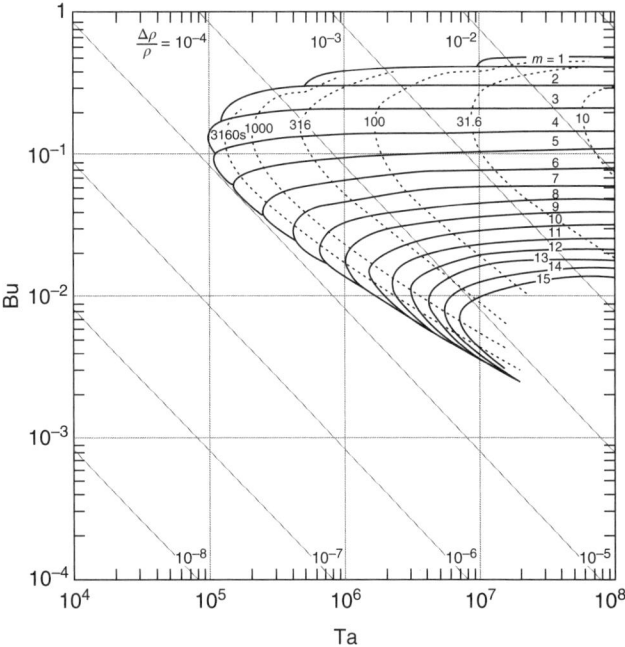

Figure 1.4. Regime diagram based on the extension of Eady's baroclinic instability theory to include Ekman layers and flat, horizontal boundaries. The wave number of maximum instability is indicated by integer numbers and the transition curves and contours of e-folding time are given on a Burger number (Bu ∼ Θ; see *Hide and Mason* [1975]) against $\mathcal{T}$ plot. (Adapted from *Mason* [1975] by permission of the Royal Society).

Taylor number (see Figure 1.4), supporting the hypothesis [*Hide and Mason*, 1975] that the "lower symmetric transition" is frictionally dominated.

The structure of the most rapidly growing instability has certain characteristic features in terms of, for example, phase tilts with height. In the thermal annulus, steady baroclinic waves are also seen to exhibit many of these features, as determined from experiment and numerical simulation. The extent to which Eady theory actually provides a complete theoretical description of the instability problem in annulus experiments, however, is a somewhat more complicated question than it at first appears. The dominant instability in the Eady model relies on the existence of horizontal temperature gradients on horizontal boundaries for the required change of sign in the lateral gradient of quasi-geostrophic potential vorticity, $\partial \overline{q}/\partial y$, for instability [e.g., *Charney and Stern*, 1962]. Elsewhere, the flow is constructed such that $\partial \overline{q}/\partial y = 0$. In practice, however, strong horizontal mass transports in the Ekman layers result in almost no horizontal temperature gradients at the boundaries; in reality $\partial \overline{q}/\partial y$ changes sign smoothly in the interior (e.g., see Figure 1.17c later). Thus, instability of an internal baroclinic jet is arguably a more appropriate starting point, preferably including a consideration of lateral shears. This was considered by *Bell and White* [1988], who examined the stability of an idealized internal zonal jet flow in a straight, rectangular channel of the form

$$U = \frac{1}{2}(1 - a_s + a_s - \sin \pi y) \sin \pi z, \quad (1.5)$$

where a_s is a constant that determines the degree of horizontal barotropic shear in the otherwise baroclinic jet. If full account is taken of lateral shear in such an internal jet (by varying a_s), however, the critical Burger number for the onset of waves is found to vary by a factor of $O(10)$. The precisely applicable value is likely dependent upon subtle details of the shape of the zonal flow and the imposed lateral boundary conditions, since the true boundary conditions at the sides of the geostrophic interior ought really to take proper account of the complex viscous boundary layer structures (e.g., Stewartson layers), although impermeable, free-slip boundaries have typically been employed (for mathematical convenience) in most theoretical studies to date.

Recent exceptions to this include the two-layer studies by *Mundt et al.* [1995a, 1995b] and the analysis of the full thermal annulus problem by *Lewis and Nagata* [2004]. *Mundt et al.* [1995a] examined the linear (and nonlinear) stability of a quasi-geostrophic, two-layer jet in a rectilinear channel in which internal viscosity was included in deriving the zonally symmetric basic state. This led to the formation of viscous (Stewartson) boundary layers adjacent to the sidewalls of the channel, within which strong zonal shear developed as the flow adjusted to the nonslip condition at each boundary. This was then shown to modify the critical Froude number for instability by a factor $O(1)$ for the gravest modes. Similar results were obtained by *Mundt et al.* [1995b] in cylindrical geometry, for which improved agreement with experimental measurements was shown compared with stability calculations assuming a free-slip outer boundary. The most sophisticated approach applied so far for the thermal annulus configuration was by *Lewis and Nagata* [2004], who used numerical continuation techniques to solve for the linear stability boundary (as a function of Θ and $\mathcal{T}$) of an axisymmetric baroclinic zonal jet in cylindrical geometry using the full Navier-Stokes equations for a viscous, Boussinesq fluid. The results indicated good agreement with the location of both the upper and lower symmetric transitions as found in laboratory experiments. They also indicated the influence of centrifugal buoyancy in modifying the stability boundary at the lower symmetric transition. These calculations all serve to demonstrate the quantitative success of linear stability theory in accounting quantitatively for the onset of the principal mode of baroclinic instability in both two-layer and continuously stratified rotating tank experiments as a supercritical global bifurcation.

1.3.3. Steady Waves and Equilibration: Weakly Nonlinear Theory

As waves grow in strength from an initial zonal flow, they typically equilibrate to either a steady or periodically varying amplitude ("amplitude vacillation")or even to a weakly chaotic flow. The linear models of baroclinic instability cannot account for equilibration and vacillation, and so we must consider the effects of nonlinearity in the interaction between the growing wave and the basic zonal flow. Weakly nonlinear theory was developed in the late 1960s as a means of introducing nonlinearity into linear instability problems while keeping the mathematics analytically tractable. The basic assumption of this approach is that the flow in which the wave grows is only weakly supercritical, and so only a small range of wave numbers is unstable and grows relatively slowly. More detailed discussions can be found [e.g., *Drazin*, 1978; *Hocking*, 1978; *Ghil and Childress*, 1987; *Pedlosky*, 1987].

Consider a zonal flow under conditions just inside the stability threshold with weak supercriticality Δ. Because the stability boundary is then asymptotically quadratic in zonal wave number k about the wave number of the first mode to go unstable k_c (see Figure 1.5), a small range of $k \sim \Delta^{1/2}$ is destabilized. In a periodic x domain where k is discretized, this may permit only one unstable wave number. We introduce a "slow" time scale τ defined by

$$\tau = \Delta^{1/2} t, \quad \Delta \ll 1, \quad (1.6)$$

(in the sense that it advances more slowly than "normal time" t) and solve for normal modes of the following wavelike form in the zonal (x) direction:

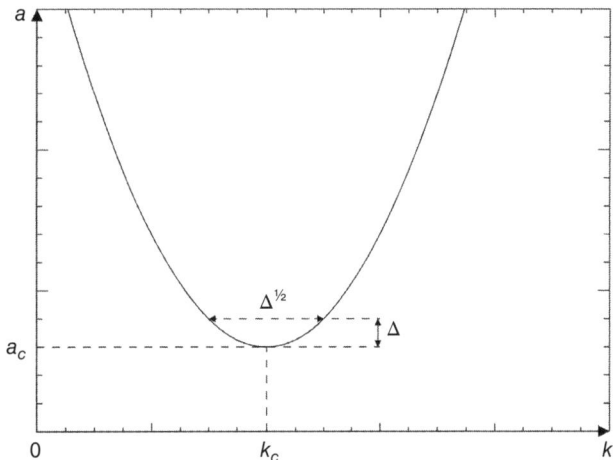

Figure 1.5. Schematic stability diagram showing the assumed (quadratic) form of the critical curve $a(k)$ as a function of wave number k in the vicinity of the first unstable mode with wave number k_c.

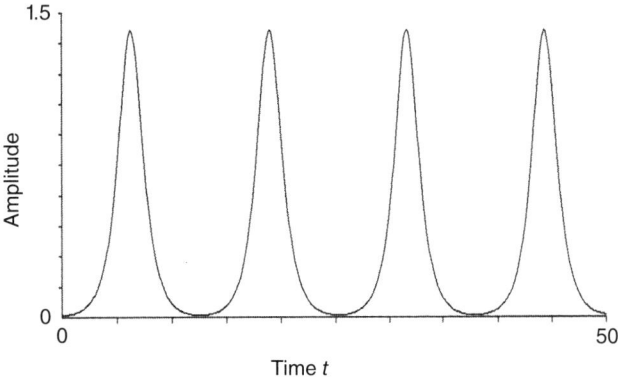

Figure 1.6. A typical solution to equations (1.9) and (1.10) showing sustained amplitude oscillations.

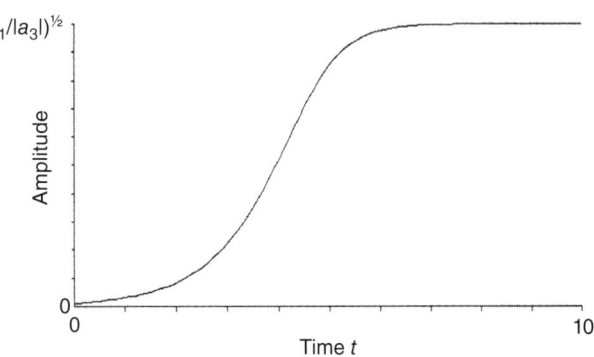

Figure 1.7. Solution to equation (1.11) and (1.12) showing the approach to a steady equilibrium.

$$\tau = \mathcal{R}\left\{A(\tau)F(y,z)\exp k(x - ct)\right\}, \quad (1.7)$$

where y is the meridional coordinate and z is in the vertical in the presence of a zonal flow of the form

$$U = U(y,z) + V(\tau)G(y,z). \quad (1.8)$$

Here A and V are respectively the slowly varying amplitudes of the wave and the correction to the zonal flow due to the nonlinear self-interaction of the wave, whose spatial structure is represented by $G y, z)$. The resulting evolution equations for A and V depend upon the relative magnitude of viscous dissipation:

(i) Weak Dissipation ($\mathcal{E}^{1/2}/\text{Ro} \ll O(\Delta^{1/2})$)

Examples include the Eady or two-layer Phillips problems with no Ekman layers [e.g., *Pedlosky*, 1987]. It can be shown that the problem then reduces to coupled ordinary differential equations (ODEs) of the form

$$\frac{d^2A}{d\tau^2} = a_1 A - |a_3|A^3, \quad (1.9)$$

$$\frac{dV}{d\tau} = a_4 \frac{dA}{d\tau}, \quad (1.10)$$

typically resulting in a sustained amplitude modulation, or "vacillation," associated with the exchange of potential energy between wave and zonal flow (e.g., see Figure 1.6).

(ii) Stronger Dissipation ($\mathcal{E}^{1/2}/\text{Ro} = O(1)$)

Examples include the Eady and Phillips problems with Ekman damping. It can be shown that the amplitude equations reduce to the well-known Landau equation [e.g., see *Pedlosky*, 1971; *Romea*, 1977; *Drazin*, 1978; *Hocking*, 1978]

$$\frac{dA}{d\tau} = a_1 A - |a_3|A^3, \quad (1.11)$$

$$V = a_4|A|^2, \quad (1.12)$$

resulting typically in an asymptotic equilibration toward a steady amplitude $A = (a_1/|a_3|)^{1/2}$ (see Figure 1.7).

These models, of course, represent a considerable oversimplification of the real equilibration processes in fully developed baroclinic instability. Indeed, *Boville* [1981] noted that *Pedlosky*'s [1970] approach led to significant inaccuracies in predicting the amplitude behavior of baroclinic waves and failed to observe the predicted amplitude oscillations close to minimum critical shear. Subsequent work [*Pedlosky*, 1982a, 1982b; *Warn and Gauthier*, 1989; *Esler and Willcocks*, 2012] suggests that the dynamical equilibration is more typically dominated by the behavior of a nonlinear critical layer that develops within the flow, leading to the wrapping up and eventual homogenization of potential vorticity. The equilibrated state may then result in convergence toward a steady wave state via a series of damped amplitude oscillations. The analytical solutions of *Warn and Gauthier* [1989] even produced periodic amplitude oscillations under certain conditions, resembling an amplitude vacillation, but in which potential vorticity is reversibly mixed and unmixed. This is only strictly applicable under conditions of weak friction close to marginal instability. The recent work of *Willcocks and Esler* [2012] also suggests that the mode of equilibration via a Landau equation, predicted in the models of *Pedlosky* [1971] and *Romea* [1977], probably applies mainly to the dissipatively destabilized instability that occurs for shears less than the critical shear in the inviscid problem in the presence of Ekman friction [*Holopainen*, 1961; *Boville*, 1981]. The full applicability of any of these models, however, still remains to be verified in detail in laboratory experiments or fully nonlinear numerical simulations in continuously stratified flows.

1.3.4. Wave Number Selection

Within the regular baroclinic wave regime, the flow tends to equilibrate typically (in the absence of a strong β-effect, e.g., associated with topographically sloping boundaries) to a state dominated by a single azimuthal wave number and its harmonics, which may be steady, quasi-periodic, or chaotic. The mechanisms by which rotating annulus waves select which wave number to favor at fully nonlinear equilibration are still not fully understood but seem likely to share some aspects in common with mechanisms identified in simple, weakly nonlinear, spectrallytruncated models of baroclinic instability. Such models [e.g., see above and *Pedlosky*, 1970; *Drazin*, 1970; *Pedlosky*, 1971] represent only the leading order nonlinear interactions between a single mixed baroclinic-barotropic traveling wave and the background ($m = 0$) zonal flow (i.e., suppressing quadratic and higher order wave-wave interactions). The nonlinear self-interaction of a growing, linearly unstable wave generates a correction to the $m = 0$ zonal flow (at second order in wave amplitude) that feeds back on the growth rate, eventually reducing it to zero (a steady wave state, for which the modal amplitude equations may asymptotically reduce to a set of Landau equations in the presence of some frictional damping), or with a more complicated, quasi-periodic or chaotic time dependence [e.g., see *Lovegrove et al.*, 2001, 2002], for which the modal amplitude equations may reduce asymptotically to the classical real or complex Lorenz equations.

When more than one distinct wave number mode is able to grow from infinitesimal amplitude on a given zonal flow, weakly nonlinear models do not provide a unique answer as to what mechanism will act to select the dominant mode. However, one commonly found factor is for the flow to preferentially select the mode that is capable of releasing the most available potential energy (APE) from the initial flow [*Hart*, 1981]. In practice, this may correspond to the mode that can reach the largest barotropic amplitude [*Hart*, 1981; *Appleby*, 1988] provided nonlinear wave-wave interactions are absent. Where wave-wave interactions are permitted, the mode selection may become hysteretic such that a nonoptimal wave mode (i.e., one that does not release the maximum possible APE) may persist as the dominant mode if it was previously dominant under more favorable conditions at an earlier time. This is found to manifest itself within the regular flow regime as intransitivity (i.e., multiple equilibrium states), in which two or more alternative flows with differing azimuthal wave number m can occur for a given set of parameters [e.g., *Hide*, 1970; *Hide and Mason*, 1975]. The state obtained in any particular experiment implicitly depends upon the initial conditions. In addition, transitions between different states in the regular regime, achieved by slowly changing the external parameters, often exhibit hysteresis [e.g., *Hide and Mason*, 1975; *Sitte and Egbers*, 2000; *von Larcher and Egbers*, 2005] in that the location of a transition in parameter space depends upon the direction from which that transition is approached (e.g., transitions from $m = 3 \to 4$ do not occur at the same point as $m = 4 \to 3$). In a situation where the forcing that maintains the background zonal state is varied cyclically with time over a range that crosses the boundary between two or more optimal modes, this can lead to complex and chaotic behavior as the flow pattern flips erratically from one dominant mode to another [e.g., see *Buzyna et al.*, 1978].

Another issue is how an initially dominant wave flow may retain its dominance and remain indefinitely stable? The presence of wave-wave and higher order nonlinear interactions might be expected to permit the possibility of secondary instabilities of the primary dominant wave mode, at least in principle, thereby preventing the sustained dominance of a single baroclinic wave mode. *Hide* [1958] and *Hide and Mason* [1975] showed empirical evidence from a range of early experiments that, depending upon the radius ratio between inner and outer cylinders, there was a maximum azimuthal wave number of a stable and persistent dominant wave such that its azimuthal wavelength always seemed to exceed roughly 1.5 times the radial extent of the wave, i.e.,

$$m_{\max} \gtrsim \frac{2\pi(b+a)}{3(b-a)}. \qquad (1.13)$$

The prototypical idealized model for such a situation considers the stability of the basic Rossby-Haurwitz (RH) mode on the sphere to wavelike barotropic perturbations [*Lorenz*, 1972; *Hoskins*, 1973; *Baines*, 1976], although this has also been generalized to investigate baroclinic perturbations and instabilities of the basic RH wave [e.g., see *Kim*, 1978; *Grotjahn*, 1984a, 1984b]. The principal criterion for barotropic stability of the RH wave can be interpreted in relation to Fjørtoft's theorem for energy transfer in a quasi-geostrophic flow [*Fjørtoft*, 1953], for which both energy and squared vorticity must be conserved in non dissipative nonlinear interactions. This essentially requires that a given wave mode must lose energy simultaneously to both a higher *and a lower* wave number mode. Thus, the longest wave number modes capable of fitting into the domain tend to be relatively stable because of the unavailability of longer wavelength modes to which they can lose energy in an instability. Such an interpretation appears to be consistent with the criterion in equation (1.13) found by *Hide* [1958] and *Hide and Mason* [1975] for the maximum azimuthal wave number that can sustain a persistent dominant wave flow.

Recent work by *Young and Read* [2013] suggests that barotropic instability may not be the only possible mechanism for breakdown of regular baroclinic wave flows.

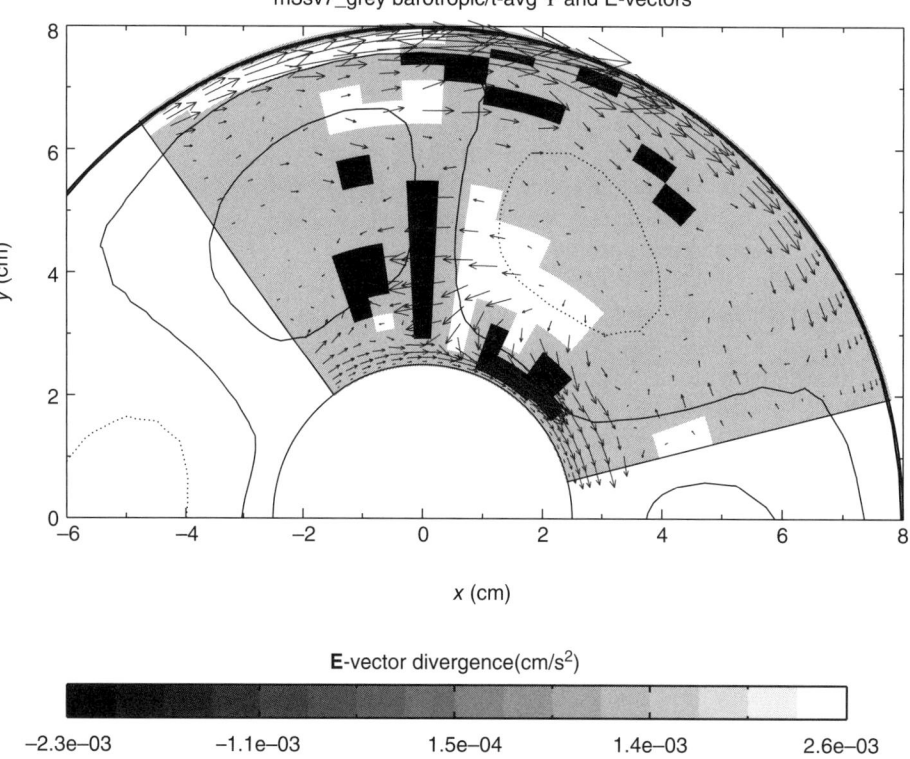

Figure 1.8. Evectors for chaotic flow in the annulus. The black contour lines show the assimilated barotropic time-averaged horizontal stream function (contours below the middle of the range are dotted), and the grey vectors are the barotropic time-averaged **E** vectors. The shading shows the **E**-vector divergence: black is up to -5×10^{-4} cm/s^2, grey is between -5×10^{-4} cm/s^{-2} and $+5 \times 10^{-4}$ cm/s^2, and white is above $+5 \times 10^{-4}$ cm/s^2. The flow is at $\Omega = 3.1$ rad/s with $T_b - T_a \approx 4.02°$C. (Adapted from *Young and Read* [2013] with permission of John Wiley & Sons, Inc.)

Based on sequences of laboratory measurements assimilated into a Boussinesq Navier-Stokes numerical model of the annulus, *Young and Read* [2013] found that localized, small-scale eddies shed from the cyclonic troughs of a large-scale baroclinic wave mode may be consistent with a localized baroclinic instability. Figure 1.8 shows the **E**vectors [*Hoskins et al.*, 1983; *James*, 1994] for these measurements at the highest rotation rate investigated. The barotropic **E**vector is a horizontal vector defined from correlations between the x–y components of the horizontal velocity (u, v) by

$$\mathbf{E} = (E_x, E_y) = (\overline{v'^2 - u'^2}, -\overline{u'v'}), \quad (1.14)$$

where the overbar represents a timeaverage and primed quantities are deviations from the time-mean flow. Its divergence provides a measure of the interaction between the time-mean flow and transient eddies such that $\nabla \cdot \mathbf{E} > 0$ implies a tendency for eddies to strengthen the mean flow [*Hoskins et al.*, 1983]. For **E**vectors pointing in the positive azimuthal direction, there is anticyclonic cyclogenesis between the divergent region and the outer cylinder, cyclonic cyclogenesis between the convergent region and the outer cylinder, and vice versa toward the inner cylinder. For the baroclinic annulus flows considered, *Young and Read* [2013] found the **E**vectors became more strongly convergent/divergent as the rotation rate increased. This acted to reinforce the main cyclone but weaken the part extending into the anticyclonic region, associated with the shedding of small-scale vortices, and doing so more and more as the rotation rate increased. The main baroclinic wave was found to be barotropically stable according to *Bell*'s [1989] criterion, but the instability was consistent with *Kim*'s [1989] observation that baroclinic Rossby waves may be baroclinically unstable if the internal Rossby deformation radius L_D is much smaller than a characteristic horizontal length scale L representative of the large-scale wave, in this case comparable with the annular gap width $b - a$. In this flow $L \gg L_D$ with increasing supercriticality (smaller values of Θ) as the rotation rate increased (as $L_D \propto 1/\Omega$), consistent with an interpretation of the chaotic vortex-shedding phenomenon discussed above as a secondary baroclinic instability of the large-scale wave.

1.3.5. Vacillating Waves and Wave-Zonal Flow Interactions

Baroclinic waves in the regular wave regime may be either steady (apart from a slow drift) or "vacillating" (i.e., with a periodic or nearly periodic time dependence; see Chapter 3 for a more detailed discussion). Laboratory observations of "amplitude vacillation" indicate that (for fluids with Pr $\gg$ 1) it occurs close to the upper stability threshold of its wave number m, around where a transition from m to $m-1$ is observed, at moderate-high Taylor number. At lower values of Pr, however, it appears that this transition sequence is reversed, with transitions, for example using air as the working fluid (with Pr $\simeq$ 0.7), from steady waves into the vacillating regime as $\mathcal{T}$ is increased [e.g., *Randriamampianina et al.*, 2006; *Castrejón-Pita and Read*, 2007]; see also Chapter 16 in this volume. The "vacillating" state then comprises the periodic modulation of both the amplitude and drift frequency of the wave on a time scale $\sim 10-100$ "days". Figure 1.9 shows a sequence of streak images taken from a typical $m=3$ flow undergoing an amplitude vacillation cycle at a level around $0.8d$ above the annulus base. This clearly shows the wave amplitude growing, reaching a maximum in amplitude in Figure 1.9d and then decaying before the cycle repeats. The wave is modulated in both amplitude and drift rate (phase speed) during the cycle, indicating a nonlinear interaction with the background zonal flow.

More detailed diagnostics show periodic variations in total heat transport and in potential energy exchanges between the wave and zonal flow [see *Pfeffer et al.*, 1980; *Hignett et al.*, 1985]. In particular variations in the slope of the azimuthal mean isotherms (see Figure 1.10) clearly show modulations in the potential energy stored in the azimuthal mean flow. The zonal flow structure (see Figure 1.10a and 1.10b) is seen to oscillate between a single jet pair at minimum wave amplitude and two double jets at maximum amplitude.

In this regard, it appears that nonlinear interactions between the dominant wave and the azimuthal mean flow are critically important for the phenomenon of amplitude vacillation. In practice, however, it may not be easy to distinguish this behavior from interference arising from a quasi-linear superposition of two wave components with the same azimuthal wave number and differing vertical structure and drift frequencies ω_1 and ω_2. Apparent "vacillation" then takes place at the difference frequency of the two components $|\omega_1 - \omega_2|$. If the two components cross-interact with the zonal flow, effects such as phase locking and zonal flow modulation may occur, reproducing several aspects of the observed flows. Some observers claim to have identified this mechanism in measurements in the laboratory [*Lindzen et al.*, 1982], though the more general relevance of this mechanism remains controversial. However, other forms of nonlinear interference vacillation, for example, involving the superposition of two modes with differing (but adjacent) azimuthal wave numbers but similar radial and vertical structures [*Ohlsen and Hart*, 1989b], may also manifest themselves as periodic modulations of baroclinic eddy variance while also modulating the azimuthal mean flow through nonlinear triad interactions of harmonics with the mean zonal flow.

Although the basic amplitude vacillation (AV) regime is typically a quasi-periodic flow characterized by two independent frequencies associated with (a) the azimuthal drift of a monochromatic wave number pattern and (b) its periodic modulation in amplitude, transitions to more chaotic states have also been observed, still apparently within the "regular" wave regime. These include transition sequences via period-doubling bifurcations to chaotic amplitude vacillations [*Hart*, 1985, 1986] and routes involving more complex transitions directly to chaos from doubly periodic "modulated" amplitude vacillation [*Farmer et al.*, 1982; *Read et al.*, 1992; *Früh and Read*, 1997]. The resulting flows were apparently consistent with the interaction of a relatively small number of spatial modes but differed in their basic azimuthal symmetry properties. Period doubling was observed as a typical route to chaos in the two-layer, open cylinder experiments of *Hart* [1985] and *Ohlsen and Hart* [1989a]. This does not seem to be typical for thermal annulus experiments, however, which tend to be dominated by higher wave number baroclinic modes ($m \gtrsim 3$). In the latter, the more typical route involves the development of a third period through the emergence of an additional wave mode that is not harmonically related to the initial dominant wave [*Read et al.*, 1992; *Früh and Read*, 1997].

However, *Young and Read* [2008] did observe a sequence of period doublings from a wave number $m = 2$ AV flow in a set of numerical simulations which led to chaotic states consistent with the endpoint of a period-doubling cascade over limited regions of parameter space. Figure 1.11 shows two examples of such flows, illustrating chaotic ((a), (b)) and period 3 ((c), (d)) vacillations. In this regime, the amplitude modulations vary in strength, alternating between two intensities in the period 2 state with successive doublings as $\mathcal{T}$ was increased until chaotic vacillation ensued. The whole sequence shows a sequence of bifurcations as successive period doublings lead to chaotic behavior followed by an indication of period 3 "periodic windows". Figure 1.12 illustrates such a sequence at even higher values of $\mathcal{T}$ showing the maximum and minimum wave amplitudes of the equilibrated baroclinic flow at various values of $\mathcal{T}$ while keeping Θ at a fixed value of $\Theta \simeq 1.75$. This shows clear evidence of a series of period-doubling transitions with chaotic regions near points A, B, and D interspersed with regions characterized by quasi-periodic vacillations.

20 MODELING ATMOSPHERIC AND OCEANIC FLOWS

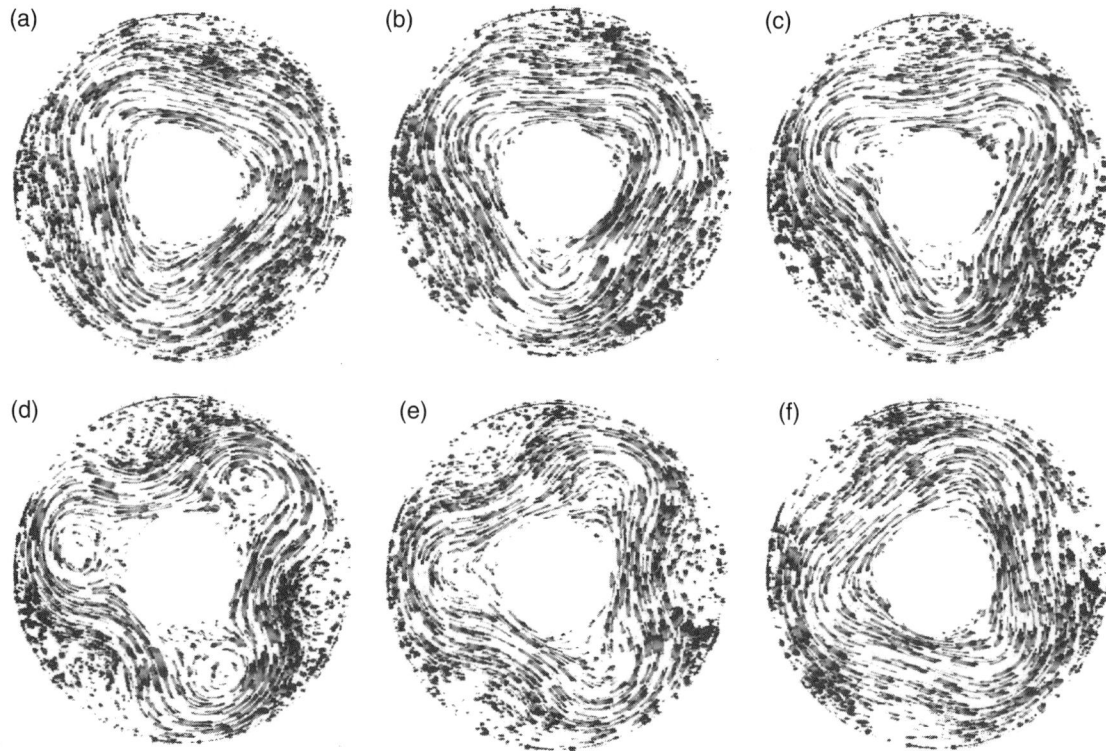

Figure 1.9. Typical horizontal flow fields (streak photographs) during an "amplitude vacillation" cycle of the rotating annulus in the same system as in Figures 1.8–1.13. For color detail, please see color plate section.

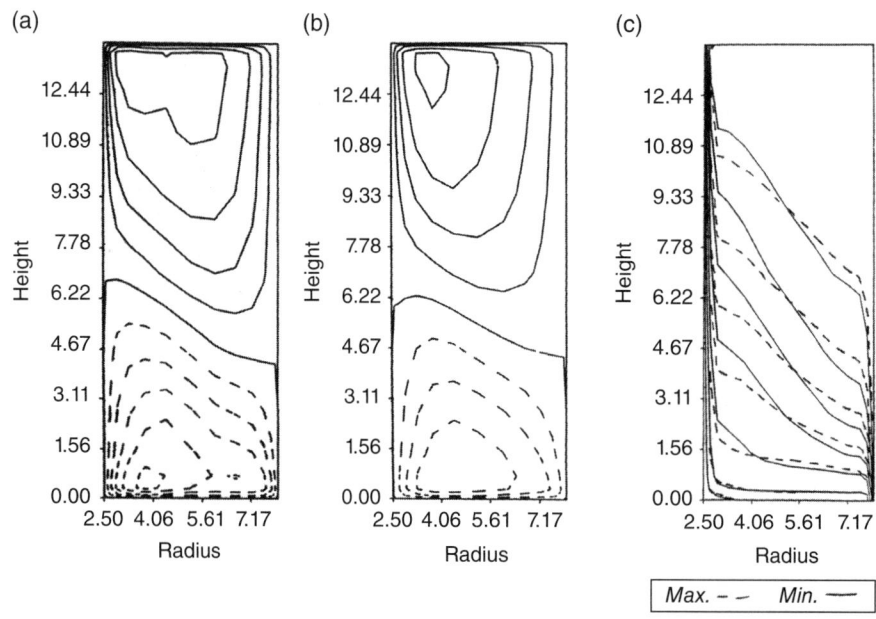

Figure 1.10. Results from numerical simulations of a baroclinic amplitude vacillation in a rotating annulus similar to that of *Hignett et al.* [1985] showing (a) azimuthal mean azimuthal flow at maximum wave amplitude, (b) azimuthal mean azimuthal flow at minimum wave amplitude, and (c) azimuthal mean temperature fields at minimum (solid) and maximum (dashed) wave amplitude during the vacillation cycle.

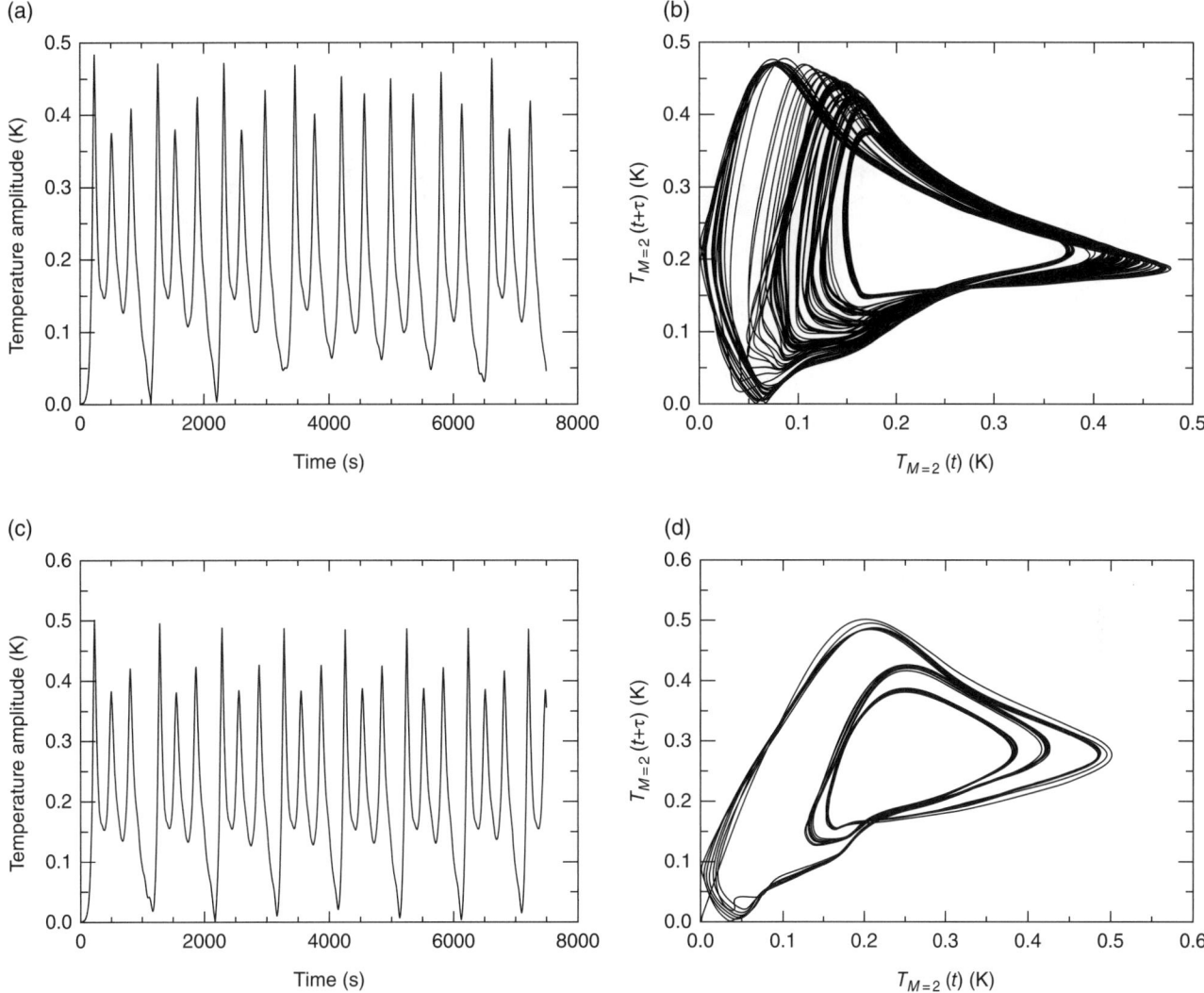

Figure 1.11. Temperature $m = 2$ amplitude time series [(a) and (c)] and delay coordinate reconstructions [(b) and (d)] in the $m = 2$AV-d period-doubled amplitude vacillation regime, obtained by *Young and Read* [2008] in Boussinesq Navier-Stokes simulations of rotating annulus flows. $\tau = 100$ s in (b) and $\tau = 85$ s in (d). Adapted from *Young and Read* [2008]. Copyright 2008, with permission from Elsevier.

Such a sequence is strongly reminiscent of the period-doubling route to chaos found in the two-layer experiments of *Hart* [1985, 1986], who showed sequences of period doublings from an $m = 1$AV flow at fixed Froude number. This would therefore appear to be a generic route to chaos in baroclinic wave flows at low enough wave numbers that sideband instabilities do not dominate the dynamics, and the main nonlinear interaction is between a single wave and the zonal flow.

This kind of bifurcation sequence has also been obtained in various studies invoking weakly nonlinear baroclinic instability theory, such as by *Pedlosky and Frenzen* [1980] for the two-layer model [see *Klein*, 1990, for a review] and *Weng et al.* [1986] for the continuously stratified Eady problem. These and other studies [e.g, see *Ghil and Childress*, 1987] have shown that the endpoint of such period-doubling sequences, even in very simple models representing the interactions of single mixed baroclinic-barotropic waves with a zonal flow, can be chaotic states of low dimension. In certain limits, several workers have shown that the single wave/zonal flow equilibration problem may reduce to a set of three coupled ODEs:

$$\frac{dX}{d\tau} = \sigma(Y - X), \qquad (1.15)$$

$$\frac{dY}{d\tau} = XZ + r_a X - Y, \qquad (1.16)$$

$$\frac{dZ}{d\tau} = XY - bZ, \qquad (1.17)$$

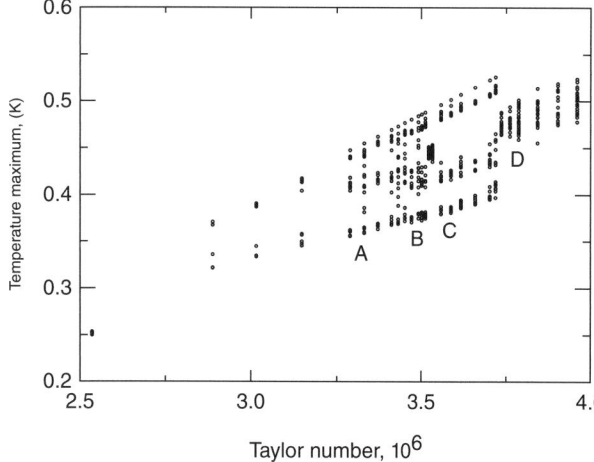

Figure 1.12. Bifurcation diagram showing the maxima of $m = 2$ temperature amplitudes at successive values of $\mathcal{T}$ keeping Θ constant at $\Theta \simeq 1.75$. Adapted from *Young and Read* [2008]. Copyright 2008 with permission from Elsevier.

where σ, r_a, and b are constants, X is related to $A(\tau)$, Y is related to $V(\tau)$ and $Z \sim F(A, V)$ [e.g., *Brindley and Moroz*, 1980; *Gibbon and McGuinness*, 1980; *Pedlosky and Frenzen*, 1980; *Klein*, 1990], which is the famous set of equations that can result in the Lorenz attractor [*Lorenz*, 1963a]. In the presence of a "planetary vorticity gradient" or β effect, the wave-zonal flow interaction problem may reduce to a set analogous to the *complex* Lorenz equations [e.g., *Gibbon and McGuinness*, 1980; *Fowler et al.*, 1982; *Lovegrove et al.*, 2001, 2002]:

$$\frac{dX}{d\tau} = \sigma(Y - X), \tag{1.18}$$

$$\frac{dY}{d\tau} = XZ + r_a X - aY, \tag{1.19}$$

$$\frac{dZ}{d\tau} = \frac{1}{2}(X^*Y + XY^*) - bZ, \tag{1.20}$$

where X and Y are now complex variables and r_a and a are complex parameters. The onset of chaos in these models as parameters are smoothly varied is characterized by a particular sequence of transitions typically involving either a sudden "snap-through" bifurcation from an initially steady wave as dissipation is reduced or a period-doubling cascade from an "amplitude vacillation" state as dissipation is increased [*Klein*, 1990].

In the thermal annulus, the situation seems more complicated, with the possibility of at least two distinct routes to chaotic behavior. In one case, periodic AV gives way to an azimuthally asymmetric, chaotically modulated vacillation in which two or more adjacent wave numbers occur in irregular competition. The final "chaotic" state appears to comprise at least three independent frequencies together with a "noisy" component associated with

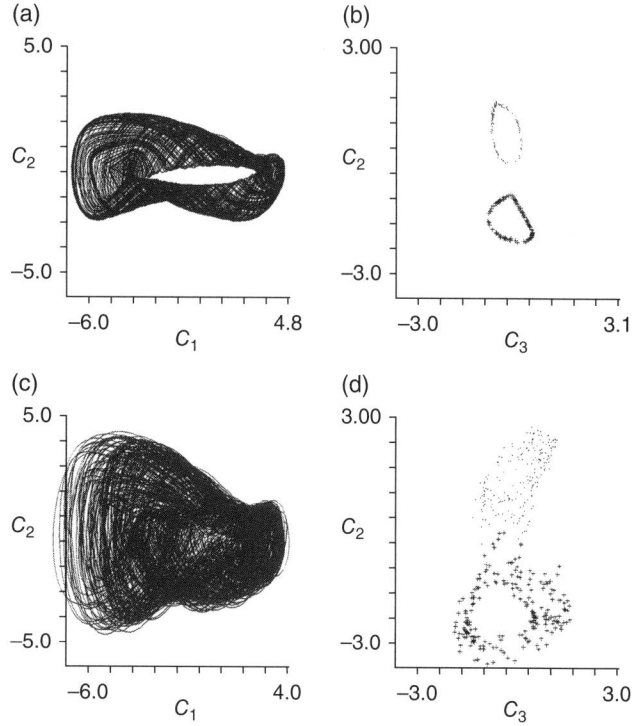

Figure 1.13. Phase portraits [(a) and (c)] and Poincaré sections [(b) and (d)] obtained from measurements of temperature in a rotating annulus showing a transition from amplitude vacillation (top) to a chaotic "modulated amplitude vacillation" (bottom). Adapted from *Read et al.* [1992] with permission.

the observed "chaos" [*Read et al.*, 1992; *Früh and Read*, 1997]. The transition may be illustrated in reconstructed phase portraits derived from time series of temperature measurements, for example. Examples are illustrated in Figure 1.13 from the experiments reported by *Read et al.* [1992]. The other main route may be via the period-doubling sequence found in model simulations by *Young and Read* [2008] although, as mentioned above, this route has so far proved elusive in real experiments.

1.3.6. Structural Vacillation and Transition to Geostrophic Turbulence

"Structural vacillation" (also known as "shape" or "tilted-trough vacillation" or SV [*White and Koschmieder*, 1981; *Buzyna et al.*, 1984]) occurs as the irregular flow transition is approached and, in its purest expression, is characterized by a nearly periodic, horizontal tilting of the radial axes of wave peaks and troughs [*Weng et al.*, 1986; *Weng and Barcilon*, 1987]. However, in practice it takes many different forms, depending upon a variety of factors, including how close the dominant wave number m may be to Hide's maximum stable wave number $m_{\max}$ [cf. equation (1.13)]. This becomes more pronounced as

Ω is increased, until the regular flow pattern breaks down into fully irregular flow [*Pfeffer et al.*, 1980; *Buzyna et al.*, 1984].

In its "purest" form, SV has been observed as the periodic tilting back and forth of the main wave troughs. In association with this observation, the lateral distribution of eddy energy within the wave was observed to shift back and forth between the inner and outer sides of the channel. This observation led some to suggest a kinematic form for the wave as the superposition of dominant waves with the same zonal wave number and with lateral structures

$$\phi_1(y) = A_1 \sin \pi y, \quad (1.21)$$
$$\phi_2(y) = A_2 \sin 2\pi y, \quad (1.22)$$

both propagating azimuthally at different phase speeds.

This type of behavior has been reproduced in a class of simple, low-order numerical models in which a small number of wave modes are allowed to interact through mutual advection in a quasi-geostrophic model. *Weng and Barcilon* [1987], for example, followed a much earlier approach pioneered by *Lorenz* [1963b] and applied it to a nonlinear version of the Eady model including the first two lateral modes (cf. equations (1.21) and (1.22)) to obtain solutions in which eddy energy oscillated in y through nonlinear interference between the two gravest y modes with the same x wave number (e.g., see Figure 1.14).

In practice, however, observed 'structural vacillations' are often more complicated than this picture would suggest, for example, with transient small-scale features growing and decaying within a large-scale pattern dominated by a single azimuthal wave number. Oscillations often appear to be strongly intermittent and irregular, and the phenomenon suggests the growth of small-scale instabilities within the large-scale pattern (either barotropic or baroclinic) that do not reach sufficient amplitude to disrupt the main pattern. *Read et al.* [1992] found that the onset of SV occurs quite suddenly at a well-defined point in parameter space, again with evidence of intermittency in time. The irregular character of the oscillations becomes steadily more apparent as Ω is increased, and the large-scale pattern becomes gradually more distorted until it begins to break up into irregular flow. This does not seem to be readily consistent with notions of chaos in the formal sense, and its precise nature is still not fully understood (see *Read et al.* [1992] for further discussion). SV is frequently regarded as an intermediate state prior to the full onset of irregular wave flow or "geostrophic turbulence."

At the highest rotation rates, the wave number spectrum is observed to fill up to become a broadband continuum, though a limited band of wave numbers still tends to dominate the spectrum. The most detailed laboratory measurements of the transition to irregular flow with

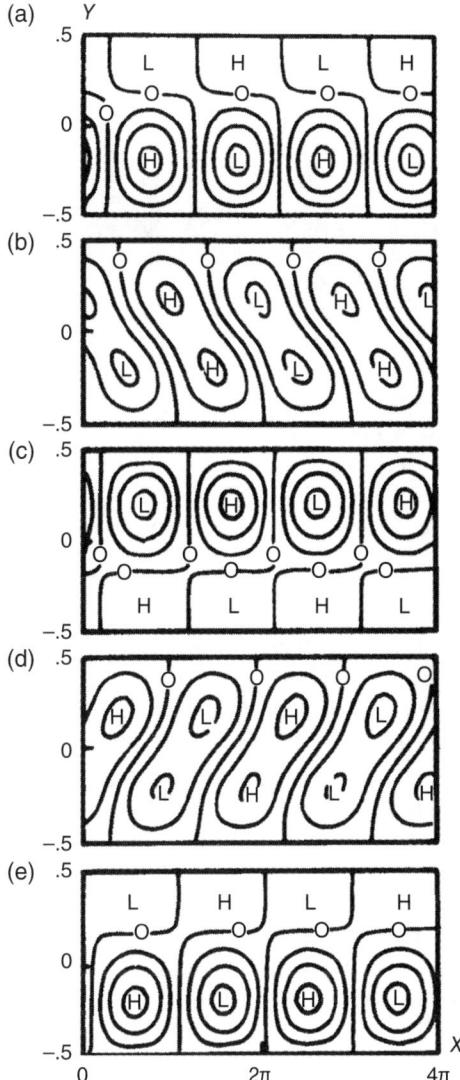

Figure 1.14. Stream function fields from a wave number 6 flow at mid-depth in the x–y plane of a zonal channel (where x is the zonal direction and y the lateral or meridional direction) during a structural vacillation cycle, as obtained in a low-order quasi-geostrophic model. Adapted from *Weng and Barcilon* [1987] with permission of John Wiley & Sons, Inc.

increasing Ω were carried out by *Buzyna et al.* [1984] [see also *Pfeffer et al.*, 1980] and *Hide et al.* [1977], in the former case using the large annulus at Florida State University. Both studies showed the gradual broadening of the wave number spectrum and increasing significance of nonharmonically related azimuthal components as the fully developed irregular regime was entered (some examples from the study by *Buzyna et al.* [1984] are shown in Figure 1.15). At extreme parameter values, the time-averaged spectrum does not display strong peaks at any particular individual wave number but appears as a broad

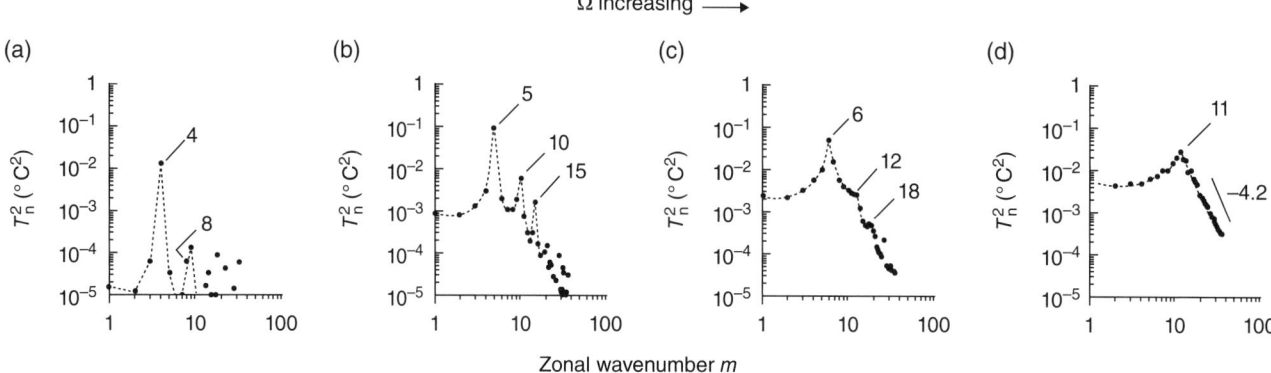

Figure 1.15. Azimuthal wave number power spectra obtained from measurements of temperature in a rotating annulus as Ω is increased through the regular wave regime toward fully developed "geostrophic turbulence". Experiments were carried out at the following points in $(\Theta, \mathcal{T})$ parameter space: (a) 1.44, 1.30×10^8, (b) 0.344, 5.40×10^8, 0.230, 8.07×10^8, and (d) 0.086, 2.16×10^9. Adapted from *Buzyna et al.* [1984] with permission.

continuum with maximum power over a range of relatively low wave numbers dominated by the main baroclinic wave activity with a characteristic quasi-power law decay toward the highest wave numbers that commonly approached $k^{-3} - k^{-4}$ [*Hide et al.*, 1977; *Pfeffer et al.*, 1980; *Buzyna et al.*, 1984] and even steeper in some cases [*Pfeffer et al.*, 1980; *Buzyna et al.*, 1984]. Simple heuristic arguments based on the Kolmogorov-Kraichnan theory [*Kraichnan*, 1967, 1971] predicts energies to decay as k^{-3} at high wave numbers in an enstrophy-dominated inertial range, but this is not the only possible explanation. The (possibly transient) formation of sharp fronts and vorticity filaments within geostrophically turbulent flows may lead to a kinetic energy spectrum with a slope as steep as k^{-4} [e.g., *Saffman*, 1971; *Brachet et al.*, 1988]. The formation of persistent, stable vortex structures within geostrophically turbulent flows can also perturb the simple Kolmogorov-Kraichnan scaling arguments by introducing some spatiotemporal intermittency to the flow. This effect can also apparently lead to spectral slopes of $k^{-3} - k^{-4}$ or even steeper [e.g., *Basdevant et al.*, 1981; *McWilliams*, 1984], although the presence of imposed vorticity gradients (e.g., due to a β effect or topography) may weaken such eddies and the flow reverts toward a k^{-3} energy spectrum.

A further question regarding these highly turbulent flows with strong background rotation is whether they also exhibit any evidence for an inverse energy cascading inertial range. Early theoretical work and models [e.g., *Salmon*, 1978; *Rhines*, 1979; *Salmon*, 1980] suggested that such inverse cascades would be relatively common in stratified, quasi-geostrophic flows, with energy being injected into the barotropic mode at around the internal Rossby deformation radius via baroclinic self-interactions, leading to a $K_3^{-5/3}$ inertial range (where K_3 is the three-dimensional total wave number) if a sufficient scale separation exists between energy injection and large-scale dissipation and forcing. The existence or otherwise of an inverse energy cascade in Earth's atmosphere or oceans continues to be an area of active controversy, with recent work suggesting that energy cascades upscale from around the first internal baroclinic deformation radius toward larger scales in the oceans [*Scott and Wang*, 2005; *Scott and Arbic*, 2007], though without a $k^{-5/3}$ inertial range apparent (at least in the sea surface height signature in satellite altimetric measurements). The development of coherent structures at large scales in the atmosphere, however, has been suggested to lead to a suppression of spectrally local nonlinear interactions and consequently the suppression of any significant inverse energy cascade at scales larger than the main energy-containing baroclinic scales in the atmosphere [e.g. *Schneider and Walker*, 2006; *O'Gorman and Schneider*, 2007].

Such issues were not considered in earlier experimental studies [*Hide et al.*, 1977; *Pfeffer et al.*, 1980; *Buzyna et al.*, 1984; *Bastin and Read*, 1997, 1998], which only presented and discussed basic temperature variance spectra and synoptic structures from their measurements. As is evident from Figure 1.15, such spectra show little obvious evidence of the $K_3^{-5/3}$ inertial range that might indicate an energy-dominated cascade at low wave numbers. *Wordsworth et al.* [2008] did explore aspects of spectral energy transfers in a set of rotating annulus experiments, both with and without sloping horizontal boundaries, even though such experiments also showed no obvious sign of the classical $K_3^{-5/3}$ inertial range. Nevertheless, these experiments did show evidence for a weak, spectrally local (eddy-eddy) upscale kinetic energy cascade in the presence of sloping boundaries and a much stronger direct (spectrally nonlocal) upscale energy transfer into the zonally symmetric component of the flow, in extreme cases deforming the large-scale baroclinic zonal flow into a pattern of multiple parallel jets and baroclinic zones. Such patterns had been anticipated in several earlier studies [*Mason*, 1975; *Bastin and Read*, 1997, 1998], which also noted the tendency to form flows with complex radial structures, but these did not examine the energy exchanges in such detail.

However, this whole subject area relating to energy exchanges in geostrophically turbulent or chaotic flows is substantially underexplored and deserves further intensive study in future research. The respective roles of baroclinic and barotropic energy exchanges (both kinetic and potential) remain uncertain in the general case, and how such roles scale and vary with key control parameters is virtually unknown. The study by *Wordsworth et al.* [2008] only considered quasi-barotropic energy exchanges in detail in the cases they were able to measure, though they attempted to infer some aspects of the baroclinic-barotropic exchanges indirectly. But a more detailed study utilizing combinations of velocity and temperature measurements with good spatial coverage and high spatial resolution are really needed to elucidate some of the most challenging outstanding questions, that are currently major issues for understanding the global circulation of both the atmosphere and oceans [cf *Schneider and Walker*, 2006; *Zurita-Gotor and Lindzen*, 2007; *Jansen and Ferrari*, 2012].

1.4. HEAT TRANSPORT AND ROLE OF BAROCLINIC WAVES

It is often asserted [e.g., *Hide and Mason*, 1975] that baroclinic annulus waves develop from a baroclinically unstable background state and serve primarily to transfer heat upward and horizontally from hot regions toward cooler ones. But heat transfer within the baroclinic rotating annulus is actually significantly more complicated than this would suggest.

If the annulus were filled with a conducting solid instead of a fluid, heat would only be transferred by molecular conduction. Hence, with isothermal vertical boundaries the equilibrated isotherms would lie on cylindrical surfaces coaxial with the axis of symmetry. Such a configuration in a fluid would be very unstable, however, since large amounts of potential energy would be stored compared with the lowest energy state that could be obtained by an adiabatic rearrangement of the fluid to place the denser fluid as low as possible (and vice versa). Where the boundary temperatures are actively maintained, however,

the complete adiabatic rearrangement of the fluid to its lowest potential energy state is not sustainable but results in a residual buoyancy-driven overturning circulation, mainly confined to thin boundary layers adjacent to the vertical boundaries (provided the Rayleigh number is large enough; see below). Figure 1.16 shows an example from a numerical simulation of the equilibrated flow in a nonrotating fluid annulus. This clearly shows the temperature field assuming a bottom-heavy configuration away from the vertical boundaries but with complex thermal structure adjacent to the sidewalls as the temperature adjusts to the isothermal boundary conditions within complex sidewall boundary layers. Some of this structure is associated with the strong overturning circulation within the sidewall boundary layers, where conductive heating/cooling is balanced by (mainly vertical) advection of heat, with a return flow in the (quasi-inviscid) interior. The interior flow redistributes hot and cold fluid until it reaches a state of minimum potential energy, i.e., with as much as possible of the cold dense fluid at the bottom of the cavity. Hence, isotherms in the interior become virtually horizontal and stably stratified in the vertical, adjusting to become vertical in thin conduction-dominated boundary layers adjacent to the sidewalls. In this state, advective heat transport (by laminar flow) is

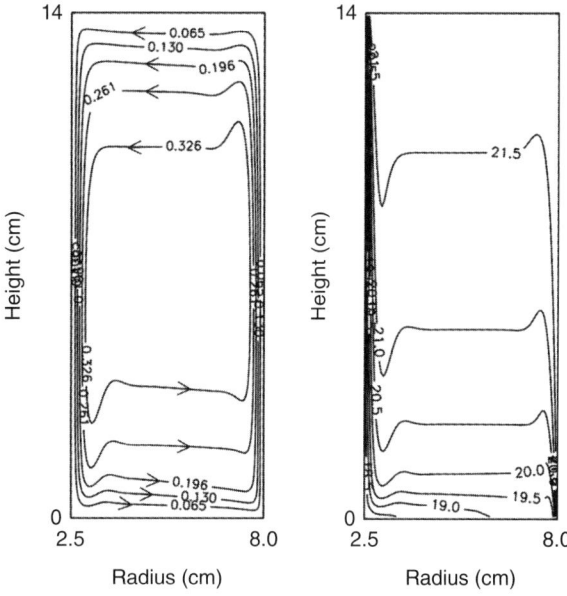

Figure 1.16. Numerical simulation of the equilibrated axisymmetric flow in a rotating annulus experiment with sidewalls maintained at different temperatures but with $\Omega = 0$: (a) stream function for the meridional circulation in the (r, z) plane; (b) corresponding temperature field with thin thermal boundary layers adjacent to each sidewall.

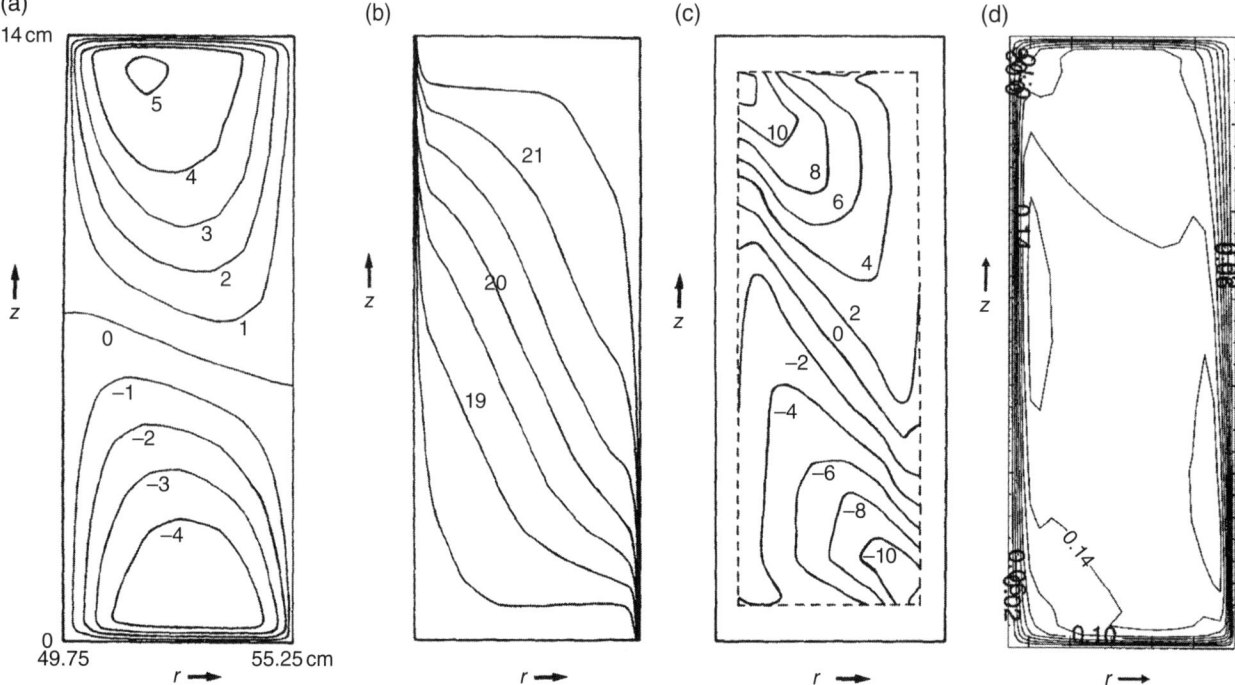

Figure 1.17. Cross sections in the (r, z) plane of (a) azimuthal velocity (mm/s), (b) temperature (°C), (c) radial gradient of QG potential vorticity (s^{-1}) (computed by *Bell and White* [1988]), and (d) meridional stream function (cm^2/s^1) in the axisymmetric regime of a rotating annulus subject to differential heating at the sidewalls. (Adapted from *Read* [2003] with permission.)

apparently as efficient as possible as measured by the nondimensional Nusselt ($\mathcal{N}$) or Péclet (Pe) numbers

$$\mathcal{N} = \frac{\text{Total heat transport}}{\text{Heat transport by conduction alone}}, \quad (1.23)$$

$$\text{Pe} = \frac{\text{Advective heat transport}}{\text{Heat transport by conduction alone}}. \quad (1.24)$$

When the system is rotated, Coriolis accelerations begin to influence the circulation, deflecting horizontal radial motion into the azimuthal direction. The axisymmetric flow at low values of Ω is therefore similar to the nonrotating case, except (a) an azimuthal component of flow is induced, producing jets antisymmetric about middepth (for identical upper and lower boundary conditions), prograde at the top (where radial flow is inward) and retrograde below (where radial flow is outward), and (b) radial flow becomes largely confined to Ekman layers adjacent to the horizontal boundaries (for Ω sufficiently large). When Coriolis accelerations dominate the interior flow, any $O(1)$ radial flow has to be geostrophic, requiring an azimuthal pressure gradient. Such a gradient cannot occur in an axisymmetric circulation unless a rigid meridional barrier is present, so radial flow strong enough to carry significant amounts of heat energy across the annulus becomes largely confined to the Ekman layers. This is clearly seen in the temperature and stream function fields of the rotating, axisymmetric flow illustrated in Figure 1.17a and 1.17d. Meanwhile, the azimuthal flow assumes a form where geostrophic balance applies except in the Ekman layers, within which viscous accelerations dominate.

Since the radial flow becomes confined to Ekman layers when $\Omega \neq 0$, the efficiency of advective heat transport is governed primarily by the mass transport, which can be accommodated within an Ekman layer. For a given geostrophic zonal flow in the interior, the advective Ekman transport is proportional to the depth of the Ekman layer (i.e., proportional to $V\mathcal{E}^{1/2}$, where $\mathcal{E}$ is the Ekman number and V the azimuthal velocity scale; see below). Since $\mathcal{E}$ is inversely proportional to Ω, the efficiency of advective heat transport must decrease with Ω given a constant ΔT, V is proportional to Ω^{-1}, and Ekman transport is proportional to $\Omega^{-1}\mathcal{E}^{1/2} \propto \Omega^{-3/2}$. Thus, as Ω increases, advective heat transport must decrease. Such a decrease is clearly apparent in Figure 1.18, which shows a compilation of both laboratory measurements of Nusselt number and from numerical simulations [*Read*, 2003; *Pérez*, 2006]. The variation of Nusselt number $\mathcal{N}$ in a pure axisymmetric flow in which baroclinic waves are suppressed is indicated by diamonds, which broadly confirm the decay as $\Omega^{-3/2}$, though note that this scaling applies more precisely to the Péclet number Pe ($=\mathcal{N}-1$; see equation (1.24)), which decays more rapidly with Ω toward zero than $\mathcal{N}$ (which decays toward $\mathcal{N}=1$). Since conductive heat transport is always present and is unaffected by rotation, the resultant thermal field becomes increasingly dominated by conduction. Hence, the isotherm structure will tend toward the vertical alignment characteristic of the conductive state as Ω increases. It is reasonable, therefore, for waves to develop in such a flow provided they are able to transfer heat radially in the interior, which is possible in a geostrophically balanced rotating fluid since waves are associated with a periodically varying azimuthal pressure gradient to balance a radial geostrophic flow.

Laboratory experiments show that such wavelike disturbances with these properties will frequently develop under many circumstances, with the primary role of transferring heat and releasing "available" potential energy. When such waves are present, experiments have consistently demonstrated that they increase the effectiveness of advective heat transfer quite significantly. The experimental measurements and simulations represented in Figure 1.18 show this effect very clearly, more or less maintaining the total Nusselt number of the whole system at a value close to its nonrotating value. Such an effect has been noted since the early work of *Bowden* [1961] and *Bowden and Eden* [1965], though a fully quantitative, theoretical understanding of this has remained elusive.

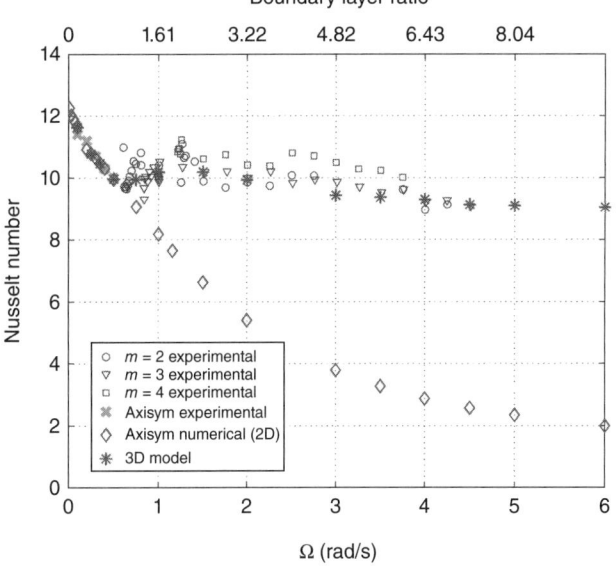

Figure 1.18. Experimental and numerically simulated total heat transport (Nusselt number) in a rotating annulus experiment using data from *Read* [2003] and *Pérez* [2006]. Both 2D and 3D models were used for the numerical simulations. (See key inside the figure for a description of each measurement.)

1.4.1. Eddy Heat Transfer in Oceans and Atmospheres

Note that the above arguments will apply in a qualitative sense to a planetary atmosphere save that the state to which the thermal structure relaxes in the absence of (large-scale) fluid motion is one not of conductive equilibrium but of radiative (or radiative-convective) equilibrium [e.g., *Pierrehumbert*, 2010]. This would suggest that parameters measuring the efficiency of atmospheric heat transport, such as the Nusselt number $\mathcal{N}$, need to be redefined, for example, with respect to a radiative-convective equilibrium state rather than a pure conductive one. In the case of Earth, pure radiative-convective equilibrium would result in a temperature contrast between equator and poles of $\sim$150 K, instead of $\sim$60 K observed on average [e.g., *Andrews et al.*, 1987], indicating the dominant role of dynamical advective heat transport in the atmosphere. Similar considerations may apply in the oceans, for which parameterization of baroclinic eddy transports are of particular importance. This is because the scales of baroclinic instability are so much smaller than the domain scale of an ocean basin that they are extremely difficult and/or expensive to resolve adequately in ocean circulation numerical models.

Various approaches toward the parameterization of eddy transports have been developed over many years for this purpose [e.g., *Plumb and Mahlman*, 1987; *Gent and McWilliams*, 1990; *Gent et al.*, 1995; *Treguier et al.*, 1997; *Killworth*, 1997; *Visbeck et al.*, 1997; *Marshall and Adcroft*, 2010]. A major advance in the development of such parameterizations that forms the basis of many contemporary schemes in current use in ocean models derives from the work of *Gent and McWilliams* [1990] and *Gent et al.* [1995]. Here an "eddy-induced" or "bolus" velocity is introduced that advects temperature and other tracers in such a manner as to flatten density surfaces. The original scheme proposed by *Gent and McWilliams* [1990] derived such an eddy-induced velocity from a parameterization of buoyancy fluxes that were assumed to act diffusively down-gradient with respect to the ambient (e.g., zonal mean) buoyancy field in the form (suitable for a zonally reentrant domain):

$$u^* = -\frac{\partial}{\partial z}\left(\frac{\overline{u'\rho'}}{\partial\overline{\rho}/\partial z}\right) \quad (1.25)$$

$$= -\frac{\partial \chi^*_{GM}}{\partial z}, \quad (1.26)$$

$$w^* = \frac{\partial}{\partial y}\left(\frac{\overline{u'\rho'}}{\partial\overline{\rho}/\partial z}\right) \quad (1.27)$$

$$= \frac{\partial \chi^*_{GM}}{\partial y}. \quad (1.28)$$

Here the term in parentheses on the right-hand sides of each of equations (1.25) and (1.27) is effectively the stream function χ_{GM}^* for the eddy-induced azimuthal mean flow (u^*, w^*) in the meridional (r, z) plane. *Gent and McWilliams* [1990] proposed that the buoyancy flux in this definition of χ^* be parameterized as a down-gradient diffusion of zonallyaveraged buoyancy

$$\chi_{GM}^* = \left(\frac{\overline{u'\rho'}}{\partial \overline{\rho}/\partial z}\right) = -\left(\mathcal{K}\frac{\partial \overline{\rho}/\partial y}{\partial \overline{\rho}/\partial z}\right), \quad (1.29)$$

where $\mathcal{K}$ is a suitably defined eddy diffusion coefficient that needs to be completed with a suitable closure model. The latter is commonly assumed to take the general form

$$\mathcal{K} = \alpha L_{eddy} U_{eddy}, \quad (1.30)$$

where L_{eddy} and U_{eddy} are characteristic scales for eddy length and velocity scales and α is a dimensionless constant, found empirically to require a value $O(10^{-2})$ [e.g., *Visbeck et al.*, 1997; *Marshall and Adcroft*, 2010].

There remains significant uncertainty, however, as to the physical basis for choosing L_{eddy} and U_{eddy}. L_{eddy} represents a prescribed "mixing length", suggestions for which have included either the so-called-Rhines scale $L_{Rhines} = (U_{rms}/\beta)^{1/2}$ [*Larichev and Held*, 1995; *Treguier et al.*, 1997], the width of the baroclinic zone [*Green*, 1970; *Visbeck et al.*, 1997], or the Rossby deformation radius [*Stone*, 1972]. U_{eddy} has been taken variously as either a "typical" thermal wind scale related to the zonal mean horizontal thermal gradient [*Green*, 1970] or setting $U_{eddy} \sim L_{eddy}/\tau_{eddy}$, where τ_{eddy} is a "typical" eddy overturning time scale that might be derived, for example, from linear baroclinic instability theory [*Stone*, 1972; *Haine and Marshall*, 1998] or weakly nonlinear theory [*Pfeffer and Barcilon*, 1978; *Read*, 2003]. In addition, problems may arise if a parameterization scheme fails to respect key conservation principles, especially energy and potential vorticity [e.g., *Marshall and Adcroft*, 2010]. The parameterization of the eddy-driven "bolus" velocity from zonal mean fields is another controversial issue since it is not always clear that the horizontal eddy buoyancy flux necessarily acts diffusively down gradient with respect to the zonal mean buoyancy field [e.g., *Treguier et al.*, 1997; *Marshall and Adcroft*, 2010]. *Treguier et al.* [1997] and *Killworth* [1997] suggested an alternative approach based on assuming that potential vorticity is more generally diffused down gradient than pure buoyancy such that the eddy-induced velocity (u^*, w^*) is defined as

$$u_{THL}^* = -\frac{\partial}{\partial z}\left(\frac{\overline{u'\rho'}}{\partial \overline{\rho}/\partial z}\right) \quad (1.31)$$

$$\simeq -\frac{\overline{u'q'}}{f} \simeq \frac{\mathcal{K}(y,z)\partial \overline{q}/\partial r}{f} \quad (1.32)$$

$$= -\frac{\partial \chi_{THL}^*}{\partial z}, \quad (1.33)$$

$$w_{THL}^* = \frac{\partial \chi_{THL}^*}{\partial y}, \quad (1.34)$$

where $\mathcal{K}_q(y, z)$ is again a suitably defined eddy diffusion coefficient, this time for potential vorticity q, and is assumed here to be variable in space.

A complete understanding of all of these issues therefore remains elusive, and there remains a continuing problem of how to verify any scheme of parameterization with the desired degree of rigor. In this respect, laboratory experiments such as the rotating, thermally driven annulus ought to have something important to contribute. Experimental techniques have been available for some time to measure both the total heat transport across the annular cavity (e.g., via calorimetric methods to determine the total heat transport across a given sidewall boundary) *and* the interior eddy variances and fluxes of heat, momentum, and vorticity associated with baroclinic waves. The quantitative interpretation of these measurements, however, requires a clear understanding of the various mechanisms at work within rotating annulus circulations to transport heat energy across the annular channel. These include direct thermal conduction and direct overturning circulations (mainly in boundary layers) as well as macroturbulent transports by baroclinic eddies themselves. In subsequent sections, therefore, we examine and review the main boundary layer and eddy processes that contribute to heat transport in the annulus, culminating in some preliminary attempts to apply an analogue of ocean baroclinic eddy parameterization schemes within an axisymmetric numerical annulus model in which baroclinic instability is artificially suppressed.

1.4.2. Regimes of Axisymmetric Flow: Heat and Momentum Transport

Although the description of the axisymmetric flow in the introduction to this section gave a plausible explanation for the observed axisymmetric and wave regimes in the annulus, it is a highly simplified discussion that glosses over more subtle aspects of the problem. In this section, we take a more quantitative view of the axisymmetric flow in the annulus to put the above discussion onto a stronger theoretical footing and as an illustration of the use of scale analysis and boundary layer theory.

Early analyses [*McIntyre*, 1968; *Sugata and Yoden*, 1992] followed the scaling approach developed by *Gill* [1966], which *Hignett et al.* [1981] and *Hignett* [1982] further extended for an incompressible Boussinesq fluid in a rotating annulus of vanishingly small relative curvature $(2[b-a]/[b+a] \ll 1$, so one may use Cartesian geometry) and neglected centrifugal accelerations. It is convenient to define a meridional stream function χ such that

$$u = \frac{\partial \chi}{\partial z}, \qquad v = -\frac{\partial \chi}{\partial x}. \qquad (1.35)$$

The steady-state equations for momentum, continuity, and heat then reduce to a zonal momentum equation

$$\nu \nabla^2 v = f \frac{\partial \chi}{\partial z} + J(v, \chi), \qquad (1.36)$$

where $f = 2\Omega$ and the Jacobian is defined as

$$J(c, d) = \frac{\partial c}{\partial x}\frac{\partial d}{\partial z} - \frac{\partial c}{\partial z}\frac{\partial d}{\partial x}; \qquad (1.37)$$

the azimuthal vorticity equation is

$$\nu \nabla^4 \chi = g\alpha \frac{\partial T}{\partial x} - f \frac{\partial v}{\partial z} - J(\chi, \nabla^2 \chi), \qquad (1.38)$$

where T is the temperature and α the volumetric expansion coefficient and vorticity ζ is defined as

$$\zeta = \frac{\partial u}{\partial z} - \frac{\partial v}{\partial x} = \nabla^2 \chi; \qquad (1.39)$$

and the temperature equation is

$$\kappa \nabla^2 T + J(\chi, T) = 0. \qquad (1.40)$$

We consider a container of aspect ratio ϵ defined by

$$\epsilon = \frac{H}{L} \qquad (1.41)$$

(where H and L are the vertical and horizontal length scales, respectively, of the domain) and apply boundary conditions

$$\chi = \frac{\partial \chi}{\partial z} = v = \frac{\partial T}{\partial z} = 0, \quad z = 0, H, \qquad (1.42)$$

$$\chi = \frac{\partial \chi}{\partial x} = v = T - T_0 = 0, \quad x = -L/2, +L/2. \qquad (1.43)$$

We make use of dimensionless parameters such as the Ekman number $\mathcal{E}$ defined by

$$\mathcal{E} = \frac{\nu}{\Omega H^2}, \qquad (1.44)$$

the Prandtl number $\Pr (= \sigma/\kappa)$, and the Rayleigh number

$$\mathrm{Ra} = \frac{g\alpha \Delta T L^3}{\kappa \nu}. \qquad (1.45)$$

It is also convenient to define Nusselt ($\mathcal{N}$) and Péclet (Pe) numbers as measuring the ratios of total heat transport and advective heat transport, respectively, to that due to conduction, which we take to be

$$\mathcal{N} = \frac{\Xi}{\kappa} + 1 = \mathrm{Pe} + 1 \qquad (1.46)$$

(where Ξ is a characteristic scale for χ). We then carry out a scale analysis with the aim of deriving the dominant dynamical balances in the interior and principal boundary layers and obtain the dependence of $\mathcal{N}$ and the zonal velocity scale on external parameters over as wide a range as possible. Initial assumptions are restricted as follows:

(i) Aspect ratio ϵ is not too different from unity.
(ii) Single thickness scales are assumed, ℓ for the side and h for the horizontal boundary layers.
(iii) Outside the boundary layers there is a distinct interior flow with length scales L and H such that $(\ell, h) \ll (L, H)$.
(iv) Prandtl number $\Pr \gg 1$.

1.4.2.1. Nonrotating Problem. We assume the flow to comprise an advective interior and thin sidewall boundary layers and nondimensionalize in the thin sidewall layer of thickness ℓ using

$$\Delta x = \ell \Delta x^*, \quad \Delta z = H\Delta z^*, \quad T - T_o = \Delta T T^*,$$

$$\chi = \Xi \chi^*. \qquad (1.47)$$

Thus equation (1.40) becomes

$$\nabla^2 T^* + \frac{\Xi \ell}{\kappa H} J(\chi^*, T^*). \qquad (1.48)$$

For advective/diffusive balance, we require

$$\Xi = \frac{\kappa H}{\ell}. \qquad (1.49)$$

For this case, (1.38) becomes

$$\nabla^4 \chi^* = \mathrm{Ra}\left(\frac{\ell}{L}\right)^4 \frac{\partial T^*}{\partial x^*} - \frac{1}{\Pr}J(\chi^*, \nabla^2\chi^*). \qquad (1.50)$$

If $\Pr \gg 1$, we obtain a buoyancy/viscous balance in the sidewall boundary layer, implying that

$$\ell = \mathrm{Ra}^{-1/4}\epsilon^{1/4}L, \qquad (1.51)$$

which was the result obtained by *Read* [1992] [see also *Fein*, 1978; *Friedlander*, 1980] for the principal boundary layer scale when $N^2\Pr/f^2 \gg \mathcal{E}^{2/3}$. In this case, the Nusselt or Péclet number is obtained from equation (1.49) as

$$\mathcal{N} - 1 = \mathrm{Pe} = \frac{\Xi}{\kappa} = O\left[\mathrm{Ra}^{1/4}\epsilon^{3/4}\right]. \qquad (1.52)$$

1.4.2.2. Effects of Rotation. In considering the relative impact of rotation on the circulation, it seems intuitive that the Ekman layer will be of importance. It is therefore convenient to follow an approach due to *Hignett et al.* [1981] and recently applied to good effect in the context of convection in rotating systems by *King et al.* [2009] and *King et al.* [2012] in defining a parameter $\mathcal{P}$ measuring the (square of the) ratio of the thickness of the Ekman layer and sidewall buoyancy layer. Thus

$$\mathcal{P} = \mathrm{Ra}^{-1/2}\mathcal{E}^{-1}, \qquad (1.53)$$

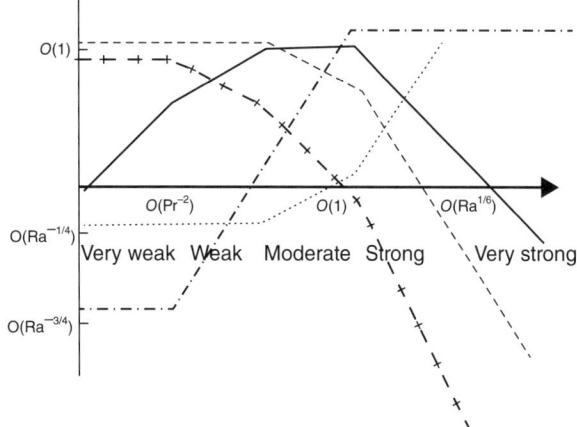

Figure 1.19. Schematic diagram showing the dependence of derived parameters on internal parameters in the various axisymmetric regimes defined in terms of $\mathcal{P}$ assuming $\epsilon = O(1)$. Quantities represented are $VL/(\kappa \mathrm{Ra}^{1/2})$ (solid line); $\Xi/(\kappa \mathrm{Ra}^{1/4})$ (dashed line); γ (dash-dotted line); ℓ/L (dotted line), and Ro (dash-crossed line). (Adapted from *Read* [1986] with permission).

(assuming hereafter for simplicity that $\epsilon = 1$) which is proportional to Ω. Based on a consideration of the full range of $\mathcal{P}$, we can effectively identify up to six distinct regimes of axisymmetric flow (see also Figure 1.19):

(i) No rotation, $\mathcal{P} = 0$.
(ii) Very weak rotation, $0 \ll \mathcal{P} \ll \mathrm{Pr}^{-2}$.
(iii) Weak rotation, $\mathrm{Pr}^{-2} \ll \mathcal{P} \ll 1$.
(iv) Moderate rotation, $\mathcal{P} \simeq 1$.
(v) Strong rotation, $1 \ll \mathcal{P} \ll \mathrm{Ra}^{1/6}$.
(vi) Very strong rotation, $\mathcal{P} \gg \mathrm{Ra}^{1/6}$.

We now briefly outline their characteristics:

(i) *No rotation*: This has already been discussed above in Section 1.4.2.1, with consequent scales for ℓ and Pe. Note that we can obtain an estimate of isotherm slope $\gamma = \Delta T_h/\Delta T$ (where ΔT_h is the horizontal temperature contrast) from a consideration of the balances in the zonal vorticity equation. Provided $\mathrm{Pr} \gg 1$, a buoyancy/viscous balance holds in the interior, so that

$$g\alpha\, \partial T/\partial x = O(g\alpha \Delta T \gamma / L) \qquad (1.54)$$
$$= \nu \nabla^4 \chi (= O(\nu \Xi L^{-4})). \qquad (1.55)$$

Hence $\gamma < \mathrm{Ra}^{-3/4} \ll 1$ and isotherms are quasi-horizontal.

(ii) *Very weak rotation*: When f is no longer zero, (1.36) is coupled to (1.38) and gyroscopic torques render v non-zero. We obtain an estimate for the zonal velocity scale V by scaling (1.38) in the Ekman layer using its characteristic depth $h = \mathcal{E}^{1/2}L$. Hence, (1.36) becomes

$$\mathrm{Pr}\mathcal{P}^{1/2}\nabla^2 v^* = \frac{fL}{V}\frac{\partial \chi^*}{\partial z^*} + J(v^*, \chi^*). \qquad (1.56)$$

Thus, for $\mathcal{P} \ll \mathrm{Pr}^{-2}$ we have an inertial/Coriolis balance in the Ekman layer (i.e., there is no proper Ekman layer), and the entire flow is characterized by local conservation of angular momentum [hence $V = O(fL)$, which is proportional to $\mathcal{P}$; see Figure 1.19].

(iii) *Weak rotation*: For $\mathcal{P} \gg \mathrm{Pr}^{-2}$, the viscous term in (1.56) becomes dominant in the Ekman layer (i.e., normal Ekman layers exist), thus rescaling V to $O(\kappa \mathrm{Ra}^{1/2}\mathcal{P}^{1/2}/L)$. This balance extends into the interior, while the previous balance in the sidewall layer is unchanged from regime (i). Despite the new scaling for V, the dominant balances (and scaling for Pe) in (1.40) also remain unchanged from (i). The rescaling of V does, however, affect the interior balance in the azimuthal vorticity equation, from a buoyancy/viscous balance to a buoyancy/Coriolis balance characteristic of the "thermal wind" balance typical of geostrophic flow. The reason why the (now geostrophic) scale for V does not go as Ω^{-1} typical of a thermal wind scale is because γ is now increasing rapidly with Ω ($\gamma = O(\mathcal{P}^{3/2})$), which more than outweighs the $\mathcal{P}^{-1}$ dependence of V for constant γ. Note also the zonal Rossby number Ro $= V/fL = O(\mathrm{Pr}^{-1}\mathcal{P}^{-1/2})$ and is therefore $\ll 1$ (see Figure 1.19).

(iv) *Moderate rotation*: In this regime, the Ekman layer thickness is comparable with that of the sidewall buoyancy layer and so is expected to begin to exert a strong influence on the meridional circulation and transport. Anticipating that V will eventually tend toward the thermal wind scale $O(\mathcal{P}^{-1})$, this range of $\mathcal{P}$ delineates the regime where V reaches a maximum $V_o = O(\kappa \mathrm{Ra}^{1/2}/L)$. If the Ekman layer exercises dominant control over the radial mass transport, Ξ will be rescaled to $O(V_o L \mathcal{E}^{1/2}) = O(\kappa \mathrm{Ra}^{1/4}\mathcal{P}^{-1/2})$, implying a slow broadening of the sidewall advective/diffusive boundary layer from ℓ to $\ell\mathcal{P}^{1/2}$.

(v) *Strong rotation*: As $\mathcal{P}$ is increased beyond 1, the Ekman layers fully dominate the meridional circulation. By this point, the isotherm slope γ has become $O(1)$ and so cannot increase any further. Then V rescales to the familiar thermal wind scale $V = O(\kappa \mathrm{Ra}^{1/2}L^{-1}\mathcal{P}^{-1})$. The expansion of the advective/diffusive sidewall layer accelerates to $\ell' = O(\mathrm{Ra}^{-1/4}L\mathcal{P}^{3/2})$, extending the influence of thermal diffusion further into the interior. The heat transport Pe is rescaled to $O(\mathrm{Ra}^{1/4}\mathcal{P}^{-3/2})$, though it remains $\gg 1$ (see Figure 1.19).

(vi) *Very strong rotation*: In this final regime, the diffusive thermal sidewall layer expands to fill the interior and no separate thermal boundary layer and interior can be distinguished (though Stewartson $\mathcal{E}^{1/3}$ layers may exist in this limit). The critical value for $\mathcal{P}$ distinguishing regimes (v) and (vi) simply arises from equating ℓ' [see (v) above] with L so that $\mathcal{P} > \mathrm{Ra}^{1/6}$. All other balances remain unchanged from (v), i.e., the geostrophic interior and strong Ekman layers. Heat transport in this regime, however, is dominated by thermal conduction so Pe → 0 and $\mathcal{N} \to 1$.

1.4.2.3. Experimental Verification. The axisymmetric regimes discussed in Section 1.4.2.2 are capable (at least in principle) of existing in real systems given an experimental system operating in an appropriate parameter range. In practice, however, regimes (iv) – (vi) are not usually obtainable because of the development of nonaxisymmetric baroclinic waves within regime (iv) and beyond. This is consistent with the notion that baroclinic waves develop when Ekman layers begin to inhibit meridional heat transport.

An exception was provided by *Hignett* [1982], who made heat transport measurements in a rotating annulus with parallel sloping upper and lower endwalls that sloped strongly in the same sense as the isotherms. Such a configuration tends to inhibit the development of baroclinic waves (by constraining fluid trajectories away from the "wedge of instability"; [*Hide and Mason*, 1975; *Mason*, 1975]). As a result, Hignett was able to show the effect of almost the full range of behavior from zero to very strong rotation on the total heat transport by axisymmetric flow in a rotating annulus. His results are shown in Figure 1.20. The dependence of $\mathcal{N}$ and V on $\mathcal{P}$ in a rotating annulus subject to internal heating was also investigated by *Read* [1986], that also confirmed the above analysis provided the definition of $\mathcal{N}$ was modified appropriately to measure heat transport efficiency in terms of the temperature contrast obtained with a given heat flux; the results are shown in Figure 1.21. These clearly show the linear scaling of V with $\mathcal{P}$ in the weak rotation regime with a transition toward $\mathcal{P}^{1/2}$ as the moderate rotation regime is entered while the Péclet number also begins to reduce toward a $\mathcal{P}^{-1/2}$ dependence in the moderate rotation regime.

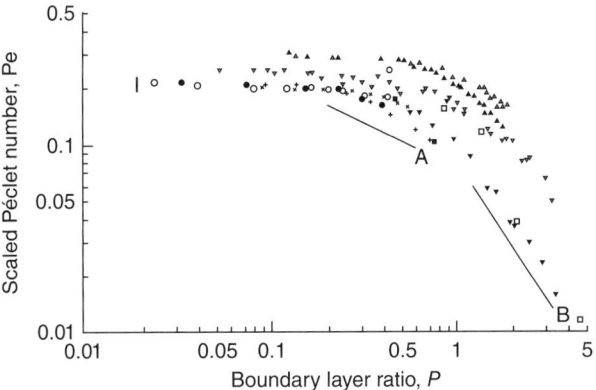

Figure 1.20. Scaled measurements of total heat transport in the axisymmetric regime of a rotating annulus as a function of $\mathcal{P}$. Adapted from *Hignett* [1982] by permission of Taylor & Francis Ltd., http://www.tandf.co.uk/journals.

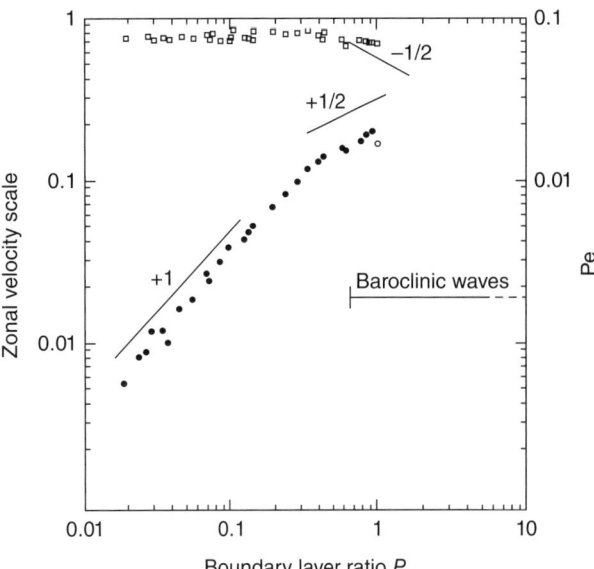

Figure 1.21. Schematic dependence of zonal (azimuthal) velocity scale on $\mathcal{P}$ in a rotating annulus subject to internal heating. Adapted from *Read* [1986] with permission.

1.4.3. Quantifying Baroclinic Eddy Transport

When baroclinic waves are not suppressed, as shown in Figure 1.18, heat transport evidently remains close to its nonrotating value as a result of eddy-induced transports. *Read* [2003] suggested that the latter can be viewed as adding to and enhancing the heat transport occurring in the axisymmetric boundary layer circulation, and the strength of this eddy-induced transport can therefore be diagnosed directly from measurements or simulations of total heat transport as the difference in Nusselt number between that of the fully three-dimensional flow and $\mathcal{N}$ obtained in a purely axisymmetric flow under the same experimental conditions. *Read* [2003] diagnosed this from a combination of numerical simulations of axisymmetric flows and experimental measurements. The results are shown in Figure 1.22, (a) as a function of both Ω and boundary layer ratio $\mathcal{P}$ and (b) as a function of the "supercriticality" of the flow defined with respect to a supercritical rotational Froude number $\mathcal{F}_s$, defined as

$$\mathcal{F}_s = \mathcal{F} - \mathcal{F}_{0m} = \frac{1}{\Theta} - \frac{1}{\Theta_{0m}}. \quad (1.57)$$

Here $\mathcal{F}$ is defined as $\mathcal{F} = 1/\Theta$ and $\Theta_{0m} = 1/\mathcal{F}_{0m}$ represents the critical values of Θ and $\mathcal{F}$ for the onset of baroclinic instability of azimuthal wave number m.

In this figure, the difference in Nusselt number represents an additional or "excess" Péclet number Pe_{xs} due to the presence of baroclinic waves. The effectiveness of baroclinic eddy heat transport grows rapidly with Ω from the first onset of baroclinic instability, rising to a value

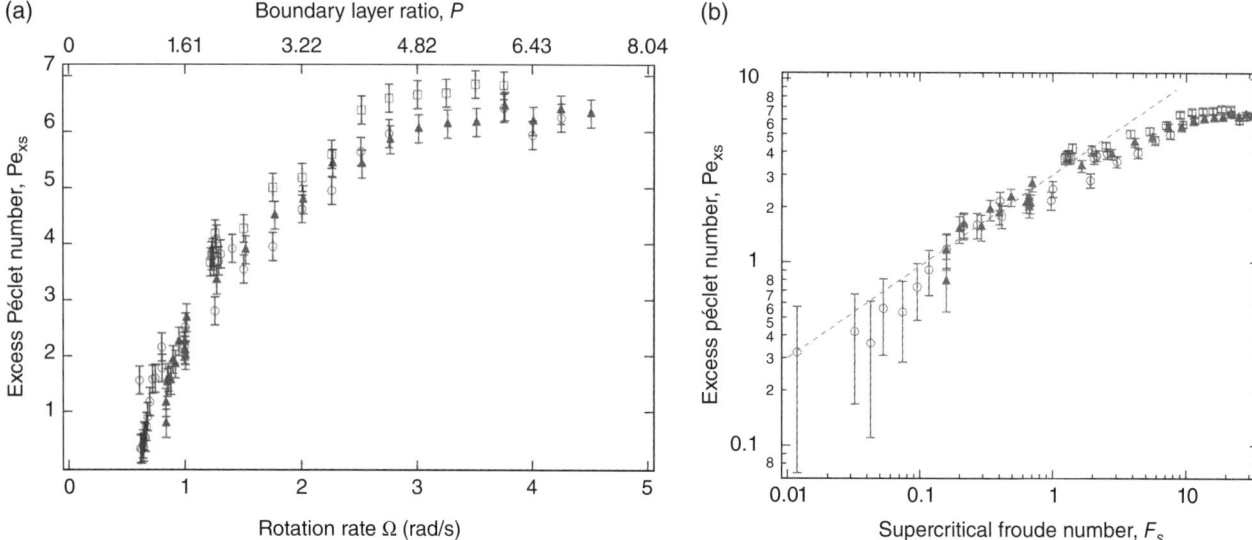

Figure 1.22. Experimental measurements of dimensionless integrated heat transport (excess Péclet number) attributable to baroclinic eddies in a rotating annulus experiment derived from data of *Read* [2003] and presented (a) as a function of Ω and $\mathcal{P}$ and (b) as a function of "supercritical" Froude number $\mathcal{F}_s$ (see text). Pe_{xs} may be compared with the nonrotating total Péclet number of around 10. Open circles are for $m = 2$ flows, filled triangles for $m = 3$, and open squares for $m = 4$ dominated flows. For color detail, please see color plate section.

(at least in the experiments discussed by *Read* [2003]) of around 6–7 at around $\Omega = 3$ rad/s, after which Pe_{xs} levels off and may even start to decrease at the highest values of Ω. This would appear to indicate that eddy-induced heat transport tends to saturate at a finite level once the instability becomes fully supercritical. Such an effect is also evident in Figure 1.22b, which shows the same data plotted against a measure of baroclinic supercriticality in terms of $\mathcal{F}_s$. This clearly indicates an initial dependence of Pe_{xs} close to $\mathcal{F}_s^{1/2}$ for $\mathcal{F}_s \lesssim O(1)$ (indicated by the dashed line in Figure 1.22b) but with a weaker dependence as $\mathcal{F}_s$ increases further. However, measurements penetrating further into the supercritical (irregular) regime would be desirable to confirm this saturation effect. The numerical simulations of *Pérez* [2006]; and *Pérez et al.* [2010] would seem to suggest that Pe_{xs} may continue to increase weakly as $\mathcal{P}$ is increased toward 10 (so the axisymmetric share of heat transport becomes relatively small), but these high Ω simulations probably had insufficient resolution to handle the increasingly turbulent wave flows at these high rotation rates, so the values of $\mathcal{N}$ may not represent accurately the heat transport of the real fluid system.

1.4.4. Testing Local Closures for Baroclinic Eddy Transport

Direct numerical simulations (DNSs), as discussed by *Pérez et al.* [2010], enable various diagnostics of complex baroclinic wave flows to be obtained that can be used to investigate quantitatively the validity of some of the assumptions underlying various approaches to eddy parameterization. *Read* [2003], *Pérez* [2006] and *Pérez et al.* [2010] have investigated various aspects of the diffusive approach to parameterizing heat transport in the context of the rotating annulus by making use of diagnostics of DNS model simulations of baroclinic annulus wave flows over a wide range of conditions with reference to the experimental measurements reported by *Read* [2003]. As mentioned above in the first part of this section, the notion that radial eddy heat flux should act down gradient with respect to its zonal mean field is implicit in various proposed parameterizations of eddy transport in the oceans and atmosphere, although this is notoriously difficult to deduce directly from theory or to verify in observations. However, many of these formulations are derived from simplified analyses based on quasi-geostrophic theory, so they should apply equally well to both geophysical situations and in the laboratory, at least within a limited range of parameters where quasi-geostrophy is reasonably valid.

Despite the well-established applicability of quasi-geostrophy to laboratory systems such as the rotating annulus [e.g., *Williams et al.*, 2010], however, there has been surprisingly little work done to investigate the parameterization of heat transfer in stably stratified flow in the laboratory and thereby to test the kinds of schemes proposed for use in models of oceans and atmospheres. In particular, *Pérez et al.* [2010] examined the extent to which baroclinic eddy fluxes of heat or vorticity (potential or

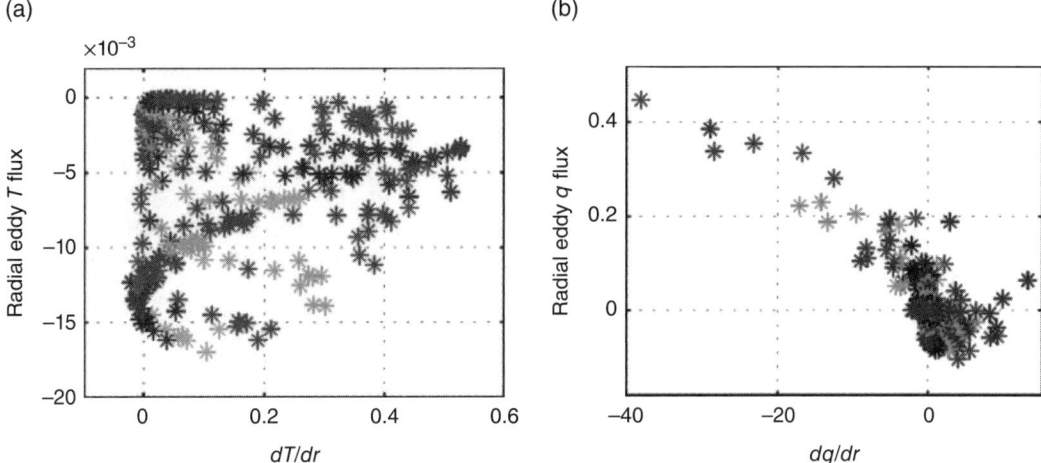

Figure 1.23. Scatter plots of eddy fluxes of heat and potential vorticity against their respective zonal mean gradient fields in numerical simulations by *Pérez* [2006]; and *Pérez et al.* [2010] of fully three-dimensional, time-dependent baroclinic waves flows under moderately supercritical conditions ($\Theta = 0.15$, $\mathcal{T} = 1.30 \times 10^7$). Plots were obtained by plotting pointwise values of fluxes and the respective radial gradient of the zonally averaged quantity across the whole meridional plane (outside boundary layers). (a) Correlation of meridional heat flux against zonal mean temperature gradient. (b) Corresponding correlation for eddy fluxes and zonal mean variations of QG potential vorticity. Adapted from *Pérez et al.* [2010]. Copyright 2010 with permission from Elsevier.

relative) act directly down-gradient with respect to various zonal mean fields in the simplest (right-cylindrical, axisymmetric annular channel with flat horizontal boundaries) configuration of the annulus experiment.

They found that, contrary to the commonly held assumption in many approaches that follow *Gent and McWilliams* [1990] (hereafter referred to as GM90), the horizontal eddy heat flux is only poorly correlated with the lateral gradient of zonal mean temperature. Figure 1.23a shows an example from Pérez et al.'s simulations in which the local eddy heat flux $\overline{(u'T')}$ is plotted against $\partial \overline{T}/\partial r$ in an equilibrated baroclinic wave flow under moderately super-critical conditions across the whole (r, z) plane of the annulus. Although some structure is evident, the dependence of $\overline{(u'T')}$ on $\partial \overline{T}/\partial r$ is clearly a lot more complicated than a simple, Fickian diffusive relationship would suggest. This appeared to be typical of most fully developed baroclinic wave simulations investigated by *Pérez et al.* [2010], with |correlation coefficients| $\lesssim 0.2$ in most cases except either under marginally unstable conditions or transiently during the initial growth of the instability, when correlation coefficients as large as -0.7 were found [*Pérez et al.*, 2010].

In contrast, fluxes of (potential or relative) vorticity were found to act quite closely down-gradient in most cases investigated. An example is shown in Figure 1.23b for quasi-geostrophic potential vorticity, plotted in the same way (and for the same case) as Figure 1.23a over the whole annular domain. In this case a strong anticorrelation is clearly evident, indicating that quasi-geostrophic potential vorticity is diffused horizontally by baroclinic eddies with respect to its zonal mean field to quite a good approximation. This behavior was found to be quite generic for almost all cases investigated, with correlation coefficients between $\overline{u'q'}$ and $\partial \overline{q}/\partial r$ ranging from -0.75 to -0.9 for both equilibrated and transient growing wave flows [*Pérez et al.*, 2010]. Similar behavior was also found for relative vorticity, in fact with even larger (negative) correlation coefficients than for potential vorticity.

Given such a clear correlation between eddy fluxes and mean gradients, *Pérez et al.* [2010] were able to deduce an effective eddy diffusivity $\mathcal{K}_q$ from a simple regression of $\overline{(u'q')}$ against $\partial \overline{q}/\partial r$ in their model simulations. A straightforward linear regression led to the remarkable result that $\mathcal{K}_q$ varied by less than a factor of 2 across the whole range of parameters investigated. Figure 1.24 shows the variation of the value of $\mathcal{K}_q$ obtained by *Pérez et al.* [2010] as a function of boundary layer ratio $\mathcal{P}$ (also cf Figure 1.18), indicating that, at least for these experiments, $\mathcal{K}_q$ was found to vary slowly between 1–2×10^{-2} cm^2/s. The largest values of $\mathcal{K}_q$ seemed to occur close to conditions of marginal instability, with $\mathcal{K}_q$ gradually reducing toward a roughly constant value $\sim 10^{-2}$ cm^2/s for all $\mathcal{P} \gtrsim 5$.

Given these results, *Pérez et al.* [2010] further tried to determine whether one or more previously proposed closures for $\mathcal{K}$ were sufficient to represent the variations found in their simulations based on an assumed form akin

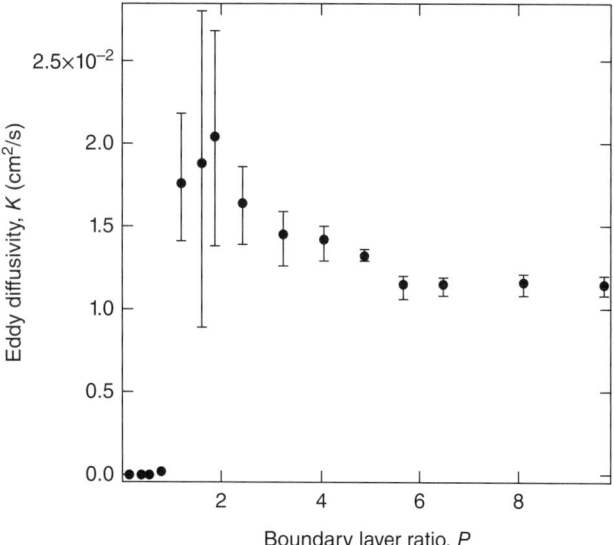

Figure 1.24. Dependence of eddy diffusivity for quasi-geostrophic potential vorticity on boundary layer ratio $\mathcal{P}$ (proportional to Ω) derived using data from correlating $\overline{(u'q')}$ against $\partial \overline{q}/\partial r$ in the rotating annulus model simulations of *Pérez et al.* [2010].

to equation (1.30) with L represented by the width of the baroclinic zone (i.e., the annulus gap width) and U_{eddy} given by a thermal wind scale

$$U_{\text{eddy}} \simeq \frac{LM^2}{N}, \quad (1.58)$$

where N is the Brunt-Väisälä frequency,

$$M^2 = \frac{g}{\overline{\rho}} \frac{\partial \overline{\rho}}{\partial r}, \quad (1.59)$$

such that $U_{\text{eddy}} \simeq 2\Omega L/\sqrt{\text{Ri}} \sim L/\tau_{\text{Eady}}$, τ_{Eady} is the linear growth time scale for the Eady model of baroclinic instability, and Ri is the characteristic shear Richardson number $\text{Ri} = N^2/(\partial u/\partial z)^2$ for the flow. In the event, no single closure seemed to apply across the whole parameter range. Such a result is not unduly surprising, since existing closures generally make assumptions based on either linear instability theory (which one might expect to hold close to marginal instability) or weakly nonlinear theory [e.g., *Pfeffer and Barcilon*, 1978]. Their results led to the conclusion that the observed variation of $\mathcal{K}_q$ with rotation was broadly consistent with weakly nonlinear theory close to conditions of marginal instability, with efficiency parameter α increasing roughly linearly with $\Theta - \Theta_c$ as suggested by *Pfeffer and Barcilon* [1978]. Under more strongly supercritical conditions, α appeared to converge to a roughly constant value that was consistent with the value obtained for example, by *Visbeck et al.* [1997], even to the extent of close quantitative agreement (0.013, cf. Visbeck et al.'s value of 0.015).

Thus, for strongly supercritical baroclinic flows, the annulus results obtained so far would seem to support a parameterization approach based on the potential vorticity diffusion hypothesis proposed by *Treguier et al.* [1997] (hereafter referred to as THL97) and *Killworth* [1997], with a closure for eddy diffusivity that is consistent with *Visbeck et al.* [1997]. Closer to marginal instability, however, a different closure would seem to be preferred that results in an increasing efficiency parameter α though not entirely following the simple, weakly nonlinear recipe of *Pfeffer and Barcilon* [1978]. Therefore, there would still seem, to be a number of unresolved issues underlying this somewhat unexpected behavior close to marginal instability. In addition, the approach of *Visbeck et al.* [1997] is overtly based on an application of linear instability theory in a regime that is far from where linear theory should be valid. The theoretical basis for this clearly deserves more attention in future work.

1.4.5. Implementing Eddy Parameterizations in an Annulus Model

The diagnostic approach discussed above using DNS is useful for investigating some of the underlying assumptions behind various approaches proposed for eddy transport parameterizations, especially those relating to the family of parameterizations following *Gent and McWilliams* [1990]. But in some respects the ultimate test of any given approach to this problem is actually to implement the parameterization in a numerical model. Although this has been common practice in generations of ocean circulation models for many years [e.g., *Danabasoglu et al.*, 1994], this is a relatively novel approach in the context of rotating annulus experiments and model simulations. Parameterizing turbulent transfers in rotating flows is, of course, of major importance for many engineering problems, e.g., in turbomachinery, where it has been customary for many years to employ large eddy simulation (LES) methods coupled with turbulence models based on rotational modifications to the classical Reynolds Averaged Navier-Stokes (RANS) model [e.g., see *Cazalbou et al.*, 2005, and references therein] to represent the effects of shear instabilities in the presence of background rotation. But the transformed Eulerian mean approach underlying the Gent-McWilliams family of parameterizations does not yet seem to have been adopted within the engineering community to represent unresolved eddy transports in stably stratified turbulence in rotating cavities.

Recently, however, *Pérez* [2006] has taken the first preliminary steps toward investigating this approach by implementing two forms of Gent-McWilliams parameterization in his 2D (axisymmetric), Boussinesq Navier-Stokes model of thermally driven flow in a fluid annulus rotating at angular velocity Ω. As with his earlier diagnostic work, Perez's model was based on the conservative, finite-difference model described in detail by *Hignett et al.* [1985], which solves the full Boussinesq Navier-Stokes equations together with continuity and temperature advection-diffusion equations in cylindrical annular geometry using an exponentially stretched mesh in r and z to ensure adequate resolution of boundary layers. Nonslip, impermeable boundary conditions were applied at each boundary of the cavity, with fixed isothermal conditions at the inner and outer sidewalls and thermally insulating conditions on the horizontal boundaries. Fairly coarse resolution was adopted (32×32 points in (r, z)) in all cases. Each parameterized simulation was initialized from an isothermal state at rest in the rotating frame and then first run to equilibrium while holding the boundary conditions fixed. The eddy parameterization was then activated by adding parameterized vertical and radial velocity components $\mathbf{u}^*$, representing the transformed Eulerian mean (TEM) velocity field [e.g., *Andrews et al.*, 1987] induced by the presence of baroclinic eddies, to the 2D axisymmetric velocity field that was used to advect momentum and temperature, and the model was then integrated to its modified equilibrium.

The $\mathbf{u}^*$ parameterization was derived from the instantaneously computed velocity and temperature fields within the model either (a) using the original Gent-McWilliams method based on equations (1.25) and (1.27) or (b) derived from the zonal mean quasi-geostrophic potential vorticity field following THL97 based on equations (1.32) and (1.34). Because the zonally averaged isotherms become very steep as the sidewall boundary layers are approached, it was necessary to place limits on the isotherm gradient utilized in the eddy parameterization. *Pérez* [2006] used the method of *slope tapering* as advocated by *Danabasoglu and McWilliams* [1995] for use in ocean circulation models. This method was also used to control the PV gradients used in the THL97 parameterization, especially close to the boundaries of the domain. The closure used for $\mathcal{K}$ was either based on the formulation by *Visbeck et al.* [1997] or (for potential vorticity) a constant diffusivity equivalent to the value diagnosed from the fully 3D model simulations.

In practice, the original GM90 method was found to be capable of matching the total Nusselt number of the fully 3D simulated flows across the full range of parameters. However, apart from at the lowest rotation speeds (close to marginal instability), the resulting temperature fields did not match closely the zonally averaged fields obtained in the full 3D eddy-resolving model. This was almost certainly a reflection of the relatively weak correlation found by *Pérez et al.* [2010] between $\overline{(u'T')}$ and $\partial \overline{T}/\partial r$ in the full 3D flow, so that the parameterized eddy-induced circulation did not accurately reflect the real TEM circulation in the 3D flow. This actually led to the development of spurious numerical instabilities within the 2D parameterized model in the most strongly super critical simulations, even though the total heat transport of the 3D flow was reproduced quite accurately in most cases.

Although the alternative THL97 PV-based parameterization was not able to match the total Nusselt number of the full 3D simulations as accurately as the GM90 scheme, it did result in much more realistic zonally averaged temperature fields in the 2D parameterized model. Figure 1.25 shows examples of (a) the eddy-induced TEM stream function χ_q^*, (b) the resulting parameterized TEM radial velocity u^*, and (c) the eddy-induced TEM radial velocity diagnosed from the corresponding fully 3D, eddy-resolving model simulation at $(\Theta, \mathcal{T}) = (1.22, 1.6 \times 10^6)$. χ_q^* takes the form of a simple overturning aligned along the principal direction of the isotherms and in the sense required to advect them toward the horizontal. The corresponding parameterized u^* resembles the diagnosed TEM radial velocity quite closely except close to the boundaries of the domain, where isotherm and PV gradients in the main fields become very large and quasi-geostrophic theory is no longer valid.

Examples of some simulated temperature fields are shown in Figure 1.26, which shows (a) the zonally averaged equilibrated temperature field in (r, z) from a fully 3D eddy-resolving simulation of moderately supercritical flow at $(\Theta, \mathcal{T}) = (0.599, 3.26 \times 10^6)$, (b) the equilibrated temperature field from a 2D axisymmetric simulation under the same conditions as in (a), and (c) the corresponding equilibrated temperature field from a 2D simulation using the THL97 eddy parameterization implemented by *Pérez* [2006]. Under these conditions, the axisymmetric isotherms (b) are much more steeply sloped than obtained in the eddy-resolving 3D model (a), where fully developed baroclinic instability acts to release a lot of stored available potential energy. This is well reflected in the parameterized simulation, where the additional eddy-induced component of the meridional circulation has strengthened the advection of temperature sufficiently to relax the isotherm slope toward the horizontal in a way that emulates quite accurately the effects of baroclinic eddies on the zonal mean flow in the 3D eddy-resolving simulation, even to the point of retaining the static stability structure. The total Nusselt number in the parameterized simulation was 9.5 compared with a time-mean value of 10.1 in the 3D eddy-resolving simulation, indicating a tendency for parameterized simulations to underestimate eddy heat transfer by around 20%.

This tendency becomes more pronounced in more strongly supercritical conditions, with a parameterized Nusselt number of 7.5 compared with a 3D Nusselt number of 9.1 at the most extreme conditions investigated by *Pérez* [2006] at $(\Theta, \mathcal{T}) = (0.017, 1.17 \times 10^8)$. As remarked earlier, however, at these more extreme parameters the

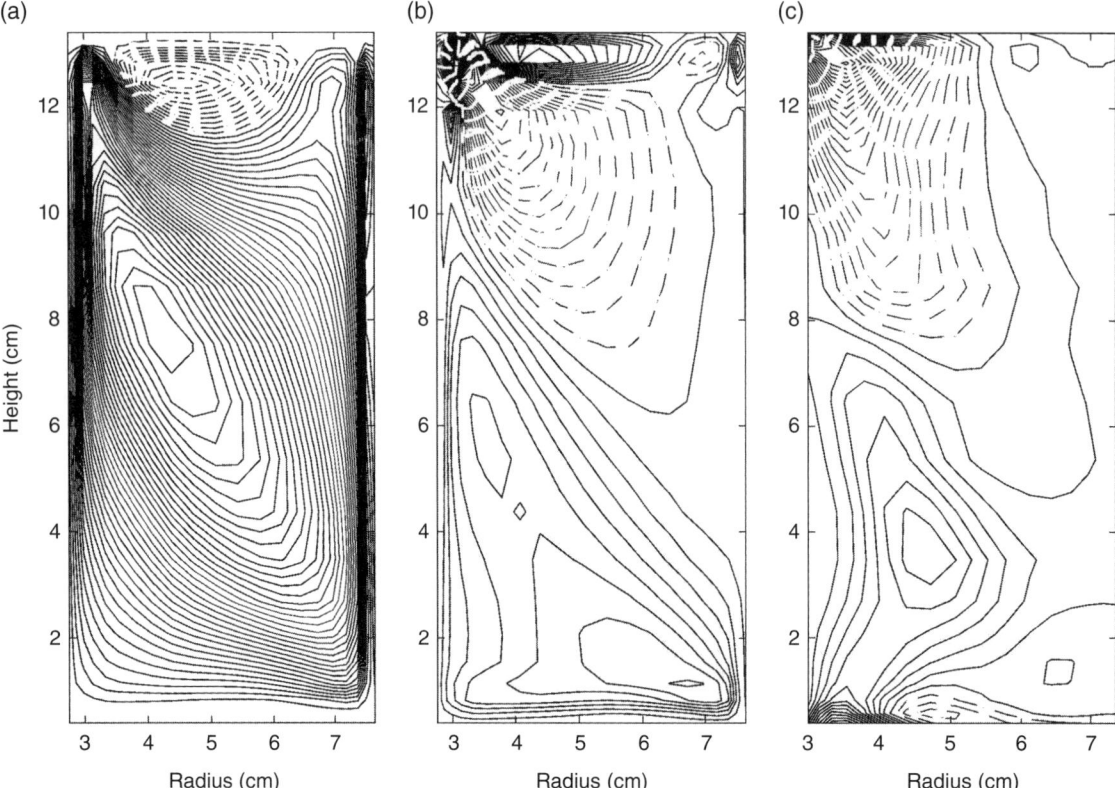

Figure 1.25. Maps in the (r, z) plane of (a) the eddy-induced transport stream function χ_q^* derived from a parameterization based on the zonal mean QG potential vorticity field [*Treguier et al.*, 1997], (b) the corresponding parameterized radial eddy-induced velocity $(u_* = 1/r \partial \chi_q^*/\partial z = (\mathcal{K}_q/f)\partial(\overline{q_{2D}})/\partial r)$, and (c) radial eddy-induced velocity diagnosed from the fully 3D simulation $(u_* = -(\overline{u'q'})/f)$ as obtained by *Pérez* [2006]. Note the remarkable degree of resemblance between (b) and (c). In each case negative values are shown with dashed contours.

simulated zonal mean temperature field begins to appear less obviously realistic, with apparent reversals of horizontal thermal gradient in the geostrophic interior, for example, that might reflect artifacts in the flow due to inadequate spatial resolution in the model. This needs to be investigated further in future work, which should include proper verification of the heat transfer and flow structure against laboratory measurements and the use of a higher resolution model.

1.5. DISCUSSION

In this chapter we have argued for the continued value and utility of rotating annulus laboratory experiments in the context of wider studies of the global circulation of planetary atmospheres and oceans. Despite many advances in the numerical modeling of atmospheres and oceans in the past 30 years, such approaches still have many limitations, particularly with regard to uncertainties associated with the use of finite resolution in space and time, the use of (often ad hoc) parameterization techniques to represent unresolved scales of motion, especially concerning the difficulty of accurately validating model simulations against measurements (which generally have incomplete and patchy coverage in space and time).

Laboratory studies help to focus attention on what factors may be fundamental to processes affecting the climate of an Earth-like planet, particularly under changing parametric conditions, in contrast to factors that may be incidental and/or specific to a particular system and may therefore be generalized across whole classes of system. The results presented above on quantifying heat transport in the thermally driven annulus system provide a prime example, in which we show how fundamental ideas on how the efficiency of heat transfer by baroclinic eddies appear to apply with equal validity both to laboratory flow systems and in the oceans (and in atmospheres too) provided the contribution to heat transfer in the laboratory due to the boundary layer circulation can be separated from that due to the baroclinic eddies themselves. This leads naturally to the use of laboratory experimental

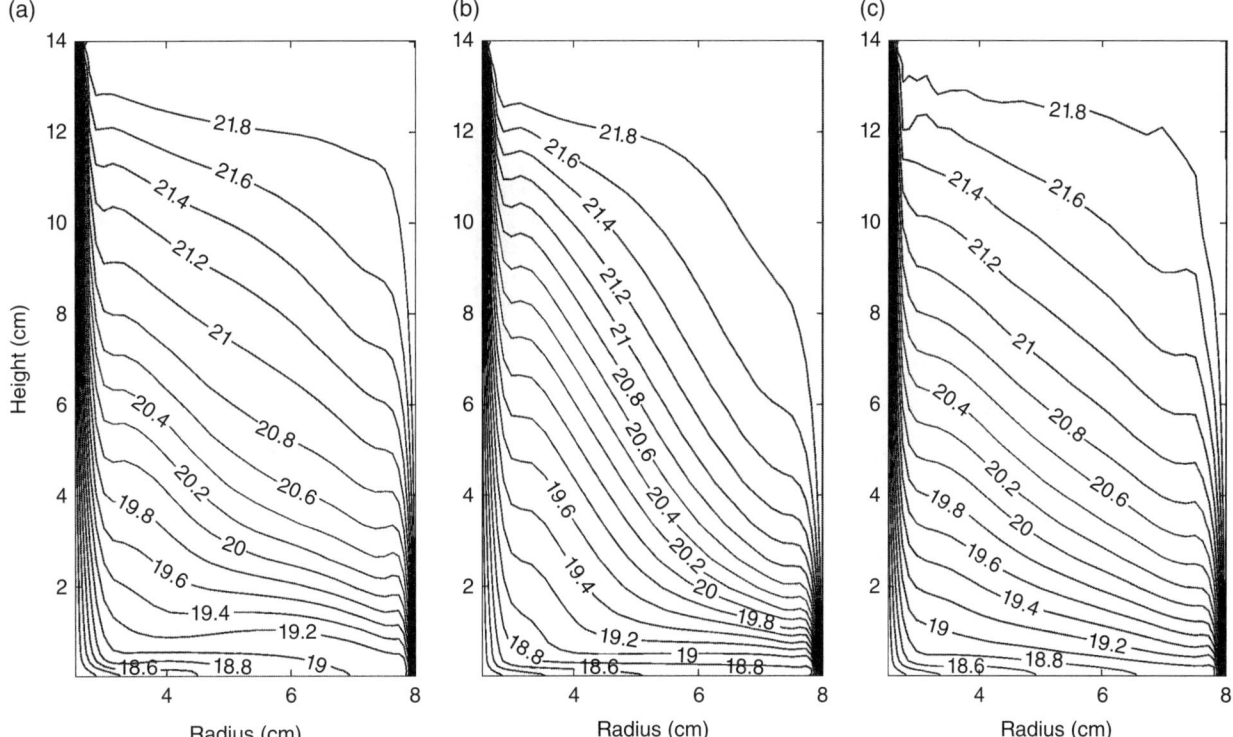

Figure 1.26. Maps in the (r, z) plane of (a) the zonal mean temperature field (in °C) derived from the equilibrated simulation using the 3D eddy-resolving numerical model (i.e., $\overline{T_{3D}}$), (b) the corresponding 2D axisymmetric temperature field T_{2D}, and (c) the equilibrated temperature field in a 2D axisymmetric model simulation using the THL97 eddy parameterization as obtained by *Pérez* [2006] for $\Delta T = 4$ K and $\Omega = 1.0$ rad/s ($\Theta = 0.599, \mathcal{T} = 3.26 \times 10^6$).

studies as rigorous means of testing theoretical hypotheses and understanding of heat transfer in geophysical problems.

A particular strength of the rotating annulus is its ability to achieve some degree of dynamical similarity with atmospheric and oceanic phenomena where background rotation is a dominating factor. In seeking to generalize results from the laboratory to geophysical systems, however, it is just as important to take account of the differences between experimental and natural systems as their similarities. The difference in geometry between cylindrical and spherical configurations is one obvious factor that must be taken into account, especially with regard to quantitative comparisons between experimental and geophysical systems. In addition, unlike in a planetary atmosphere or ocean, for example, diffusive boundary layers play major roles in the thermally driven annulus system in maintaining the mean stratification and horizontal thermal contrast in the quasi-geostrophic interior. This may make it difficult to use results from laboratory circulation systems to address mechanisms for setting the stratification in an ocean or atmosphere, for example, unless the experiment can be specifically reconfigured to reduce or allow for the influence of boundary layers.

The work described in Section 1.4 provides a powerful example of how combining insights and results from both real experiments and numerical model simulations can help to unravel the quantitative effects of boundary layer and quasi-geostrophic circulations within laboratory flows, thereby assisting in generalizing results from the latter to other systems. This methodological approach in intertwining laboratory measurements with numerical simulation offers the prospect of greatly increasing the scientific value of laboratory-based studies in the future

(a) by utilizing laboratory measurements to directly validate and compare numerical modeling techniques and to investigate e.g. convergence properties of model simulations with increasing resolution,

(b) by enabling simulations to be run that can test hypotheses under conditions (e.g., by artificially suppressing key instabilities) that may be difficult to realize directly in the laboratory, and

(c) ultimately to allow direct deterministic model predictions from initial states obtained using statistical-dynamical assimilation methods that combine model simulations with laboratory measurements.

The latter directly emulates the operational practice of numerical weather and climate prediction for Earth's

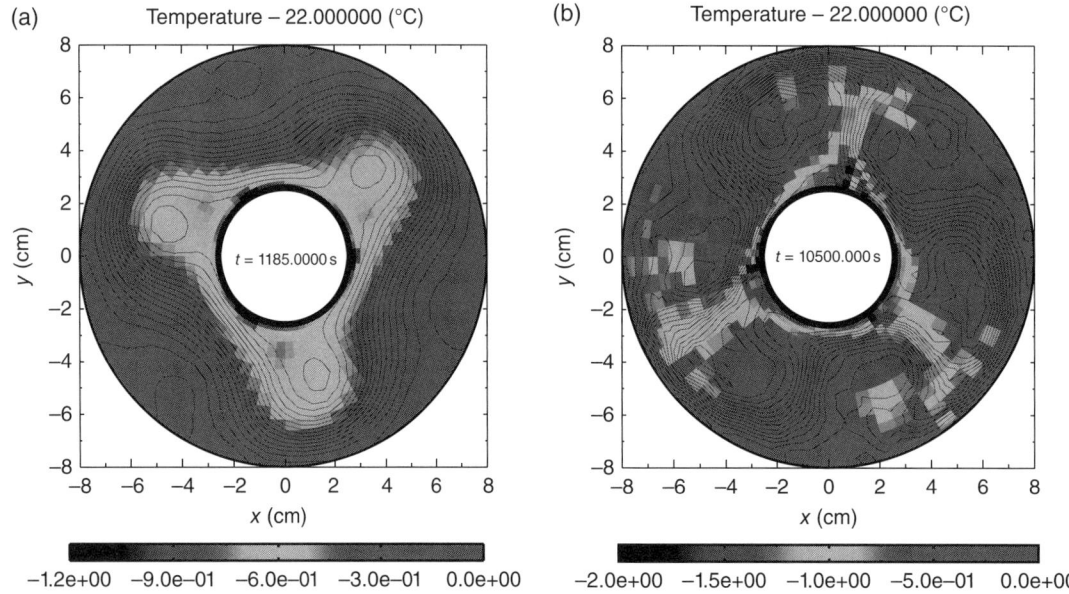

Figure 1.27. Representative temperature fields (colors) and horizontal stream function (contours) produced from assimilated horizontal velocity observations obtained in the same system as shown in Figures 1.8–1.13 and 1.21–1.26. Fields are plotted for regular (a) $\Omega = 0.875$ rad/s, $T_b - T_a \approx 4.07°C$) and chaotic flow (b) $\Omega = 3.1$ rad/s, $T_b - T_a \approx 4.02°C$) at $z = 9.7$ cm above the base of the annulus. Temperatures are relative to 22°C. Adapted from *Young and Read* [2013] with the permission of John Wiley & Sons, Inc. For color detail, please see color plate section.

atmosphere and oceans, offering the same potential uses to (a) obtain analyses of complete fields in the presence of incomplete and noisy measurements, (b) enable deterministic model predictions from assimilated measurements to quantify predictability and sensitivity to initial conditions, and (c) identify, characterize, and quantify systematic model errors.

The work by *Young and Read* [2013] applying data assimilation to the rotating annulus experiment in the form of analysis correction [*Lorenc et al.*, 1991] began to address some of these points. They demonstrated that it is possible to take methods developed for meteorological analysis and prediction and use them in the context of the laboratory experiment toward a useful end. In particular, they addressed the problem of incomplete measurements using the analysis correction procedure with a Boussinesq Navier-Stokes model to recover unobserved variables such as temperature (Figure 1.27) solely from irregularly distributed horizontal velocity observations at five vertical levels. The diagnostics required to shed light on the secondary instabilities at high rotation rate described in Section 1.3.4 were only obtainable because unobserved variables and vertically averaged quantities were retrieved via the assimilation procedure.

Although they did not address any outstanding problems with the analysis correction method itself (it has since been superseded by newer methods), this work laid the foundations to do so with newer methods not yet fully established in operational meteorological practice. Potential methods of interest include the various flavors of the ensemble Kalman filter (a version of which *Ravela et al.* [2010] have applied in this context) and other experimental methods that have been tested thus far primarily using low-dimensional systems [e.g., *Stemler and Judd*, 2009; *van Leeuwen*, 2010]. Laboratory experiments bridge the gap between these low dimensional systems and geophysical systems such as the atmosphere and, by using laboratory experiments, methods can be tested under laboratory conditions using a real fluid, a nonidealized model, and incomplete and noisy observations.

1.5.1. Planetary Circulation Regimes

An important question that still deserves a lot more attention than has been evident in the literature to date is the extent to which the rich and complex diversity of different flow regimes and bifurcations exhibited in the laboratory are shared, even qualitatively, by a full scale planetary atmosphere. The inability to carry out controlled experiments on real atmospheres is a major obstacle to progress in this regard (although of course such an approach would have other undesirable consequences for the inhabitants of such a planetary system!). The solar system provides a small sample of around eight planetary bodies with substantial atmospheres that occupy very different positions in parameter space. But this samples the parameter space too sparsely to address the problem in much detail [e.g., see *Showman et al.*, 2010; *Read*, 2011, for reviews]. The growing number of discoveries of planets around other stars (e.g., see http://exoplanet.eu/ and *Schneider et al.* [2011]) offers the eventual prospect of sampling parameter space much more densely, but the available measurements are as yet much too crude to be able to provide quantitative characterization of circulation regimes. So at the present time (and for the foreseeable future) the only way of addressing and characterizing the diversity of planetary circulation regimes and bifurcations is through the use of numerical model simulations.

To date, however, relatively little has been done to define and sample an appropriate parameter space for planetary circulations that comes anywhere near matching what has been achieved in the laboratory, at least in terms of breadth and detail with respect to the dominant dimensionless parameters. The early work of *Geisler et al.* [1983] laid some of the foundations for this approach in using a stripped-down version of an atmospheric global circulation model (GCM) to investigate a range of prototypical circulations with an imposed equator-pole thermal contrast and varying rotation speeds of an Earth-like planetary atmosphere. By explicit analogy with laboratory rotating annulus experiments, they presented their results with respect to two dimensionless parameters:

(i) A thermal Rossby number defined as

$$\Theta_S = \left(\frac{gH}{f_0^2 L^2}\right) \frac{\Delta T}{T_r}, \quad (1.60)$$

where H is a pressure scale height ($= RT_r/g$), R is the gas constant, $f_0 = 2\Omega \sin\phi_0$ (ϕ_0 was taken to be a latitude of 45°), L is a horizontal length scale, ΔT is the imposed equator-pole temperature contrast at the surface, and T_r is a reference temperature.

(ii) A Taylor or inverse squared Ekman number, defined as

$$\mathcal{T} \simeq \mathcal{E}^{-2} = \frac{\Omega^2 H^4}{K_v^2}, \quad (1.61)$$

where K_v is a vertical "eddy viscosity" coefficient.

For various practical reasons *Geisler et al.* [1983] only studied cases equivalent to an Earth-like planet rotating at the same speed as or slower than Earth itself. But this did enable them to demonstrate the existence of a "lower symmetric" regime at relatively small ΔT, where wavy flows gave way to axisymmetric circulations, the boundary of which was found close to the line defined by $\Theta_S \simeq 10^5 \mathcal{E}^2$. This roughly emulates the lower symmetric regime boundary found in rotating annulus experiments using relatively high Prandtl number fluids [*Fein*, 1973]. They also found evidence for a regular baroclinic wave regime at higher values of Θ_S than for Earth itself ($\Theta_S \gtrsim 0.05$), where the flow was dominated by near-monochromatic waves, peaking in amplitude at midlatitudes and drifting in longitude at a roughly steady rate. These waves were either steady in amplitude or apparently undergoing periodic oscillations reminiscent of the amplitude vacillation behavior seen in the annulus.

A significant difference from the laboratory systems was found, however, at the highest values of Θ_S, which would lie above the corresponding upper symmetric transition in the laboratory and would therefore be expected to exhibit axisymmetric flow. Instead, the flow in the spherical shell was found to transition from a predominantly baroclinic wave flow to a barotropically unstable flow, also with zonally propagating waves of relatively low zonal wave number $m \simeq 2$ drifting around an intense polar vortex. This kind of behavior has since been confirmed in more recent work [e.g., *Mitchell and Vallis*, 2010] in which such barotropically unstable flow at low planetary rotation speeds may also be associated with strongly superrotating zonal flow at low latitudes. Such a flow appears be consistent with the strongly superrotating circulations found on very slowly rotating planets such as Venus and Titan.

Since Geisler et al.'s early study, there has been a steady trickle of other work [e.g., *Williams and Holloway*, 1982; *Del Genio and Suozzo*, 1987; *Williams*, 1988a, 1988b; *Jenkins*, 1996; *Navarra and Boccaletti*, 2002; *Barry et al.*, 2002; *Schneider and Walker*, 2006] exploring other areas of parameter space, including cases corresponding to even faster rotation speeds (lower values of Θ_S) than Earth. Another early pioneer of this kind of modeling study was Gareth Williams [*Williams and Holloway*, 1982; *Williams*, 1988a, 1988b], who presented results from an Earth-like GCM for which the planetary rotation rate was varied between $\Omega_E/16$ and $\Omega_E \times 8$ (where Ω_E is Earth's rotation speed). At higher rotation speeds than that of Earth, Williams' model simulations suggested that the dominant scale of baroclinic instability would continue to decrease with increasing Ω, but with a tendency (at $\Omega \gtrsim 2\Omega_E$) for the subtropical zonal jet stream to break up into a set of two or more parallel jets associated with parallel trains of baroclinically unstable eddies. At the highest rotation speeds, up to seven or eight parallel jets were obtained in each hemisphere, resulting in a circulation pattern that bore a strong resemblance to that of Jupiter's or Saturn's cloud bands. Williams did not attempt to locate his simulations in a dimensionless parameter space, but *Read* [2011] computed approximate values of Θ_S and dissipation parameters to locate these experiments retrospectively. In common with some more recent work, the results appear to suggest that the multiple jets organize themselves on a scale comparable with the Rhines scale and are largely generated and controlled by the nonlinear interactions between eddies and the zonal flow.

This tendency to form multiple, parallel wave trains and zonal flows is reminiscent of the kind of flow regimes obtained in thermally driven annulus experiments with oppositely sloping end walls to generate a topographic β effect [e.g., *Mason*, 1975; *Bastin and Read*, 1997, 1998; *Wordsworth et al.*, 2008]. As argued by *Read* [2011], however, in contrast to laboratory experiments with boundaries of variable end wall slope, the global Rhines length scale ($\sim (U_{\rm rms}/\beta)^{1/2} \sim (U_{\rm rms}a/(2\Omega))^{1/2}$) is not independent of the thermal Rossby number in a planetary circulation. It is largely set by the spherical geometry and, in simple cases, may scale roughly as $\sqrt{\Theta_S}$ (though the full situation may be more complicated than this; e.g., see *Jansen and Ferrari* [2012] for further discussion). Thus, provided an analog of the planetary vorticity gradient is present in the laboratory, there appear to be strong parallels between the principal sequences of regime transitions in both cylindrical annular laboratory experiments and planetary atmospheres in spherical shells across much of the parameter space. But many gaps in our understanding of these parallels remain to be explored in detail.

However, the ability to run many experiments in the laboratory in order to sample parameter space densely offers the possibility of testing various theoretical scalings for circulations that operate on a planetary scale in atmospheres and oceans. Numerical models will continue to struggle to match this, especially at high planetary rotation rates where the range of dynamically significant scales of motion demands the use of very high resolution models.

Acknowledgments. Several aspects of this chapter originated in a series of graduate lectures given at the University of Oxford. It is a pleasure to thank the many colleagues and collaborators with whom I have worked on this problem for a number of years. Particular thanks are due to Drs. R. Hide, P. Hignett, M. J. Bell, and A. A. White of the UK Met. Office for their many insights and to D. W. Johnson, R. M. Small, W.-G. Früh, P. D. Williams, and A. A. Castrejón-Pita in connection with some of the experimental work discussed herein. We are also grateful to two anonymous referees whose comments greatly assisted in improving the presentation of this chapter.

REFERENCES

Abbe, C. (1907), Projections of the globe appropriate for laboratory methods of studying the general circulation of the atmosphere, *B. Am. Math. Soc.*, *13*, 502–506.

Andrews, D. G., J. R. Holton, and C. B. Leovy (1987), *Middle Atmosphere Dynamics*, Academic Press, Orlando, Fla.

Appleby, J. C. (1982), Comparative theoretical and experimental studies of baroclinic waves in a two-layer system, Ph.D. thesis, Univ. of Leeds.

Appleby, J. C. (1988), Selection of baroclinic waves, *Quart. J. R. Meteor. Soc.*, *114*, 1173–1179.

Baines, P. G. (1976), The stability of planetary waves on a sphere, *J. Fluid Mech.*, *73*, 193–213.

Barry, L., G. C. Craig, and J. Thuburn (2002), Poleward heat transport by the atmospheric heat engine, *Nature*, *415*, 774–777.

Basdevant, C., B. Legras, R. Sadourney, and M. Beland (1981), A study of barotropic model flows: Intermittency, waves and predictability, *J. Atmos. Sci.*, *38*, 2305–2326.

Bastin, M. E., and P. L. Read (1997), A laboratory study of baroclinic waves and turbulence in an internally heated rotating fluid annulus with sloping endwalls, *J. Fluid Mech.*, *339*, 173–198.

Bastin, M. E., and P. L. Read (1998), Experiments on the structure of baroclinic waves and zonal jets in an internally heated, rotating, cylinder of fluid, *Phys. Fluids*, *10*, 374–389.

Bell, M. J. (1989), Theoretical investigations prompted by experiments with baroclinic fluids, Ph.D. thesis, Imperial College London.

Bell, M. J., and A. A. White (1988), The stability of internal baroclinic jets: Some analytical results, *J. Atmos. Sci.*, *45*, 2571–2590.

Bernadet, P., A. Butet, M. Deque, M. Ghil, and R. Pfeffer (1990), Low-frequency oscillations in a rotating annulus with topography, *J. Atmos. Sci.*, *47*, 3023–3043.

Boville, B. A. (1981), Amplitude vacillation on a β-plane, *J. Atmos. Sci.*, *38*, 609–618.

Bowden, M. (1961), An experimental investigation of heat transfer in rotating fluids, Ph.D. thesis, Durham Univ., UK.

Bowden, M., and H. F. Eden (1965), Thermal convection in a rotating fluid annulus: Temperature, heat flow and flow field observations in the upper symmetric regime, *J. Atmos. Sci.*, *22*, 185–195.

Brachet, M. E., M. Meneguzzi, H. Politano, and P.-L. Sulem (1988), The dynamics of freely-decaying two-dimensional turbulence, *J. Fluid Mech.*, *194*, 333–349.

Brindley, J., and I. Moroz (1980), Lorenz attractor behaviour in a continuously stratified baroclinic fluid, *Phys. Lett.*, *77A*, 441–444.

Buzyna, G., R. L. Pfeffer, and R. Kung (1978), Cyclic variations of the imposed temperature contrast in a thermally driven rotating annulus of fluid, *J. Atmos. Sci.*, *35*, 859–881.

Buzyna, G., R. L. Pfeffer, and R. Kung (1984), Transition to geostrophic turbulence in a rotating differentially heated annulus of fluid, *J. Fluid Mech.*, *145*, 377–403.

Castrejón-Pita, A. A., and P. L. Read (2007), Baroclinic waves in an air-filled thermally driven rotating annulus, *Phys. Rev. E*, *75*, 026,301.

Cazalbou, J.-B., P. Chassaing, G. Dufour, and X. Carbonneau (2005), Two-equation modeling of turbulent rotating flows, *Phys. Fluids*, *17*, 055,110.

Charney, J. G., and M. E. Stern (1962), On the stability of internal baroclinic jets in a rotating atmosphere, *J. Atmos. Sci.*, *19*, 159–172.

Danabasoglu, G., and J. C. McWilliams (1995), Sensitivity of the global ocean circulation to parameterizations of mesoscale tracer transports, *J. Clim.*, *8*, 2967–2987.

Danabasoglu, G., J. C. McWilliams, and P. Gent (1994), The role of mesoscale tracer transports in the global ocean circulation, *Science*, *264*, 1123–1126.

Davies, T. V. (1959), On the forced motion due to heating of a deep rotating liquid in an annulus, *J. Fluid Mech.*, *5*, 593–621.

Del Genio, A., and R. J. Suozzo (1987), A comparative study of rapidly and slowly rotating circulation regimes in a terrestrial general circulation model, *J. Atmos. Sci.*, *44*, 973–986.

Drazin, P. G. (1970), Non-linear baroclinic instability of a continuous zonal flow, *Quart. J. R. Meteor. Soc.*, *96*, 667–676.

Drazin, P. G. (1978), Variations on a theme of eady, in *Rotating Fluids in Geophysics*, edited by P. H. Roberts and A. M. Soward, pp. 139–169, Academic Press, London and New York.

Esler, J. G., and B. T. Willcocks (2012), Nonlinear baroclinic equilibration at finite supercriticality, *Geophys. Astrophys. Fluid Dyn.*, *106*, 320–350.

Exner, F. M. (1923), Uber die bildung von windhosen und zyklonen, *Sitzungsber der Akad. der Wiss. Wien, Abt. IIa*, *132*, 1–16.

Farmer, D., J. Hart, and P. Weidman (1982), A phase space analysis of baroclinic flow, *Phys. Lett.*, *91A*, 22–24.

Fein, J. (1973), An experimental study of the effects of the upper boundary condition on the thermal convection in a rotating cylindrical annulus of water, *Geophys. Fluid Dyn.*, *5*, 213–248.

Fein, J. S. (1978), *Boundary Layers in Homogeneous and Stratified-Rotating Fluids*, Univ. Presses of Florida, Tallahasee, Fla.

Fein, J. S., and R. L. Pfeffer (1976), An experimental study of the effects of prandtl number on thermal convection in a rotating, differentially heated cylindrical annulus of fluid, *J. Fluid Mech.*, *75*, 81–112.

Fjørtoft, R. (1953), On the changes in the spectral distribution of kinetic energy for two-dimensional nondivergent flow, *Tellus*, *5*, 225–230.

Fowler, A. C., J. D. Gibbon, and M. J. McGuinness (1982), The complex Lorenz equations, *Physica D*, *4*, 139–163.

Friedlander, S. (1980), *An Introduction to the Mathematical Theory of Geophysical Fluid Dynamics*, North Holland, Amsterdam, The Netherlands.

Früh, W. G., and P. L. Read (1997), Wave interactions and the transition to chaos of baroclinic waves in a thermally driven rotating annulus, *Phil. Trans. Roy. Soc. London*, *A355*, 101–153.

Fultz, D. (1951), Experimental analogies to atmospheric motions, in *Compendium of Meteorology*, edited by T. F. Malone, Am. Meteorol. Soc., New York.

Geisler, J. E., E. J. Pitcher, and R. C. Malone (1983), Rotating-fluid experiments with an atmospheric general circulation model, *J. Geophys. Res.*, *88*, 9706–9716.

Gent, P. R., and J. C. McWilliams (1990), Isopycnal mixing in ocean circulation models, *J. Phys. Oceanogr.*, *20*, 150–155.

Gent, P. R., J. Willebrand, T. J. Mcdougall, and J. C. McWilliams (1995), Parameterizing eddy-induced tracer transports in ocean circulation models, *J. Phys. Oceanogr.*, *25*, 463–474.

Gibbon, J. D., and M. J. McGuinness (1980), A derivation of the Lorenz equations for some unstable dispersive physical systems, *Phys. Lett.*, *77A*, 295–299.

Gill, A. E. (1966), The boundary-layer regime for convection in a rectangular cavity, *J. Fluid Mech.*, *26*, 515–536.

Green, J. S. A. (1970), Transfer properties of the large-scale eddies and the general circulation of the atmosphere, *Quart. J. R. Met. Soc.*, *96*, 157–185.

Grotjahn, R. (1984a), Baroclinic instability in a long wave environment. Part i. Review, *Quart. J. R. Meteor. Soc.*, *110*, 663–668.

Grotjahn, R. (1984b), Baroclinic instability in a long wave environment. Part ii. Ageostrophic energy conversions, *Quart. J. R. Meteor. Soc.*, *110*, 669–693.

Haine, T. W. N., and J. C. Marshall (1998), Gravitational, symmetric and baroclinic instability of the ocean mixed layer, *J. Phys. Oceanogr.*, *28*, 534–658.

Harlander, U., T. von Larcher, Y. Wang, and C. Egbers (2011), PIV- and LDV-measurements of baroclinic wave interactions in a thermally driven rotating annulus, *Exp. Fluids*, *51*, 37–49.

Hart, J. E. (1972), A laboratory study of baroclinic instability, *Geophys. Fluid Dyn.*, *3*, 181–209.

Hart, J. E. (1979), Finite amplitude baroclinic instability, *Ann. Rev. Fluid Mech.*, *11*, 147–172.

Hart, J. E. (1980), An experimental study of nonlinear baroclinic instability and mode selection in a large basin, *Dyn. Atmos. Oceans*, *4*, 115–135.

Hart, J. E. (1981), Wavenumber selection in nonlinear baroclinic instability, *J. Atmos. Sci.*, *38*, 400–408.

Hart, J. E. (1985), A laboratory study of baroclinic chaos on the f-plane, *Tellus*, *37A*, 286–296.

Hart, J. E. (1986), A model for the transition to baroclinic chaos, *Physica D*, *20*, 350–362.

Hide, R. (1958), An experimental study of thermal convection in a rotating fluid, *Phil. Trans. R. Soc. Lond.*, *A250*, 441–478.

Hide, R. (1970), Some laboratory experiments on free thermal convection in a rotating fluid subject to a horizontal temperature gradient and their relation to the theory of the global atmospheric circulation, in *The Global Circulation of the Atmosphere Joint Conference, 25–29 August 1969*, edited by G. A. Corby, Royal Meteorol. Soc., London.

Hide, R., and P. J. Mason (1970), Baroclinic waves in a rotating fluid subject to internal heating, *Phil. Trans. R. Soc. Lond.*, *A268*, 201–232.

Hide, R., and P. J. Mason (1975), Sloping convection in a rotating fluid, *Adv. Phys.*, *24*, 47–100.

Hide, R., P. J. Mason, and R. A. Plumb (1977), Thermal convection in a rotating fluid subject to a horizontal temperature gradient: Spatial and temporal characteristics of fully developed baroclinic waves, *J. Atmos. Sci.*, *34*, 930–950.

Hignett, P. (1982), A note on the heat transfer by the axisymmetric thermal convection in a rotating fluid annulus, *Geophys. Astrophys. Fluid Dyn.*, *19*, 293–299.

Hignett, P., A. Ibbetson, and P. D. Killworth (1981), On rotating thermal convection driven by non-uniform heating from below, *J. Fluid Mech.*, *109*, 161–187.

Hignett, P., A. A. White, R. D. Carter, W. D. N. Jackson, and R. M. Small (1985), A comparison of laboratory measurements and numerical simulations of baroclinic wave flows in a rotating cylindrical annulus, *Quart. J. R. Meteor. Soc.*, *111*, 131–154.

Hocking, L. M. (1978), Theory of hydrodynamic stability, in *Rotating Fluids in Geophysics*, edited by P. H. Roberts and A. M. Soward, pp. 437–469, Academic Press, London and New York.

Holopainen, E. O. (1961), On the effect of friction in baroclinic waves, *Tellus*, *13*, 363–367.

Holton, J. R. (1972), *An Introduction to Dynamic Meteorology*, Academic Press, New York.

Hoskins, B. J. (1973), Stability of the Rosby-Haurwitz wave, *Quart. J. R. Meteor. Soc.*, *99*, 723–745.

Hoskins, B. J. (1983), Dynamical processes in the atmosphere and the use of models, *Quart. J. R. Meteor. Soc.*, *109*, 1–21.

Hoskins, B. J., I. N. James, and G. H. White (1983), The shape, propagation and mean-flow interaction of large-scale weather systems, *J. Atmos. Sci.*, *40*, 1595–1612.

Jacoby, T. N. L., P. L. Read, P. D. Williams, and R. M. B. Young (2011), Generation of inertia-gravity waves in the rotating thermal annulus by a localised boundary layer instability, *Geophys. Astrophys. Fluid Dyn.*, *105*, 161–181.

James, I. N. (1994), *Introduction to Circulating Atmospheres*, Cambridge Univ. Press. Cambridge, UK.

Jansen, M., and R. Ferrari (2012), Macroturbulent equilibration in a thermally forced primitive equation system, *J. Phys. Oceanogr.*, *69*, 695–713.

Jenkins, G. S. (1996), A sensitivity study of changes in Earth's rotation rate with an atmospheric general circulation model, *Glob. Plan. Change*, *11*, 141–154.

Jnosi, I. M., P. Kiss, V. Homonnai, M. Pattantyús-Ábrahám, B. Gyüre, and T. Tél (2010), Dynamics of passive tracers in the atmosphere: Laboratory experiments and numerical tests with reanalysis wind fields, *Phys. Rev. E*, *82*, 046,308.

Killworth, P. D. (1997), On the parameterization of eddy transfer. Part I: Theory, *J. Mar. Res.*, *55*, 1171–1197.

Kim, K. (1978), Instability of baroclinic Rossby waves: Energetics in a two-layer ocean, *Deep Sea Res.*, *25*, 795–814.

King, E. M., S. Stellmach, J. Noir, U. Hansen, and J. M. Aurnou (2009), Boundary layer control of rotating convection systems, *Nature*, *457*, 301–304.

King, E. M., S. Stellmach, and J. M. Aurnou (2012), Heat transfer by rapidly rotating Rayleigh-Bénard convection, *J. Fluid Mech.*, *691*, 568–582.

King, J. C. (1979), Instabilities and nonlinear wave interactions in a two-layer rotating fluid, Ph.D. thesis, Univ. of Leeds.

Klein, P. (1990), Transition to chaos in unstable baroclinic systems: A review, *Fluid Dyn. Res.*, *5*, 235–254.

Kraichnan, R. H. (1967), Inertial ranges in two-dimensional turbulence, *Phys. Fluids*, *10*, 1417–1423.

Kraichnan, R. H. (1971), Inertial range transfer in two- and three-dimensional turbulence, *J. Fluid Mech.*, *47*, 525–535.

Larichev, V. D., and I. M. Held (1995), Eddy amplitudes and fluxes in a homogeneous model of fully developed baroclinic instability, *J. Atmos. Sci.*, *25*, 2285–2297.

Leach, H. (1981), Thermal convection in a rotating annulus: effects due to bottom topography, *J. Fluid Mech.*, *109*, 75–87.

Lewis, G. M., and W. Nagata (2004), Linear stability analysis for the differentially heated rotating annulus, *Geophys. Astrophys. Fluid Dyn.*, *98*, 129–152.

Li, G., R. Kung, and R. Pfeffer (1986), An experimental study of baroclinic flows with and without two-wave bottom topography, *J. Atmos. Sci.*, *43*, 2585–2599.

Lindzen, R. S., B. Farrell, and D. Jacqmin (1982), Vacillation due to wave interference: Applications to the atmosphere and to annulus experiments, *J. Atmos. Sci.*, *39*, 14–23.

Lorenc, A. C., R. S. Bell, and B. Macpherson (1991), The Meteorological Office analysis correction data assimilation scheme, *Q. J. Roy. Meteorol. Soc.*, *117*, 59–89.

Lorenz, E. N. (1963a), Deterministic non-periodic flow, *J. Atmos. Sci.*, *20*, 130–141.

Lorenz, E. N. (1963b), The mechanics of vacillation', *J. Atmos. Sci.*, *20*, 448–464.

Lorenz, E. N. (1967), *The Nature and Theory of the General Circulation of the Atmosphere*, No. 218, T. P. 115, World Metesral. Assoc., Geneva, Switzerland.

Lorenz, E. N. (1972), Barotropic instability of Rossby wave motion, *J. Atmos. Sci.*, *29*, 258–264.

Lovegrove, A. F. L., P. L. Read, and C. J. Richards (2000), Generation of inertia-gravity waves in a baroclinically unstable fluid, *Quart. J. R. Meteor. Soc.*, *126*, 3233–3254.

Lovegrove, A. F. L., I. M. Moroz, and P. L. Read (2001), Bifurcations and instabilities in rotating, two-layer fluids: I. f-plane, *Nonlin. Proc. Geophys.*, *8*, 21–36.

Lovegrove, A. F. L., I. M. Moroz, and P. L. Read (2002), Bifurcations and instabilities in rotating, two-layer fluids: Ii. β-plane, *Nonlin. Proc. Geophys.*, *9*, 289–309.

Marshall, D. P., and A. J. Adcroft (2010), Parameterization of ocean eddies: Potential vorticity mixing, energetics and Arnol'd's first stability theorem, *Ocean Modelling*, *32*, 188–204.

Mason, P. J. (1975), Baroclinic waves in a container with sloping endwalls, *Phil. Trans. Roy. Soc. London*, *278*, 397–445.

Maubert, P., and A. Randriamampianina (2002), Transition vers la turbulence géostrophique pour un écoulement d'air en cavité tournante différentiellement chauffée, *C. R. Méchanique*, *330*, 365–370.

McIntyre, M. E. (1968), The axisymmetric convective regime for a rigidly bounded rotating annulus, *J. Fluid Mech.*, *32*, 625–655.

McWilliams, J. C. (1984), The emergence of isolated coherent vortices in turbulent flow, *J. Fluid Mech.*, *146*, 21–43.

Mitchell, J. L., and G. K. Vallis (2010), The transition to superrotation in terrestrial atmospheres, *J. Geophys. Res.*, *115*, E12008.

Mundt, M. D., N. H. Brummell, and J. E. Hart (1995a), Linear and nonlinear baroclinic instability with rigid sidewalls, *J. Fluid Mech.*, *291*, 109–138.

Mundt, M. D., J. E. Hart, and D. R. Ohlsen (1995b), Symmetry, sidewalls, and the transition to chaos in baroclinic systems, *J. Fluid Mech.*, *300*, 311–338.

Navarra, A., and C. Boccaletti (2002), Numerical general circulation experiments of sensitivity to Earth rotation rate, *Climate Dyn.*, *19*, 467–483.

O'Gorman, P. A., and T. Schneider (2007), Recovery of atmospheric flow statistics in a general circulation model without nonlinear eddy-eddy interactions, *Geophys. Res. Lett.*, *34*, L2280.

Ohlsen, D. R., and J. E. Hart (1989a), The transition to baroclinic chaos on the β-plane, *J. Fluid Mech.*, *203*, 23–50.

Ohlsen, D. R., and J. E. Hart (1989b), Nonlinear interference vacillation, *Geophys. Astrophys. Fluid Dyn.*, *45*, 213–235.

Pedlosky, J. (1970), Finite-amplitude baroclinic waves, *J. Atmos. Sci.*, *27*, 15–30.

Pedlosky, J. (1971), Finite-amplitude baroclinic waves with small dissipation, *J. Atmos. Sci.*, *28*, 587–597.

Pedlosky, J. (1982a), Finite-amplitude baroclinic waves at minimum critical shear, *J. Atmos. Sci.*, *39*, 555–562.

Pedlosky, J. (1982b), A simple model for nonlinear critical layers in an unstable baroclinic waves, *J. Atmos. Sci.*, *39*, 2119–2127.

Pedlosky, J. (1987), *Geophysical Fluid Dynamics*, Springer, Berlin.

Pedlosky, J., and C. Frenzen (1980), Chaotic and periodic behavior of finite-amplitude baroclinic waves, *J. Atmos. Sci.*, *37*, 1177–1196.

Pérez, E. P. (2006), Heat transport by baroclinic eddies: Evaluating eddy parameterizations for numerical models, Ph.D. thesis, Univ. of Oxford.

Pérez, E. P., P. L. Read, and I. M. Moroz (2010), Assessing eddy parameterization schemes in a differentially heated rotating annulus experiment, *Ocean Modelling*, *32*, 118–131.

Pfeffer, R. L., and A. Barcilon (1978), Determination of eddy fluxes of heat and eddy temperature variances using weakly nonlinear theory, *J. Atmos. Sci.*, *35*, 2099–2110.

Pfeffer, R. L., G. Buzyna, and R. Kung (1980), Time dependent modes of behavior of thermally-driven rotating fluid, *J. Atmos. Sci.*, *37*, 2129–2149.

Phillips, N. A. (1954), Energy transformations and meridional circulation associated with simple baroclinic waves in a two-level quasi-geostrophic model, *Tellus*, *6*, 273–286.

Pierrehumbert, R. T. (2010), *Principles of Planetary Climate*, Cambridge Univ. Press, Cambridge, UK.

Plumb, R., and J. Mahlman (1987), The zonally-averaged transport characteristics of the GFDL general circulation/transport model, *J. Atmos., Sci.*, *44*, 298–327.

Randriamampianina, A., W.-G. Früh, P. L. Read, and P. Maubert (2006), Direct numerical simulations of bifurcations in an air-filled rotating baroclinic annulus, *J. Fluid Mech.*, *561*, 359–389.

Ravela, S., J. Marshall, C. Hill, A. Wong, and S. Stransky (2010), A realtime observatory for laboratory simulation of planetary flows, *Exp. Fluids*, *48*, 915–925.

Read, P. L. (1986), Regimes of axisymmetric flow in an internally heated rotating fluid, *J. Fluid Mech.*, *168*, 255–289.

Read, P. L. (1988), On the scale of baroclinic instability in deep, compressible atmospheres, *Quart. J. R. Meteor. Soc.*, *114*, 421–437.

Read, P. L. (1992), Dynamics and instabilities of Ekman and Stewartson layers, in *Rotating Fluids in Geophysical and Industrial Applications*, edited by E. J. Hopfinger, pp. 49–84, Springer-Verlag, Vienna and New York.

Read, P. L. (2003), A combined laboratory and numerical study of heat transport by baroclinic eddies and axisymmetric flows, *J. Fluid Mech.*, *489*, 301–323.

Read, P. L. (2011), Dynamics and circulation regimes of terrestrial planets, *Plan. Space Sci.*, *59*, 900–914.

Read, P. L., and S. H. Risch (2011), A laboratory study of global-scale wave interactions in baroclinic flow with topography i: Multiple flow regimes, *Geophys. Astrophys. Fluid Dyn.*, *105*, 128–160.

Read, P. L., M. J. Bell, D. W. Johnson, and R. M. Small (1992), Quasi-periodic and chaotic flow regimes in a thermally driven, rotating fluid annulus, *J. Fluid Mech.*, *238*, 599–632.

Read, P. L., N. P. J. Thomas, and S. H. Risch (2000), An evaluation of Eulerian and semi-Lagrangian advection schemes in simulations of rotating, stratified flows in the laboratory. Part I: Axisymmetric flow, *Mon. Weather Rev.*, *128*, 2835–2852.

Rhines, P. B. (1979), Geostrophic turbulence, *Ann. Rev. Fluid Mech.*, *11*, 401–441.

Richardson, L. F. (1922), *Weather Prediction by Numerical Process*, Dover Publications, New York.

Romea, R. D. (1977), The effects of friction and β on finite amplitude baroclinic waves, *J. Atmos. Sci.*, *34*, 1689–1695.

Rossby, C. G. (1926), On the solution of problems of atmospheric motion by means of model experiments, *Mon. Weather Rev.*, *54*, 237–240.

Saffman, P. G. (1971), On the spectrum and decay of random two-dimensional vorticity distributions of large Reynolds number, *Stud. Appl. Math.*, *50*, 377–383.

Salmon, R. (1980), Baroclinic instability and geostrophic turbulence, *Geophys. Astrophys. Fluid Dyn.*, *15*, 167–211.

Salmon, R. S. (1978), Two-layer quasi-geostrophic turbulence in a simple special case, *Geophys. Astrophys. Fluid Dyn.*, *10*, 25–52.

Schneider, J., Dedieu, C., Le Sidaner, P., Savalle, R., and Zolotukhin, I. (2011), Defining and cataloging exoplanets: The exoplanet.eu database, *Astron. Astrophys.*, *532*, A79. doi:10.1051/0004-6361/201116713.

Schneider, T., and C. C. Walker (2006), Self-organization of atmospheric macroturbulence into critical states of weak nonlinear eddy-eddy interactions, *J. Atmos. Sci.*, *63*, 1569–1586.

Scott, R. B., and B. K. Arbic (2007), Spectral energy fluxes in geostrophic turbulence: Implications for ocean energetics, *J. Phys. Oceanogr.*, *37*, 673–688.

Scott, R. B., and F. Wang (2005), Direct evidence of an oceanic inverse energy cascade from satellite altimetry, *J. Phys. Oceanogr.*, *35*, 1650–1666.

Showman, A. P., J. Cho, and K. Menou (2010), Atmospheric circulation of extrasolar planets, in *Exoplanets*, edited by S. Seager, pp. 471–516, Univ. of Arizona Press, Tueson, Ariz.

Sitte, B., and C. Egbers (2000), Higher order dynamics of baroclinic waves, in *Physics of Rotating Fluids. Proc. 11th Int. Couette-Taylor Workshop, July 20–23, 1999, Bremen, Germany*. C. Egbers and G. Pfister (Eds), LNP 549, pp. 355–375. Springer-Verlag, Berlin & Heidelberg.

Spence, T. W., and D. Fultz (1977), Experiments on wave-transition spectra and vacillation in an open rotating cylinder, *J. Atmos. Sci.*, *34*, 1261–1285.

Stemler, T., and K. Judd (2009), A guide to using shadowing filters for forecasting and state estimation, *Phys. D*, *238*, 1260–1273.

Stone, P. (1972), A simplified radiative-dynamical model for the static stability of rotating atmospheres, *J. Atmos. Sci.*, *29*, 405–418.

Sugata, S., and S. Yoden (1992), Steady axisymmetric flow due to differential heating in a rotating annulus and its dependence on experimental parameters, *J. Met. Soc. Japan*, *70*, 1005–1017.

Sugata, S., and S. Yoden (1994), Chaotic Lagrangian motion and heat transport in a steady, baroclinic annulus wave, *J. Met. Soc. Japan*, *72*, 569–587.

Tajima, T., T. Nakamura, and T. Kuroda (1995), Laboratory experiments of Lagrangian motions in a steady baroclinic wave - internal structures of vortices, *J. Met. Soc. Japan*, *73*, 37–46.

Tajima, T., T. Nakamura, and K. Kurosawa (1999), Experimental observations of 3D Lagrangian motions in steady baroclinic waves: II, *J. Met. Soc. Japan*, *77*, 17–29.

Tamaki, K., and K. Ukaji (2003), An experimental study of wave dispersion in a differentially heated rotating fluid annulus with a radially sloping bottom, *J. Met. Soc. Japan*, *81*, 951–962.

Treguier, A. M., I. M. Held, and V. D. Larichev (1997), Parametrization of quasigeostrophic eddies in primitive equation ocean models, *J. Phys. Oceanogr.*, *27*, 567–580.

Ukaji, K., and K. Tamaki (1989), A comparison of laboratory experiments and numerical simulations of steady baroclinic waves produced in a differentially heated rotating annulus, *J. Met. Soc. Japan*, *67*, 359–373.

Vallis, G. K. (2006), *Atmospheric and Oceanic Fluid Dynamics: Fundamentals and Large-Scale Circulation*, Cambridge Univ. Press, Cambridge, UK.

van Leeuwen, P. J. (2010), Nonlinear data assimilation in geosciences: An extremely efficient particle filter, *Q. J. Roy. Meteorol. Soc.*, *136*, 1991–1999.

Vettin, F. (1857), Uber den aufsteigen Luftström, die Entstehung des Hagels und der Wirbel-Stürme, *Ann. Physik Chemie*, *102*, 246–255.

Vettin, F. (1884), Experimentale darstellung von luftbewegungen unter dem einflusse von temperatur-unterschieden und rotations-impulsen, *Meteorolol. Z.*, *1*, 227–230, 271–276.

Visbeck, M. J., J. Marshall, and T. W. N. Haine (1997), Specification of eddy transfer coefficients in coarse-resolution ocean circulation models, *J. Phys. Oceanogr.*, *27*, 381–402.

von Larcher, T., and C. Egbers (2005), Experiments on transitions of baroclinic waves in a differentially heated rotating annulus, *Nonlin. Proc. Geophys.*, *12*, 1033–1041.

von Larcher, T., A. Fournier, and R. Hollerbach (2013), The influence of a sloping bottom endwall on the linear stability in the thermally driven rotating annulus with a free surface, *Theor. Comput. Fluid Dyn.*, *27*, 433–451.

Warn, T., and P. Gauthier (1989), Potential vorticity mixing by marginally unstable baroclinic disturbances, *Tellus*, *41A*, 115–131.

Weng, H.-Y., and A. Barcilon (1987), Wave structure and evolution in baroclinic flow regimes, *Quart. J. R. Meteor. Soc.*, *113*, 1271–1294.

Weng, H.-Y., A. Barcilon, and J. Magnan (1986), Transitions between baroclinic flow regimes, *J. Atmos. Sci.*, *43*, 1760–1777.

White, A. A. (1988), The dynamics of rotating fluids: numerical modelling of annulus flows, *Met. Mag.*, *117*, 54–63.

White, H. D., and E. L. Koschmieder (1981), Convection in a rotating, laterally heated annulus. Pattern velocities and amplitude oscillations, *Geophys. Astrophys. Fluid Dyn.*, *18*, 301–320.

Willcocks, B. T., and J. G. Esler (2012), Nonlinear baroclinic equilibration in the presence of Ekman friction, *J. Phys. Oceanogr.*, *42*, 225–242.

Williams, G. P. (1988a), The dynamical range of global circulations. I, *Clim. Dyn.*, *2*, 205–260.

Williams, G. P. (1988b), The dynamical range of global circulations. II, *Clim. Dyn.*, *3*, 45–84.

Williams, G. P., and J. L. Holloway (1982), The range and unity of planetary circulations, *Nature*, *297*, 295–299.

Williams, P. D., T. W. N. Haine, and P. L. Read (2005), On the generation mechanisms of short-scale unbalanced modes in rotating two-layer flows with vertical shear, *J. Fluid Mech.*, *528*, 1–22.

Williams, P. D., T. W. N. Haine, and P. L. Read (2008), Inertia-gravity waves emitted from balanced flow: Observations, properties, and consequences, *J. Atmos. Sci.*, *65*, 3543–3556.

Williams, P. D., P. L. Read, and T. W. N. Haine (2010), Testing the limits of quasi-geostrophic theory: Application to observed laboratory flows outside the quasi-geostrophic regime, *J. Fluid Mech.*, *649*, 187–203.

Wordsworth, R. D., P. L. Read, and Y. H. Yamazaki (2008), Turbulence, waves, and jets in a differentially heated rotating annulus experiment, *Phys. Fluids*, *20*, 126,602.

Young, R. M. B., and P. L. Read (2008), Flow transitions resembling bifurcations of the logistic map in simulations of the baroclinic rotating annulus, *Physica D*, *237*, 2251–2262.

Young, R. M. B., and P. L. Read (2013), Data assimilation in the laboratory using a rotating annulus experiment, *Quart. J. R. Meteor. Soc.*, p. in press.

Zurita-Gotor, P., and R. S. Lindzen (2007), Theories of baroclinic adjustment and eddy equilibrations, in *The Global Circulation of the Atmosphere*, edited by T. Schneider and A. H. Sobel, pp. 22–46, Princeton University Press, Princeton, N.I.

2
Primary Flow Transitions in the Baroclinic Annulus: Prandtl Number Effects

Gregory M. Lewis, Nicolas Périnet, and Lennaert van Veen

2.1. INTRODUCTION

The baroclinic annulus experiments, also called differentially heated rotating fluid annulus experiments, have been used extensively to study baroclinic instability, which is an important mechanism in the process of cyclogenesis in the midlatitudes [*Holton*, 2004]. The experiments consist of observing the flow of a differentially heated fluid that is contained between two coaxial cylinders that are both rotated at rate Ω. The differential heating can be generated by holding the inner cylinder at a different temperature, usually colder, than the outer cylinder, but other mechanisms have been used, for example, internal heating. See, e.g., *Read* [2001] for a review. The differential heating and rotation provide the necessary conditions to observe baroclinic instability; in fact, it has been shown that over a wide range of parameters, it is the primary mechanism of instability [see, e.g., *Williams*, 1971; *Hide and Mason*, 1975].

For small values of the forcing (rate of rotation and differential heating), a steady axisymmetric flow is observed. However, as the parameters are varied, the steady flow becomes unstable to baroclinic perturbations, and a steadily rotating wave equilibrates. These waves are often referred to as steady waves [*Hide and Mason*, 1975] or baroclinic waves because they result as a consequence of baroclinic instability [*Williams*, 1971]. A wide variety of other complex wave flows may also be observed in different regions of the parameter space, for example, modulated waves (e.g., amplitude vacillation [*Hignett*, 1985]), mixed azimuthal mode flows (also called wave dispersion [*Pfeffer and Fowlis*, 1968] or interference vacillation [*Früh and Read*, 1997]), as well as more complicated flows such as modulated amplitude vacillation [*Read et al.*, 1992;

Institute of Technology, University of Ontario, Oshawa, Ontario, Canada.

Früh and Read, 1997] and geostrophic turbulence [*Read*, 2001].

The Taylor number $\mathcal{T}$ and the thermal Rossby number $\mathcal{R}$ are generally regarded as the two most important dimensionless parameters that determine the nature of the observed flow [*Hide and Mason*, 1975], where the Taylor number is given by

$$\mathcal{T} = \frac{4\Omega^2 R^4}{\nu^2}, \quad (2.1)$$

where Ω is the rate of rotation, $R = r_b - r_a$ is the difference between the outer radius r_b and inner radius r_a of the annulus, and ν is the kinematic viscosity of the fluid, and the thermal Rossby number is given by

$$\mathcal{R} = \frac{\alpha g D \Delta T}{\Omega^2 R^2}, \quad (2.2)$$

where ΔT is the imposed horizontal temperature gradient, α is the coefficient of thermal expansion, D is the depth of the fluid, and g is the gravitational acceleration. Indeed, if all other parameters are held fixed, the thermal Rossby number and Taylor number together have a one-to-one relationship with the rotation rate Ω and differential heating ΔT, which are the physical parameters that are varied in a particular experiment. Thus, experimental results are usually presented as a regime diagram on a log-log graph of Taylor number versus thermal Rossby number. However, the dynamics of the fluid in fact depend on no less than four other dimensionless parameters. One possible set of dimensionless parameters consists of the thermal Rossby number, the Taylor number, the vertical aspect ratio $\delta = D/R$, the horizontal aspect ratio $\eta = r_a/R$, a rotational Froude number $F = \Omega^2 R/g$, and the Prandtl number $\mathrm{Pr} = \nu/\kappa$, where κ is the thermal diffusivity. This set is not unique, and other complete sets with different dimensionless parameters can be constructed.

Modeling Atmospheric and Oceanic Flows: Insights from Laboratory Experiments and Numerical Simulations, First Edition. Edited by Thomas von Larcher and Paul D. Williams.
© 2015 American Geophysical Union. Published 2015 by John Wiley & Sons, Inc.

Different experiments have been performed using annuli of different dimensions. A comparison of these experiments can be used to determine the dependence of the flow on the aspect ratio δ. For example, it is observed that an increase in δ leads to an onset of steady waves with smaller wave number [see, e.g., *Fein*, 1973; *Hignett et al.*, 1985]. The effects of the centrifugal buoyancy have been discussed by *Lewis and Nagata* [2004] and can be related to the dependence on the rotational Froude number F, which measures the relative strength of centrifugal and gravitational buoyancy.

Perhaps the most interesting parameter, other than the thermal Rossby number and Taylor number, is the Prandtl number Pr. Various experiments have been performed using fluids of different Prandtl number. Most experiments have used fluids of moderate to high Prandtl number, e.g., water with Pr $\approx$ 7 (*Fowlis and Hide* [1965], *Fein* [1973], and others), a water-glycerol mixture with Pr $\approx$ 13 [*Hignett et al.*, 1985] or higher [*Read et al.*, 1992; *Früh and Read*, 1997], or silicone oil with Pr $\approx$ 63 [*Fein and Pfeffer*, 1976]. However, *Fein and Pfeffer* [1976] also used mercury as the working fluid, which has a low Prandtl number (Pr $\approx$ 0.025), and *Castrejon-Pita and Read* [2007] performed experiments using air, which has a near-unity Prandtl number. See also Chapter 3. Many theoretical studies have looked at fluids of moderate to high Prandtl number, and some theoretical results exist for Prandtl number near unity [*Randriamampianina et al.*, 2006; *Read et al.*, 2008; *Lewis*, 2010]; however, lower Prandtl number fluids have not been studied extensively.

In this chapter, we focus on the effects that the Prandtl number has on the dynamics near the primary transition, i.e., the transition from the steady axisymmetric solution to rotating waves. In particular, we follow previous work [*Lewis and Nagata*, 2003; *Lewis*, 2010] and use a dynamical systems approach that uses linear stability analysis, center manifold reduction, and normal forms to determine the behavior of the model equations near the primary transition. The methods will enable us to extend the results of *Lewis and Nagata* [2003] and *Lewis* [2010] and compute results over a large range of Prandtl number, covering values ranging from Pr = 0.025 to Pr = 13. Results show that for smaller values of the Prandtl number it is possible to observe a hysteretic primary transition, and there are regimes in parameter space where stable mixed-mode flows exist.

In the next section, we describe the model equations. In Section 2.3, we describe the methods involved in computing the primary transition curve, i.e., the curve in the parameter space that indicates where baroclinic instability sets in, and present an example of a regime diagram for an experiment that uses water as the working fluid. The nonlinear analysis and results for a wide range of Prandtl number are presented in Section 2.4. We conclude with a discussion of how the analysis can be extended to study the flow far from the transition, in particular, in the wave and vacillation regimes.

2.2. MODEL

As in many previous theoretical investigations of the annulus experiments, we take the governing equations of the fluid flow to be the Navier-Stokes equations in the Boussinesq approximation [see, e.g., *Williams*, 1971; *Lu and Miller*, 1997; *Lewis and Nagata*, 2003, 2004; *Lewis*, 2010; *Randriamampianina et al.*, 2006; *Miller and Butler*, 1991]. Here, we take this to mean that all fluid properties are constant except for the fluid density ρ, whose variation itself is assumed negligible, except for when it is multiplied by buoyancy forces such as gravity or the centrifugal force. The equation of state of the fluid is taken to be

$$\rho = \rho_0 \left[1 - \alpha \left(T - T_0\right)\right], \quad (2.3)$$

where ρ_0 is the fluid density at a reference temperature T_0 and α, called the coefficient of thermal expansion, is assumed to be constant and small. Given these assumptions, the fluid, to first order, can be considered as incompressible. The heat equation with an advection term is used to model the evolution of the fluid temperature.

For many fluids, such as water, the Boussinesq approximation is an accurate model for reasonable thermal forcing. For air in the context of the annulus, the Boussinesq approximation is still a reasonable approximation; with a differential heating of 40 K and using a reference temperature $T_0 = 313$ K, the error in the kinematic viscosity ν and the thermal diffusivity κ is approximately $\pm 11\%$. However, over this same range, the Prandtl number Pr = ν/κ varies by less than 1%. In some investigations [e.g., *Hignett et al.*, 1985; *Young and Read*, 2008], an extension to the Boussinesq approximation is made. In particular, the kinematic viscosity ν, thermal diffusivity κ, and coefficient of thermal expansion α are assumed to vary linearly or quadratically with temperature. In this case as well, to first order the fluid is assumed to be incompressible. Such an approximation allows for a more accurate representation at larger differential heating.

It is convenient to write the governing equations in a frame of reference rotating with the annulus at angular velocity Ω and in cylindrical polar coordinates, where the radial, azimuthal, and vertical (or axial) coordinates are denoted r, φ, and z, respectively. As such, the Navier-Stokes equations in the Boussinesq approximation describing the fluid flow can be written as

$$\frac{\partial \mathbf{u}}{\partial t} = \nu \nabla^2 \mathbf{u} - \frac{1}{\rho_0} \nabla p - 2\Omega \mathbf{e}_z \times \mathbf{u} \\ + \left[g\mathbf{e}_z - \Omega^2 r \mathbf{e}_r\right] \alpha \left(T - T_0\right) - (\mathbf{u} \cdot \nabla)\mathbf{u}, \quad (2.4)$$

$$\frac{\partial T}{\partial t} = \kappa \nabla^2 T - (\mathbf{u} \cdot \nabla) T, \qquad (2.5)$$

$$\nabla \cdot \mathbf{u} = 0, \qquad (2.6)$$

where $\mathbf{u}(r, \varphi, z, t) = u(r, \varphi, z, t)\mathbf{e}_r + v(r, \varphi, z, t)\mathbf{e}_\varphi + w(r, \varphi, z, t)\mathbf{e}_z$ is the fluid velocity vector that has been decomposed into components that lie along the standard cylindrical polar basis vectors $\mathbf{e}_r$, $\mathbf{e}_\varphi$, and $\mathbf{e}_z$; $\mathbf{\Omega} = \Omega \mathbf{e}_z$ is the rotation vector; $\Omega = |\mathbf{\Omega}|$ is the rate of rotation about the axis of the annulus; p is the pressure deviation from $p_0 = \rho_0 g(D - z) + \rho_0 \Omega^2 r^2/2$; ν is the kinematic viscosity; κ is the coefficient of thermal diffusivity; g is the gravitational acceleration; T_0 is the reference temperature; and ∇ is the usual gradient operator in cylindrical coordinates. The domain is

$$r_a \leq r \leq r_b, \qquad 0 \leq \varphi < 2\pi, \qquad 0 \leq z \leq D.$$

The boundary conditions are chosen to mimic the ideal experimental apparatus. In particular, for experiments using a rigid, flat top and bottom of the annulus, the boundary conditions on the fluid velocity are taken to be nonslip at all boundaries. The differential heating is imposed through the boundary conditions on the temperature; that is, the inner wall of the annulus is held fixed at $T = T_a$ and the outer wall at $T = T_b$. We take $\Delta T = T_b - T_a > 0$, as in most studies of the annulus, although some studies have considered $\Delta T < 0$ [*Koschmieder and White*, 1981]. It is assumed that the top and bottom of the annulus are perfect thermal insulators. Thus, the boundary conditions are

$$\mathbf{u} = \mathbf{0} \quad \text{on} \quad r = r_a, r_b \quad \text{and} \quad z = 0, D, \qquad (2.7)$$

$$T = T_a \quad \text{on} \quad r = r_a,$$

$$T = T_b \quad \text{on} \quad r = r_b,$$

$$\frac{\partial T}{\partial z} = 0 \quad \text{on} \quad z = 0, D, \qquad (2.8)$$

with 2π periodicity in φ for $\mathbf{u}$, T, and p.

For the situation of a free upper surface instead of a rigid flat top, *Williams* [1971] and *Miller and Butler* [1991] assumed stress-free boundary conditions at $z = D$. The differences in boundary conditions induce some small but interesting differences in the observations. In particular, a hysteretic primary transition is seen at parameter values for which this did not occur in the rigid-lid case. We do not investigate this further here.

The effects due to centrifugal buoyancy are included via the term $\Omega^2 r \alpha (T - T_0) \mathbf{e}_r$. These effects are generally included in models because they are not insignificant at higher rotation rates and are necessary to obtain a quantitative comparison with experiments. See *Lewis and Nagata* [2004] for a discussion of this term's quantitative effects on the primary transition.

Nondimensionalization is usually not performed, essentially because little benefit is gained by doing so. Unlike in some other classical fluid dynamics problems, the introduction of dimensionless parameters does not significantly reduce the dimension of the parameter space. In particular, there are a total of seven dimensional parameters: $T_b - T_a \equiv \Delta T$, Ω, ν, κ, α, $r_b - r_a \equiv R$, and D. Many choices of nondimensionalization are possible; however, in all cases, the resulting equations depend on no less than six dimensionless parameters (see Section 2.1).

2.3. TRANSITION CURVE

Most of the numerical studies of the annulus use a numerical experimentation approach. This essentially consists of performing a series of numerical integrations of the governing equations from a specific initial state for a small set of parameter values; this is similar to the procedure used in many laboratory experiments. This can be a computationally efficient method for determining the basic features of the flow in various regions in the parameter space, including approximations of the extent of the various flow regimes. However, such methods have difficulty pinpointing the boundaries between the various flow regimes and in uncovering certain details of the regimes. For instance, regions in which two flows are stable and the mechanism by which these bistable flows are generated cannot be determined. Here we implement an alternative approach that uses numerical linear stability analysis [*Lewis and Nagata*, 2003; *Lewis*, 2010]. This not only provides a method for precisely determining the location of the transition curve but also is the first step in a weakly nonlinear analysis (center manifold reduction and normal forms) that can provide important information about the flow transitions. Although computationally more challenging than numerical experimentation, the analysis supplies complementary information regarding the primary transition. We will describe the nonlinear analysis in the next section, while in this section we will provide some details of the linear stability analysis. The methods and presentation follow the work of *Lewis and Nagata* [2003], where more details can be found.

In the linear stability analysis, we seek the parameter values at which the steady axisymmetric solution (basic state) loses stability to small (baroclinic) perturbations. These locations in parameter space will correspond to the primary transition, i.e., the axisymmetric-to-wave transition. We find the steady axisymmetric solution by looking for solutions of equations (2.4)–(2.6) subject to the boundary conditions (2.7) and (2.8) in the form

$$\mathbf{u} = \mathbf{u}^{(0)}(r, z), \quad p = p^{(0)}(r, z), \quad T = T^{(0)}(r, z),$$

i.e., solutions that are independent of time t and the azimuthal variable φ. The axisymmetric solution also depends on the parameters, but we do not indicate this dependence explicitly. The steady axisymmetric solution can be found directly using a stream function approach. See *Lewis and Nagata* [2003] for details. Analytical solutions cannot be found, and therefore numerical approximations are used. We discuss the numerical methods briefly below. The solution corresponds to a toric convection cell in which the fluid rises at the outer cylinder and falls at the inner while being deflected to the right as it passes from the outer to the inner cylinder in the top of the annulus and from the inner to the outer cylinder in the bottom.

It is convenient to compute the linearized stability of the steady axisymmetric solution from the perturbation equations, which can be obtained by substituting

$$\mathbf{u} = \mathbf{u}^{(0)} + \hat{\mathbf{u}}, \quad p = p^{(0)} + \hat{p}, \quad T = T^{(0)} + \hat{T} \quad (2.9)$$

into equations (2.4)–(2.6) and forming equations for the hatted variables, i.e., the perturbations. The trivial solution $\hat{\mathbf{u}} = \mathbf{0}$, $\hat{p} = 0$, $\hat{T} = 0$ then satisfies the perturbation equations and corresponds to the steady axisymmetric solution. In what follows, we drop the hats.

It is also convenient to represent the perturbation equations in the following (abstract) form:

$$\dot{U} = \mathbf{L}\,U + \mathbf{N}(U), \quad (2.10)$$

where U is a vector of the functions $\mathbf{u}$ and T, $\mathbf{L}$ is a (partial differential) linear operator such that $\mathbf{L}\,U$ is the linear part of the equation, $\mathbf{N}(U)$ is the nonlinear part, and the velocity components of U are divergence free. Written as such, some important theoretical properties have been established that enable the analysis described below [*Henry*, 1981].

Assuming that the perturbations are sufficiently small that the nonlinear part may be neglected, the linear stability of the steady axisymmetric solution can be determined from the spectrum of the linearization $\mathbf{L}$. In particular, we assume that the perturbations may be written as

$$\mathbf{u} = \mathbf{u}(r, \varphi, z, t) = e^{\lambda t}\tilde{\mathbf{u}}_m(r,z)e^{im\varphi}, \quad (2.11)$$

with m an integer and likewise for T and p, where the azimuthal dependence of these functions can be assumed due to periodicity in the azimuthal variable. Consequently, we obtain a linear eigenvalue problem for the eigenvalues λ and the eigenfunctions $\left[\tilde{\mathbf{u}}_m(r,z), \tilde{T}_m(r,z)\right]e^{im\varphi}$ for each azimuthal wave number m. The eigenfunctions will determine the form of the bifurcating solutions, and thus we anticipate that they will be azimuthal waves of some integer azimuthal wave number m.

If $m \neq 0$, it is possible to eliminate the pressure and azimuthal velocity. The resulting three equations in the three remaining unknowns $\tilde{u}_m(r,z)$, $\tilde{w}_m(r,z)$, and $\tilde{T}_m(r,z)$ may be written in the form of a generalized linear eigenvalue problem

$$\lambda \mathbf{A}_m \tilde{U}_m = \mathbf{L}_m \tilde{U}_m, \quad (2.12)$$

where

$$\tilde{U}_m = \begin{pmatrix} \tilde{u}_m \\ \tilde{w}_m \\ \tilde{T}_m \end{pmatrix}$$

and $\mathbf{A}_m$ and $\mathbf{L}_m$ are 3×3 matrices of linear operators. If $m = 0$, a stream function method can be used in exactly the same manner as in the calculation of the axisymmetric solution. Again numerical approximations are implemented.

If, for a given set of parameter values, all eigenvalues λ have negative real parts, then all (small) perturbations from the axisymmetric solution will decay, see (2.11), and thus, the steady axisymmetric solution is asymptotically stable. If any eigenvalue λ has a positive real part, then a perturbation in the direction of the corresponding eigenfunction will grow, and the steady axisymmetric solution is unstable. The steady solution is called neutrally stable if there are a finite number of eigenvalues with zero real part while all other eigenvalues have negative real part. In a two-dimensional parameter space, there is a curve along which the steady axisymmetric solution is neutrally stable. On one side of this curve the steady solution is linearly stable, while on the other side the solution is linearly unstable. Thus, we expect that a transition will occur as parameters are varied from the stable side to the unstable side.

Procedurally, because the eigenvalue problem takes the form (2.12), a neutral stability curve is constructed for each azimuthal wave number m, where a neutral stability curve is a curve in the space of parameters along which at least one eigenvalue corresponding to a given wave number m has zero real part while all other eigenvalues for that wave number have negative real part. Thus, to one side of the neutral stability curve for wave number m, the axisymmetric solution is stable to all perturbations of azimuthal wave number m, while to the other side it is unstable to some perturbation with that wave number and thus is unstable. Consequently, the portions of the neutral stability curves, for which the eigenvalues corresponding to all other wave numbers have negative real part, form a segment of the transition curve.

In Figure 2.1 is a transition curve for the case where water is the working fluid; specifically, the experimental results of *Fein* [1973] are compared to the linear stability analysis of *Lewis and Nagata* [2003]. It can be seen that the linear stability analysis is successful in predicting the location of the primary transition. The approximations are not as accurate at large Rossby number; this may be a consequence of discretization error, which becomes larger at higher differential heating.

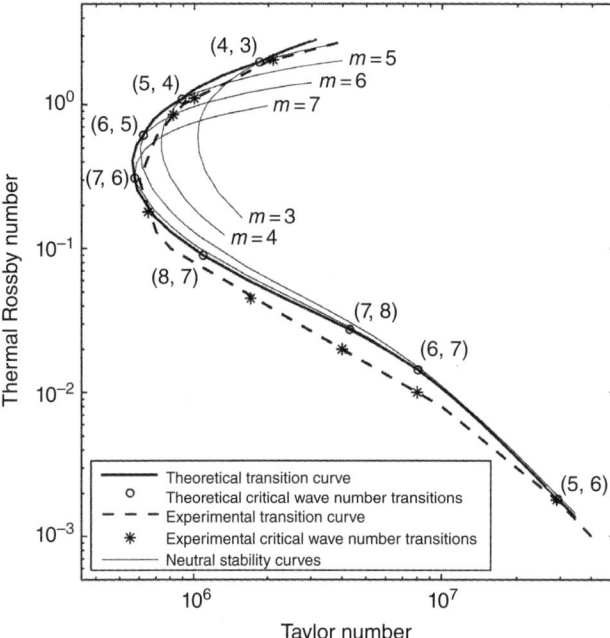

Figure 2.1. Primary transition curve with neutral stability curves for moderate Prandtl number, Pr ≡ 7.1 (water). Reproduced following *Lewis and Nagata* [2003] with the numerical linear stability analysis results of *Lewis and Nagata* [2003] and the experimental results of *Fein* [1973].

The linear stability analysis only indicates the stability of the axisymmetric solution and does not determine to what the perturbations from an unstable solution will equilibrate. The nonlinear analysis discussed in the next section may provide some insight into this.

2.4. NONLINEAR ANALYSIS

Due to the rotational symmetry of the annulus, all eigenvalues for nonzero wave numbers will come in complex conjugate pairs (see, e.g., *Hennessy and Lewis* [2012]). Except at isolated points along the transition curve, there is a single complex conjugate pair of eigenvalues with zero real part, and thus we expect that the transition corresponds to a Hopf bifurcation. Consequently, we expect that a periodic orbit is born at the transition. Whether this periodic orbit is stable or unstable can be determined by the nonlinear analysis. Due to the form of the eigenfunctions, we expect that this periodic orbit corresponds to a rotating wave (i.e., a baroclinic wave) with azimuthal wave number corresponding to that of the eigenfunction of the eigenvalue with zero real part at the transition.

At isolated points along the transition curve, specifically at the intersection of the neutral stability curves of different wave number, there is what we call a wave number transition. That is, we expect that the wave number of the bifurcating wave will be different for the portion of the transition curve on either side of these points. At such points, there are two complex conjugate pairs of eigenvalues with zero real part, while all other eigenvalues have negative real part, i.e., two rotating waves of different wave number bifurcate simultaneously as parameters are varied through this point. These critical values define the locations of mode interaction points (also called double Hopf bifurcation points). The nonlinear interaction of the two bifurcating waves will produce interesting dynamics that will be observed as parameters are varied near these critical points.

Suppose that the mode interaction points have been isolated at the critical parameter values $\Omega = \Omega_0$ and $\Delta T = \Delta T_0$ so that for Ω near Ω_0 and ΔT near ΔT_0 there are eigenvalues

$$\lambda_1 = \mu_1 + i\omega_1, \quad \bar{\lambda}_1, \quad \lambda_2 = \mu_2 + i\omega_2, \quad \bar{\lambda}_2, \quad (2.13)$$

where $\mu_j = \mu_j(\Omega, \Delta T)$ and $\mu_j(\Omega_0, \Delta T_0) = 0$ for $j = 1, 2$, i.e., at $\Omega = \Omega_0$ and $\Delta T = \Delta T_0$, there are two complex conjugate pairs of eigenvalues with zero real part. Also, assume that all the other eigenvalues have negative real parts.

The eigenfunctions corresponding to the above eigenvalues are

$$\Phi_1, \quad \bar{\Phi}_1, \quad \Phi_2, \quad \bar{\Phi}_2,$$

where they have the form

$$\Phi_j = \left[\tilde{\mathbf{u}}_{m_j}(r, z), \tilde{T}_{m_j}(r, z)\right] e^{im_j \varphi},$$

with m_j ($j = 1, 2$, $m_1 \neq m_2$) being the azimuthal wave number corresponding to Φ_j. The adjoint eigenfunctions corresponding to the Φ_j are denoted by Φ_j^*, where the Φ_j^* are found from the adjoint eigenvalue problem.

We write U as

$$U = z_1 \Phi_1 + \bar{z}_1 \bar{\Phi}_1 + z_2 \Phi_2 + \bar{z}_2 \bar{\Phi}_2 + \Psi, \quad (2.14)$$

where $z_1 \Phi_1 + \bar{z}_1 \bar{\Phi}_1 + z_2 \Phi_2 + \bar{z}_2 \bar{\Phi}_2 \in E^c$ and $\Psi \in E^s$, the center eigenspace E^c the span of the eigenfunctions corresponding to the eigenvalues with zero real parts when $\Omega = \Omega_0$ and $\Delta T = \Delta T_0$ and the stable eigenspace E^s the span of all the other eigenfunctions, which are the eigenfunctions that correspond to eigenvalues with negative real parts.

Given certain technical assumptions on (2.10), we can invoke the center manifold theorem, which enables us to determine the dynamics of the full partial differential model equations (PDEs) close to the bifurcation point from a low-order ordinary differential equation (ODE) [*Henry*, 1981; *Lewis and Nagata*, 2003]. The theorem states that there exists an exponentially attracting manifold that is tangent to the center eigenspace E^c, and thus, close to the bifurcation point, the equations governing the dynamics on the manifold are topologically equivalent to those of the full PDEs. In particular, the existence and stability of solutions of the PDEs can be determined from the center manifold equations, the dimension of which is the dimension of E^c, in this case 4. Practically, the assumptions allow us to express Ψ of (2.14) on the center manifold in terms of the center space variables, i.e., the complex wave amplitudes z_j; thus the equations on the center manifold are decoupled from the stable space equations. Because we are interested in results close to the bifurcation point, it is convenient to work in Taylor series for the nonlinear part $\mathbf{N}(U)$ and the function that describes the center manifold, the coefficients of which can be computed order by order. The underlying assumption is that the contribution of the nonlinear terms are small but important. The analysis is sometimes called weakly nonlinear and is similar to the method of separation of time scales.

The four-dimensional ODE for the complex wave amplitudes z_j that describe the dynamics on the center manifold is

$$\dot{z}_1 = \lambda_1 z_1 + G_{11} z_1^2 \bar{z}_1 + G_{12} z_1 z_2 \bar{z}_2 + O(4),$$
$$\dot{z}_2 = \lambda_2 z_2 + G_{21} z_1 \bar{z}_1 z_2 + G_{22} z_2^2 \bar{z}_2 + O(4), \quad (2.15)$$

where $\lambda_j = \lambda_j(\Omega, \Delta T)$, $O(4)$ indicates terms of order 4 and higher, and the normal form coefficients G_{kl} are complex numbers that depend on the nonlinear part $\mathbf{N}(U)$ of (2.10), the eigenfunctions, the adjoint eigenfunctions, and certain Taylor coefficients of the center manifold function. Equations (2.15) are often referred to as the amplitude equations. The details are suppressed here for space considerations; however, all details, in the present context, can be found in the literature [*Lewis and Nagata*, 2003]. This normal form requires the nonresonance condition that the imaginary parts of the eigenvalues, ω_1 and ω_2, satisfy $n_1 \omega_1 + n_2 \omega_2 \neq 0$ for all integers n_1 and n_2 with $|n_1| + |n_2| \leq 4$ at the critical parameter values $\Omega = \Omega_0$ and $\Delta T = \Delta T_0$.

We write $z_1 = \rho_1 e^{i\theta_1}/\sqrt{|G_{11}^r|}$ and $z_2 = \rho_2 e^{i\theta_2}/\sqrt{|G_{22}^r|}$, where G_{ij}^r is the real part of the normal form coefficients G_{ij} and substitute these expressions into (2.15). In these scaled polar coordinates, the truncated normal form equations are

$$\dot{\rho}_1 = \rho_1 \left(\mu_1 + a\rho_1^2 + b\rho_2^2\right), \quad (2.16)$$
$$\dot{\rho}_2 = \rho_2 \left(\mu_2 + c\rho_1^2 + d\rho_2^2\right), \quad (2.17)$$
$$\dot{\theta}_1 = \omega_1, \quad (2.18)$$
$$\dot{\theta}_2 = \omega_2, \quad (2.19)$$

where

$$a = \frac{G_{11}^r}{|G_{11}^r|} = \pm 1, \quad b = \frac{G_{12}^r}{|G_{22}^r|}, \quad c = \frac{G_{21}^r}{|G_{11}^r|},$$
$$d = \frac{G_{22}^r}{|G_{22}^r|} = \pm 1, \quad (2.20)$$

and $\lambda_j = \mu_j + i\omega_j$. The $O(|\rho_1, \rho_2|^4)$ terms are ignored in the $\dot{\rho}_j$ equations, and the $O(|\rho_1, \rho_2|^2)$ terms are ignored in the $\dot{\theta}_j$ equations. Ignoring these terms does not affect the local dynamics, except for fine details of the dynamics.

Thus, given m_1 and m_2, the coefficients of the scaled normal form equations a, b, c, d can be written in terms of the following functions: the axisymmetric solution, the eigenfunctions and adjoint eigenfunctions, and certain Taylor coefficients of the center manifold function, which are all only functions of two spatial variables r and z. The eigenfunctions are found from the eigenvalue problem (2.12), and the coefficients of the center manifold function are found from linear PDEs that are derived using certain properties of the center manifold. See *Hassard et al.* [1981] for a general discussion; the equations for this problem are written out by *Lewis and Nagata* [2003].

The axisymmetric solution, eigenfunctions, and coefficients of the center manifold cannot be computed analytically, and therefore we use numerical approximations. Upon discretization of the spatial derivatives using second-order finite differences, the axisymmetric solution is found from a nonlinear system of equations; the linear stability, including the eigenfunctions, is found from a generalized matrix eigenvalue problem; and the coefficients of the center manifold are found from systems of linear equations. A nonuniform grid is used in order to better resolve the boundary layers that occur in the flow. The discretization of the PDEs leads to large, sparse systems, for which sparse matrix computations are required in order to achieve sufficient grid resolution. For more details and a discussion of the implementation and convergence, see *Lewis and Nagata* [2003] and *Lewis* [2010]. In particular, in [*Lewis*, 2010], computations with grid size up to $N = 150$ are performed, where N is the number of grid points in each of the independent variables. It is shown that $N = 50$, the value used in the results presented below, is sufficient resolution to obtain a good approximation of the normal form coefficients.

Given the values of the coefficients a, b, c, and d and the value of $A = ad - bc$, the dynamics close to the mode interaction points can be determined by an analysis of the truncated normal form equations (2.16)–(2.19). Because ρ_j and θ_j correspond to a scaled amplitude and phase, respectively, of the complex amplitude z_j, these can be reinterpreted in terms of the full PDEs through (2.14).

The analysis of (2.16)–(2.19) in a general context is described in detail by *Guckenheimer and Holmes* [1983] and *Kuznetsov* [2004]. Here we present a short summary in the context of the present application.

The last two equations of the normal form, (2.18) and (2.19), indicate that to first order the variables θ_1 and θ_2 are simply linear functions of time t and therefore represent a constant rotation. Thus, we need only consider the dependence on the radial variables ρ_1 and ρ_2, i.e., the dynamics can be found from considering only (2.16) and (2.17).

For the reduced system (2.16)–(2.17), the following fixed points exist when the quantities inside the square root signs are positive:

1. $\rho_1 = \rho_2 = 0$
2. $\rho_2 = 0$ and $\rho_1 = \rho_p = \sqrt{\dfrac{\mu_1}{-a}}$
3. $\rho_1 = 0$ and $\rho_2 = \rho_q = \sqrt{\dfrac{\mu_2}{-d}}$
4. $\rho_1 = \rho_1^{(T)} = \sqrt{\dfrac{-d\mu_1 + b\mu_2}{A}}$ and
$\rho_2 = \rho_2^{(T)} = \sqrt{\dfrac{c\mu_1 - a\mu_2}{A}}$

Fixed point 1 corresponds to the trivial fixed point of the full normal form equations (2.16)–(2.19) and corresponds to the steady axisymmetric solution of the model equations. For fixed point 2, because we have $\rho_2 = 0$, the constant rotation due to the θ_1 dependence implies that there exists a periodic orbit when $\rho_1 = \rho_p > 0$. This periodic orbit corresponds to a wave number m_1 rotating wave of the full model equations. Likewise fixed point 3 corresponds to a rotating wave of wave number m_2. Fixed point 4 has $\rho_1 > 0$ and $\rho_2 > 0$, and thus, the θ_1 and θ_2 dependence implies that the corresponding orbit of (2.16)–(2.19) lies on a 2-torus. If ω_1 and ω_2 are not rationally related, then the orbit is quasi-periodic on the torus. This is the case for all bifurcation points considered here. Fixed point 4 corresponds to a mixed-mode flow for the full model equations. A straightforward analysis can determine the dependence of the stability of the fixed points on the coefficients [*Guckenheimer and Holmes*, 1983; *Kuznetsov*, 2004]. The correspondence between the fixed points 1–4 and the fluid flows described here holds for the entire discussion below, i.e. we determine how the stability of these flows, and the extent in parameter space in which they exist, change as the Prandtl number is varied.

2.4.1. Results Over a Range of Prandtl Numbers

In this section, we investigate the effects of Prandtl number on the dynamics that are observed close to the mode interaction points. Specifically, as the Prandtl number is varied, we use a pseudo-arc length continuation method [see, e.g., *Govaerts* [2000]] to track the location of the mode interaction points in parameter space, and at each location, we compute the normal form coefficients from which the dynamics are determined. For further discussion of pseudo-arc length continuation see Section 2.5. The normal form coefficients have previously been calculated for two different annuli with fluids of different Prandtl number: *Lewis and Nagata* [2003] compared to experiments of *Fein* [1973], who used water as the working fluid, and *Lewis* [2010] compared to numerical experiments of *Randriamampianina et al.* [2006], in which the fluid parameters are taken to be those of air. Here we connect and extend these results by varying the kinematic viscosity ν and thermal diffusivity κ such that for high Prandtl number they correspond to the values for a gycerol-water mixture as in the experiments of *Hignett et al.* [1985], for moderate Prandtl number the values correspond to those of water [*Fein*, 1973], for near unity Prandtl number they correspond to the values of air [*Castrejon-Pita and Read*, 2007], while at low Prandtl number they correspond to mercury [*Fein and Pfeffer*, 1976]. We do not consider variations in the coefficient of thermal expansion α of the various fluids, because a change in α can be unambiguously compensated for with a change in ΔT, as they both only appear in the thermal Rossby number. Thus, although a change in α will change the specific value of ΔT at which the transition occurs, the change in α would have no effect on the critical values of the nondimensional parameters, nor would it affect the predicted dynamics.

The parameters for the geometry of the annulus are taken to be

$$r_a = 3.48 \text{ cm}, \qquad r_b = 6.02 \text{ cm}, \qquad D = 10 \text{ cm}.$$

These values are similar to those used by *Randriamampianina et al.* [2006] and *Pfeffer and Fowlis* [1968]. We also fix

$$\alpha = 3.30 \cdot 10^{-3} \text{ K}^{-1}, \qquad g = 980 \text{ cm/s}^2.$$

Results for three mode-interaction points are presented in Figures 2.2–2.4, which display the values of the normal form coefficients a, b, c, and d and of $A = ad - bc$ as a function of Prandtl number Pr. Other mode interaction points exist but for only some values of the Prandtl number; these are not included here. Also, due to spatial resonance, the $(m_1, m_2) = (2, 1)$ mode interaction point involves a different, more complicated analysis and is a topic of future research. The regions of qualitatively different behavior are separated with vertical lines, and the corresponding dynamics are labeled according to the cases identified by *Guckenheimer and Holmes* [1983], where the borders of these regions correspond to a change in sign of one of the coefficients a, b, c, and d or of A. For a table which relates the signs of the coefficients to the given dynamic region, see the work of *Guckenheimer and Holmes* [1983]. The prime in the case labels indicates that it is the "time-reverse" case of the corresponding case of

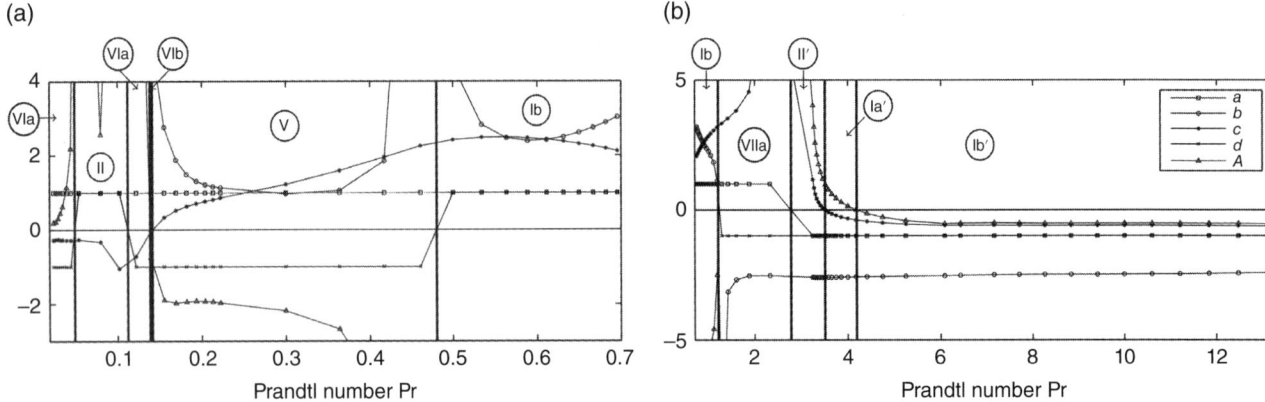

Figure 2.2. Results for the $(m_1, m_2) = (3, 2)$ mode interaction point. Computed values of the normal form coefficients a, b, c, and d and $A = ad - bc$ are shown as a function of the Prandtl number Pr. (a) Results for Pr between 0.025 and 0.7 and (b) results for Pr between 0.7 and 13, where at Pr = 0.025 the parameters correspond to those of mercury, at Pr = 0.7 to those of air, and at Pr = 13 to those of a water-glycerol mixture. In order to ensure that the values of ν and κ correspond to those of these fluids at the given Prandtl number, it is necessary to change the magnitude of the incrementation of ν and κ as the Prandtl number is varied; discontinuous changes in the incrementation result in kinks in the graphs of b, c, and A. Vertical lines indicate transitions between qualitatively different dynamics that are labeled according to the case numbers of *Guckenheimer and Holmes* [1983].

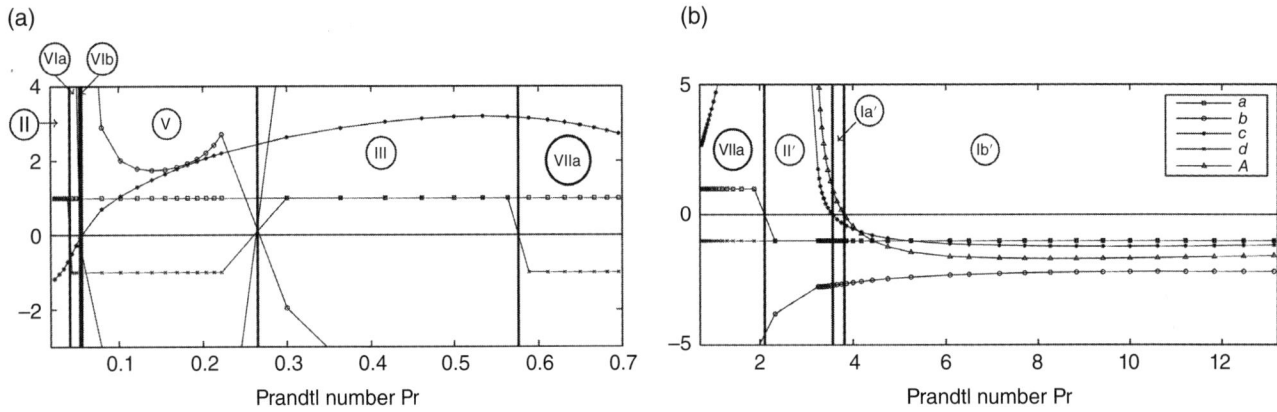

Figure 2.3. Results for the $(m_1, m_2) = (4, 3)$ mode interaction point. See caption of Figure 2.2 for description.

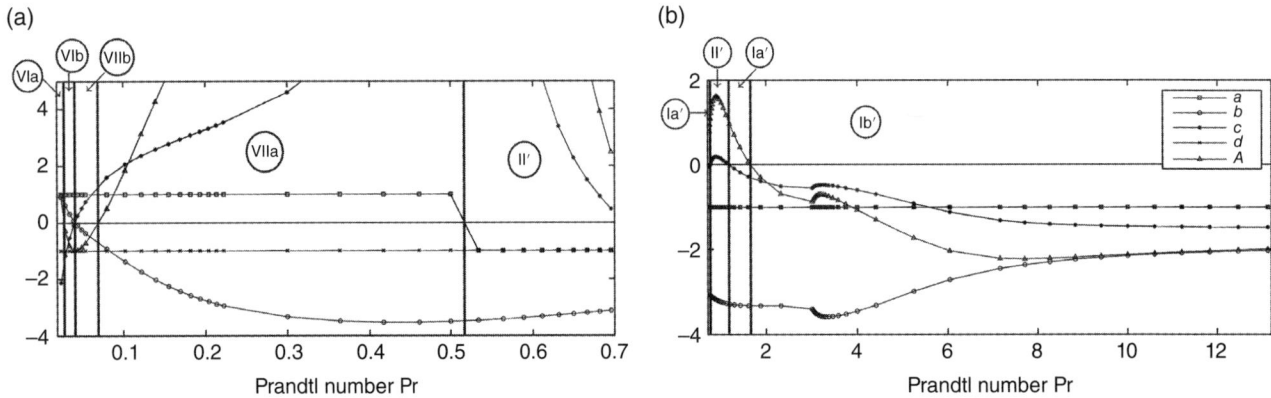

Figure 2.4. Results for the $(m_1, m_2) = (5, 4)$ mode interaction point. See caption of Figure 2.2 for description.

Guckenheimer and Holmes [1983]; e.g., case II′ is the time reverse of case II.

The results indicate that case Ib′ applies for all mode interaction points for all Prandtl numbers greater than 4. This case corresponds to the "simple" case 1 of *Kuznetsov* [2004]. The results of *Lewis and Nagata* [2003], in which a fluid of Prandtl number 7.1 is studied, fall into this category; the details of the dynamics of all high Prandtl number are thus qualitatively the same. This is consistent with the experimental results reported in [*Fowlis and Hide*, 1965; *Pfeffer and Fowlis*, 1968].

This is an important case because it indicates the mechanism by which hysteresis of rotating waves occurs in the annulus. In this case, all normal form coefficients are negative and A is also negative. The negative a and d indicate that supercritical Hopf bifurcations occur at $\mu_1 = 0$ and $\mu_2 = 0$, respectively; i.e., there exist periodic orbits corresponding to fixed points 2 and 3 for $\mu_1 > 0$ and $\mu_2 > 0$, respectively. Fixed point 4 exists in the wedge defined by $\mu_2 > -\mu_1/b$ and $\mu_2 < -c\mu_1$. Linear stability analysis reveals that fixed point 2 is stable for $\mu_2 < -c\mu_1$ and unstable elsewhere (where it exists) and 3 is stable for $\mu_2 > -\mu_1/b$ and unstable elsewhere. Thus, both fixed points 2 and 3 are stable inside the wedge, and thus, there is bistability of the corresponding wave solutions. Fixed point 4 is always unstable, and thus, the corresponding mixed-mode flow would not be observable in experiment.

The corresponding two-parameter bifurcation diagram with corresponding phase portraits is presented in Figure 2.5. It is presented with μ_1 and μ_2 as parameters with the understanding that these depend directly on the physical parameters Ω and ΔT and thus on the nondimensional parameters, the thermal Rossby number $\mathcal{R}$, and the Taylor number $\mathcal{T}$, with which most regime diagrams are presented. This figure and similar ones that follow should be interpreted as follows: The line $\mu_j = 0$ corresponds to the neutral stability curve of wave number m_j, $j = 1, 2$. Thus, the four quadrants in Figure 2.5 correspond to the four regions created by the intersection of two neutral stability curves. The relative location of the curves in the $\mathcal{T}$-$\mathcal{R}$ plane corresponding to the lines $\mu_2 = -\mu_1/b$ and $\mu_2 = -c\mu_1$ are indicated in Figure 2.5, but the exact form they take depends on the values of b and c and on how μ_1 and μ_2 depend on Ω and ΔT.

The hysteresis between the wave solutions may occur on a one-parameter path through parameter space such as the one parameterized by s that is indicated in Figure 2.5. See Figure 2.6 for the bifurcation diagram that would be observed along the path s. Thus, it would be expected that the transition from the rotating wave of dominant wave number m_1 to one of wave number m_2 would occur at different parameter values depending on whether you followed a path of increasing or decreasing s.

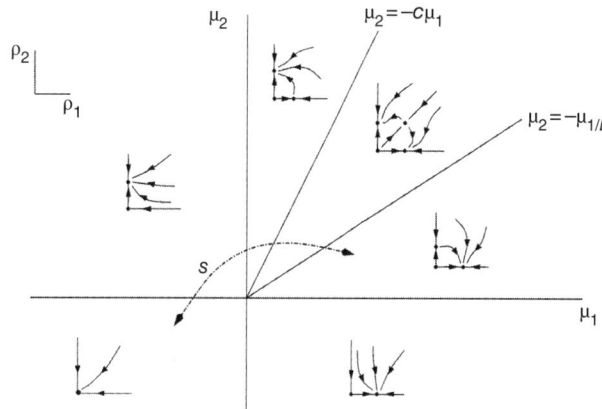

Figure 2.5. Two-parameter bifurcation diagram for case Ib′ (reproduced following [*Lewis and Nagata*, 2003]). The diagram is displayed using the real parts of the critical eigenvalues μ_1, μ_2 as the bifurcation parameters. The regions of different character are separated by solid lines. In each region, the corresponding phase portrait is drawn, where the phase portraits are presented in (ρ_1, ρ_2) coordinates. Fixed points at the origin correspond to the steady axisymmetric solutions of the full model equations; fixed points along the horizontal (vertical) axis correspond to wave number m_1 (m_2) rotating waves, while fixed points for which $\rho_1 > 0$ and $\rho_2 > 0$ correspond to mixed-mode flows for the model equations.

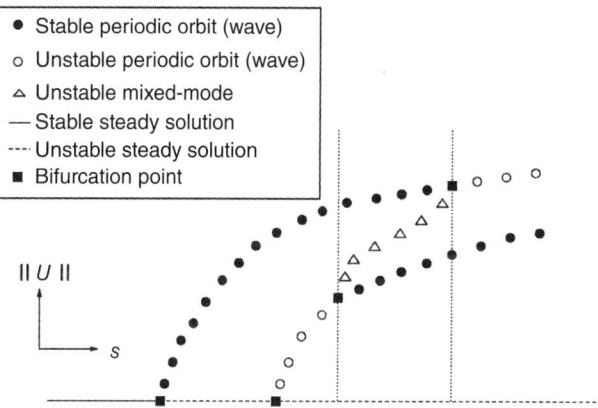

Figure 2.6. One-parameter bifurcation diagram that may be observed along the path parameterized by s that is indicated in the two-parameter bifurcation diagram for case Ib′ plotted in Figure 2.5.

For all mode interaction points, as the Prandtl number is decreased, there is a transition from case Ib′ to case Ia′, equivalently from simple case 1 to simple case 2 of *Kuznetsov* [2004]. Case Ia′ is another important case; the change in case occurs via a change in sign of A while all other coefficients remain of the same sign. The change in

A corresponds to a relative change in value of b and c such that the line $\mu_2 = -\mu_1/b$ now lies above $\mu_2 = -c\mu_1$. Because the fixed points 2 and 3 are stable under the same conditions, they are both now *unstable* in the wedge defined by $\mu_2 < -\mu_1/b$ and $\mu_2 > -c\mu_1$. Linear stability analysis of fixed point 4 reveals that it is now always *stable* when it exists. Thus, the mixed-mode flow corresponding to fixed point 4 may be observable in experiment. The two-parameter bifurcation diagram is presented as simple case 2 by *Kuznetsov* [2004]; it has a similar structure as in Figure 2.5, except with the change in relative orientation of the lines $\mu_2 = -\mu_1/b$ and $\mu_2 = -c\mu_1$ and the change in stability of the fixed points as indicated.

Subsequently, for all mode interaction points, a decrease in Prandtl number leads to a change in sign of the coefficient c, which implies that $A > 0$. Thus, we have case II′, which is the time reverse of case II of *Guckenheimer and Holmes* [1983]. *Kuznetsov* [2004] refers to this as simple case 3. See Figure 2.7. This is similar to case Ia′ except that the line $\mu_2 = -c\mu_1$ now lies below the line $\mu_2 = 0$. As above, the negative a and d indicate that there exist periodic orbits corresponding to fixed points 2 and 3 for $\mu_1 > 0$ and $\mu_2 > 0$, respectively, and fixed point 4 exists in the wedge defined by $\mu_2 < -\mu_1/b$ and $\mu_2 > -c\mu_1$. Linear stability analysis reveals that fixed points 2 and 3 are stable (unstable) outside (inside) this wedge, when they exist, and that fixed point 4 is always stable. Thus, again the mixed-mode flow corresponding to fixed point 4 may be observable in experiment.

Again, for all mode interaction points, as the Prandtl number is decreased, there is a transition to case VIIa. In this case, we have $a = 1$, $b < -1$, $c > 1$, $d = -1$, and $A > 0$. For most instances when this case applies, we also have that $b < -1$ and $c > 1$, which corresponds to subcase (b) of case VIIa [*Guckenheimer and Holmes*, 1983]. This case is presented as "difficult" case I by *Kuznetsov* [2004]. In this case, there is a supercritical Hopf bifurcation at $\mu_2 = 0$ and a subcritical Hopf bifurcation at $\mu_1 = 0$. The subcritical Hopf bifurcation indicates that there is an unstable periodic orbit for negative values of μ_1. Such a bifurcation is associated with a hysteretic primary transition. However, the hysteresis occurs via a different mechanism than the hysteresis observed in the wave regime. In particular, if it is assumed that there is a smooth transition between the supercritical and subcritical cases, we can assume that the branch of stable wave solutions of wave number m_j that exists in the wave regime (i.e., for positive values of μ_j) in the supercritical case does not disappear. Thus, the branch of unstable wave solutions that emanate from the subcritical Hopf bifurcation will link up with the stable branch in a saddle node bifurcation at some negative value of μ_j, creating a region in parameter space in which the stable axisymmetric solution and stable wave solutions coexist. This creates the necessary conditions to observe hysteresis. The analysis presented here, i.e., the computation of only the third-order coefficients, is insufficient to prove whether this actually occurs or to determine the extent of the hysteresis (i.e., the parameter value at which the saddle node bifurcation occurs). The higher order coefficients may be able to determine these. However, even the computation of the coefficients of the next order is prohibitively difficult, and so we do not compute it here. In Section 2.5, we discuss alternative methods that are capable of computing both the stable and unstable branches and therefore of showing the existence and extent of the hysteresis. Hysteretic transitions have been observed in the numerical experiments of *Randriamampianina et al.* [2006] in the case of a near-unity Prandtl number fluid. However, the range of hysteresis is very small, which may account for why it is not observed in the experiments of *Castrejon-Pita and Read* [2007]. Hysteresis is also observed in the case of higher Prandtl number fluid when the upper surface of the fluid is free [*Fein*, 1973]. Figures 2.2–2.4 all indicate that such hysteretic transitions become more common at lower Prandtl number, first appearing at Prandtl number less than 3, in the $(m_1, m_2) = (3, 2)$ mode interaction. However, the figures do not indicate the largest value of the Prandtl number at which hysteresis occurs because the mode interaction points only provide information at a single point on the transition curve. The other mode interaction points show a hysteretic primary transition at much lower values of the Prandtl number. We also note that Case VIIa, subcase (b) occurs in the $(m_1, m_2) = (4, 3)$ mode-interaction point at Prandtl number corresponding to that of air. However, this nor any of the other Cases observed for these parameters explain the occurrence of the experimentally observed weak waves, see [*Castrejon-Pita and Read*, 2007; *Lewis*, 2010].

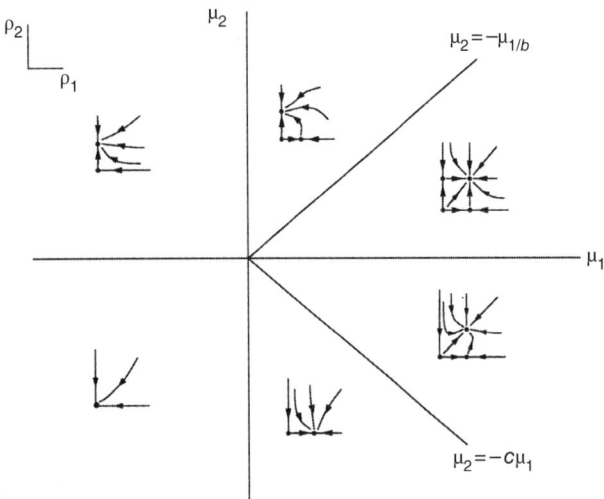

Figure 2.7. Two-dimensional bifurcation diagram for case II′. See caption of Figure 2.5 for description.

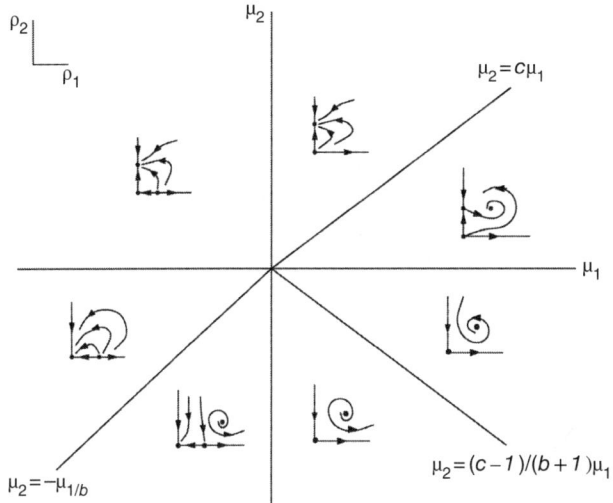

Figure 2.8. Two-dimensional bifurcation diagram for case VIIa, subcase (b). See caption of Figure 2.5 for description.

The fixed point 4 of case VIIa, subcase (b) exists in the wedge defined by $\mu_2 < -\mu_1/b$ and $\mu_2 < c\mu_1$, where the borders of the wedge, given by $\mu_2 = -\mu_1/b$ and $\mu_2 = c\mu_1$, are in the lower left quadrant and upper right quadrant, respectively, of the parameter space. A linear stability analysis reveals that the eigenvalues of fixed point 4 are a complex conjugate pair with real parts positive for $\mu_2 < (c-1)/(b+1)\mu_1$ and with real parts negative for $\mu_2 > (c-1)/(b+1)\mu_1$, i.e.; fixed point 4 is a stable focus in the later case. Thus, there is a stable mixed-mode solution in the wedge defined by $\mu_2 < c\mu_1$ and $\mu_2 > (c-1)/(b+1)\mu_1$. The phase portraits in all regions of the space of parameters are presented in Figure 2.8.

Along the curve $\mu_2 = (c-1)/(b+1)\mu_1$, a Hopf bifurcation of fixed point 4 occurs. This would imply that there is a periodic orbit for the reduced system (2.16)–(2.17) near this line. In the full normal form equations (2.16)–(2.19), this bifurcation corresponds to a bifurcation from the 2-torus, and the periodic orbit of the reduced system would correspond to a 3-torus. In general, the corresponding flow that would be observed in the annulus would have three non-rationally-related frequencies. The stability of this solution (i.e., whether the bifurcation is subcritical or supercritical) would depend on the fifth-order terms in the normal form equations. These would be formidable to compute in this context, and so we do not attempt it here. It would be interesting to see if evidence of such a solution could be seen in either experiment or numerical simulation.

Similarly, in case VIa, a Hopf bifurcation of fixed point 4 occurs, and again, depending on the higher order terms, it may be possible to observe a three-frequency flow. In this case, however, the wedge in which the mixed-mode solution may be observed may be small. For a bifurcation diagram see *Guckenheimer and Holmes* [1983], where an example of a situation in which a stable 3-torus exists is also given. Also, see *Kuznetsov* [2004] in which this case corresponds to difficult case 6.

Case VIa is observed at low Prandtl number, in particular, at the Prandtl number of mercury for the $(m_1, m_2) = (5, 4)$ and $(3, 2)$ mode interaction points and near this for the third point. These results are not inconsistent with the experimental results of *Fein and Pfeffer* [1976], although they do not explain some key experimental observations, in particular, the absence of the upper symmetric regime. In the current analysis, the primary transition occurs very near to the experimentally observed transition from non-geostrophic turbulence to the regular wave regime, which suggests that the observed turbulence may be related to this bifurcation. In particular, case VIa predicts the occurrence of 3-frequency flow, near which chaotic motion may be found.

Preceding the appearance of case VIa, for all mode interaction points, is case VIb. However, for intermediate values of the Prandtl number, we observe some differences in the dynamics for the different mode interaction points. The remaining cases that may be observed are discussed below.

In Cases Ib, II, and III[1], the analysis can say little about the flow that may be observed. In particular, no stable periodic or mixed-mode solution in any region of parameter space is predicted. See *Guckenheimer and Holmes* [1983] for the bifurcation diagram with corresponding phase portraits of case Ib. In all these cases, both Hopf bifurcations are subcritical, and therefore without the computation of higher order coefficients, it is not possible to determine the dynamics that would be observed. The mixed-mode solution corresponding to fixed point 4 is also unstable and thus would not be physically observable near these bifurcation points, although characteristics of the solution may be observed in transient flows.

In cases V, VIIb, and VIb,[2] we have $a = +1$ and $d = -1$, i.e., a subcritical Hopf bifurcation occurs at $\mu_1 = 0$ while a supercritical Hopf occurs at $\mu_2 = 0$. Thus, we would expect to observe a stable rotating wave of wave number m_2 as parameters are varied across $\mu_2 = 0$ such that $\mu_1 < 0$. However, all other solutions are unstable, and therefore, the analysis cannot determine the dynamics that will be observed in other regions. For example, for case V, the two-parameter bifurcation diagram with corresponding phase portraits for all regions of the space of parameters is

---

[1] These cases correspond to the time reverse of simple case 1, the time reverse of simple case 2, and the time reverse, with index switch, of simple case 3, respectively, of *Kuznetsov* [2004].

[2] These cases correspond to the time reverse, with index switch, of difficult case 4, difficult case 3, and difficult case 5, respectively, of *Kuznetsov* [2004].

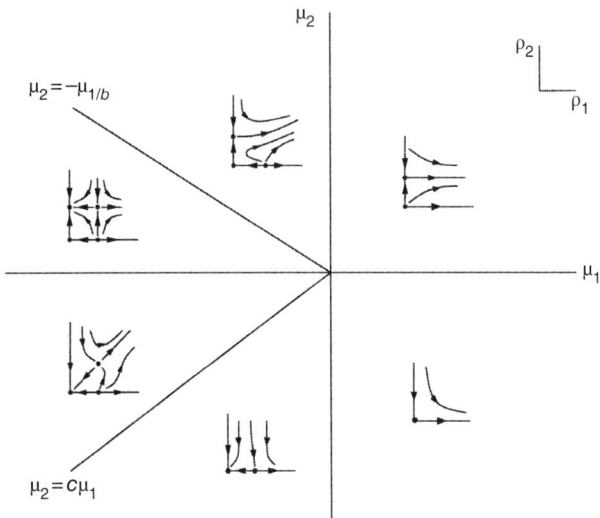

Figure 2.9. Two-parameter bifurcation diagram for case V. See caption of Figure 2.5 for description.

presented in Figure 2.9. In this case, the fixed point 4 exists in the wedge defined by $\mu_2 < -\mu_1/b$ and $\mu_2 > c\mu_1$, where the borders of the wedge given by $\mu_2 = -\mu_1/b$ and $\mu_2 = c\mu_1$ are in the upper left quadrant and lower left quadrant, respectively, of the parameter space. In (2.16)–(2.19), a bifurcation from a periodic orbit to a 2-torus, i.e., a Neimark-Sacker bifurcation, occurs as a parameter is varied across these borders. A linear stability analysis of the fixed points [*Guckenheimer and Holmes*, 1983; *Kuznetsov*, 2004] reveals that fixed point 3 is stable when it exists inside this wedge and unstable outside while fixed points 2 and 4 are always unstable. Thus, the mixed-mode solution corresponding to fixed point 4 would not be physically observable.

For cases VIIb and VIb, see *Guckenheimer and Holmes* [1983] or *Kuznetsov* [2004] for a bifurcation diagram.

2.5. CONCLUSION AND FUTURE WORK

Using methods from dynamical systems theory, we performed a detailed study of the effects of the Prandtl number on the primary transition. The analysis indicates that, for Prandtl number between approximately 4 and 13, bistability of rotating waves is observed, i.e., there are regions within the wave regime where the wave number of the rotating wave depends on the initial conditions and there is hysteresis in the transition between these waves. As the Prandtl number is decreased, quasi-periodic mixed azimuthal mode flows can be observed. In certain cases, depending on the values of the higher order normal form coefficients, it may also be possible to observe three-frequency flows. However, these coefficients are not computed here. Also, it is determined that for smaller values of the Prandtl number the primary transition may become hysteretic. Similar dynamics are observed for all mode interaction points.

In future work, we will extend the analysis beyond the primary transition and study other types of flow such as vacillation. However, the above analysis consisting of the reduction to the normal form equations (2.16)–(2.19) is based on weakly nonlinear theory and thus only predicts the qualitative behavior of the full problem (2.4)–(2.8) for parameter values and flow structures close to those of the primary bifurcation points. Further away from these points, we need to consider the fully nonlinear problem. Moreover, in some scenarios, such as the one in which the axisymmetric solution loses stability in a subcritical Hopf bifurcation, the third-order coefficients that are computed in this chapter are not sufficient to determine to what the flow will equilibrate, and the computation of the necessary higher order terms is a monumental task.

The dynamical systems approach can still be applied by implementing numerical continuation methods to the fully nonlinear problem. Using such methods, we are able to compute both stable and, in contrast to numerical and laboratory experiments, unstable special solutions of dynamical systems, such as equilibria and travelling waves. To conclude this chapter, we briefly describe the methods that we are implementing to extend the results discussed here to parameter values away from the primary transition.

Continuation methods aim to track special solutions as a parameter of the system is varied; that is, they aim to compute *branches* of special solutions in the given parameter. In the cases described in Section 2.4, these solutions are expected to be stationary (fixed points of type 1), periodic (fixed points of types 2 and 3) or quasi-periodic (fixed points of type 4). The method we use is called pseudo-arclength continuation, where the qualifier *pseudo-arclength* indicates that the computed branch of solutions is parameterized by the approximate arc length, rather than by the system parameter. The pseudo-arclength approach can be explained independently of the type of solution under consideration. Thus, we will describe it in the context of steady solutions, and modifications for computing periodic solutions will be mentioned later.

In the following, we will discuss the computation of solutions to a spatial discretization of model equations (2.4)–(2.8). The discretization can be the result of a finite-difference approximation of the spatial derivatives, as we used to compute the approximations presented in Section 2.4, or a spectral decomposition of the variables. In either case, the discretized model consists of a large set of ordinary differential equations with a highly dissipative structure due to the diffusive and viscous terms in the continuous model. The applications of pseudo-arclength continuation we describe are essentially the same as those that would be applied to low-dimensional dynamical systems, except that any dense linear problems that arise must be solved by iterative methods because direct methods are prohibitively memory and time intensive.

For the application of continuation methods to a branch of steady solutions, we could proceed, as we did in Section 2.3, by setting the time derivative equal to zero. Instead, we elect to compute steady solutions as fixed points of the flow over some finite time t. This makes the extension to periodic orbits straightforward and, as explained below, yields linear problems directly amenable to iterative solvers.

We define the map Φ to be the flow of the discretized system, i.e., the solution $\mathbf{X}(t)$ of the discretized model equations with initial condition $\mathbf{X}(0)$, for fixed values of the continuation parameter α:

$$\frac{d}{dt}\Phi(\mathbf{X}(0), t, \alpha) = \mathbf{F}(\Phi(\mathbf{X}(0), t, \alpha)), \quad (2.21)$$

where $\mathbf{X}(t)$ is a vector formed from the discretized velocity vector $\mathbf{u}$ and the temperature T, $\mathbf{F}$ corresponds to the discretization of equations (2.4)–(2.8), and the continuation parameter α could be, for instance, the rotation rate Ω or the differential heating ΔT. The stationarity condition can then be written as:

$$\Phi(\mathbf{X}, t, \alpha) - \mathbf{X} = \mathbf{0} \quad (2.22)$$

for arbitrary time t.

Generically, the solutions lie on curves, i.e., the solution branches. To proceed, it is necessary to know an initial point $(\mathbf{X}_0, \alpha_0)$ on the solution branch. This can be a stable fixed point computed by numerical integration or the discretization of a theoretically known solution. Given a known point $(\mathbf{X}_i, \alpha_i)$ along a branch, the continuation method is composed of two parts: (1) a prediction step, in which an initial guess $(\mathbf{X}^0_{i+1}, \alpha^0_{i+1})$ of a new point $(\mathbf{X}_{i+1}, \alpha_{i+1})$ along the branch is formed, and (2) a correction step, in which the initial guess is iteratively refined until it is sufficiently close to an exact solution.

The prediction $(\mathbf{X}^0_{i+1}, \alpha^0_{i+1})$ is generated from the previous (known) point $(\mathbf{X}_i, \alpha_i)$ and from an approximation of the tangent vector to the branch, $\mathbf{t}_i$. Specifically, we choose

$$(\mathbf{X}^0_{i+1}, \alpha^0_{i+1}) = (\mathbf{X}_i, \alpha_i) + \Delta \mathbf{t}_i, \quad (2.23)$$

where Δ is the step size, which approximates the increment of the arc length along the computed solution branch.

The correction step is based on Newton-Raphson iteration. However, the linear equation to be solved in each Newton-Raphson iteration is underdetermined, reflecting the fact that the next fixed point is not uniquely defined but lies on a curve. To render the next fixed point unique, we seek a solution in the hyperplane orthogonal to the tangent vector $\mathbf{t}_i$. Thus, instead of solving (2.22) alone, we solve the extended system:

$$\begin{aligned}\Phi(\mathbf{X}_{i+1}, t, \alpha_{i+1}) - \mathbf{X}_{i+1} &= 0, \\ (\mathbf{X}_{i+1} - \mathbf{X}^0_{i+1}) \cdot \mathbf{t}_{i\mathbf{X}} + (\alpha_{i+1} - \alpha^0_{i+1}) t_{i\alpha} &= 0.\end{aligned} \quad (2.24)$$

The second subscript attributed to the tangent vector denotes its component. The linear system that is to be solved at the kth Newton-Raphson iteration takes the form

$$\begin{bmatrix} \nabla_\mathbf{X}\Phi - \mathbb{I} & \nabla_\alpha\Phi \\ \mathbf{t}_{i\mathbf{X}} & t_{i\alpha} \end{bmatrix} \begin{bmatrix} \Delta\mathbf{X} \\ \Delta\alpha \end{bmatrix} = -\begin{bmatrix} \Phi(\mathbf{X}^{k-1}_{i+1}, t, \alpha_{i+1}) - \mathbf{X}^{k-1}_{i+1} \\ (\mathbf{X}^{k-1}_{i+1} - \mathbf{X}^0_{i+1}, \alpha^{k-1}_{i+1} - \alpha^0_{i+1})^t \mathbf{t}_i \end{bmatrix} \quad (2.25)$$

and leads to the kth approximation to the $(i+1)$st solution on the branch: $(\mathbf{X}^k_{i+1}, \alpha^k_{i+1}) = (\mathbf{X}^{k-1}_{i+1}, \alpha^{k-1}_{i+1}) + (\Delta\mathbf{X}, \Delta\alpha)$.

This linear system is large and dense and can, in general, not be solved by direct methods. Instead, we can use an iterative solver like GMRES [*Saad and Schultz*, 1986]. This method constructs an approximate solution to the linear systems based on a sequence of matrix-vector products. Each product can be computed from an integration of the linearized equations. Thus, each GMRES iteration requires one integration of the discretized model, along with its linearization, over time t. The number of iterations necessary to obtain a solution with sufficient accuracy to ensure convergence of the Newton-Raphson iteration depends strongly on the spectrum of the matrix. Generally speaking, the more clustered the spectrum, the faster the convergence of GMRES. Because of the dissipative nature of the discretized model, the spectrum will be more clustered for longer integration time t, since the majority of the eigenvalues of $\nabla\Phi$ correspond to strongly damped perturbations and accumulate around zero. Consequently, the choice of the integration time is a trade-off between the time taken by each GMRES iteration on one hand and the required number of iterations on the other.

The algorithm for tracking periodic solutions is largely the same as that for stationary states. The main difference is that the integration time is now the period τ of the orbit, which is unique but unknown and therefore becomes a new variable of the problem. An additional scalar equation is needed for closure. Given a solution $(\mathbf{X}_i, \alpha_i, \tau_i)$ and a prediction $(\mathbf{X}^0_{i+1}, \alpha^0_{i+1}, \tau^0_{i+1})$ for the next solution, the equations to solve are

$$\begin{aligned}\Phi(\mathbf{X}_{i+1}, \tau_{i+1}, \alpha_{i+1}) - \mathbf{X}_{i+1} &= 0, \\ (\mathbf{X}_{i+1} - \mathbf{X}^0_{i+1}) \cdot \mathbf{t}_{i\mathbf{X}} + (\alpha_{i+1} - \alpha^0_{i+1})t_{i\alpha} & \\ + (\tau_{i+1} - \tau^0_{i+1})t_{i\tau} &= 0, \\ f(\mathbf{X}_{i+1}) &= 0,\end{aligned} \quad (2.26)$$

where f is a suitable smooth function that defines a Poincaré plane of intersection. With this method, integration over time τ is necessary for each GMRES iteration. Unfortunately τ is often much higher than the time t optimized for GMRES. This makes the numerical solution of (2.26) very costly.

By using the hypotheses that the periodic solutions are waves traveling with constant phase speed ω and these waves do not deform (e.g., the fixed points of types 2 and 3 discussed in Section 2.4), the time consumption is rendered comparable to that of the method for finding stationary solutions described above. Instead of (2.26), we solve

$$\Phi(\mathbf{X}_{i+1}, t, \alpha_{i+1}) - \mathbf{R}(\omega_{i+1} t)\mathbf{X}_{i+1} = 0,$$
$$(\mathbf{X}_{i+1} - \mathbf{X}^0_{i+1}) \cdot \mathbf{t}_{i\mathbf{X}} + (\alpha_{i+1} - \alpha^0_{i+1})t_{i\alpha}$$
$$+ (\omega_{i+1} - \omega^0_{i+1})t_{i\omega} = 0, \quad (2.27)$$
$$(\mathbf{X}_{i+1} - \mathbf{X}^0_{i+1}) \cdot \mathbf{r} = 0.$$

Here, $\mathbf{R}(\omega t)\mathbf{X}_{i+1}$ represents the discretization of the velocity and temperature fields rotated over an angle ωt in the azimuthal direction and $\mathbf{r}$ is the generator of these rotations for $\mathbf{X}^0_{i+1}$. The phase speed ω replaces the period τ in the set of unknowns. The third equation imposes the condition that the consecutive Newton-Raphson update steps do not shift the azimuthal phase of the traveling wave. The rotated fields are efficiently computed in Fourier space as an element-wise multiplication by phase factors. Using forward and backward fast Fourier transforms for the angular coordinate ϕ, this is done at a computational cost that is negligible compared to the time stepping.

The combination of pseudo-arclength continuation with GMRES was first introduced by *Sánchez et al.* [2004] and is called *Newton-Krylov continuation*. *Viswanath* [2007] later adapted the method to the computation of equilibrium and modulated traveling waves. A third application worth mentioning here, in the light of the normal form analysis presented in Section 2.4, is that to the computation of invariant tori. *Sánchez et al.* [2010] cast this problem into a form suitable for Newton-Krylov continuation. The computation of quasi-periodic solutions is much less straightforward than that of equilibria and periodic solutions, though, and might fail if the quasi-periodic solution is strongly unstable or has a high ratio of frequencies. Nevertheless, it could be applied to examine quasi-periodic flows in regions of the parameter space where stable fixed points of type 4 exist or regions further from the transition where amplitude or other vacillation is observed.

Armed with these methods, a detailed characterization of the parameter space within the nonaxisymmetric regime can be determined. In particular, coupled with the computation of the stability of the solutions, the transitions between the various complex flows can be found. This will lead to a better understanding of how the instabilities lead to the observed dynamics, which in turn will shed light on the dynamics of all differentially heated rotating fluids.

Acknowledgments. The author would like to acknowledge the support of the Natural Sciences and Engineering Research Council of Canada and SHARCNET.

REFERENCES

Castrejon-Pita, A., and P. Read (2007), Baroclinic waves in an air-filled thermally driven rotating annulus, *Phys. Rev. E*, *75*(2), 026301. doi: http://dx.doi.org/10.1103/PhysRevE.75.026301.

Fein, J. (1973), An experimental study of the effects of the upper boundary condition on the thermal convection in a rotating cylindrical annulus of water, *Geophys. Fluid Dyn.*, *5*, 213–248.

Fein, J., and R. Pfeffer (1976), Experimental study of effects of Prandtl number on thermal convection in a rotating, differentially heated cylindrical annulus of fluid, *J. Fluid Mech.*, *75*, 81–112.

Fowlis, W., and R. Hide (1965), Thermal convection in a rotating annulus of liquid: Effect of viscosity on the transition between axisymmetric and non-axisymmetric flow regimes, *J. Atmos. Sci.*, *22*, 541–558.

Früh, W.-G., and P. Read (1997), Wave interactions and the transition to chaos of baroclinic waves in a thermally driven rotating annulus, *Philos. Trans. R. Soc. London, A*, *355*, 101–153.

Govaerts, W. (2000), *Numerical Methods for Bifurcations of Dynamical Equilibria*, SIAM, Philadelphia, Penna.

Guckenheimer, J., and P. Holmes (1983), *Nonlinear Oscillations, Dynamical Systems and Bifurcations of Vector Fields, Applied Mathematical Sciences*, vol. 42, Springer-Verlag, New York.

Hassard, B., N. Kazarinoff, and Y.-H. Wan (1981), *Theory and Applications of Hopf Bifurcation, London Mathematical Society Lecture Note Series*, vol. 41, Cambridge Univ. Press, Cambridge, UK.

Hennessy, M., and G. Lewis (2013), The primary flow transition in a differentially heated rotating channel of fluid with O(2) symmetry, *J. Computat. Appl. Math.*, *254*, 116–131.

Henry, D. (1981), *Geometric Theory of Semilinear Parabolic Equations, Lecture Notes in Mathematics*, vol. 840, Springer-Verlag, Berlin.

Hide, R., and J. Mason (1975), Sloping convection in a rotating fluid, *Adv. Geophys.*, *24*, 47–100.

Hignett, P. (1985), Characteristics of amplitude vacillation in a differentially heated rotating fluid annulus, *Geophys. Astrophys. Fluid Dynam.*, *31*, 247–281.

Hignett, P., A. White, R. Carter, W. Jackson, and R. Small (1985), A comparison of laboratory measurements and numerical simulations of baroclinic wave flows in a rotating cylindrical annulus, *Q. J. R. Meteorol. Soc.*, *111*, 131–154.

Holton, J. (2004), *An Introduction to Dynamic Meteorology*, 4th ed., Elsevier, Burlington, Mass.

Koschmieder, E., and H. White (1981), Convection in a rotating, laterally heated annulus: The wave number transitions, *Geophys. Astrophys. Fluid Dynam.*, *18*, 279–299.

Kuznetsov, Y. (2004), *Elements of Applied Bifurcation Theory*, 3rd ed., Springer-Verlag, New York.

Lewis, G. (2010), Mixed-mode solutions in an air-filled differentially heated rotating annulus, *Phys. D*, *239*, 1843–1854.

Lewis, G., and W. Nagata (2003), Double Hopf bifurcations in the differentially heated rotating annulus, *SIAM J. Appl. Math.*, *63*(3), 1029–1055.

Lewis, G., and W. Nagata (2004), Linear stability analysis for the differentially heated rotating annulus, *Geophys. Astrophys. Fluid Dynam.*, *98*(2), 129–152.

Lu, H.-I., and T. Miller (1997), Wave dispersion in a rotating, differentially-heated fluid model, *Dynam. Atmos. Oceans*, *27*, 505–526.

Miller, T., and K. Butler (1991), Hysteresis and the transition between axisymmetric flow and wave flow in the baroclinic annulus, *J. Atmos. Sci.*, *48*(6), 811–823.

Pfeffer, R., and W. Fowlis (1968), Wave dispersion in a rotating, differentially heated cylindrical annulus of fluid, *J. Atmos. Sci.*, *25*, 361–371.

Randriamampianina, A., W.-G. Früh, P. Read, and P. Maubert (2006), Direct numerical simulations of bifurcations in an air-filled rotating baroclinic annulus, *J. Fluid Mechan.*, *561*, 359–389.

Read, P. (2001), Transition to geostrophic turbulence in the laboratory, and as a paradigm in atmospheres and oceans, *Surv. Geophys.*, *22*, 265–317.

Read, P., M. Bell, D. Johnson, and R. Small (1992), Quasi-periodic and chaotic flow regimes in a thermally driven, rotating fluid annulus, *J. Fluid Mechan.*, *238*, 599–632.

Read, P., P. Maubert, A. Randriamampianina, and W. Früh (2008), DNS of transitions towards structural vacillation in an air-filled, rotating baroclinic annulus, *Phys. Fluids*, *20*(4), 044107. doi: http://dx.doi.org/10.1063/1.2911045.

Saad, Y., and M. H. Schultz (1986), GMRES: A generalized minimal residual algorithm for solving nonsymmetric linear systems, *SIAM J. Sci. Statist. Comput.*, *7*, 856.

Sánchez, J., M. Net, B. García-Archila, and C. Simó (2004), Newton-Krylov continuation of periodic orbits for Navier-Stokes flows, *J. Comput. Phys.*, *201*, 13–33.

Sánchez, J., M. Net, and C. Simó (2010), Computation of invariant tori by Newton–Krylov methods in large-scale dissipative systems, *Phys. D*, *239*, 123–133.

Viswanath, D. (2007), Recurrent motions within plane Couette turbulence, *J. Fluid Mechan.*, *580*, 339–358.

Williams, G. (1971), Baroclinic annulus waves, *J. Fluid Mechan.*, *49*, 417–449.

Young, R., and P. Read (2008), Flow transitions resembling bifurcations of the logistic map in simulations of the baroclinic rotating annulus, *Phys. D*, *237*(18), 2251–2262.

3

Amplitude Vacillation in Baroclinic Flows

Wolf-Gerrit Früh

3.1. PHENOMENOLOGY OF AMPLITUDE VACILLATION

The first reference to the term *vacillation* in reference to baroclinic flows, with a qualitative description, can be found in a brief note from January 1953 by *Hide* [1953] on observations in a rotating baroclinic annulus: "One cycle of this phenomenon, which has been termed 'vacillation' begins (say) with a symmetrical wave pattern with its continuous 'jet'. Some seconds later there is a distinct leaning backward of the troughs and a decrease of their width. This is followed by the troughs returning to N.-S. orientation and then leaning forward in preparation of the stage when the 'jet' stream is actually interrupted and intense cyclones are formed in the position of the wave troughs. The cyclones decay and the 'jet' is re-established; the wave pattern returns to the initial stage and the cycle starts again. The period corresponds to a few 'weeks'." A full description of his observations can be found in *Hide* [1958], and two typical snapshots of a vacillating wave number 3 are shown in Figure 3.1. While it will become apparent in this chapter that this excerpt describes structural vacillation, rather than amplitude vacillation, it initiated detailed research into vacillating flows in many places.

This term vacillation was subsequently taken up as a technical term and its definition refined, distinguishing between amplitude vacillation and shape vacillation [*Hide and Mason*, 1975]. *Fowlis and Pfeffer* [1969] characterized amplitude vacillation based on an array of thermistors in a large baroclinic annulus. Since then, amplitude vacillation has been investigated in laboratory experiments with a range of thermally driven baroclinic rotating annulus experiments, e.g., by *White and Koschmieder* [1981], *Tamaki and Ukaji* [1985], and *Sitte and Egbers* [2000], in

Figure 3.1. Two flow stages within an amplitude vacillation cycle. Courtesy of Peter Read, published in *Lappa* [2012], Copyright © John Wiley & Sons. Used with permission.

addition to those by Hide and Pfeffer. While these experiments generated the baroclinic flow by thermal forcing of the sidewalls, it can also be mechanically forced by a differentially rotating lid in contact with the upper layer of a two-layer fluid. This system has also been investigated in detail by *Hart* [1972] and successors. Like the thermally driven annulus, this two-layer experiment has also shown amplitude vacillation in experiments with both two immiscible fluids [*Hart*, 1973, 1976] and salt-stratified water [*Flór et al.*, 2011].

Another example of flow observations referred to as amplitude vacillation is from a thermally driven annulus rotated so rapidly that the centrifugal term outweighs terrestrial gravity [*Azouni et al.*, 1986; *Or and Busse*, 1987; *Schnaubelt and Busse*, 1992]. In that case, the fluid is no longer stably stratified, and the resulting flow is closer to rotationally constrained Rayleigh-Bénard convection than to baroclinic instability. The term vacillation is also used to describe atmospheric phenomena such as tropospheric wave-zonal flow fluctuations [*Koo et al.*, 2002; *Feliks et al.*, 2011], the stratospheric vacillation

---

School of Engineering & Physical Sciences, Heriot-Watt University, Edinburgh, United Kingdom.

Modeling Atmospheric and Oceanic Flows: Insights from Laboratory Experiments and Numerical Simulations, First Edition. Edited by Thomas von Larcher and Paul D. Williams.
© 2015 American Geophysical Union. Published 2015 by John Wiley & Sons, Inc.

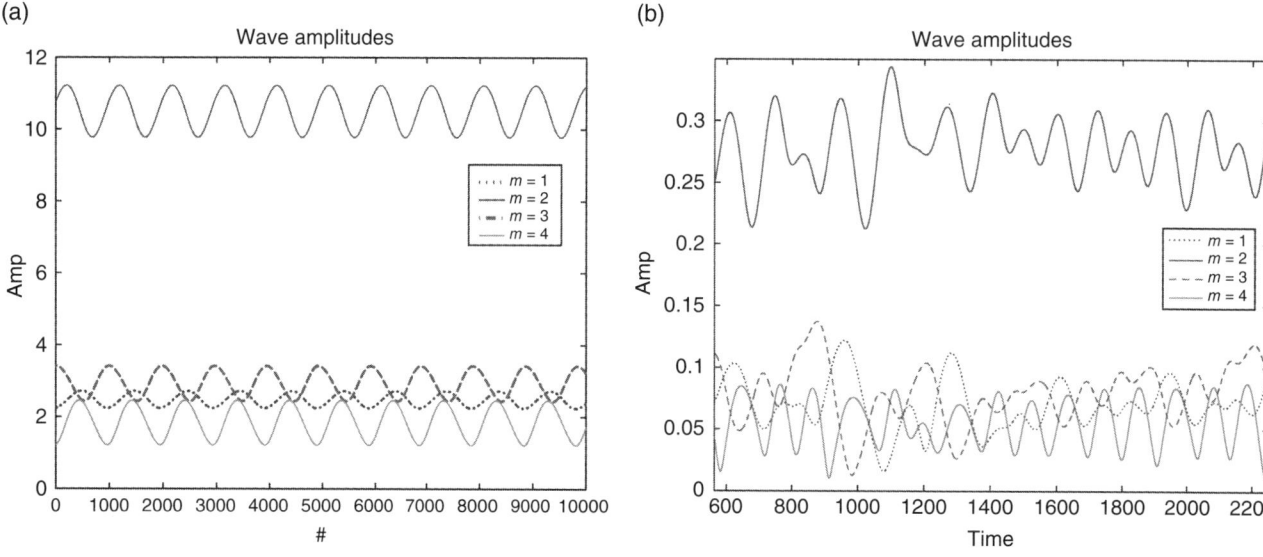

Figure 3.2. Typical time series of wave amplitudes for (a) amplitude vacillation and (b) a modulated amplitude vacillation. Quantities shown are Fourier amplitudes from the temperature fields calculated by the model described by *Randriamampianina and Crespo Del Arco* [2014].

[*Christiansen*, 2000; *Pogoreltsev et al.*, 2009; *Studer et al.*, 2012], or seasurface temperature (SST) fluctuations [*Watterson*, 2001], and for climate fluctuations [*Son and Lee*, 2006].

A widely accepted definition of amplitude vacillation (AV) is now that it is a (fairly) regular oscillation of the magnitude of a well-defined wave mode while the spatial structure remains (essentially) unchanged. A more complex form of vacillation is modulated amplitude vacillation (MAV) [*Read et al.*, 1992; *Früh and Read*, 1997], which frequently displays chaotic oscillations and usually involves fluctuations of several wave modes of different spatial structure. A periodic amplitude vacillation of a wave number 2 and a chaotic modulated amplitude vacillation are illustrated in Figure 3.2, both taken from the direct numerical simulations of an air-filled annulus described by *Randriamampianina et al.* [2006] and *Randriamampianina and Crespo Del Arco* [2014] in this book. A special case of a flow that appears like a modulated amplitude vacillation is the superposition of two steady wave modes; this has been termed *interference vacillation* [*Ohlsen and Hart*, 1989; *Lindzen et al.*, 1982].

For laboratory experiments of baroclinic flows, these types of amplitude vacillations are contrasted to "structural vacillation", "tilted-trough vacillation" or "shape vacillation" [*Hide and Mason*, 1975; *Tamaki and Ukaji*, 1993; *Pfeffer et al.*, 1997; *Früh et al.*, 2007; *Hide*, 2011; *Read et al.*, 1992; *Früh and Read*, 1997]. These flows are mainly characterized by distinct changes in the shape but little changes in the power of the waves. While these vacillations can occur at a distinct time scale, they tend to be much less regular than amplitude vacillations, and the time scale is shorter than that of the typical amplitude vacillation. *Hart* [1972] also reported the phenomenon of "frequency vacillation" in the mechanically driven two-layer experiment that appeared to involve an oscillation of the wave speed independently of the wave amplitude though a similar time series was later interpreted by *Hart* [1979] as "wave number vacillation" where the flow structure vacillated between mode 1 and mode 2.

3.1.1. Nondimensional Parameters

In the thermally driven baroclinic annulus, the two principal non–dimensional parameters are usually the Taylor number and the thermal Rossby number. The Taylor number,

$$\text{Ta} = \left(\frac{fL^2}{\nu}\right)^2 \frac{L}{d} = \frac{4\Omega^2(b-a)^5}{\nu^2 d}, \quad (3.1)$$

is essentially the ratio of the Coriolis term to viscous dissipation, where $\Omega = f/2$ is the angular velocity of the annulus, L and d the horizontal and vertical length scales ($L = b - a$ with a the radius of the inner cylinder and b that of the outer), and ν the kinematic viscosity. Using an aspect ratio $\gamma = L/d$, the Taylor number can be equated to the Ekman number as $\text{Ta} = \gamma E^{-2}$.

The thermal Rossby number,

$$\Theta = \frac{g\alpha d \Delta T}{\Omega^2 (b-a)^2} \quad (3.2)$$

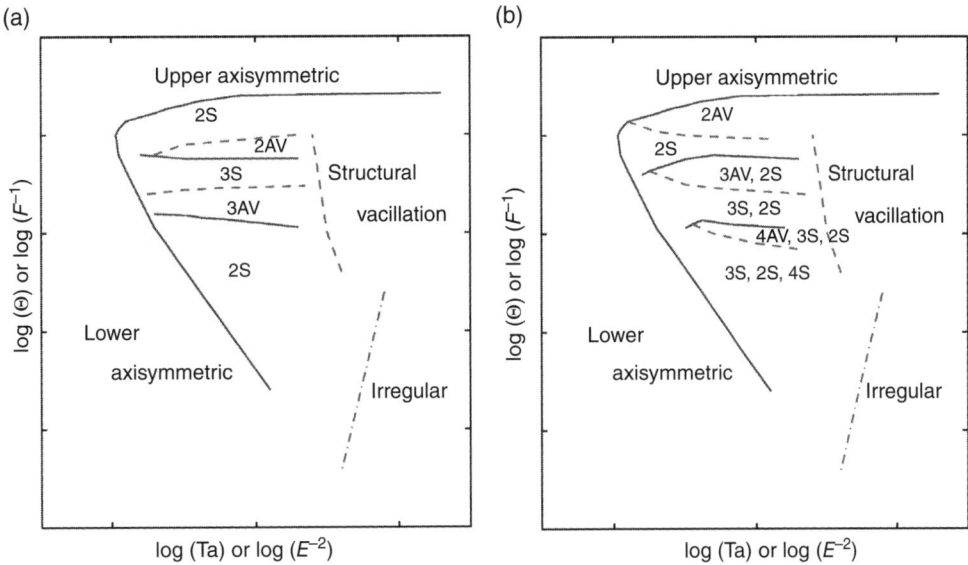

Figure 3.3. Typical regime diagrams for flow regimes in the thermally driven baroclinic annulus; (a) filled with air (after *Castrejon-Pita and Read* [2007], *Randriamampianina et al.* [2006], and *Read et al.* [2008]) and (b) filled with a water-glycerol fluid (after *Hide and Mason* [1975] and *Hignett* [1985]). The numbers refer to the dominant wave number and S indicates a steady wave while AV indicates amplitude vacillations including complex vacillations such as MAV. Common ranges of the values explored in such regime diagrams are $O(10^{-3}) \lesssim \Theta \lesssim O(1)$ and $O(10^5) \lesssim \mathrm{Ta} \lesssim O(10^{10})$.

is a measure of the vertical stratification through a ratio of the buoyancy term over the Coriolis term. The term *thermal Rossby number* originates from the standard Rossby number, $\mathrm{Ro} = U/(fL)$, where the scaling velocity is defined through the thermal wind balance $U_{th} = g\alpha \Delta T d/(fL)$. The parameter is also related to the stratification through the buoyancy frequency, $N^2 = -(\partial \rho/\partial z)g/\rho$ (commonly also referred to as the Brunt-Väisälä frequency), and by that to the rotational Froude number $F = (fL/(Nd))^2$ as $\Theta = 4/F$. In either case, the thermal Rossby number is a measure of the thermal forcing of the system whereas the Taylor number is a measure of the dissipation. The mechanically driven two-layer experiments tend to use the Rossby number defined by the mechanically imposed velocity as one of the two principal parameters and then either the Froude number or a dissipation parameter defined as $r = \sqrt{E}/\mathrm{Ro}$.

The flow observations can then be summarized in regime diagrams in the parameter space defined by the two parameters. While the illustrations of the regime diagrams in Figure 3.3 were derived from the thermally driven annulus, the equivalent regime diagrams for the mechanically driven systems will show similar features and structures. Since the regime diagrams intend to highlight the typical occurrence of amplitude vacillation-type flows, they were compiled from a wide range of experimental observations across different systems and hence do not show either more complex flows such as MAV flows or specific values of the nondimensional parameters since they depend on other parameters as well, such as aspect ratios and the Prandtl number. Common ranges of the values explored in these regimes are $O(10^{-3}) \lesssim \Theta \lesssim O(1)$ and $O(10^5) \lesssim \mathrm{Ta} \lesssim O(10^{10})$. Irrespective of precise values, a common observation is that amplitude vacillations, including AV and MAV, are found within the parameter space occupied by the regular and steady waves, whereas the less regular structural vacillation types are found at higher Taylor numbers or lower values of the dissipation parameter as the flow moves toward turbulent flow.

The classic Reynolds number as the ratio of the dissipation to the advection term is linked to the Taylor and Rossby numbers through $\mathrm{Re} = UL/\nu = (U/fL)(fL^2/\nu) = \mathrm{Ro}/E$. Identifying the thermal Rossby number with the Rossby number and $\mathrm{Ta} = \gamma E^{-2}$, the Reynolds number can be written as $\mathrm{Re} = \Theta\sqrt{\mathrm{Ta}/\gamma}$, and lines of constant Reynolds number in Figure 3.3 are lines with a slope of $-1/2$ with low-Reynolds-number values in the lower left corner and high Reynolds numbers in the upper right corner of the regime diagrams. Inserting the typical ranges for Θ and Ta results in a typical range of the Reynolds number as $O(1) \lesssim \mathrm{Re} \lesssim O(5)$. As will be seen later in Section 3.3.5, equation (3.4), the Ekman or Taylor number quantifies the dissipation through the Ekman layers, which is proportional to the potential vorticity. The Reynolds number, on the other hand, quantifies the horizontal viscous diffusion and is proportional to the second derivative of the potential vorticity and hence highly dependent on the length scales of the flow structures.

3.1.2. Transition to Amplitude Vacillation

Amplitude vacillations tend to develop from their corresponding steady wave flow through a supercritical bifurcation as precursor to a mode transition to a different wave number. While the occurrence of structural vacillation appears to be determined more by dissipation, the onset of amplitude vacillation and mode transitions are more determined by the thermal forcing. Other factors known to affect the occurrence of amplitude vacillation are the tank geometry and the fluid's Prandtl number. An impressionistic synthesis of the various experimental reports by, in particular, *Hide et al.* [1977], *Jonas* [1981], *Hignett* [1985], *Pfeffer et al.* [1973], *Pfeffer et al.* [1980b], *Buzyna et al.* [1984], *Sitte and Egbers* [2000], and *von Larcher and Egbers* [2005] suggests the following generalization: *Vacillations seem to be more prevalent in a wider and deeper annulus filled with a higher Prandtl number fluid.*

The fact that vacillation appears more easily in a wider gap could be a different phrasing of another observation, namely that amplitude vacillation tends to be seen more when the baroclinic wave has a relatively low wave number. *Hide and Mason* [1970] showed that only a finite range of wave number can be observed, given by the ratio of the zonal wave length to the gap width,

$$\alpha = \frac{m\pi(a+b)}{(b-a)} \quad \text{as } 0.25\alpha \leq m \leq 0.75\alpha. \quad (3.3)$$

The key difference between the regime diagram for a low-Prandtl-number fluid (Pr $\lesssim$ 1 in Figure 3.3a) and for a high-Prandtl-number fluid (10 $\lesssim$ Pr $\lesssim$ 80 in Figure 3.3b) is the relative position of the vacillating regime. For a lower Prandtl number, a steady wave can develop an amplitude vacillation as the thermal forcing is reduced, prior to a transition to a flow with a higher wave number, while the onset of amplitude vacillation in a fluid with a higher Prandtl number is usually found when the thermal forcing and stratification are increased. While there is no experimental evidence for a systematic trend in low-Prandtl-number fluids, it has been observed in many experiments that vacillation is rare in water but becomes more widely observed at higher Prandtl numbers, to a degree where steady waves become rare as the Prandtl number reaches values in excess of 40. For example, the water-filled annulus of *von Larcher and Egbers* [2005] only exhibited flow resembling amplitude vacillation in the region between the $m = 2$ and $m = 3$ dominated range in the narrow-gap annulus, whereas the annulus filled with a silicone fluid of *Pfeffer et al.* [1980b] appears to show always vacillating flows in the regular wave range.

Bifurcation studies in a high-Prandtl-number fluid by *Read et al.* [1992] have suggested that the onset of vacillation on increasing Θ is consistent with a Hopf bifurcation, and similarly *Randriamampianina et al.* [2006] showed the same for the onset of amplitude vacillation in direct numerical simulations of a low-Pr fluid on decreasing Θ. The intermediate case was covered by *Sitte and Egbers* [2000], who were able to show that both existed, a Hopf bifurcation from a steady wave 2 to a 2AV on decreasing Θ toward the $m = 3$ region and a Hopf bifurcation from a steady wave 3 on increasing Θ toward the $m = 2$ region. The region between these two bifurcations showed secondary bifurcations to chaotic modulated vacillations, each involving both modes, $m = 2$ and $m = 3$, but dominated by their respective original mode. While the hysteresis in the transition between modes involving only steady waves and amplitude vacillation is substantial, the transition between the 2-dominated and the 3-dominated chaotic flows seen by *Sitte and Egbers* [2000] had little hysteresis and is more gradual, similar to the transition between complex amplitude vacillation flow observed by *Früh and Read* [1997]. It appears that the modulated vacillations always involve activity in other modes, especially the sidebands of the dominant mode, and thereby facilitate the transition from one dominant mode to the next lower or higher mode.

In the corresponding two-layer experiment, *Hart* [1972] observed that amplitude vacillation emerged from a steady wave when the driving of the lid and consequently the Rossby number were decreased. If the vertical velocity shear, either driven by the lid or through the thermal wind balance, is taken as the "forcing" of the system, then the Rossby number defined by the lid rotation takes the equivalent role as the thermal Rossby number defined by the thermal wind. In that case, the observed transition from a steady wave to the amplitude vacillation in the two-layer system corresponds to the thermally driven annulus filled with a low-Prandtl-number fluid.

3.2. MECHANICS OF AMPLITUDE VACILLATION

In this section, some possible processes resulting in amplitude vacillation will be presented. As all observations suggest that AV is a global modulation of a finite-amplitude steady wave mode and that the steady wave originates from a global mode instability of a zonal flow, the processes are usually expressed in terms of energy transfer between modes and the underlying zonal flow. In this framework, the energy is described in such terms as *zonal kinetic energy*, *zonal potential energy*, *eddy kinetic energy*, and *eddy potential energy* [*Lorenz*, 1955]. Following *Hart* [1976], the possible transfer routes can be illustrated as in Figure 3.4. In the mechanically driven two-layer system, the lid injects kinetic energy into the two fluids, which are then converted to a distortion of the interface, while in the thermally driven annulus, the imposed horizontal temperature difference sets up the sloped isotherms and the vertical shear flow. In both

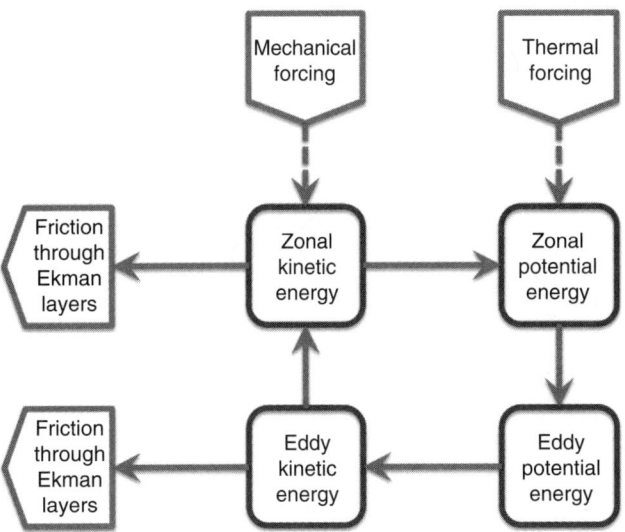

Figure 3.4. Routes of energy transfer from the forcing of the vertical shear in the two-layer experiment or from the imposed baroclinicity in the thermally driven annulus to dissipation of the kinetic energy and incorporating the thermal forcing relevant for *Pfeffer et al.* [1973]. Adapted from Fig. 6 of *Hart* [1976]. Copyright © American Meteorological Society. Used with permission.

cases, the baroclinic instability releases the zonal potential energy stored in the sloped isotherms or isopycnals and transfers this to the eddy potential energy and then the kinetic energy of growing wave modes. These lose energy through friction from the Ekman layers and through horizontal viscous diffusion but also feed back into the zonal flow. The feedback which can lead to an equilibration to a steady wave arises from the fact that the energy transfer from the eddies to the zonal flow reduces the baroclinicity until a balance between the energy supply from the forcing is balanced by energy loss through Ekman friction and diffusion.

Amplitude vacillation can set in when this balancing point of forcing and dissipation becomes unstable, and a slight increase in wave amplitude does not lead to a sufficient reduction in zonal potential energy and vice versa. *Pfeffer et al.* [1973] used experimental data to test a suggestion by *Pfeffer and Chiang* [1967] that the main energy conversion resulting in amplitude vacillation would be the two routes between zonal and eddy potential energy and between eddy potential and kinetic energy. The observations showed that the two potential energy terms were shifted by a quarter vacillation period, or phaseshifted by $\pi/2$, with the zonal potential energy leading. This means that the time of maximum zonal potential energy coincided with increasing eddy potential while the maximum eddy potential energy coincided with decreasing zonal potential energy. The results for the kinetic energy terms were noisier but suggested that they were coinciding with their respective potential energy terms or very slightly delayed. An idealization of Figures 17–19 from *Pfeffer et al.* [1973] in our Figure 3.5 illustrates the various energy terms in (a) and the main energy conversion terms in (b). A positive value in Figure 3.5b corresponds to an energy flow in the direction given by the arrow in the annotation and a negative value indicates a reverse energy flow. In particular, Figure 3.5b shows that the energy transfer between the potential energy forms appears to be always from the zonal to the eddy potential energy and the transfer between the two eddy energies appears to be always from the eddy potential energy to the eddy kinetic energy. In contrast, the transfer between the two zonal energy forms changes sign: The zonal kinetic energy receives energy from the zonal potential energy during minimum energy transfer from the zonal potential energy to eddy potential and then to eddy kinetic energy. This then changes to a drain from the zonal kinetic energy to the zonal kinetic energy at times when the energy transfer from this zonal kinetic energy to the other forms of energy is large. The transfer between the two kinetic energy terms was always very small but appeared to be in the direction from eddy kinetic to zonal kinetic energy. The reversal of the energy transfer between the kinetic energy terms is illustrated in Figure 3.6.

3.3. MODELLING APPROACHES

3.3.1. Computational Fluid Dynamics

Hignett et al. [1985] succeeded in reproducing a realistic amplitude vacillation in the finite-difference Navier-Stokes model for the Oxford annulus filled with a fluid of Prandtl number of around Pr ≈ 13 from *James et al.* [1981] even at a relatively low resolution of 16 grid cells in the radial and vertical, respectively, and 64 in the azimuthal direction using a stretched grid to resolve the boundary layers adequately. A later version of this model, now known as MORALS, was used by *Young and Read* [2008] to construct a more detailed regime diagram for this apparatus and found very good agreement in the structure of the regime diagram, similar to that of Figure 3.3b.

Lu et al. [1994] developed a numerical model using finite–difference discretization in the radial and vertical directions but a spectral representation in the azimuthal direction to model the larger Florida annulus with a relatively narrow gap and filled with a viscous fluid of Prandtl number 73 (experiment B in *Pfeffer et al.* [1980b]) and found good agreement, in particular the fact that vacillating flows were extremely common and steady waves very rare. *Lu and Miller* [1997] then analyzed two particular vacillating cases, one classified as amplitude vacillation and the other as structural vacillation. In particular, they

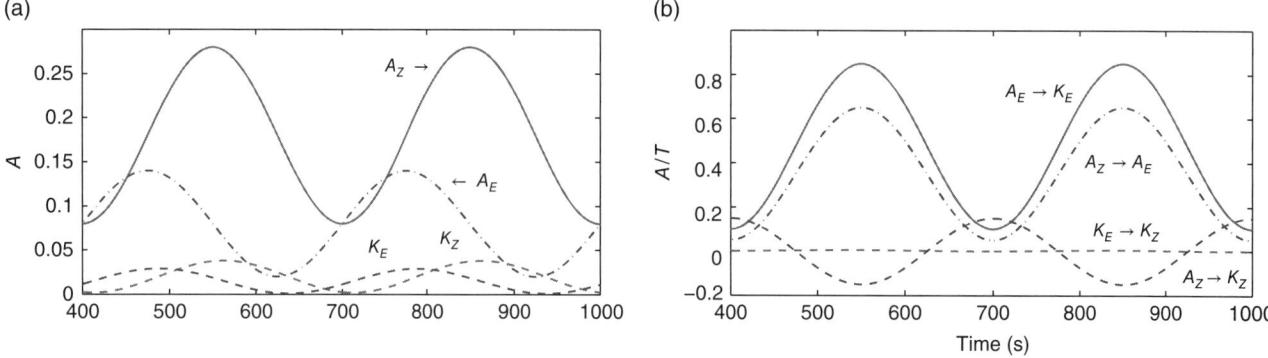

Figure 3.5. Idealization of the energetics for an amplitude vacillation cycle (after *Pfeffer et al.* [1973]) using A_Z for zonal potential energy, A_E for eddy potential energy, K_E for eddy kinetic energy, and K_Z for zonal kinetic energy: (a) energy contained in that type and (b) rate of energy transfer from one type to another. Adapted from Figures 17, 18, and 19 of *Pfeffer et al.* [1973]. Copyright © American Meteorological Society. Used with permission.

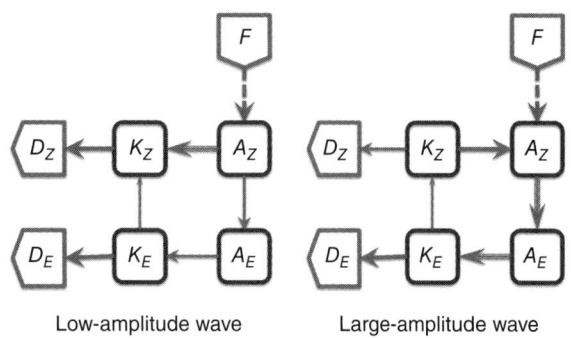

Figure 3.6. Schematic of the energy flow during the two extreme stages of an amplitude vacillation as calculated by *Pfeffer et al.* [1973] using F for forcing, A_Z for zonal potential energy, A_E for eddy potential energy, K_E for eddy kinetic energy, K_Z for zonal kinetic energy, and D for dissipation. Adapted from Figures 20 and 21 of *Pfeffer et al.* [1973]. Copyright © American Meteorological Society. Used with permission.

observed that the amplitude vacillation showed a clear oscillation of the relative phase of the wave in the lower part of the annulus compared to that of the upper part. Steady baroclinic waves have long been associated with a clear westward tilt of the temperature field and associated vertical heat transport [e.g., *Hide and Mason*, 1975]. With this in mind, the strong variation in the westward tilt is associated with the transfer between eddy kinetic energy (little tilt) and eddy potential energy (strong tilt) from the basic energy transfer model given in Figure 3.4. In contrast, the case classified as structural vacillation shows no such vertical tilt of the flow features but is essentially barotropic.

A high-resolution spectral Fourier-Chebyshev model of the thermal annulus filled with air was used by *Maubert and Randriamampianina* [2002] with the then-surprising observation that vacillating flows occurred on increasing the rotation rate (or the forcing toward favoring higher wave numbers). At the time this was surprising as all previous vacillation studies had been carried out in liquids where vacillation is found on decreasing the rotation rate. *Randriamampianina et al.* [2006] then traced a full bifurcation sequence from the initial instability to a chaotic modulated amplitude vacillation that was subsequently confirmed experimentally by *Castrejon-Pita and Read* [2007].

3.3.2. Quasi-Geostrophic Approximation

In the quasi-geostrophic approximation, the momentum equations are scaled against the Coriolis term and then ordered in a series of terms of increasing power of the Rossby number, where the Rossby number is the ratio of the advection term to the Coriolis term, $\mathrm{Ro} = U/(fL)$ [e.g., *Pedlosky*, 1987]. If the Rossby number is small, the leading balance of forces is the Coriolis force to the horizontal pressure gradient, which leads to the definition of the geostrophic stream function. The terms of $O(\mathrm{Ro}^2)$ then give an equation for the evolution, advection, and diffusion of this geostrophic stream function.

Based on the quasi-geostrophic approximation, a variety of models have been developed, all of which center around wave mode perturbations for the horizontal motion around an idealized baroclinic basic state. The vertical structure of this baroclinic basic state could be continuous, such as the Eady model [*Eady*, 1949] or the Charney model [*Charney*, 1949], or discrete, such as Phillips' two-layer model [*Phillips*, 1951]. The model can be used for high-resolution modeling for a systematic truncation to a low order or for investigating the evolution of a specific perturbation.

3.3.3. Low-Order Models

The picture of a flow with a regular spatial structure has led to a number of low-order dynamical systems models of amplitude vacillation in which the components are as follows:

   1. A constant forcing, often represented as a constant vertical shear velocity, applied positively to a pair of wave mode amplitude equations

   2. A mean-flow correction equation, coupled to the wave amplitude (the larger the wave, the stronger the correction), which counteracts the forcing (applied negatively to the wave amplitude equations)

   3. Dissipation applied negatively to both the wave amplitude and the mean-flow correction equations

This suggests that the system requires at least three dimensions, but the energy transfer routes indicated in Figure 3.4 suggests that four components are needed. Translating the amount of necessary information to a normal mode decomposition, this would suggest a traveling barotropic wave (consisting of two modes or amplitude and phase), a traveling baroclinic wave (also two degrees of freedom), and a mean-flow correction. While this adds up to five degrees of freedom, it is recognized that one of the phases can be eliminated by a suitable coordinate transformation, leaving four degrees of freedom. Reducing this to only three degrees of freedom would only be possible if either the relative phase or the relative amplitude between the barotropic and baroclinic mode is constant.

3.3.4. Eady-Type Models

The basic instability as developed by *Eady* [1949] led to the formulation of the nonlinear dynamics of finite-amplitude waves driven by a linear vertical shear. *Weng et al.* [1986] expressed the flow through a (nondimensional) stream function $\phi = -yz + \varphi' + \bar{\phi}$, where $-yz$ is the basic Eady profile, φ' the wave field, and $\bar{\phi}$ the mean-flow correction, which they then expressed in modes using $e^{ikx} \sin l\pi y$ for the horizontal component of the fields and $\sinh \mu z$ and $\cosh \mu z$ for the vertical structure. Using this, *Weng et al.* [1986] followed a bifurcation scenario from the initial instability to a steady wave with the lowest radial ($\sin \pi y$) wave number, which then underwent a period doubling bifurcation, followed eventually by the growth of the second radial mode, $\sin 2\pi y$, which is referred to as structural vacillation. However, while period doubling has been observed in the two-layer experiment, the only well-documented period doubling in the thermally driven annulus was associated with strong stationary forcing due to an imperfection in the apparatus [*Früh and Read*, 1997]. *Weng and Barcilon* [1987] suggested that, while structural vacillation is due to the interference of two radial modes of the same zonal wave, amplitude vacillation is due to the interference of two vertical modes of the same zonal wave. This is in accord with the CFD results from *Lu and Miller* [1997], who identified the vertical transport of energy associated with periodic changes in the vertical tilt of the wave structure during the vacillation cycle. *Weng and Barcilon* [1988] added more zonal wave modes to the model, but in a way that did not allow for wave-wave interactions. With this they demonstrated that wave-mean flow interactions alone are sufficient to give rise to wave number vacillation as well as amplitude vacillation and structural vacillation.

3.3.5. Two-Layer Models

Two-layer quasi-geostrophic models are a standard tool in geophysical fluid dynamics [e.g., *Pedlosky*, 1987], and can be formulated for the stream function in each layer or for the barotropic and baroclinic components. For example, on a β plane in the layer formulation with ψ_i, with $i = 1, 2$ for the upper and lower layers, respectively, the equations can be written as

$$\left\{\frac{\partial}{\partial t} + J(\psi_i, \cdot)\right\} q_i = -\frac{r}{2}\nabla^2 \psi_i + \frac{1}{\text{Re}}\nabla^2 q_i \quad (3.4)$$

with

$$q_i = \nabla^2 \psi_i + \beta y + (-1)^{-i} F_i (\psi_1 - \psi_2),$$

$i = 1$ for upper layer, 2 for lower layer,

$$\nabla = \frac{\partial^2}{\partial x^2} + \frac{\partial^2}{\partial y^2},$$

$$J(\psi, q) = \frac{\partial \psi}{\partial x}\frac{\partial q}{\partial y} - \frac{\partial \psi}{\partial y}\frac{\partial q}{\partial x},$$

$$r = \frac{\sqrt{E}}{\text{Ro}} = \sqrt{\frac{\nu f_0}{D^2}}\frac{L}{U}, \text{ the dissipation parameter,}$$

$$F_i = \frac{\rho_0 f_0^2 L^2}{g(\rho_2 - \rho_1)D_i}, \text{ the Froude number,}$$

$$\text{Re} = \frac{UL}{\nu} = \frac{1}{r^2 \text{Ro}}, \text{ the Reynolds number.}$$

The equivalent form for the barotropic component, $\psi_s = (\psi_1 + \psi_2)/2$, and the baroclinic component, $\psi_d = (\psi_2 - \psi_1)/2$, is for the barotropic component (s = sum)

$$\frac{\partial}{\partial t}\nabla^2 \psi_s + \beta\frac{\partial \psi_s}{\partial x} + u_s\frac{\partial}{\partial x}\nabla^2 \psi_s + u_d\frac{\partial}{\partial x}\nabla^2 \psi_d$$
$$+ J\left(\psi_s, \nabla^2 \psi_s\right) + J\left(\psi_d, \nabla^2 \psi_d\right)$$
$$= -r\nabla^2 \psi_s + \text{Re}^{-1}\nabla^4 \psi_s \quad (3.5)$$

and for the baroclinic component

$$\frac{\partial}{\partial t}\left(\nabla^2 - 2F\right)\psi_d + \beta\frac{\partial \psi_d}{\partial x}$$
$$+ u_s\frac{\partial}{\partial x}\nabla^2\psi_d + u_d\frac{\partial}{\partial x}\nabla^2\psi_s + 2Fu_d\frac{\partial \psi_s}{\partial x}$$
$$+ J\left(\psi_s, \nabla^2\psi_d\right) + J\left(\psi_d, \nabla^2\psi_s\right) - 2FJ(\psi_s, \psi_d)$$
$$= -r\nabla^2\psi_d + Re^{-1}\left(\nabla^2 - 2F\right)\nabla^2\psi_d \quad (3.6)$$

followed by a suitable spectral expansion and truncation, for example, Fourier modes for a straight channel [*Früh*, 1996],

$$\psi_{s,d} = \sum_{n=1}^{N}\phi(t)_{s,d}^{n}\cos n\pi y$$
$$+ \sum_{m,n=1}^{M,N}\left(\chi(t)_{s,d}^{mn}\cos\frac{2m\pi}{\alpha}x + \sigma(t)_{s,d}^{mn}\sin\frac{2m\pi}{\alpha}x\right)\sin n\pi y. \quad (3.7)$$

To satisfy the lateral boundary conditions, only some of the cross-channel Fourier modes are possible, but the nonlinear interactions in the equations result in terms of the other modes, and this energy must be projected onto those that do satisfy the boundary conditions. If the product of two wave terms has a zero zonal wave number ($m = 0$), it has a mean-flow structure of the form $\sin n\pi y$ while the modes satisfying the boundary conditions are of the form $\cos \ell\pi y$. This means that each term of radial mode n has to be expressed by a series of zonal flow correction terms ℓ with

$$c_{n\ell} = \frac{2}{\pi}\left[1 - (-1)^{n+\ell}\right]\frac{n}{n^2 - \ell^2}. \quad (3.8)$$

The Reynolds number term was originally omitted in the low-order models as it was assumed that the Stewartson layers at the side boundaries were "passive" while the relevant dissipation occurred through Ekman suction from the Ekman layers as the horizontal boundaries. However, *Smith* [1974] showed that the side boundaries are involved in the energy balance for the fluid interior and, in particular, that their absence resulted in a nonphysical energy source of mean-flow kinetic energy.

One of the earliest applications by *Lorenz* [1963b] of a truncation of a two-layer model to investigate specifically amplitude vacillation arrived at a 14-dimensional system, capturing a barotropic zonal flow, a baroclinic zonal flow, and two different radial modes of a wave with a common zonal wave number, each represented by a cosine and a sine component of the stream function as well as a temperature component. Depending on the parameter values, this system produced steady wave solutions, periodic vacillations, as well as aperiodic flow that appeared to arise from a homoclinic bifurcation.

Further studies have successively reduced the dimension of the system to the absolute minimum required for vacillation to isolate a simple sufficient mechanism for vacillation. For example, *Pedlosky and Frenzen* [1980] derived from the quasi-geostrophic two-layer equations a set of ordinary differential equations of the form

$$\frac{dA}{dt} = B - \gamma A, \quad (3.9)$$
$$\frac{dB}{dt} = -\frac{\gamma}{2}(B - \gamma A) + A - c\left(A^2 + V_k\right), \quad (3.10)$$
$$\frac{dV_k}{dt} = \gamma\left(A^2 - \alpha V_k\right), \quad (3.11)$$

where A is the amplitude of the represented baroclinic wave, $B = dA/dt + \gamma A$, and V_k is a set of $k = 1, \ldots, M$ cross-channel modes of the mean-flow correction to capture the mean-flow correction adequately; cf. equation (3.8). In this model, only the wave amplitude is represented explicitly but not the phase speed.

Pedlosky and Frenzen [1980] showed that this system can be reduced to a form equivalent to the classic Lorenz equations, originally derived as the simplest model for Rayleigh-Bénard convection [*Lorenz*, 1963a]. The relationship between the Lorenz equations and the two-layer model equations was subsequently analyzed and discussed by *Lovegrove et al.* [2001].

3.4. WAVE INTERACTIONS

3.4.1. Wave Triad Interactions

A common description of the underlying processes is to identify the transfer to and from the eddy kinetic and potential energies through nonlinear wave interactions which arise explicitly in the advection term $(\mathbf{u}\cdot\nabla)\mathbf{u}$ of the Navier-Stokes or "primitive" equations or the Jacobian $J(\nabla^2\psi, \psi)$ in the vorticity-stream function form of the momentum equations when the flow field is expanded into Fourier modes. The product of sine and cosine terms leads to contributions to the equations for the modes with mode numbers of the sum and difference of the two terms in the product. This leads to the notion of wave triads: two waves with zonal and radial wave numbers (m, n) and (m', n') combine in the multiplication to structures with wave numbers $(m'', n'') = (m \pm m', n \pm n')$, which then appear in the evolution equation for those respective modes. These possibilities are constrained for two-dimensional and nondivergent flow such that energy has to flow to both larger and smaller scales in such a way that both kinetic energy and entrophy are conserved [*Fjørtoft*, 1953]. In addition, *Hasselmann* [1967] pointed out that within a triad only the wave with the highest frequency can support energy transfer to the other two members of this triad, which was confirmed experimentally by *McEwan et al.* [1972]. Finally, the energy transfer to a mode will be most effective if the frequency of the forcing is equal or close to that of the wave itself ("resonance"). All these together then lead to the concept of resonant triads [*Bretherton*, 1964; *Plumb*, 1977], in which triads can interact if their frequencies align to maximize energy transfer form one scale to others, which is expressed in the selection criteria for the zonal and radial wave numbers as

$$m \pm m' \pm m'' = 0, \quad n \pm n' \pm n'' = 0, \quad (3.12)$$

respectively, and resonance condition for their frequencies

$$\omega - \omega' - \omega'' \ll \langle \omega \rangle, \quad (3.13)$$

where $\langle \omega \rangle$ is the average drift frequency of the three modes. True resonance is achieved if the left-hand side is equal to zero but energy transfer can also take place at nonzero but small values. In a steady wave case, the nonlinear coupling would lead to entrainment of the frequencies such that they do add up to zero, but in cases with varying mode amplitudes the strength of entrainment may also fluctuate. If the left-hand side is nonzero when the coupling is weak, the waves may drift apart, but if that drift is slow, i.e., $\ll \langle \omega \rangle$, their relative phases will still be close enough to reestablish entrainment when the amplitude of the driving mode becomes strong enough again.

The basic form of nonlinear interactions through a resonant triad is illustrated in Figure 3.7 in a diagram following *Ablowitz and Segur* [1981, Section 4.2.b] using the dispersion relation for linear Rossby waves against the zonal wave number for the first three cross-channel modes. Here, the selection criteria are satisfied by choosing two wave modes and calculating the difference between the wave modes to identify candidates for triads. To determine whether there is the possibility for resonance, the difference in the frequency of the chosen pair of waves is calculated. The wave number difference and the frequency difference are then used to place the circle in the dispersion diagram. If a mode is found within that circle, it satisfies the condition and can participate in the resonant triad interaction. As the selection rule applies to both the zonal and cross-channel wave number, the graph is in fact a projection of a three-dimensional graph with axes m, n, ω, where the lines for the different cross-channel modes are displaced in the third direction onto the plane of the zonal modes only. So, in this picture one has to ensure that the mode within the circle also satisfies the second of the selection rules in equation (3.12). The example shown is the triad of zonal wave numbers 4, 3, and 1, where the two selection rules are satisfied. The resonance condition is not fully satisfied but still within a range allowing for some energy transfer. For this case, linear Rossby waves were used for illustration purposes. However, finite-amplitude baroclinic waves have a substantially modified frequency. For one, the strongly nonlinear shape of a finite-amplitude wave leads to the fact that a single wave mode is not represented by a single mode but by a superposition of the fundamental mode and its harmonics, all moving with the same group velocity. In addition, the frequency or angular velocity of a wave depends on the wave amplitude. As a result, the resonance condition may be satisfied for a certain range of wave amplitudes but not for another.

3.4.2. Harmonic Forcing and Zonal Mean-Flow Correction

One special case of the selection rules is where $m' = m$, in which case the "triads" are to feed energy to the first harmonic of the wave, $m'' = 2m$, and to the zonal flow, i.e., $m'' = 0$. To satisfy Fjørtoft's constraint of transfer to larger and smaller scales, the energy transfer to the harmonic requires the flow of energy from $(m, 2)$ to $(m, 1)$ and $(2m, 1)$. Since finite-amplitude waves are never sinusoidal, there is always strong energy transfer between a mode and its harmonic.

Similarly, for the mean-flow correction, the transfer requires $(m, 1)$ and $(m, 2)$, though with the complication that the self-interaction is of the form of $\sin n\pi y$ while the modes satisfying the boundary conditions are of the form $\cos \ell \pi y$; cf. equation (3.8). Since the resonance condition is irrelevant for the zonal mean flow, this route for energy transfer is always possible and only depends on the wave amplitude, whereas the energy transfer through resonant

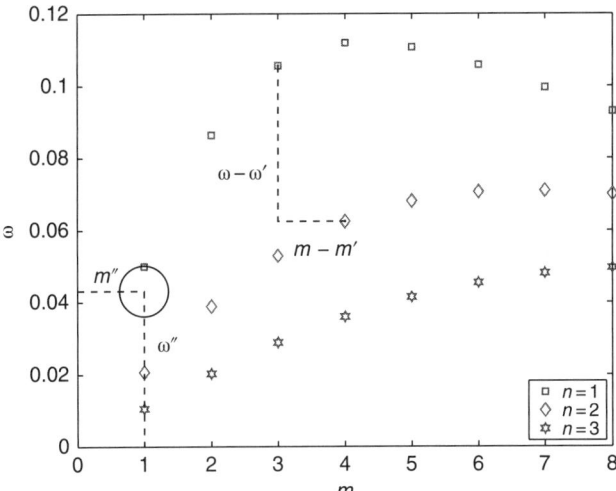

Figure 3.7. Basic routes of energy transfer through an almost resonant triads involving the first radial/cross-channel modes of wave numbers $(m, n) = (3, 1)$ and $(m', n') = (4, 1)$ and the second radial mode of wave number $(m'', n'') = (1, 2)$.

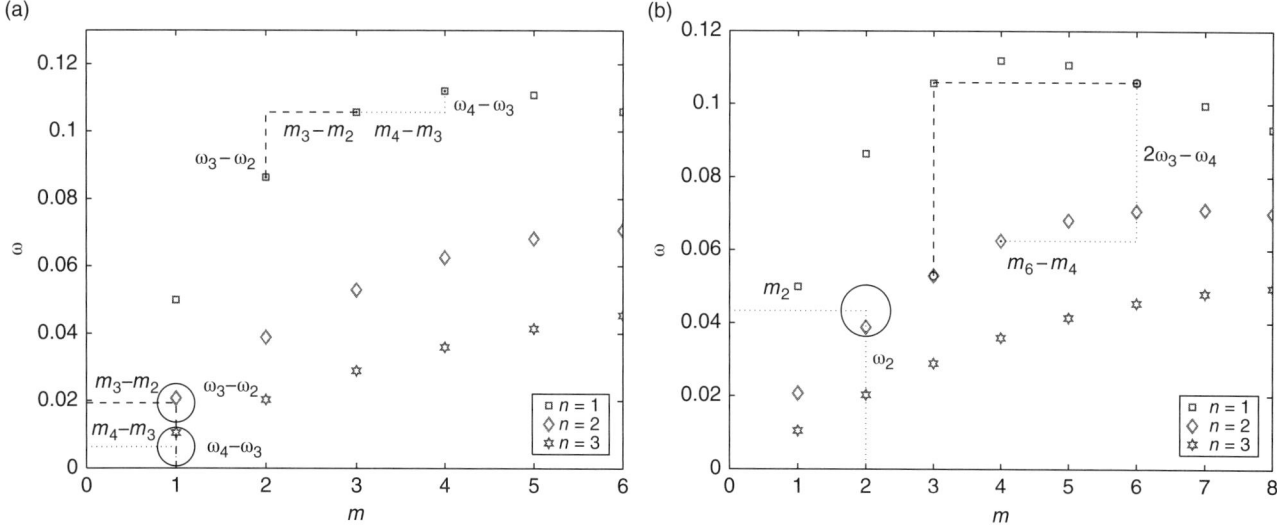

Figure 3.8. Two possible routes of energy transfer to the sidebands, either (a) involving two triads coupled by the common long wave with $m' = 1$ or (b) through the harmonic of the dominant mode, $m' = 2m$.

triads depends on the wave amplitudes and resonance conditions.

3.4.3. Higher-Order Wave Interactions

If no triad is fully resonant, higher-order interaction scenarios can affect the baroclinic wave. One classic example of this is the Benjamin-Feir instability [*Benjamin and Feir*, 1967] where a monochromatic surface wave with wave vector **k** in a channel develops a slow modulation through the rise of a long wave of wave vector δ**k** and "sideband" waves with wave vectors **k**$\pm\delta$**k**. As with the resonance condition for the resonant triads (equation (3.13)), a condition for the sideband instability can be written as [*Zakharov*, 1968]

$$2\omega_k - \omega_{k-\delta k} - \omega_{k+\delta k} = 0. \quad (3.14)$$

In the rotating annulus or two-layer experiments, the possible wave numbers are a discrete set, $k = 2\pi m/L$, and the longest possible wave is that with the wave number $m = 1$, i.e., $\delta k = 2\pi/L$. With this, we can propose an illustration of how this sideband instability can occur through a coupled set of triads; one option invokes this long wave, $m' = 1$, while the other possible route involves the first harmonic of the main wave mode, as illustrated in Figure 3.8.

The questions that arise from this framework are: Is any particular set of possible nonlinear interaction the essential process to destabilize a steady baroclinic wave and lead to amplitude vacillation? Do a range of interaction possibilities allow for all or specific types of amplitude vacillation.

3.4.4. Wave Interactions in Experiments and CFD

Hide et al. [1977] developed a method to quantify the degree to which sidebands interact from spatially resolved temperature measurements in the thermally driven rotating annulus. A Fourier analysis gave the amplitude A and phases ϕ of the zonal modes (but not resolving radial modes). Integrating equation (3.14) suggests

$$2\phi_m - \phi_{m-1} - \phi_{m+1} \approx \text{const},$$

which lead to the definition of a sideband phase-locking function

$$\Phi_m = 2\phi_m - \phi_{m-1} - \phi_{m+1}, \quad (3.15)$$

and *Hide et al.* [1977] observed that this phase-locking function was indeed fluctuating around a constant value of $\Phi_m \approx \pi$ for fully developed steady waves and amplitude vacillations. This did not hold for irregular flow or for flows with a noticeable structural vacillation. The fluctuation around a constant value implies that the resonance would only be nearly satisfied and that nonlinear interactions couple the waves when they are strong enough, that the waves start to drift apart when that coupling becomes weaker as the main mode decays, and that they become reentrained when the wave grows again in the vacillation cycle. While the sideband phase locking confirmed the presence of nonlinear wave interactions, it does not distinguish between the two possible interaction routes illustrated in Figure 3.8. A theoretical study by *Plumb* [1977] suggested the route through the long wave (Figure 3.8a), while an analysis of a numerical simulation by *James et al.* [1981] suggested the route through the harmonic (Figure 3.8b).

3.4.5. Wave Interaction Scenarios in Low-Order Models

Numerous studies have investigated the onset of vacillation in a range of low-order models, each of them isolating a few or even a single possible route by which a steady, equilibrated wave starts to develop a vacillation. The most basic of them, for example, demonstrated that the wave-mean flow interaction between a single zonal wave and the mean flow is able to render a finite-amplitude steady wave unstable to vacillation if the forcing as quantified by the Froude number is large enough or if the dissipation parameter r is small enough. As a comprehensive review of the earlier two-layer models by *Klein* [1990] has shown, the "interesting" behavior of vacillating and chaotic flows in the simplest models with a single unstable wave of wave number (k, l) was mostly found at an intermediate balance of forcing and dissipation, as quantified by $r/\Delta^{1/2} = O(1)$, where $\Delta = F - (k^2 + l^2)/2$. *Klein* [1990] also found that the inclusion of more wave modes into the models tends to stabilize the flow but it does not fundamentally alter the types of flows observed.

Früh [1996] analyzed the various possible wave interaction scenarios in a set of low-order models where the included wave modes were carefully chosen to allow or suppress specific wave-wave interaction routes based on the selection criteria (equation (3.12)). For the analysis, the sideband phase-locking function in equation (3.14) was adapted to the triad resonance condition in equation (3.13) to define a triad phase-locking function. In the full model, they observed a sequence of bifurcations that, at least superficially, resembled the types of transitions found in the experiment, from a steady wave through an amplitude vacillation to some forms of chaotic modulated amplitude vacillations, all of which involved substantial energy transfer between the different zonal wave modes, and finally to fairly irregular flow within the constraints of the dimensions of the system. All the flows involving more than one zonal mode showed clear resonant triad interactions, where the strength depended on the relative mode amplitudes. Removing specific triads from these models resulted only in moderate changes of the observed flows, which suggested that the dynamics would make use of preferred triads if they are available but that they could make use of alternative routes for energy transfer. The results changed more substantially if all triads were removed and only wave-mean flow interactions were retained. In that case, the preferred route was through a competition between different zonal modes. This can be understood through the fact that the most unstable zonal mode is not usually the mode to which a flow would eventually equilibrate. The final steady wave regime would usually be dominated by a lower zonal wave number [*Hart*, 1973, 1981; *Pedlosky*, 1981]. For this reason, there is an amplitude dependence of the growth rates of the waves such that a higher mode can grow preferably during a stage of strong zonal flow but that this mode suffers stronger damping and reduced growth compared to a longer wave when the zonal flow is reduced through the original growth of the shorter wave, leading to an alternation of which mode received more energy from the zonal flow.

3.5. PRANDTL NUMBER EFFECTS

3.5.1. Observations

As discussed by *Lewis* [2013], the Prandtl number as the ratio of the kinematic viscosity over the thermal diffusivity,

$$\text{Pr} = \frac{\nu}{\kappa}, \qquad (3.16)$$

affects the first transition, from axisymmetric flow to regular waves. We have also already mentioned in Section 3.1 that the Prandtl number affects the transition to amplitude vacillation strongly in a way that can be summarised as follows: bifurcation to AV from a steady wave on decrease of Θ if $\text{Pr} \lesssim 1$, none or very little AV in water, onset of AV on increase of Θ if $\text{Pr} \gtrsim 10$, and amplitude vacillation prevalent if $\text{Pr} \gg 10$. One key characteristic defined largely by the Prandtl number alone is the relative thickness of the momentum and temperature boundary layers.

3.5.2. Possible Role of Boundary Layers

As the low-order models are, so far, all based on the quasi-geostrophic equations that do not solve explicitly the heat equations or the boundary layers, they rely on capturing the effect of boundary layers implicitly through Ekman suction from horizontal boundary layers and horizontal diffusion from vertical boundary layers. However, the relative thickness of the thermal and velocity boundary layers affects the relative contribution to the heat transport through the boundary layers and through the fluid interior, respectively. A linear analysis by *Barcilon and Pedlosky* [1967] identified that two parameters, namely the Ekman number and the product of the thermal Rossby number and Prandtl number, organize the relative contribution from different boundary layers into three scenarios,

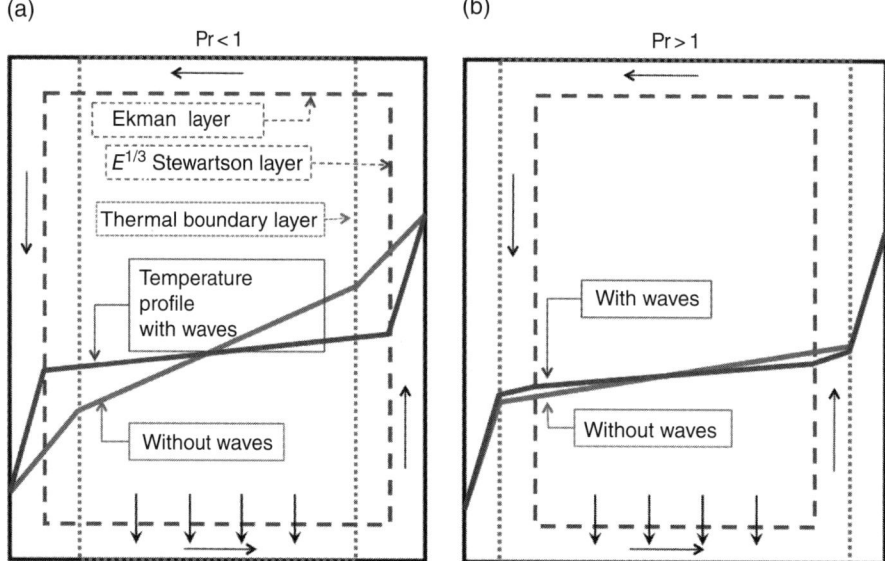

Figure 3.9. Schematic diagram of the effect of baroclinic waves on the mean radial temperature profile for (a) Pr < 1 and (b) Pr > 1. Shown are sections in the r–z plane with cooling at the right wall, heating at the left, and insulated rigid upper and lower boundaries.

Homogenous Fluid, Pr Θ < $E^{1/2}$ The fluid interior is homogeneous and constrained by the Taylor-Proudman theorem. Dissipation is through Ekman suction, and the sidewall boundary layers are the two Stewartson layers of thickness $E^{1/3}$ and $E^{1/4}$.

Weakly Stratified, $E^{1/2}$ < Pr Θ < $E^{2/3}$ The Taylor-Proudman theorem is still strong but thermal stratification is increasingly noticeable. The $E^{1/4}$ Stewartson layer is largely unaffected but buoyancy very close to the wall affects the $E^{1/3}$ Stewartson layer and two new boundary layers develop: one very thin layer of thickness $E^{1/2}/(\text{Pr}\,\Theta)^{1/4}$ in which the viscous stresses balance buoyancy and an outer, hydrostatic layer of thickness $(\text{Pr}\,\Theta)^{1/2}$.

Strongly Stratified, Pr Θ > $E^{2/3}$ The fluid is strongly stratified, the hydrostatic $(\text{Pr}\,\Theta)^{1/2}$ layer merges with the interior, and only the $E^{1/2}/(\text{Pr}\,\Theta)^{1/4}$ buoyancy layer remains. Ekman suction is very weak and the interior is controlled by viscous diffusion.

Read [2003] analyzed the heat transfer calculated using the MORALS model (introduced above in Section 3.3.1) as a function of a boundary layer ratio, defined as the squared ratio of the thermal sidewall boundary layer thickness to the Ekman layer thickness. This demonstrated that the Ekman layer was the limiting factor when the thermal boundary layer was wider than the Ekman layer but that the heat transport by the axisymmetric flow became constant when the thermal boundary layer became thinner than the Ekman layer.

3.5.3. Low-Order Model of Wave-Boundary Layer Interaction

Based on the scaling by *Barcilon and Pedlosky* [1967] and the observations by *Read* [2003] that the thinner of the two vertical boundary layers in a way determines the behavior of the heat transfer in the steady wave through the Ekman layers and fluid interior, respectively, we can propose a simple conceptual model of the interaction between baroclinic waves and the thermal forcing of the Ekman circulation and the thermal wind in the interior using the following argument, which is also illustrated schematically in Figure 3.9.

Since the analysis by *Barcilon and Pedlosky* [1967] suggests a change from the $E^{1/3}$ Stewartson layer to thermal boundary layers scaling with Pr Θ, we use the $E^{1/3}$ Stewartson layer as our reference layer in the following argument, which is based on the relative thickness of the vertical velocity and thermal boundary layers.

To determine the relative thicknesses and put these in the context of the nondimensional parameters, a set of axisymmetric solutions of the MORALS code for a range of Prandtl numbers but otherwise fixed parameters was generated (see Appendix Appendix A: for model specifications). Applying the classification from *Barcilon and Pedlosky* [1967], all solutions presented here are nominally strongly stratified, with the transition from weak stratification to strong stratification at Pr ~ 0.1. Since the Ekman number was a constant in all computations, the $E^{1/3}$ Stewartson layer thickness was used as our reference layer. Figure 3.10 shows the distances of the theoretical and

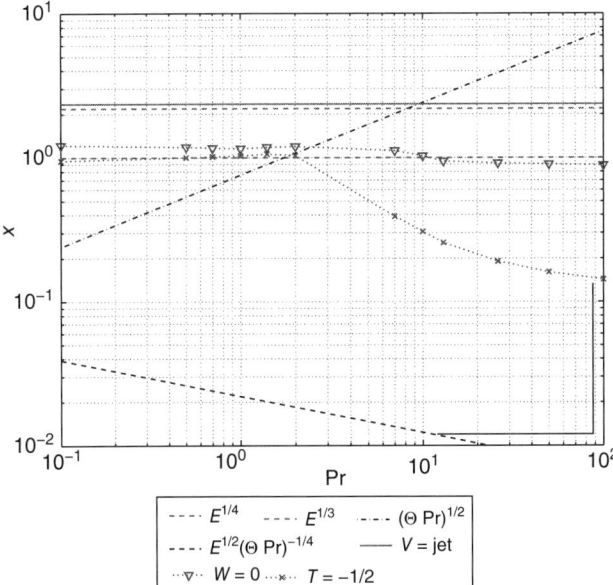

Figure 3.10. Comparison of boundary layer thicknesses obtained from axisymmetric solutions of the MORALS code with the theoretical Stewartson layer thicknesses and thermal layer thicknesses against Prandtl number. The thickness variable is scaled against the $E^{1/3}$ Stewartson layer.

computed boundary layer thicknesses at the inner cold wall. There the theoretical two Stewartson layers are shown as the horizontal dashed lines and the location of the maximum jet velocity in the interior is shown as the solid line, while the outer hydrostatic thermal boundary layer is the dash-dotted line and the inner buoyancy layer is the dashed line. The extent of interior baroclinic flow is measured as the location of the maximum velocity of the baroclinic jet. This is located just outside the outer $E^{1/4}$ Stewartson layer and did not appear to depend on the Prandtl number at all. Since the axisymmetric solutions always showed a strong downwelling at the cold sidewall adjacent to a small upwelling (and vice versa at the warm wall), the extent of the vertical flow is measured as the location of zero vertical velocity and shown as the triangles with the dotted line. This is clearly limited by the inner $E^{1/3}$ Stewartson layer. To obtain a measure of the thermal boundary layer, the distance from the wall where the temperature had increased from $-\Delta T/2$ to $-\Delta T/4$ relative to the mean temperature was found. This is shown by the crosses and dotted line. For $\mathrm{Pr} \lesssim 2$ this thermal layer has a constant thickness that appears to be limited by the vertical convection since it is very close to the edge of the velocity boundary layer. Once the Prandtl number increases beyond $\mathrm{Pr} \sim 2$, the temperature gradient near the walls increases noticeably and follows a decay similar to that of the buoyancy boundary layer. The changeover from a solution where the heat transfer is limited by the velocity boundary layer to that where the thermal boundary layer thickness affects the heat transfer occurs at a point where the theoretically derived hydrostatic $(\mathrm{Pr}\,\Theta)^{1/2}$ layer becomes thicker than the $E^{1/3}$ Stewartson layer. With this, we can now state the principle of the Prandtl number dependent coupling between the boundary layer and the baroclinic waves:

1. The horizontal temperature profile across the entire domain is characterized by high-temperature gradients within the thermal boundary layers and smaller gradients outside of them.
2. The velocity behavior is split into velocity boundary layers and the "fluid interior." The interior is defined by the horizontal velocities and is therefore terminated by the $E^{1/3}$ Stewartson layers.
3. The effective forcing of the thermal wind in the fluid interior is determined by the horizontal temperature difference across the fluid interior.
4. Therefore the temperature at the top of the $E^{1/3}$ Stewartson layers determines the thermal wind. The ratio of the effective temperature contrast over the imposed temperature contrast as a function of the Prandtl number for axisymmetric flow derived from MORALS results is shown in Figure 3.11.
5. Finite-amplitude baroclinic waves, at least those with a wave number preferred by Hide's geometric constraint, i.e., equation (3.3), tend to fill the gap up to the edges of the Stewartson layers.
6. Finite-amplitude waves enhance the heat transfer through the interior and thereby act to reduce the temperature gradient within the fluid interior.
7. If the thermal boundary layers are thinner than the Stewartson layers, the temperature at the edges are

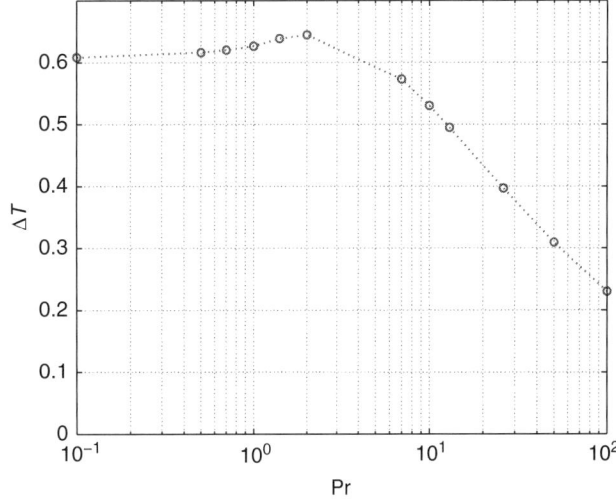

Figure 3.11. Effective temperature contrast across the fluid interior against Prandtl number obtained from axisymmetric solutions of the MORALS code.

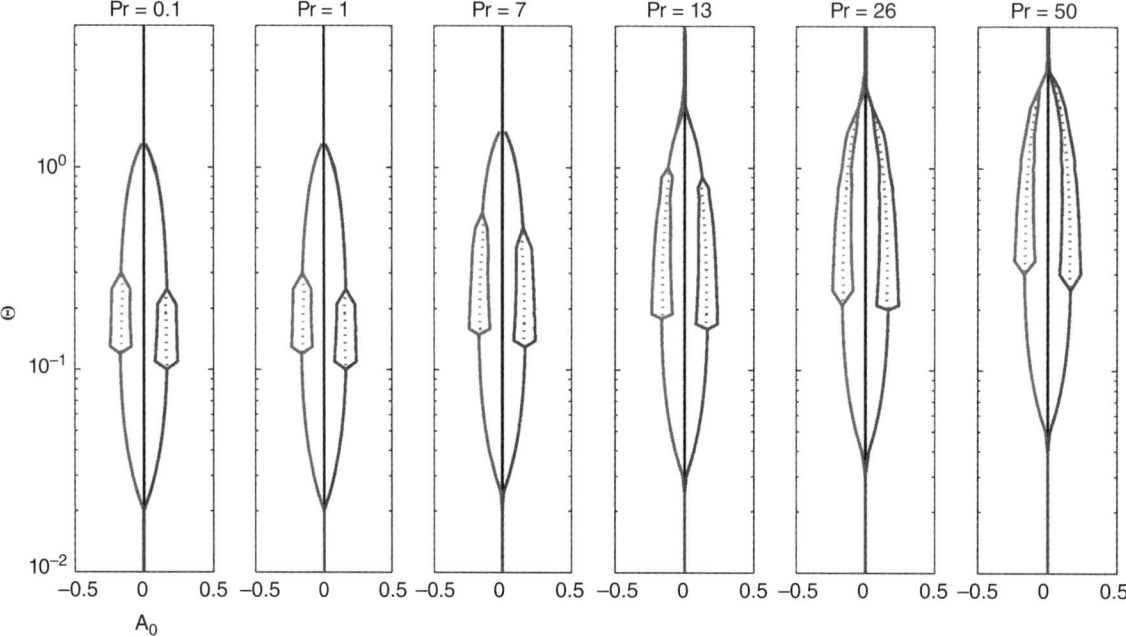

Figure 3.12. Bifurcation diagrams for Ta = 10^5 and various Prandtl numbers with the thermal Rossby number Θ as the bifurcation parameter. The quantity plotted is the amplitude of the mean-flow correction, on the left of the central axis for increasing Θ and on the right of the axis for decreasing Θ. The solid lines indicate the range of the amplitude and the dotted lines the mean amplitude.

affected only slightly since most of the imposed temperature contrast is taken by the thermal boundary layers. Hence the feedback between sidewall forcing and baroclinic waves is minor

8. If, on the other hand, the thermal boundary layers are thicker than the Stewartson layers, the temperature at the edges of the boundary layers is affected substantially because the baroclinic waves extend to within the thermal boundary layers and thus provide an additional heat transfer route across the annulus directly from one thermal boundary to that at the opposite sidewall.

This principle can be formalized in a one-dimensional model for the temperature at the interface between the Stewartson and the fluid interior based on the vertical heat convection through the Ekman circulation, the radial heat conduction, and the radial heat convection, which is a function of the wave amplitude. If the radial extent of the fluid interior is taken to extend from $y = \pm 1$ with $y = 0$ at the center of the annulus gap, the $E^{1/3}$ layer has a thickness of δ, and the nondimensional temperatures at the sidewalls are $T = \pm 1$, then the energy equation

$$\frac{\partial T}{\partial t} = -\mathbf{u}\cdot\nabla T + \frac{1}{\Pr}\nabla^2 T$$

can be discretized for the temperature at the interface, $T(y) = \theta$, as

$$\frac{\partial \theta}{\partial t} = -v_w\frac{\theta}{1} - w_s\frac{1-(-1)}{\gamma_v-(-\gamma_v)} + \frac{1}{\Pr}\frac{(1-\theta)/\delta - \theta/1}{1+\delta/2 - 1/2}$$

where $v_w\theta$ is the heat convection into the interior through the waves with a radial fluid velocity v_w and w_s/γ_v the vertical heat convection through the Stewartson layer with a vertical velocity w_s. The parameter γ_v is the vertical aspect ratio of the annulus. The last term is the horizontal heat conduction from the wall (at $y = 1 + \delta$) to the edge of the boundary layer (at $y = 1$), and from that edge to the center of the annulus (at $y = 0$). One implicit assumption here is that the flow and temperature fields are symmetric around the center line. This can then be rearranged to a differential equation with a constant term, a term proportional to the effective temperature, θ, and a term proportional to the strength of the baroclinic waves as quantified by v_w,

$$\frac{\partial \theta}{\partial t} = \frac{2}{\Pr(1+\delta)\delta} - \left(\frac{2}{\Pr\delta} - v_w\right)\theta - w_s\frac{1}{\gamma_v}. \quad (3.17)$$

Coupling this equation to a low-order two-layer model can then simulate the effect of the now wave-dependent effective thermal forcing from the edge of the Stewartson layer. An implementation of this into the minimal two-layer model of a single wave and a single mean-flow correction term by *Lovegrove et al.* [2002] is presented in Appendix B. Some initial results of this are shown as a set of bifurcation diagrams for a selection of Prandtl numbers in Figure 3.12, where the Taylor number was kept fixed at Ta = 10^5 and the bifurcation parameter was the thermal Rossby number. The diagrams show the mean and extreme values of the wave amplitude of the barotropic component, $\sqrt{\chi_s^2 + \sigma_s^2}$ with χ_s the barotropic cosine mode and σ_s the sine mode, following the convention used in equation (3.7) for either increasing Θ (left branch) or decreasing as a test for hysteresis. In general, the axisymmetric flow develops into a steady wave on increasing Θ, then develops an amplitude vacillation before returning to the axisymmetric flow. The main feature of Figure 3.12 is that the location of the steady waves and the amplitude vacillation shifts toward higher Θ as the Prandtl number is increased. Furthermore, the relative extent of the steady over the vacillating flows shifts toward more prevalent vacillation for higher Prandtl numbers, which is consistent with experimental observations. The clear reversal of the bifurcation order is not captured in this very simple model, although a very small steady wave regime can be observed for the lowest Prandtl number above the vacillating regime.

In conclusion, there is some evidence that the proposed feedback between wave amplitude and the effective thermal forcing can contribute to the vacillation, as seen in experiments, though this very simple model is far from complete.

3.6. OTHER FORMS OF VACILLATION

3.6.1. Interference and Wave Number Vacillations

Though not strictly speaking an "amplitude vacillation," the processes explaining the interference vacillation found in the two-layer experiment are worth mentioning. Following on from the concept of the onset of a wave through a Hopf bifurcation, *Ohlsen and Hart* [1989] used the two-layer model developed for the two-layer experiment to investigate the dynamics following a double-Hopf bifurcation where two modes become unstable simultaneously. They found that the two modes can both grow to a finite amplitude. These two coexisting modes can then generate zonal flow oscillations through interactions of a mode with its sidebands. Interference vacillation was also found by *Harlander et al.* [2011] in the thermally driven annulus with a free surface. In contrast to the two-layer experiment, there was no indication that the two waves were coupled through nonlinear interactions, but they appeared to be a linear superposition of two modes of different zonal wave number drifting at different speeds.

Related to amplitude vacillation is the wave number vacillation as it arises from the interaction between two or three wave modes and the zonal flow. *Weng and Barcilon* [1988] suggested that the two-wave wave number vacillation arises from an imbalance in the forcing of the wave from the baroclinic zonal flow and viscous damping of the two participating unstable modes.

3.6.2. Structural Vacillation

Over the years, many processes have been invoked to understand the nature of "structural vacillation" and the transition to it. Key features of structural vacillation compared to amplitude vacillation is that smaller horizontal scales are involved besides the dominant wave mode and that the vertical structure is much less baroclinic. The more barotropic nature of this flow was observed experimentally by *Pfeffer et al.* [1980a] and subsequently confirmed through CFD for those cases by *Lu and Miller* [1997] and for the air-filled annulus by *Read et al.* [2008]. Furthermore, the time scale associated with the structural vacillation is usually considerably shorter than that of an amplitude vacillation, though still longer than the rotation period of the apparatus. A systematic experimental study by *Tamaki and Ukaji* [1993] suggested that the onset of structural vacillation occurs at a fairly well-defined place in parameter space irrespective of which wave mode dominates the flow, which is in distinct contrast to the amplitude vacillation where its onset depends on the wave number of the flow. *Read et al.* [1992] and *Früh and Read* [1997] noticed that the overall heat transfer fluctuates only slightly compared to that of amplitude vacillations, another hint that the processes driving the onset of structural vacillation are fundamentally different from those leading to amplitude vacillation.

A closer analysis of the structural vacillation found by *Read et al.* [2008] and *Früh et al.* [2007] is presented here in terms of the temporal and radial spectra of the fluctuations relative to the steady wave. The azimuthally averaged temporal spectra are shown in Figure 3.13 as a contour plot in the radial frequency plane, where the

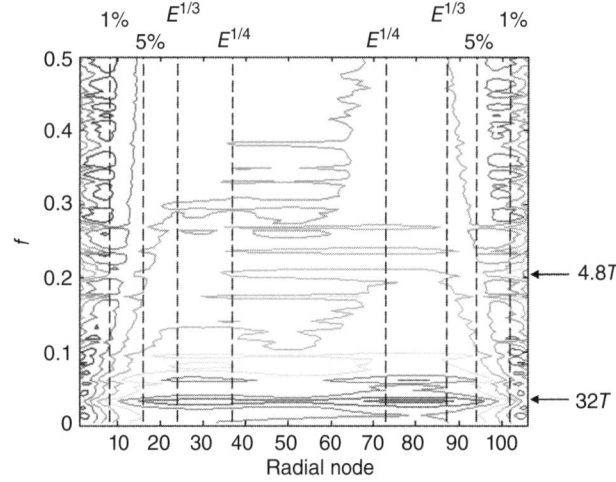

Figure 3.13. Radius-frequency contour plot of azimuthally averaged power spectral magnitude for structural vacillation in air-filled annulus. For color detail, please see color plate section.

radial coordinate is the node number i from the computational grid with $r_i = \cos(i\pi/N), i = 1,\ldots,N$, as described by *Randriamampianina and Crespo Del Arco* [2014] in this book, which stretches the radial coordinate to resolve some scales near the boundary. To show the key distances, the dashed lines indicated the near-wall region, within 1 and 5% of the gap width and the inner $E^{1/3}$ and outer $E^{1/4}$ Stewartson layers. The key regions of variability are located in the fluid interior but even more so in the outer Stewartson layers at the longer time scale of 32 nondimensional time units ($t = 2\Omega t^*$). This period, of around 10 rotation periods of the apparatus, is consistent with the usually observed vacillation periods of structural vacillation. Additionally, there is a source of much faster fluctuations within the near-wall regions at a time scale of only three to five time units of the same magnitude as the rotation period of the annulus. This suggests that a relatively fast boundary layer process might be involved in the onset of structural vacillation whereas amplitude vacillation was understood as a global instability of the fluid interior. The azimuthal spatial spectrum in Figure 3.14 shows the clear peaks of the dominant mode 2 and its harmonics superimposed on a general decay with a decay rate between $p_\theta \sim m^{-2.2}$ and m^{-3}, which would be consistent with a quasi-geostrophic turbulence spectrum [*Charney*, 1971; *Waite and Bartello*, 2006]. The radial spectrum in Figure 3.15, on the other hand, has a spectrum with radial wave number as $p_r \sim k^{-5/3}$, which is closer to a mesoscale energy spectrum in terms of the horizontal wave number [*Nastrom and Gage*, 1985] or strongly stratified turbulence where strong small-scale static instability is present [*Lindborg*, 2006].

One suggested route, the instability of a higher radial mode of the dominant wave growing to a finite-amplitude modulation of that higher mode superimposed on the

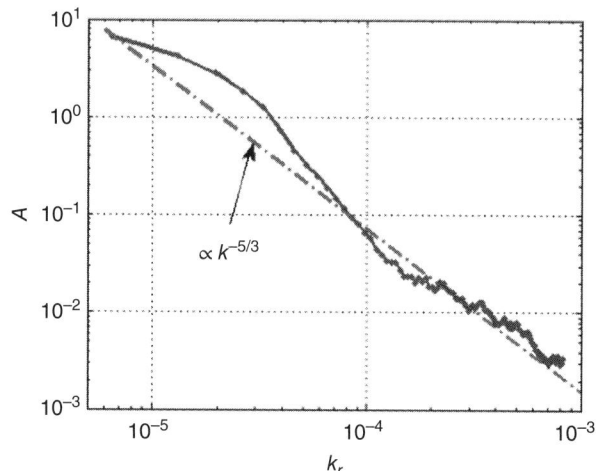

Figure 3.15. Radial spatial spectrum averaged over time and azimuth for structural vacillation.

still steady fundamental radial mode of that wave, is supported by the experiments of *Früh and Read* [1997] and explained in a low-order Eady-type model by *Weng et al.* [1986]. This explanation would suggest that there is a clear bifurcation route from the steady wave to the AV, which then bifurcates to a SV in some way. However, this bifurcation remains elusive, and calculations of the Grassberger-Procaccia dimension of amplitude vacillations and structural vacillations have repeatedly shown that amplitude vacillations are well behaved and appear to follow low-dimensional dynamics whereas the dimension estimates for measurements from structural vacillations do not converge to a reliable estimate [*Guckenheimer and Buzyna*, 1983; *Read et al.*, 1992; *Früh and Read*, 1997]. Another explanation might be a localized instability of the large-amplitude wave resulting in possibly a barotropic instability of the type of a detached shear layer [*Früh and Read*, 1999] but localized in space along the edges of individual wave lobes or in the form of a breaking wave leading to internal gravity waves such as those described by *Jacoby et al.* [2011]. Yet another option might be a boundary layer instability as the large-amplitude wave impinges on the sidewalls, such as seen by *Read et al.* [2008] and *Früh et al.* [2007].

3.7. AMPLITUDE VACILLATION AS STEP TOWARD CHAOS AND TURBULENCE

In this context, the distinction between "chaos" and "turbulence" is based on the assumption that chaotic flow is governed by deterministic equations that can be modeled by a finite (hopefully small) number of degrees of freedom whereas turbulence requires so many dimensions

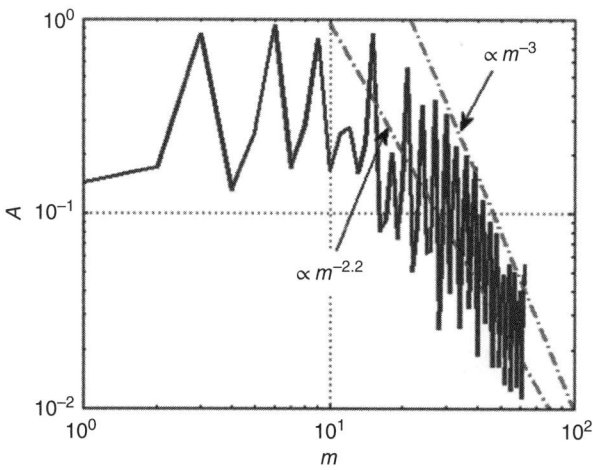

Figure 3.14. Azimuthal spatial spectrum averaged over time and radius for structural vacillation.

that it might, from a practical point of view, be as well an infinite-dimensional system or a nondeterministic system.

To distinguish these two cases, the attractor dimension reconstructed from experimental or numerical data can be used as a guide. So far, chaotic modulated amplitude vacillation and similarly complex forms of amplitude vacillation have always appeared to follow fairly low-dimensional dynamics when their Grassberger-Procaccia dimension was estimated, as demonstrated for the thermally driven annulus by *Guckenheimer and Buzyna* [1983], *Read et al.* [1992], *Früh and Read* [1997], and *Sitte and Egbers* [2000]. From all studies, it is clear that amplitude vacillation is a key candidate to explore a number of standard bifurcation scenarios through secondary Hopf bifurcations, period-doubling cascades, and intermittency-type bifurcation, to name but a few. The majority of the evidence points to a picture whereby the chaotic flow is in some way the result of global mode instabilities or through attractor crises arising from two coexisting attractors associated with different zonal global wave modes. Either type of transition was always found to lead to strictly low-dimensional behavior where the Grassberger-Procaccia dimension tended to be less than 4. Ultimately, the chaotic flow is normally terminated by steady wave flow again, usually of a lower wave number if $Pr > 1$ and a higher wave number if $Pr < 1$.

Another possible progression is to a structural vacillation. The evidence, however, points to an understanding that AV and SV are fundamentally different types of flow and that a transition from AV to SV occurs more by accident than through a systematic transition. This is because the transition is only found at high Prandtl numbers where AV is so ubiquitous in the regular wave regime that it is virtually the only possible regular wave flow on which an SV can develop, whereas SV does also develop on a steady wave through an as-yet poorly understood mechanism.

The transition to turbulence, on the other hand, seems to be closely linked to either structural vacillation through the emergence of possibly localized flow structures, as the Taylor number is increased, or the progressive emergence of higher wave modes as the rotation rate is increased (simultaneous increase of Ta and Θ). A reevaluation of the dimension estimates for structural vacillation from *Guckenheimer and Buzyna* [1983] by *Pfeffer et al.* [1997] supports the proposition that structural vacillation represents a secondary instability on top of the remaining stable baroclinic wave, which gradually gains predominance as the flow becomes turbulent. Their argument is based on the observation that, at the onset of SV, the dimension estimates suggest a Grassberger-Procaccia dimension of 1.6, a number that persists as "the answer." At the same time a second scaling region develops at smaller scales in phase space. That second scaling region suggests a dimension of between 7 and 10, with an estimated dimension of full geostrophic turbulence of 11. These observations are consistent with those of *Sitte and Egbers* [2000] and *Früh and Read* [1997], who observed two scaling ranges for their weak structural vacillation, one suggesting a dimension of 1.3, the other 4.5. Complementary dimension estimates for the integrated total heat flux measured simultaneously with the temperature measurements in the same experiment gave inconsistent results with a suggested dimension of 5.8. Usually the heat transfer dimension D_Q would be related to that from the temperature data, D_T, as $D_Q = D_T - 1$ since the total heat transfer does not resolve the spatial structure within the flow. To overcome the difficulties presented by dimension estimates and spurious Lyapunov exponents, *Pfeffer et al.* [1997] used Lorenz analog diagrams to visualize the degree of chaos by presenting the phase space distance between subsequent states to show how the apparently stable global flow structure is being broken up by spatially separated fluctuations.

3.8. CONCLUSIONS

This review of amplitude vacillation has attempted to introduce a range of methods to investigate this phenomenon from careful experimentation in a thermally or a mechanically driven apparatus complemented by high-resolution CFD as well as targeted low-order models. By combining the findings from the various approaches, it has been possible to build up a fairly comprehensive picture of the processes leading to and involved in amplitude vacillation. The main processes remain nonlinear wave-wave interactions and wave-mean flow interactions but also feedback mechanisms between the fluid interior and the boundary layers.

This survey has reiterated the fact that the baroclinic annulus and the two-layer experiment are key fluid experiments to investigate a rich variety of nonlinear dynamics, including chaotic flow and geostrophic turbulence. The success, but also the challenges in modeling the observed flows successfully in CFD models, makes this system a good candidate for model development and validation. For straightforward code validation it is possible to find relatively simple flows that are (or should be!) easy to model, and for model development there is the option to model slightly more complex flows that either involve a higher resolution or combine new processes such as gravity waves. It is also possible to push the experimental conditions to truly complex flows that are likely to remain a serious challenge to computational modeling.

From a practical point of view, a frequently asked question is how this experiment can possibly help to understand real atmospheric flows, let alone help to predict weather and climate more accurately. This is a valid question, especially as the boundary layers are much more important in the laboratory than in the atmosphere. Furthermore, atmospheres do not have the Stewartson layers to contend with and therefore should not be affected by the Prandtl number in the way as described in Section 3.5. However, through developing some understanding of how the Prandtl number affects the annulus flow, it is possible to disentangle the laboratory-specific dynamics from those that are relevant to atmospheric or oceanic dynamics and to the atmospheric modeler.

Acknowledgments. The author wishes to thank in particular Raymond Hide, Peter Read, Patrice Klein, Christoph Egbers, and Thomas von Larcher for many inspiring discussions on the baroclinic annulus and the concept of vacillation.

Appendix A: MORALS CODE SETUP

The two-dimensional (2D) solver of MORALS was set up for axisymmetric flow integration with a grid resolution of 24—24 and 32—32 without any appreciable difference, where the grid was stretched using a hyperbolic tangent function. As a result, no further grid refinement was carried out.

The dimensions of the annulus were an inner radius $a = 2.5$ cm, outer radius $b = 8$ cm, and depth $d = 14$ cm, and the temperatures were $18°C$ at the inner wall and $22°C$ at the outer wall.

The fluid properties at the reference temperature $T_0 = 22°C$ were a density $\rho_0 = 1.043$ g/cm^3, kinematic viscosity of $\nu_0 = 0.0162$ g/cm^3, and thermal diffusivity κ_0 calculated to set the Prandtl number as follows:

Pr	0.1	0.5	0.7	1	1.4	2
κ_0	0.162	0.0324	0.0231	0.0162	0.01157	0.00810
Pr	7	10	13	26	50	100
κ_0	0.00231	0.00162	0.00129	0.000623	0.000324	0.000162

The variation of the fluid properties with temperature was a volume expansion coefficient for the fluid of $\alpha = 3.07 \times 10^{-4}$ K^{-1} and quadratic approximations as

$$\rho = \rho_0 \left[1 - \alpha(T - T_0) - 7.83 \times 10^{-6}(T - T_0)^2\right],$$

$$\nu = \nu_0 \left[1 - 2.79 \times 10^{-2}(T - T_0) - 6.73 \times 10^{-4}(T - T_0)^2\right],$$

$$\kappa = \kappa_0 \left[1 - 2.33 \times 10^{-3}(T - T_0)\right].$$

Appendix B: LOW-ORDER MODEL OF BOUNDARY LAYER FEEDBACK

The model to couple the one-dimensional boundary layer model with a single zonal baroclinic wave mode with zonal mode m and radial mode $n = 1$ was formulated to simulate the behavior for annulus parameters given by

γ_h, the horizontal aspect ratio, $\pi(a+b)/(b-a)$

γ_v, the vertical aspect ratio, $d/(b-a)$

n, the zonal wave number

Pr, the Prandtl number, equation (3.16)

Ta, the Taylor number, equation (3.1)

Θ, the thermal Rossby number, equation (3.2)

$\beta = df/dy$, the β effect

Defining equivalences between the thermally driven annulus and the two-layer system by using the thermal wind as the definition for both, the Rossby number of the two-layer system, and the shear forcing, we can associate the two-layer terms on the left-hand side with thermal annulus parameters on the right-hand side as

$$\begin{aligned} \text{Ro} &= \theta/2, \\ U_d &= \theta/2, \\ U_s &= 0, \\ F &= 8/\theta, \\ r &= \left[4/\left(\gamma_v^3 \text{Ta}\right)\right]^{1/4} 2/\theta. \end{aligned} \quad \text{(B.1)}$$

Furthermore, the thickness of the $E^{1/3}$ Stewartson layer can be described through the vertical aspect ratio of the annulus and the Taylor number as

$$\delta = \gamma_v^{-1/2} \text{Ta}^{-1/6}. \quad \text{(B.2)}$$

Using the correspondences, thermal annulus conditions can be converted to those of the minimal baroclinic two-layer model analyzed by *Lovegrove et al.* [2002] as

$$\begin{aligned} k_1 &= 2m\pi/\gamma_h, \\ K^2 &= k_1^2 + \pi^2, \\ \Delta_s &= r\left(1 + r\,\text{Ro}\,K^2\right), \\ \Delta_d &= rK^2/\left(K^2 + 2F\right) + r^2\,\text{Ro}\,K^2, \\ \Delta_b &= r\pi^2/\left(\pi^2 + 2F\right) + r^2\,\text{Ro}\,\pi^2, \\ \beta_s &= \left(\beta/K^2 - U_s\right)k_1, \end{aligned}$$

$$\beta_d = \left(\beta/\left(K^2 + 2F\right) - U_s\right) k_1,$$
$$v_s = U_d k_1,$$
$$v_d = U_d \left(K^2 - 2F\right)/\left(K_1^2 + 2F\right) k_1,$$
$$\gamma_s = 16 k_1^3/(16 K^2),$$
$$\gamma_d = 16 k_1 \left(k_1^2 - 2F\right)/\left(6\left(K^2 + 2F\right)\right),$$
$$\gamma_b = 32 F k_1/\left(3\left(\pi^2 + 2F\right)\right). \quad \text{(B.3)}$$

The model by *Lovegrove et al.* [2002] consists of four equations for the baroclinic wave of zonal and radial mode numbers m and $n = 1$, respectively, in terms of the cosine component C and sine componen S for the barotropic and baroclinic vertical modes using subscript s for sum or barotropic and d for difference or baroclinic, as well as an equation for a single mean-flow correction term, A. The coupling between the standard model by *Lovegrove et al.* [2002] and the boundary layer adds equations for the temperature at the interface between the interior and the Stewartson layer θ in the forms

$$\dot{C}_s = -\Delta_s C_s + \beta_s S_s - (v_s + \gamma_s A) S_d,$$
$$\dot{S}_s = -\beta_s C_s - \Delta_s S_s + (v_s + \gamma_s A) C_d, \quad \text{(B.4)}$$
$$\dot{C}_d = -(v_s + \gamma_d A) S_s - \Delta_d C_d + \beta_d S_d, \quad \text{(B.5)}$$
$$\dot{S}_d = (v_s + \gamma_d A) C_s - \beta_d C_d - \Delta_d S_d, \quad \text{(B.6)}$$
$$\dot{A} = \gamma_b S_d C_s - \gamma_b C_d S_s - \Delta_b A, \quad \text{(B.7)}$$
$$\dot{\theta} = -\frac{\Delta_b}{\delta} A, \quad \text{(B.8)}$$
$$-\left(\frac{\Delta_b}{\delta} + \frac{1}{\Pr} + v_n \exp(-\Pr)\right) \theta + \frac{1}{\Pr(1 + gS)}. \quad \text{(B.9)}$$

While the effect of the waves on the interface temperature is explicit, the reverse effect is implicit in the fact that the forcing parameter as defined equation (B.1) depends on the interface temperature.

REFERENCES

Ablowitz, M., and H. Segur (1981), *Solitons and the Inverse Scattering Transform*, SIAM Studies in Applied Mathematics, SIAM, Philadelphia, Penna.

Azouni, A., E. W. Bolton, and F. H. Busse (1986), Experimental study of convection columns in a rotating cylindrical annulus, *Geophys. Astrophys. Fluid Dyn.*, 34, 301–317.

Barcilon, V., and J. Pedlosky (1967), A unified linear theory of homogeneous and stratified rotating fluids, *J. Fluid Mech.*, 29, 609–621.

Benjamin, T. B., and J. E. Feir (1967), The disintegration of wave trains on deep water. Part I: Theory, *J. Fluid Mech.*, 27, 417–430.

Bretherton, F. (1964), Resonant interactions between waves. The case of discrete oscillations, *J. Fluid Mech.*, 20, 457–479.

Buzyna, G., R. L. Pfeffer, and R. Kung (1984), Transitions to geostrophic turbulence in a rotating differentially heated annulus of fluid, *J. Fluid Mech.*, 145, 377–403.

Castrejon-Pita, A. A., and P. L. Read (2007), Baroclinic waves in an air-filled thermally driven rotating annulus, *Phys. Rev. E*, 75, 026,301.

Charney, J. G. (1949), On a physical base for numerical prediction of large-scale motions in the atmosphere, *J. Meteor.*, 6, 371–385.

Charney, J. G. (1971), Geostrophic turbulence, *J. Atmos. Sci.*, 28, 1087–1095.

Christiansen, B. (2000), Chaos, quasiperiodicity, and interannual variability: Studies of a stratospheric vacillation model, *J. Atmos. Sci.*, 57(18), 3161–3173.

Eady, E. T. (1949), Long waves and cyclone waves, *Tellus*, 1, 33–52.

Feliks, Y., M. Ghil, and A. W. Robertson (2011), The atmospheric circulation over the North Atlantic as induced by the SST field, *J. Clim.*, 24(2), 522–542, doi:10.1175/2010JCLI3859.1.

Fjørtoft, R. (1953), On changes in the spectral distribution of kinetic energy in two-dimensional, non-divergent flow, *Tellus*, 5, 225–230.

Flór, J.-B., H. Scolan, and J. Gula (2011), Frontal instabilities and waves in a differentially rotating fluid, *J. Fluid Mech.*, 685, 532–542.

Fowlis, W. W., and R. L. Pfeffer (1969), Characteristics of amplitude vacillation in a rotating, differentially heated fluid annulus determined by a multi-probe technique, *J. Atmos. Sci.*, 26, 100–108.

Früh, W.-G. (1996), Low-order models of wave interactions in the transition to baroclinic chaos, *Nonlin. Proc. Geophys.*, 3, 150–165.

Früh, W.-G., and P. L. Read (1997), Wave interactions and the transition to chaos of baroclinic waves in a thermally driven rotating annulus, *Phil. Trans. R. Soc. Lond. A*, 355, 101–153.

Früh, W.-G., and P. L. Read (1999), Experiments in a barotropic rotating shear layer. I: Instability and steady vortices, *J. Fluid Mech.*, 383, 143–173.

Früh, W.-G., P. Maubert, P. Read, and A. Randriamampianina (2007), DNS of structural vacillation in the transition to geostrophic turbulence, in *Advances in Turbulence XI*, edited by J. M. L. M. Palma and A. Silva Lopes, No. 117 in Springer Proceedings in Physics, pp. 432–434, Springer Verlag, Porto, Portugal.

Guckenheimer, J., and G. Buzyna (1983), Dimension measurements for geostrophic turbulence, *Phys. Rev. Lett.*, 51, 1438–1441.

Harlander, U., T. von Larcher, Y. Wang, and C. Egbers (2011), PIV- and LDV-measurements of baroclinic wave interactions in a thermally driven rotating annulus, *Exp. Fluids*, 51(1), 37–49.

Hart, J. E. (1972), A laboratory study of baroclinic instability, *Geophys. Fluid Dyn.*, 3, 181–209.

Hart, J. E. (1973), On the behaviour of large amplitude baroclinic waves, *J. Atmos. Sci.*, *30*, 1017–1034.

Hart, J. E. (1976), The modulation of an unstable baroclinic wave field, *J. Atmos. Sci.*, *33*, 1874–1889.

Hart, J. E. (1979), Finite amplitude baroclinic instability, *Ann. Rev. Fluid Mech.*, *11*, 147–172.

Hart, J. E. (1981), Wavenumber selection in nonlinear baroclinic instability, *J. Atmos. Sci.*, *38*, 400–408.

Hasselmann, K. (1967), A criterion for nonlinear wave stability, *J. Fluid Mech.*, *30*, 737–739.

Hide, R. (1953), Some experiments on thermal convection in a rotating liquid, *Q. J. Roy. Met. Soc.*, *79*, 161.

Hide, R. (1958), An experimental study of thermal convection in a rotating liquid, *Phil. Trans. R. Soc. Lond. A*, *250*, 441–478.

Hide, R. (2011), Regimes of sloping thermal convection in a rotating liquid annulus, *Geophys. Astrophys. Fluid Dyn.*, *105*(2–3), 117–127.

Hide, R., and P. J. Mason (1970), Baroclinic waves in a rotating fluid subject to internal heating, *Phil. Trans. R. Soc. Lond. A*, *268*, 201–232.

Hide, R., and P. J. Mason (1975), Sloping convection in a rotating fluid, *Adv. Phys.*, *24*, 47–99.

Hide, R., P. J. Mason, and R. A. Plumb (1977), Thermal convection in a rotating fluid subject to a horizontal temperature gradient: Spatial and temporal characteristics of fully developed baroclinic waves, *J. Atmos. Sci.*, *34*, 930–950.

Hignett, P. (1985), Characteristics of amplitude vacillation in a differentially heated rotating fluid annulus, *Geophys. Astrophys. Fluid Dyn.*, *31*, 247–281.

Hignett, P., A. A. White, R. D. Carter, W. D. N. Jackson, and R. M. Small (1985), A comparison of laboratory measurements and numerical simulations of baroclinic wave flows in a rotating cylindrical annulus, *Q. J. Roy. Met. Soc.*, *111*, 131–154.

Jacoby, T., P. Read, P. Williams, and R. Young (2011), Generation of inertia-gravity waves in the rotating thermal annulus by a localised boundary layer instability, *Geophys. Astrophys. Fluid Dyn.*, *105*(203), 161–181, doi:10.1080/03091929.2011.560151.

James, I. N., P. R. Jonas, and L. Farnell (1981), A combined laboratory and numerical study of fully developed steady baroclinic waves in a cylindrical annulus, *Q. J. Roy. Met. Soc.*, *107*, 51–78.

Jonas, P. R. (1981), Some effects of boundary conditions and fluid properties on vacillation in thermally driven rotating flow in an annulus, *Geophys. Astrophys. Fluid Dyn.*, *18*, 1–23.

Klein, P. (1990), Transitions to chaos in unstable baroclinic systems: A review, *Fluid Dyn. Res.*, *5*, 235–254.

Koo, S., A. Robertson, and M. Ghil (2002), Multiple regimes and low-frequency oscillations in the Southern Hemisphere's zonal-mean flow, *J. Geophys. Res. Atmos.*, *107*(D21), doi:10.1029/2001JD001353.

Lappa, M. (2012), *Rotating Thermal Flows in Natural and Industrial Processes*, John Wiley & Sons, Hoboken, N.J.

Lewis, G. (2013), Effects of Prandtl number on the axisymmetric to nonaxisymmetric transition in the differentially heated rotating annulus, in *Modelling Atmospheric and Oceanic Flow: Insights from Laboratory Experiments and Numerical Simulations.*, edited by T. von Larcher and P. D. Williams, AGU, Washington, D.C.

Lindborg, E. (2006), The energy cascade in a strongly stratified fluid, *J. Fluid Mech.*, *550*, 207–242.

Lindzen, R. S., R. B. Farrell, and D. Jacqmin (1982), Vacillations due to wave interference: Applications to the atmosphere and to annulus experiments, *J. Atmos. Sci.*, *39*, 14–23.

Lorenz, E. N. (1955), Available potential energy and the maintenance of the general circulation, *Tellus*, *7*, 157–167.

Lorenz, E. N. (1963a), Deterministic nonperiodic flow, *J. Atmos. Sci.*, *20*, 130–141.

Lorenz, E. N. (1963b), The mechanics of vacillation, *J. Atmos. Sci.*, *20*, 448–464.

Lovegrove, A., I. Moroz, and P. Read (2001), Bifurcations and instabilities in rotating two-layer fluids: I. f-Plane, *Nonlin. Proc. Geophys.*, *8*, 21–36.

Lovegrove, A., I. Moroz, and P. Read (2002), Bifurcations and instabilities in rotating two-layer fluids: II. β-Plane, *Nonlin. Proc. Geophys.*, *9*, 280–309.

Lu, H., and T. L. Miller (1997), Characteristics of annulus baroclinic flow structure during amplitude vacillation, *Dyn. Atmos. Oceans*, *27*, 485–503.

Lu, H. I., T. L. Miller, and K. A. Butler (1994), A numerical study of wave-number selection in the baroclinic annulus flow system, *Geophys. Astrophys. Fluid Dyn.*, *75*(1), 1–19.

Maubert, P., and A. Randriamampianina (2002), Transition vers la turbulence géostrophique pour un écoulement d'air en cavité tournante différentiellement chauffée, *C. R. Méchanique*, *330*, 365–370.

McEwan, A., D. Mander, and R. Smith (1972), Forced resonant second-order interaction between damped internal waves, *J. Fluid Mech.*, *55*, 589–608.

Nastrom, G. D., and K. S. Gage (1985), A climatology of atmospheric wavenumber spectra observed by commercial aircraft, *J. Atmos. Sci.*, *42*, 950–960.

Ohlsen, D. R., and J. E. Hart (1989), Nonlinear interference vacillation, *Geophys. Astrophys. Fluid Dyn.*, *45*, 213–235.

Or, A. C., and F. H. Busse (1987), Convection in a rotating cylindrical annulus: Part 2. Transitions to asymmetric and vacillating flow, *J. Fluid Mech.*, *174*, 313–326.

Pedlosky, J. (1981), The nonlinear dynamics of baroclinic wave ensembles, *J. Fluid Mech.*, *102*, 169–209.

Pedlosky, J. (1987), *Geophysical Fluid Dynamics*, 2nd ed., Springer-Verlag, Berlin, Heidelberg, and New York.

Pedlosky, J., and C. Frenzen (1980), Chaotic and periodic behaviour of finite-amplitude baroclinic waves, *J. Atmos. Sci.*, *37*, 1177–1196.

Pfeffer, R. L., and Y. Chiang (1967), Two kinds of vacillation in rotating laboratory experiments, *Monthly Weather Rev.*, *95*, 75–82.

Pfeffer, R. L., G. Buzyna, and W. W. Fowlis (1973), Synoptic features and energetics of wave-amplitude vacillation in a rotating, differentially-heated fluid, *J. Atmos. Sci.*, *31*, 622–645.

Pfeffer, R. L., G. Buzyna, and R. Kung (1980a), Time-dependent modes of thermally driven rotating fluids, *J. Atmos. Sci.*, *37*, 2129–2149.

Pfeffer, R. L., G. Buzyna, and R. Kung (1980b), Relationships among eddy fluxes of heat, eddy temperature variances and basic-state temperature parameters in thermally driven rotating fluids, *J. Atmos. Sci.*, *37*, 2577–2599.

Pfeffer, R. L., S. Applequist, R. Kung, C. Long, and G. Buzyna (1997), Progress in characterizing the route to geostrophic turbulence and redesigning thermally driven rotating annulus experiments, *Theor. Comput. Fluid Dyn.*, *9*, 253–267.

Phillips, N. A. (1951), A simple three-dimensional model for the study of large-scale extratropical flow patterns, *J. Meteor.*, *8*, 381–394.

Plumb, R. A. (1977), The stability of small amplitude Rossby waves in a channel, *J. Fluid Mech.*, *80*, 705–720.

Pogoreltsev, A., A. Kanukhina, E. Suvorova, and E. Savenkova (2009), Variability of planetary waves as a signature of possible climatic changes, *J. Atmos. Solar-Terrestrial Phys.*, *71* (14–15), 1529–1539, doi:10.1016/j.jastp.2009.05.011.

Randriamampianina, A., and E. Crespo Del Arco (2014), High resolution method for direct numerical simulation of the instability and transition in a baroclnic cavity, in *Modelling Atmospheric and Oceanic Flow: Insights from Laboratory Experiments and Numerical Simulations*, edited by T. von Larcher and P. D. Williams, AGU, Washington, D.C.

Randriamampianina, A., W.-G. Früh, P. L. Read, and P. Maubert (2006), Direct numerical simulations of bifurcations in an air-filled rotating baroclinic annulus, *J. Fluid Mech.*, *561*, 359–389.

Read, P. (2003), A combined laboratory and numerical study of heat transport by baroclinic eddies and asixymmetric flows, *J. Fluid Mech.*, *489*, 301–323, doi:10.1017/S002211200300524X.

Read, P. L., M. J. Bell, D. W. Johnson, and R. M. Small (1992), Quasi-periodic and chaotic flow regimes in a thermally-driven, rotating fluid annulus, *J. Fluid Mech.*, *238*, 599–632.

Read, P. L., P. Maubert, A. Randriamampianina, and W.-G. Früh (2008), Direct numerical simulation of transitions towards structural vacillation in an air-filled, rotating, baroclinic annulus, *Phys. Fluids*, *20*, 044107, 1–17.

Schnaubelt, M., and F. H. Busse (1992), Convection in a rotating cylindrical annulus: Part 3. Vacillating and spatially modulated flows, *J. Fluid Mech.*, *245*, 155–173.

Sitte, B., and C. Egbers (2000), Higher order dynamics of baroclinic waves, in *Physics of Rotating Fluids*, edited by C. Egbers and G. Pfister, pp. 355–375, Springer Verlag, Berlin, Heidelberg, and New York.

Smith, R. K. (1974), On limit cycles and vacillating baroclinic waves, *J. Atmos. Sci.*, *31*, 2008–2011.

Son, S., and S. Lee (2006), Preferred modes of variability and their relationship with climate change, *J. Climate*, *19*(10), 2063–2075, doi:10.1175/JCLI3705.1.

Studer, S., K. Hocke, and N. Kämpfer (2012), Intraseasonal oscillations of stratospheric ozone above Switzerland, *J. Atmos. Solar-Terrestrial Phys.*, *74*, 180–198, doi:10.1016/j.jastp.2011.10.020.

Tamaki, K., and K. Ukaji (1985), Radial heat transport and azimuthally averaged temperature fields in a differentially heated rotating fluid annulus undergoing amplitude vacillation, *J. Met. Soc. Japan*, *63*, 168.

Tamaki, K., and K. Ukaji (1993), Characteristics of tilted-trough vacillation in a differentially heated rotating fluid annulus, *J. Met. Soc. Japan*, *71*, 553–566.

von Larcher, T., and C. Egbers (2005), Experiments on transitions of baroclinic waves in a differentially heated rotating annulus, *Nonlin. Proc. Geophys.*, *12*(6), 1033–1041.

Waite, M., and P. Bartello (2006), The transition from geostrophic to stratified turbulence, *J. Fluid Mech.*, *568*, 89–108.

Watterson, I. (2001), Zonal wind vacillation and its interaction with the ocean: Implications for interannual variability and predictability, *J. Geophys. Res. Atmos.*, *106*(D20), 23,965–23,975, doi:10.1029/2000JD000221.

Weng, H.-Y., and A. Barcilon (1987), Wave structure and evolution in baroclinic flow regimes, *Q. J. R. Met. Soc.*, *113*, 1271–1294.

Weng, H.-Y., and A. Barcilon (1988), Wavenumber transition and wavenumber vacillation in Eady-type baroclinic flows, *Q. J. R. Met. Soc.*, *114*, 1253–1269.

Weng, H.-Y., A. Barcilon, and J. Magnan (1986), Transitions between baroclinic flow regimes, *J. Atmos. Sci.*, *43*, 1760–1777.

White, H. D., and E. L. Koschmieder (1981), Convection in a rotating, laterally heated annulus. Pattern velocities and amplitude oscillations, *Geophys. Astrophys. Fluid Dyn.*, *18*, 301–320.

Young, R. M. B., and P. L. Read (2008), Flow transitions resembling bifurcations of the logistic map in simulations of the baroclinic rotating annulus, *Phys. D Nonlin. Phenom.*, doi:10.1016/j.physd.2008.02.014.

Zakharov, V. (1968), Stability of periodic waves of finite amplitude on the surface of a deep fluid, *J. Appl. Mech. Tech. Phys.*, *9*, 190–194.

Section II: Balanced and Unbalanced Flows

4

Rotation Effects on Wall-Bounded Flows: Some Laboratory Experiments

P. Henrik Alfredsson[1] and Rebecca J. Lingwood[1,2]

4.1. INTRODUCTION

Rotation of a solid boundary with respect to a fluid or rotation of a fluid system occurs in both constructed and naturally occurring systems. An example of the former is rotating machinery, e.g., rotating compressors, turbines, propellers, and centrifuges, and examples of the latter are flows in the atmosphere and the oceans.

When studying such systems, one may choose from two principally different approaches, namely to view the system either from a nonrotating frame of reference or from the rotating system itself. Depending on the flow system under investigation, one approach may be more appropriate than the other. In the former case the rotation enters through rotating boundary conditions, in the other through the body forces set up by the rotation. In the latter case, the rotation is felt through two well-known physical effects, namely the centrifugal and Coriolis forces. In the rotating frame of reference, the continuity equation is unchanged from its nonrotating equivalent and is given as

$$\nabla \cdot \mathbf{u} = 0, \quad (4.1)$$

where $\mathbf{u}$ is the velocity vector in the rotating system, whereas the Navier-Stokes equation becomes

$$\frac{\partial \mathbf{u}}{\partial t} + (\mathbf{u} \cdot \nabla)\mathbf{u} = -\frac{1}{\rho}\nabla p + \nu \nabla^2 \mathbf{u} - \Omega \times (\Omega \times \mathbf{r}) - 2\Omega \times \mathbf{u}, \quad (4.2)$$

where Ω is the system rotation rate, which is assumed to be constant, and the boundary conditions are cast in the rotating frame. The centrifugal term $\Omega \times (\Omega \times \mathbf{r})$ can be incorporated into the pressure term and is not

[1] *Royal Institute of Technology, Linné FLOW Centre, KTH Mechanics, Stockholm, Sweden.*

[2] *Institute of Continuing Education, University of Cambridge, Cambridge, United Kingdom.*

dynamically important (however, it may be problematic from a practical experimental point of view), whereas the Coriolis term $2\Omega \times \mathbf{u}$ may completely change and dominate the flow behavior compared with a nonrotating counterpart.

4.1.1. Categorization of Different Rotating Flow Systems

One may categorize rotating flows in many different ways, but here we restrict ourselves to some cases where there have been laboratory experiments of canonical flows reported in the literature. For the present undertaking, we focus on three different categories (see also Figure 4.1):

(a) System rotation vector parallel to mean-flow vorticity

(b) Flows set up by the rotation of one or more boundaries

(c) System rotation aligned with the mean-flow direction

System rotation of shear flows in category (a) occurs in many types of rotating machinery as well as in geophysical flows. A canonical flow of this type is flow in the x direction through a two-dimensional channel with rotation along the z axis (Ω_z). In this case the Coriolis acceleration gives rise to a body force that can be "unstably stratified" and, if it is large enough (compared with restoring viscous forces), can therefore destabilize the flow. So, for example, for pressure-driven Poiseuille flow, half of the channel will be stabilized and half of the channel destabilized, whereas for wall-driven Couette flow, the full channel will be either stabilized (cyclonic rotation, i.e., the vorticity of the mean flow is in the same direction as the rotation vector) or destabilized (anticyclonic rotation). Analogously, for flow over a flat plate rotating about a spanwise axis, the boundary layer on the leading side will be

---

Modeling Atmospheric and Oceanic Flows: Insights from Laboratory Experiments and Numerical Simulations, First Edition. Edited by Thomas von Larcher and Paul D. Williams.
© 2015 American Geophysical Union. Published 2015 by John Wiley & Sons, Inc.

86 MODELING ATMOSPHERIC AND OCEANIC FLOWS

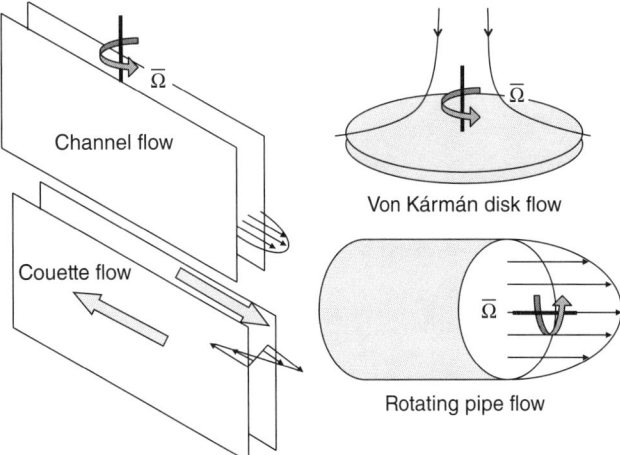

Figure 4.1. Schematics of different flow fields where rotation effects are of importance.

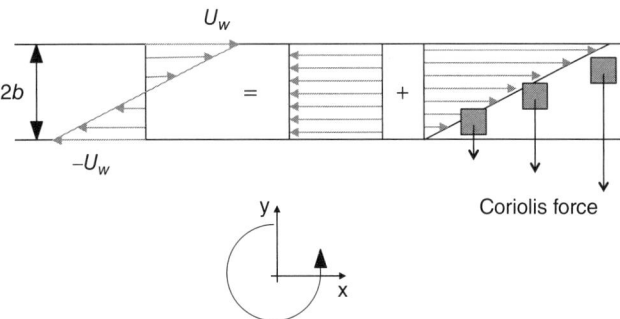

Figure 4.2. Effects of the Coriolis force on rotating plane Couette flow.

destabilized and flow on the trailing side will be stabilized. (The terms *leading* and *trailing* originate from turbomachinery where the "leading" side of a turbine or compressor blade is the one approaching the flow.) The stabilization and destabilization occur in both the laminar and turbulent regimes for all these examples.

In this category of rotating flows, experiments on pressure-driven channel flow have been carried out by *Johnston et al.* [1972], *Lezius and Johnston* [1976], and *Alfredsson and Persson* [1989], and experiments on wall-driven plane Couette flow have been performed by *Alfredsson and Tillmark* [2005], *Hiwatashi et al.* [2007], and *Tsukahara et al.* [2010a]. Some experiments on flat-plate boundary layers with system rotation have also been undertaken by *Watmuff et al.* [1985], *Masuda and Matsubara* [1989], *Matsubara and Masuda* [1991], and *Nickels and Joubert* [2000].

As an example, the effect of the Coriolis force on a rotating flow is illustrated by Figure 4.2, which shows rotating plane Couette flow. By dividing the plane Couette mean flow into a superposition of a constant velocity and a linearly varying flow across the channel, it is clear from the Coriolis term $2\Omega \times \mathbf{u}$ that the constant-velocity profile only gives a constant acceleration across the channel and a corresponding constant (opposing) Coriolis force. However, the linearly varying profile gives a linearly varying Coriolis acceleration across the channel and a corresponding linearly varying Coriolis force, which for sufficiently high Reynolds number leads to instability in the case of anticyclonic rotation, as in Figure 4.2. If the rotation is cyclonic, then the Coriolis force is in the opposite direction and the flow is stabilized instead.

Taylor-Couette flow may fall into category (a) (and also into category (b)). In this case streamline curvature also affects the flow through the introduction of centrifugal forces, and this is usually the dominant factor. However, there are certain parameter ranges where the rotation effects may be dominant, i.e., for corotating cylinders. Another example where centrifugal effects may compete with Coriolis effects is the pressure-driven *curved* channel flow. Pressure-driven curved channel flow is usually denoted *Dean flow*, and in that case the outer half of the channel (concave-wall side) is unstable due to centrifugal forces, and the instability may result in streamwise vortices. Depending on the sign of the spanwise rotation vector, the Coriolis force may either enhance or counteract the centrifugal force [*Matsson and Alfredsson*, 1990, 1994]. A similar centrifugal instability occurs for a boundary layer flow over a concave wall, in that case denoted as Görtler instability.

Category (b) includes a number of different flow classes where, for example, the flow is set up by one or two rotating disks or cylinders with the fluid in between (see, e.g., the recent special volume edited by *Healey* [2007] where studies of different disk flows are presented). One such class is the so-called BEK (Bödewadt, Ekman, and von Kármán) family of boundary layer flows (as defined by *Lingwood* [1997]). One of the first observations of this type of flow was made by the Norwegian Arctic explorer Fridtjof Nansen, who observed that the drift of surface ice was angled at $20° - 40°$ to the right of the wind direction (in the Northern Hemisphere) and correctly attributed this effect to the fact that the Coriolis force introduced by the rotation of Earth is not negligible compared with the forces from the wind and current, which together produce the slow drift velocity. Based on this suggestion, the Swedish scientist Vagn Walfrid Ekman (as his Ph.D. research) studied the influence of Earth's rotation on wind-driven ocean currents, resulting from balanced pressure gradient, Coriolis, and frictional forces. *Ekman* [1905] showed that the oceanic flow system, bounded by the ocean surface, has a boundary layer structure within which the mean velocity can be represented by a vector that changes length exponentially with depth below the surface and changes angle linearly with depth, the so-called Ekman spiral. Although boundary layer velocity profiles that approximate the Ekman layer occur in the atmospheric boundary layer and in wind-driven surface layers of the ocean, turbulence always plays a role in atmospheric and oceanic boundary layers because of the high Reynolds numbers involved. Unsteadiness of the mean-flow and thermal effects may also be important.

From Ekman's Ph.D. work, the flow induced when the boundary and the fluid at a large distance from the boundary approach the same rotation rate has become known as the Ekman layer. Later, *Bödewadt* [1940] studied the boundary layer flow created near a stationary plate in a body of fluid rotating with constant rotation rate at large distances from the plate (think of the boundary layer at the bottom of a cup of stirred tea!). For the Bödewadt layer the centrifugal and radial pressure gradient forces are in equilibrium in the fluid rotating at large distances above the plate but the centrifugal forces are reduced within the boundary layer, due to viscous action, and the axially independent radial pressure gradient causes a radial flow that is predominantly inward (that is why the tea leaves settle in the center of the cup) and a consequent (from continuity arguments) upward axial flow.

The *von Kármán* [1921] layer is in some senses the reverse of the Bödewadt layer. It is induced by a rotating boundary in otherwise quiescent fluid and sets up a flow where fluid is thrown radially outward at the boundary due to the action of centrifugal forces and is replaced by a downward axial flow. The BEK boundary layers are particular examples of a family of rotating flows and are sketched in Figure 4.3. In addition to the BEK family, one can envisage other rotation-induced flows, such as boundary layer flows over rotating cones or spheres, with or without added coflow (see, e.g., *Kohama* [2000]; *Garrett and Peake* [2007]; *Hussain et al.* [2011]).

In the laboratory, the Bödewadt and von Kármán flows are set up on plates, that are stationary and rotating, respectively, in the laboratory frame. The Ekman flow is usually set up between two disks with slightly different rotation rates ($\Omega_z = \Omega_0 \pm \Delta\Omega$), where boundary layers are formed on both disks and the flow spirals radially inward on the slower moving disk and vice versa on the faster moving disk. In the region between the disks and outside the boundary layers, the fluid rotates with Ω_0 and there is an axial flow exchange between the two boundary layers.

In category (c) we have, for example, the pressure-driven axially rotating pipe or channel flows. For an axially rotating pipe, i.e., with the rotation axis parallel to the mean-flow direction, the rotation has no direct influence on the axial laminar mean flow and the basic parabolic axial mean-flow profile of the nonrotating case is retained. Early linear stability analysis and experiments have shown that rotation has a destabilizing effect [*Pedley*, 1969; *Imao et al.*, 1992]. In the turbulent case, the azimuthal reflection symmetry is broken and the cross-flow Reynolds stress term (which is zero for the nonrotating case) gives rise to a deviation from solid-body rotation (first observed by *Kikuyama et al.* [1983]). The streamwise velocity distribution becomes less full in this case. Only one experiment of channel flow with streamwise rotation has been reported, where *Recktenwald et al.* [2007] measured the velocity field with particle image velocimetry (PIV). Comparisons were made with direct numerical simulations (DNSs) and in that case, which is more complicated than the rotating pipe flow, a secondary flow is set up perpendicular to the main flow direction, and the mean-flow profiles were found to become fuller than for the nonrotating case.

4.1.2. Layout of Chapter

The flows in the different categories above differ with respect to their geometry but, more importantly, in how rotation affects them. In the following, we focus on three different flows that are relatively amenable to laboratory investigation, one from each category described above: One is plane Couette flow undergoing system rotation about an axis normal to the mean flow, another is the von Kármán boundary layer flow, and the third is axially rotating pipe flow. These flows are associated with interesting flow features that may be relevant to both technical and naturally occurring fluid systems. In Section 4.2, we define the important nondimensional parameters that govern them and discuss some of their interesting flow features in various parameter ranges. Various experimental realizations of the three different flow systems are described in Section 4.3 and considerations and limitations regarding the laboratory systems are discussed. Some intriguing results are described in Section 4.4, and in Section 4.5

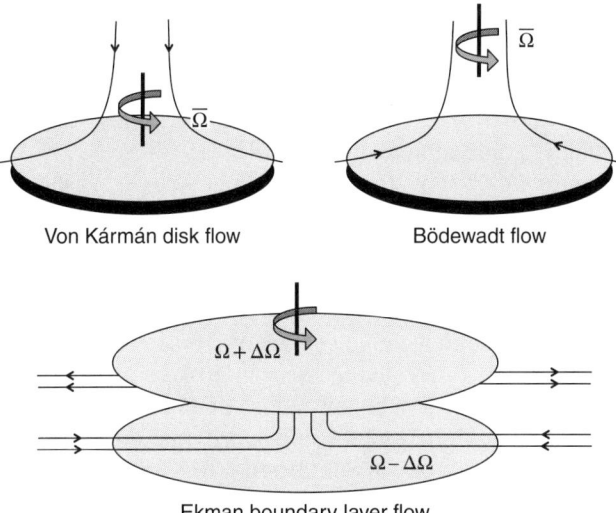

Figure 4.3. Three different examples of the BEK family of flows.

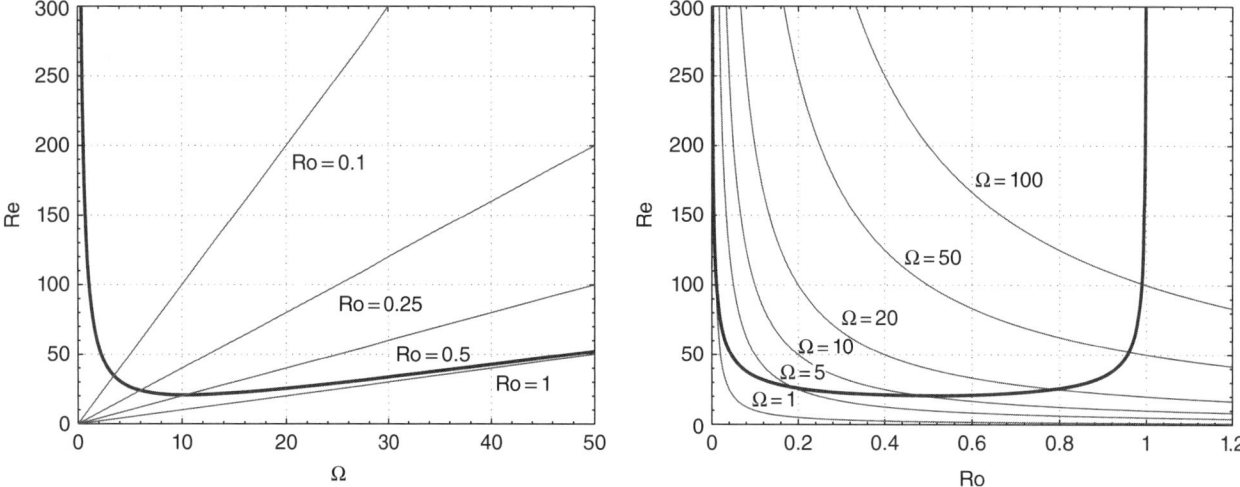

Figure 4.4. Stability diagram for RPCF in the ΩRe plane (left) and RoRe plane (right). The thick line is the neutral stability curve, and the thin lines show constant values of Ro and Ω, respectively. Note that the flow is linearly stable for Ro > 1.

we comment on the implications for experiments of model systems in the age of large-scale computations.

4.2. FLOW PARAMETERS AND FEATURES OF SOME ROTATING FLOWS

4.2.1. Rotating Plane Couette Flow

Rotating plane Couette flow (RPCF) consists of a two-dimensional Couette flow channel with two walls at $y = \pm b$, with flow in the x direction and rotation along the z axis (Ω_z). In principle, this flow is determined by two control parameters, namely the Reynolds number and a suitably defined nondimensional parameter that reflects the rotation rate. Here the Reynolds number Re is conveniently defined using half the velocity difference between the walls (U_w) and half the channel width (b). In the engineering literature, the rotation is usually given as a rotation number Ro = $\Omega_z b/U_w$, but from an experimental point of view, it is more convenient to use the nondimensional rotation rate Ω defined as $\Omega = 2\Omega_z b^2/\nu$ since then it is possible to change Re by varying U_w without affecting Ω. However, these parameters are, of course, connected through the relation Ro = $\Omega/$Re. In the geophysical literature, the preferred nondimensional rotation parameter is the Rossby number, which is the inverse of the rotation number. In this chapter, we use the definitions of Ro and Ω given above since these definitions are used in much of the laboratory work on these flows.

The RPCF is characterized by a large number of different flow states, which have been explored by *Tsukahara et al.* [2010a]. For anticyclonic rotation, the minimum Reynolds number for instability given by linear stability theory is Re = 20.6 at a rotation number of Ro = 0.5 or

Figure 4.5. Flow visualization using reflective flakes of 2D roll cells in RPCF at Re = 50 and Ω = 2.6 obtained by *Tsukahara et al.* [2010a]. Mean flow is in the horizontal direction and the length and width of the photograph is approximately $60h$ and $30h$, respectively. Copyright © Cambridge University Press, 2011. Reprinted with permission.

$\Omega = 10.3$. in fact, it is possible to write the neutral stability curve in analytical form as $\text{Re}_{\text{crit}} = \Omega + 107\Omega^{-1}$. Figures 4.4a,b show the neutral stability curve in the ΩRe plane and RoRe plane together with the corresponding lines for some constant values of Ro and Ω, respectively. The primary linear instability takes the form of streamwise independent roll cells (see Figure 4.5) that reach across the channel with a spanwise size close to the channel width (i.e., $2b$), giving them a cross-sectional aspect ratio close to 1. For higher values of Ω and Re, the roll cells develop into more complicated three-dimensional nonstationary types. An interesting feature is the three-dimensional nonlinear *steady* states that were theoretically predicted by *Nagata* [1998] and also observed in experiments by *Hiwatashi et al.* [2007], and *Tsukahara et al.*

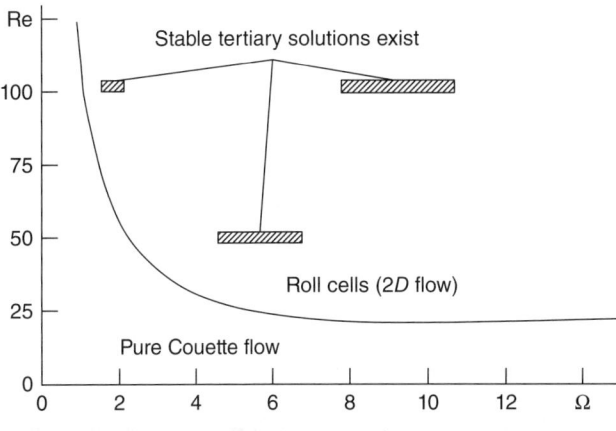

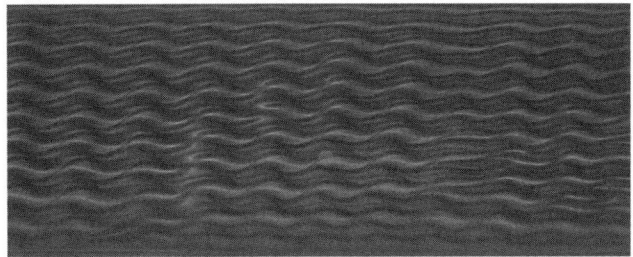

Figure 4.6. Predicted nonlinear steady states by *Nagata* [1998] (upper) and experimental realization by *Tsukahara et al.* [2010a] (lower) at Re = 100, Ω = 8.7. Copyright © Cambridge University Press, 2011. Photograph reprinted with permission.

[2010a] (see Figure 4.6). For even higher Reynolds numbers, the roll cells themselves become turbulent.

For the nonrotating case, plane Couette flow (PCF) starts to develop turbulent spots around Re = 325 to Re = 360 [*Tillmark and Alfredsson*, 1991, 1992; *Daviaud et al.*, 1992] and turbulent stripes [*Prigent et al.*, 2002; *Barkley and Tuckerman*, 2005, 2007] are observed at intermediate Reynolds numbers in both experiments and simulations. For cyclonic rotation, formation of turbulent spots is suppressed until higher Reynolds numbers and are followed by turbulent stripes if the Reynolds number is further increased, as observed both in experiments and in DNS [*Tsukahara et al.*, 2010b]. The first DNSs of a few cases of cyclonic rotation at low Reynolds numbers were reported already by *Komminaho et al.* [1996] and showed the stabilization effect. At sufficiently high Re the flow becomes fully turbulent, as can be seen in the mapping of the parameter space done by *Tsukahara et al.* [2010a] (see their Figure 2).

4.2.2. von Kármán Boundary Layer Flow

The laminar von Kármán boundary layer flow can be expressed as an exact solution of the Navier-Stokes equations, which is an attractive feature for theoretical analyses. The laminar boundary layer also has a constant thickness (proportional to $\sqrt{\nu/\Omega_z}$) and is therefore parallel in the physical sense (z is the coordinate normal to the disk surface). In the von Kármán flow, there is only one nondimensional parameter, namely a Reynolds number based on the angular rotational speed, the radius, and the viscosity. Usually the Reynolds number is taken as Re = $r\sqrt{\Omega_z/\nu}$, where r is the (dimensional) radius and hence Re increases linearly with r.

One of the most interesting features of the rotating disk boundary layer flow is that the onset of transition to turbulence is observed experimentally within an unusually limited range of Reynolds numbers (500 < Re < 520). Reported transition Reynolds numbers slightly outside of this range can be largely attributed to the use of different definitions for the onset of transition rather than fundamental variations in the physical phenomena; see *Imayama et al.* [2012b]. This degree of reproducibility across different experimental facilities contrasts with the transition behavior observed for most other boundary layer flows. This was first pointed out by *Lingwood* [1995], and she also offered a hypothesis that this is due to the rotating disk boundary layer being absolutely unstable. Figure 4.7 shows a visualization from *Cederholm and Lundell* [1998] of the temperature of a heated disk, where the transition to turbulence is seen as a distinct change in temperature at a specific radial position (i.e., specific Reynolds number).

The von Kármán flow is a three-dimensional boundary layer flow that is susceptible to an inviscid cross-flow instability (often referred to as type 1 or B instability) due to an inflectional mean radial velocity component, giving rise to so-called cross-flow vortices. Figure 4.7 shows

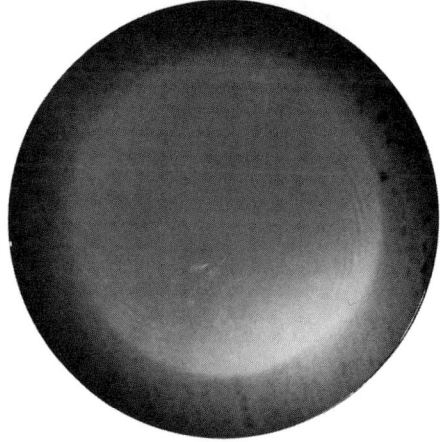

Figure 4.7. Temperature distribution of a heated rotating disk shown by temperature-sensitive floating crystals [*Cederholm and Lundell*, 1998]. The photograph is obtained as a long-time exposure using stroboscopic light. Copyright © Cederholm & Lundell, 1998. Reprinted with permission. For color detail, please see color plate section.

(in the blue-green region) the footprints of such vortices, which are stationary with respect to the rotating disk, vortices that are predicted by linear stability theory to form part of the convectively unstable cross-flow mode. The initial discovery of the cross-flow instability was by *Smith* [1947], was further substantiated by *Gregory et al.* [1955], and was visualized in the now classical study of *Kohama* [1984]. A second convectively unstable mode, which is viscous in nature and often referred to as type 2 (or type A) instability, that has a lower critical Reynolds number for linear instability also exists. While the von Kármán flow is induced by the disk rotation, acting like a centrifugal fan, a similar three-dimensional boundary layer is established in nonrotating configurations, e.g., on swept wings, where cross-flow instability has been found to be of importance for understanding transition (see the review by *Saric et al.* [2003]).

In the 1980s and 1990s several studies of the linear instability of rotating disk flow were made to investigate the type 1 and type 2 convectively unstable modes. However, the theoretical discovery by *Lingwood* [1995] of an absolute instability turned the research in a new direction. As discussed by *Huerre and Monkewitz* [1990], the response of a flow to impulsive forcing shows whether it is convectively or absolutely unstable. If the response to the transient disturbance grows with time at a fixed location in space, then the flow is absolutely unstable. *Lingwood* [1995] showed that above a critical Reynolds number there is an absolute instability of the von Kármán boundary layer produced by a coalescence of the inviscidly unstable (type 1) mode and a third mode that is spatially damped and inwardly propagating. The local absolute instability was found to occur above Re = 507, and this value was confirmed by *Lingwood* [1996] in a concurrent experimental study. Her work suggested that this instability mechanism is responsible for the onset of nonlinear behavior and thereby possibly for the onset of laminar-turbulent transition.

Davies and Carpenter [2003] performed direct numerical simulations solving the linearized Navier-Stokes equations and suggested that the convective behavior eventually prevails even for strongly locally absolutely unstable regions and concluded that the absolute instability does not produce a linear amplified global mode. This work has been followed by several recent theoretical (e.g., *Pier* [2003]; *Healey* [2010]) and experimental (e.g., *Corke et al.* [2007]; *Imayama et al.*, [2012a, 2012b]) studies that have focused on detailed investigations of the absolute instability and the transition to turbulence and to determine whether *Davies and Carpenter's* [2003] finding of linear global stability is a product of the linear approximation and/or neglect of the inwardly traveling disturbances from the outer radial boundary. *Pier* [2003] proposed a subcritical mechanism where finite-amplitude disturbances with secondary instability lead to a nonlinear steep-fronted global mode at the location of the onset of absolute instability. *Healey* [2010] proposed a supercritical mechanism where inclusion of the finite radius of the disk in the theoretical analysis creates global instability leading directly to a steep-fronted global mode regardless of how low the background disturbance level is.

Only relatively few studies have so far investigated the turbulent disk flow [*Cham and Head*, 1969; *Erian and Tong*, 1971; *Itoh and Hasegawa*, 1994; *Imayama*, 2012], in part because of experimental difficulties associated with achieving spatially well-resolved measurements. The existing measurements show both similarities and dissimilarities with a turbulent flat-plate boundary layer and will be further discussed in Section 4.4.2.

4.2.3. Axially Rotating Pipe Flow

In the axially rotating pipe flow, two nondimensional flow parameters arise, the Reynolds number based on the average flow velocity and pipe radius (R) and the angular rotation speed of the pipe (Ω_x). The rotation rate can be given as the azimuthal velocity of the pipe inner wall, $V_w = R\Omega_x$, normalized with the bulk velocity such that $S = V_w/U_b$, where S is sometimes called the swirl rate.

As already mentioned, the fully developed laminar axial mean flow is not affected by the axial rotation since the fluid undergoes solid-body rotation. However, theoretical studies show that rotation destabilizes the flow, so that it is no longer linearly stable at all Reynolds numbers. Furthermore, the study of rotating Hagen-Poiseuille flow by *Shrestha et al.* [2013] shows the first experimental evidence (via flow visualization) of absolute instability of a confined spatially developing forward flow. Their results build on theoretical linear stability analyses of the convective versus absolute instability of the flow (e.g., *Fernandez-Feria and del Pino* [2002]; *del Pino et al.* [2003]) and the three-dimensional numerical simulations of *Sanmiguel-Rojas and Fernandez-Feria* [2005] and show that the axially rotating pipe flow can be absolutely unstable for moderate values of Reynolds numbers and low values of the swirl rate. *Shrestha et al.*'s [2013] flow visualizations show transitions, with increased Reynolds number and swirl rate, from stable cases, where the rotating boundary layer develops with an axisymmetric conical shape in the inlet region with no disturbance field downstream, to convectively unstable cases, where the inlet region is unchanged but with sinusoidal waveforms in the downstream region, to absolutely unstable cases, where the azimuthal symmetry in the inlet region is broken by a spiral structure overlaid on the conical boundary layer development. Further analysis of nonlocal and nonlinear effects is required to elucidate the laminar-turbulent transition process.

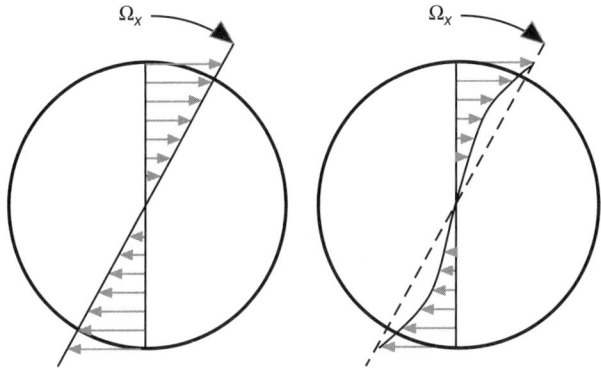

Figure 4.8. Azimuthal velocity profile in laminar (left) and turbulent (right), axially rotating pipe flow.

In the turbulent case, rotation breaks the azimuthal reflection symmetry. This means that, in the turbulent case, a nonzero azimuthal/radial Reynolds stress component is established. This makes the pressure drop (or rather skin friction) decrease and the mean velocity profile tends to become more parabolic, i.e., closer to the laminar profile. This also influences the azimuthal mean velocity distribution such that it no longer increases linearly with the radius, as it would if the fluid were in solid-body rotation, but rather has a parabolic distribution as illustrated by Figure 4.8. This has been observed in several experiments (for further discussion of this, see Section 4.4.3) but also in the numerical simulation by *Orlandi and Fatica* [1997].

4.3. LABORATORY FACILITIES

4.3.1. Rotating Plane Couette Flow

The rotating plane Couette flow has so far only been studied in the Fluid Physics Laboratory at KTH and results have been presented in a few papers [*Tillmark and Alfredsson*, 1996; *Alfredsson and Tillmark*, 2005; *Hiwatashi et al.*, 2007; *Tsukahara et al.*, 2010a, 2010b]. Even without rotation the plane Couette flow channel is a rather complicated mechanical system and several variations have been tried. An innovative idea for the construction of an experimental plane Couette flow channel was introduced by *Tillmark and Alfredsson* [1991, 1992] (who also provide a review of earlier concepts). This new concept involves an "infinite" transparent belt moving in the horizontal direction sliding against two vertical glass walls, thereby giving a counter movement of the two walls. The belt mechanism is immersed in an open water tank giving a free surface through which the belt is slightly protruding. This latter fact proved to be crucial since the surface tension between the belt and the glass plates keeps the belt well attached to the walls. By using a transparent belt, the channel is optically accessible both for flow visualization and, for example, PIV or laser Doppler velocimetry (LDV). This design has been subsequently exploited in at least four other Couette flow apparatuses around the world (France, Norway, the United States, and Japan, by *Daviaud et al.* [1992], *Malerud et al.* [1995], *Zettner and Yoda* [2001], and *Hagiwara et al.* [2002], respectively). However, so far only the original apparatus at KTH has been used to study the flow with system rotation.

A drawing of the RPCF apparatus is shown in Figure 4.9. The full length of the apparatus is 2500 mm and the test section height is 360 mm and its length is 1500 mm. The total weight of the filled channel is around 300 kg. The distance between the belt walls is adjustable and is typically 10 or 20 mm but can accommodate distances up to 70 mm. The larger the wall distance, the larger the rotation rate that can be achieved since $\Omega \sim b^2$: doubling the channel width gives a fourfold increase in the maximum rotation rate. However, the larger the distance between the walls, the worse the aspect ratio of the test channel.

Since the upper water surface is free under rotation it will have a parabolic shape, and this is a limiting factor on the rotational speed to about $\Omega_z = 0.6$ rad/s. In order to avoid a sloshing motion of the water in the channel, the rotating table supporting the channel needs to be accurately balanced.

The research so far on the RPCF has mainly been focused on flow visualization, using reflective flakes, of the various flow regimes for $\Omega = \pm 30$. However, recently some new results including two-dimensional (2D) PIV measurements in the streamwise-spanwise plane have been reported, also for higher Ω [*Suryadi et al.*, 2012, 2013].

4.3.2. von Kármán Boundary Layer Flow

When designing a rotating disk experiment there are certain limitations that need to be taken into account. If the laminar flow, its stability, and subsequent transition to turbulence are the objects of the study, it is important to note that the thickness of the laminar boundary layer is approximately $\delta = 5\sqrt{\nu/\Omega_z}$ whereas the nondimensional radius (or Reynolds number) needed to capture the transition region is at least Re = 600, giving $r/\delta > 120$. Therefore, for a boundary layer thickness of 2 mm, which may be the smallest practical thickness if boundary layer measurements are to be made, the diameter of the rotating disk would need to be at least 480 mm. This is independent of the fluid used, although most studies of the von Kármán flow have had a disk rotating in air. To reach Re = 600 in air at $r = 240$ mm, the angular speed needs to be $\Omega_z = 94$ rad/s, or about 15 rev/s (900 rpm). This relatively high rotation rate means that balancing the disk is crucial and that other vibrational sources need to be minimized.

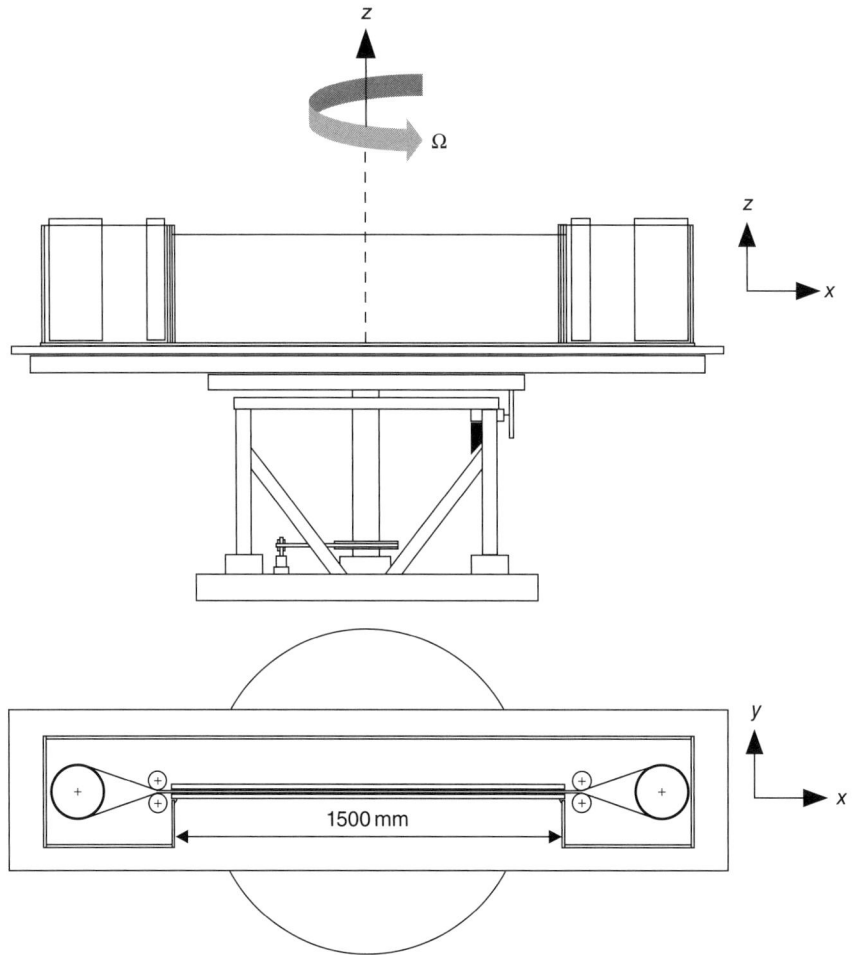

Figure 4.9. RPCF apparatus designed and constructed by *Tillmark and Alfredsson* [1996]: (a) side view; (b) top view.

Table 4.1 provides an extensive list of rotating disk experimental studies. As can be seen, most disks have a diameter around 500 mm, with three exceptions with larger diameters in the range 900–1000 mm. These latter cases were for studies of the turbulent boundary layer. The viscous length scale ($\ell_* = \nu/u_\tau$, where u_τ is the friction velocity) of the turbulent boundary layer can be estimated as

$$\ell_* = \frac{\nu}{u_\tau} = \sqrt{\frac{2}{c_f}}\frac{\nu}{\Omega_z r} = \sqrt{\frac{2}{c_f}}\mathrm{Re}^{-2}r. \qquad (4.3)$$

The skin friction coefficient $c_f[= 2u_\tau^2/(\Omega_z r)^2]$ is expected to be a weak function of the Reynolds number, and then it is clear from equation (4.3) that the larger the radius at a given Reynolds number, the larger the viscous length scale.

Most studies listed in Table 4.1 were made in air, but there are also some (e.g., flow visualization studies) that were made in water. The stability studies used hot-wire anemometry, whereas for the studies of the turbulent boundary layer both Pitot tubes and hot wires were used.

It is noteworthy that there are no studies using LDV or PIV. One reason for this is that the particles required to seed the flow would tend to stick to the disk surface, thereby creating stationary cross-flow vortices in the laminar region. Another problem is spatial resolution in the wall-normal direction. The thinness of the boundary layer would require very small measurement volumes (for LDV) or thin laser sheets (for PIV). For a disk experiment in air it may also be problematic to have particles that follow the flow accurately since the centrifugal force acting on the particles close to the disk may be substantial (this would be less problematic in a water experiment).

For accurate hot-wire measurements, a good calibration of the anemometer system is necessary. It is usually best to do the calibration in situ, that is, in the flow where the measurements will be taken. In a wind tunnel this is done in the free stream against a Pitot tube reading, where the Pitot tube is placed close to the hot wire. Here, the main problem is to obtain good readings from a Pitot tube at low velocities. In contrast to measurement of boundary layers on flat plates or wing profiles in wind tunnels, the rotating

Table 4.1. Examples of different rotating disk experimental facilities.[a]

Reference	D (mm)	N_h (rpm)	Orientation	Type of Study[b]
Smith [1947]	610	2120	Vertical	Stab
Gregory et al. [1955]	305	3000	Horizontal	Stab, Trans
Cobb and Saunders [1956]	457	2000	Vertical	Heat transfer
Gregory and Walker [1960]	457	2000	Horizontal	Stab, Trans
Cham and Head [1969]	900	1900	Horizontal	Turb
Chin and Litt [1972]	150	~1500	Horizontal	Trans
Fedorov et al. [1976]	100–200	5–$25 \cdot 10^3$	—	Trans
Clarkson et al. [1980]	610	50	Horizontal	Trans
Malik et al. [1981]	457	870	Vertical	Stab
Kobayashi et al. [1980]	400	1880	Vertical	Stab, Trans
Kohama [1984]	400	5000	Horizontal	Stab, Trans
	600	1200		Stab, Trans
Wilkinson and Malik [1985]	456	934	Horizontal	Stab
Littell and Eaton [1994]	1000	1500	Horizontal	Turb
Itoh and Hasegawa [1994]	1000	720	Vertical	Turb
Lingwood [1996]	475	1400	Horizontal	Stab
Jarre et al. [1996a, 1996b]	500	60	Horizontal	Stab
Cederholm and Lundell [1998]	400	2000	Horizontal	Stab, Heat transfer
Corke and Knasiak [1998]	457	1000	Horizontal	Stab
Othman and Corke [2006]	457	1000	Horizontal	Stab
Imayama et al. [2012a, 2012b]; Imayama [2012]	474	1542	Horizontal	Stab, Trans, Turb

[a] All experiments were made in air except Chin and Litt [1972], Clarkson et al. [1980], and Jarre et al. [1996a, 1996b]. N_h denotes the highest rotational speed reported in each study.
[b] Three main categories can be noted: Stability (Stab), Transition (Trans), and Turbulence (Turb).

disk flow has no free stream where the probe can be calibrated. However, the theoretical mean azimuthal velocity profile for the laminar flow is well known and may be used to calibrate the hot wire if its distance to the disk is known accurately and if the disk apparatus is well balanced and runs at a constant speed. Figure 4.10 shows an example of how the distance of the hot-wire sensing element can be estimated with an accuracy of about 10 μm by a comparative measurement with an accurate gauge block using a highly magnifying lens system and a high-resolution camera [Imayama, 2012]. Since the azimuthal velocity is much larger than the radial velocity in most of the boundary layer, the probe can be reliably calibrated if the sensing element is oriented radially, thereby effectively only sensing the azimuthal component. Different velocities can be obtained by moving the probe normal to the wall for a given rotational speed and radius, but it is also possible to change the velocity "seen" by the hot wire by varying the radial position of the probe or the rotational speed.

For the rotating disk flow the largest velocity is at the wall surface itself, which has the advantage that heat transfer to the wall from the hot wire becomes negligible compared with the convective heat transfer. This is useful both for the hot-wire calibration and, of course, for measurements close to the wall. However, there are the normal practical constraints that positioning the hot wire close to the wall risks damaging it (and the disk itself) if there is any imbalance in the disk rotation or any error in positioning. A disadvantage is that the hot wire cannot be calibrated against the mean-velocity distribution above a Reynolds number of about 500 where nonlinearity and transition begin to modify the mean flow compared with the theoretical mean flow. Furthermore,

$$u_\theta(z=0) = \mathrm{Re}\sqrt{\nu \Omega_z},$$

and it is clear that in order to obtain high azimuthal velocities in the laminar flow (i.e., for Re < 500), high rotation speeds are required. However, this also leads to a thin boundary layer since $\delta \sim \sqrt{\nu/\Omega_z}$, which makes it harder to determine the probe position accurately and to position it sufficiently close (in a nondimensional sense) to the rotating disk in order to reach high velocities for calibration purposes. Imayama [2012] partially circumvented this problem by extrapolating the calibration voltage to the wall where the velocity is known, and all the results

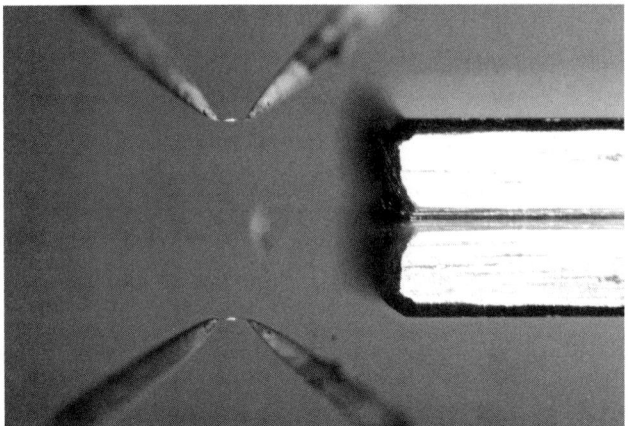

Figure 4.10. Hot-wire sensor and its reflection close to the wall for comparative measurement with a 1 mm thick gauge block in order to determine its position relative to the wall. The obtainable accuracy is estimated to better than 10 μm. The length of the hot-wire sensor is approximately 0.3 mm. From *Imayama* [2012]. Copyright © Imayama, 2012. Reprinted with permission.

based on that calibration method showed satisfactory consistency.

Another issue related to studies of turbulent boundary layers over rotating disks is the determination of the skin friction or, equivalently, the friction velocity (u_τ). Since this quantity is essential for scaling the boundary layer variables in the turbulent case, a direct measure is necessary and correlations obtained for the well-established two-dimensional turbulent flat-plate boundary layer cannot be relied upon. To date, the skin friction has been obtained through hot-wire measurements in the viscous sublayer close to the disk surface ($z^+ = zu_\tau/\nu = z/\ell_* < 6$). As an example the thickness of the viscous sublayer is about 100 μm at Re = 800 and a radius of $r = 400$ mm (independent of the fluid), corresponding to a viscous length scale of about 15 μm, emphasizing the need for very small probes and accurate positioning. So far, direct measurement of the friction velocity has only been reported by *Itoh and Hasegawa* [1994], who determined the skin friction in both the azimuthal and radial directions, and by *Imayama* [2012]. In the case of hot-wire measurements of turbulent fluctuations, spatial averaging of turbulent fluctuations along the sensor may seriously affect the results (see, e.g., *Segalini et al.* [2011]). Even a sensor length of $20\ell_*$ can give reductions of the turbulence intensity of the order of 10% in the near-wall region, which means that very small sensor lengths need to be used in order for the measurements of turbulence quantities not to be affected by spatial averaging.

4.3.3. Axially Rotating Pipe Flow

There have only been a few different experimental studies of the axially rotating pipe flow. The experimental challenge is to have a long enough pipe so that both the rotational effects and the pipe flow itself can become fully developed. One of the first studies was a stability experiment by *Nagib et al.* [1971], who used a fairly short pipe ($L/D \approx 23$, where L is the length of the pipe and D its diameter) so the parabolic profile was not fully developed; however, the rotation was obtained by the letting the fluid (water) pass through a porous material inside the rotating pipe, thereby efficiently bringing the fluid into rotation. They observed from flow visualization and hot-film measurements that the transitional Reynolds number decreased from 2500 to 900 when the swirl rate increased from 0 to 3. More recently, *Imao et al.* [1992] used a $300D$ long pipe (with water) to study the flow stability at low Reynolds numbers (500–1000) and up to rather high swirl rates ($S = 0-11$). They described details of the instabilities through both LDV measurements and flow visualizations and demonstrated that the instability takes the form of spiral waves.

There have also been a number of experimental investigations of turbulent pipe flow. One of the first was that of *White* [1964], followed by *Kikuyama et al.* [1983], *Imao et al.* [1996], and *Facciolo et al.* [2007]. Only Facciolo et al. used air as the fluid. One reason why water is preferable to air is that its lower kinematic viscosity makes it possible to keep both the Reynolds and swirl numbers high, without an excessive rotation rate, since

$$V_w = S \operatorname{Re} \frac{\nu}{D}.$$

Water may also be preferable for studies using LDV or flow visualization. An advantage of using air is that the pipe can be open at the end and the flow easily accessed by probes or LDV.

White [1964] measured the pressure drop in nonrotating parts upstream and downstream of the rotating pipe and found that it, and hence the skin friction, decreased when rotation was applied. *Kikuyama et al.* [1983] showed through LDV measurements that the streamwise velocity profile became less flat and thereby the velocity gradient at the wall decreased. They also showed that the fluid was not in solid-body rotation but lagged behind, giving a nearly parabolic profile (illustrated in Figure 4.8). Similar results were shown by *Imao et al.* [1996] but were also supplemented by measurements of five of the shear stresses. The normal stresses were lower for the cases with rotation.

In the study of *Facciolo et al.* [2007] an airflow facility was used and a schematic of it is shown in Figure 4.11, with the major components described in the caption. The

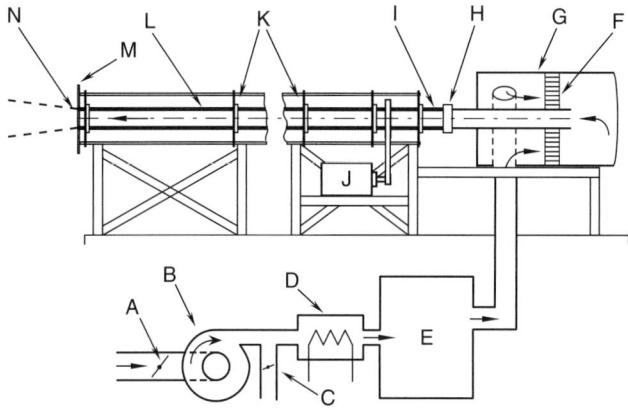

Figure 4.11. Rotating pipe apparatus used by *Facciolo et al.* [2007]: (A) throttle valve, (B) centrifugal fan, (C) valve regulated bypass, (D) electrical heater, (E) distribution chamber, (F) honeycomb, (G) stagnation chamber, (H) coupling between stationary and rotating pipe, (I) honeycomb (J) DC motor, (K) ball bearings, (L) rotating pipe, (M) circular and plate, (N) pipe outlet. Adapted from *Ferro* [2012]. Copyright © Ferro, 2012. Reprinted with permission.

rotating part of the honed steel pipe has a diameter of 60 mm and is 6000 mm long, giving a ratio $L/D = 100$. After the settling chamber, there is a 1000 mm long nonrotating pipe that is connected to the rotating part through a rotating connection. At the entrance of the pipe a 100 mm long honeycomb is mounted to bring the air into rotation where it can develop freely. The pipe is supported by four ball bearings distributed along its length. The maximum rotational speed obtainable was 900 rpm.

4.4. SOME INTRIGUING RESULTS IN ROTATING FLOWS

4.4.1. Rotating Plane Couette Flow

There are a number of interesting observations related to RPCF. The theoretical discovery of stationary tertiary states by *Nagata* [1998] shows that the Navier-Stokes equations may have multiple steady states, and at least some of these states seem to have been verified in experiments [*Hiwatashi et al.*, 2007; *Tsukahara et al.*, 2010a]. The flow also has a large number of possible flow states within the parameter ranges spanned by experiments, revealing various types of instabilities and also robust turbulent structures. The stabilization of turbulence for cyclonic rotation is also quite remarkable for anybody who has observed it in the laboratory. Even a small rotation rate of the order of one revolution in 20s can dramatically wipe out all turbulence within less than one full revolution of the channel.

The stabilizing effect on turbulence can be understood by scrutinizing the turbulent energy equation. For unidirectional flows (e.g., channel flows) there is no variation of mean quantities in the $x_1 = x$ and $x_3 = z$ directions, and the mean velocity components satisfy $U_2 = U_3 = 0$. The equations for the different components can be written as

$$\frac{\partial}{\partial t}\left(\overline{u^2}\right) = -2\overline{uv}\left(\frac{dU}{dy} - 2\Omega_z\right) + \Pi_{11} - \epsilon_{11} + D_{11}, \quad (4.4)$$

$$\frac{\partial}{\partial t}\left(\overline{v^2}\right) = -4\overline{uv}\,\Omega_z + \Pi_{22} - \epsilon_{22} + D_{22}, \quad (4.5)$$

$$\frac{\partial}{\partial t}\left(\overline{w^2}\right) = \Pi_{33} - \epsilon_{33} + D_{33}, \quad (4.6)$$

$$\frac{\partial}{\partial t}(-\overline{uv}) = \overline{v^2}\frac{dU}{dy} + 2\left(\overline{u^2} - \overline{v^2}\right)\Omega_z + \Pi_{12} - \epsilon_{12} + D_{21}. \quad (4.7)$$

Here, Π_{ij} is the pressure transport, ϵ_{ij} the dissipation, and D_{ij} the viscous diffusion (for a detailed discussion of these equations see *Johnston et al.* [1972]). Depending on the sign of Ω_z, turbulent energy will be transferred to or from the streamwise component, equation (4.4), and vice versa for the wall-normal one, equation (4.5), through the term $\pm 4\overline{uv}\Omega_z$, which appears with opposite sign in the two equations. For anticyclonic rotation (note that the mean-flow vorticity ω_z for PCF is equal to $\omega_z = -dU/dy$), $\overline{u^2}$ will decrease and $\overline{v^2}$ increase since $-\overline{uv}$ is positive where dU/dy is positive and vice versa. It is not obvious what the overall result will be on the turbulence, except that it will tend towards an equalization of the two components. On the other hand, negative (cyclonic) rotation transfers energy to $\overline{u^2}$ from $\overline{v^2}$. The normal component is important for turbulence production since it directly influences $\overline{uv}$ (see equation (4.7)) and it can be intuitively understood that negative rotation will lead to a decrease in $-\overline{uv}$ and hence in a decrease in the production of turbulent energy.

This type of reasoning can be generalized to other types of shear flows and shows the competition between the mean-flow vorticity and the background vorticity imposed by system rotation. It is not hard to imagine that similar reasoning may be applied to flows on a planetary scale.

4.4.2. von Kármán Boundary Layer Flow

One of the most important and interesting aspects of the rotating disk flow is the absolute instability of the boundary layer above a critical Reynolds number (i.e., outward from a critical radial position). While the convectively unstable nature of the rotating disk boundary layer was well established, it was (as far as the authors are aware) the first of any boundary layer flow found to exhibit an absolute instability [*Lingwood*, 1995, 1996], and the first example to link the onset of nonlinearity and transition to turbulence to such a well-defined critical Reynolds number, explaining the observed highly reproducible transition Reynolds number. Subsequently, other members of the BEK family have been shown to exhibit absolute instability as well [*Lingwood*, 1997]. The BEK flows are "semiclosed" in the sense that a disturbance that grows in time at a fixed radial position while continuing to convect azimuthally will not convect out of the domain of interest but cycle round and build on itself. It is pertinent to note that it is the azimuthal periodicity of the rotating disk configuration, namely its semiclosed nature, that means the unidirectional (radial) absolute instability is particularly relevant to the physical phenomena observed.

In a similar three-dimensional boundary layer without azimuthal periodicity, e.g., a swept-wing boundary layer, a disturbance that grows in time at a fixed chordwise position while continuing to convect in the spanwise direction (i.e., as would apply analogously to a unidirectional absolute instability) could well pass out of the domain of interest, e.g., off the end of the wing, before triggering nonlinearity.

The onset of transition for the rotating disk flow was identified by *Lingwood* [1995] as being highly reproducible, which was the motivation for looking for absolute instability of the boundary layer flow. The flow was studied using linear stability theory assuming a locally spatially invariant boundary layer and found to be locally absolutely unstable above a Reynolds number of 507. She hypothesized that the growth in time at a fixed radial position would lead to nonlinearity and the onset of transition and corroborated this experimentally [*Lingwood*, 1996]. However, for a complete and mathematically consistent theoretical stability analysis, it is necessary to include the spatial variation of the flow (i.e., variations in Reynolds number with radius), and therefore it is the *global* stability characteristics that are relevant to the real flow. If there is a global instability associated with laminar-turbulent transition, it would imply that the onset of transition should be highly repeatable across different experimental facilities. While it has previously been shown that local absolute instability does not necessarily lead to linear global instability, *Pier* [2003] showed that finite-amplitude traveling waves are subject to a secondary instability leading to a steep-fronted nonlinear global mode situated at the onset of primary local absolute instability. Furthermore, recently *Healey* [2010] has shown, using the linearized complex Ginzburg-Landau equation, that if the finite nature of the flow domain (finite radius) is accounted for, then local absolute instability can give rise to linear global instability, leading directly to a nonlinear global mode located at the convective-absolute transition boundary. *Imayama et al.* [2012b] investigated experimentally the effect of the disk edge on transition and compared with others' results for the onset of nonlinearity/transition, which is shown to lie at approximately 500–520. They found that there is in fact even greater reproducibility in the transition Reynolds number for smooth/clean disks than might have previously been thought when a consistent definition for transition is applied to others' results. This supports the hypothesis of *Lingwood* [1995, 1996] that the local absolute instability is the trigger for transition as well as the suggestion of *Healey* [2010] that it is the finite disk (present, of course, in all experiments) that allows the absolute instability to lead to (supercritical) linear global instability (and then to a nonlinear steep-fronted global mode).

The *turbulent* rotating disk boundary layer flow is arguably more relevant to geophysical flows than the laminar one; however, it is yet relatively underinvestigated. Recent experimental results by *Imayama* [2012] for the turbulent boundary layer compare the mean azimuthal velocity and variance, as well as higher moments, with results obtained for a (nonrotating) two-dimensional turbulent flat-plate boundary layer. The mean azimuthal velocity profile shows a strong similarity to the mean velocity for the two-dimensional flat-plate boundary layer, especially in the near-wall region and in the logarithmic region. However, the so-called wake region, i.e., the region where the flow deviates from the logarithmic distribution, is clearly less pronounced for the turbulent rotating disk flow. *Imayama* [2012] has been able to determine the skin friction (or friction velocity) accurately by direct measurement of the velocity distribution close to the disk and shows that the turbulence intensities in the rotating disk experiment are qualitatively similar, although they have lower values near the wall compared to measurements in two-dimensional boundary layers. This can partly be explained by the spatial averaging of the hot-wire probe, which for the experiments of *Imayama* [2012] has a length of almost $30\ell_*$ despite the fact that the sensing length is only 0.3 mm long. While the higher moments, i.e., the skewness and flatness factors, also show similar behavior in the near-wall and logarithmic regions as compared to the two-dimensional boundary layer, differences are apparent in the outer region.

An intriguing aspect of *Imayama's* [2012] experimental results as compared to two-dimensional flat-plate boundary layers [*Örlü and Schlatter*, 2013] is the difference between the spectral maps of the azimuthal and streamwise turbulence fluctuations for the two cases. First, the spectral peak corresponding to the maximum in the rms distribution near the wall is located at higher wave numbers (smaller wavelengths) for the rotating disk boundary layer. Second, the energy distribution across the boundary layer differs. It is suggested that the differences observed between these two canonical turbulent boundary layer flows may reflect the influence of rotation on the inclination and dynamics of the large-scale turbulent structures.

4.4.3. Axially Rotating Pipe Flow

One of the most surprising results for the mean-flow field of a rotating pipe flow is the deviation from solid-body rotation. In a cylindrical coordinate system the equation for the mean circumferential velocity can be reduced to (for a derivation see *Wallin and Johansson* [2000])

$$\nu \left(\frac{d^2 V}{dr^2} + \frac{1}{r}\frac{dV}{dr} - \frac{V}{r^2} \right) = \frac{d}{dr}(\overline{vw}) + 2\frac{\overline{vw}}{r}. \quad (4.8)$$

Equation (4.8) can be integrated twice, first from 0 to r and thereafter from r to R, to give

$$V(r) = V_w \frac{r}{R} - \frac{r}{\nu} \int_r^R \frac{\overline{vw}}{r} dr. \quad (4.9)$$

The first term on the right-hand side gives the solid-body rotation, whereas the second term gives a contribution from the Reynolds stress $\overline{vw}$ that results in the deviation from the solid-body rotation. As was discussed earlier and sketched in Figure 4.8, the mean circumferential velocity distribution can be described accurately by a parabola, which then can be used to obtain an expression from equation (4.9) of $\overline{vw}$ such that

$$\frac{\overline{vw}}{U_b^2} = \frac{2S}{Re}\frac{r}{R}, \quad (4.10)$$

i.e., the distribution is a straight line. As can be seen, the resulting cross-stream Reynolds stress is obtained through a delicate balance between viscous and rotation effects and is due to the fact that, by rotating the pipe, the azimuthal reflection symmetry of the flow is broken. The fact that rotation also influences the mean streamwise velocity such that the wall shear stress decreases and maximum velocity in the center of the pipe increases is also rather perplexing.

Finally, the theoretical modeling of the turbulent flow in axially rotating pipes by *Oberlack* [1999] should be mentioned. He used a Lie group approach to the Reynolds-averaged Navier-Stokes equations and was able to derive new scaling laws for both the azimuthal and streamwise mean velocities for the rotating pipe flow. For instance, the theory gives the azimuthal mean velocity as

$$\frac{V}{V_w} = \xi \left(\frac{r}{R} \right)^\psi,$$

which with $\xi = 1$ and $\psi = 2$ corresponds to the parabolic velocity distribution observed in experiments and simulations. Some other results from that analysis are also substantiated by the experiments of *Facciolo et al.* [2007].

4.5. CONCLUSIONS

It is clear that rotation plays an important, if not dominant, role in many geophysical situations, whether it be in the atmosphere or the oceans. However, for these flows there are also other competing effects, such as temperature and density stratification. In this review, we have highlighted some cases where rotation drastically changes or is responsible for setting up the flow field and where the resulting flow fields are preferably studied through laboratory experiments. The examples reviewed here do not have in mind any specific geophysical analogies (or for that matter technical applications) but are intended, through some general flow cases, to describe various aspects of rotation and how it affects stability, transition, and turbulence.

At this point, one may wish to consider the continuing reasons for undertaking laboratory experiments when simulations are becoming more and more powerful, especially in the case of system rotation where it is easy to add the Coriolis term to a working simulation of the corresponding nonrotating case (to obtain a solution in the rotating frame). Some simulations of RPCF and axially rotating pipe flow have been performed, but they are still limited to a few parameter values and low Reynolds numbers. In the case of RPCF, *Tsukahara et al.* [2010a] were able to explore the parameter plane consisting of more than 400 combinations of Re and Ω, a task that would be too time consuming to contemplate using numerical simulations. Also, for both RPCF and axially rotating pipe flow, simulations still cannot (and maybe never will) fully resolve small-scale turbulence at high Reynolds numbers. Such explorations would therefore require the use of experimental facilities for the foreseeable future.

For the single-rotating-disk case, full nonlinear simulations have not been reported so far. Here, one would expect the treatment of the boundary conditions to be crucial given the role upstream (i.e., radially inwardly) traveling modes have in absolute instability. An advantage that rotating disk simulations have over some other flow cases is that there is only one nondimensional parameter that defines the flow. In the Linné FLOW Centre, simulations of the rotating disk boundary layer flow are presently underway and we hope will give interesting results in the future to compare directly with our experimental results (e.g., *Imayama et al.* [2012a, 2012b]; *Imayama* [2012]). Moreover, our interest is also directed toward other types of rotating body induced flows, for instance, the boundary layer flows over a rotating sphere and rotating cones where simulations would be even harder to carry out due to geometric complications.

Acknowledgments. This review is based on both the authors' own work, and that of several former and present Ph.D. students and postdoctoral researchers. They are all acknowledged for their contributions. A special thanks to Dr. Nils Tillmark, who has been the main designer of both the Couette flow apparatus and the rotating pipe facility at the Fluid Physics Laboratory of KTH Mechanics.

REFERENCES

Alfredsson, P. H., and H. Persson (1989), Instabilities in channel flow with system rotation, *J. Fluid Mech.*, *202*, 543–557.

Alfredsson, P. H., and N. Tillmark (2005), Experimental observations of instabilities in rotating plane Couette flow, *Fluid Mech. Appl.*, *77*, 173–193.

Barkley, D., and L. S. Tuckerman (2005), Turbulent-laminar patterns in plane Couette flow, *Fluid Mech. Appl.*, *77*, pp. 107–127.

Barkley, D., and L. S. Tuckerman (2007), Mean flow of turbulent-laminar patterns in plane Couette flow, *J. Fluid Mech.*, *576*, 109–137.

Bödewadt, U. T. (1940), Die Drehströmung über festem Grund, *ZAMM*, *20*, 241–253.

Cederholm, A., and F. Lundell (1998), A study of the velocity and temperature boundary layers over a heated rotating disk, M.S. thesis, KTH Mechanics, Royal Inst. Tech., Stockholm, Sweden.

Cham, T.-S., and M. R. Head (1969), Turbulent boundary-layer flow on a rotating disk, *J. Fluid Mech.*, *37*, 129–147.

Chin, D.-T., and M. Litt (1972), An electrochemical study of flow instability on a rotating disk, *J. Fluid Mech.*, *54*, 613–625.

Clarkson, M. H., S. C. Chin, and P. Shacter (1980), Flow visualization of inflexional instabilities on a rotating disk, AIAA, Washington, D.C.

Cobb, E. C., and O. A. Saunders (1956), Heat transfer from a rotating disk, *Proc. R. Soc. Lond. A. Math. Phys. Sci.*, *236*, 343–351.

Corke, T. C., and K. F. Knasiak (1998), Stationary travelling cross-flow mode interactions on a rotating disk, *J. Fluid Mech.*, *355*, 285–315.

Corke, T. C., E. H. Matlis, and H. Othman (2007), Transition to turbulence in rotating-disk boundary layers: Convective and absolute instabilities, *J. Eng. Math.*, *57*, 253–272.

Daviaud, F., J. Hegseth, and P. Bergé (1992), Subcritical transition to turbulence in plane Couette flow, *Phys. Rev. Lett.*, *69*, 2511–2514.

Davies, C., and P. W. Carpenter (2003), Global behaviour corresponding to the absolute instability of the rotating-disc boundary layer, *J. Fluid Mech.*, *486*, 287–329.

del Pino, C., J. Ortega-Casanova, and R. Fernandez-Feria (2003), Nonparallel stability of the flow in an axially rotating pipe, *Fluid Dyn. Res.*, *69*, 261–281.

Ekman, V. W. (1905), On the influence of the Earth's rotation on ocean currents, *Arkiv för Matematik Astronomi och Fysik*, *2*, 1–52.

Erian, F. F., and Y. H. Tong (1971), Turbulent flow due to a rotating disk, *Phys. Fluid*, *14*, 2588–2591.

Facciolo, L., N. Tillmark, A. Talamelli, and P. H. Alfredsson (2007), A study of swirling turbulent pipe and jet flows, *Phys. Fluids*, *19*, 035,105.

Faller, A. J. (1991), Instability and transition of disturbed flow over a rotating disk, *J. Fluid Mech.*, *230*, 245–269.

Fedorov, B. I., G. Z. Plavnik, I. V. Prokhorov, and L. G. Zhukhovitskii (1976), Transitional flow conditions on a rotating disk, *J. Eng. Phys. Thermophys.*, *31*, 1448–1453.

Fernandez-Feria, R., and C. del Pino (2002), The onset of absolute instability of rotating Hagen-Poiseuille flow: A spatial stability analysis, *Phys. Fluids*, *14*, 3087–3097.

Ferro, M. (2012), Experimental study on turbulent pipe flow, M.S. thesis, KTH Mechanics, Royal Inst. Tech., Stockholm, Sweden.

Garret, S. J., and N. Peake (2007), The absolute instability of the boundary layer on a rotating cone, *Eur. J. Mech. B-Fluids*, *26*, 344–353.

Gregory, N., and W. S. Walker (1960), Experiments on the effect of suction on the flow due to a rotating disk, *J. Fluid Mech.*, *9*, 225–234.

Gregory, N., J. T. Stuart, and W. S. Walker (1955), On the stability of three-dimensional boundary layers with application to the flow due to a rotating disk, *Phil. Trans. R. Soc. Lond.*, *248*, 155–199.

Hagiwara, Y., S. Sakamoto, M. Tanaka, and K. Yoshimura (2002), PTV measurement on interaction between two immiscible droplets and turbulent uniform shear flow of carrier fluid, *Exp. Therm. Fluid Sci.*, *26*, 245–252.

Healey, J. J. (2007), Instabilities of flows due to rotating disks: Preface, *J. Eng. Math.*, *57*, 199–204.

Healey J. J. (2010), Model for unstable global modes in the rotating-disk boundary layer, *J. Fluid Mech.*, *663*, 148–159.

Hiwatashi, K., P. H. Alfredsson, N. Tillmark, and M. Nagata (2007), Experimental observations of instabilities in rotating plane Couette flow, *Phys. Fluids*, *19*, 048,103.

Huerre, P., and P. A. Monkewitz (1990), Local and global instabilities in spatially developing flows, *Annu. Rev. Fluid Mech.*, *22*, 473–537.

Hussain, Z., S. J. Garrett, and S. O. Stephen (2011), The instability of the boundary layer over a disk rotating in an enforced axial flow, *Phys. Fluids*, *23*, 114,108.

Imao, S., M. Itoh,, Y. Yamada, and Q. Zhang (1992), The characteristics of spiral waves in an axially rotating pipe, *Exp. Fluids*, *12*, 277–285.

Imao, S., M. Itoh, and H. Takeyoshi (1996), Turbulent characteristics of the flow in an axially rotating pipe, *Int. J. Heat Fluid Flow*, *17*, 444–451.

Imayama, S. (2012), Experimental study of the rotating-disk boundary-layer flow, Licentiate thesis, KTH Mechanics, Royal Inst. Tech., Stockholm, Sweden.

Imayama, S., P. H. Alfredsson, and R. J. Lingwood (2012a), A new way to describe the transition characteristics of a rotating-disk boundary-layer flow, *Phys. Fluids*, *24*, 031,701.

Imayama, S., P. H. Alfredsson, and R. J. Lingwood (2012b), An experimental study of edge effects on rotating-disk transition, *J. Fluid Mech.*, *716*, 638–657.

Itoh, M., and I. Hasegawa (1994), Turbulent boundary layer on a rotating disk in infinite quiescent fluid, *JSME Int. J.*, B *37*, 449–456.

Jarre, S., P. Le Gal, and M. P. Chauve (1996a), Experimental study of rotating disk instability. I. Natural flow, *Phys. Fluids*, 8, 496–508.

Jarre, S., P. Le Gal, and M. P. Chauve (1996b), Experimental study of rotating disk instability. II. Forced flow, *Phys. Fluids*, 8, 2985–2994.

Johnston, J. P., R. M. Halleen, and D. K. Lezius (1972), Effects of spanwise rotation on the structure of two-dimensional fully developed turbulent channel flow, *J. Fluid Mech.*, 56, 533–557.

Kikuyama, K., M. Murakami, K. Nishibori, and K. Maeda (1983), Flow in an axially rotating pipe, *Bull. JSME*, 26, 506–513.

Kobayashi, R., Y. Kohama, and C. Takamadate (1980), Spiral vortices in boundary layer transition regime on a rotating disk, *Acta Mech.*, 35, 71–82.

Kohama, Y. (1984), Study on boundary layer transition of a rotating disk, *Acta Mech.*, 50, 193–199.

Kohama, Y. P. (2000), Three-dimensional boundary layer transition study, *Curr. Sci.*, 79, 800–807.

Komminaho J., A. Lundbladh, and A. V. Johansson (1996), Very large structures in plane Couette flow, *J. Fluid Mech.*, 320, 259–285.

Lezius, D. K., and J. P. Johnston (1976), The structure and stability of turbulent boundary layers in rotating channel flow, *J. Fluid Mech.*, 77, 153–175.

Lingwood, R. J. (1995), Absolute instability of the boundary layer on a rotating disk, *J. Fluid Mech.*, 199, 17–33.

Lingwood, R. J. (1996), An experimental study of absolute instability of the rotating-disk boundary-layer flow, *J. Fluid Mech.*, 314, 373–405.

Lingwood, R. J. (1997), Absolute instability of the Ekman layer and related rotating flows, *J. Fluid Mech.*, 331, 405–428.

Littell, H. S., and J. K. Eaton (1994), Turbulence characteristics of the boundary layer on a rotating disk, *J. Fluid Mech.*, 266, 175–207.

Malerud, S., K. J. Måløy, and W. I. Goldburg (1995), Measurements of turbulent velocity fluctuations in a planar Couette cell, *Phys. Fluids, 7*, 1949–1955.

Malik, M. R., S. P. Wilkinson, and S. A. Orszag (1981), Instability and transition in rotating disk flow, *AIAA J.*, 19, 1131–1138.

Masuda, S., and M. Matsubara (1989), Visual study of boundary layer transition on rotating flat plate, in *Laminar-Turbulent Transition*, IUTAM Symposium Toulouse/France September 1989, edited by D. Arnal and R. Michel, pp. 465–474, Springer Berlin Heidelberg.

Matsson, O. J. E., and P. H. Alfredsson (1990), Curvature and rotation induced instabilities in channel flow, *J. Fluid Mech.*, 210, 537–563.

Matsson, O. J. E., and P. H. Alfredsson (1994), The effect of spanwise system rotation on Dean vortices, *J. Fluid Mech.*, 274, 243–265.

Matsubara, M, and S. Masuda (1991), Turbulent spots in rotating Blasius boundary layer, in *Advances in Turbulence 3, Proc. Third Eur. Turbulence Conf., Stockholm July 1990*, edited by A. V. Johansson and P. H. Alfredsson, pp. 204–210, Springer Berlin Heidelberg.

Nagata, M. (1998), Tertiary solutions and their stability in rotating plane Couette flow, *J. Fluid Mech.*, 358, 357–378.

Nagib, H. M., Z. Lavan, A. A. Fejer, and L. Wolf (1971), Stability of pipe flow with superposed solid body rotation, *Phys. Fluids*, 14, 766–768.

Nickels, T. B., and P. N. Joubert (2000), The mean velocity profile of turbulent boundary layers with system rotation, *J. Fluid Mech.*, 408, 323–345.

Oberlack, M. (1999), Similarity in non-rotating and rotating turbulent pipe flows, *J. Fluid Mech.*, 379, 1–22.

Orlandi, P., and M. Fatica (1997), Direct simulations of turbulent flow in a pipe rotating about its axis, *J. Fluid Mech.*, 343, 43–72.

Örlü, R., and P. Schlatter (2013), Comparison of experiments and simulations for zero pressure gradient turbulent boundary layers at moderate Reynolds numbers. *Exp. Fluids*, 54, 1547.

Othman, H., and T. C. Corke (2006), Experimental investigation of absolute instability of a rotating-disk boundary layer, *J. Fluid Mech.*, 565, 63–94.

Pedley, T. J. (1969), On the instability of viscous flow in a rapidly rotating pipe, *J. Fluid Mech.*, 35, 97–115.

Pier, B. (2003), Finite-amplitude crossflow vortices, secondary instability and transition in the rotating-disk boundary layer, *J. Fluid Mech.*, 487, 315–343.

Prigent, A., G. Grégoire, H. Chaté, O. Dauchot, and W. van Saarloos (2002), Large-scale finite-wavelength modulation within turbulent shear flows, *Phys. Rev. Lett.*, 89, 014,501.

Recktenwald, I., T. Weller, W. Schroeder, and M. Oberlack (2007), Comparison of direct numerical simulations and particle-image velocimetry data of turbulent channel flow rotating about the streamwise axis, *Phys. Fluids*, 19, 085,114.

Sanmiguel-Rojas, E., and R. Fernandez-Feria (2005), Nonlinear waves in the pressure driven flow in a finite rotating pipe, *Phys. Fluids*, 17, 014,104.

Saric, W. S., H. L. Reed, and E. B. White (2003), Stability and transition of three-dimensional boundary layers, *Annu. Rev. Fluid Mech.*, 35, 413–440.

Segalini, A., R. Örlü, P. Schlatter, P. H. Alfredsson, J.-D. Rüedi, and A. Talamelli (2011), A method to estimate turbulence intensity and transverse Taylor microscale in turbulent flows from spatially averaged hot-wire data, *Exp. Fluids*, 51, 693–700.

Shrestha, K., L. Parras, C. Del Pino, E. Sanmiguel-Rojas, and R. Fernandez-Feria (2013), Experimental evidence of convective and absolute instabilities in rotating Hagen-Poiseuille flow, *J. Fluid Mech.*, 716, R12.

Smith, N. H. (1947), Exploratory investigation of laminar-boundary-layer oscillations on a rotating disk, NACA TN-1227 (can be downloaded from http://www.ntrs.nasa.gov/).

Suryadi, A., N. Tillmark, and P. H. Alfredsson (2012), Rotating plane Couette flow at high rotation number, paper presented at 65th Annual Meeting of the APS Division of Fluid Dynamics, http://meetings.aps.org/link/BAPS.2012.DFD.R14.6.

Suryadi, A., N. Tillmark, and P. H. Alfredsson (2013), Velocity measurements of streamwise roll cells in rotating plane Couette flow, *Exp. Fluids*, 54, 1617.

Tillmark, N., and P. H. Alfredsson (1991), An experimental study of transition in plane Couette flow, in *Advances in Turbulence 3, Proc. Third Eur. Turbulence Conf., Stockholm July 1990*, edited by A. V. Johansson and P. H. Alfredsson, pp. 235–242, Springer Berlin Heidelberg.

Tillmark, N., and P. H. Alfredsson (1992), Experiments on transition in plane Couette flow, *J. Fluid Mech.*, *235*, 89–102.

Tillmark, N., and P. H. Alfredsson (1996), Experiments on rotating plane Couette flow, in *Advances in Turbulence*, vol.4, edited by S. Gavrilakis, L. Machiels, and P. A. Monkewitz, pp. 391–394, Kluwer.

Tsukahara, T., N. Tillmark, and P. H. Alfredsson (2010a), Flow regimes in a plane Couette flow with system rotation, *J. Fluid Mech.*, *648*, 5–33.

Tsukahara, T., Y. Kawaguchi, H. Kawamura, N. Tillmark, and P. H. Alfredsson (2010b), Turbulence stripe in transitional channel flow with/without system rotation, in *IUTAM Bookseries vol. 18, 7th IUTAM Symposium on Laminar-Turbulent Transition*, Stockholm June 2009, edited by P. Schlatter and D. S. Henningson, pp. 421–426, Springer Netherlands.

von Kármán, T. (1921), Über laminare und turbulent Reibung, *Z. Angew. Math. Mech.*, *1*, 233–252.

Wallin, S., and A. V. Johansson (2000), An explicit algebraic Reynolds stress model for incompressible and compressible turbulent flow, *J. Fluid Mech.*, *403*, 89–132.

Watmuff, J. H., H. T. Witt, and P. N. Joubert (1985), Developing turbulent boundary layers with system rotation, *J. Fluid Mech.*, *157*, 405–448.

White, A. (1964), Flow of a fluid in an axially rotating pipe, *J. Mech. Eng. Sci.*, *6*, 47–52.

Wilkinson, S. P., and M. R. Malik (1985), Stability experiments in the flow over a rotating disk, *AIAA J.*, *23*, 588–595.

Zettner, C. M., and M. Yoda (2001), The circular cylinder in simple shear at moderate Reynolds numbers: An experimental study, *Exp. Fluids*, *30*, 246–353.

5

Altimetry in a GFD Laboratory and Flows on the Polar β-Plane

Yakov D. Afanasyev

5.1. INTRODUCTION

Rotating table experiments are essential in the study of geophysical fluid dynamics (GFD) where planetary rotation is important. They are a valuable research tool that allows one to gain insight into the dynamics of oceanic and atmospheric flows. They can be especially useful for studying three-dimensional (3D) flows with relatively high Reynolds numbers. Indeed, at Reynolds numbers of the order of 10^3–10^5, laboratory experiments can be even more time effective than current 3D numerical models. Although numerical simulations provide more flexibility in the choice of the physical setup, including initial and boundary conditions, the range of control parameters, and type of forcing, experiments are indispensable for the purpose of testing and validating numerical models. A numerical simulation run in parallel with the experiment, especially with some form of data assimilation from the experiment [e.g., *Ravela et al.*, 2010], can be very beneficial for supplementing the experimental data and obtaining complete information about the flow.

One limiting factor of an experiment is a measurement problem. It is often desirable to resolve all significant scales from the viscous scale of an order of 1 mm up to the scale of the experimental tank (typically 1–2 m). Particle image velocimetry (PIV) is widely used by researchers for measurements of the velocity field in the entire flow domain. PIV is a well-developed method applicable for virtually all possible flows. However, there is a resolution problem with the PIV technique. By design, PIV uses interrogation windows of size from 12^2 to 24^2 or more pixels. Given the total size of the imaging array used in a video camera, the window size ultimately determines the spatial resolution, or rather the range of scales that PIV can resolve at the same time.

In this chapter we describe a different technique for global measurements of dynamic fields. This technique is based on optical altimetry and allows one to measure the slope of the surface elevation in every pixel of the image of the flow. With modern megapixel cameras this makes the problem of spatial (and temporal) resolution a thing of the past. The slope of the surface elevation can be converted to barotropic velocity using relations suitable for a rotating fluid. Laboratory altimetry is not unlike satellite radar altimetry, which has provided global observations of the sea surface elevation since the TOPEX/Poseidon satellite was launched in 1992. The altimetry technique is somewhat less versatile than PIV, but it is well suited for rotating GFD experiments. Because of its high spatial and temporal resolution, laboratory altimetry can be used effectively for detailed comparison and testing of numerical models [*Slavin and Afanasyev*, 2012].

In a two-layer fluid this technique can be used in combination with a traditional optical density method that allows one to measure the thickness of one of the layers almost simultaneously with the surface elevation measurements [*Afanasyev et al.*, 2009]. Thus the baroclinic component of velocity can also be resolved. The combination of barotropic and baroclinic components provides complete information about the system. The amount of data obtained in such an experiment is comparable with that in a numerical simulation such that one can think of a laboratory tank as of an "analogue computer" for GFD modeling.

This chapter is organized as follows. Section 5.2 gives a brief review of the general features of a rotating layer with a free surface, including the so-called polar β-plane approximation. We then proceed in Section 5.3 to the description of the experimental technique, including the methods of conversion of the slope of the surface

---

Department of Physics and Physical Oceanography, Memorial University of Newfoundland, St. John's, Newfoundland, Canada.

elevation into velocity. A brief theoretical description of inertial and Rossby waves is given in Section 5.4. Examples of different flows studied with the altimetric imaging velocimetry (AIV) technique are described in Section 5.5.

5.2. POLAR β-PLANE

In GFD, an ocean or the atmosphere is often considered as thin layers of fluid if one is concerned about large-scale motions. In order to avoid using a spherical coordinate system, a local Cartesian system (x, y) tangent to the surface of the planet at some latitude φ_0 is often employed together with the β-plane approximation to describe the variation of the Coriolis parameter in the meridional direction. The Coriolis parameter is defined as $f = 2\Omega \sin \varphi$, where Ω is the rotation rate of the planet and φ the latitude. The β-plane approximation gives a linear dependence on y of the form $f = f_0 + \beta y$, where the distance $y = a(\varphi - \varphi_0)$ increases toward the north. Here a is the radius of the planet. This linear dependence is the result of the expansion of the Coriolis parameter about a reference latitude φ_0,

$$f \approx 2\Omega \sin \varphi_0 + \frac{2\Omega \cos \varphi_0}{a} y, \quad (5.1)$$

such that $f_0 = 2\Omega \sin \varphi_0$ and $\beta = 2\Omega \cos \varphi_0/a$. The validity of this approximation has been extensively discussed in the literature [*Phillips* 1963, 1966; *Veronis*, 1963]. However, the β-plane cannot be centered at the pole of the planet. A different approximation can be used instead for the domains which include the pole. The Coriolis parameter can be written as $f = 2\Omega \cos \phi$, where ϕ is the colatitude. Expanding near the pole, $\phi = 0$, we obtain

$$f \approx 2\Omega \left(1 - \frac{\phi^2}{2}\right) = 2\Omega - \frac{\Omega}{a^2} r^2 = f_0 - \gamma r^2, \quad (5.2)$$

where r is the radial distance from the pole and $f_0 = 2\Omega$. We have now adopted the polar coordinate system (r, θ) and introduced the parameter $\gamma = \Omega/a^2$. This approximation is called the polar β-plane or γ-plane approximation. Note that the Coriolis parameter f varies quadratically with distance.

In the laboratory, we do not reproduce the variation of the Coriolis parameter with distance. Instead, the depth of the fluid is varied. The depth of a rotating fluid with a free surface varies quadratically with the distance r from the axis of rotation. In a cylindrical tank of radius R the depth h is given by

$$h(r) = H_0 + \frac{\Omega^2}{2g} \left(r^2 - \frac{R^2}{2}\right), \quad (5.3)$$

where H_0 is the depth of the layer in the absence of rotation, Ω is the rotation rate of the tank, and g is the gravitational acceleration. The dynamical equivalence of the varying depth to the quadratically varying Coriolis parameter in the polar β-plane approximation results from the conservation of the potential vorticity (PV), defined as $q = (\zeta + 2\Omega)/h$. Here ζ is the vertical component of the relative vorticity and h is given by (5.3). Assuming that the percentage change in h is small and that ζ is much smaller than the background vorticity 2Ω (i.e., the Rossby number Ro $= |\zeta|/2\Omega \ll 1$), we obtain the following approximate expression for PV:

$$q \approx \frac{1}{H_0} \left\{ \zeta + 2\Omega \left[1 - \frac{\Omega^2}{2gH_0} \left(r^2 - \frac{R^2}{2}\right)\right] \right\}. \quad (5.4)$$

Comparing the term in curly brackets in the above formula with (5.2), we define the laboratory polar β-plane with parameter γ given by

$$\gamma = \frac{\Omega^3}{gH_0}. \quad (5.5)$$

Within this laboratory framework we can also introduce a local β-plane if desired. Choosing r_0 to be the reference distance from the pole ("midlatitudes" of the tank), we obtain an approximate expression for the depth as follows:

$$h(r) = H_0 + \frac{\Omega^2}{2g} \left((r_0 - y)^2 - \frac{R^2}{2}\right)$$

$$\approx H_0 + \frac{\Omega^2}{2g} \left(r_0^2 - 2r_0 y - \frac{R^2}{2}\right) = h(r_0) - \frac{\Omega^2 r_0}{g} y.$$

Here the y axis of the local Cartesian system with the origin at r_0 is directed toward the pole. The PV can then be written in the form

$$q \approx \frac{1}{h(r_0)} \left[\zeta + 2\Omega \left(1 + \frac{\Omega^2 r_0}{gh(r_0)} y\right)\right] \quad (5.6)$$

such that the β-parameter is defined as

$$\beta = \frac{2\Omega^3 r_0}{gh(r_0)}. \quad (5.7)$$

5.3. ALTIMETRIC IMAGING VELOCIMETRY

One can think of a paraboloidal surface of a liquid when in a solid-body rotation as an undisturbed reference surface similar to a geoid in satellite altimetry. A flow generated in the rotating layer creates perturbations of the reference surface. Let η be the elevation of the surface measured with respect to the reference paraboloidal surface. Unlike satellite altimetry where the distance between the satellite and the sea surface is measured, laboratory altimetry measures the angles, namely the gradient of the surface elevation, $\nabla \eta = (\partial \eta/\partial x, \partial \eta/\partial y)$, in the horizontal

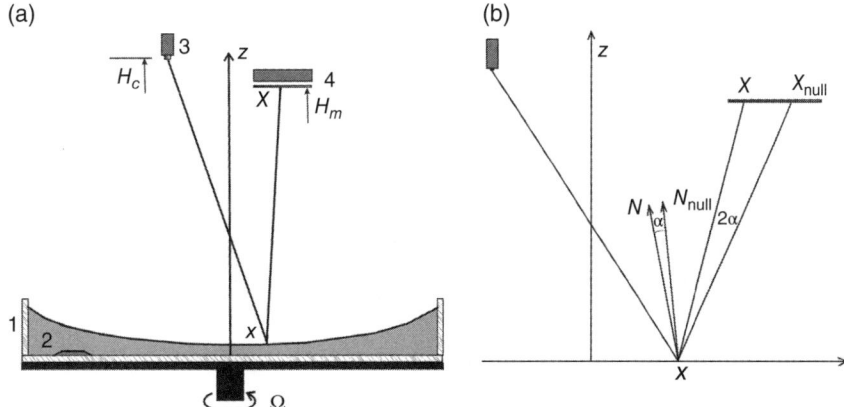

Figure 5.1. (a) Experimental setup and (b) geometry of the reflection from the perturbed surface. In (a) the numbers refer to (1), rotating tank (2) mountain for experiments with bottom topography, (3) photo/video camera, and (4) light source with color mask. In (b) the angle α is the angle between the normal to the surface at null point N_{null} and the normal to the perturbed surface N.

plane (x, y). The two components of the slope of η are color coded and the images of the surface obtained by this method are color maps of the surface. They show the flows of different scales, including small/mesoscale eddies, gravity and inertial waves, and global scale currents and Rossby waves. We call this laboratory technique altimetric imaging velocimetry because it allows us not only to visualize the flow field but also to calculate the velocity field from measured slopes. Before we describe the conversion of the slopes into velocities, we give in what follows some more technical details of AIV.

Figure 5.1a shows a sketch of a typical experimental setup. A video or photo camera observes from above the reflection of light from the free surface of the fluid. The light source is also located above the water and contains a color mask, where the color varies in two directions across the mask. The mask resembles a color wheel used by painters. A rainbow of bright saturated colors converges toward the center where the color mix results in a grayish tone. Thus each point of the mask is characterized by a unique color that can be measured using three indices. The images are usually coded in values of red, green, and blue (RGB) color. It is convenient to convert the RGB color into a so-called *Luv* color space where the first index L stands for lightness and u and v denote chromaticity values. Ultimately, only two indices are required to code the color in the x and y directions across the mask such that we can ignore the lightness. This also makes this method less vulnerable to errors due to the variation of lightness/brightness across the experimental images.

The AIV has evolved from the (gray-scale) optical altimetry introduced by *Rhines et al.* [2007] by incorporating color coding. The principle of optical altimetry can be briefly described as follows. If a paraboloidal surface is illuminated by a light source of small but finite size located at a height of about twice the focal length of the paraboloid and slightly off the axis of rotation, the surface appears to be lit almost uniformly when observed from a certain distance and angle. Under these conditions (see below) any point of the surface observed by the camera adjacent to the light source corresponds to a small area (point) of the light source. When the surface is perturbed, the slopes of the surface elevation field change the angles of reflection. As a result an observer sees different areas of the light source. If there is a gradient of brightness across the surface of the light source such as that which occurs naturally in the light bulb, which is always brighter in the center where the filament is, the entire (perturbed) surface of water will be visualized by variations of intensity. When, instead of a light bulb a color mask is used, the observer will see the reflections of the different areas of the mask.

The optical setup is simple and in fact does not require any optical elements since the parabolic surface of the rotating fluid is used as if it were a mirror in a Newtonian telescope. The setup can include an inexpensive video or photo camera and can be used in a teaching laboratory. A more specialized camera that does not use color compression is necessary when more accurate results are required for research purposes.

Using the reflection law of optics, we can describe the path of the ray originating from a certain point at the mask (X, Y) reflecting from the point (x, y) at the surface of water and coming to the camera. The ray paths can be calculated using the vector reflection law:

$$\mathbf{R} = \mathbf{I} - 2(\mathbf{I} \cdot \mathbf{N})\mathbf{N} \qquad (5.8)$$

where $\mathbf{R}$ and $\mathbf{I}$ are reflected and incident unit vectors and $\mathbf{N}$ is the normal to the surface. The vectors can be written as follows:

$$\mathbf{I} = \frac{(x - X_c, -H_c)}{\sqrt{(x - X_c)^2 + H_c^2}}, \quad \mathbf{R} = \frac{(-x + X, H_m)}{\sqrt{(-x + X)^2 + H_m^2}},$$

$$\mathbf{N} = \frac{(-\Omega_0^2 x/g, 1)}{\sqrt{\Omega_0^2 x^2/g^2 + 1}}, \quad (5.9)$$

where x, X_c, X are the x coordinates of the point on the surface, the camera lens, and the point on the slide and H_m and H_c are the heights of the color mask and the camera above the surface of water (see Figure 5.1). The equations resulting from the reflection law give in the first approximation that at certain rotation rate of the tank, Ω_0, the rays coming to the camera from all points of the surface originate from one point of the color mask, namely its center (for details see *Afanasyev et al.*, [2009]). The rotation rate determines the focal distance of the paraboloid such that at this particular value of the rotation rate the camera sees the surface uniformly illuminated by nearly one color (that of the center of the mask). We call this rotation rate a null point and all altimetric measurements are performed at this rate. The value of Ω_0 is determined by the geometry of the setup:

$$\Omega_0^2 = \frac{g(H_c + H_m)}{2 H_c H_m}. \quad (5.10)$$

Usually the camera and the mask are at approximately the same height such that $\Omega_0^2 = g/H_m$. Thus the lower the mask/camera, the faster the rotation rate must be.

A more accurate solution of the reflection law equations shows that a color does vary across the surface at the null point (in the absence of perturbations). Although an analytical expression of the solution exists, it is more convenient to use measured values. In the beginning of each experiment before the flow is started, one image showing this color distribution is recorded. A color-matching procedure implemented in a numerical code allows us to relate the color observed in each point of the surface to its position on the mask (X, Y) where the particular color originates. Thus the null-point image gives the values of $(X_{\text{null}}, Y_{\text{null}})$ across the surface of water. When the flow is started, a pressure field created by the flow perturbs the surface of water. The surface slope changes with respect to that of the null field. As a result a ray coming to the camera originates from a point on the mask (X, Y) that is different from that of the null point $(X_{\text{null}}, Y_{\text{null}})$. Figure 5.1b shows the geometry of the reflection from the perturbed surface. The slope $\partial \eta / \partial x = \alpha$ due to the perturbation can be found from the geometry in the form

$$\frac{\partial \eta}{\partial x} = \frac{1}{2} \frac{(X - X_{\text{null}}) H_m}{(X_{\text{null}} - x)^2 + H_m^2}. \quad (5.11)$$

The y component of the slope is determined in a similar manner. The mask coordinates (X, Y) are determined by color matching similar to the color matching for the null field. This procedure gives the gradient of the surface elevation, $\nabla \eta = (\partial \eta / \partial x, \partial \eta / \partial y)$.

Integrating $\nabla \eta$ over x and y, we can obtain the field of surface elevation η and hence the barotropic pressure $p = \rho g \eta$. It is also important to obtain the velocity field of the flow. Fortunately, the gradient of the surface elevation can be easily converted into a velocity field in a rapidly rotating fluid where geostrophic and quasi-geostrophic approximations can be employed. Consider the shallow-water equation

$$\frac{\partial \mathbf{V}}{\partial t} + (\mathbf{V} \cdot \nabla) \mathbf{V} + \left(f_0 - \gamma r^2 \right) (\mathbf{k} \times \mathbf{V}) = -g \nabla \eta, \quad (5.12)$$

where $\mathbf{V}$ is the horizontal velocity vector and $\mathbf{k}$ is the vertical unit vector. The relative importance of the unsteady term (the first term) and the nonlinear term (second term) is determined by the temporal Rossby number $\text{Ro}_T = 1/(\Omega T)$ and the Rossby number $\text{Ro} = U/(\Omega L)$. Here T is the time scale of the flow evolution, while U and L are velocity and length scales of the flow. A third dimensionless number, γ^*, is given by the ratio of the γ-term to the Coriolis parameter, $\gamma^* = \gamma L^2 / f_0$. If all three dimensionless numbers are small, we may neglect the corresponding terms in (5.12) and find that the main balance is between the Coriolis term and the pressure gradient term, which gives the geostrophic equation

$$f_0 (\mathbf{k} \times \mathbf{V}_g) = -g \nabla \eta. \quad (5.13)$$

The geostrophic velocity $\mathbf{V}_g$ is then obtained by taking the cross product with vector $\mathbf{k}$ of both sides of the above equation,

$$\mathbf{V}_g = \frac{g}{f_0} (\mathbf{k} \times \nabla \eta). \quad (5.14)$$

Substituting (5.14) into the small terms of equation (5.11) such that the total velocity $\mathbf{V}$ remains only in the Coriolis term $f_0 (\mathbf{k} \times \mathbf{V})$, we then solve for $\mathbf{V}$ to obtain

$$\mathbf{V} = \mathbf{V}_g - \frac{g}{f_0^2} \frac{\partial}{\partial t} \nabla \eta - \frac{g}{f_0^3} J(\eta, \nabla \eta) + \frac{\gamma r^2}{f_0} \mathbf{V}_g, \quad (5.15)$$

where

$$J(A, B) = \frac{\partial A}{\partial x} \frac{\partial B}{\partial y} - \frac{\partial B}{\partial x} \frac{\partial A}{\partial y}$$

is the Jacobian operator. Note that we assumed in fact that $\mathbf{V} = \mathbf{V}_g + \mathbf{V}_a$, where the ageostrophic part of the total velocity is relatively small, $\mathbf{V}_a < \mathbf{V}_g$. The right-hand side (RHS) of equation (5.15) contains only the components of $\nabla \eta$ and their time derivatives. These quantities are measured by AIV. Note that in order to measure $\nabla \eta$, we need only one image of the flow, while to measure the time derivative, two successive images are required. Thus, (5.15) allows one to calculate the (barotropic) velocity of the flow in the quasi-geostrophic approximation. Note that the fact that we use quasi-geostrophic velocity conversion does not mean of course that the flows in the tank are necessarily quasi-geostrophic. In the experiments we measure a "true" field of the gradient of the surface elevation (in fact the pressure gradient), but the calculated velocity is more accurate when the dimensionless numbers Ro_T, Ro, and γ^* are small.

Yet another important application of AIV is to study baroclinic flows. In a baroclinic flow AIV gives only the surface (barotropic) velocity. In some cases this may be enough for an adequate description of the flow, especially if the baroclinic component of velocity is relatively weak. However, both barotropic and baroclinic components can be measured in the special case of a two-layer fluid. In this case AIV is used in combination with the optical density method. Here we describe this method, referring to *Afanasyev et al.* [2009] for more details. In order to measure the thickness of one of the layers, we dye this layer with a red food dye. The layer does not have to cover the entire area of the domain. In some experiments the dyed fluid is supplied from a source to form boundary currents, lenses, or plumes that have a distinct boundary separating the dyed fluid from the clear ambient water. The thickness h_1 of the dyed layer can be measured by relating the value of the chromaticity u or v to the thickness of the dyed fluid. This approach differs from that used by others [e.g., *Linden and Simpson*, 1994; *Cenedese and Dalziel*, 1998], who used the brightness L for this purpose. Chromaticity is chosen because it is less affected by background variations of brightness. For calibration purposes a wedge-shaped cuvette filled with the fluid with the same concentration of dye as that used in the experiment is placed above the tank. The profile of the chromaticity a across the cuvette gives the relation between the depth of the fluid in the cuvette and its color intensity. This relation is used to calculate h_1 in every pixel of the image of the flow. The baroclinic component of the flow can then be calculated using (5.15), where the layer thickness h_1 is used instead of the surface elevation η and the reduced gravity $g' = g\,\Delta\rho/\rho$, where $\Delta\rho$ is the density difference between the layers, is used instead of g. The optical density method requires a uniform background light. The method can be used either in a separate experiment or in combination with AIV. Almost simultaneous measurements can be performed by switching back and forth from the background light to the color mask light. The time difference between the velocity fields measured by the optical density method and AIV will be determined by the frame rate of the camera.

5.4. INERTIAL AND PLANETARY/TOPOGRAPHIC WAVES

A rotating fluid system sustains inertial waves with frequencies below the value of the Coriolis parameter $\omega \leq f_0$ [e.g., *Batchelor*, 1967]. Planetary or topographic waves in a rotating layer occur due to variation of the Coriolis parameter or due to variation of the depth of the layer $h(r)$, respectively. Here we call both planetary and topographic waves "Rossby waves," assuming their dynamic equivalence via the conservation of PV. These waves are of relatively low frequency, with the frequency being proportional to the parameter β or γ, depending on which approximation is used. In fact, they constitute a subset of inertial waves. *Phillips* [1965] showed that Rossby waves in a rotating annulus with quadratically varying depth can be obtained from a set of inertial wave modes when appropriate boundary conditions are applied. The theory of both inertial waves and Rossby waves is well known, and the details can be found in textbooks. Here we describe only briefly the surface elevation and velocity fields of both types of waves for the purpose of showing their surface signature. AIV allows one to observe and reconstruct the structure of the waves by their surface signature.

Consider first inertial waves in a rotating layer of water. For simplicity, we assume that the Coriolis parameter and the depth of the layer are constant. Let us start with linearized equations of motion for a layer of depth H_0

$$\frac{\partial u}{\partial t} - f_0 v = -\frac{1}{\rho}\frac{\partial p}{\partial x}, \qquad \frac{\partial v}{\partial t} + f_0 u = -\frac{1}{\rho}\frac{\partial p}{\partial y},$$
$$\frac{\partial w}{\partial t} = -\frac{1}{\rho}\frac{\partial p}{\partial z}, \qquad \frac{\partial u}{\partial x} + \frac{\partial v}{\partial y} + \frac{\partial w}{\partial z} = 0 \quad (5.16)$$

with boundary conditions

$$w(0) = \frac{\partial \eta}{\partial t}, \qquad p(0) = \rho g \eta, \qquad w(-H_0) = 0. \quad (5.17)$$

We look for a solution in the form of horizontally propagating modes with vertical structure determined by their z-dependent amplitudes,

$$(u, v, w, p, \eta) = (u_0(z), v_0(z), w_0(z), p_0(z), \eta_0)$$
$$\times \exp[i(\omega t - kx - ly)]. \quad (5.18)$$

The system of equations (5.16) can then be reduced to a single equation for pressure, which yields a solution of the form

$$p = \rho g \eta \frac{\cos \gamma_n (z + H_0)}{\cos \gamma_n H_0} \quad (5.19)$$

with dispersion relation

$$-\gamma_n \tan \gamma_n H_0 = \frac{\omega^2}{g}, \quad (5.20)$$

where

$$\gamma_n = \sqrt{\frac{\omega^2 (k^2 + l^2)}{(f_0^2 - \omega^2)}}. \quad (5.21)$$

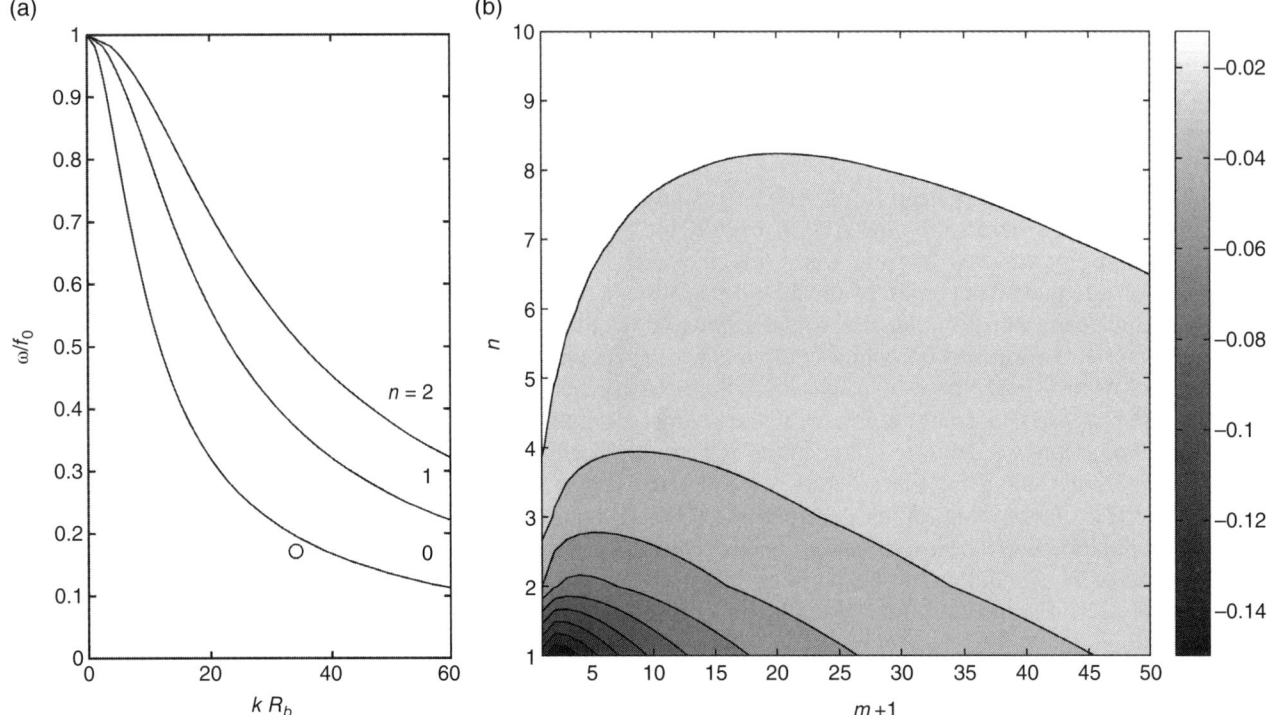

Figure 5.2. Dispersion relations for (a) inertial waves and (b) Rossby waves. The frequencies of inertial wave modes $n = 0, 1, 2$ are calculated by numerically solving (5.20). The wave vector k is nondimensionalized by the barotropic radius of deformation R_b. The frequencies of the Rossby waves are calculated from (5.26) for different azimuthal, m, and radial, n, wave numbers. The dimensional parameters are $\gamma = 1.2 \times 10^{-3} \text{s}^{-1} \text{cm}^{-2}$, $R_b = 21.6 \text{ cm}$, $R = 55 \text{ cm}$, and $f_0 = 4.58 \text{ s}^{-1}$. The gray scale shows the dimensionless frequency ω/f_0.

Note that the value of the product $\gamma_n H_0$ should lie between $(1 + n)\pi/2$ and $(1 + n)\pi$. The dispersion relation (5.20) is shown in Figure 5.2 a for modes $n = 0, 1, 2$.

Once the vertical structure is resolved, we can easily write the velocity components in terms of surface elevation:

$$u = -g \left[i\omega \frac{\partial \eta}{\partial x} + f_0 \frac{\partial \eta}{\partial y} \right] \frac{\cos \gamma_n (z + H_0)}{(f_0^2 - \omega^2) \cos \gamma_n H_0},$$

$$v = g \left[f_0 \frac{\partial \eta}{\partial x} - i\omega \frac{\partial \eta}{\partial y} \right] \frac{\cos \gamma_n (z + H_0)}{(f_0^2 - \omega^2) \cos \gamma_n H_0}, \quad (5.22)$$

$$w = -g i \gamma_n \eta \frac{\sin \gamma_n (z + H_0)}{\omega \cos \gamma_n H_0}.$$

The relations (5.19) and (5.22) allow us to obtain all of the characteristics of the inertial wave from its signature on the surface determined by the surface elevation η. Note that in order to obtain the velocity field the frequency of the wave has to be measured. This requires a relatively long set of observations (for the duration of one or more inertial periods). A Fourier transform can then yield the frequency.

Inertial waves are nongeostrophic and nonhydrostatic motions. Within a context of modes propagating horizontally in the fluid of constant depth, these waves are described by equations 5.20–5.22. However, if the depth of the layer is not constant, the modal description can be used only approximately. Alternatively, without the assumption of constant depth, inertial waves can be considered as transverse oscillations with ray paths such that the angle θ between the direction of the wave and the rotation axes is determined by the frequency of the wave, $\theta = \cos^{-1}(\omega/f_0)$ (e.g., *Greenspan* [1968]).

Consider next Rossby waves in a rotating system where either the Coriolis parameter or the depth of the layer varies with distance from the pole. While the theory of Rossby waves on the β-plane is very familiar, it is instructive to consider an alternate version of this theory on the polar β-plane [e.g., *Rhines*, 2007; *Afanasyev et al.*, 2012] as follows. The shallow water equation (5.12) can be completed with a continuity equation [*Gill*, 1982] of the form

$$\frac{\partial}{\partial t}\eta + (\mathbf{V}_g \cdot \nabla)\eta + H_0 (\nabla \cdot \mathbf{V}_a) = 0. \quad (5.23)$$

Substituting (5.13) and (5.14) into (5.23) and then linearizing and changing to polar coordinates (r, θ), we obtain a single equation for surface elevation,

$$\frac{\partial}{\partial t}\left(\nabla^2 \eta - \frac{1}{R_b^2}\eta\right) + 2\gamma \frac{\partial \eta}{\partial \theta} = 0, \quad (5.24)$$

where $R_b = \sqrt{gH_0}/f_0$ is the barotropic radius of deformation. The solution of (5.24) can be found in the form of wave modes,

$$\eta = \eta_0 \exp(im\theta - i\omega t) J_m\left(\frac{\alpha_{mn}r}{R}\right) \quad (5.25)$$

with dispersion relation

$$\omega_{mn} = \frac{-2m\gamma R_b^2}{\alpha_{mn}^2 R_b^2/R^2 + 1}. \quad (5.26)$$

Here α_{mn} is the nth root of the Bessel function of the mth order and R is the radius of the domain (the radius of the tank). Figure 5.2b shows frequency given by (5.26) for different values of the wave numbers n and m.

5.5. OBSERVATIONS

Here we give several examples of experiments where flows were observed using AIV. Consider first flows over bottom topography. These flows are rich in phenomena, including inertial and Rossby waves, vortices, and jets. In our experiment a mountain with a flat top (plateau) was fixed to the bottom of the tank rotating initially with "null" rate $\Omega_0 = 2.29\,\text{s}^{-1}$ (Figure 5.1). The mountain was located at "midlatitudes" of the tank. The height of the mountain was $h_m = 0.1H_0 = 1\,\text{cm}$ and its diameter was $d = 0.11R = 6\,\text{cm}$. Here $R = 55\,\text{cm}$ is the radius of the tank. To generate a flow above the mountain, the rotation rate of the table was changed abruptly to a slightly lower value, $\Omega = 0.98\Omega_0$, which imposed a solid-body cyclonic (eastward) flow. The tank rotates counterclockwise, modeling the Northern Hemisphere with the North Pole in the center.

This mean zonal flow, which is initially of rotation rate $\Omega_0 - \Omega$, then gradually relaxes on the Ekman time scale $T_E = H_0/\sqrt{\nu\Omega} = 65\,\text{s}$, where ν is the kinematic viscosity of water. The perturbation resulting from the interaction of the mean zonal flow with the bottom topography is a lee wave pattern downstream of the mountain shown in Figure 5.3. The velocity vector field calculated from the measured gradient of the surface elevation using equation (5.22) is superimposed on the altimetry image in Figure 5.3a. Note that the color altimetry image is converted to gray scale and only every 40th velocity vector in each direction is displayed here to avoid overcrowding. Figure 5.3b shows the isolines of the surface elevation η calculated by integrating $\nabla\eta$ over x and y. The scale of η indicates that the amplitude of the wave is about $75\,\mu\text{m}$. To gain some physical insight into the interaction of the incoming flow with the mountain, consider a simple

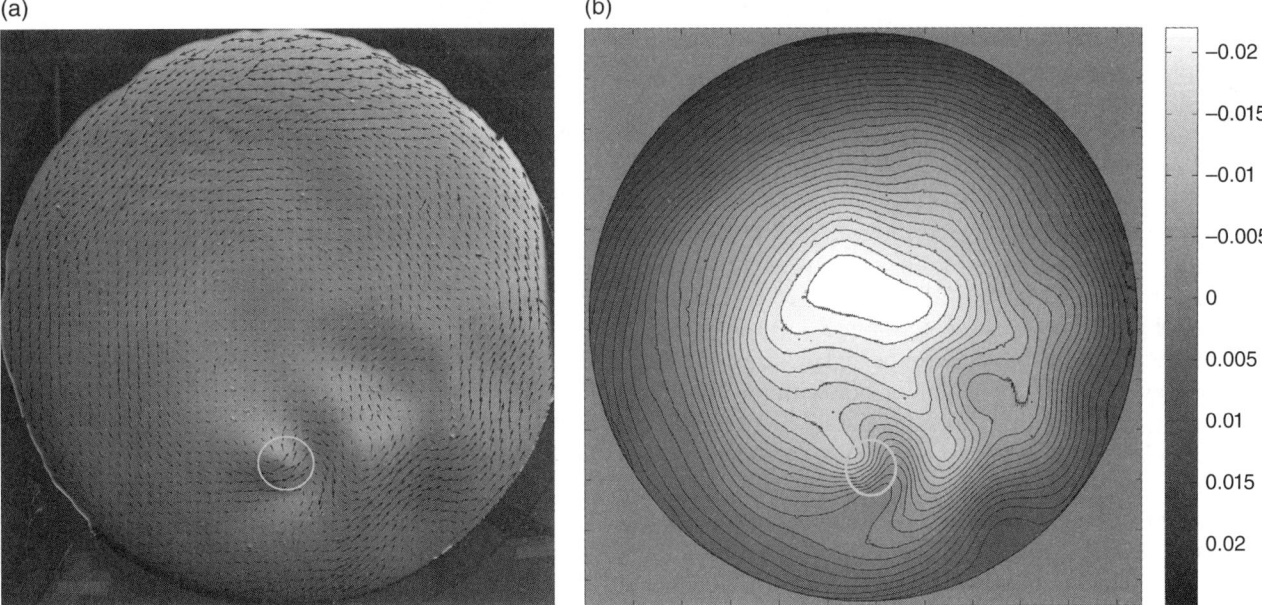

Figure 5.3. A lee wave behind a small mountain in eastward flow (counterclockwise) at $t = 25$ s. The mountain (white circle) is positioned at "6 o'clock." (a) Velocity vectors are superimposed on the altimetric image of the surface slope which is converted from color to gray scale. Only every 40th velocity vector in each direction is displayed here. (b) Isolines show surface topography in centimeters. A stationary Rossby wave is clearly seen as a pattern of hills and valleys above and downstream of the mountain. The amplitude of the first wave crest in the lee of the mountain is about 75 μm.

argument based on the conservation of PV. In this experiment the flow is ageostrophic enough for the incoming fluid parcels to climb over the plateau. In order to conserve PV, the decrease in depth should be compensated by decreasing either the Coriolis parameter or the relative vorticity. The Coriolis parameter is lower southward such that the flow can turn toward the equator (topographic steering). However, in our experiment topographic steering is not obvious and the main effect is the generation of anticyclonic relative vorticity. The parcel with induced circulation experiences the Coriolis force, which is greater at the northern side of the parcel due to the larger value of the Coriolis parameter. The resulting total force is a restoring force that drives the (anticyclonic) parcel northward. The parcel moving northward then acquires cyclonic vorticity and is moved southward and the process repeats itself. This is the basic mechanism of the generation of a Rossby wave.

It is useful to check if the observed wave pattern is consistent with the dispersion relation (5.26) which can be simply modified to include an incoming stream of angular velocity Ω':

$$\omega_{mn} - m\Omega' = \frac{-2m\gamma R_b^2}{\alpha_{mn}^2 R_b^2/R^2 + 1}. \quad (5.27)$$

The angular velocity of the mean flow is such that $\Omega' = \Omega_0 - \Omega$ initially. For stationary waves $\omega_{mn} = 0$. A small mountain, not unlike a point source, excites an entire spectrum of waves with corresponding wave numbers m and n. A dominant mode, however, can be easily identified in Figure 5.3 with a zonal wavelength (in angular measure) of about 30°, which corresponds to $m = 12$. The first zero of the Bessel function is at $\alpha_{12,1} = 16.7$, which gives the zonal phase speed $\Omega_{ph} = \omega/m = -0.024 \, s^{-1}$. Here the minus sign indicates westward propagation. This value is close (and of opposite sign) to the mean rotation rate of the fluid in the tank, $\Omega' = 0.021 \, s^{-1}$ measured at the same time as when the wave pattern in Figure 5.3 was observed.

The reader should also refer to the high contrast images of flows over bottom topography made with the original gray scale altimetry by *Rhines et al.* (2007). They show in particular an effect of upstream blocking caused by very long Rossby waves that are fast enough to propagate upstream from the mountain. These waves are described by a simple solution of (5.27) for stationary waves given by $m = 0$. Their speed is somewhat limited by dispersion, which is represented by the term with α_{mn} such that the fastest phase speed is achieved for the minimum value $\alpha_{01} = 2.4$. These waves have zonal crests and form a pattern of almost steady zonal currents such that there is a stagnant region upstream of the mountain. These currents are in fact β-plumes, not unlike those generated by spatially localized sources/forcing on the β-plane [*Stommel*, 1982; *Davey and Killworth*, 1989; *Rhines*, 1994]. Experimental demonstration of β-plumes on the polar β plane was provided in the work of *Afanasyev et al.* [2012]. An important dimensionless parameter in the problem of flows over topography is $R_\beta = \beta d^2/U$ [*Lighthill*, 1967; *Rhines*, 2007], which is the square of the ratio of the width (north-south) of the mountain to the zonal length scale of the stationary Rossby wave speed. To calculate R_β, we have to consider the flow in the β-plane context rather than in the context of the polar β-plane as we have been doing so far. In our experiment, $R_\beta = 0.5$ is relatively small, which indicates that the mountain is not wide enough and the current is too strong for blocking to occur. A classic lee Rossby wave is observed instead.

The flow is strikingly different if the mean flow is westward (rather than eastward as in the previous example). The westward mean flow is induced by accelerating the tank slightly above the null rate to $\Omega = 1.02\Omega_0$. The perturbation in the lee of the mountain is now in the form of a train of inertial waves (indicated by 1 in Figure 5.4). The Inset in Figure 5.4a shows the relative vorticity of the flow in the lee of the mountain while Figure 5.4b shows isolines of the surface elevation η. The displacement of the isolines of η indicates that the amplitude of the inertial waves is about 2 μm, which is quite small. The wavelength can be easily estimated by measuring the distance between the crests of the wave. It is noticeably shorter than that of the Rossby wave in the eastward flow. Just as we did with Rossby waves, we can perform a consistency check of the observations with the appropriate dispersion relation. For stationary waves we assume again that their frequency is $\omega = kU$. For measured values of the wavelength of the wave of about 4 cm and the velocity of the mean flow at the latitude of the mountain of 0.5 cm/s, we obtain the dimensionless frequency $\omega/f_0 = 0.17$. This value is shown by the circle in Figure 5.2a and is in good agreement with the dispersion relation (5.20). Note that we can also calculate the vertical wave number γ_n using (5.21). The value of the product $\gamma_n H_0 = 2.7 < \pi$ indicates that this is mode $n = 0$, which has the most simple vertical structure.

The flow shown in Figure 5.4 is actually more complicated than a simple lee wave. Wave breaking can be observed in the region indicated by 2 in Figure 5.4. This region is a shear layer formed by a relatively high velocity current at the northern flank of the mountain. This current is indicated by the concentrated isolines in Figure 5.4b and most likely occurs due to the effect of topographic steering. Indeed, the flow climbing the mountain is deflected northward where the Coriolis parameter is larger (or equivalently the depth is smaller) to conserve PV. Inertial waves break in the shear layer and create disturbances that propagate downstream. Careful observation shows that these disturbances, in turn, emit inertial

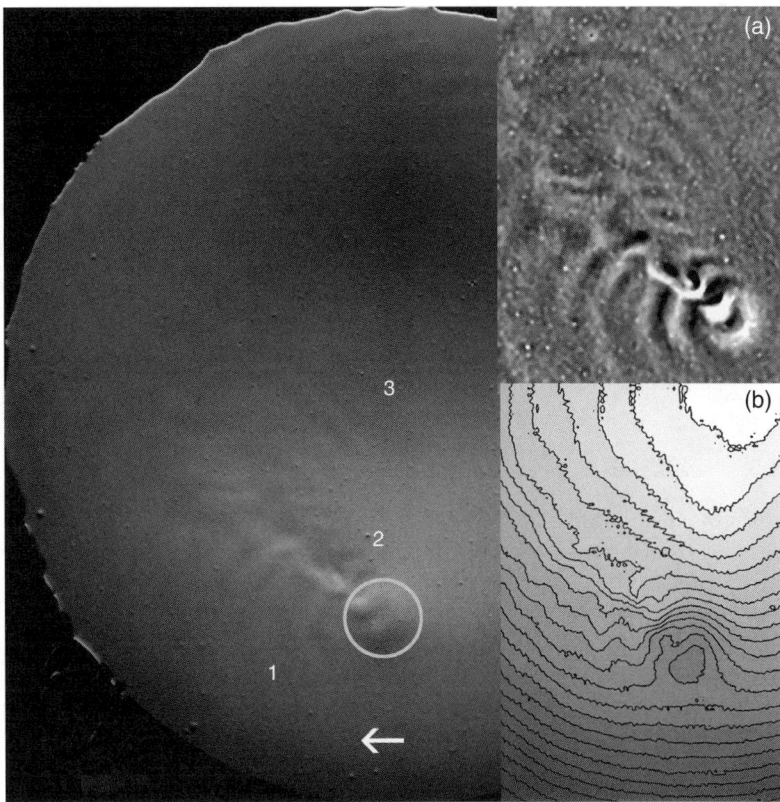

Figure 5.4. Altimetric image of the westward flow above a small mountain (white circle). The direction of the mean flow (clockwise solid-body rotation) is indicated by the arrow. The train of inertial waves in the lee of the mountain is indicated by 1 while 2 indicates the region where the instability and wave breaking occurs. Secondary emission of the inertial waves by the instability is indicated by 3. (a) Vertical component of the relative vorticity in the lee of the mountain. The gray scale shows the dimensionless vorticity ζ/f_0 in the range between -0.5 and 0.5. (b) Isolines of surface topography with intervals of 2 μm between the isolines.

waves that are visible in region 3 (Figure 5.4). Note that the emission of these secondary waves and the fact that they originate from the breaking primary wave are more evident from the video sequences of the flow rather than from a still image.

Another interesting and geophysically relevant example is the flows around an obstacle in the rotating fluid. Here we give only a brief description that illustrates the application of AIV. Further details can be found in *Afanasyev et al.* [2008a]. Altimetric images from our experiments are shown in Figure 5.5. A cylindrical obstacle of diameter $d = 5.9$ cm was fixed to the bottom of the tank (at 3 o'clock in Figure 5.5). The same technique we used in the previous experiment to create a mean flow over the mountain was used here. The rotation rate of the tank was slightly changed up or down from the null rotation to create either westward or eastward mean flow. Note that in *Afanasyev et al.* [2008a] a different method was used: The cylinder was towed in the azimuthal direction while the rotation rate of the tank was kept constant. The disadvantage of the simpler method we use here is that the mean flow decays, albeit on a relatively slow Ekman time scale.

In the experiments with an obstacle we observe again the difference between the westward and eastward flows. When the mean flow is westward, vortex shedding from the cylinder forms a regular vortex street (see Figures 5.5a, 5.6a, and 5.6c) similar to the von Karman–Benard vortex street in classical fluid dynamics. Inertial waves, even if they are emitted, are not easily visible there. When the mean flow is eastward, we observe a familiar Rossby wave in the lee of the cylinder (as shown in Figure 5.5 b), and surprisingly, no vortex street is observed at all. To understand the physics, we can consider again the fluid parcels in the PV context. The difference between the flows over the mountain and the flows around the obstacle is that in the latter case the relative vorticity is changed not only due to displacement of parcels northward or southward as they travel around the cylinder but also due to generation of vorticity in the boundary layer at the wall of the cylinder. The main dimensionless parameter that controls the boundary layer dynamics is the Reynolds number

110 MODELING ATMOSPHERIC AND OCEANIC FLOWS

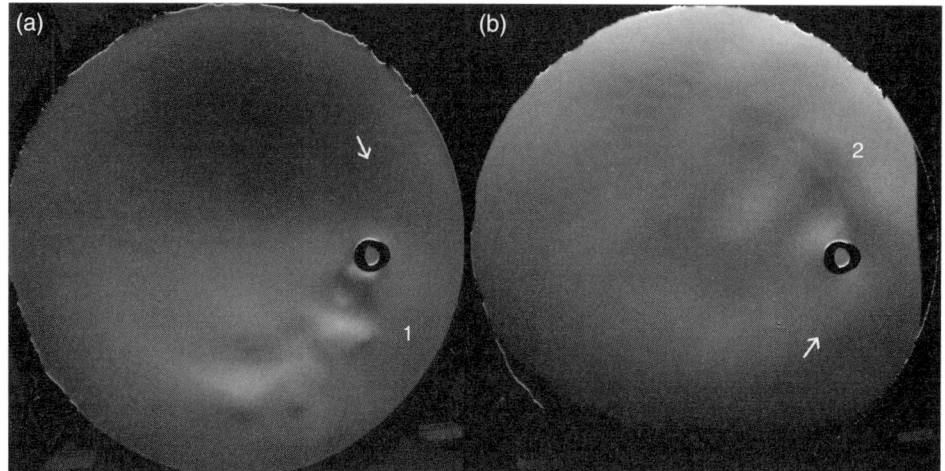

Figure 5.5. Altimetric images of (a) westward and (b) eastward flow around a circular cylinder (at 3 o'clock). A perturbation in the form of the vortex street in the lee of the cylinder is indicated by 1 while a stationary Rossby wave is indicated by 2.

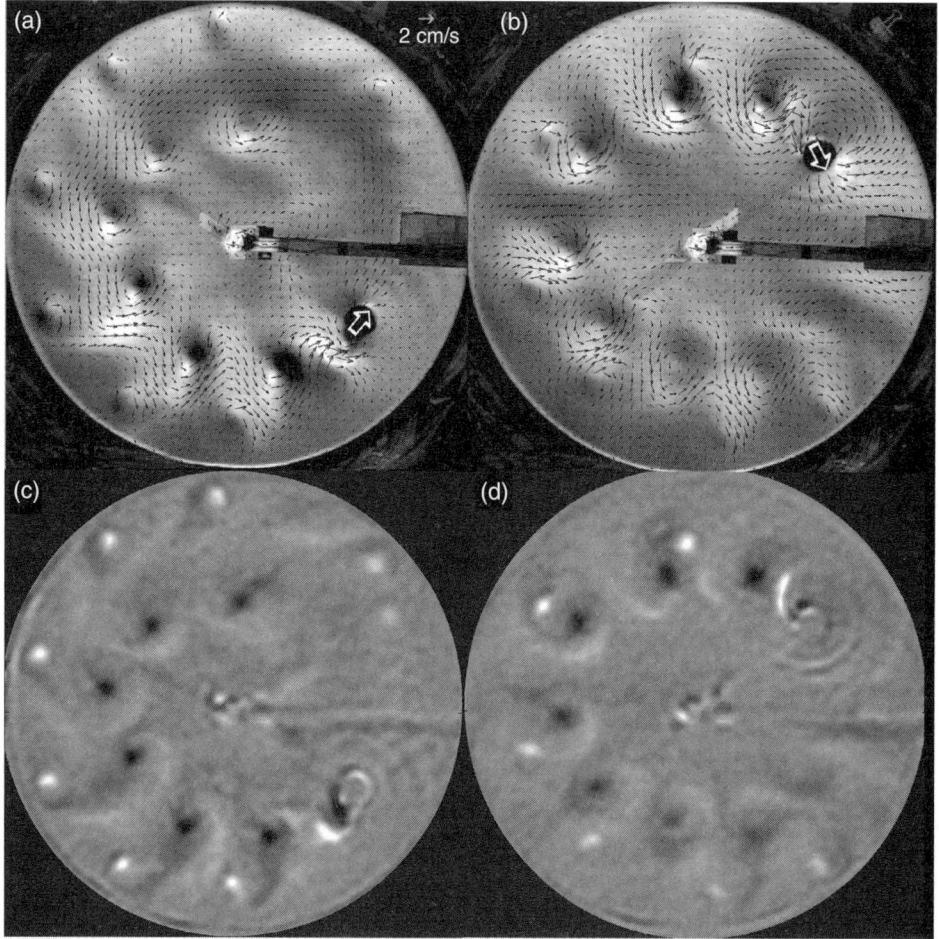

Figure 5.6. Flows behind a cylinder towed in a rotating fluid. The direction of motion of the cylinder is indicated by the large arrows: In (a) and (c) the cylinder is transported eastward while in (b) and (d) it is westward. Velocity vectors in (a) and (b) are superposed on the altimetric images while (c) and (d) show relative vorticity for the same flows, respectively. Vorticity is normalized by the Coriolis parameter and varies from -0.5 to 0.5. Cyclonic vortices are black while anticyclones are white in vorticity maps (c) and (d). The control parameters for both flows are Re = 1730 and R_β = 0.6.

Re = Ud/ν. At Re above about 50 one would expect the formation of the vortex street behind the cylinder. In our experiment the Reynolds number is quite large, Re ≈ 600, yet the vortex street does not form. Another relevant control parameter in this problem is again $R_\beta = \beta d^2/U$, which characterizes the relative importance of the β-effect on the scale of the cylinder. Exploration of the parameter space (Re, R_β) reveals that the vortex street does form in the eastward flow when the Reynolds number is sufficiently high. In fact, the boundary between the Rossby wave regime and the vortex street regime is described by an empirical linear relation

$$\text{Re} = 1100 R_\beta + 50. \qquad (5.28)$$

The vortex street regime is then predicted to occur above the line given by (5.28). Another surprising effect of this flow is that the observed vortex street is different from its classical counterpart: The order of vortices in the street is opposite to that in the regular (classical) street.

In this flow anticyclonic vorticity is generated on the northern side of the cylinder while cyclonic vorticity is generated on its southern side. In a regular vortex street the vortices are expected to maintain their relative positions. However, in this flow, immediately after being shed from the cylinder, the vortices exchange their positions such that the anticyclones move to the southern side while the cyclones move to the northern side. As a result of this "castling," the vortex street resembles a so-called reverse von Karman–Benard vortex street, which is normally formed behind a self-propelled body (a body providing a thrust force on the fluid). The reverse vortex street can be clearly seen in Figures 5.6b and 5.6d, which show the velocity and relative vorticity in the flow induced by a cylinder moving westward (this is equivalent to eastward mean flow). This anomalous vortex street can be compared with the regular vortex street shown in Figures 5.6a and 5.6c where the cylinder moves eastward (westward mean flow).

Another interesting and geophysically relevant effect is the interaction between vortex streets and Rossby waves. In fact, each vortex moves fluid parcels in the north-south direction changing their relative vorticity and thus creating Rossby waves. The generation of Rossby waves by individual vortices and eddies has been widely discussed in the literature [e.g., *Korotaev*, 1997; *Reznik et al.*, 2001; *Flor and Eames*, 2002; *McWilliams*, 2006]. Multiple vortices that are arranged in a certain organized pattern can be highly synchronized with Rossby waves. As a result, the dynamics can be something in between that of a vortical flow and of a wave-dominated flow. The boundary between these two regimes is provided by the Rhines scale [*Rhines*, 1975], which can be written as

$$L_\beta = \left(\frac{U_{\text{rms}}}{\beta}\right)^{1/2}, \qquad (5.29)$$

where U_{rms} is the root mean square (rms) velocity of the flow. This scale was obtained by finding a length scale at which a field of vortices (eddies) in two-dimensional isotropic turbulence is synchronized with linear Rossby waves on the β-plane. We equate the frequency of a vortex (its rotation rate) to the frequency of the Rossby wave obtained from the dispersion relation. The wavelength of the Rossby wave is determined by the distance between the vortices in the vortex arrangement. The original Rhines scale was obtained for β-plane flow, and it can be modified for the polar β-plane as follows (see also *Slavin and Afanasyev* [2012]). Consider vortices arranged in an approximately checkerboard pattern such that the oppositely signed vortices are located next to each other. This arrangement specifies the wave number $k = \pi/d$, where d is a typical diameter of a vortex. Taking the rotation rate of a vortex to be Ω_v, we require the synchronization between the vortex arrangement and the Rossby wave to be such that $\omega = \Omega_v$, where ω is the frequency of the wave obtained from the dispersion relation (5.26). We further require a match between the wave number of the vortex arrangement and that of the Rossby wave. The quantity α_{mn}/R in (5.26) represents the isotropic wave number and is analogous to the wave number $k = (k_x^2 + k_y^2)^{1/2}$ defined for a two-dimensional Fourier transform in Cartesian coordinates. We take $\alpha_{mn}/R = k$ and also take the azimuthal wave number m at "midlatitudes" of the domain (at $R/2$) to be equal to $kR/2$. Then solving (5.26) for k, we obtain

$$k = \frac{\gamma R}{f_0 \text{Ro}} + \sqrt{\frac{(\gamma R)^2}{f_0 \text{Ro}^2} - \frac{1}{R_b^2}}. \qquad (5.30)$$

Here Ro = $2\Omega_v/f_0$ is the Rossby number of a vortex. Note that R is the radius of the domain, which is the radius of the tank for laboratory flows or the radius of a planet for oceanic/atmospheric flows. The (modified) Rhines scale can then be introduced as a wavelength of the vortex arrangement/wave pattern $L_\gamma = 2\pi/k$. When the radius of deformation is relatively large, $R_b > f_0 \text{Ro}/(\gamma R)$, we can neglect the second term under the square root in (5.30), which gives

$$L_\gamma = \frac{\pi f_0 \text{Ro}}{\gamma R}. \qquad (5.31)$$

Note that Eq. (5.31) is analogous to the classic form of the Rhines scale (5.29) but written for the polar β-plane. Thus, when the typical size of the vortices created either by the flow around an obstacle or by some other process (a typical oceanographic example is the field of eddies created by a baroclinic instability) reaches L_γ, the emission of Rossby waves becomes very significant. For larger scales, the flow is wave dominated, in many cases exhibiting a pattern of zonal flows (jets). Some examples of baroclinic flows where jets are developed are given by *Slavin and Afanasyev* [2012].

Let us now return to the problem of emission of inertial waves. Inertial waves provide a mechanism of adjustment of rotating flows. They can also be viewed in the context of so-called spontaneous emission. The idea behind this is that a balanced flow should exist without any significant emission of inertial waves. Balanced flow is the vortical flow that can be described by an appropriate balance relation such as a geostrophic or quasi-geostrophic relation. Inertial waves are filtered out from these relations and assumed to be negligible. By relaxing the original restrictions of the model, one can obtain source terms on the RHS of the balanced equations. *Lighthill* [1952] derived the source term for acoustic sound generation by turbulence in a compressible fluid. *Ford* [1994] derived a similar term for the more geophysically relevant problem of emission of gravity waves in a rotating shallow water system. However, it is not yet completely clear what the most favorable circumstances for wave emission are and how much energy is radiated in the emission events. A few laboratory experiments reported in the literature demonstrate that wave emission does occur in different circumstances, although the percentage of radiated energy is quite small. For example, *Lovegrove et al.* [2000] and *Williams et al.* [2005] reported observations of inertia-gravity modes in baroclinically unstable rotating two-layer flows.

The emission of interfacial gravity waves by interacting vortex dipoles in a nonrotating two-layer fluid was demonstrated by *Afanasyev* [2003]. It was shown that for an effective emission a match between the speed of the translating vortex structure (dipole) and the phase speed of the wave as well as a match between the size of the dipole and the wavelength of the wave is required. Here we briefly review the results of experiments where the emission of pure inertial waves by baroclinically unstable flows was observed in the rotating fluid [*Afanasyev et al.*, 2008b]. In those experiments, the flows were induced by a small source of fresh water located on the surface of a saline layer at the wall of the tank. The fresh water from the source forms a coastal current that is wedge shaped in cross section. The current is approximately in geostrophic balance. A jet of width equal to the baroclinic radius of deformation forms at the outer edge of the current where the density interface intersects the surface forming a density front. The current is leaning on the wall to its right such that the Coriolis force is directed toward the wall. For further details on the dynamics of coastal density currents in the rotating fluid, the reader is referred to the review by *Griffiths* [1986].

The freshwater coastal current however rapidly succumbs to the frontal/baroclinic instability, which results in the formation of typical meanders along the current (Figure 5.7). Meandering coastal currents are a familiar phenomenon; they can often be observed in the satellite images of the ocean. The dynamics of finite-amplitude meanders is quite complicated. The meanders grow and

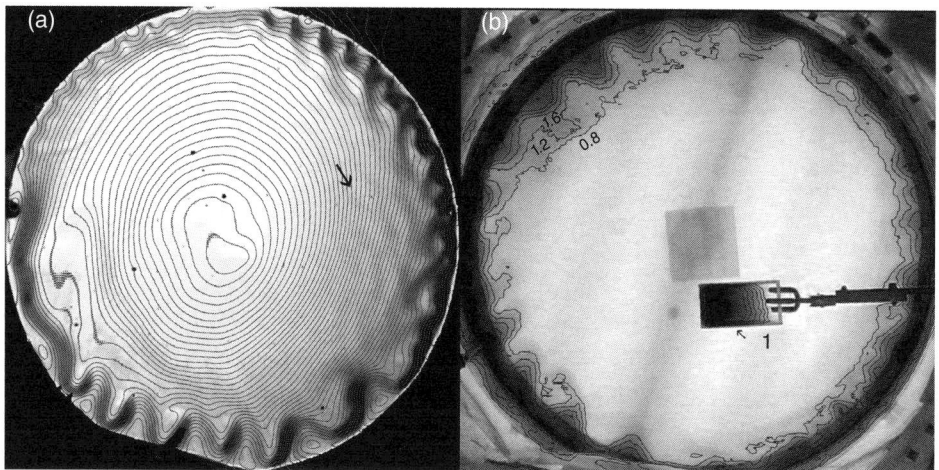

Figure 5.7. Meandering coastal current induced by a source of fresh water on the surface of a saline layer. (a) Isolines of surface topography superimposed on the altimetric image of the flow. The height difference between the adjacent isolines is 80 μm. Inertial waves are emitted by the baroclinic meanders. They are visible in the area between 2 and 4 o'clock (indicated by the arrow) and propagate in the unstratified interior of the tank away from the coastal current. (b) Isolines of density interface depth obtained by optical density method in a separate experiment. The contours of depth are from 0.8 to 3 cm with the interval of 0.4 cm. The current is colored by red food dye. The original color image is converted to gray scale. A wedge-shaped cuvette (1) is used for calibration of the optical density measurements.

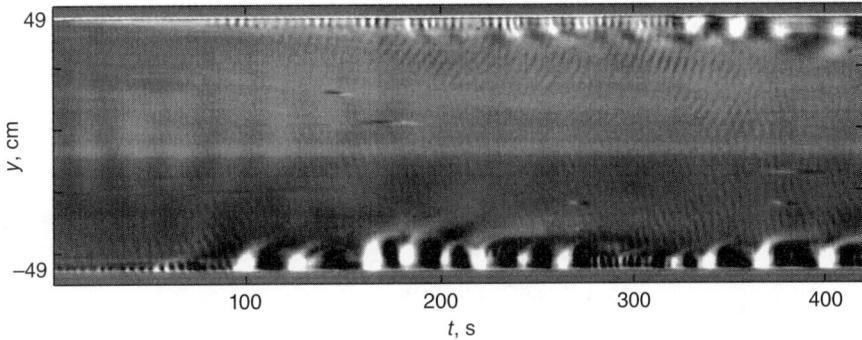

Figure 5.8. Hovmöller diagram showing the time line of the x component of velocity of the flow shown in Figure 5.5 measured along the y axis across the tank. The gray scale shows velocity in the range from -0.8 to 0.8 cm/s. The y axis crosses the wall of the tank at $y = -49$ cm and $y = 49$ cm (6 o'clock and 12 o'clock, respectively). The meanders of the coastal current passing through these two points are clearly visible along the bottom and the top of the diagram. Inertial waves emitted by the meanders can be seen in the interior in the form of lighter and darker bands.

interact with each other by pairing. Video sequences of this flow reveal that spontaneous bursts of wave radiation occur in this flow. The waves are radiated by the evolving meanders and then propagate toward the center of the tank. Fluid in the interior of the tank away from the stratified coastal current at the wall is unstratified.

The waves propagating in the interior have frequency below the inertial frequency and are pure inertial waves. Their amplitude is quite low, and it is difficult to see them in still images. The area where the waves can be seen is indicated by the arrow in Figure 5.7a. To visualize the waves more clearly and to follow their evolution in time, a Hovmöller diagram can be used (see Figure 5.8). The diagram shows the time evolution of velocity measured along a line across the tank. The dynamical range of velocities is intentionally reduced in the diagram to reveal low-amplitude waves. The bursts of inertial waves in the interior can be traced to meanders of the coastal current at the wall of the tank (top and bottom of the diagram). It is clear that the waves are mostly emitted by small-scale meanders (see top half of Figure 5.8, where most of the inertial waves are visible and where the meanders are smaller than on the bottom), which indicates that a certain match in scale and synchronization in time between meanders and waves must be realized for the effective emission. One might expect that the size of the meanders determines the wavelength of the wave while the translation velocity of the meander is matched to the phase speed of the wave, similar to that described by *Afanasyev* [2003].

Figure 5.9 shows velocity profiles measured across the coastal current and across the wave at three different times separated by small time intervals. One can see that the crest of the wave (indicated by the arrows in Figure 5.9) propagates toward the wall. This might seem counterintuitive because the source of the wave is the meander at the wall so the wave might be expected to propagate away from the source (similar to the way that gravity waves on the surface of a pond propagate away from a thrown rock). However, this is in fact one of the peculiarities of inertial waves. Their phase speed is directed *toward* the source while the group velocity is of course directed away from the source such that the energy propagates away.

Another example of the interesting interplay between vortices and waves is provided by our recent experiments with rotating (nearly) two-dimensional turbulence [*Afanasyev and Craig*, 2013; *Zhang and Afanasyev*, 2014]. Rotating turbulence is relevant to a wide range of applications, including geophysical, astrophysical, and engineering flows. It has been the subject of theoretical and experimental studies for quite a long time [e.g., *Ibbetson and Tritton*, 1974; *Hopfinger et al.*, 1982; *Jacquin et al.*, 1990; *Morize et al.*, 2005; *Bewley et al.*, 2007], yet it is not completely understood. In our experiments it was desirable to avoid the β-effect by using the f-plane setup, namely a uniformly rotating fluid of constant depth. Since it was also desirable to take advantage of the AIV technique, which requires the paraboloidal surface, a paraboloidal bottom was employed. The null rotation of the table was such that the form of the free surface of water was approximately the same as that of the bottom, thus providing a layer of constant depth. The container with paraboloidal bottom was of diameter $D = 90$ cm and was inserted into the main tank.

The flow was forced electromagnetically [e.g., *Marteau et al.*, 1995; *Afanasyev and Wells*, 2005]. For this purpose over 300 permanent rare earth magnets (size $2.5 \times 2.5 \text{ cm}^2$) of alternating polarity were attached under the bottom of the container such that the distance between the magnets was 4.5 cm. The water in the tank was saline to allow electric current to flow between two electrodes placed at the sides of the container. The Lorentz force on the ions occurs due to the vertical magnetic

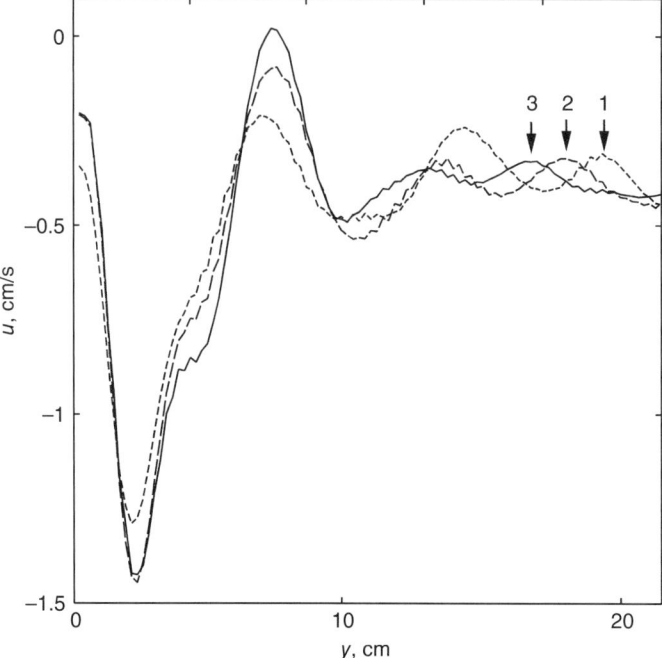

Figure 5.9. Profiles of the x component of the velocity measured along the y axis near the wall of the tank. The time interval between the profiles is 0.5 s. The high values of velocity near the wall are due to the coastal current, and the low-amplitude peaks away from the wall are indicative of inertial waves. Arrows 1–3 indicate the propagation of the crest of the wave toward the wall.

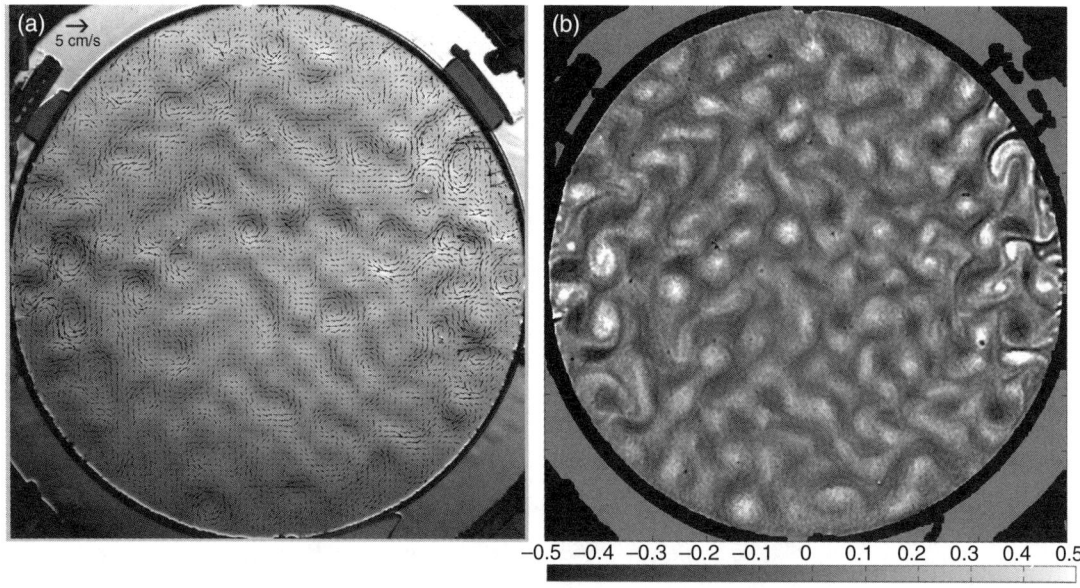

Figure 5.10. Continuously forced turbulent flow on the f-plane: (a) velocity vectors superposed on the altimetric image of the flow and (b) the relative vorticity field. The gray scale shows dimensionless vorticity ζ/f_0.

field and horizontal electric current and drives the ions in the horizontal direction perpendicular to the electric current. Thus, each magnet provides a localized force on the fluid in the horizontal direction, which generates a vortex dipole. The dipoles induced by individual magnets interact with each other and go through the process of adjustment to the background rotation. A typical image of the resulting turbulent flow is shown in Figure 5.10. The evolution of turbulence comprises vortex formation, translation, interaction, and decay by shear.

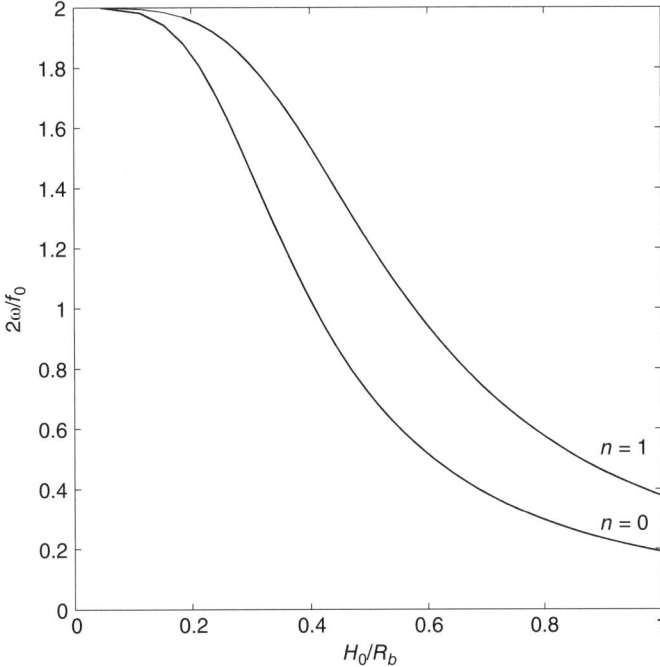

Figure 5.11. Frequency in the form of the Rossby number Ro = $2\omega/f_0$ of inertial waves synchronized with the vortex array vs. the depth of the layer normalized by the radius of deformation. The solid lines show two modes, $n = 0$ and $n = 1$. The frequency was found by solving (5.20) and (5.21), where the wave number k is determined by the spacing between the magnets $d = 4.5$ cm.

The turbulent vortical motion created by the electromagnetic forcing in the rotating layer of water should coexist and interact with inertial waves. It is interesting to see if inertial waves can be identified in this turbulent flow and also to discuss the condition when they can be emitted most effectively by vortices. The Rhines scale was determined by synchronizing/matching the field of vortices to the Rossby waves. Let us perform similar analysis for turbulence on the f-plane by matching vortices and inertial waves instead of Rossby waves. Assume again that the vortices are arranged in a checkerboard pattern with wave number $k = \pi/d$, where d is a diameter of a vortex. We require the synchronization between the vortex arrangement and inertial wave equating the rotation rate of a vortex, Ω_v, and the frequency of the wave, ω, obtained from the dispersion relation (5.20). We also match the wave number of the vortex arrangement to that of the inertial wave such that (5.21) is written in the form $\gamma_n = k\sqrt{\omega^2/(f_0^2 - \omega^2)}$. Solving then the dispersion relation numerically, we can obtain the frequency of the wave/vortex array for different parameters of the system. Figure 5.11 shows the values of frequency for different values of the depth of the layer. Here the spacing between the magnets used in our experiments determines the typical size of the vortex d and hence the value of the wave number k for inertial waves. Note that the graph in Figure 5.11 indicates that in order for the synchronization to occur when the Rossby number is below unity, the layer should be relatively deep. This depth will decrease if the size of the vortices/spacing between the magnets is decreased. Typical values of the Rossby number for the flow shown in Figure 5.10 are lower than 0.5 (except in a few large vortices). With depth $H_0 = 10$ cm such that $H_0/R_b = 0.45$, this flow falls below the line in Figure 5.11, which indicates that the condition for the most efficient emission of inertial waves is not achieved.

In an attempt to identify the waves in this turbulent flow, we performed a double Fourier transform (in time and space) of the velocity components of the flow. The transform was performed in one direction across the tank and for the time interval of about 90 inertial periods with seven samples per period. The resulting energy spectrum is shown in Figure 5.12 in frequency and wave number space. A line showing the dispersion relation (5.20) for inertial waves of the lowest mode $n = 0$ is also shown on the diagram. One can see that a peak of energy is located at the (dimensionless) wave number 10 (the forcing wave number) at zero frequency (since the forcing is constant). Note that (steady) geostrophic motions lie along the horizontal axis ($\omega = 0$). Further examination of the diagram in Figure 5.12 reveals that the peak at wave number 10 extends toward higher frequencies up to the limiting curve given by the dispersion relation. There is also noticeable energy at somewhat lower frequencies (about 0.1–0.2)

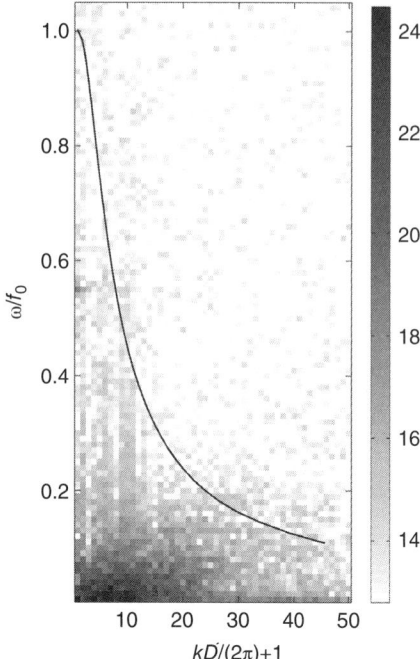

Figure 5.12. Energy spectrum of rotating turbulence shown in frequency and wave number space. The gray scale shows energy in logarithmic scale. The frequency is normalized by the Coriolis parameter while the wave number is normalized by the diameter of the container, $D = 90$ cm. The solid line shows the dispersion relation (5.20) for the inertial waves.

but at higher wave numbers (20–40). Thus it is quite clear that inertial waves are detected in this flow, although their energy is quite small compared to that of the vortices.

5.6. SUMMARY

This chapter has reviewed a new laboratory technique of optical altimetry designed to visualize rotating flows with a free surface and to measure their surface elevation and velocity fields. An additional benefit of the rotating system, where the depth of the layer varies quadratically with the distance from the center, is that it models the planetary polar β-plane. The theoretical background for Rossby and inertial waves is given as well as the method for calculating the velocity field from the slopes of the surface elevation. This chapter gives several examples of experiments that illustrate the advantages of laboratory altimetry in studying global patterns of GFD flows, including Rossby and inertial waves and vortices. Typical features of the β-plane dynamics are demonstrated in experiments with flows over a bottom topography and around an obstacle where striking differences are observed between the eastward and westward flows. The interesting and geophysically relevant issue of emission of waves by vortices/eddies is discussed in the context of emission of Rossby waves by vortex arrays on the polar β-plane and emission of inertial waves by vortices in quasi two-dimensional rotating turbulence on the f-plane. The Rhines scale, which was originally derived using the idea of synchronization between waves and vortices of matching wave number, is relevant in the former case. A modification of the original Rhines scale to the polar β-plane is discussed. It allows one to predict at what scale the emission of Rossby waves becomes significant and as a result a change in the regime of the flow can occur. A similar consideration of synchronization/matching between vortices on the f-plane gives a prediction of when the most efficient emission of inertial waves can occur.

Acknowledgments. I would like to thank J. Craig for his help with expriments on two-dimensional turbulence and S. H. Curnoe for thorough proofreading of the draft of this chapter. The support of the Natural Sciences and Engineering Research Council of Canada is gratefully acknowledged.

REFERENCES

Afanasyev, Y. D. (2003), Spontaneous emission of gravity waves by interacting vortex dipoles in a stratified fluid: laboratory experiments, *Geophys. Astrophys. Fluid Dynamics*, 97(2), 79–95.

Afanasyev, Y. D., and J. D. C. Craig (2013), Rotating shallow water turbulence, *Phys. Fluids*, 25, 106603.

Afanasyev, Y. D., and J. Wells (2005), Quasi-two-dimensional turbulence on the polar beta-plane: Laboratory experiments, *Geophys Astrophys Fluid Dyn.*, 99, 1.

Afanasyev, Y. D., P. B. Rhines, and E. G. Lindahl (2008a), Laboratory experiments with altimetric imaging velocimetry: Cylinder wakes on the polar beta-plane, *Phys. Fluids*, 20, 086604.

Afanasyev, Y. D., P. B. Rhines, and E. G. Lindahl (2008b), Emission of inertial waves by baroclinically unstable flows: Laboratory experiments with altimetric imaging velocimetry, *J. Atmos. Sci.*, 65, 250.

Afanasyev, Y. D., P. B. Rhines, and E. G. Lindahl (2009), Velocity and potential vorticity fields measured by altimetric imaging velocimetry in the rotating fluid, *Exp. Fluids.*, 47, 913.

Afanasyev, Y. D., S. O'Leary, P. B. Rhines, and E. G. Lindahl (2012), On the origin of jets in the ocean, *Geophys Astrophys Fluid Dyn.*, 106(2), 113–137.

Batchelor, G. K. (1967), *An Introduction to Fluid Dynamics*, Cambridge University Press, Cambridge.

Bewley, G. P., D. P. Lathrop, L. R. M. Maas, and K.R. Sreenivasan (2007), Inertial waves in rotating grid turbulence, *Phys. Fluids*, 19, 071701, 1–4.

Cenedese, C., and S. B. Dalziel (1998), Concentration and depth fields determined by the light transmitted through a dyed solution, Paper 061, in *Proceedings of the 8th International*

Symposium on Flow Visualization, Eds. G. M. Carlomagno and I. Grant, ISBN 0953399101, Sorrento.

Davey, M. K., and P. D. Killworth (1989), Flows produced by discrete sources of buoyancy, *J. Phys. Ocean.*, *19*, 1279.

Flor, J.-B., and I. Eames (2002), Dynamics of monopolar vortices on a topographic beta-plane, *J. Fluid Mech.*, *456*, 353.

Ford, R. (1994), Gravity wave radiation from vortex trains in rotating shallow water, *J. Fluid Mech.*, *281*, 81–118.

Gill, A. E. (1982), *Atmosphere-Ocean Dynamics*, Academic, London.

Greenspan, H. (1968), *The Theory of Rotating Fluids*, Cambridge University Press, Cambridge.

Griffiths, R. W. (1986), Gravity currents in rotating systems, *Annu. Rev. Fluid Mech.*, *18*, 59–89.

Hopfinger, E. J., F. K. Browand, and Y. Gagne (1982), Turbulence and waves in a rotating tank, *J. Fluid Mech.*, *125*, 505.

Ibbetson, A., and D. J. Tritton (1974), Experiments on turbulence in a rotating fluid, *J. Fluid Mech.*, *68*, 639.

Jacquin, L., O. Leuchter, C. Cambon, and J. Mathieu (1990), Homogeneous turbulence in the presence of rotation, *J. Fluid Mech.*, *220*, 1.

Korotaev, G. K. (1997), Radiating vortices in geophysical fluid dynamics, *Surv. Geophys.*, *18*, 567–619.

Lighthill, J. M. (1952), On sound generated aerodynamically. Part I: General theory, *Proc. Roy. Soc.*, *21*, A211, 564–587.

Lighthill, J. M. (1967), On waves generated in dispersive systems by travelling forcing effects, with application to the dynamics of rotating fluids, *J. Fluid Mech.*, *27*(4), 725–752.

Linden, P. F, and J. E. Simpson (1994), Continuous releases of dense fluid from an elevated point source in a crossflow, in *Mixing and Transport in the Environment*, edited by K. J. Beven, P. C. Chatwin, and J. H. Millbank, pp. 401–418, Wiley, New York.

Lovegrove, A. F., P. L. Read, and C. J. Richards (2000), Generation of inertia-gravity waves in a baroclinically unstable fluid, *Q. J. R. Meteorol. Soc.*, *126*, 3233–3254.

Marteau, D., O., Cardoso, and P. Tabeling (1995), Equilibrium states of two-dimensional turbulence: An experimental study, *Phys. Rev. E, 51*, 5124–5128.

McWilliams, J. C. (2006), *Fundamentals of Geophysical Fluid Dynamics*, Cambridge University Press, Cambridge.

Morize, C., F. Moisy, and M. Rabaud (2005), Decaying grid-generated turbulence in a rotating tank, *Phys. Fluids*, *17*, 095105.

Phillips, N. A. (1963), Geostrophic motion, *Rev. Geophys.*, *1*, 123–176.

Phillips, N. A. (1965), Elementary Rossby waves, *Telus*, *17*(3), 295–301.

Phillips, N. A. (1966), The equations of motion for a shallow rotating atmosphere and the traditional approximation, *J. Atmos. Sci.*, *23*, 626–628.

Ravela, S., J. Marshall, C. Hill, A. Wong, and S. Stransky (2010), A realtime observatory for laboratory simulation of planetary flows, *Exp. Fluids*, *48*, 915.

Reznik, G. M., and R. Grimshaw (2001), Ageostrophic dynamics of an intense localized vortex on a β-plane, *J. Fluid Mech.*, *443*, 351.

Rhines, P. B. (1975), Waves and turbulence on a beta-plane, *J. Fluid Mech.*, *69*(3), 417.

Rhines, P. B. (1994), Jets, *Chaos 706*, *4*, 313.

Rhines, P. B. (2007), Jets and orography: Idealized experiments with tip-jets and Lighthill blocking, *J. Atmos. Sci.*, *64*, 3627.

Rhines, P. B., E. G. Lindahl, and A. J. Mendez (2007), Optical altimetry: A new method for observing rotating fluids with application to Rossby waves on a polar beta-plane, *J. Fluid Mech.*, *572*, 389.

Slavin, A. G., and Y. D. Afanasyev (2012), Multiple zonal jets on the polar beta plane, *Phys. Fluids*, *24*, 016603, doi:10.1063/1.3678017.

Stommel, H. M. (1982), Is the South Pacific helium-3 plume dynamically active?" *Earth Planet. Sci. Lett.*, *61*, 63.

Veronis, G. (1963), On the approximations involved in transforming the equations of motion from a spherical surface to the β-plane. I. Barotropic systems. *J. Marine Res.*, *21*, 110–124.

Williams P. D., T. W. N. Haine, and P. L. Read, (2005), On the generation mechanisms of short-scale unbalanced modes in rotating two-layer flows with vertical shear, *J. Fluid Mech.*, *528*, 1–22.

Zhang, Y., and Afanasyev, Y. D. (2014), Beta-plane turbulence: experiments with altimetry. Phys. Fluids 26, 026602.

6
Instabilities of Shallow-Water Flows with Vertical Shear in the Rotating Annulus

Jonathan Gula[1] and Vladimir Zeitlin[2]

6.1. INTRODUCTION

There is a long tradition of experiments in differentially rotating annuli in order to understand the baroclinic instability and, more generally, the instabilities of fronts in geophysical fluid dynamics (GFD) [*Hide*, 1958; *Fultz et al.*, 1959; *Hide and Fowlis*, 1965; *Hart*, 1972]. Recently the interest in such experiments was revived in the context of the so-called spontaneous emission of inertia-gravity waves by balanced flows (see *Ford* [1994], *O'Sullivan and Dunkerton* [1995], and the references in the special collection of *Journal of Atmospheric Sciences* on this subject, *Dunkerton et al.* [2008]). Thus, short-wave patterns coupled to the baroclinic Rossby waves were observed in independent experiments [*Lovegrove et al.*, 2000; *Williams et al.*, 2005; *Flór*, 2007; *Flór et al.*, 2011] on instabilities of the two-layer rotating flows in the annulus at high enough Rossby numbers.

On the other hand, the classical experiments on unstable density (coastal) currents by *Griffiths and Linden* [1982] also used annular geometry and a two-layer setting, with lighter fluid overflowing the denser one in the rotating tank. Recently, similar experiments, but with sloping bottom, were performed by *Pennel et al.* [2012].

Motivated by all these experiments, we undertook a thorough stability analysis of a two-layer shallow-water system in the rotating annulus both with the rigid lid and with a free surface and outcropping interface. Our main goal in the rigid-lid configuration was to check to what extent the ageostrophic short-wave instabilities in shallow water may account for experimental observations. In the outcropping configuration it was instructive to see to what extent the simple two-layer shallow water-theory reproduces the experiment.

The experiments mentioned above are not strictly speaking shallow-water ones, although no pronounced vertical structure was observed, as to our knowledge. The results we present below may serve, nevertheless, to understand the vertically averaged behavior of the full system. Moreover, *Williams et al.* [2005] interpreted their experiments in terms of shallow-water dynamics, referring to *Ford* [1994]. As to the density currents, their instabilities are traditionally studied with shallow-water models, as in the classical paper by *Griffiths et al.* [1982]. Being standard in GFD, the two-layer shallow-water approximation is a reasonable compromise between the realistic representation of the observed fluid flow and the computational effort (and amount of resources) necessary for a full stability analysis. It is, in addition, self-consistent and universal, as, for example, the fine vertical structure of the flow may vary from one experiment to another.

In Section 6.2 we present our results for the superrotating rigid-lid configurations (following *Gula et al.* [2009b], where most of them were published). In Section 6.3 we give new results for the free surface configuration with outcropping, and in Section 6.4 we analyze the influence of bathymetry on the instabilities.

6.2. STABILITY OF FRONTS UNDER RIGID LID

A typical configuration used in laboratory experiments by *Williams et al.* [2005] and *Flór et al.* [2011] is presented in Figure 6.1. The annulus has an inner vertical sidewall of radius r_1, an outer vertical sidewall of radius r_2, and a

---

[1] *Institute for Geophysics and Planetary Physics, University of California, Los Angeles, California, United States of America.*

[2] *Laboratoire de Météorologie Dynamique, École Normale Supérieure; and Université Pierre et Marie Curie, Paris, France.*

Modeling Atmospheric and Oceanic Flows: Insights from Laboratory Experiments and Numerical Simulations, First Edition. Edited by Thomas von Larcher and Paul D. Williams.
© 2015 American Geophysical Union. Published 2015 by John Wiley & Sons, Inc.

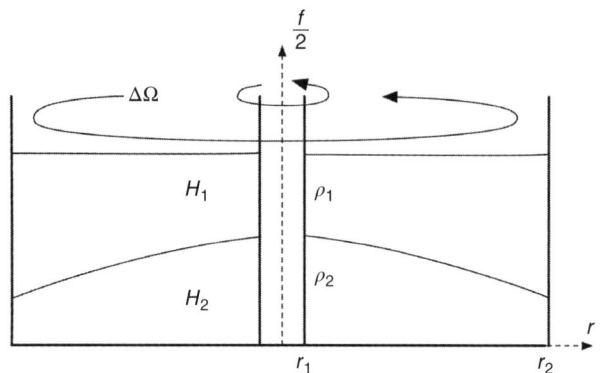

Figure 6.1. Schematic representation of a two-layer flow in the annulus with a superrotating lid.

total depth $2H_0$. The radial width of the annulus is therefore $r_2 - r_1$, and the two layers have equal depths H_0 at rest. The base and the lid are both horizontal and flat. The angular velocity about the axis of symmetry is Ω, and the upper lid is superrotating at $\Omega + \Delta\Omega$. This differential rotation provides a vertical velocity shear of the balanced basic state that is close to solid-body rotation of each fluid layer with different angular velocities. Such a state will be represented in the stability analysis that follows by a cyclogeostrophic equilibrium in each layer, with linear radial profile of the azimuthal velocity, within the rotating two-layer shallow-water model. In order to fulfil a complete linear stability analysis, we use below the collocation method.

Our analysis is purely inviscid; however, in the experiment the mean axisymmetric flow is controlled by friction. As is well known [see, e.g., *Hart*, 1972], Ekman, Stewartson, and shear boundary layers are present in the two-layer rotating fluid in the tank, and the related torques are acting upon the quasi-inviscid interior. Moreover, the interfacial layer has an internal structure depending on whether two fluids are immiscible or not (see Chapter 11). All this internal structure will be neglected in what follows, and the layers will be considered to be in solid rotation.

6.2.1. Equations of Motion, Basic States, and Linear Stability Problem

Consider the two-layer rotating shallow-water model on the plane rotating with constant angular velocity Ω. The momentum and continuity equations are written in polar coordinates as

$$D_j u_j - \left(f + \frac{v_j}{r}\right) v_j - r\Omega^2 = -\partial_r \Pi_j,$$
$$D_j v_j + \left(f + \frac{v_j}{r}\right) u_i = -\frac{\partial_\theta \Pi_j}{r}, \quad (6.1)$$
$$D_j h_j + h_j \nabla \cdot \mathbf{v_j} = 0,$$

where $\mathbf{v_j} = (u_j, v_j)$, h_j, and Π_j are velocity (radial, azimuthal), thickness, and pressure normalized by density (geopotential), respectively, in the jth layer (counted from the top), $j = 1, 2$; f is the Coriolis parameter, $f = 2\Omega$; and D_j denote Lagrangian derivatives in respective layers. The boundary conditions are $u = 0$ at $r = r_1, r_2$.

By introducing the time scale $1/f$, the horizontal scale $r_0 = r_2 - r_1$, the vertical scale H_0, and the velocity scale $V_0 = fr_0$, we use nondimensional variables from now on without changing notation. By linearizing about a steady state with constant azimuthal velocities $V_1 \neq V_2$, we obtain the following nondimensional equations (the ageostrophic version of the Phillips model in cylindrical geometry):

$$\partial_t u_j + \frac{V_j}{r}\partial_\theta u_j - v_j - 2\frac{V_j v_j}{r} = -\partial_r \pi_j,$$
$$\partial_t v_j + u_j \partial_r V_j + \frac{V_j}{r}\partial_\theta v_j + u_j + \frac{V_j u_j}{r} = -\frac{\partial_\theta \pi_j}{r}, \quad (6.2)$$
$$\partial_t h_j + \frac{1}{r}(rH_j u_j)_r + \frac{1}{r}H_j \partial_\theta v_j + \frac{V_j}{r}\partial_\theta h_j = 0,$$

where the pressure perturbations in the layers, π_j, are related through the interface perturbation η as usual,

$$\pi_2 - \pi_1 + s(\pi_2 + \pi_1) = \text{Bu}\,\eta, \quad (6.3)$$

and $s = (\rho_2 - \rho_1)/(\rho_2 + \rho_1)$ is the stratification parameter, $\text{Bu} = (R_d/r_0)^2$ is the Burger number, $\text{Ro} = \Delta\Omega/(2\Omega)$ is the Rossby number (as used in experiments, cf. *Flór et al.* [2011]), $R_d = (g'H_0)^{\frac{1}{2}}/(2\Omega)$ is the Rossby deformation radius, and $g' = 2\Delta\rho g/(\rho_1 + \rho_2) = 2sg$ is the reduced gravity.

The depth profiles $H_j(r)$ and respective velocities $V_j(r)$ in (6.2) correspond to a steady cyclogeostrophically balanced state of the two-layer system that obeys the nondimensional equations.

$$V_j + \frac{V_j^2}{r} + \frac{r}{4} = \partial_r \Pi_j, \quad (6.4)$$

where the $r/4$ term corresponds to the centrifugal effect at the interface, while the other terms correspond to the classical cyclogeostrophic equilibrium.

The rotation rates of the layers lie in the interval between the rotation rate of the base (0 in the rotating frame) and that of the upper lid (Ro in the rotating frame). Therefore, in general,

$$V_2 = \alpha_2 r, \quad V_1 = \alpha_1 r \quad (6.5)$$

and we get the following expressions for the heights of the layers in the state of cyclogeostrophic equilibrium for such mean flow:

$$H_j = H_j(0) + (-1)^j[\alpha_2 + \alpha_2^2 - \alpha_1 - \alpha_1^2]\frac{r^2}{2\,\text{Bu}}$$
$$+ (-1)^j s[\alpha_2 + \alpha_2^2 + \alpha_1 + \alpha_1^2 + 1/2]\frac{r^2}{2\,\text{Bu}}. \quad (6.6)$$

Hart [1972] considered the top, bottom, and interfacial friction layers and found that the rotation rates are $\alpha_1 = (2 + \chi) \text{Ro}/2(1 + \chi)$ and $\alpha_2 = \text{Ro}/2(1 + \chi)$, where $\chi = (\nu_2/\nu_1)^{1/2}$ is the viscosity ratio between the two layers. If the two layers have close viscosities $\chi = 1$, it leads to $(\alpha_1, \alpha_2) = (0.75 \text{Ro}, 0.25 \text{Ro})$.

A calculation based on a layerwise balance of the torques in *Williams et al.* [2004] gives values for (α_1, α_2) of the same order but depending on the turntable angular velocity. The direct measurements of the radial velocity profiles by *Flór* [2007] are closer to $(\alpha_1, \alpha_2) \approx (0.9 \text{Ro}, 0.1 \text{Ro})$. We will therefore keep these last values throughout the chapter, but this particular choice means no loss of generality, as changing the relative rotation rate just means rescaling the Rossby number.

Supposing a harmonic form of the solution in the azimuthal direction,

$$(u_j(r, \theta), v_j(r, \theta), \pi_j(r, \theta))$$
$$= (\tilde{u}_j(r), \tilde{v}_j(r), \tilde{\pi}_j(r)) \exp[ik(\theta - ct)] + \text{c.c.}, \quad (6.7)$$

where k is the azimuthal wave number ($k \in \mathbb{N}$), and omitting tildes we get, from (6.2),

$$k(V_j - rc)iu_j - (r + 2V_j)v_j + r\partial_r \pi_j = 0,$$
$$-(r + V_j + r\partial_r(V_j))iu_j + k(V_j - rc)v_j + k\pi_j = 0,$$
$$-\partial_r(rH_j iu) + kH_j v + k(V_j - rc)(-1)^j \eta = 0,$$
$$\pi_2 - \pi_1 + s(\pi_2 + \pi_1) = \text{Bu}\,\eta\,.$$
$$(6.8)$$

The system (6.8) is an eigenproblem of order 6 that can be solved by applying the spectral collocation method [*Trefethen*, 2000].

The dispersion diagrams we thus obtain show that the branches of dispersion relation corresponding to different modes can intersect, leading to linear wave resonances and thus creating instabilities of various nature [*Cairns*, 1979; *Sakai*, 1989].

Following *Cairns* [1979] and *Ripa* [1983], the flow with velocity U_0 is unstable if there exists a pair of waves with intrinsic frequencies $\tilde{\omega}_1$ and $\tilde{\omega}_2$ that satisfy the following conditions: The waves propagate in the opposite directions with respect to the basic flow $\tilde{\omega}_1 \tilde{\omega}_2 < 0$, meaning that they have opposite energy anomalies, and have almost the same Doppler-shifted (absolute) frequencies ($\tilde{\omega}_1 + kU_0 \sim \tilde{\omega}_2 - kU_0$) and thus can phase lock and resonate. The interpretation of the unstable modes as resonances between the neutral waves provides a classification of different instabilities and corresponding regions of parameter space.

Namely, we will display below the instabilities resulting from the resonance between Rossby waves in upper and lower layers (the baroclinic instability), the resonance between Rossby and Kelvin or Poincaré waves in respective layers (Rossby-Kelvin instability), and the resonances between two Poincaré, or Kelvin and Poincaré, or two Kelvin modes (Kelvin-Helmholtz shear instability). We should recall at this point the physical nature of different waves in the two-layer shallow-water system: Rossby waves propagate due to potential vorticity gradients, whatever their origin, Kelvin waves propagate due to (and along) the boundaries in the rotating systems, and Poincaré waves are inertia-gravity waves propagating due to the density jump at the interface or at the free surface. Although each instability occupies its proper domain in the parameter space, we will see that there exist crossover regions where two different instabilities coexist and may compete.

6.2.2. Instabilities and Growth Rates

We first present the overall stability diagram in the space of parameters of the model and then illustrate different parts of this diagram by displaying the corresponding unstable modes and dispersion curves. The stability diagram was obtained by calculating the eigenmodes and the eigenvalues of the problem (6.5), (6.8) for about 50,000 points in the space of parameters (there are typically 200–300 points along each axis in the figures below) and then interpolating. Only discrete azimuthal wave numbers correspond to realizable modes. We nevertheless present the results as if the spectrum of wave numbers were continuous for better visualization. They are synthesized in Figures 6.2 and 6.3 displaying the growth rates and the wave numbers, respectively, of the most unstable modes. Both figures represent the plane of parameters Ro-Bu (Figures 6.2 and 6.3). We also show in Figure 6.4 how the dispersion diagrams evolve while changing parameters and approaching the instability band spreading from low left to upper right in Figures 6.2 and 6.3. One clearly sees how the initially stable flow without imaginary eigenvalues of c develops instabilities of various nature as parameters change. Thus, as shown in the left column of Figure 6.4, decreasing the Burger number leads to distortion of the dispersion curves of Rossby modes and their reconnection leading to Rossby-Rossby (RR) resonance, i.e., the baroclinic instability. Different distortion of dispersion curves of Rossby modes takes place if Ro increases at constant Bu, leading to reconnection with (a) a Kelvin-mode curve and Rossby-Kelvin (RK) resonance with corresponding instability and (b) a Poincaré-mode curve and Rossby-Poincaré (RP) resonance and corresponding instability. Further increase in Ro leads to reconnection of Kelvin-mode curves and Kelvin-Kelvin (KK) resonance and related shear instability with features

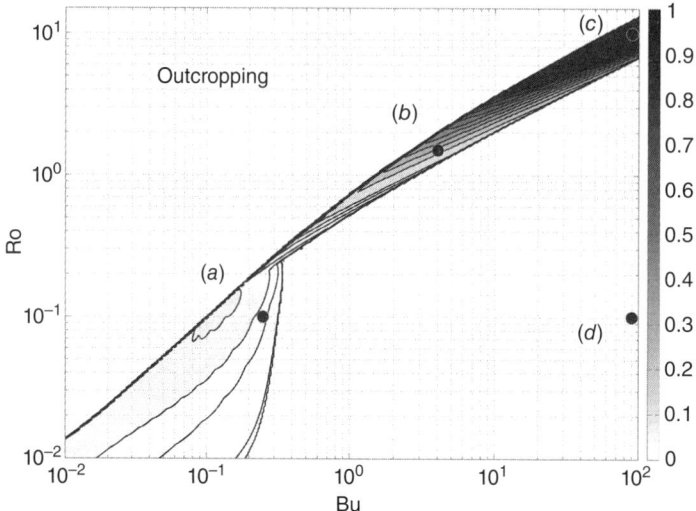

Figure 6.2. Growth rate of most unstable modes in (Ro, Bu) space. Darker zones correspond to higher growth rates. Contours displayed are 0.001, 0.01, 0.02 and further interval at 0.02. The thick upper frontier line marks the outcropping limit when the interface between the two layers intersects the bottom or the top. Adapted from *Gula et al.* [2009b].

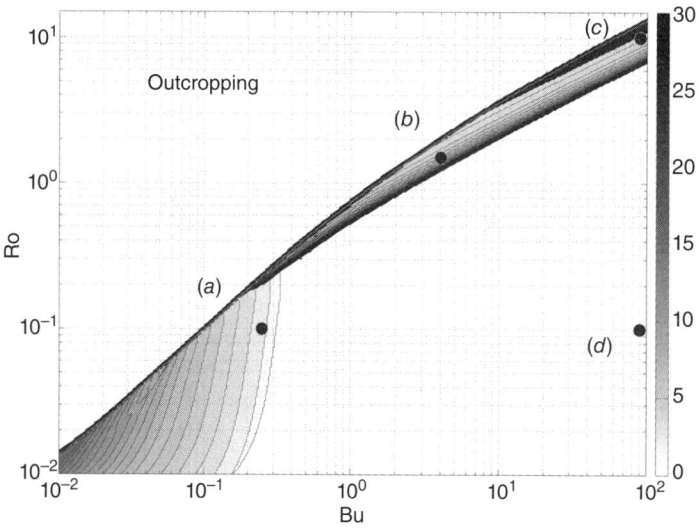

Figure 6.3. Wave number of most unstable modes in (Ro, Bu) space corresponding to Figure 6.2. Darker zones correspond to higher wave numbers. The interval between subsequent contours is 1. Adapted from *Gula et al.* [2009b].

similar to the classical Kelvin-Helmholtz (KH) instability [*Paldor and Ghil*, 1991]. Note that although KK, Kelvin-Poincaré (KP), and Poincaré-Poincaré (PP) resonances are physically different, they are frequently confused in the literature and appear under the general name of KH instability. Similarly RK and RP instabilities are often both called RK [cf. *Sakai*, 1989]. We follow this simplified convention.

In the context of wave resonances, there are three essential parameters in the problem: $V = \Delta\Omega r_0$, the velocity (or velocity shear) of the basic flow; $C_R = \Omega \Delta H / H_0 r_0$, the characteristic phase velocity of the Rossby waves; and $C_G = \sqrt{g' H_0}$, the characteristic phase velocity of the gravity waves. The interpretation of the results may be done on the basis of the alternative set of nondimensional parameters defined as

$$F^* = \frac{V}{C_G} = \frac{\Delta\Omega r_0}{\sqrt{g' H_0}},$$

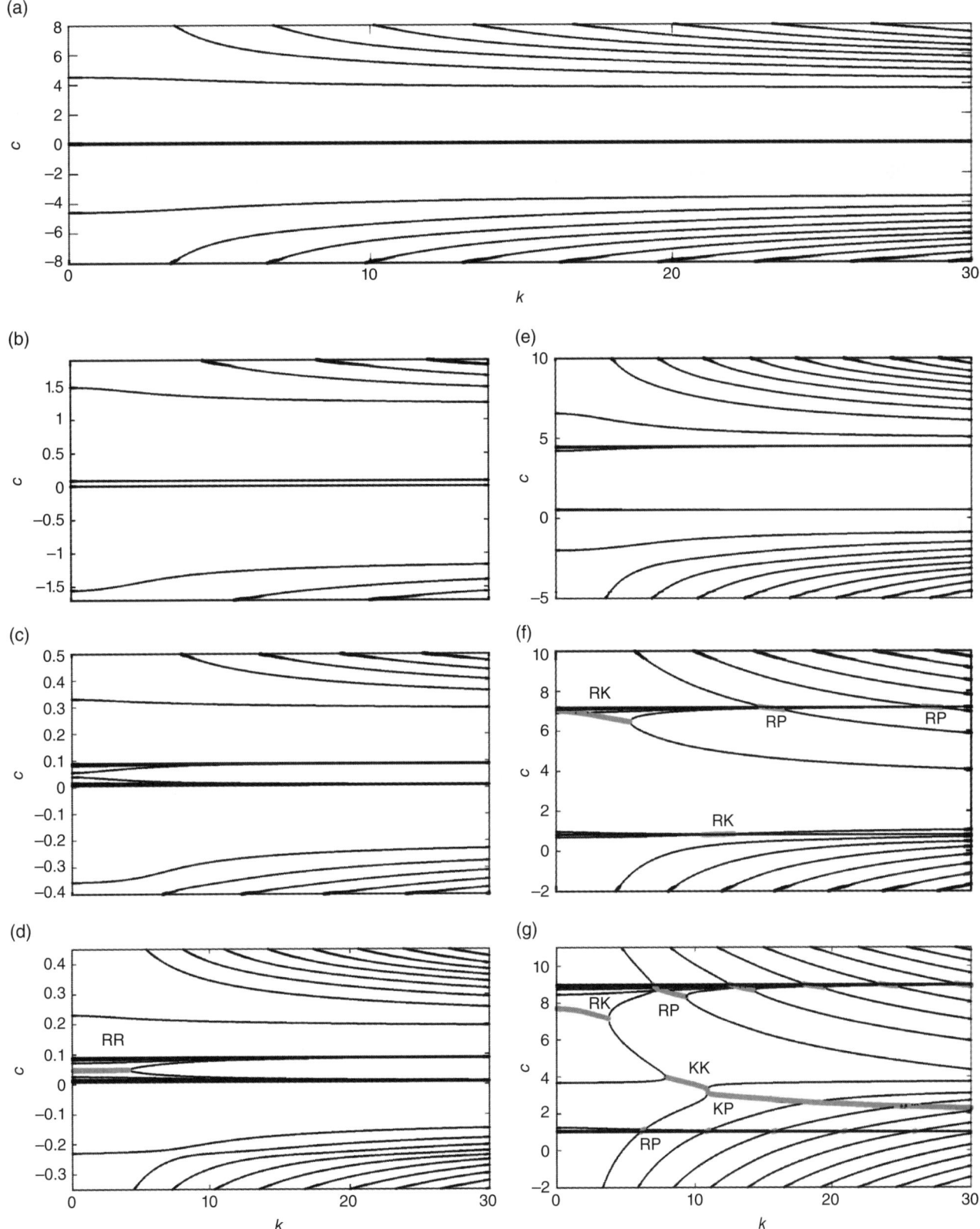

Figure 6.4. Dispersion diagram $c-k$ for the stable configuration corresponding to point (d) in Figure 6.2 (upper panel) and its evolution with the change of parameters: Ro = 0.1, Bu decreasing from top to bottom. Left panel: (a) Bu = 90, (b) Bu = 10, (c) Bu = 0.5, and (d) Bu = 0.25. Bu = 90, Ro increasing from top to bottom. Right panel: (a) Ro = 0.1, (e) Ro = 5, (f) Ro = 8, and (g) Ro = 10. Thick gray lines correspond to unstable regions; nonzero Im(c). Adapted from *Gula et al.* [2009b].

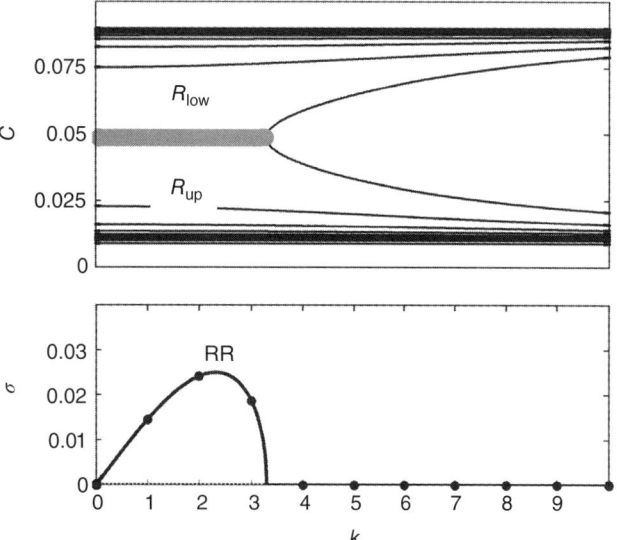

Figure 6.5. Dispersion diagram (upper panel) and growth rate (lower panel) of the eigenmodes of the superrotating lid configuration for Ro = 0.1 and Bu = 0.25 (see Figure 6.2a). Thick gray line in the upper panel corresponds to the RR resonance and respective unstable modes. Here and sunsequent figures the subscripts "low" and "up" indicate a wave in the lower and upper layer, respectively.

the Froude number, and

$$R^* = \frac{V}{C_R} = \frac{g'H}{2\Omega^2 r_0^2},$$

a new Rossby number. With these definitions one finds the baroclinic instability at small R^* and KH instabilities at large F^*, which matches the traditional view of these instabilities. However, to keep a closer link with experimental results of *Williams et al.* [2005] and *Flór et al.* [2011], the discussion below is based on Bu and Ro. (For more details, see *Gula et al.* [2009b].)

Thus, as for the ageostrophic Phillips model in a straight channel [cf. *Sakai*, 1989; *Gula et al.*, 2009a], several types of instabilities are present, namely, (a) the baroclinic instability for small values of Bu and Ro (RR resonance), (b) the Rossby-Kelvin instability (RK or RP resonance) for intermediate values of Bu and Ro, and (c) the Kelvin-Helmholtz instability (KK or KP resonance) for high values of Bu and Ro. As usual, the KH instability is characterized by highest growth rates and shortest wavelengths, the baroclinic instability is long-wave and low growth-rate, and RK instability is intermediate, although spanning a wide range of wave numbers.

In Figures 6.5–6.7 we give the dispersion diagrams corresponding to different values of (Ro, Bu) referring to typical cases (a), (b), (c), respectively, in Figures 6.2 and 6.3.

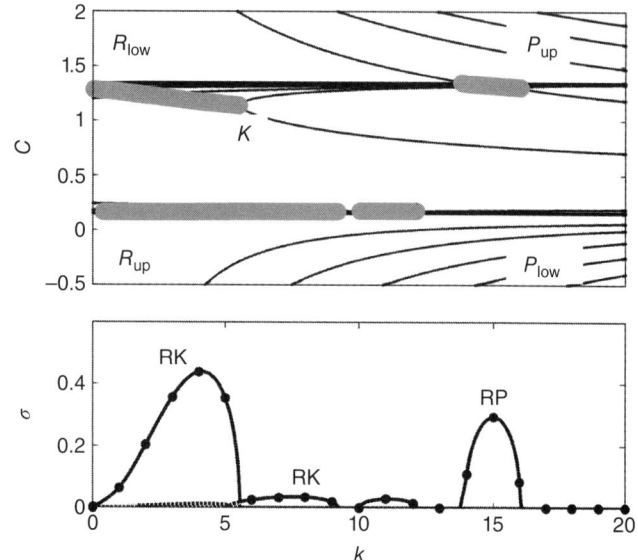

Figure 6.6. Dispersion diagram (upper panel) and growth rate (lower panel) of the eigenmodes of the superrotating lid configuration for Ro = 1.5 and Bu = 3.5 (see Figure 6.2b). Thick gray lines on the upper panel correspond to the RK and RP resonances and respective unstable modes.

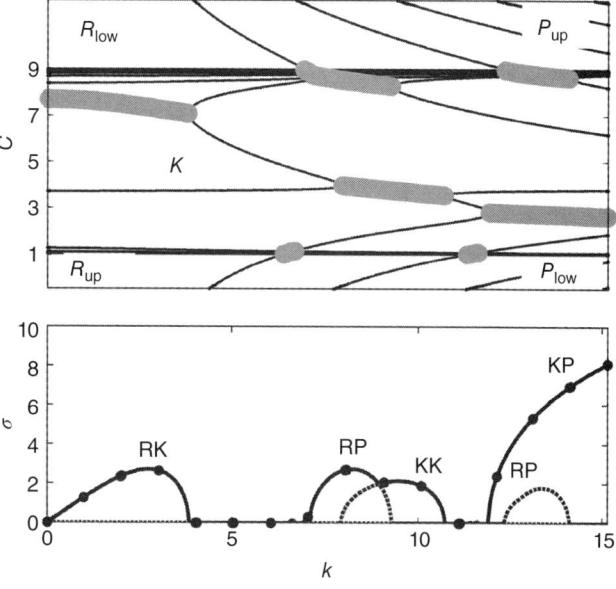

Figure 6.7. Dispersion diagram (upper panel) and growth rate (lower panel) of the eigenmodes of the superrotating lid configuration for Ro = 10 and Bu = 90 (see Figure 6.2c). Thick gray lines on the upper panel correspond to the RK, KK, RP, and KP resonances and respective unstable modes.

We present also in Figure 6.8 the structure of unstable modes in both layers and the corresponding maps of the interface deviation, which is an often measured quantity in experiments.

Figure 6.5 shows a dispersion diagram in the zone of baroclinic instability. Two Rossby waves, one propagating in each layer, are in resonance having the same Doppler-shifted phase speed and giving rise to a baroclinic instability, as explained, e.g., in *Hoskins et al.* [1985]. The structure of the unstable mode is shown in Figure 6.8a.

Figure 6.6 shows a dispersion diagram in the Rossby-Kelvin instability area. The gravest radial Rossby mode propagating in the upper layer resonates with a Kelvin wave propagating in the lower layer and gives rise to the dominant RK instability; see *Sakai* [1989] and *Gula et al.* [2009a]. Resonances of higher Rossby modes with the Kelvin wave give weaker RK instabilities, and the resonance of a lower-layer Rossby mode with a Poincaré wave gives the RP instability. The structure of the most unstable RK and RP modes is shown in Figures 6.8b and 6.8c.

Figure 6.7 shows a dispersion diagram in a KH instability area. A Kelvin wave propagating in the upper layer resonates with another Kelvin wave propagating in the lower layer and gives rise to a KH instability. For these values of parameters we can see that RP and RK instabilities are also present but with lower growth rates. The structure of an unstable KK mode is shown in Figure 6.8d.

Thus RK and KH instabilities coexist for large Bu and Ro having comparable growth rates although different characteristic wave numbers. As follows from Figure 6.7 and from the comparison of Figures 6.2 and 6.3, in general, close values of the growth rates may correspond to essentially different wavelengths of the most unstable modes. This means that different instabilities may coexist and compete.

6.3. STABILITY OF OUTCROPPING BUOYANCY-DRIVEN BOUNDARY CURRENTS

6.3.1. Equations of Motion, Basic States, Linearization and Boundary Conditions

Another configuration used in experiments with the rotating annulus is the free surface-outcropping one [*Griffiths and Linden*, 1982; *Pennel et al.*, 2012]. Note that outcropping was excluded in the analysis of the previous

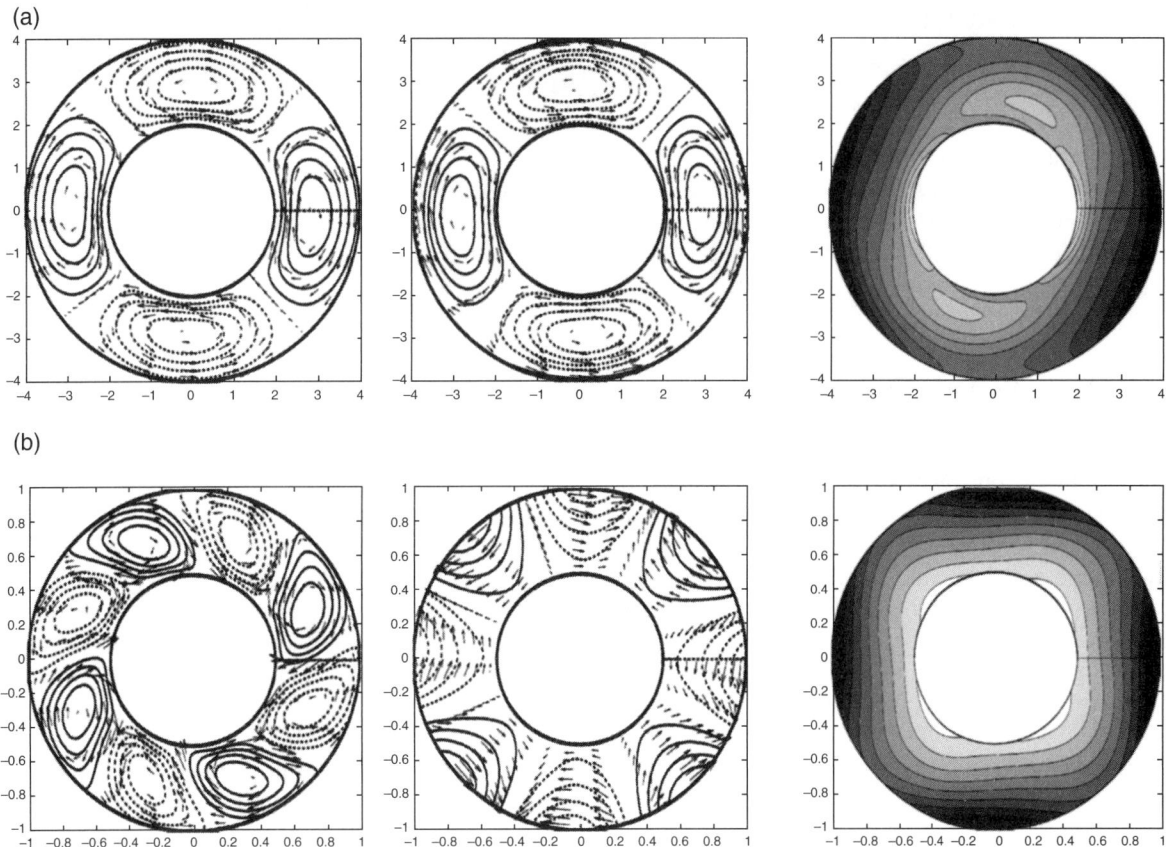

Figure 6.8. Pressure and velocity fields in the upper (left) and lower (middle) layers and interface height (right) of (a) the baroclinically unstable RR mode at $k = 2$ ($kR_d = 0.9$, see Figure 6.5), (b) the unstable RK mode at $k = 4$ ($kR_d = 7.5$, see Figure 6.6),

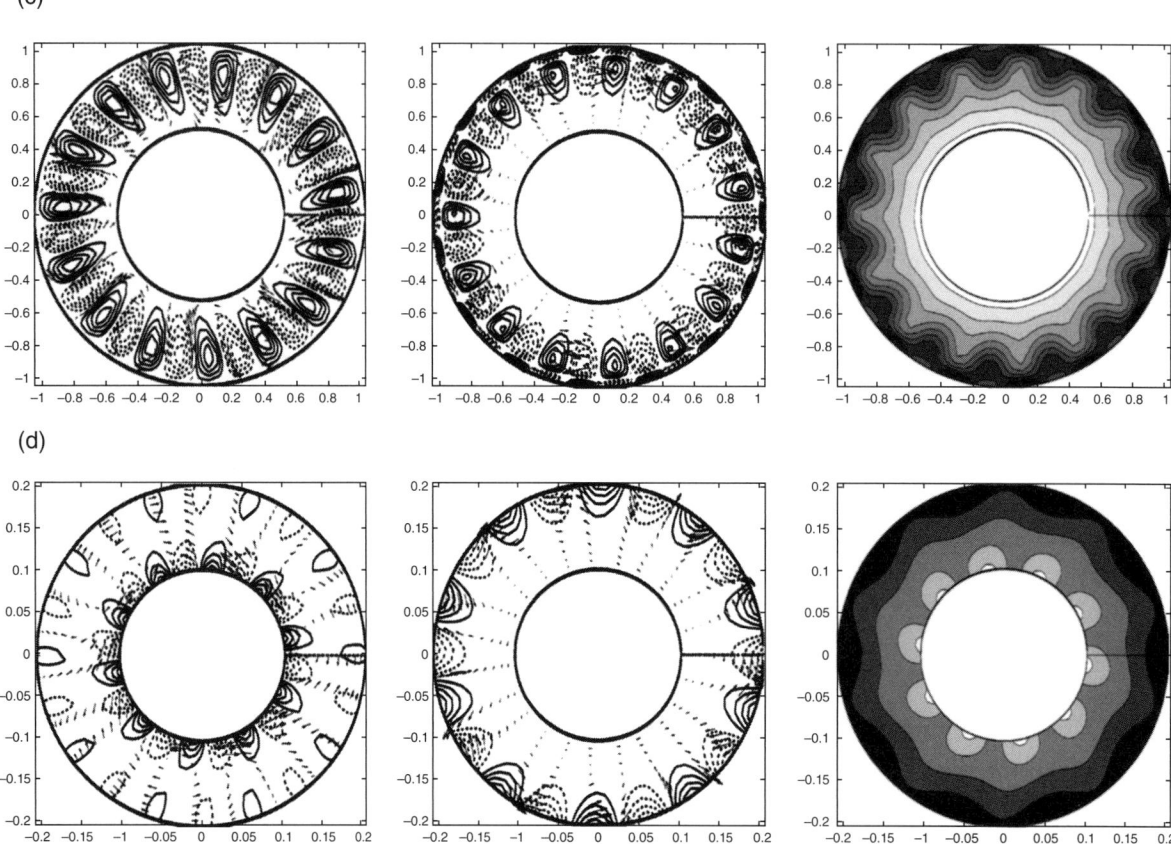

Figure 6.8 Continued. (c) the unstable RP mode at $k = 15$ ($kR_d = 28$, see Figure 6.6), and (d) the unstable KK mode at $k = 10$ ($kR_d = 95$, see Figure 6.7). The full lines correspond to positive and the dotted lines to negative values. (a) Both fields are typical of a Rossby mode. (b) The field in the upper layer is typical of a Rossby mode while the field in the lower layer is typical of a Kelvin mode. (c) The field in the upper layer is typical of a Rossby mode while the field in the lower layer is typical of a Poincaré mode. (d) Both fields are typical of a Kelvin mode.

section. So we consider now the situation where the interface between the layers joins the free surface forming a surface front, as shown in Figure 6.9. This is an idealized configuration of a buoyancy-driven coastal current in a circular basin. In the classical experiments by *Griffiths and Linden* [1982], a volume of lighter salty water of density ρ_1 flows above a denser water of density ρ_2 and is confined between the surface front and the internal cylinder. In the work of *Thivolle-Cazat and Sommeria* [2004] and *Pennel et al.* [2012], the lighter fluid flows along the external cylinder. In the following we consider an upper layer of lighter fluid of density ρ_1 with a free surface terminating at a point $r = r_0 = r_1 + L$ with mean velocity $U_1(r)$ and a lower layer of density $\rho_2 > \rho_1$ with a mean velocity $U_2(r)$.

We work with the two-layer shallow-water equations in the cylindrical geometry, as in the previous section, and perform a cylindrical equivalent of the stability analysis of *Gula and Zeitlin* [2010] and *Gula et al.* [2010] for coastal currents. Another difference with the previous section is that we now consider a free surface instead of a rigid lid for the comparison with experiments. In this section the slope of the bottom, γ, is set to be zero, its influence to be studied in the next section.

By introducing the time scale $1/f$, the horizontal scale L, which is the unperturbed width of the density current, the vertical scale $H_0 = H_1(r_1)$, and the velocity scale fL, we use nondimensional variables from now on without changing notation. Note that with this scaling the characteristic value of the velocity gives the Rossby number. By linearizing about a steady state in cyclogeostrophic equilibrium, we obtain nondimensional equations identical to equations (6.2), where the pressure perturbations in the layers π_j are now related through the layers' heights h_j via the hydrostatic relations as follows:

$$\nabla \pi_j = \frac{\mathrm{Bu}}{2s} \nabla (\delta_s^{j-1} h_1 + h_2). \qquad (6.9)$$

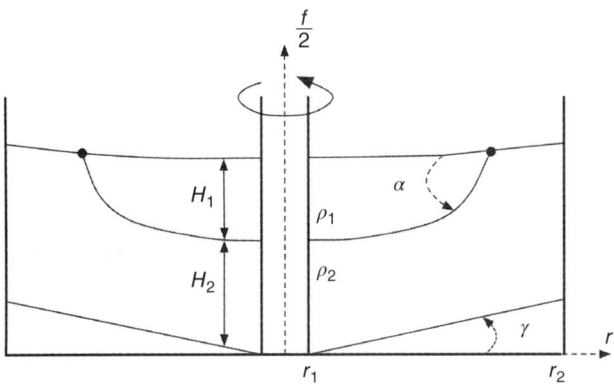

Figure 6.9. Schematic representation of a two-layer outcropping flow in the annulus with linearly sloping bottom.

Here $\delta_s = \rho_1/\rho_2$ is the density ratio, $s = (\rho_2 - \rho_1)/(\rho_2 + \rho_1)$ is the stratification parameter, and $Bu = (R_d/r_0)^2$ the Burger number.

The depth profiles $H_j(r)$ and respective velocities $V_j(r)$ in (6.2) correspond to steady cyclogeostrophically balanced states in each layer:

$$V_j + \frac{V_j^2}{r} + \frac{r}{4} = \frac{Bu}{2s} \partial_r (\delta_s^{j-1} H_1 + H_2). \quad (6.10)$$

We look for solutions harmonic in the azimuthal direction:

$$(u_j(r,\theta), v_j(r,\theta), \pi_j(r,\theta))$$
$$= (\tilde{u}_j(r), \tilde{v}_j(r), \tilde{\pi}_j(r)) \exp[ik(\theta - ct)] + c.c. \quad (6.11)$$

The boundary condition of no normal flow at the coast is the same as in the previous case for both layers, $u_j(r_1) = 0$. The boundary conditions at the front for the upper layer are

$$H_1(r) + h_1(r,\theta,t) = 0, \quad D_t L_R = v \quad \text{at } r = L_R(\theta), \quad (6.12)$$

where $r = r_1 + L$ is the location of the free streamline of the basic state, $L_R(\theta, t)$ is the position of the perturbed free streamline, and $D_t = \partial_t + u\partial_r + v/r\partial_\theta$ is the Lagrangian derivative. Physically, they correspond to the conditions that the fluid terminates at the boundary, which is a material line. The linearized boundary conditions give (1) the relation between the perturbation of the position of the free streamline and the value of the height perturbation,

$$L_R = -\frac{h_1}{H_{1r}}\bigg|_{r=r_1+L}, \quad (6.13)$$

and (2) the continuity equation evaluated at $r = r_1 + L$. Hence, the only constraint to be imposed at the front for the upper layer is the regularity of $(u_1, v_1, h_1 + h_2)$.

We also have to ensure the continuity of pressure of the lower layer across the front. In the region $r > r_1 + L$ with no upper layer, the lower layer obeys the one-layer rotating shallow-water equations with (hydrostatic) pressure proportional to the height of the fluid column. In what follows we consider an outer cylinder to be far enough from the front ($r_2 \gg r_0$) so that its influence is negligible. Moreover, below we will limit ourselves, for technical simplicity, only by the *balanced* component of π_2, which, in the leading order, in polar coordinates, satisfies the equation [cf., e.g., *Reznik et al.*, 2001]

$$\frac{1}{r}\partial_r(r\partial_r\pi_2) + \left(k^2 - \frac{1}{R_{d2}^2} - \frac{k^2}{r^2}\right)\pi_2 = 0, \quad (6.14)$$

where $R_{d2} = \sqrt{gH_2}/f$ is the Rossby deformation radius of the lower layer. We thus impose the continuity of the full solution for π_2 in the inner region, $r < r_1 + L$, with the decaying balanced solution in the outer region, $r > r_1 + L$, at $r = r_1 + L$. By this choice an unbalanced part of the one-layer flow beyond the front, consisting of freely propagating surface inertia-gravity waves, is discarded. We thus loose possible resonances of the eigenmodes of the inner flow with the outer inertia-gravity wavefield and related radiative instabilities. For small to moderate Rossby numbers, which is the case of existing experiments, and our case below, these instabilities are weak [cf. *Zeitlin*, 2008]. As we will see later, the stability analysis under these assumptions reproduces the experiments well, which gives an a posteriori justification.

Injecting (6.11) into (6.2) and (6.9), we obtain an eigenvalue problem of order 6 that can be solved by applying the spectral collocation method along the same lines as in the previous section. In what follows, we will first consider the simplest case of a bottom layer initially at rest ($U_2 = 0$) and an upper flow with a constant rotation rate $U_1 = \alpha r$.

6.3.2. Resonances and Instabilities

As in the configuration of Section 6.2, the instabilities in the outcropping case originate from resonances between the eigenmodes of the linearized problem. As in the previous section, the wave species are Poincaré (inertia-gravity) modes, Rossby modes (if PV gradients are present), and unidirectional Kelvin modes trapped at the boundary. Additional ingredients in the outcropping configuration are the frontal modes trapped in the vicinity of the free streamlines (outcropping lines). These modes are described in *Iga* [1993] as mixed Rossby-gravity waves

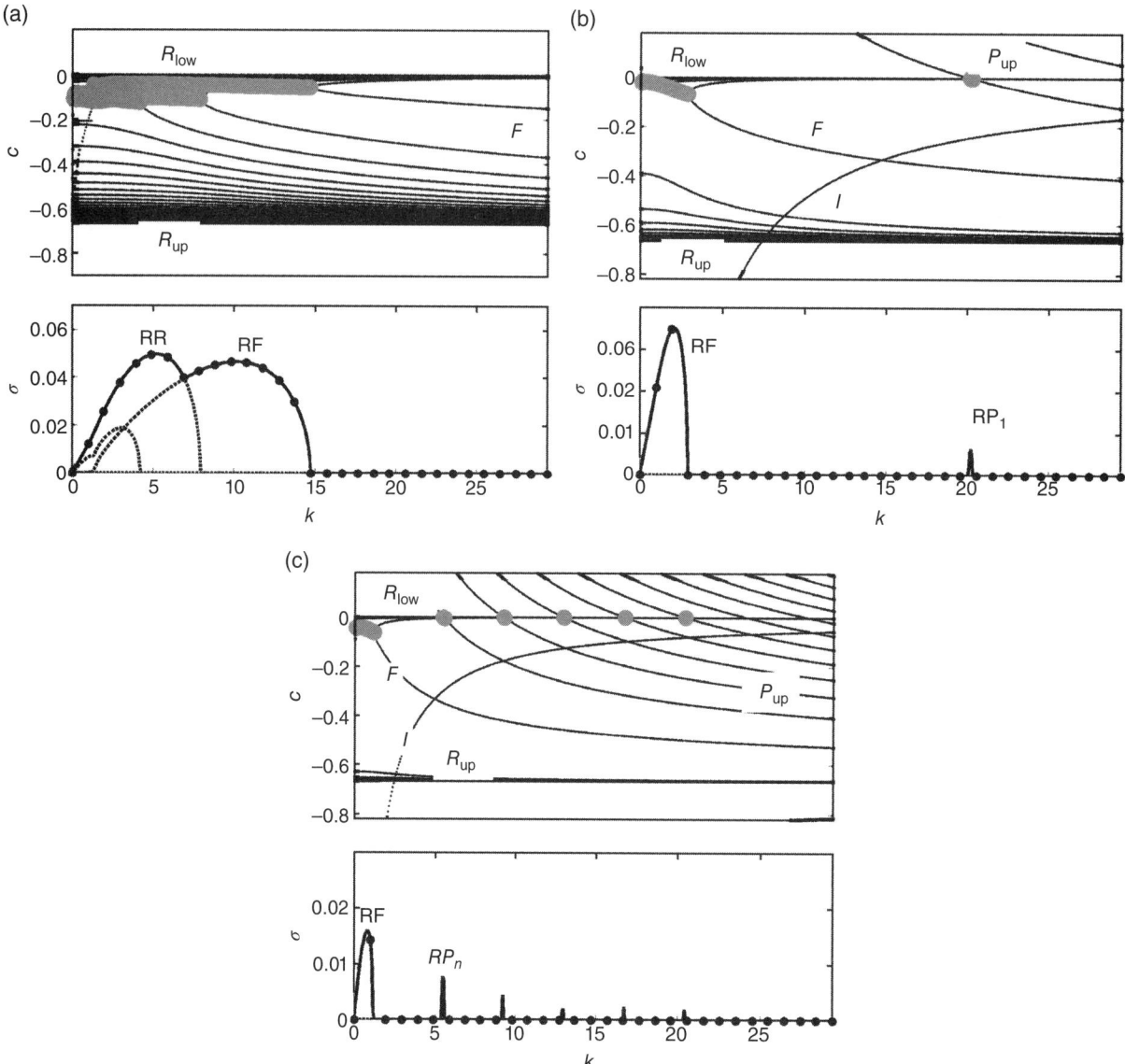

Figure 6.10. Dispersion diagram (upper panels) and growth rate (lower panels) of the eigenmodes of the outcropping configuration for (a) Ro = 0.02, (b) Ro = 0.2, and (c) Ro = 0.6 with $\delta_H = 0.1$. Thick gray lines on the upper panel correspond to the unstable modes.

in the sense that they behave like Rossby modes as long as the wave number is small and like gravity modes when the wave number becomes large. Note that *Hayashi and Young* [1987] and other authors refer to these modes as Kelvin waves, as they propagate along the (internal) boundary. More generally, the frontal mode can be interpreted as a vortical mode, as in the work of *Meacham and Stephens* [2001], *Gula and Zeitlin* [2010], and *Gula et al.* [2010], in a sense of a wave that exists due to the PV gradient at the outcropping point, because this point may be interpreted as a point connecting the finite-depth layer with a layer of infinitesimal thickness [*Boss et al.*, 1996]. We use the denomination "frontal" for such modes in what follows.

Figure 6.10 shows the dispersion diagram and corresponding growth rates of the eigenmodes of the outcropping coastal flow with a depth ratio $\delta_H = H_1(r_1)/[H_1(r_1) + H_2(r_1)] = 0.1$ and a density ratio $\delta_s = \rho_1/\rho_2 = 0.99$ as a function of k for increasing values of vertical shear: Ro = 0.02, Ro = 0.2, and Ro = 0.6.

For low Rossby and Burger numbers (Figure 6.10a), Rossby modes in the lower layer with $c \approx U_2 = 0$ can resonate with Rossby modes in the upper layer with $c \approx -U_0$. This is the standard mechanism of the baroclinic instability, as explained in the previous section, which occurs for wave numbers $kR_d < 1$. The corresponding pressure and velocity fields in both layers are plotted in Figure 6.11a. Both fields are typical of a Rossby mode. Rossby modes in the lower layer can also resonate with the frontal mode in the upper layer (RF interaction). The frontal mode has the characteristics of a Rossby wave for low wave numbers, and the unstable mode under consideration is therefore very similar to the classical baroclinic instability (RR). The corresponding pressure and velocity fields in both layers are plotted in Figure 6.11b.

For higher Rossby and Burger numbers (Figures 6.10b and 6.10c), the Rossby-Rossby interaction is not allowed anymore as the horizontal extension of the surface current is too small compared to the Rossby deformation radius. The RF mode is the primary unstable mode with wave numbers $kR_d \approx 0.5 \div 1$. The second instability in Figure 6.10b corresponds to the first Poincaré mode in the upper layer resonating with a Rossby wave in the lower layer. The pressure and velocity fields for this mode are plotted in Figure 6.11c, and confirm this interpretation. Note that the same instability appears at higher k for Poincaré modes of higher order with decreasing growth rates (Figure 6.10c). This is the RP instability.

A new dispersion curve $kc = 1$ denoted by I in the Figures 6.10b and 6.10c also appears. It corresponds to inertial motion in the lower layer, with the quiescent upper layer. The absence of pressure variations is typical for inertial oscillations. This mode was already discussed by *Paldor and Ghil* [1991] and *Gula et al.* [2010]. In spite of intersections of this curve with other branches of the dispersion diagram, no resonances and hence no instabilities between the inertial motion and other modes arise due to its pressureless character. Indeed, pressure fluctuations are required for the instability to arise [*Cairns*,

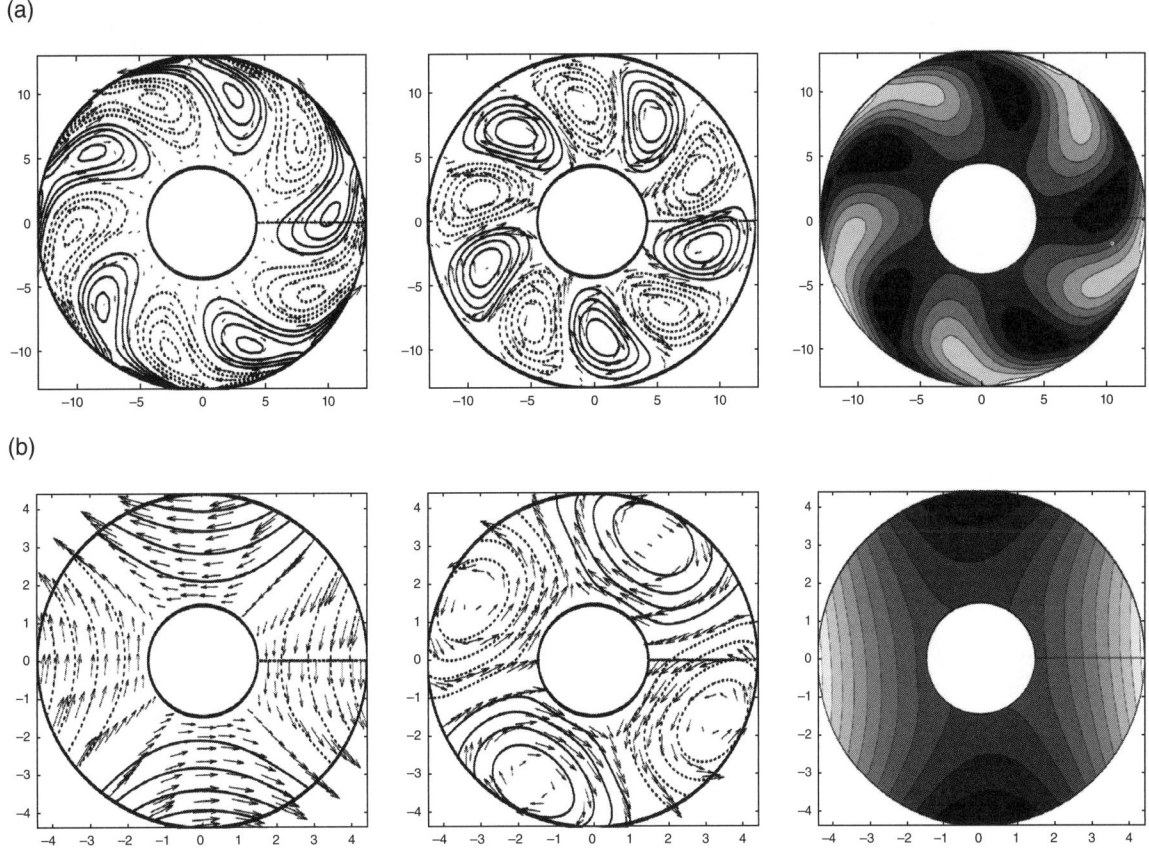

Figure 6.11. Pressure and velocity fields of the upper (left) and lower (middle) layers and interface height (right) of (a) the unstable RR mode for Ro = 0.02 and $k = 5$ ($kR_d = 0.6$, see Figure 6.10a), (b) the unstable Rossby-frontal (RF) mode for Ro = 0.2 and $k = 2$ ($kR_d = 0.6$, see Figure 6.10b),

130 MODELING ATMOSPHERIC AND OCEANIC FLOWS

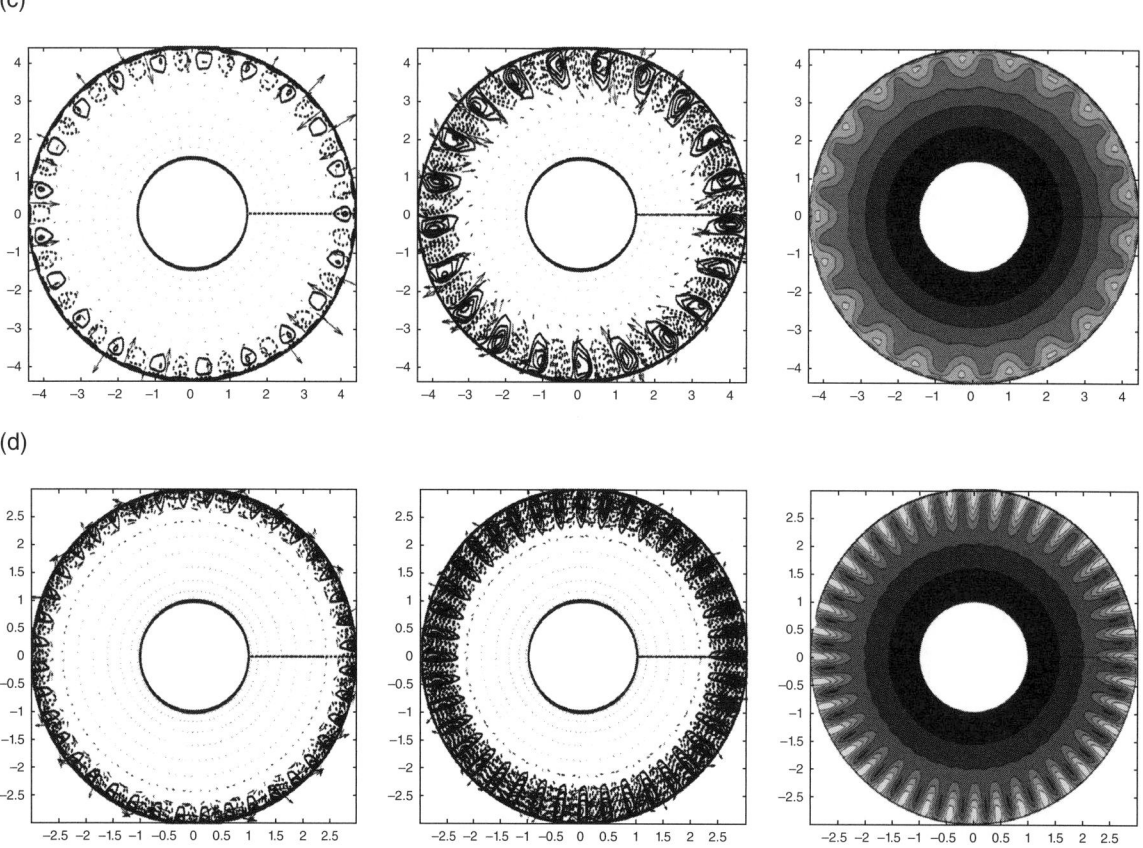

Figure 6.11 Continued. (c) the unstable RP mode at $k = 21$ ($kR_d = 7.5$, see Figure 6.12b), and (d) the unstable frontal-Poincaré mode at $k = 40$ ($kR_d = 23$, see Figure 6.10c). Full lines: positive; dotted lines: negative values.

1979] as the coupling between the layers, necessary for the resonance between upper and lower layer waves, is achieved through pressure (cf. the instability criterion in *Sakai* [1989]).

It should be emphasized that resonances involving the Kelvin mode at the inner boundary, in contradistinction with the previous section and the results of *Gula and Zeitlin* [2010] and *Gula et al.* [2010], are never of significant impact in the present configuration, where the vertical shear at the inner cylinder is small compared to the shear at the location of the front.

Figure 6.12 shows the dispersion diagram and corresponding growth rates for a larger depth ratio $\delta_H = H_1(r_1)/[H_1(r_1) + H_2(r_1)] = 0.5$ and a density ratio $s = \rho_1/\rho_2 = 0.99$. Comparison of Figures 6.10 and 6.12 shows that the unstable modes become more and more vigorous when the depth ratio increases. For high Rossby numbers, a new zone of instabilities with high growth rates appears at very high wave numbers. They are due to the interaction of the frontal mode in the upper layer with various Poincaré modes in the lower layer (FP_1). Such short-wave instability is analogous to the one described by *Paldor and Ghil* [1991] for the zero-PV case and by *Gula et al.* [2010] for the constant-PV case in the planar geometry. The frontal mode having the characteristics of a gravity wave for high wave numbers, this instability is therefore very similar to the vertical shear instabilities, which we have seen previously (KK, KP, or PP). The pressure and velocity fields for the first Poincaré mode in the lower layer and the frontal mode in the upper layer are plotted in Figure 6.11d.

Barotropic interactions in the upper layer, as studied by *Gula and Zeitlin* [2010] in the rectilinear case, are not present in this analysis due to the absence of both horizontal potential vorticity gradient and current reversal in the upper layer, which would allow Rossby-Frontal or Kelvin-Frontal barotropic interactions, respectively. These interactions are usually absent in experimental studies, as mentioned above, due to the lack of surface forcing, which would allow for stronger horizontal shear in the equilibrium state. Experiments realized through geostrophic adjustment of a

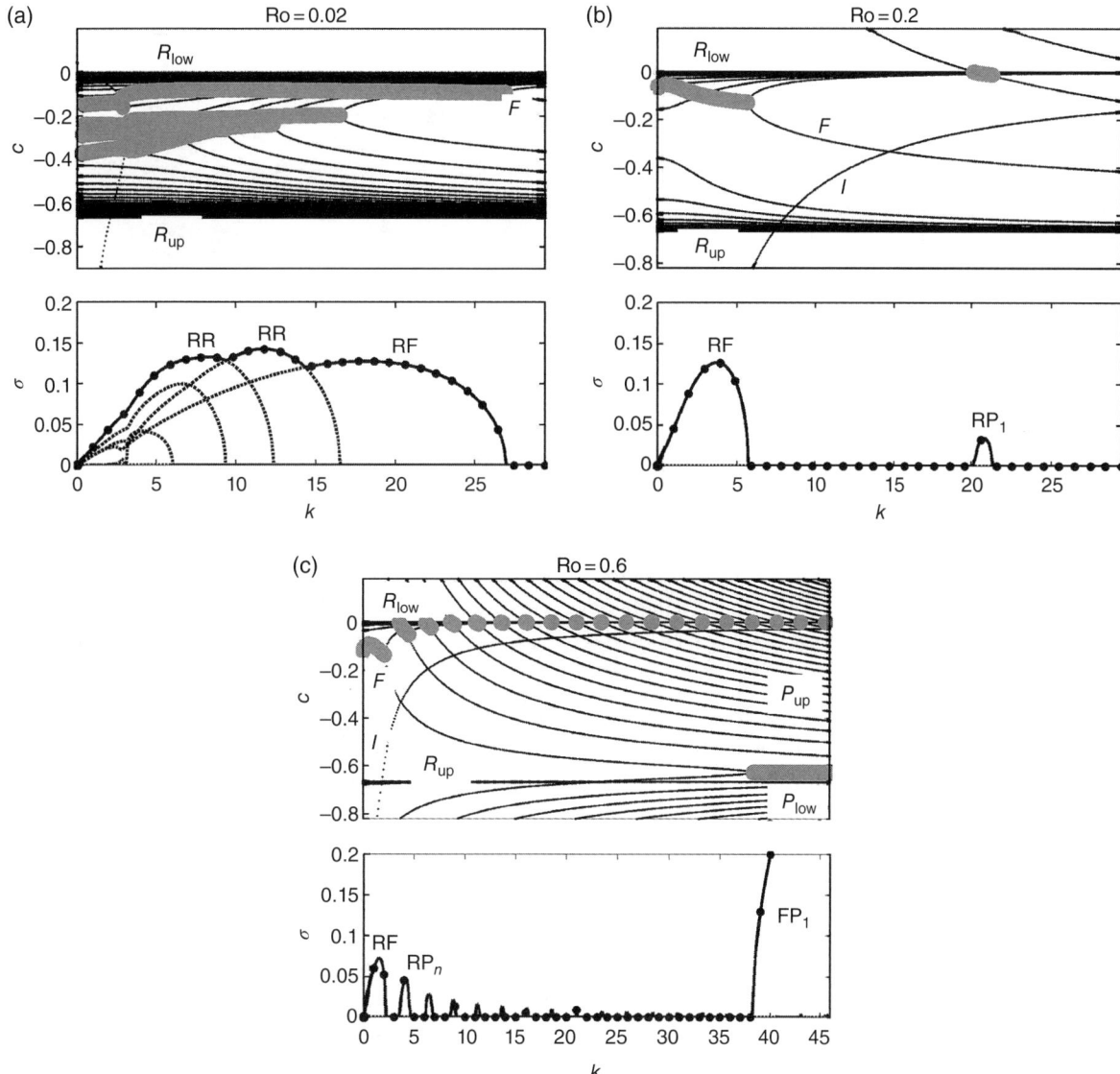

Figure 6.12. As in Figure 6.10, but with $\delta_H = 0.5$. Thick gray lines in the upper panel correspond to the unstable modes.

constant-PV layer will, in general, have low Rossby and Burger numbers and, therefore, correspond to the RF regime.

6.3.3. Comparison with Experiments

The experiments of *Griffiths and Linden* [1982] were conducted in a circular tank mounted on a rotating turntable, as in Figure 6.9, filled with a solution of density ρ_2. The boundary current was created by injecting a lighter solution of density ρ_1 between the inner cylinder and a bottomless cylinder of radius r_c such as $r_1 < r_c < r_2$. The experiment was then initiated by vertically withdrawing this cylinder and allowing the upper layer (height h_0 and width $L_0 = r_c - r_1$) to move under the influence of buoyancy, Coriolis, and centrifugal forces.

The upper layer is stationary before the geostrophic adjustment and therefore has constant potential vorticity. Under the assumption of no diabatic mixing taking place during the collapse, potential vorticity in the upper layer should be conserved in the final balanced state and is then written as

$$Q_1 = \frac{f + \partial_r V_1 + V_1/r}{H_1} = \frac{f}{h_0}. \qquad (6.15)$$

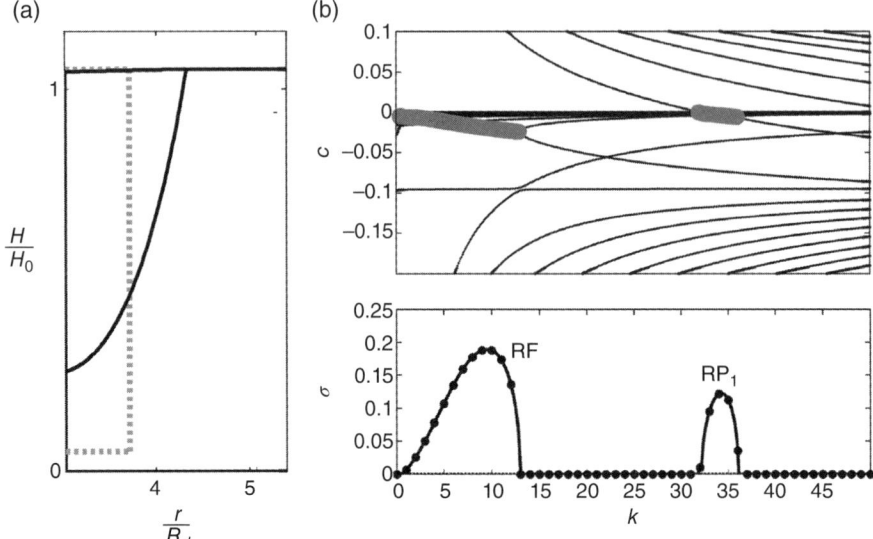

Figure 6.13. (a) Initial experimental state (dashed line) and basic state height after reaching cyclo-geostrophic equilibrium (thick line) for $F_0 = 1.4$ and $\delta_0 = 0.95$. (b) Dispersion diagram (upper panel) and growth rate (lower panel) of the eigenmodes. Thick gray lines in the upper panel correspond to the RF and RP resonances and respective unstable modes.

The steady cyclogeostrophically balanced state in each layer is given by (6.10). Assuming a bottom layer initially at rest in the rotating frame, the basic state velocity in the upper layer is then given by the solution of the following ordinary differential equation (ODE):

$$V_{1rr} + \frac{V_{1r}}{r} - \frac{V_1}{r^2} - \frac{f}{h_0 g(1-\delta_s)}\left(fV_1 + \frac{V_1^2}{r}\right) = 0,$$
(6.16)

with

$$V_{1r}(r_1) = 0, \qquad H_1(r_1 + L) = 0. \qquad (6.17)$$

We have solved (6.16) numerically using Runge-Kutta method. The results of these calculations for two sets of parameters are shown in Figures 6.13a and 6.15a.

We use the parameters of *Griffiths and Linden* [1982], $F_0 = f^2 L_0^2/g' h_0$, the Froude number, and $\delta_0 = h_0/H$, the initial depth ratio. We plot for comparison two cases corresponding to (a) $F_0 = 1.4$ and $\delta_0 = 0.95$ (the dispersion diagram is shown in Figure 6.13 and the structure of the most unstable mode in Figure 6.14) and (b) $F_0 = 14.4$ and $\delta_0 = 0.19$ (the dispersion diagram is shown in Figure 6.15 and the structure of the most unstable mode in figure 6.16). Photographs of the corresponding laboratory experiments (Figures 3 and 4, respectively in [*Griffiths and Linden*, 1982]) are reproduced in Figures 6.14d and 6.16d, and one can see a perfect agreement with the observed wave numbers for these experiments.

In both cases the most unstable wave number corresponds to the RF mode. The Rossby eigenmodes are absent in the upper layer owing to the uniformity of PV. As seen from the comparison with equation (6.10), the corresponding set of dispersion curves $c \approx -U_0$ is absent in Figures 6.13 and 6.15. However, as was already discussed in the previous section, the frontal mode has the characteristics of a Rossby wave for low wave numbers, and the RF mode is therefore very similar to the classical baroclinic instability.

Photographs of the laboratory experiments (Figure 6.14d and 6.16d) show the instability at a later nonlinear stage as compared to the initial linearly growing stage computed by linear analysis. The nonlinear evolution of the RF instability in the rectilinear two-layer shallow-water model was simulated by *Gula et al.* [2010]. The frontal disturbances were observed to evolve in agreement, modulo rectilinear geometry, with the sequence of photographs of *Griffiths and Linden* [1982], and ultimately led to the formation and detachment of outward propagating cyclone-anticyclone vortex pairs, as observed in the work of *Griffiths and Linden* [1982] and *Thivolle-Cazat and Sommeria* [2004].

The stability analysis has been repeated for different values of the inner cylinder radius r_1 (not shown) and reproduces the results of *Griffiths and Linden* [1982], demonstrating a small influence of this parameter and hence of the wall in such a configuration. It is interesting to note that the RF instability was interpreted, even for

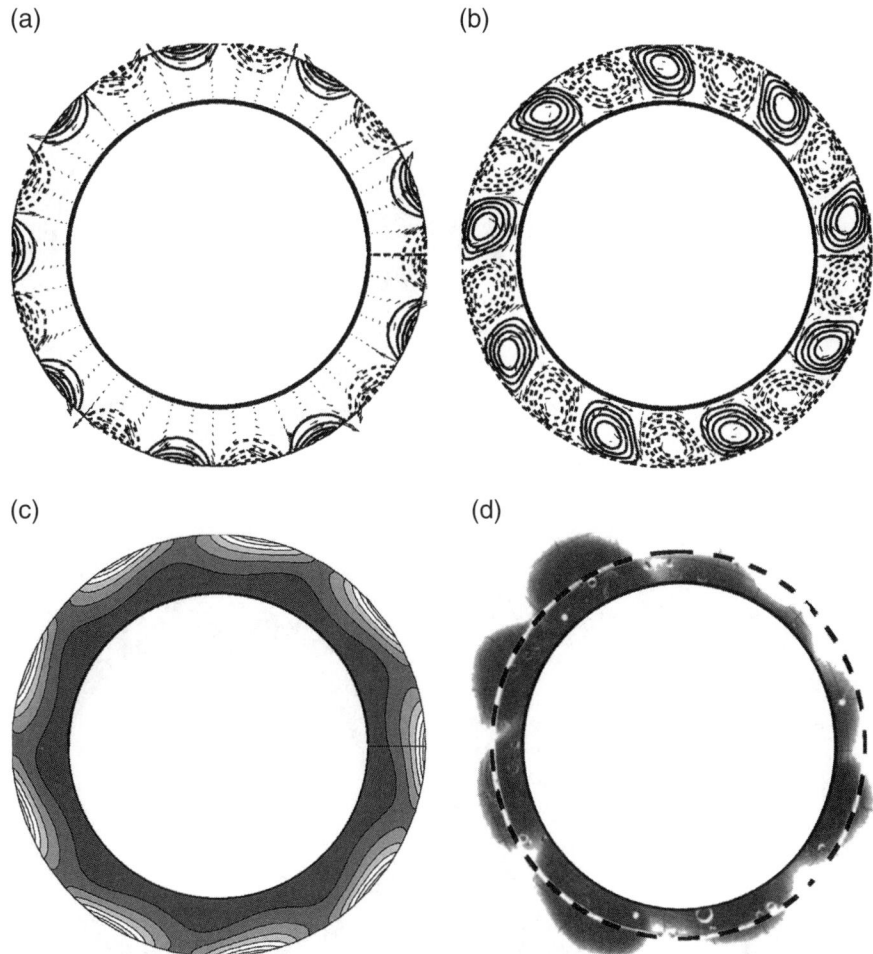

Figure 6.14. Pressure and velocity fields of (a) upper and lower (b) layers and (c) interface height of the baroclinically unstable mode at $k = 9$ ($kR_d = 14.5$, see Figure 6.13). Full lines: positive; dotted lines: negative values. The field in the upper layer is typical for a frontal mode, while the field in the lower layer is typical for a Rossby mode. (d) Photograph of the corresponding experiment adapted from *Griffiths and Linden* [1982].

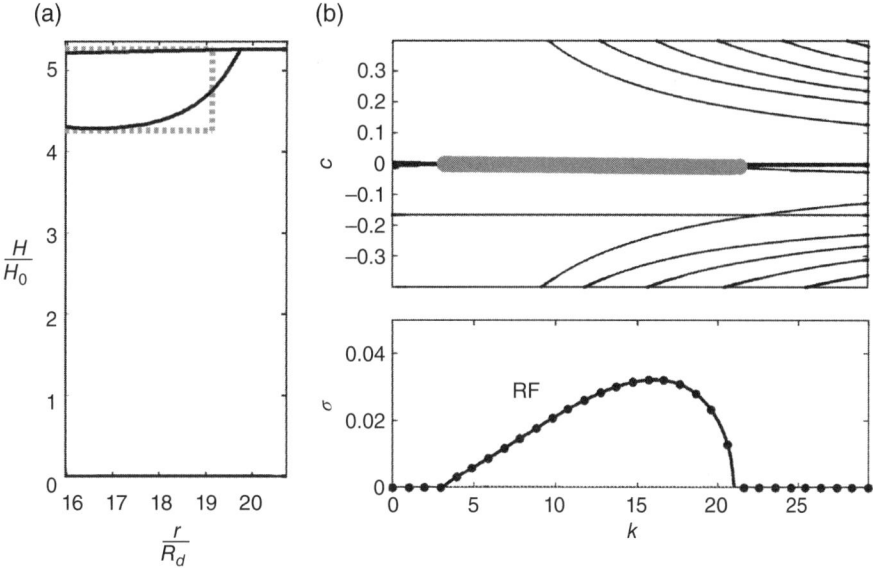

Figure 6.15. (a) Initial experimental state (dashed line) and basic state height after reaching cyclo-geostrophic equilibrium (thick line) for $F_0 = 14.4$ and $\delta_0 = 0.19$. (b) Dispersion diagram (upper panel) and growth rate (lower panel) of the eigenmodes. Thick gray lines in the upper panel correspond to the RF and RP resonances and respective unstable modes.

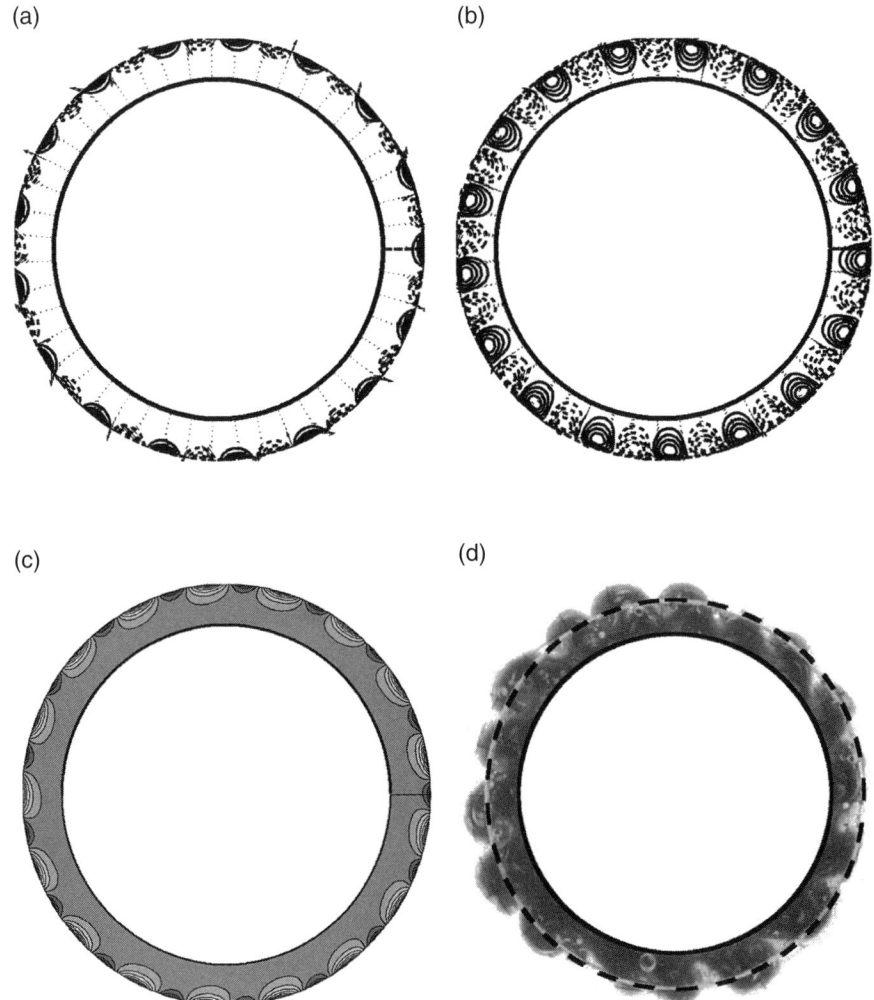

Figure 6.16. Pressure and velocity fields of the upper (a) and lower (b) layers and (c) interface height of the baroclinically unstable mode at $k = 17$ ($kR_d = 5$, see Figure 6.15). Full lines: positive; dotted lines: negative values. The field in the upper layer is typical for a frontal mode, while the field in the lower layer is typical for a Rossby mode. (d) Photograph of the corresponding experiment adapted from *Griffiths and Linden* [1982].

small Rossby numbers, as ageostrophic by many authors, because the uniform PV in the upper layer does not satisfy the Charney-Stern theorem of the PV gradient inversion between the two layers. Yet, it is still possible to interpret it as generalized quasi-geostrophic instability; see a discussion of this point by *Boss et al.* [1996].

6.4. IMPACT OF BATHYMETRY ON INSTABILITIES

In this section we study the impact of bathymetry in a form of a constant-slope shelf on the stability of the flow. We use the same set of equations and parameters as in the previous section with the addition of a bottom topography with height $H_t(r)$. Following *Pennel et al.* [2012], we define a topography parameter To as the ratio of the shelf slope γ to the isopycnal slope α,

$$\text{To} = \frac{\gamma}{\alpha}, \quad (6.18)$$

where α is defined as the slope of the interface between the layers at the location of the front. The parameter To has been found relevant for quantifying the shelf impact on the surface current (*Pennel et al.* [2012]), as was previously suggested by works in the quasi-geostrophic Phillips model. Positive values of To, as in Figure 6.9, correspond to isopycnal and shelf slopes in the same direction, which is typical of upwelling events along the coast of western boundary currents. Negative values of To correspond to

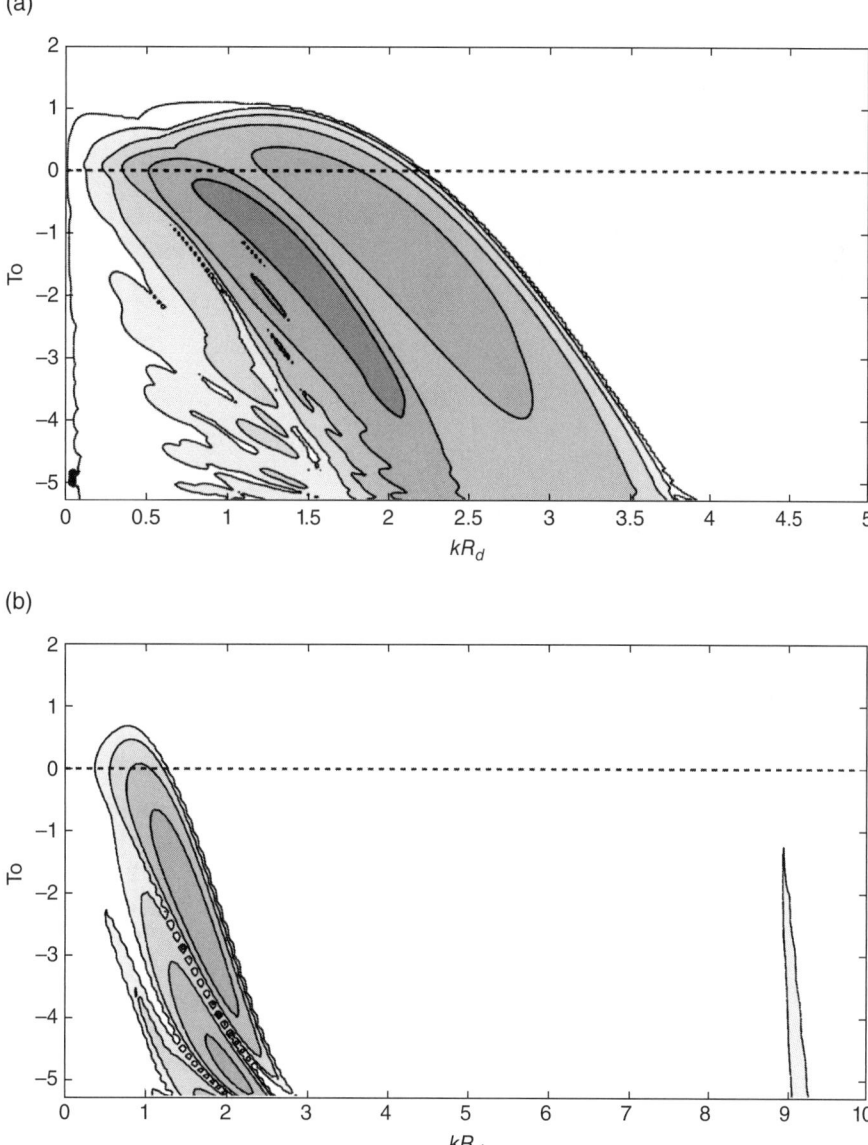

Figure 6.17. Growth rate of the most unstable modes as a function of the topography parameter To for (a) Ro = 0.02 and (b) Ro = 0.2 with $\delta_H = 0.1$

isopycnal and shelf slopes in the opposite directions and are typical of buoyant coastal currents.

In order to study the influence of the topography on the stability of the current and to vary the parameter To without changing other parameters, we will define the aspect ratio between the two layers as $\delta H = H_1(r_1)/H_2(r_1 + L)$ and keep it constant while varying the topography height.

The novelty of the configuration with nontrivial topography is that the latter allows for specific topographic waves, in addition to frontal, Kelvin, Rossby, and Poincaré waves discussed above. These waves may resonate with other types of waves and thus lead to new instabilities. On the other hand, topography changes the propagation speed of these waves and may thus "detune" the resonances, leading to stabilization of the flow. We observe both effects, depending on To.

Figure 6.17 shows the growth rates of the most unstable modes as a function of To and nondimensional wave number for the set of parameters used in the dispersion diagrams of Figures 6.10a, 6.10b, and 6.12c. In all cases there is a strong stabilization of the flow for a positive To, with growth rates vanishing for To close to 1

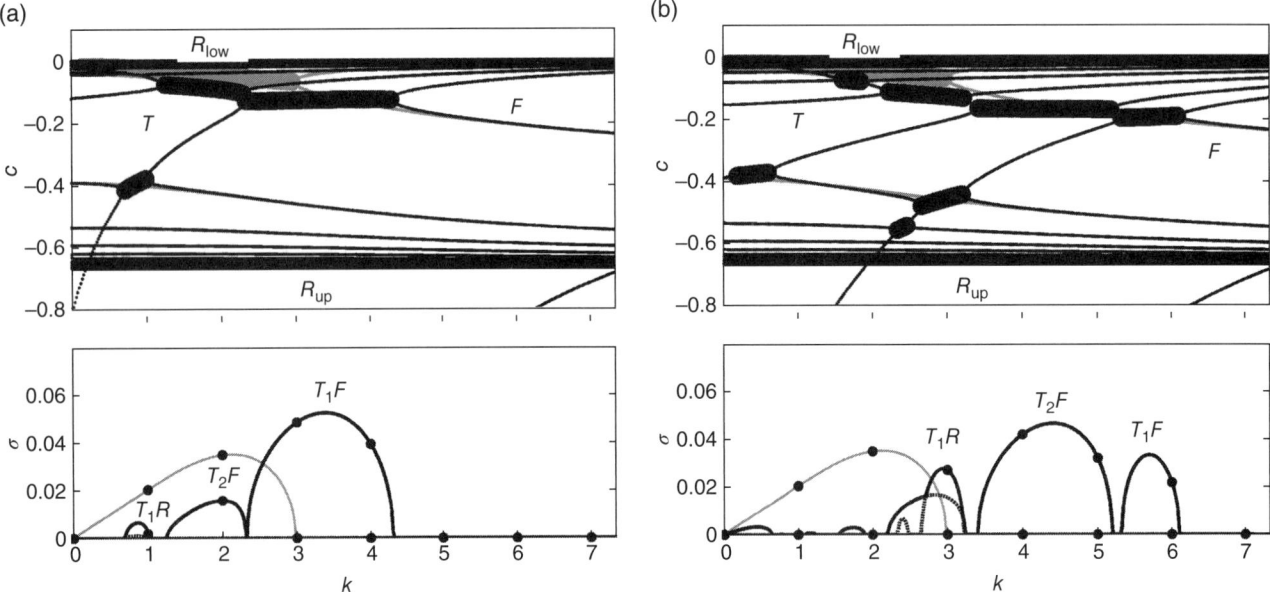

Figure 6.18. Dispersion diagram (upper panel) and growth rate (lower panel) of the eigenmodes for: (a) Ro = 0.2 and To = −2 and (b) To = −5. Thick black lines in the upper panel correspond to the TF and TR resonances and respective unstable modes. Gray lines in both panels correspond to the same case with no topography (Figure 6.10b).

(topography almost parallel to the interface). It is clear from the expressions for the PV gradient in the lower layer,

$$\partial_r Q_2 = -f \frac{\partial_r H_2}{H_2^2}, \quad (6.19)$$

and for the topography parameter,

$$\text{To} = \left. \frac{\partial_r H_t}{\partial_r H_2 + \partial_r H_t} \right|_{r_1+L}, \quad (6.20)$$

that To → 1 will imply $\partial_r Q_2$ → 0, which explains the stabilization.

In Figure 6.18 we present the stability diagrams at two different negative values of To at Ro = 0.2 showing that the most unstable modes in such configurations are due to the resonances of a frontal wave with a topographic wave (either the first T_1 or the second T_2 radial mode), although the resonances of the first topographic mode with a Rossby wave are also observed. Figure 6.19 displays the corresponding unstable modes. Thus, topography plays a crucial role in the destabilization of the flow in this regime.

6.5. SUMMARY AND DISCUSSION

We thus performed a stability analysis of the shear flows in the rotating annulus in the framework of the two-layer shallow-water model in the wide range of parameters, both in the superrotating rigid-lid and free-surface outcropping configurations, including topography effects in the latter case. We got a detailed structure of the unstable modes to be compared with the experimental results in these configurations. Such a comparison shows a good agreement with the density-current experiments. The experiments with two-layer fluid with superrotating lid do show the RK instability as follows from the analysis of *Flór et al.* [2011], with good quantitative agreement with two-layer shallow-water results, while short-wave structures on the background of Rossby waves, which were observed in experiments of *Williams et al.* [2005] and *Flór* [2007]), do not find a direct explanation in terms of dominant unstable modes that we found. This means that these short-wave structures are probably due to the fine vertical structure of the interface between the layers, which does not exist in the shallow-water approximation (see a discussion by *Flór et al.* [2011] on this subject as well as Chapter 11 in this volume). Our results on the influence of topography on the instabilities show that in the case of opposite orientations of the isopycnal and shelf slopes the destabilization of density currents is due to resonances of Rossby and topographic waves, and thus the influence of topography may be crucial.

Acknowledgments. The work of V. Zeitlin was supported by the French ANR grant SVEMO.

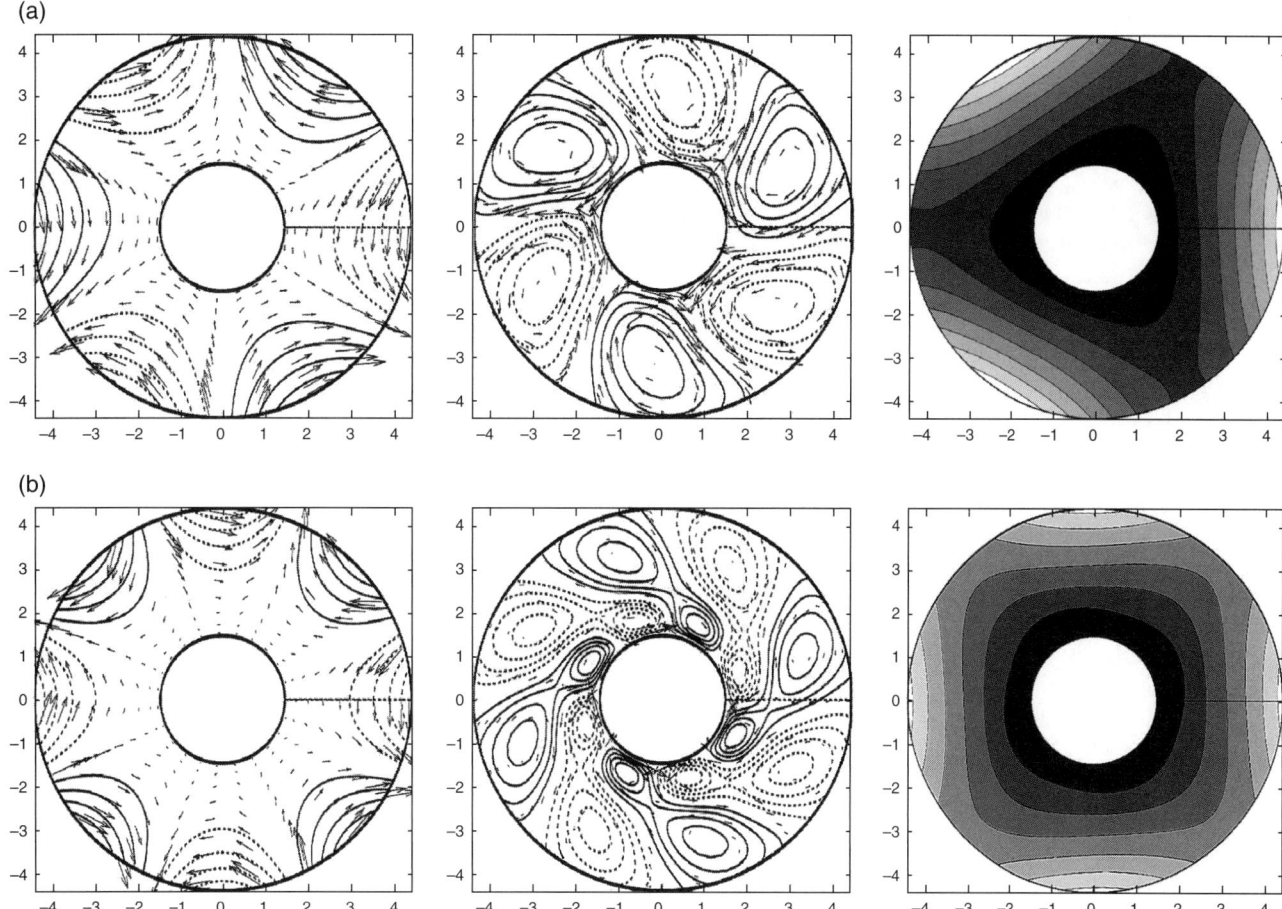

Figure 6.19. Pressure and velocity fields of the upper (left) and lower (middle) layers and interface height (right) of the most unstable mode for Ro = 0.2 and (a) To = −2 (see Figure 6.18a) and (b) To = −5 (see Figure 6.18b). Full lines: positive; dotted lines: negative values.

REFERENCES

Boss, E., N. Paldor, and L. Thompson (1996), Stability of a potential vorticity front: From quasi-geostrophy to shallow water, *J. Fluid Mech.*, *315*, 65–84.

Cairns, R. A. (1979), The role of negative energy waves in some instabilities of parallel flows, *J. Fluid Mech.*, *92*, 1–14.

Dunkerton, T., P. Lelong, and C. Snyder (Eds.) (2008), Spontaneous imbalance, *J. Atmos. Sci.*, special collection.

Flór, J.-B. (2007), Frontal instability, inertia-gravity wave radiation and vortex formation, in *Proceedings of the 18th CFM Conference, Grenoble, France*.

Flór, J.-B., H. Scolan, and J. Gula (2011), Frontal instabilities and waves in a differentially rotating fluid, *J. Fluid Mech.*, *685*, 532–542.

Ford, R. (1994), Gravity wave radiation from vortex trains in rotating shallow water, *J. Fluid Mech.*, *281*, 81–118.

Fultz, D., R. R. Long, G. V. Owens, W. Bohan, R. Kaylor, and I. Weil (1959), Studies of thermal convection in a rotating cylinder with some implications for large-scale atmospheric motions, *Met. Monogr.*, *4*, 1–104.

Griffiths, R. W., and P. F. Linden (1982), Part I: Density-driven boundary currents. *Geophys. Astrophys. Fluid Dyn.*, *19*, 159–187.

Griffiths, R. W., P. D. Killworth, and M. E. Stern, (1982), Ageostrophic instability of ocean currents, *J. Fluid Mech.*, *117*, 343–377.

Gula, J., and V. Zeitlin (2010), Instabilities of buoyancy driven coastal currents and their nonlinear evolution in the two-layer rotating shallow water model. Part I. Passive lower layer, *J. Fluid Mech.*, *659*, 69–93.

Gula, J., R. Plougonven, and V. Zeitlin, (2009a), Ageostrophic instabilities of fronts in a channel in the stratified rotating fluid, *J. Fluid Mech.*, *627*, 485–507.

Gula, J., V. Zeitlin, and R. Plougonven (2009b), Instabilities of two-layer shallow-water flows with vertical shear in the rotating annulus, *J. Fluid Mech.*, *638*, 27–47.

Gula, J., V. Zeitlin, and F. Bouchut (2010), Instabilities of buoyancy driven coastal currents and their nonlinear evolution in the two-layer rotating shallow water model. Part II. Active lower layer, *J. Fluid Mech.*, *665*, 209–237.

Hart, J. E. (1972), A laboratory study of baroclinic instability, *Geophys. Astrophys. Fluid Dyn,.* *3*, 181–209.

Hayashi, Y.-Y. and W. R. Young (1987), Stable and unstable shear modes of rotating parallel flows in shallow water, *J. Fluid Mech., 184*, 477–504.

Hide, R. (1958), An experimental study of thermal convection in a rotating liquid, *Phil. Trans. R. Soc. London Ser. A Math. Phys. Sci. 250*(983), 441–478.

Hide, R. and W. W. Fowlis (1965), Thermal convection in a rotating annulus of liquid: Effect of viscosity on the transition between axisymmetric and non-axisymmetric flow regimes, *J. Atmos. Sci. 22*, 541–558.

Hoskins, B. J., M. E. McIntyre, and A. W. Robertson (1985), On the use and significance of isentropic potential vorticity maps, *Q. J. R. Meteorol. Soc., 111*(470), 877–946.

Iga, K. (1993), Reconsideration of Orlanski's instability theory of frontal waves, *J. Fluid Mech., 255*, 213–236.

Lovegrove, A. F., P. L. Read, and C. J. Richards, (2000), Generation of inertia-gravity waves in a baroclinically unstable fluid, *Q. J. R. Meteorol. Soc., 126*, 3233–3254.

Meacham, S. P., and J. C. Stephens (2001), Instabilities of gravity currents along a slope, *J. Phys. Oceanogr., 31*, 30–53.

O'Sullivan, D., and T. J. Dunkerton (1995), Generation of inertia-gravity waves in a simulated life cycle of baroclinic instability, *J. Atmos. Sci., 52*(21), 3695-3716.

Paldor, N., and M. Ghil, (1991), Shortwave instabilities of coastal currents, *Geophys. Astrophys. Fluid Dyn., 58*, 225–241.

Pennel, R., A. Stegner, and K. Beranger (2012), Shelf impact on buoyant coastal current instabilities, *J. Phys. Oceanogr., 42*, 39–61.

Reznik, G. M., V. Zeitlin, and M. Ben Jelloul (2001), Nonlinear theory of geostrophic adjustment. Part 1. Rotating shallow-water model, *J. Fluid Mech., 445*, 93–120.

Ripa, P. (1983), General stability conditions for zonal flows in a one-layer model on the β-plane or the sphere, *J. Fluid Mech., 126*, 463–489.

Sakai, S. (1989), Rossby-Kelvin instability: A new type of ageostrophic instability caused by a resonance between Rossby waves and gravity waves, *J. Fluid Mech., 202*, 149–176.

Thivolle-Cazat, E., and J. Sommeria (2004), in *Shallow Flows*, edited by G. H. Jirka and W. S. J. Uijttewaal, pp. 23–30, An experimental investigation of a boundary current instability, Taylor and Francis Group, London.

Trefethen, L. N. (2000), *Spectral Methods in Matlab.*, SIAM, Philadelphia, Penna.

Williams, P. D., P. L. Read, and T. W. N. Haine (2004), A calibrated, non-invasive method for measuring the internal interface height field at high resolution in the rotating, two-layer annulus, *Geophys. Astrophys. Fluid Dyn., 98*, 453–471.

Williams, P. D., T. W. N. Haine, and P. L. Read (2005), On the generation mechanisms of short-scale unbalanced modes in rotating two-layer flows with vertical shear, *J. Fluid Mech., 528*, 1–22.

Zeitlin, V. (2008), Decoupling of balanced and unbalanced motions and inertia-gravity wave emission: Small versus large Rossby numbers, *J. Atmos. Sci., 65*, 3528–3542.

7
Laboratory Experiments on Flows Over Bottom Topography

Luis Zavala Sansón[1] and Gert-Jan van Heijst[2]

7.1. INTRODUCTION

The motions of mesoscale and large-scale geophysical flows in the oceans and the atmosphere (with sizes of the order of tens to hundreds of kilometers) are essentially affected by Earth's rotation. At these scales geophysical flows can be also strongly influenced by bottom topography. Thus, the essential dynamical ingredients to be discussed here are topography effects in a rotating system. More specifically, we shall present recent experimental studies on homogeneous flows simulating geophysical situations, in which the dynamics are dominated by the combination of rotation and topography. We will also discuss the two-dimensionality of shallow, nonrotating fluid systems. The presentation and discussion of the topic are based on the physical scales of different experimental facilities, in a similar vein to the study of *von Arx* [1957]. In contrast with that study, we focus the discussion on numerous examples that underline the role of topography effects in laboratory experiments with horizontal scales L ranging from some centimeters to a few meters and vertical scales H from some milimeters to about 1 m. In addition, we ignore stratification effects, gravity waves, and, to a great extent, free surface effects. It will be shown that laboratory experiments performed in large containers are useful to study rather different problems than those performed within containers with smaller dimensions. In other words, we aim at pointing out advantages and disadvantages of experiments performed within the available range of length scales in typical laboratories.

Several examples presented in subsequent sections address specific oceanographic or atmospheric problems. For instance, the presence of a weak, linear topographic slope in a rotating tank simulates the planetary β-effect (the latitudinal gradient of the Coriolis parameter [see, e.g., *van Heijst*, 1994]. Using this effect, it is possible to study the self-propagation mechanism of monopolar vortices in the laboratory. The drift of vortices under the β-plane approximation is characteristic of mesoscale eddies in the ocean (see *Vukovich* [2007] for loop current eddies in the Gulf of Mexico) and in the atmosphere (e.g., *Chan* [2005] for tropical cyclones). During their drift, vortices may encounter topographic features that modify their structure or trajectory [see, e.g., *Zehnder*, 1993; *de Steur and van Leeuwen*, 2009]. Another important topic is the vertical advection of chemical and biological material due to mesoscale features in the ocean. Several mechanisms introducing or extracting nutrients into/from the upper ocean surface have been recently discussed by several authors [*Lévy*, 2008], and in particular due to topographic effects [*Genin*, 2004]. Some other relevant geophysical phenomena over topography, such as the propagation of topographic Rossby waves, the self-organization properties of quasi-two-dimensional turbulence, and the dynamics of shallow flows, can be further studied by means of laboratory experiments.

Previous reviews discussing flow-topography phenomena are the studies of *Hopfinger and van Heijst* [1993], *Boyer and Davies* [2000], and *van Heijst and Clercx* [2009]. Those works provide a modern and complete description of several flows subject to different topographic scenarios. They also describe the similarity parameters that provide a physical justification to compare experimental results with oceanographic or atmospheric observations as well as a review of different experimental methods for flow measurement and visualization. A classical reference to the dynamics of rotating flows is the book of *Greenspan* [1968].

In Section 7.2 we briefly review some typical characteristics of the modeling of geophysical flows in rotating fluid containers as well as some aspects of (nonrotating)

[1]*Departamento de Oceanografía Física, CICESE, Ensenada, Baja California, México.*
[2]*Department of Applied Physics, Eindhoven University of Technology, Eindhoven, The Netherlands.*

Modeling Atmospheric and Oceanic Flows: Insights from Laboratory Experiments and Numerical Simulations,
First Edition. Edited by Thomas von Larcher and Paul D. Williams.
© 2015 American Geophysical Union. Published 2015 by John Wiley & Sons, Inc.

shallow-layer flows. Sections 7.3–7.5 contain several examples of laboratory experiments performed in facilities with very different horizontal and vertical sizes. Finally, Section 7.6 presents some concluding remarks.

7.2. THEORY AND EXPERIMENTAL BACKGROUND

7.2.1. Essential Balance in a Rotating System

The motion of a homogeneous fluid in a steadily rotating tank presents a remarkable characteristic: When the rotation axis is aligned with gravity, fluid motion is predominantly horizontal, i.e., in a plane perpendicular to the rotation axis. This implies that fluid motion takes place in the form of vertical columns that remain always parallel to the angular velocity of the system. This phenomenon was predicted by S. S. Hough since 1897 [*Gill*, 1982, p. 506] and by *Proudman* [1916], and it has been observed and reported in numerous laboratory studies since the early experiments of *Taylor* [1917, 1921].

In order to illustrate this behaviour, consider a homogeneous, incompressible fluid with density ρ and kinematic viscosity ν moving in a steadily rotating system with constant angular velocity Ω. Fluid motion is governed by the Navier-Stokes equations and by the continuity equation representing conservation of mass:

$$\frac{\partial \mathbf{u}}{\partial t} + \mathbf{u} \cdot \nabla \mathbf{u} + 2\Omega \times \mathbf{u} = -\frac{1}{\rho}\nabla P + \nu \nabla^2 \mathbf{u}, \quad (7.1)$$

$$\nabla \cdot \mathbf{u} = 0, \quad (7.2)$$

where $\mathbf{u}$ and P are the velocity field and pressure, respectively. (The pressure is actually the reduced pressure $p - \rho\Phi$, with p the mechanical pressure and Φ an effective potential representing conservative forces per unit mass. This potential usually contains both the gravitational potential Φ_g and the centrifugal potential Φ_c whose gradient is the centrifugal acceleration $\nabla\Phi_c \equiv -\Omega \times (\Omega \times \mathbf{r})$, with $\mathbf{r}$ being the position vector.)

Consider the system's angular velocity as $\mathbf{\Omega} = \mathbf{k}\Omega$, with the rotation axis along direction $\mathbf{k}$. The governing equations can be written in nondimensional form by introducing a length L, a time L/U, with U a characteristic velocity, and considering the pressure scale as $2\Omega UL$:

$$\epsilon\left(\frac{\partial \mathbf{u}'}{\partial t'} + \mathbf{u}' \cdot \nabla \mathbf{u}'\right) + \mathbf{k} \times \mathbf{u}' = -\nabla P' + E\nabla^2 \mathbf{u}', \quad (7.3)$$

$$\nabla \cdot \mathbf{u}' = 0. \quad (7.4)$$

The nondimensional numbers in (7.3) are the Rossby number $\epsilon = U/fL$, where $f = 2\Omega$ is the Coriolis parameter, and the Ekman number $E = \nu/fL^2$. The Rossby number measures the importance of the advective acceleration U^2/L compared with the Coriolis acceleration fU. Alternatively, the Rossby number can be interpreted as the ratio of the inertial period and the advective time scale $\epsilon = f^{-1}/(L/U)$; in this sense, a small Rossby number refers to "slow" motions with respect to the rotation period. The Ekman number E estimates the importance of viscous effects with respect to Coriolis accelerations. The relative vorticity of the velocity field is defined as $\boldsymbol{\omega} = \nabla \times \mathbf{u}$ and scales as U/L. The nondimensional vorticity equation is

$$\epsilon\left(\frac{\partial \boldsymbol{\omega}'}{\partial t} + \mathbf{u}' \cdot \nabla \boldsymbol{\omega}'\right) = (\mathbf{k} + \epsilon\boldsymbol{\omega}') \cdot \nabla \mathbf{u}' + E\nabla^2 \boldsymbol{\omega}'. \quad (7.5)$$

Since rotation effects are fundamental in mesoscale and large-scale geophysical flows, rotating tank experiments are usually designed in such a way that the Rossby and the Ekman numbers are typically very small, i.e., $\epsilon \ll 1$ and $E \ll 1$. Considering quasi-steady motions and ignoring nonlinear and viscous effects, it is verified from (7.3) that the flow is nearly in geostrophic balance,

$$\mathbf{k} \times \mathbf{u}' \sim -\nabla P'. \quad (7.6)$$

For this particular balance the vorticity equation (7.5) is reduced to

$$\mathbf{k} \cdot \nabla \mathbf{u}' \sim 0, \quad (7.7)$$

which is the celebrated Taylor-Proudman theorem. To facilitate the physical understanding of (7.6) and (7.7), consider a laboratory fluid tank steadily rotating about the z' axis of a Cartesian coordinate frame (x', y', z'). Both the angular velocity vector and the constant-gravity vector are aligned in the vertical direction. It is straightforward to show that the horizontal momentum equations indicate that the horizontal accelerations are balanced by the pressure gradients, and the z' equation expresses that the flow is so slow that it can be considered to remain in hydrostatic balance. The vorticity equation (7.7) implies

$$\frac{\partial \mathbf{u}'}{\partial z'} \sim 0, \quad (7.8)$$

that is, each component of the velocity (u', v', w') is independent of the coordinate parallel to the axis of rotation.

The most striking consequence of the Taylor-Proudman theorem is that the horizontal divergence is zero, because $\partial w'/\partial z' = 0$ in the continuity equation. This condition states that there is no divergence or convergence of fluid in any plane (x', y') perpendicular to the axis of rotation. If the vertical velocity component is zero at some level, for instance the solid bottom tank, then it is zero for all z'. In this case the flow is purely two-dimensional, and the motion takes place in the form of columns, always parallel to the rotation axis. Such a visually attractive effect is easily observed in rotating tank experiments by adding dye to the flow. What happens when these fluid columns experience a change of depth?

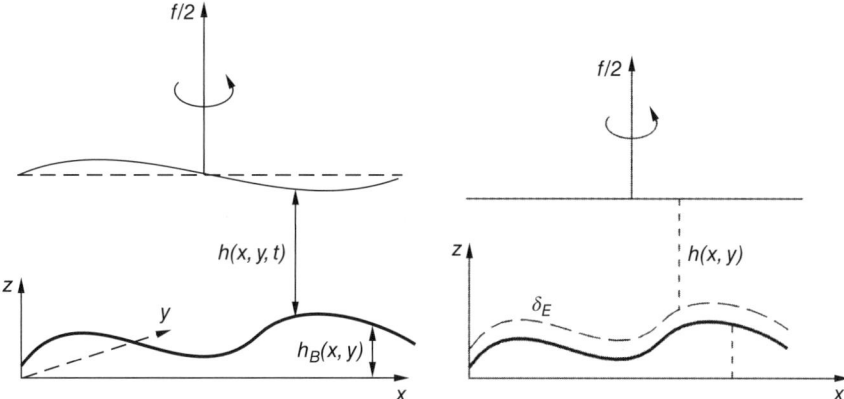

Figure 7.1. Left: Schematic view of a homogeneous fluid layer over spatially variable topography in a rotating system. Right: Bottom Ekman layer with thickness δ_E and the geostrophic interior with depth $h(x, y)$ under the rigid-lid approximation. Typically $\delta_E \ll h(x, y)$ in laboratory experiments.

7.2.2. Topography Effects

The principal effect of topography on the dynamics of rotating flows is by stretching or squeezing of vertical fluid columns. It is assumed that the horizontal flow field (u, v) is independent of the vertical coordinate z, in line with the discussion in the previous section. In dimensional terms it means that $u \equiv u(x, y, t)$ and $v \equiv v(x, y, t)$. In addition, the flow remains in hydrostatic balance in the vertical direction. These assumptions allow the formulation of the flow dynamics in terms of the vertical component of the relative vorticity $\omega = \partial v/\partial x - \partial u/\partial y$. Taking the curl of the equations of motion (7.1), the evolution equation for ω is

$$\frac{\partial \omega}{\partial t} + u\frac{\partial \omega}{\partial x} + v\frac{\partial \omega}{\partial y} + \left(\frac{\partial u}{\partial x} + \frac{\partial v}{\partial y}\right)(\omega + f) = \nu \nabla^2 \omega. \quad (7.9)$$

Now it is required to find the horizontal velocity components in terms of the relative vorticity in order to close the system. For this purpose the continuity equation (7.2) is integrated in the vertical direction over the full layer depth, i.e., over $h_B \leq z \leq h + h_B$, where $h(x, y, t)$ is the layer depth and $h_B(x, y)$ describes the spatially variable bottom topography (Figure 7.1, left). Note that h contains the free surface elevation associated with the flow itself, which may be time dependent. In the discussion below it is necessary to consider the rigid-lid approximation, which consists of neglecting the temporal variations of h when compared with changes associated with topographic variations. This approximation filters out gravity waves as if the top surface was flat, and the surface elevation is no longer a dependent variable. In addition, a thin Ekman layer of thickness $\delta_E \equiv (2\nu/f)^{1/2}$ is considered at the solid bottom (Figure 7.1, right). Depending on the boundary conditions imposed, the z integration of the continuity equation may lead to different formulations of quasi-two-dimensional flows over topography. Before presenting a complete dynamical model, we discuss first the main inviscid and viscous contributions separately.

1. *Inviscid Topography Effects.* Ignoring the Ekman layer, the vertical integration of the continuity equation (using kinematical conditions at the surface and at the bottom) implies that the horizontal divergence is given by material changes of fluid columns:

$$\frac{\partial u}{\partial x} + \frac{\partial v}{\partial y} = -\frac{1}{h}\frac{Dh}{Dt}. \quad (7.10)$$

Inserting this expression in the inviscid form of (7.9) yields the material conservation of potential vorticity $q = (\omega + f)/h$:

$$\frac{Dq}{Dt} = 0. \quad (7.11)$$

This property indicates that the relative vorticity of a fluid column will change when experiencing depth variations due to stretching/squeezing effects on fluid columns. As a column moves toward a deep region, it is stretched and gains positive vorticity. In contrast, fluid motions toward shallower regions imply squeezing effects and the production of negative vorticity. Since most of laboratory experiments are characterized by weak viscous effects, this dynamical behavior is fundamental when studying the effects of variable topography.

2. *Viscous Topography Effects.* The main effect of the viscous Ekman boundary layer at the bottom is to slow down the flow motion. This behavior arises as a consequence of the Ekman pumping-suction condition between the thin boundary layer and the rest of the fluid column. Essentially, the Ekman theory states that the flow inside the Ekman layer induces a nonzero vertical velocity [see, e.g., *Pedlosky*, 1987]. By means of this mechanism, fluid is exchanged between the Ekman layer and

the interior geostrophic region. The Ekman condition is expressed by the vertical velocity on top of the Ekman layer, which is proportional to the relative vorticity of the interior flow: $w|_{z=\delta_E} = \delta_E \omega/2$. Using this condition as the bottom boundary condition when the continuity equation is vertically integrated (together with the rigid-lid approximation), it is easily verified that

$$\frac{\partial u}{\partial x} + \frac{\partial v}{\partial y} = \frac{1}{2} E^{\frac{1}{2}} \omega, \quad (7.12)$$

with $E^{\frac{1}{2}} = \delta_E/H$. Thus, the horizontal divergence is produced by the entrainment or detrainment of fluid from the Ekman layer and is proportional to ω. When the flow has positive relative vorticity, the Ekman layer pumps fluid into the geostrophic interior domain, whereas fluid columns with negative vorticity imply a flow into the Ekman layer. In both cases the relative vorticity ω in the interior decays by stretching and squeezing effects, respectively. By using this result in the vorticity equation (7.9), and neglecting all nonlinear terms and lateral friction (in order to isolate the bottom damping effects), it is found that

$$\frac{\partial \omega}{\partial t} = -\frac{1}{2} E^{\frac{1}{2}} f \omega. \quad (7.13)$$

Thus the relative vorticity decay induced by the Ekman layer is exponential, $\omega \propto \exp(-t/T_E)$, where the decay rate defines the characteristic Ekman time scale

$$T_E = \frac{2}{f E^{\frac{1}{2}}} \equiv \frac{H}{(\nu\Omega)^{\frac{1}{2}}}. \quad (7.14)$$

7.2.3. Quasi-Two-Dimensional Models

Inviscid and viscous topography effects can be incorporated in a single formulation as derived by *Zavala Sansón and van Heijst* [2002]. Considering both effects, the z integration of the continuity equation gives the horizontal divergence as the sum of their contributions:

$$\frac{\partial u}{\partial x} + \frac{\partial v}{\partial y} = -\frac{1}{h} \frac{Dh}{Dt} + \frac{\delta_E}{2h} \omega, \quad (7.15)$$

where the kinematic boundary condition is used at the free surface and the Ekman pumping-suction condition is used at the lower boundary. The key point is to use the rigid-lid approximation $\partial h/\partial t \sim 0$, which implies that now $h \equiv h(x,y)$. This allows the definition of a transport function ψ from (7.15) such that, up to $O[(\delta_E/h)^2]$, the horizontal velocity components are

$$u = \frac{1}{h}\left(\frac{\partial \psi}{\partial y} - \frac{\delta_E}{2h}\frac{\partial \psi}{\partial x}\right), \quad v = \frac{1}{h}\left(-\frac{\partial \psi}{\partial x} - \frac{\delta_E}{2h}\frac{\partial \psi}{\partial y}\right). \quad (7.16)$$

Inserting these expressions together with the horizontal divergence (7.15) in the evolution equation (7.9) yields:

$$\frac{\partial \omega}{\partial t} + J(q,\psi) - \frac{\delta_E}{2h}\nabla\psi \cdot \nabla q = \nu\nabla^2\omega - \frac{\delta_E}{2h}\omega(\omega+f) \quad (7.17)$$

where J is the Jacobian operator. The relative vorticity is written in terms of the transport function:

$$\omega = -\frac{1}{h}\nabla^2\psi + \frac{1}{h^2}\nabla h \cdot \nabla\psi + \frac{\delta_E}{h^3}J(h,\psi). \quad (7.18)$$

This model is essentially a shallow-water formulation with a rigid lid. The inviscid version (omitting all viscous terms) and its properties are clearly explained by *Grimshaw et al.* [1994]. A more conventional formulation consists of considering topographic variations much smaller than the total fluid depth. This is the quasi-geostrophic approximation model, which can be derived by writing the fluid depth as $h(x,y) = H - \Delta h(x,y)$, where H is the mean depth, and small deviations are such that $|\Delta h(x,y)| \ll H$. The vorticity equation has the form

$$\frac{\partial \omega}{\partial t} + J(q^{qg},\psi^{qg}) = \nu\nabla^2\omega - \frac{1}{2}E^{1/2}f\omega, \quad (7.19)$$

where now the potential vorticity is defined as $q^{qg} = \omega + f\Delta h/H$ and the stream function as $\psi^{qg} = \psi/H$. Note that units of these two fields are different from their counterparts in the shallow-water model. Another difference is that the horizontal velocity has zero divergence, and therefore the velocity components are $u = \partial\psi^{qg}/\partial y$ and $v = -\partial\psi^{qg}/\partial x$. The corresponding expression of the relative vorticity in terms of the stream function is the Poisson equation $\omega = -\nabla^2\psi^{qg}$. In most studies, only linear Ekman terms are considered.

We make a short digression here: For a geophysical flow, all models above apply under the so-called f-plane approximation, which consists of a plane tangent to Earth's surface centered at a reference midlatitude ϕ_0. The difference is that the geophysical Coriolis parameter is now given in terms of the angular velocity component perpendicular to the plane, $f \equiv f_0 = 2\Omega_e \sin\phi_0$, with Ω_e the angular speed of the planet. The approximation is valid for flow motions up to order $L \sim 100$ km. For larger scales, of order $L \sim 1000$ km, but keeping the plane approximation, corrections due to the curvature of Earth's surface must be included. Such corrections are of the form $f = f_0 + \beta y$, where y is the latitudinal direction and the corresponding variation of f is given by the parameter $\beta = 2\Omega_e \cos\phi_0/R_e$, with R_e the mean radius of Earth. The so-called β-effect implies profound consequences on the evolution of geophysical flows, and it can be simulated in laboratory experiments with topography, as shall be explained below.

7.2.4. Experimental Considerations

Experiments on geophysical flows are usually performed within a container rotating about a vertical axis aligned with gravity. The depth scale H is usually smaller than the horizontal scale L, but not necessarily. This is an important point to keep in mind when discussing the use of the shallow-water equations versus the quasi-geostrophic approximation. The experimental tank is set to rotate steadily for a certain time until the fluid inside reaches a state of solid-body rotation. How long is this time? The *spin-up* process is essentially viscous: When the container is set in motion, the fluid takes some time to adjust to the rotation by viscous stresses exerted by the lateral boundaries (through Stewartson layers) and the solid bottom (Ekman layer). An additional Ekman layer may be present at the upper boundary, due either to the presence of a solid lid or to stresses generated at the free surface (e.g., by wind). However, the effects associated with this top Ekman layer will be ignored in the rest of the chapter. The main damping contribution during the spin up is due to bottom friction effects, which become effective after a few Ekman periods T_E (7.14). This is the appropriate time scale for adjustment to solid-body rotation, and it has to be taken into account before starting an experiment.

Once the fluid has reached a state of rest in the rotating system (i.e., a state of solid-body rotation), the actual experiment is started: The fluid is set in motion by generating vortices, currents, or turbulent flows or by any other desired initial flow. Several methods are carefully described by *van Heijst and Clercx* [2009] and references therein. In the following sections we shall give further details of the initial forcing in different experiments. We should stress here the requirement of generating a flow with a low to moderate Rossby number in order to ensure that the main horizontal balance in the fluid is geostrophic.

During the experiment, viscous topographic effects are expected to become important at times of the order of T_E, again. Thus, experiments devoted to study flows in the absence of bottom friction should have a duration shorter than the Ekman period. In contrast, when bottom friction effects are the main subject of study, it is necessary to perform experiments during one or more Ekman periods. The conventional way to include Ekman damping is by considering only the dominant linear term in the vorticity equation (7.17) and neglecting the nonlinear Ekman terms. This is an acceptable approximation for most laboratory experiments. However, weak nonlinear Ekman friction effects might be important in some cases, for instance, when considering weak asymmetries between cyclonic and anticyclonic regions (*Zavala Sansón and van Heijst* [2000a]).

In the presence of variable topography, the flow response will strongly depend on the specific configuration used at the bottom of the container. Regardless of the shape of the topography, it is fundamental to have an estimation of the effects associated with depth variations before starting an experiment. Such an estimation is obtained by means of the *topographic β-effect*, which is dynamically equivalent to the planetary β-effect. In order to understand this equivalence, it is useful to compare the potential vorticity of the following two cases: (i) the laboratory with variable topography and (ii) the planetary β-plane assuming a constant depth. In the inviscid limit, conservation of potential vorticity in these two cases is expressed by

$$q_{\text{lab}} \equiv \frac{\omega + f}{h} = \text{const.}, \qquad q_{\text{pla}} \equiv \omega + \beta y = \text{const.} \quad (7.20)$$

These relationships show that increasing (decreasing) depth in the laboratory is equivalent to decreasing (increasing) y in the ocean. Now assume the fluid depth in the container to vary linearly along an arbitrary direction, say y, such that $h = H - \delta_B y/W$, where H is the maximum fluid depth, δ_B is the height of the topography over a horizontal distance W, and the topographic slope is such that $\delta_B/W \ll 1$. As the Rossby number of the flow is assumed small enough, conservation of potential vorticity for fluid parcels in the rotating laboratory tank implies that $q_{\text{lab}} \approx \omega + \beta_t y = \text{const.}$, with

$$\beta_t = \frac{f \delta_B}{WH}. \quad (7.21)$$

Thus, the first-order effect of the bottom topography in the laboratory is equivalent to the planetary β-effect along the meridional y direction. Therefore, it is usually referred to as the topographic β-effect. Note that the "meridional" direction in the laboratory depends on the orientation of the topography.

In order to write the topography parameter β_t in nondimensional terms, we must multiply by the horizontal scale of the flow L and divide by the Coriolis parameter: $\beta'_t = L\delta_B/WH$. Different values can be estimated as the topography varies in different locations. It is important to point out that the appropriate parameter to compare the influence of topographic variations in experiments and in geophysical situations is the nondimensional parameter β'_t, and not the topographic slope δ_B/W.

These arguments can be applied to the parabolic free surface deformation of a rotating system, which also produces an equivalent β-effect. In this case, the free surface elevation $\eta \approx f^2 W^2/8g$ plays the role of δ_B, and W is about half the length of the container. Thus, $\beta_\eta \approx f^3 W/8gH$. In many experiments the stretching and squeezing effects due to the parabolic deformation of the free surface are undesirable and are usually neglected. In order to justify it, the experimental setup must be designed to have $\beta_\eta \ll \beta_t$. In some experiments, a parabolic bottom plate was introduced for a specific Ω value having the same shape as the free surface in order to eliminate any topographic effects associated with the free surface deformation [e.g., *Trieling et al.*, 2010].

7.2.5. Shallow Flows

Flows in shallow fluid layers, with the layer depth H significantly smaller than the horizontal scales L, i.e., $\delta = H/L \ll 1$, are anisotropic because of the difference in scales: On the basis of the continuity equation it is commonly argued that the horizontal and vertical velocity components U and W, respectively, scale as $W \sim U\delta \ll U$. In the ideal case of a stress-free bottom and upper surface, one would have $w = 0$ and $\partial/\partial z = 0$, so that the vorticity vector is $\boldsymbol{\omega} = (0,0,\omega)$ and the motion is purely 2D and governed by

$$\frac{\partial \omega}{\partial t} + J(\omega, \psi) = \nu \nabla^2 \omega, \qquad (7.22)$$

with ψ the stream function.

The presence of a no-slip bottom is usually taken into account by adopting a parabolic (Poiseuille) profile in the z direction. This quasi-2D (planar) flow is then assumed to be governed by

$$\frac{\partial \omega}{\partial t} + J(\omega, \psi) = \nu \nabla^2 \omega - \lambda \omega \qquad (7.23)$$

with $\lambda = \nu(\pi/2H)^2$ the Rayleigh friction coefficient [see, e.g., *Clercx and van Heijst*, 2009]. In recent studies by *Akkermans et al.* [2008a, 2008b], *Cieślik et al.* [2009b], and *Kamp* [2012], however, it was found that significant vertical motions may occur in shallow-layer flows even for $\delta \to 0$ and that a correlation exists between the vertical motion and the rotation/strain of the primary horizontal flow field.

Experimentally, shallow flows have been studied in laboratory configurations of varying sizes. Large-scale experiments on turbulent mixing layers, for example, have been carried out in tanks of horizontal dimensions ranging from 1 m to tens of meters [see, e.g., *Jirka and Uijttewaal*, 2004]. Such flows are usually characterized by larger Re values, and hence are turbulent. Shallow flow experiments in much smaller geometries (with a water depth $H \approx 10$ mm and a typical horizontal size $L \approx 50$ cm) have been performed by a number of researchers [e.g., *Tabeling et al.*, 1991; *Xia et al.*, 2009; *Figueroa et al.*, 2009], with the purpose of studying the characteristics of 2D turbulence. The motion in these experiments is typically generated by electromagnetic forcing, as will be described in Section 7.5.

7.3. LARGE-SCALE EXPERIMENTS $O(10\ M)$

The very first attempts to perform experiments with a rotating tank were rather modest, as the pioneering work by *Taylor* [1921]. During the 1950s and 1960s the study of geophysical flows by means of laboratory models showed a trend toward the construction of large-scale apparatus. Part of the story of these developments is narrated by *von Arx* [1957], who described the 4 m diameter rotating tank constructed at the Woods Hole Oceanographic Institution. In 1960 the Coriolis platform, a 13 m diameter device, was built by the University of Grenoble and a group of French science agencies in Grenoble, France. The original purpose was to model tidal motions in English Channel (La Manche) affected by the rotation of Earth. The platform was dismantled in 2010 and has recently been rebuilt with numerous technological improvements. Another well-known large-scale facility is the 5 m diameter platform at the Norwegian University of Science and Technology in Trondheim, Norway.

With a large-scale platform it is possible to generate flows with very low Rossby and Ekman numbers, as required for cases where rotation effects are fundamental. For modeling fluid phenomena over variable topography, a large-scale tank is very convenient because a topographic feature can be constructed with great detail given the range of typical depths (around $H \sim 0.5$ to 1 m). Another advantage is that the Ekman time scale, proportional to H, can be relatively large in comparison with smaller rotating tanks, which makes possible to perform long experiments without the influence of Ekman damping (T_E can be as long as 30 min, whereas in a medium-size tank it is typically 3–5 min).

7.3.1. The β-Drift of Monopolar Vortices

On a planetary β-plane in the Northern Hemisphere, cyclonic vortices move northwestward, while anticyclones drift in the southwestern direction [see, e.g., *McWilliams and Flierl*, 1979]. This effect is due to the redistribution of relative vorticity in the vortex core, as fluid columns change latitude while conserving potential vorticity. As a result, vortices preserve their quasi-circular shape, superimposed by a dipolar component causing the vortex motion.

Laboratory experiments of planetary flows cannot simulate this phenomenon directly, because the first-order effects of the curvature of Earth are not present in a uniformly rotating fluid tank. However, the planetary β-plane can be simulated by taking advantage of the dynamical equivalence (7.20) shown above: Using a uniform weak sloping topography over the length of a rotating tank, a nearly uniform topographic β_t is obtained over the whole domain. Then, the shallow part corresponds to the north and the deep part to the south. The experimental β_t value

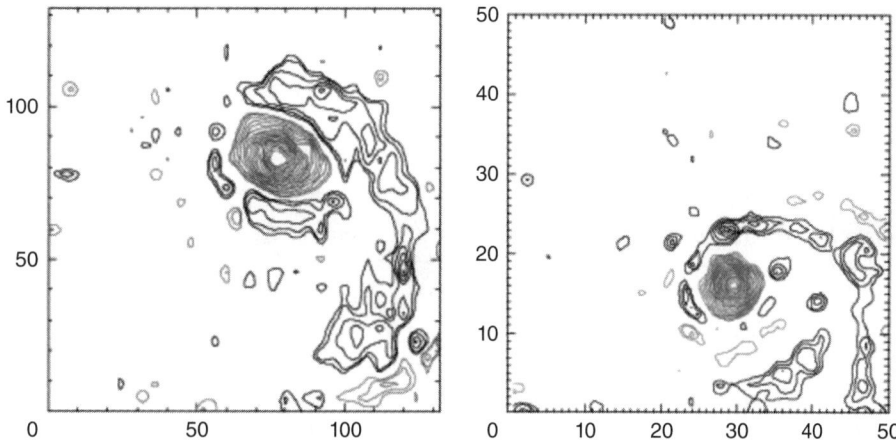

Figure 7.2. Drift of monopolar cyclonic vortices over a sloping bottom topography in a rotating tank. Adapted from *Flór and Eames* [2002]. See also Table 7.1. Left: Relative vorticity contours of a stirring vortex drifting in "northwest" direction. The central core contains positive vorticity contours, whereas the surrounding negative vorticity is characteristic of this type of vortices. Right: Same for a sink vortex, but now the negative vorticity around the vortex is generated only by the topographic β-effect. The vortices are generated at the "southeast" (lower right) corner. The scale of both domains is in cm.

is given by (7.21). The use of a sloping bottom to mimick the β-effect was originally developed by *Pedlosky and Greenspan* [1967], and it was widely used in experiments on the wind-driven ocean circulation or the intensification and separation of western currents, among other applications [*Griffiths and Kiss*, 1992]. More recently, the topographic β-plane has been used to study the motion of vortices. A detailed derivation and discussion are given by *van Heijst* [1994].

There have been numerous experimental studies on the motion of monopolar vortices on a topographic β-plane [e.g., *Masuda et al.*, 1990; *Carnevale et al.*, 1991]. In particular, *Flór and Eames* [2002] studied the motion of barotropic vortices on the Coriolis platform. Following *Carnevale et al.* [1991], the study of Flór and Eames examined the drift of two different cyclonic vortices, namely, "sink" and "stirring" vortices. These names are associated with the generation methods and have been widely explained in the reviews by *Hopfinger and van Heijst* [1993] and *van Heijst and Clercx* [2009]. Both structures are initially axisymmetric. On a flat bottom, sink vortices have a single-signed vorticity and are hence nonisolated, while stirring vortices consist of a core surrounded by an annulus of oppositely signed vorticity in such a way that the vortex contains zero net vorticity. The authors included in their analysis an additional parameter α measuring the steepness of the azimuthal velocity profiles.

Once generated, cyclonic vortices drift toward the "northwest" direction, which corresponds with an uphill motion with a leftward component. The northwestward motion of sink and stirring cyclones depends on their different vorticity distribution and also on the steepness of their corresponding profiles. Stirring vortices travel with a pronounced northward trajectory, while sink vortices drift more in a northwest direction (Figure 7.2). This result was already known from experiments in a medium-size rotating tank by *Carnevale et al.* [1991]. The relevance of the study by Flór and Eames in a larger container is that quantitative measurements of the vortex structures, trajectories, velocity, and vorticity fields can be obtained with much higher resolution and detail. For instance, sink vortices develop an enveloping patch of negative vorticity as they drift, similar to the anticyclonic ring of stirring vortices (see Figure 7.2). This patch is associated with Rossby wave radiation due to the β-effect.

Another advantage of the large container is that it allows the use of a very gentle slope ($s \approx 0.053$), in comparison with the more abrupt topographies used in smaller containers by *Masuda et al.* [1990] and *Carnevale et al.* [1991]. A weaker slope improves the simulation of the planetary β-effect, as illustrated in Table 7.1, where the dimensional and nondimensional β values in the experimental flows and in oceanic flow conditions are presented. For the oceanic case, typical values of mesoscale vortices at midlatitude are considered, which give a nondimensional $\beta' \approx 0.01$. The effective β parameter in the studies by *Masuda et al.* [1990] and *Carnevale et al.* [1991] are overestimated by approximately a factor of 7 and 4, respectively, while the parameter obtained in the large platform (0.015) is much closer to the oceanic value.

7.3.2. Topographic Waves

For small-amplitude motions (e.g., in the wake of a moving vortex), the planetary β-effect induces radiation of so-called Rossby waves. These waves can be visualized

Table 7.1. Dimensional and nondimensional topographic β parameters in the experiments performed on a large-scale platform by Flór and Eames [2002] (column FE) compared with the corresponding values in two experiments using medium-scale containers by Masuda et al. [1990] and Carnevale et al. [1991] (columns M and C, respectively). Considering that in the ocean $f \approx 10^{-4}$ and $\beta \approx 10^{-11}$ m^{-1} s^{-1}, and a mesoscale vortex with length scale $L \approx 10^5$ m, yields a nondimensional, planetary parameter $\beta' \approx 0.01$.

	M (1990)	C (1991)	FE (2002)
s (slope)	0.333	0.133	0.053
f (s^{-1})	4	1.2	0.25
H (m)	0.24	0.171	0.35
β_t (m^{-1} s^{-1})	5.55	0.933	0.038
L (m)	0.05	0.05	0.1
$\beta'_t = L\beta_t/f$	0.069	0.039	0.015

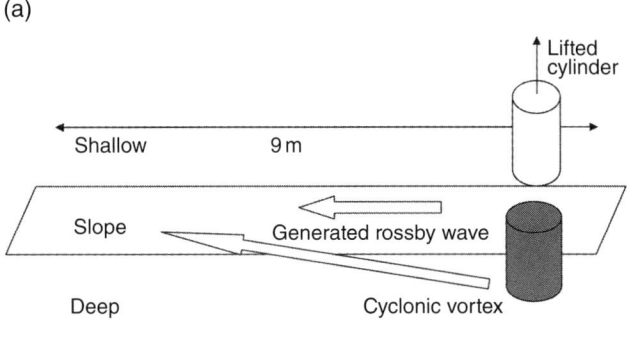

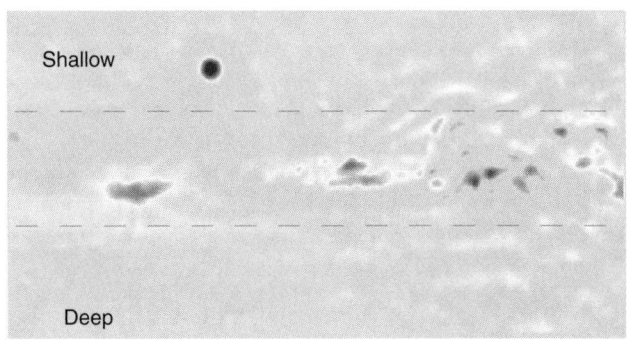

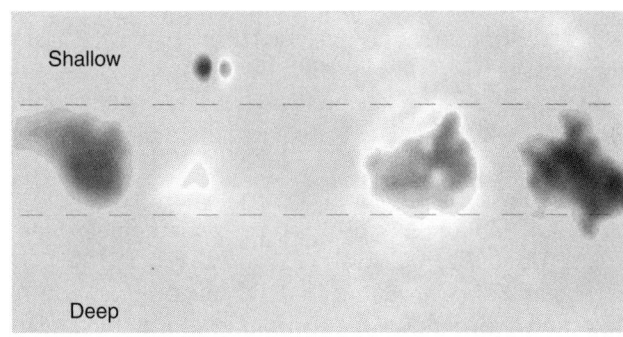

Figure 7.3. Preliminary experimental study of topographic waves excited by the passage of a cyclonic vortex (unpublished results by Zavala Sansón and van Heijst). (a) Schematic view (not to scale) of the experiments: A cyclonic vortex generated by the collapse technique drifts along the topographic slope and generates topographic Rossby waves traveling with shallow water to the right. (b) Relative vorticity surfaces measured in a horizontal plane at time $t = 14T$, with $T = 30$ s the rotation period of the Coriolis platform. The domain shown is a 3.5 m × 1.8 m rectangle. The submerged slope is 0.5 m wide (approximated separation between dashed lines) and centered in the figure. The vortex is generated near the lower right corner. The small, strong cyclonic vortex at the shallow side is a spurious vortex generated during the experiment and does not play any role on the evolution of the topographic waves over the slope. (c) Corresponding surfaces of the velocity component v, perpendicular to the topography contours. For color detail, please see color plate section.

as an alternating pattern of clockwise and anticlockwise circulation cells translating westward. Because of the dynamical equivalence (7.20), similar oscillations are also generated over a topographic slope, and they are usually referred to as topographic (Rossby) waves. Because these waves are associated with conservation of potential vorticity, their frequency is subinertial [Brink, 1991]. When generated near the coastline, the waves are also referred to as continental shelf waves, and they travel along the coast with shallow water to the right (left) in the Northern (Southern) Hemisphere. They can also travel along a submerged slope [Rhines, 1969].

Since wavelike motions are in general very weak features, fine observations are required to obtain quantitative measurements. In addition, Rossby waves rapidly spread and occupy wide areas, so a large facility is necessary to detect them. This has been done in the Coriolis platform in different studies. For instance, Pierini et al. [2002] conducted experiments in a straight channel with a bottom topography consisting of a linear slope separating two regions with depths 0.3 and 0.6 m. The length of the channel was 4.3 m, and the width was 2 m, so the slope was about $0.3/2 = 0.15$. Wave motions were generated by using a wavemaker consisting of a paddle oscillating in front of the topographic feature. The rotation periods of the platform in different experiments were between 35 and 50 s, while the period of the wavemaker was 90 s. Under these conditions, the authors measured the horizontal velocity field on the free surface and detected three alternating patches of opposite circulation along a distance of 4 m over the topography (see their Figure 7.3). The duration of the longest measurements is about 8 min. The authors concluded that in all the analyzed cases the first Rossby normal mode was generated.

Similar experiments were carried out by *Cohen et al.* [2010]. In their study, the bottom topography extended along an external annulus 4 m wide around the Coriolis platform. The maximum fluid depth at the internal circumference was 1 m, while the depth at the external wall was 0.6 m, so the slope is 0.4/4=0.1. The period of the platform was about 30 s. At some point over the topography the waves were generated by two short rods connected to a motor. The characteristics of the rods determined the type of excited wave. Field measurements were carried out over an observation area at the opposite side of the wavemaker. Using this information, the authors measured the velocity fields and the corresponding velocity time series in order to calculate the characteristic oscillations of the waves. Their purpose was to analyze the dispersion relation of the topographic waves measured in the experiments and to compare the results with the dispersion relation of a harmonic model (in which the cross-topography structures of the waves are trigonometric functions) and a proposed model based on Airy functions. Thus, this type of experiments can be useful not only to observe the waves but also to test analytical theories.

Topographic waves can be excited by the passage of a vortex over a topographic slope. This has been the subject of a series of recent experiments in the Coriolis platform by the authors of this chapter. The bottom topography consisted of a steep submarine escarpment that has approximately a tanhshape with a maximum depth of 0.8 m, a minimum depth of 0.55 m, and a width of 0.5 m, so the slope is 0.25/0.50=0.5. A cyclonic vortex was generated at the deep side of the escarpment. This vortex was generated by lifting a solid cylinder out of the fluid, which creates a low-pressure zone and a corresponding cyclonic motion. This method is referred to as the "collapse technique" by *Kloosterziel and van Heijst* [1992]. The presence of the bottom slope makes the vortex move along the topography, as sketched in Figure 7.3a. These experiments were performed over a 9 m long, straight topography that covers a great part of the platform diameter. Measurements of vorticity and horizontal velocity have revealed the structure of the topographic waves. Figures 7.3b and 7.3c show the vorticity field and the velocity component v transversal to the topography. The slope is centerd along the x axis, so the upper part of the figure corresponds to the shallow and the lower part to the deeper part of the flow region. The graphs, covering a domain of almost 4 m in the horizontal, are a composite picture obtained from two cameras mounted in the corotating superstructure of the platform. At the time the photographs were taken, the vortex had already passed the view field, and in its wake the Rossby wave is clearly observed in the v field (Figure 7.3c), rather than in the vorticity field ω (Figure 7.3b). The whole pattern drifts to the left, along the topography, which is the topographic compass direction "west." This study was designed to perform quantitative measurements of the vertical velocity component w in order to estimate the relevance of the topographic waves on the vertical transport. The results will be published elsewhere. Regarding vertical motions, we shall now discuss a similar study on vortices over submarine mountains.

7.3.3. Vertical Motions

As discussed above, a first approach in the modeling of geophysical flows consists of considering the fluid motion as organized in the form of vertical columns. However, stretching and squeezing effects triggered by a flow impinging on variable topography may drive vertical motions and material transport. Some authors have proposed these effects as a mechanism to explain the upwelling of deep waters over seamounts or continental slopes, which are nutrient-rich waters that in turn favor plankton and fish abundances over these topographic features [see, e.g., *Beckmann and Mohn*, 2002]. An equivalent debate exists about the action of planetary Rossby waves on the pumping of nutrients to the ocean surface [*Uz et al.*, 2001].

Laboratory experiments in large-scale facilities may provide new insights not only in the horizontal but also in the vertical flow fields generated by a vortex over a submarine obstacle. This was recently done by *Zavala Sansón et al.* [2012], who performed a series of experiments on the Coriolis platform on the evolution of cyclonic vortices over an axisymmetric mountain. The mountain consisted of a solid, axisymmetric structure, with a maximum height of 0.3 m above the bottom of the rotating platform and a radius of 0.5 m. The large dimensions of the experimental tank allowed to place the mountain far enough (i.e., several mountain diameters) from the lateral walls. The experiments reproduced the main results in the horizontal velocity field observed in previous studies: cyclonic vortices drift in an anticyclonic direction around a conical hill due to the topographic β-effect associated with the slope of the topography (reported by *Carnevale et al.* [1991] using a medium-scale rotating tank). Another important result is the generation of anticyclonic vortices over the summit of a seamount, as shown in the numerical simulations of *Verron and Le Provost* [1985].

How is the structure of the vertical velocity field when a cyclonic vortex rotates around the tip of the mountain? The measurements revealed that the flow in a vertical plane across the mountain has an oscillatory character associated with the orbital motion of the vortex around the topography. This behavior was measured in a number of experiments and was simulated numerically afterward. The simulations were based on the quasi-two-dimensional model (7.17) that allows to solve the

148 MODELING ATMOSPHERIC AND OCEANIC FLOWS

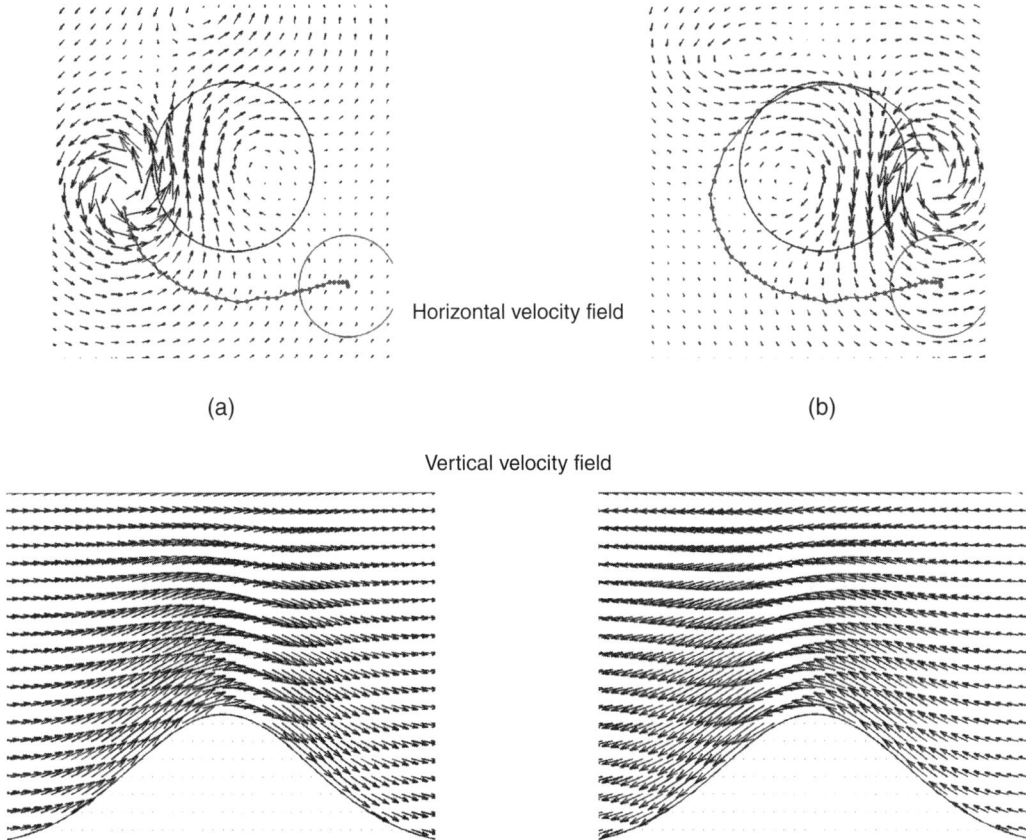

Figure 7.4. Numerical simulation of a cyclonic vortex around a submarine mountain. Adapted from *Zavala Sansón et al.* [2012]. Upper panels: Horizontal velocity field and trajectory calculated at $t=13.5T$ (left) and $t=19T$ (right), with $T=30$ s the rotation period of the Coriolis platform. The vortex is initialized at the southeast flank of the mountain (solid circle). The domain is a square of sides 2.07 m centerd on the mountain. Lower panels: Corresponding velocity fields in the vertical $y-z$ plane across the mountain with dimensions 1 m × 0.8 m. The maximum horizontal (vertical) velocity is 0.05 m/s (0.01 m/s).

horizontal, z-independent velocity field. Subsequently, the vertical velocity w can be calculated by integrating the continuity equation again, but now up to an arbitrary level z [*Pedlosky*, 1987, p. 63]. Using the definitions of h, h_B, and H, it is verified that

$$w(x,y,z;t) = -(z-H)\left(\frac{\partial u}{\partial x} + \frac{\partial v}{\partial y}\right). \quad (7.24)$$

Figure 7.4 presents the horizontal and vertical velocity fields at two different times in a typical simulation. As the vortex approaches the mountain in a spiral trajectory, the horizontal fields reveal that the flow over the mountain is due to the cyclonic vortex as it turns around the topography. For instance, when the vortex is located at the left-hand side of the mountain (Figure 7.4a), the circulation in the vertical plane is in the positive y direction. Afterward, the vortex moves to the right flank of the mountain (Figure 7.4b), and consequently the flow is directed in the negative y direction. The flow is reversed every time the vortex crosses the vertical plane.

7.4. MEDIUM-SCALE EXPERIMENTS O(1 M)

It is much easier and affordable to build, maintain, and operate a smaller rotating platform with dimensions of about 1 m than operating large-scale facilities. Rotating tanks of medium size are commonly used in several institutions all over the world since a few decades. These experimental apparatus have proven to be a great method not only for scientific research but also for educational purposes. As in the case of larger devices, medium-scale experiments are still characterized by relatively small Rossby numbers, and therefore they are suitable for experiments on flows over variable bottom topography.

7.4.1. Vortices Over Topographic Features

As mentioned above, many studies were aimed at gaining a better understanding of the evolution of vortices over topographic features such as a submarine obstacle. The vortex behavior will strongly depend on the shape and

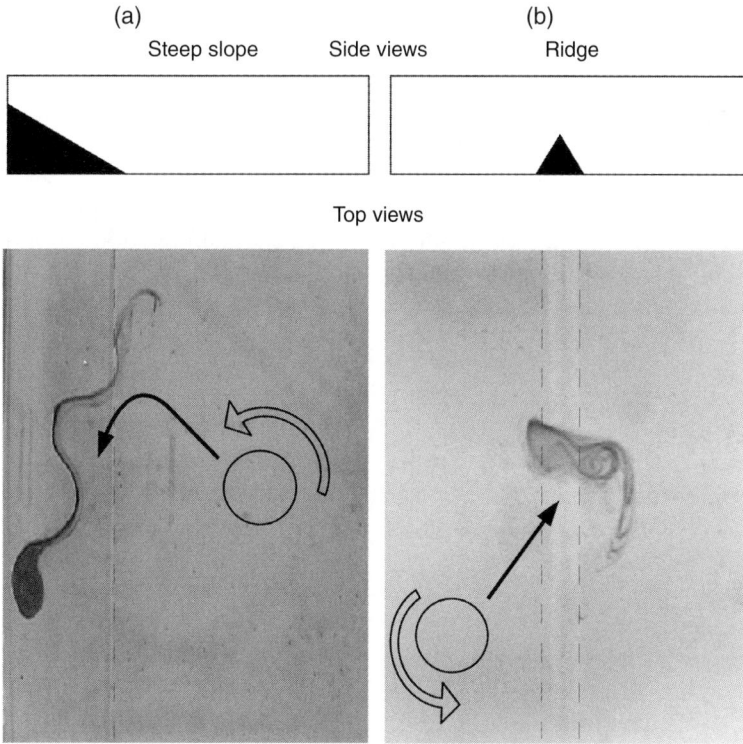

Figure 7.5. Dye photographs showing strong vortex-topography interactions in medium-scale rotating tanks. (a) Cyclonic vortex over a steep slope at the "western" wall. Adapted from *Zavala Sansón and van Heijst* [2000b]. The vortex is forced to move to the south, while meandering and being deformed over the slope. The domain is a 0.8 m × 1.2 m rectangle. The width of the slope is 0.23 m. The maximum fluid depth is 0.2 m, and the minimum depth at the western wall is 0.08 m. (b) Cyclonic vortex over a topographic ridge. Adapted form *Zavala Sansón* [2002]. The domain is a 0.8 m × 1.2 m rectangle. The width of the ridge is 0.1 m, and the height is 0.03 m. The maximum fluid depth is 0.21 m.

characteristics of the topographic feature as well as on the vortex parameters. Some authors adopt a topographic β-plane in order to produce the drift of monopolar vortices toward the obstacle, whereas in other studies the vortices are initially generated over the variable topography. Several cases were reviewed by *van Heijst and Clercx* [2009]. Here we underline two remarkable effects observed in this type of experiment.

Strong vortex deformation: An abrupt topographic feature is characterized by a typical height that is a sustantial fraction of the total fluid depth. Such a condition implies important stretching and squeezing effects on fluid columns and therefore important changes in their relative vorticity. This leads to a strong deformation of the vortices and/or to clear deviations of their initial trajectory. Consider, for instance, the fate of a cyclonic vortex impinging a pronounced "western" slope as shown in Figure 7.5a. The vortex is initialized far from the continental slope, and it was made to drift to the northwest by the topographic β-effect of an additional overall weakly sloping bottom. The photograph shows a dye visualization of the vortex, which has been deflected to the south, being strongly deformed, and leaving a meandering tail along the topography. These meanders are so intense that they may form new vortices. In fact, they are a manifestation of very intense topographic Rossby oscillations, as described in the previous section.

Another remarkable deformation is that of a cyclonic vortex over a topographic ridge (Figure 7.5b). Initially the circular vortex acquires a translating motion toward the topography due to the influence of the closest slope. Then it climbs to the top and the shape becomes elliptical, with the major axis perpendicular to the isobaths. The vortex deformation when crossing the topography leads to a remarkable event, namely, the vortex splitting in two parts, one at each side of the ridge. The vortex elliptical distortion can be understood by means of the topographic Rossby radiation over the opposite slopes of the topography, which are manifested by two anticyclonic cells that strangulate the vortex (see Figures 10 and 11 in *Zavala Sansón* [2002]). The remarkable distortions of the vortices over abrupt topographies strongly suggest that these cases should be treated under the shallow-water rather than the quasi-geostrophic approximation.

Vortex attraction toward a topographic obstacle: It is not necessary to use a topographic β-plane to study vortex-topography interactions. The sole presence of a topographic feature might be enough to set the vortex motion, even if the vortex is relatively far from the obstacle. The reason is the β-drift induced by the nearest slope. Thus, on the Northern Hemisphere, cyclones (anticyclones) are attracted toward topographic upslopes (downslopes). The important point is that the trajectory depends on the far-field structure of the vortex (given a certain topography). This effect was used in the studies described above to attract cyclonic vortices toward a ridge and a submarine mountain [*Zavala Sansón*, 2002; *Zavala Sansón et al.*, 2012]. It has been proposed by *Zehnder* [1993] that this mechanism, the effect of the vortex far field, might play a role on the trajectory of tropical cyclones under the influence of continental topography.

7.4.2. Turbulence Over Variable Topography

In the absence of external forcing, strictly two-dimensional turbulent flows are characterized by a self-organization process in which energy is transferred from small scales of motion toward larger scales. This is the concept of the inverse energy cascade proposed by *Kraichnan* [1967], based on the pioneering ideas of L. F. Richardson since 1922. A similar process is observed in homogeneous, quasi-two-dimensional turbulence, where rotation effects and bottom topography variations are taken into account. An interesting example is the spin up of fluid over a sloping bottom in a rectangular tank observed by *van Heijst et al.* [1994], who discuss the influence of the topography on the formation of a regular pattern of vortices along the container. When random variable topography is present, the inverse energy cascade is halted as the flow tends toward a steady state, aligned with topography contours with shallow water to the right. This implies anticyclonic structures over topographic bumps and cyclonic vorticity over hollows. This process was examined within the quasi-geostrophic context by *Bretherton and Haidvogel* [1976], who showed that the quasi-steady state of the flow aligned along topographic contours is characterized by a linear relationship between potential vorticity and the stream function. Using the more general shallow-water approximation, *Zavala Sansón et al.* [2010] studied numerically the long-term evolution of turbulent flows over random topography. In their results a linear relationship between potential vorticity and transport function was verified, equivalent to that found by Bretherton and Haidvogel.

We have already described some experiments using isolated topographic features (submarine obstacles, linear slopes, ridges, etc.); however, irregular topographies covering the whole container have not been sufficiently studied in the laboratory. One reason may be that bottom topography variations in typical experiments using medium-size tanks are usually not very small, which restricts the use of the quasi-geostrophic model. One of the few experimental studies on this topic was performed by *Zavala Sansón* [2007], who examined the evolution of dipolar and tripolar vortices over a sine-shaped topography in one of the horizontal directions. The qualitative results showed that in both cases the long-term evolution of the vortices was characterized by the alignment of the flow along the topographic contours. Cyclonic relative vorticity was distributed over the deep regions, while anticyclonic vorticity spreads over shallow parts of the domain.

Another important reason for the scarcity of experimental cases on rotating turbulence over topography is that the generation of an initially turbulent flow in a medium-scale facility might be difficult to reproduce. Forcing methods are usually mechanical. For instance, a disordered small-scale initial flow field was obtained by *Maassen et al.* [2002] by passing a grid of vertical bars through the fluid in nonrotating experiments with stratified fluids. In rotating tank experiments, *Morize et al.* [2005] generated a turbulent initial condition by lifting a square grid placed at the flat bottom of the container.

Using the former technique, the studies of *Tenreiro et al.* [2010, 2013] provide recent examples of rotating turbulence over a discontinuous topography. In their experiments, the turbulent flow was examined in rectangular and square containers (1.5 m × 1 m and 1 m × 1 m, respectively) in which the bottom was divided in two regions, deep and shallow. This is perhaps the most simple topographic variation that can be analyzed by experimental methods, because both regions have a flat bottom over which the self-organization process is expected to occur. However, the presence of the step that divides both sides of the tank implied rather complicated and unexpected results. Let us first describe the general characteristics of the observed phenomena: Initially the flow evolves almost as if the step was not present, generating larger structures according to the inverse energy cascade. After a few rotation periods, the step leads to a flow along the topography that always maintains the shallow region on its right. Depending on the strength of the flow and the step height, at some time the vortical structures are no longer able to cross the topography. As a result, the flow evolves almost independently in the shallow and deep regions.

Let us illustrate the long-term state of the flow by means of numerical simulations based on the vorticity equation (7.17) and omitting Ekman friction. It is important to emphasize that the results are radically different when the aspect ratio δ_C of the container is changed from rectangular ($\delta_C = 2$, *Tenreiro et al.* [2010]) to square ($\delta_C = 1$, *Tenreiro et al.* [2013]). In the rectangular case, the shallow and deep regions are themselves square regions

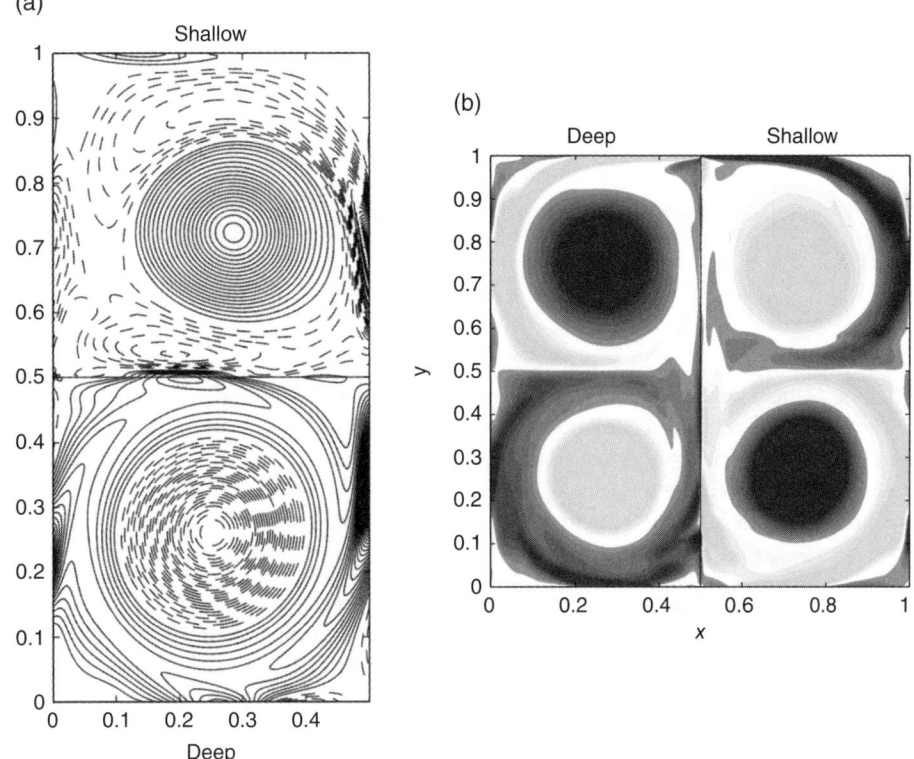

Figure 7.6. Numerical simulations of self-organizing turbulent flows in the presence of a topographic step dividing the (dimensionless) domain in shallow and deep regions. The plots represent relative vorticity distributions after several rotation periods of the system. (a) Rectangular domain: relative vorticity contours; solid (dashed) contours represent positive (negative) values. The step runs along the x direction (the step height is 0.05). The long-term state of the flow is a large vortex at each region. The sign of these structures can change with slight variations of the initial conditions, so there is no preferential final state in this system. Adapted from *Tenreiro et al.* [2010]. (b) Square domain: relative vorticity surfaces calculated from an ensemble of 12 simulations; dark (light) colors indicate positive (negative) vorticity. The step runs along the y direction. In contrast with the previous case, the long-term evolution is systematically given by the four-vortex arrangement shown in the figure, regardless of the initial conditions. Adapted from *Tenreiro et al.* [2013].

(see Figure 7.6a). As the self-organization process occurs, the final state is characterized by one or two dominant vortices in each region. One could naively expect to find an anticyclone over the shallow region and a cyclone over the deep region, according to the results of *Bretherton and Haidvogel* [1976]. However, the experiments showed that it was impossible to predict the sign of the resulting vortices: small variations in the initial conditions led the flow toward a completely different final state. This result was reinforced with a statistical analysis. On the other hand, when using a square container, the results were radically different: The long term evolution of the flow was characterized by the four vortices shown in Figure 7.6b. It was found that this arrangement was systematically obtained in numerical simulations with very different initial conditions, and this preferential vorticity distribution was associated with the topographic step and the aspect ratio of the domain. The analysis included numerical simulations using rectangular domains with larger aspect ratios (up to $\delta_C = 5$), and it was concluded that a preferential final state is always found as long as the length of the step L_S is equal to the longest side of the container L_C. When $L_S/L_C < 1$, the flow organization after long times remains uncertain.

7.4.3. Bottom Friction With Rotation

Up to now we have focused the discussion on the effects of variable topography, that is, on inviscid topography effects. In a rotating system, the viscous effects associated with the bottom topography are related with the Ekman boundary layer, as described in Section 7.2. Ekman friction is typically incorporated in the equations of motion by adding a dominant, linear damping term, and this is also the case in the vorticity formulation. However, the vorticity equation (7.17) also includes nonlinear Ekman terms. Perhaps the most important consequence of these

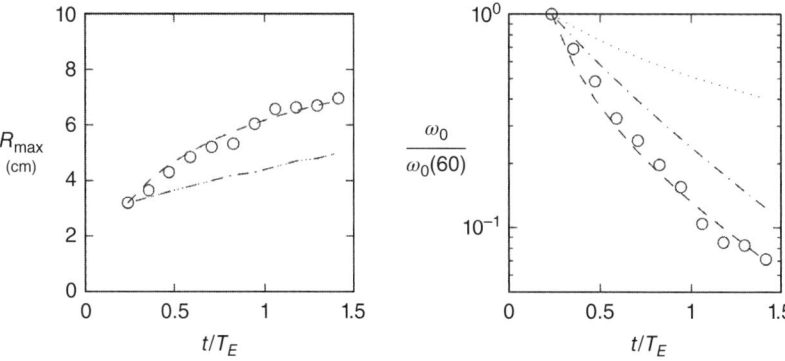

Figure 7.7. Time evolution of the sink vortex parameters while decaying over a flat bottom mainly due to Ekman friction. Adapted from *Zavala Sansón and van Heijst* [2000a]. Left: Radius of maximum velocity R_{max} (left). Right: Normalized peak vorticity ω_0. Circles denote experimental measurements. The dashed line represents a numerical simulation using the vorticity equation (7.17), the dashed-dotted line is a simulation using only linear Ekman damping, and the dotted line represents the case with no Ekman effects. (60) denotes initial parameters 60 s after the forcing was stopped. The fluid depth is 18 cm.

higher-order terms is that the decay rate of cyclonic and anticyclonic fluid columns is different, i.e., the nonlinear Ekman corrections predict an asymmetry between cyclones and anticyclones. This is better appreciated by omitting lateral viscous effects and the Jacobian term and considering only the linear and nonlinear Ekman contributions at the right-hand side of (7.17):

$$\frac{\partial \omega}{\partial t} = -\frac{1}{2}E^{\frac{1}{2}}\omega(\omega + f). \qquad (7.25)$$

The solution of this equation is

$$\omega = \frac{\omega_0 \exp(-t/T_E)}{(\omega_0/f)[1 - \exp(-t/T_E)] + 1}, \qquad (7.26)$$

which states a different decay of cyclonic and anticyclonic fluid columns. Previous studies have also taken into account the nonlinear Ekman corrections in the special case of axisymmetric flows. For instance, the spin up in a rotating cylinder was examined by *Wedemeyer* [1964]. The Wedemeyer model considers the azimuthal component of the Navier-Stokes equation, written in cylindrical coordinates while assuming axisymmetric flow. His results were later extended by a number of authors in order to study the spin-down of barotropic vortices [*Kloosterziel and van Heijst*, 1992; *Maas*, 1993].

Medium-scale rotating facilities are suitable for testing the Ekman friction theories on the decay of homogeneous flows. The reason is that the Ekman time scale is long enough to perform precise measurements of the flow decay, and short enough to avoid undesired perturbations during the flow evolution (e.g., the effect of the free surface). *Zavala Sansón and van Heijst* [2000a] discussed a series of experiments on the decay of different types of vortices. For instance, consider an axisymmetric sink vortex (see Section 7.3.1) generated in a rotating container sufficiently far away from the lateral walls. Over a flat bottom, the evolution of these structures is stable, i.e., the vortices maintain their initially circular shape while being dissipated. The decay is characterized by an increase of the radius of maximum azimuthal velocity R, while the peak vorticity ω_0 decays. These two parameters are shown in Figure 7.7, where the experimental measurements are compared with three different numerical simulations: (i) the general model (7.17) (using a flat bottom or constant depth $h \equiv H$), (ii) the same equation but only including linear Ekman friction, and (iii) with no Ekman effects. Evidently, the simulations including nonlinear Ekman corrections simulate the observations in a much more effective way than just using the conventional linear term. The increase of R in Figure 7.7a is due to two effects: lateral friction and nonlinear Ekman terms. Note that the simulation with linear friction and with no Ekman effects coincide, which indicates that linear friction does not contribute to the expansion of the vortex [*García Sánchez and Ochoa*, 1995]. The decay of the vortex in Figure 7.7b is enhanced by nonlinear Ekman terms with respect to linear damping, as suggested by (7.26).

7.5. SMALL-SCALE EXPERIMENTS O(0.1 M)

Large-scale flows encountered in many environmental situations, such as those in harbors, groyne fields of rivers, lakes, and estuaries, are often referred to as "shallow flows." The shallowness of the flow domain suggests that vertical motions in such configurations are generally much smaller than the large-scale horizontal motions. This anisotropy of shallow flows gives rise to an interesting and rich phenomenology. Examples are the meandering of shallow currents and the formation of larger circulation cells in semi-enclosed basins [see, e.g., *Jirka*, 2001; *Jirka and Uijttewaal*, 2004].

7.5.1. Experiments on Shallow-Layer Flows

Flows in shallow fluid layers have been studied quite intensively in the laboratory by a number of researchers [e.g., *Tabeling et al.*, 1991; *Xia et al.*, 2009], because they were assumed to mimic quasi-2D turbulence to a good approximation. Essentially, the experimental arrangement consists of a thin layer of electrolytic fluid (salt solution) contained in a square tank. The fluid depth H (typically 5–10 mm) is chosen much smaller than the horizontal length scale L (typically 20 cm or more), the shallowness implicitly assumed to result in planar flows with negligible vertical motions. The fluid is set in motion by electromagnetic forcing: The combination of an electrical current running through the fluid between two electrodes mounted along opposing walls and a magnetic field introduced by one or more magnets underneath the tank bottom results in a Lorentz force that acts on the fluid. For example, by placing a single disk-shaped magnet underneath the tank bottom, the locally vertically oriented magnetic field, in combination with the horizontal electrical current, gives rise to a locally horizontal Lorentz force. By applying an electrical current pulse of short duration, one may thus generate a dipolar vortex structure that propels itself in the direction of its axis. By a suitable arrangement of the magnets, more complicated initial flow conditions can be created, e.g., colliding dipoles, arrays of dipoles, and an irregular arrangement of vortex structures, leading to a turbulent field.

This generation technique has been used to create dipolar vortices that were forced to move over topographic features such as a sloping bottom or a step in the bottom [see *van Heijst et al.*, 2012]. Such a situation is observed in the case of an inhomogeneous obliquely incident breaking wave train on a beach, where two vortices of opposite circulation are generated [see, e.g., *Bühler and Jacobson*, 2001]. As a first step to analyze this problem, a single vortex over a linearly sloping bottom may be considered as being a segment of a larger 3D vortex ring. In this inviscid approach the motion of the vortex is identical to the self-propulsion speed of the ring, and in the case of a vortex ring with a uniform vorticity distribution and a circular cross section, it will thus move without change of shape in a direction parallel to the coast [see *Bühler and Jacobson*, 2001]. As a slightly better model, one could take into account the stretching/squeezing of vortex tubes in the shallow vortex, according to Kelvin's circulation theory, leading to changes in its vorticity distribution. This "modulation" of the vorticity structure implies a shape change and hence a more complicated motion of the vortex.

For the case of a vortex pair moving over a sloping bottom toward the coast, one could argue that each vortex behaves according to the "vortex-ring model" and that the oppositely signed vortices separate when approaching the coast, subsequently moving away from each other in opposite directions along the coast. This behavior is similar to that of two oppositely signed point vortices approaching a wall [see *Lamb*, 1932, Section 155.3], the motion of which is governed by the "image principle." Inclusion of viscous effects drastically changes this behavior, as shown in the numerical simulations by *Orlandi* [1990] of a 2D dipole structure colliding against a no-slip wall. The boundary layer formed at this wall contains vorticity with a sign opposite to that of the neighboring vortex core. When the dipole moves closer to the wall, the oppositely signed vorticity is removed by advection induced by the dipole and moves around the dipole cores, which leads to a widening of the primary structure, a splitting of the dipole, and eventually a rebound from the wall.

The shallow-layer experiments by *van Heijst et al.* [2012] have revealed that 3D effects may play a significant role in shallow flows with bottom topography, for both the dipolar vortex approaching an upward step and the dipole climbing a linearly sloping bottom. In the case of the step topography, it was observed that the no-slip sidewall of the step gives rise to generation of vertical vorticity ω_z, the signature of which is also clearly observed in the upper parts of the fluid column. As a result, a shield of oppositely signed vorticity is formed around the dipole, causing its arrest at the deeper side of the step. Before this takes place, a portion of the dipole in the upper part of the fluid column might cross the step, although it decays quickly because of the shallowness of the fluid layer on that side. The dipole moving uphill over the sloping bottom experiences a similar effect. The boundary layer at the no-slip bottom implies the generation of vorticity, that is strongest in the shallower part of the fluid layer. The vorticity vector pointing locally parallel to the bottom implies a vertical component ω_z that is strongest near the front of the dipole. This bottom-induced vorticity ω_z has a sign opposite to that of the primary vortex, and its signature is clearly observed even at the free surface. As in the case of step topography, the bottom-induced vorticity component becomes concentrated in a band around the vortex dipole, which causes its widening and its slowdown, as visible in the measured flow evolution shown in Figure 7.8.

This exploratory study has revealed that viscous effects, i.e., no-slip conditions at the nonhorizontal parts of the bottom, play an important role in shallow flows with bottom topography. The combination of these viscous effects and the 3D nature of the bottom topography gives rise to generation of vorticity (including a vertical component ω_z), which significantly influences the flow evolution. Clearly, such flows cannot be described simply by the two-dimensional Navier-Stokes equation with an additional "bottom friction" term.

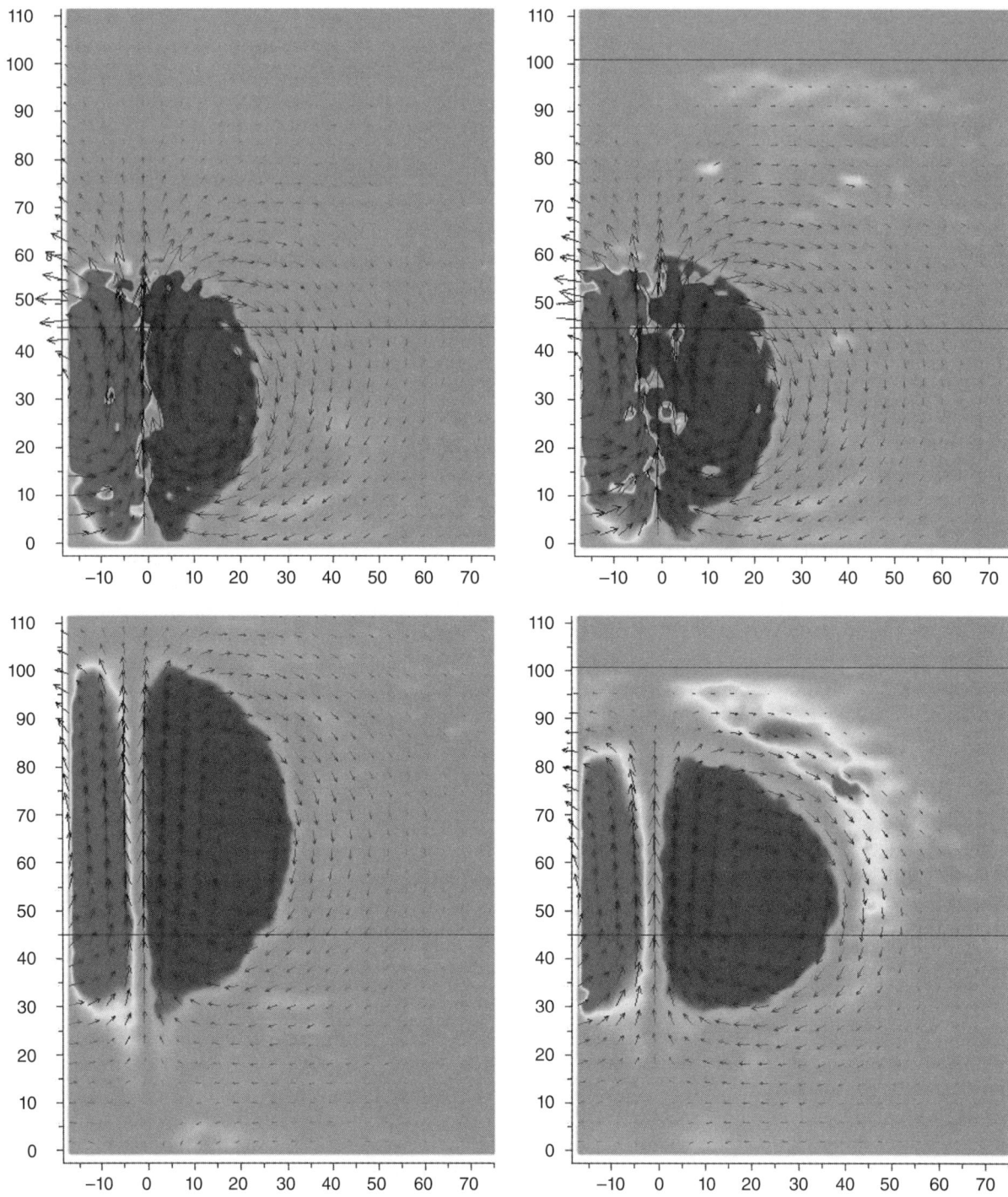

Figure 7.8. Experimentally measured flow evolution of a dipolar vortex encountering a weak slope topography with angle $\alpha = 3.4°$ (left panels) and a steeper slope with inclination $\alpha = 6.8°$ (right panels). Adapted from *van Heijst et al.* [2012]. The snapshots are taken at $t = 3$ s (top) and $t = 9$ s (bottom). The graphs show the horizontal flow field (vectors) and the ω_z vertical vorticity component (red/blue colors indicate positive/negative values) measured at the free surface. The horizontal black lines indicate the position of the edge of the slope. For color detail, please see color plate section.

Detailed 3D flow measurements by *Akkermans et al.* [2008a, 2008b, 2010], *Cieślik et al.* [2009a, 2009b, 2010], *Lacaze et al.* [2010] and *Sous et al.* [2004] on electromagnetically generated vortex structures in a shallow fluid layer have clearly revealed that such flows are generally not quasi-2D: Vertical motion is observed to occur throughout the evolution of the flow. These experimental observations were confirmed by 3D numerical flow simulations. In the case of a flow field generated by an array of magnets, large-scale meandering motions were observed instead of larger coherent vortex structures, that would emerge when the flow would be (quasi-)2D. This behavior is attributed to 3D effects: The PIV flow measurements and the numerical simulations showed that significant vertical motions were present throughout the flow evolution. Upward motion was mainly observed in areas around the elliptical points in the horizontal flow field, while downward motion was concentrated in elongated regions in strain-dominated parts of the horizontal flow field. The relation between vertical, secondary motion and the strain-vorticity characteristics of the primary horizontal flow field has been analyzed in detail by *Kamp* [2012].

Background rotation is commonly believed to promote the two dimensionality of flows by virtue of the Taylor-Proudman theorem valid for geostrophic motion. In a recent experimental and numerical study, *Duran-Matute et al.* [2012] have investigated the effects of background rotation on shallow-layer flows for the specific case of an axisymmetric swirl flow. It was found that two dimensionality was enforced only in extreme cases: for very high rotation rates and in very shallow fluid layers. In particular, two dimensionality was promoted for the Rossby number Ro = 0.2 (defined as Ro = $\omega_z/2\Omega$, with ω_z the axial vorticity of the primary vortex structure), independently of the fluid depth. Shallowness of the fluid layer only promotes two dimensionality when the layer depth is small compared to the typical (Ekman) layer thickness. *Duran-Matute et al.* [2012] have carried out a systematic study, and they report on a number of different regimes in the relevant parameter space.

7.5.2. Scaling Arguments

It is commonly argued that the shallowness of the flow domain implies that shallow flows are quasi-2D. This argument is based on the continuity equation: By using L and H as horizontal and vertical length scales and U and W as typical horizontal and vertical velocity scales, respectively, one derives

$$\frac{U}{L} \sim \frac{W}{H} \rightarrow W \sim U\frac{H}{L} = U\delta, \qquad (7.27)$$

with $\delta = H/L$ the domain aspect ratio. As $\delta \ll 1$ in shallow flows, this would imply that $W \ll U$. In a recent study, *Duran-Matute et al.* [2010a] demonstrated that a more subtle scaling argument is appropriate. These authors considered the case of a single axisymmetric shallow vortex flow, and we will here summarize their analysis. The initial state is an axisymmetric swirl flow in terms of cylindrical coordinates (r, θ, z) given by

$$t = 0: \quad \mathbf{v} = (v_r, v_\theta, v_z) = (0, v_\theta, 0) \qquad (7.28)$$

with $\partial/\partial \theta = 0$. During the subsequent evolution a secondary flow in the (r, z) plane may develop, but it is assumed that this flow remains axisymmetric. The continuity equation then becomes

$$\frac{1}{r}\frac{\partial}{\partial r}(rv_r) + \frac{\partial v_z}{\partial z} = 0, \qquad (7.29)$$

implying that no relation between the primary swirl velocity v_θ and the vertical velocity v_z can be derived. This is in contrast to argument (7.27) based on the continuity equation in Cartesian formulation. Apparently, the scaling ratio v_z/v_θ must depend on the flow dynamics, which is governed by the Navier-Stokes equation. The θ component of this equation is essentially an evolution equation for v_θ (with terms containing v_r and v_z), and the r and z components can be combined into an evolution equation for the azimuthal vorticity $\omega_\theta = \partial v_r/\partial z - \partial v_z/\partial r$, which describes the secondary circulation in the (r, z) plane. The latter equation contains the following viscous term:

$$\frac{\nu}{r}\frac{\partial v_\theta^2}{\partial z} = \frac{2\nu}{r}v_\theta\frac{\partial v_\theta}{\partial z} \qquad (7.30)$$

Apparently, a vertical gradient in the primary flow field v_θ implies a nonzero ω_θ and hence a secondary flow in the (r, z) plane. Again, the problem is nondimensionalized by adopting the length scales L and H and the velocity scales U and W. In the nondimensional equations for v_θ and ω_θ the aspect ratio δ and the Reynolds number Re = UL/ν now appear as dimensionless quantities. Under the assumption of a shallow layer, i.e., $\delta \ll 1$, the flow quantities were expanded in powers of δ. It was shown that the radial and vertical velocities scale as

$$\frac{v_r}{v_\theta} = O(\text{Re}\,\delta^2), \quad \frac{v_z}{v_\theta} = O(\text{Re}\,\delta^3). \qquad (7.31)$$

This scaling is in remarkable contrast to the commonly adopted scaling (7.27) based on the continuity equation. *Duran-Matute et al.* [2010a] have performed 3D numerical simulations of a decaying Lamb-Oseen vortex and compared these with the analytical solution for the case of zero secondary circulation (i.e., $\omega_\theta = 0$). Good agreement between the analytical and numerical results was found for $\delta\text{Re}^{1/2} \leq 3$ and $\delta\text{Re}^{1/3} \leq 1$. For these values of $\delta\text{Re}^{1/2}$ and $\delta\text{Re}^{1/3}$ the shallow swirling flow can hence be considered as quasi-2D.

In a separate study *Duran-matute et al.* [2010b] have analyzed the evolution of decaying dipolar vortices in a shallow fluid layer, both numerically and experimentally. By quantifying the magnitude of the vertical motion and the (nonzero) horizontal divergence with respect to the primary swirling motion in the vortex cores, they found that the three dimensionality of the shallow ($\delta \ll 1$) dipolar vortex flow depends on the quantity $\text{Re}\delta^2$. Three different regimes could be distinguished: (i) a quasi-2D regime for $\text{Re}\delta^2 \leq 6$, in which the flow is dominated by viscous effects. In this case the secondary flow is negligibly small; (ii) a transitional regime for $6 \leq \text{Re}\delta^2 \leq 15$, in which secondary motion inside the dipole can be observed as well as a spanwise circulation cell in front of the translating dipole. These 3D effects are rather weak, however, and they do not seriously affect the primary dipole flow; (iii) a 3D regime for $\text{Re}\delta^2 \geq 15$, in which the principal dipole flow is seriously affected by the secondary motions.

The flow behavior was examined for different initial conditions, but these resulted in rather marginal differences in the $\text{Re}\delta^2$ values of the transitional regime. The agreement found between the experiments and the numerical simulations underline the conclusion that the quantity $\text{Re}\delta^2$ is the crucial parameter characterizing the decaying dipole flow. As this quantity is apparently the relevant parameter in describing the two basic, generic flow structures (monopolar and dipolar flows), it is anticipated that it is also the crucial parameter characterizing the three dimensionalization of other types of shallow flows.

7.6. CONCLUDING REMARKS

We have described laboratory experiments on flows over bottom topography in arrangements of different scales, ranging from rotating fluid containers of more than 10 m in diameter to fluid layers of a few millimeters in thickness. The choice of the spatial dimensions depends on the particular aspects of the flow that one is interested in and also on the measurement resolution that one wishes to achieve. As has been discussed, the dimensions of different experimental facilities may imply different advantages and disadvantages. For example, nonrotating shallow-layer experiments with electromagnetic forcing in a thin layer of electrolyte are characterized by a relatively high damping rate, which puts a serious limitation on the usefulness of such a setup. In addition, recent experiments have revealed that the fluid shallowness does not guarantee the two dimensionality of the flow, and a more careful scaling argument is required. On the other hand, large-scale rotating experiments are often less easy to run (and far more expensive) than the medium-scale $O(1\text{ m})$ experiments, just because of their size. In any case, medium- and large-scale facilities are ideal to study the effects of variable topography.

Acknowledgments. The authors gratefully acknowledge their colleagues and former Ph.D. students for numerous fruitful discussions throughout the years: Leon Kamp, Herman Clercx, Ruben Trieling, Rinie Akkermans, Matias Duran-Matute, Andrzej Cieślik, Ron Theunissen, Saskia Maassen, and Miguel Tenreiro. The experimental work of the authors performed in the Coriolis platform in 2009 was funded by the 6th Framework Programme of the European Commission within the Integrated Infrastructure Initiative Hydralab III. The authors gratefully acknowledge the scientific and technical help of H. Didelle, S. Viboud, J. Sommeria, and L. Gostiaux during our stay in Grenoble.

REFERENCES

Akkermans, R. A. D., A. R., Cieślik, L. P. J., Kamp, R. R., Trieling, H. J. H. Clercx, and G. J. F. van Heijst (2008a), The three-dimensional structure of an electromagnetically generated dipolar vortex in a shallow fluid layer, *Phys. Fluids*, *20*, 116,601.

Akkermans, R. A. D., L. P. J., Kamp, H. J. H. Clercx, and G. J. F. van Heijst (2008b), Intrinsic three-dimensionality in electromagnetically driven shallow flows, *Europhys. Lett.*, *83*, 24,001.

Akkermans, R. A. D., L. P. J., Kamp, H. J. H. Clercx, and G. J. F. van Heijst (2010), Three-dimensional flow in electromagnetically driven shallow two-layer fluids, *Phys. Rev. E*, *82*, 026,314.

Beckmann, A., and C. Mohn (2002), The upper ocean circulation at Great Meteor Seamount Part II: Retention potential of the seamount-induced circulation, *Ocean Dyn.*, *52*, 194–204.

Boyer, D. L., and P. A. Davies (2000), Laboratory studies of orographic effects in rotating and stratified flows, *Annu. Rev. Fluid Mech.*, *32*, 165–202.

Bretherton, F. P., and D. B. Haidvogel (1976), Two-dimensional turbulence above topography, *J. Fluid Mech.*, *78*, 129–154.

Brink, K. H. (1991), Coastal-trapped waves and wind-driven currents over the continental shelf, *Annu. Rev. Fluid Mech.*, *23*, 389–412.

Bühler, O., and T. E. Jacobson (2001), Wave-driven currents and vortex dynamics on barred beaches, *J. Fluid Mech.*, *449*, 313–339.

Carnevale, G. F., R. C. Kloosterziel, and G. J. F. van Heijst (1991), Propagation of barotropic vortices over topography in a rotating tank, *J. Fluid Mech.*, *233*, 119–139.

Chan, J. C. L. (2005), The physics of tropical cyclone motion, *Annu. Rev. Fluid Mech.*, *37*, 99–128.

Cieślik, A. R., R. A. D., Akkermans, L. P. J., Kamp, H. J. H. Clercx, and G. J. F. van Heijst (2009a), Dipole-wall collision in a shallow fluid, *Eur. J. Mech. B/Fluids*, *28*, 397–404.

Cieślik, A. R., L. P. J., Kamp, H. J. H. Clercx, and G. J. F. van Heijst (2009b), Meandering streams in a shallow fluid layer, *Europhys. Lett.*, *85*, 54,001.

Cieślik, A. R., L. P. J., Kamp, H. J. H. Clercx, and G. J. F. van Heijst (2010), Three-dimensional structures in a shallow flow, *J. Hydr. Env. Res.*, *4*, 89–101.

Clercx, H. J. H., and G. J. F. van Heijst (2009), Two-dimensional Navier-Stokes turbulence in bounded domains, *Appl. Mech. Rev.*, *62*, 020,802/1–25.

Cohen, Y., N. Paldor, and J. Sommeria (2010), Laboratory experiments and a non-harmonic theory for topographic Rossby waves over a linearly sloping bottom on the f-plane, *J. Fluid Mech.*, *645*, 479–496.

de Steur, L., and P. J. van Leeuwen (2009), The influence of bottom topography on the decay of modeled Agulhas rings, *Deep Sea Res. I*, *56*, 471–494.

Duran-Matute, M., L. P. J., Kamp, R. R. Trieling, and G. J. F. van Heijst (2010a), Scaling of decaying shallow axisymmetric swirl flows. *J. Fluid Mech.*, *648*, 471–484.

Duran-Matute, M., J., Albagnac, L. P. J. Kamp, and G. J. F. van Heijst (2010b), Dynamics and structure of decaying shallow dipolar vortices, *Phys. Fluids*, *22*, 116,606.

Duran-Matute, M., L. P. J., Kamp, R. R. Trieling, and G. J. F. van Heijst (2012), Regimes of two-dimensionality of decaying shallow axisymmetric swirl flows with background rotation, *J. Fluid Mech.*, *691*, 214–244.

Figueroa, A., F., Demiaux, S. Cuevas, and E. Ramos (2009), Electrically driven vortices in a weak dipolar magnetic field in a shallow electrolytic layer, *J. Fluid Mech.*, *641*, 245–261.

Flór, J. B., and I. Eames (2002), Dynamics of monopolar vortices on a topographic beta-plane, *J. Fluid Mech.*, *456*, 353–376.

García Sánchez, R. F., and J. L. Ochoa (1995), Decaimiento de vórtices barotrópicos en el plano f, *Rev. Mex. Fís.*, *41*, 533–555.

Genin, A. (2004), Bio-physical coupling in the formation of zooplankton and fish aggregations over abrupt topographies, *J. Mar. Sys.*, *50*, 3–20.

Gill, A. E. (1982), *Atmosphere-Ocean Dynamics*, Academic Press, London.

Greenspan, H. P. (1968), *The Theory of Rotating Fluids*, Cambridge Univ. Press, Cambridge.

Griffiths, R. W., and A. E. Kiss (1992), Flow regimes in a wide "sliced-cylinder" model of homogeneous beta-plane circulation, *J. Fluid Mech.*, *399*, 205–236.

Grimshaw, R., Y. Tang, and D. Broutman (1994), The effect of vortex stretching on the evolution of barotropic eddies over a topographic slope. *Geophys. Astrophys. Fluid Dyn.*, *76*, 43–71.

Hopfinger, E. J., and G. J. F. van Heijst (1993), Vortices in rotating fluids, *Annu. Rev. Fluid Mech.*, *25*, 241–289.

Jirka, G. H. (2001), Large scale flow structures and mixing processes in shallow flows, *J. Hydraulic Res.*, *39*, 567–573.

Jirka, G. H. and W. S. J. Uijttewaal, (Eds.) (2004), *Shallow Flows*, Taylor & Francis Group, London.

Kamp, L. P. J. (2012), Strain-vorticity induced secondary motion in shallow flows. *Phys. Fluids*, *24*, 023,601.

Kelley, D. H., and N. T. Ouellette (2011), Onset of three-dimensionality in electromagnetically driven thin-layer flows, *Phys. Fluids*, *23*, 045,103.

Kloosterziel, R. C., and G. J. F. van Heijst (1992), The evolution of stable barotropic vortices in a rotating free-surface fluid, *J. Fluid Mech.*, *239*, 607–629.

Kraichnan, R. H. (1967), Inertial ranges transfer in two-dimensional turbulence, *Phys. Fluids*, *10*, 1417–1423.

Lacaze, L., P., Brancher, O. Eiff, and L. Labat (2010), Experimental characterization of the 3D dynamics of a laminar shallow vortex dipole, *Exp. Fluids*, *48*, 225–231.

Lamb, H. (1932), *Hydrodynamics*, 6th edn., Cambridge Univ. Press, Cambridge.

Lévy, M. (2008), The modulation of biological production by oceanic mesoscale turbulence, in *Transport and Mixing in Geophysical Flows*, edited by J. B. Weiss and A. Provenzale, pp. 219–261, Springer, Berlin.

Maas, L. R. (1993), Nonlinear and free-surface effects on the spin-down of barotropic axisymmetric vortices, *J. Fluid Mech.*, *246*, 117–141.

Maassen, S. R., H. J. H. Clercx, and G. J. F. van Heijst (2002), Self-organization of quasi-two-dimensional turbulence in stratified fluids in square and circular containers, *Phys. Fluids*, *14*, 2150–2169.

Masuda, A., K. Marubayashi, and M. Ishibashi (1990), A laboratory experiment and numerical simulation of an isolated barotropic eddy in a basin with topographic β, *J. Fluid Mech.*, *213*, 641–655.

McWilliams, J. C., and G. R. Flierl (1979), On the evolution of isolated, nonlinear vortices, *J. Phys. Oceanogr.*, *9*, 1155–1182.

Morize, C., F. Moisy, and M. Rabaud (2005), Decaying grid-generated turbulence in a rotating tank, *Phys. Fluids*, *17*, 095105.

Orlandi, P. (1990), Vortex dipole rebound from a wall, *Phys. Fluids A*, *2*, 1429–1436.

Pedlosky, J. (1987), *Geophysical Fluid Dynamics*, Springer-Verlag, New York.

Pedlosky, J., and H. P. Greenspan (1967), A simple laboratory model for the oceanic circulation, *J. Fluid Mech.*, *27*, 291–304.

Pierini, S., A.M Fincham, D. Renouard, M. R. D'Ambrosio, and H. Didelle (2002), Laboratory modeling of topographic Rossby normal modes, *Dyn. Atmos. Oceans*, *26*, 205–225.

Proudman, J. (1916), On the motion of solids in a liquid possessing vorticity, *Proc. R. Soc. A*, *92*, 408–424.

Rhines, P. B. (1969), Slow oscillations in an ocean of varying depth. Part 1. Abrupt topography, *J. Fluid Mech.*, *37*, 161–189.

Sous, D., N. Bonneton, and J. Sommeria (2004), Turbulent vortex dipoles in a shallow water layer, *Phys. Fluids*, *16*, 2886–2898.

Tabeling, P., S., Burkhart, O. Cardoso, and H. Willaime (1991), Experimental study of freely decaying two-dimensional turbulence, *Phys. Rev. Lett.*, *67*, 3772–3775.

Taylor, G. I. (1917), Motion of solids in fluids when the flow is not irrotational, *Proc. R. Soc. A*, *93*, 99–113.

Taylor, G. I. (1921), Experiments with rotating fluids, *Proc. R. Soc. A*, *100*, 114–121.

Tenreiro, M., L. Zavala Sansón, G. J. F. van Heijst, and R. R. Trieling (2010), Experiments and simulations on self-organization of confined quasi-two-dimensional turbulent flows with discontinuous topography, *Phys. Fluids*, *18*, 056,603.

Tenreiro, M., R. R. Trieling, L. Zavala Sansón, and G. J. F. van Heijst (2013), Preferential states of rotating turbulent flows in a square container with a step topography, *Phys. Fluids.* *25*, 015,109.

Trieling, R. R., G. J. F. van Heijst, and Z. Kizner (2010), Laboratory experiments on multipolar vortices in a rotating fluid, *Phys. Fluids*, *22*, 094,104.

Uz, B. M., J. A. Yoder, and V. Osychny (2001), Pumping of nutrients to ocean surface waters by the action of propagating planetary waves, *Nature*, *409*, 597–600.

van Heijst, G. J. F. (1994), Topography effects on vortices in a rotating fluid, *Meccanica*, *29*, 431–451.

van Heijst, G. J. F., and H. J.H Clercx (2009), Laboratory modeling of geophysical vortices, *Annu. Rev. Fluid Mech.*, *41*, 143–164.

van Heijst, G. J. F., L. R. M. Maas, and C. W. M. Williams (1994), The spin-up of fluid in a rectangular container with a sloping bottom, *J. Fluid Mech.*, *265*, 125–159.

van Heijst, G. J. F., L. P. J. Kamp, and R. Theunissen (2012), Shallow flows with bottom topography, in *Environmental Fluid Mechanics: Memorial Volume in Honour of Prof. G. H. Jirka*, edited by W. Rodi and M. Uhlman, pp. 73–84, CRC Press. Boca Raton, Fla.

Verron, J., and Le Provost (1985), A numerical study of quasi-geostrophic flow over isolated topography, *J. Fluid Mech.*, *154*, 231–252.

von Arx, W. S. (1957), An experimental approach to problems in physical oceanography, *in Progress in Physics and Chemistry of the Earth*, Vol. 2, pp. 1–29, Pergamon, New York.

Vukovich, F. M. (2007), Climatology of ocean features in the Gulf of Mexico using satellite remote sensing data, *J. Phys. Oceanogr.*, *37*, 689–707.

Wedemeyer, E. H. (1964), The unsteady flow within a spinning cylinder, *J. Fluid Mech.*, *20*, 383–399.

Xia, H., M. Shats, and G. Falkovich (2009), Spectrally condensed turbulence in thin layers, *Phys. Fluids*, *21*, 125,101.

Zavala Sansón, L. (2002), Vortex-ridge interaction in a rotating fluid, *Dyn. Atmos. Oceans*, *35*, 299–325.

Zavala Sansón, L. (2007), The long-time decay of rotating homogeneous flows over variable topography, *Dyn. Atmos. Oceans*, *44*, 29–50.

Zavala Sansón, L. and G. J. F. van Heijst (2000a), Nonlinear Ekman effects in rotating barotropic flows, *J. Fluid Mech.*, *412*, 75–91.

Zavala Sansón, L., and G. J. F. van Heijst (2000b), Interaction of barotropic vortices with coastal topographies: Laboratory experiments and numerical simulations, *J. Phys. Oceanogr.*, *30*, 2141–2162.

Zavala Sansón, L., and G. J. F. van Heijst (2002), Ekman effects in a rotating flow over bottom topography, *J. Fluid Mech.*, *471*, 239–255.

Zavala Sansón, L., A. González-Villanueva, and L. M. Flores (2010), Evolution and decay of a rotating flow over random topography, *J. Fluid Mech.*, *642*, 159–180.

Zavala Sansón, L., A. C. Barbosa Aguiar, and G. J. F. van Heijst (2012), Horizontal and vertical motions of barotropic vortices over a submarine mountain, *J. Fluid Mech.*, *661*, 32–44.

Zehnder, J. A. (1993), The influence of large-scale topography on barotropic vortex motion, *J. Atmos. Sci.*, *50*, 2519–2532.

8
Direct Numerical Simulations of Laboratory-Scale Stratified Turbulence

Michael L. Waite

8.1. INTRODUCTION

As already discussed in previous chapters, stable density stratification is a fundamental property of atmospheric and oceanic flows. Stratification creates buoyancy forces, which inhibit vertical motions and can have a profound effect on dynamics across a wide range of length scales. This chapter is about turbulence in such fluids, which is commonly called stratified turbulence. Stratified turbulence differs from other types of turbulence, such as isotropic three-dimensional, two-dimensional, and quasi-geostrophic (QG) turbulence, in several important ways. It is characterized by approximately horizontal velocities, thin layers of intense vertical shear, quasi-two-dimensional vortices, and interactions with internal gravity waves (for reviews of stratified turbulence, see *Lin and Pao* [1979], *Hopfinger* [1987], and *Riley and Lelong* [2000]). We will make a few common idealizations and consider stratified turbulence that is statistically homogeneous, nonrotating, and not dominated by internal gravity waves. For a recent discussion of rotating stratified turbulence, see *Bartello* [2010], and for gravity waves, see *Staquet and Sommeria* [2002].

Stratified turbulence has relevance for geophysical flows because, over large regions of the atmosphere and ocean, the Brunt–Väisälä frequency N is much larger than the Coriolis parameter f; typical midlatitude values of N/f are $O(100)$. As a result, there is a range of length scales in both fluids over which buoyancy forces are thought to dominate the Coriolis effect. This range is between larger scales where both rotation and stratification are important and smaller scales where both effects are negligible. These scales correspond roughly to the atmospheric mesoscale ($O(100)$ to $O(1)$ km) and oceanic sub-mesoscale ($O(10)$ km to $O(100)$ m). The cascade of energy through this scale range forms a key link in the global energy budget, as it connects large-scale QG turbulence with microscale isotropic turbulence and ultimately dissipation. In the case of the atmosphere, various attempts have been made to explain the observed mesoscale energy spectrum with a stratified turbulence hypothesis, which posits that the essential nonlinear dynamics of the mesoscale regime are captured by stratified turbulence [*Gage*, 1979; *Lilly*, 1983; *Lindborg*, 2006]. While its details have evolved, this hypothesis has motivated much of the research on stratified turbulence over the last 30 years. We will use primarily atmospheric terminology in this chapter (e.g., "mesoscale"), but the stratified turbulence hypothesis has also been advanced to explain the sub-mesoscale cascade in the ocean [*Riley and Lindborg*, 2008].

Since the early experiments of the 1960s and 1970s [reviewed by, e.g., *Lin and Pao*, 1979] and the pioneering computations of *Riley et al.* [1981], laboratory experiments and numerical simulations have played important roles in investigating the dynamics of stratified turbulence. But even today there are major differences between the parameter regimes of the atmosphere and ocean and those accessible numerically and in the laboratory. Stratified turbulence is characterized by the Froude and Reynolds numbers. Distinguishing between horizontal and vertical scales (denoted by subscripts h and v), the Froude numbers are

$$\mathrm{Fr}_h \equiv \frac{U}{NL_h}, \quad \mathrm{Fr}_v \equiv \frac{U}{NL_v}, \quad (8.1)$$

where U is the horizontal velocity scale and L_h and L_v are horizontal and vertical length scales. The Reynolds number is given by

Department of Applied Mathematics, University of Waterloo, Waterloo, Ontario, Canada.

Modeling Atmospheric and Oceanic Flows: Insights from Laboratory Experiments and Numerical Simulations, First Edition. Edited by Thomas von Larcher and Paul D. Williams.
© 2015 American Geophysical Union. Published 2015 by John Wiley & Sons, Inc.

$$\mathrm{Re} \equiv \frac{UL_h}{\nu}, \qquad (8.2)$$

where ν is the kinematic viscosity. Stratified turbulence requires small Fr_h (for "stratified") and large Re (for "turbulence"), but the definition of "large Re" depends on the degree of stratification. It has been recognized for some time that, at large but fixed Re, decreasing Fr_h can suppress turbulence [e.g., *Laval et al.*, 2003; *Riley and deBruynKops*, 2003; *Waite and Bartello*, 2004]. Stronger stratifications therefore require larger Reynolds numbers. Building on the work of *Smyth and Moum* [2000] and others (*Riley and deBruynKops* [2003], *Hebert and de Bruyn Kops* [2006a,b]), *Brethouwer et al.* [2007] argued that stratified turbulence requires large buoyancy Reynolds number

$$\mathrm{Re}_b \equiv \mathrm{Re}\,\mathrm{Fr}_h^2 \equiv \frac{U^3}{\nu N^2 L_h}, \qquad (8.3)$$

which implies that the criterion for "large" Re increases like Fr_h^{-2} as $\mathrm{Fr}_h \to 0$. Flows with $\mathrm{Re}_b \lesssim 1$ are strongly damped by vertical viscosity, even if $\mathrm{Re} \gg 1$ (see Section 8.2.3).

Mesoscale motions in the atmosphere lie well inside the stratified turbulence parameter regime: Typical values of $U = 1$ m/s, $L = 100$ km, $N = 10^{-2}$ s^{-1}, and $\nu = 10^{-5}$ m^2/s yield

$$\mathrm{Fr}_h = 10^{-3}, \qquad \mathrm{Re} = 10^{10}, \qquad \mathrm{Re}_b = 10^4. \qquad (8.4)$$

Geophysical values of Fr_h can be readily obtained in the laboratory and computationally, but realizable values of Re, and hence Re_b, are many orders of magnitude smaller. Contemporary laboratory experiments and direct numerical simulations (DNSs) of stratified turbulence can obtain Re as high as $O(10^4)$ [e.g., *Praud et al.*, 2005; *Bartello and Tobias*, 2013]. While this is a large value, it is not necessarily sufficient to yield large Re_b at small Fr_h, especially under decaying conditions. For example, consider the decaying grid turbulence experiments of *Praud et al.* [2005]: A representative experiment has initial Froude and Reynolds numbers of 0.086 and 9000, which give $\mathrm{Re}_b = 67$. However, as the turbulence decays, U decreases and L_h increases while N and ν stay fixed. Since $\mathrm{Re}_b \propto U^3/L_h$, the buoyancy Reynolds number decreases rapidly and falls to $O(1)$ after only a few turnover times. Indeed, *Brethouwer et al.* [2007] surveyed a number of experimental papers and found them all to have $\mathrm{Re}_b \lesssim O(1)$, except for when the stratification was very weak.

Direct numerical simulations face a similar challenge in capturing stratified turbulence with large Re_b, because such flows have a wide scale separation between the energy-containing scale L_h, the Ozmidov scale

$$L_O \equiv 2\pi \left(\frac{\epsilon}{N^3}\right)^{1/2}, \qquad (8.5)$$

and the Kolmogorov scale

$$L_d \equiv 2\pi \left(\frac{\nu^3}{\epsilon}\right)^{1/4}, \qquad (8.6)$$

where ϵ is the kinetic energy dissipation rate (length scales are defined with the factor 2π for consistency with the usual wave number definitions). The Ozmidov scale is the scale below which isotropic three-dimensional turbulence occurs [*Lumley*, 1964; *Ozmidov*, 1965], while L_d is the scale below which three-dimensional turbulence is damped by viscosity. Using the Taylor relation $\epsilon \sim U^3/L_h$, it can be shown that the ratios between these scales are

$$\frac{L_O}{L_h} \sim \mathrm{Fr}_h^{3/2}, \qquad \frac{L_d}{L_O} \sim \frac{1}{\mathrm{Re}_b^{3/4}}, \qquad (8.7)$$

so strong stratification at large Re_b necessitates

$$L_h \gg L_O \gg L_d \qquad (8.8)$$

[e.g., *Brethouwer et al.*, 2007]. Since DNS requires spatial resolutions of $\Delta x \lesssim L_d$ [e.g., *Moin and Mahesh*, 1998], the computational challenge presented by (8.8) is quite demanding. As a consequence of this difficulty, numerical studies of stratified turbulence commonly employ hyperviscosity or other ad hoc sub-grid-scale models to avoid direct resolution of L_d [e.g., *Herring and Métais*, 1989; *Waite and Bartello*, 2004; *Lindborg*, 2006; *Waite*, 2011]. Nevertheless, a number of recent DNS studies have reached modest Re_b values up to $O(100)$ at small Fr_h [*Kimura and Herring*, 2012; *Bartello and Tobias*, 2013; *Almalkie and deBruynKops*, 2012].

In this chapter, we will review and investigate the dynamics of stratified turbulence with buoyancy Reynolds numbers around unity. This is the regime of laboratory-scale stratified turbulence, though it is often employed, via experiments and simulations, as an idealization of the atmospheric mesoscale and oceanic sub-mesoscale. In keeping with both themes of this book, that is, laboratory and numerical models for atmospheric and oceanic flows, we will use DNSs to investigate the dynamics of laboratory-scale turbulence. This chapter is about the laboratory parameter regime, not a particular set of laboratory experiments; the focus will be on the use of theory and idealized simulation to better understand turbulence at these scales. We will begin in Section 8.2 with a brief review of stratified turbulence theory, including the different parameter regimes, applications to the atmospheric mesoscale, and how viscous effects become important when $\mathrm{Re}_b \sim O(1)$, even when the Reynolds number is large. In Section 8.3, new DNSs of stratified turbulence are presented, with Re_b in the range of 0.2–2. The dependence of energy spectra, transfer spectra, and related length scales on Fr_h, Re, and Re_b is discussed. Conclusions are given in Section 8.4.

8.2. BACKGROUND

8.2.1. Equations and Scale Analysis

The equations of motion for an incompressible stratified fluid subject to the Boussinesq approximation are

$$\frac{\partial \mathbf{u}}{\partial t} + \mathbf{u} \cdot \nabla \mathbf{u} = -\frac{1}{\rho_0}\nabla p + b\hat{z} + \nu \nabla^2 \mathbf{u}, \quad (8.9)$$

$$\nabla \cdot \mathbf{u} = 0, \quad (8.10)$$

$$\frac{\partial b}{\partial t} + \mathbf{u} \cdot \nabla b + N^2 w = \kappa \nabla^2 b, \quad (8.11)$$

where $\mathbf{u} = u\hat{x} + v\hat{y} + w\hat{z}$ is the velocity, b is the buoyancy, and p is the dynamic pressure. The definition of b depends on the fluid in question: It is $g\theta/\theta_0$ in a dry atmosphere and $-g\rho/\rho_0$ in the ocean, where θ and ρ are potential temperature and density perturbations, θ_0 and ρ_0 are constant reference values, and g is gravity. The mass diffusivity κ is assumed to be equal to ν (i.e., unit Schmidt number Sc), but much of the following discussion is also valid for the oceanic regime of $Sc \gg 1$. The Brunt–Väisälä frequency N is defined for the atmosphere as

$$N^2 \equiv \frac{g}{\theta_0}\frac{d\bar{\theta}}{dz}, \quad (8.12)$$

where $\bar{\theta}(z)$ is the basic state potential temperature. Here, N is assumed to be constant.

The equations of motion can be nondimensionalized in different ways depending on whether the underlying flow is vortical or wavelike. The main difference in these approaches is in the time scale, which is characterized by the advection time scale L_h/U for vortical flows and the buoyancy time scale $1/N$ for waves [*Drazin*, 1961; *Riley et al.*, 1981; *Lilly*, 1983]. Using the vortical time scale, the dimensionless equations are [following *Riley and Lelong*, 2000]

$$\frac{\partial \mathbf{u}'_h}{\partial t'} + \mathbf{u}'_h \cdot \nabla'_h \mathbf{u}'_h + Fr_v^2 w' \frac{\partial \mathbf{u}'_h}{\partial z'}$$
$$= -\nabla'_h p' + \frac{1}{Re}\left(\nabla_h'^2 + \frac{1}{\alpha^2}\frac{\partial^2}{\partial z'^2}\right)\mathbf{u}'_h, \quad (8.13)$$

$$Fr_h^2\left(\frac{\partial w'}{\partial t'} + \mathbf{u}'_h \cdot \nabla'_h w' + Fr_v^2 w' \frac{\partial w'}{\partial z'}\right)$$
$$= -\frac{\partial p'}{\partial z'} + b' + \frac{Fr_h^2}{Re}\left(\nabla_h'^2 + \frac{1}{\alpha^2}\frac{\partial^2}{\partial z'^2}\right)w', \quad (8.14)$$

$$\nabla'_h \cdot \mathbf{u}'_h + Fr_v^2 \frac{\partial w'}{\partial z'} = 0, \quad (8.15)$$

$$\frac{\partial b'}{\partial t'} + \mathbf{u}'_h \cdot \nabla'_h b' + Fr_v^2 w'\frac{\partial b'}{\partial z'} + w' = \frac{1}{Re}\left(\nabla_h'^2 + \frac{1}{\alpha^2}\frac{\partial^2}{\partial z'^2}\right)b', \quad (8.16)$$

where primes denote dimensionless variables and subscript h denotes the horizontal component. The aspect ratio of the flow is

$$\alpha \equiv \frac{L_v}{L_h} \equiv \frac{Fr_h}{Fr_v}. \quad (8.17)$$

Consider first the inviscid dynamics of (8.13)–(8.16) (viscous effects will be reviewed in Section 8.2.3). By definition, strong stratification means $Fr_h \ll 1$. But the limiting behavior of (8.13)–(8.16) is largely controlled by the size of Fr_v, which has so far not been specified. Early work on stratified turbulence was based on the premise of small Fr_v [*Riley et al.*, 1981; *Lilly*, 1983], which leads to the neglect of several terms in (8.13)–(8.16). This assumption implies that vertical advection is small, the flow is nearly in hydrostatic balance, and the horizontal velocity field is approximately nondivergent. The equations of motion reduce to vertically decoupled layers of two-dimensional turbulence in this limit, which is commonly referred to as quasi-two-dimensional, layerwise two-dimensional, or pancake turbulence. More rigorous approaches, based on averaging over high-frequency internal gravity waves, yield the same limiting equations (with the possible inclusion of advection by a vertically sheared horizontal mean flow) [*Babin et al.*, 1997; *Embid and Majda*, 1998].

There is a self-destructive paradox built into the layerwise two-dimensional turbulence picture, which was anticipated by *Lilly* [1983]. Vertical decoupling implies a collapse of vertical scale. Depending on the ultimate size of L_v after this collapse, the vertical Froude number may no longer be small and the scaling may break down. In the inviscid case, *Lilly* [1983] predicted that Kelvin–Helmholtz instabilities would ultimately develop, halting the vertical collapse and three-dimensionalizing the flow at small scales. Such instabilities were subsequently observed in the numerical simulations of *Laval et al.* [2003] and *Riley and deBruynKops* [2003], along with a number of subsequent studies.

Billant and Chomaz [2001] revisited the scale analysis of *Riley et al.* [1981] and *Lilly* [1983] and argued that their assumption of $Fr_v \ll 1$ is inappropriate for stratified turbulence. They claimed that the vertical scale would adjust naturally to keep $Fr_v \sim O(1)$, implying that L_v would be set by the buoyancy scale

$$L_b \equiv 2\pi\frac{U}{N}. \quad (8.18)$$

Note that L_b is sometimes called the overturning scale because of its association with the appearance of small-scale density overturning [e.g., *Munk*, 1981; *Waite and Bartello*, 2006], and it is distinct from the smaller Ozmidov scale (which, confusingly, is sometimes also called the buoyancy scale). The argument of *Billant and Chomaz* [2001] was based on the self-similarity of the equations of motion when $L_v \sim L_b$ as well as the finding that L_b is the dominant vertical scale of the zigzag instability [*Billant and Chomaz*, 2000]. However, this scaling is also implied by *Lilly* [1983], since L_b is the vertical scale at which the Richardson number of layerwise two-dimensional turbulence becomes $O(1)$ and, presumably, Kelvin–Helmholtz instabilities develop. *Waite and Bartello* [2004] measured L_v in numerical simulations of stratified turbulence and confirmed that $L_v \sim L_b$, and subsequent numerical studies have been consistent with this finding. The inviscid limiting dynamics of (8.13)–(8.16) at $\mathrm{Fr}_h \ll 1$ and $\mathrm{Fr}_v \sim 1$ are very different from the classical picture of layerwise two-dimensional turbulence. Such turbulence is three dimensional in the sense that horizontal and vertical advection have the same order of magnitude, but it is anisotropic because $\alpha \sim \mathrm{Fr}_h \ll 1$.

8.2.2. Cascade Theories and Application to Atmospheric Mesoscale

Observations of the atmospheric kinetic energy spectrum suggest that it has a double power law form in horizontal wave number: At synoptic scales, it has a spectral slope of -3, in agreement with QG turbulence theory [*Charney*, 1971], but in the mesoscale the slope shallows to something resembling $-\frac{5}{3}$ [*Nastrom and Gage*, 1985; *Cho et al.*, 1999]. A number of different theories were proposed in the late 1970s and early 1980s to explain the observed form of the mesoscale spectrum. *Lilly* [1983] and *Gage and Nastrom* [1986] advanced a stratified turbulence hypothesis: They argued that the layerwise two-dimensional nature of stratified turbulence with small Fr_v might support an inverse cascade of energy through the mesoscale, in analogy with two-dimensional turbulence [*Kraichnan*, 1967]. Such a cascade would require a small-scale source of kinetic energy, which *Lilly* [1983] speculated could be due to moist convection at the $O(1)$ km scale. Around the same time, a different explanation based on a direct cascade of gravity wave energy was proposed [*Dewan*, 1979; *VanZandt*, 1982]. More recent alternatives to the stratified turbulence hypothesis have included theories based on QG [*Tung and Orlando*, 2003] and surface-QG turbulence [*Tulloch and Smith*, 2009]. *Waite and Snyder* [2013] have considered the effect of direct mesoscale forcing by latent heating.

Although it was an intriguing idea, the inverse cascade theory for stratified turbulence was not supported by numerical simulations. *Herring and Métais* [1989] attempted to obtain an inverse cascade by applying small-scale forcing in simulations over a range of Froude numbers, but they were unsuccessful. *Lilly et al.* [1998] considered rotating stratified turbulence with small-scale forcing; they found an inverse cascade when strong rotation was present, but not for purely stratified turbulence. *Smith and Waleffe* [2002] found a direct transfer of energy from small-scale forcing into a vertically sheared mean flow, but there was no cascade through intermediate scales. A theoretical explanation for the lack of an inverse cascade was given by *Waite and Bartello* [2004]: The two-dimensional inverse cascade is a result of the conservation of enstrophy (along with energy) by wave number triads; the analogous quantity in stratified turbulence is the potential enstrophy, which is approximately conserved by triads for $\mathrm{Fr}_v \ll 1$. But the relationship between energy and (potential) enstrophy is weaker in stratified turbulence than in two-dimensional turbulence, because gravity waves carry some of the energy but no potential enstrophy. Even when $\mathrm{Fr}_v \ll 1$, layerwise two-dimensional turbulence eventually leaks energy into gravity waves, which cascade downscale and destroy the possibility of an inverse cascade. More recent data analysis after *Nastrom and Gage* [1985] has also contributed to the rejection of the inverse cascade theory, as it shows a downscale flux of mesoscale energy below scales of around 100 km [*Lindborg and Cho*, 2001].

The lack of an inverse cascade seemed to mark the end of the stratified turbulence hypothesis for the atmospheric mesoscale, but it was revived in a very different form by *Lindborg* [2006]. The idea that stratified turbulence naturally develops a vertical scale of L_b, and hence $\mathrm{Fr}_v \sim O(1)$, suggests that it should be anisotropic but three dimensional. *Lindborg* [2006] proposed a theory for stratified turbulence with a direct cascade of energy from large to small horizontal scales. Proceeding on dimensional grounds along the lines of the *Kolmogorov* [1941] theory for three-dimensional turbulence, *Lindborg* [2006] argued that the kinetic energy spectrum should have the form

$$E_K(k_h) \sim \epsilon^{2/3} k_h^{-5/3}, \qquad E_K(k_v) \sim N^2 k_v^{-3}, \qquad (8.19)$$

in horizontal and vertical wave numbers k_h and k_v. He suggested that the mesoscale kinetic energy spectrum could be understood as a stratified turbulence direct cascade from synoptic scales to the microscale. The same claim has also been advanced by *Riley and Lindborg* [2008] to explain the energy spectrum of the ocean sub-mesoscale.

Lindborg [2006] presented numerical simulations that showed good agreement with (8.19), at least in the horizontal. To minimize viscous effects and capture the strong anisotropy at small Fr_h, he used thin numerical grids with $\Delta z \ll \Delta x$, along with separate horizontal and vertical hyperviscosity to keep dissipation focused around the grid scale. Most other numerical studies of stratified turbulence have employed isotropic grids with $\Delta z \sim \Delta x$ and have obtained a wider range of spectral slopes. Several DNS studies [*Riley and deBruynKops*, 2003; *Brethouwer et al.*, 2007; *Almalkie and deBruynKops*, 2012; *Bartello and Tobias*, 2013; *Kimura and Herring*, 2012] are broadly consistent with *Lindborg* [2006], despite having much larger dissipation ranges. Steeper spectra, with slopes as low as −5, have been found in some studies [*Laval et al.*, 2003; *Waite and Bartello*, 2004], but as *Brethouwer et al.* [2007] pointed out, this steepening seems to result from excessive vertical dissipation. Horizontal slopes of −2 to −3 have also been found, raising some questions about the universality of (8.19) [*Waite*, 2011; *Kimura and Herring*, 2012]. The vertical spectrum in (8.19) has been more difficult to reproduce numerically, but *Kimura and Herring* [2012] have reported a clear −3 slope at small vertical scales. At vertical scales larger than L_b, the vertical spectrum tends to be flat, which is consistent with the layerwise decoupling that occurs for $L_v \gg L_b$ [*Waite and Bartello*, 2004].

Brethouwer et al. [2007] speculated that the stratified turbulence inertial subrange extends down to the Ozmidov scale, below which it transitions to isotropic three-dimensional turbulence. However, the details of this transition are not entirely clear. As envisioned by *Lilly* [1983], the layered structure of stratified turbulence can lead to the development of shear instabilities at small horizontal scales. Such instabilities have been observed in numerical simulations and are associated with bumps in the kinetic energy spectrum at large k_h [*Laval et al.*, 2003; *Brethouwer et al.*, 2007; *Waite*, 2011]. These bumps are located at horizontal scales around the buoyancy scale and appear to result from a nonlocal transfer of energy from large, quasi-horizontal vortices to small-scale Kelvin-Helmholtz billows [*Waite*, 2011; *Khani and Waite*, 2014]. These results suggest that L_b rather than L_O marks the small-scale end of the *Lindborg* [2006] inertial subrange. In practice, these scales are usually quite similar, as

$$\frac{L_O}{L_b} \sim \mathrm{Fr}_h^{1/2}, \qquad (8.20)$$

so very small Fr_h are required to get a wide separation between L_b and L_O [e.g., *Brethouwer et al.*, 2007].

8.2.3. Viscous Effects

Our discussion so far has been focused on the inviscid dynamics of (8.13)-(8.16). However, it is viscosity that ultimately distinguishes the laboratory and geophysical regimes of stratified turbulence. *Riley and deBruynKops* [2003] pointed out that viscous coupling rather than Kelvin-Helmholtz instabilities sets the vertical scale of stratified turbulence when viscous effects are sufficiently strong. Assuming a balance between the vertical viscosity and advective terms in (8.13), they estimated that the resulting vertical scale would be on the order of the viscous scale

$$L_{\mathrm{visc}} \equiv 2\pi \sqrt{\frac{L_h \nu}{U}} \qquad (8.21)$$

(note that L_{visc} is not the same as the Kolmogorov scale L_d; see also *Godoy-Diana et al.* [2004]). Indeed, numerical simulations have shown strongly layered flows with laminar coupling in the vertical scale when the stratification is increased at fixed Re (or, analogously, fixed resolution with ad hoc grid-scale dissipation) [*Laval et al.*, 2003; *Waite and Bartello*, 2004]. *Brethouwer et al.* [2007] showed that the vertical scale is set by L_{visc} at small Re_b and L_b at large Re_b, with a transition range over $1 \lesssim \mathrm{Re}_b \lesssim 10$. The laboratory experiments of *Praud et al.* [2005] also exhibit a viscous scaling of vertical scale, consistent with their relatively small values of Re_b.

Viscous effects can be important in strongly stratified turbulence, even at large Reynolds number. When $\mathrm{Re} \gg 1$, the horizontal part of the viscous term in (8.13) is small, but the vertical part has magnitude

$$\frac{1}{\mathrm{Re}\,\alpha^2} \sim \begin{cases} 1 & \text{if } \mathrm{Re}_b \lesssim 1 \quad (L_v \sim L_{\mathrm{visc}}), \\ 1/\mathrm{Re}_b & \text{if } \mathrm{Re}_b \gtrsim 1 \quad (L_v \sim L_b), \end{cases} \qquad (8.22)$$

[*Brethouwer et al.*, 2007]. So viscous effects due to vertical gradients can be significant at large Reynolds number if the stratification is strong enough to make $\mathrm{Re}_b \lesssim 1$. The apparent paradox of strong vertical viscosity at large Reynolds numbers is explained by the anisotropy of stratified turbulence. As stratification increases, the characteristic vertical scale decreases like L_b and the aspect ratio of the flow decreases like Fr_h. At large but fixed Re, L_b ultimately falls inside the vertical dissipation range, at which point the horizontal cascade is suppressed; this transition occurs around $\mathrm{Re}_b \sim 1$ [*Riley and deBruynKops*, 2003; *Brethouwer et al.*, 2007]. The cascade theory of *Lindborg* [2006] assumes an inertial subrange in k_h over which viscous effects are negligible and thus cannot be expected to hold in the laboratory regime of $\mathrm{Re}_b \lesssim 1$, where vertical viscosity may not be restricted to large k_h. Instead, simulations point to a steep -5 spectrum when vertical damping (viscous or ad hoc) is strong [*Laval et al.*, 2003; *Waite and Bartello*, 2004; *Brethouwer et al.*, 2007].

8.3. DIRECT NUMERICAL SIMULATION

8.3.1. Setup

In order to further explore the parameter regime of laboratory-scale stratified turbulence, we have performed a number of DNSs with modest buoyancy Reynolds

Table 8.1. Parameters, nondimensional numbers, and length scales for each simulation. In the run labels, A, B, C denote runs with approximately equal Re_b; 1, 2, 3, etc., denote runs with equal v and approximately equal Re. Dimensional quantities are given in cgs units.

Run	N	v	n	Δx	Δt	U	L_h	Fr_h	Re	Re_b	L_b	L_O	L_d
A1	0.1	1	128	7.4	0.25	1.6	150	0.11	240	2.7	100	32	16
A2	0.141	0.5	256	3.7	0.125	1.6	170	0.069	530	2.5	71	19	9.4
A3	0.2	0.25	512	1.8	0.0625	1.6	190	0.043	1200	2.3	51	11	5.7
A4	0.283	0.125	512	1.8	0.05	1.6	180	0.032	2400	2.4	37	6.5	3.4
A5	0.4	0.0625	1024	0.92	0.025	1.6	170	0.023	4300	2.3	25	3.8	2.0
A6	0.566	0.03125	1536	0.61	0.0167	1.7	180	0.017	9400	2.6	18	2.4	1.2
B1	0.2	1	128	7.4	0.25	1.6	170	0.046	270	0.58	50	11	16
B2	0.283	0.5	256	3.7	0.125	1.7	180	0.032	620	0.65	38	6.8	9.3
B3	0.4	0.25	512	1.8	0.05	1.7	190	0.022	1300	0.59	26	3.8	5.7
B4	0.566	0.125	512	1.8	0.05	1.7	190	0.016	2500	0.63	19	2.3	3.3
B5	0.8	0.0625	1024	0.92	0.02	1.7	180	0.011	4900	0.63	13	1.4	2.0
B6	1.13	0.03125	1536	0.61	0.0133	1.7	190	0.0081	10000	0.69	9.6	0.87	1.2
C1	0.4	1	128	7.4	0.2	1.7	210	0.020	370	0.15	27	3.9	16
C2	0.566	0.5	256	3.7	0.1	1.8	220	0.014	790	0.16	20	2.4	9.4
C3	0.8	0.25	512	1.8	0.04	1.8	230	0.0097	1600	0.15	14	1.4	5.6
C4	1.13	0.125	512	1.8	0.04	1.8	230	0.0069	3400	0.16	10	0.85	3.3
C5	1.6	0.0625	1024	0.92	0.015	2.0	270	0.0045	8600	0.17	7.7	0.51	1.9
C6	2.26	0.03125	1536	0.61	0.01	1.9	250	0.0034	15000	0.18	5.3	0.31	1.1

numbers. Simulations are designed to have roughly equal velocity and length scales, so different Fr_h, Re, and Re_b are obtained by varying N and v. Three sets of six numerical experiments are considered, corresponding to approximately equal buoyancy Reynolds numbers of ≈ 2, 0.6, and 0.2 (labeled A, B, and C, respectively). In each set of simulations, v is varied by factors of $\frac{1}{2}$ and N is varied by factors of $\sqrt{2}$ to obtain a spread of Re and Fr_h with the same Re_b. The overall ranges of Fr_h and Re are 0.003–0.1 and 200–10,000, respectively, which are realizable values in laboratory experiments. Parameters for all runs are listed in Table 8.1.

The computational domain size is $L \times L \times (L/4)$, with periodic boundary conditions on u and b in each direction. The numerical model employs a Fourier-transform-based spectral method with third-order Adams-Bashforth time stepping, and viscous terms are treated implicitly with a trapezoidal approach. The Fourier discretization uses $n \times n \times (n/4)$ wave numbers, yielding a wave number spacing of $\Delta k \equiv 2\pi/L$ in each direction. Aliasing errors are eliminated by the standard two-thirds rule [e.g., *Durran*, 2010], which is implemented by truncating wave vectors $\mathbf{k} = (k_x, k_y, k_z)$ outside the sphere,

$$|\mathbf{k}| \leq \frac{2}{3}\frac{n\pi}{L}. \quad (8.23)$$

Accounting for this truncation, the effective grid spacing is

$$\Delta x \equiv \Delta y \equiv \Delta z \equiv \frac{3}{2}\frac{L}{n}. \quad (8.24)$$

The grid spacing for a given v is chosen to resolve the Kolmogorov scale with $L_d/\Delta x \approx 2$. The time step Δt was selected to be roughly as large as possible while ensuring stability.

Forcing was applied to obtain statistically stationary turbulence and avoid the rapid decay of unforced simulations [as in, e.g., *Riley and deBruynKops*, 2003]. We have modified the deterministic forcing approach of *Sullivan et al.* [1994] to excite vortical motion at large horizontal scales (a similar approach was employed by *Bartello* [2000] to force barotropic vortical motion). The forcing maintains a fixed value E_0 of horizontally rotational kinetic energy in a cylindrical wave number shell with radius k_f, given by

$$|k_h - k_f| \leq \frac{\Delta k}{2}, \quad \text{where} \quad k_h \equiv \sqrt{k_x^2 + k_y^2}. \quad (8.25)$$

Over every time step, the energy in this shell naturally evolves away from its initial value. The forcing is applied by uniformly scaling the rotational modes in the shell at the end of every time step to return their kinetic energy to E_0. Rotational forcing is used to avoid exciting large-amplitude internal gravity waves at the forcing scale

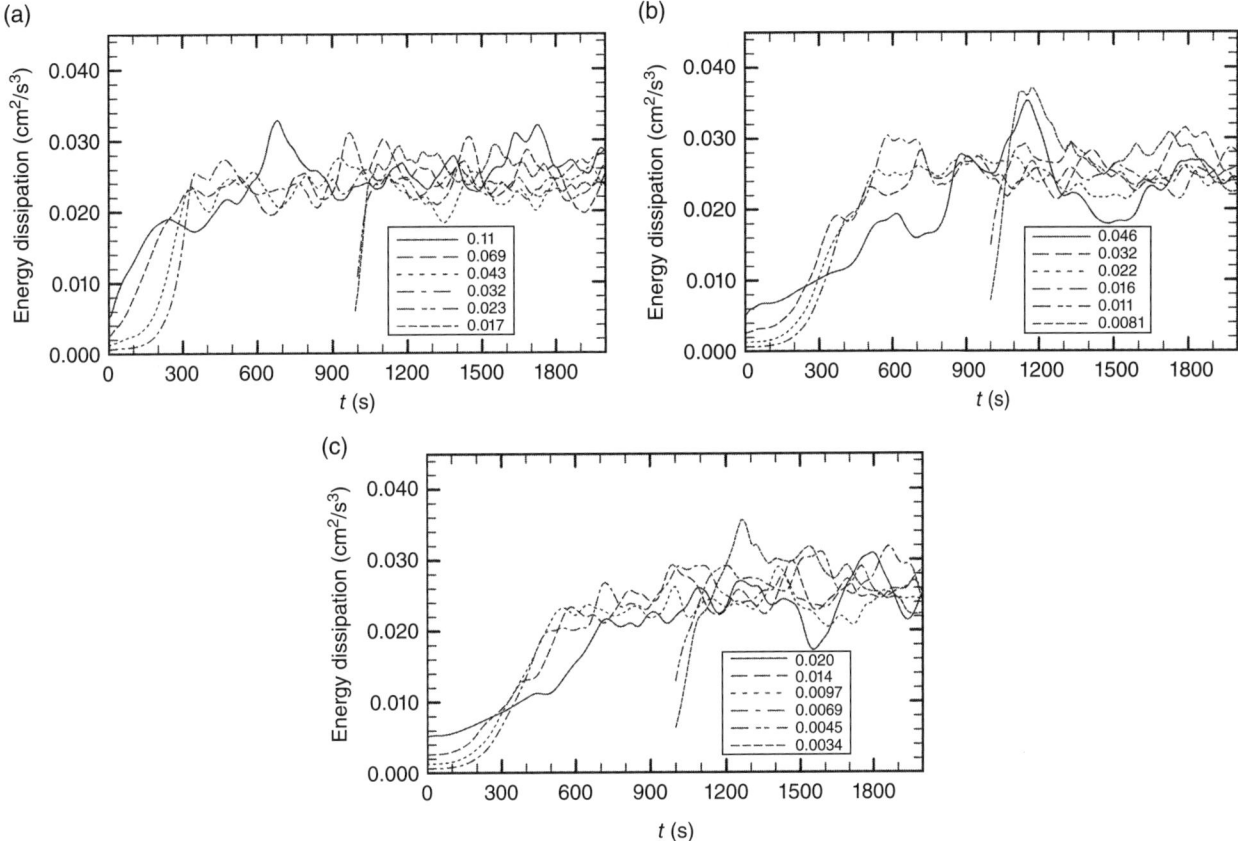

Figure 8.1. Time series of kinetic energy dissipation in simulations with $Re_b \approx$ (a) 2, (b) 0.6, and (c) 0.2 (simulation sets A, B, and C). Dash patterns denote different Fr_h.

(forcing total kinetic energy was found to produce wave-dominated flows for certain parameter values). Relatively large horizontal scales are forced by setting $k_f = 3\,\Delta k$.

The dimensional domain size L is set to 2π meters for simplicity, so the wave number spacing is $\Delta k = 1\,\text{m}^{-1}$. The forcing level E_0 is chosen to produce velocities comparable to those of laboratory experiments. We let $E_0 = 1\,\text{cm}^2/\text{s}^2$, which yields rms velocities of $U \approx 1\text{–}2$ cm/s and dissipation rates of $\epsilon \approx 0.02\text{–}0.03\,\text{cm}^2/\text{s}^3$. Characteristic horizontal scales calculated from Taylor's relation $L_h \equiv U^3/\epsilon$ are around 2 m, yielding turnover time scales L_h/U of $O(100)$ s. The Kolmogorov scales of the different simulations range between 1 and 16 cm.

Lower-resolution simulations with $n \leq 512$ are initialized with low-level noise and run for 2000 s. Higher-resolution simulations with $n \geq 1024$ are first spun up to $t = 1000$ s with $n = 512$ and correspondingly lower Re and then continued to $t = 2000$ s at full resolution. Time series of ϵ (Figure 8.1) show that the simulations started from rest take several hundred seconds to spin up, while those restarted at $t = 1000$ s adjust much more quickly. The turbulence appears reasonably stationary by $t = 1200$ s in all cases. The various quantities presented below (energy spectra, transfers, etc.) are averaged over the last 800 s of each simulation. Values of Fr_h, Re, and Re_b for each simulation, along with velocity and length scales, are given in Table 8.1. These values are computed using the rms velocity U, the time-averaged dissipation ϵ, and L_h from Taylor's relation.

8.3.2. Energy Spectra

Horizontal wave number spectra of kinetic energy $E_K(k_h)$ are computed by integrating the energy in each wave vector $\mathbf{k}$ over cylindrical wave number shells of width Δk [as in, e.g., *Waite*, 2011]. In other words, $E_K(k_h)\,\Delta k$ gives the total kinetic energy in modes with horizontal wave numbers within $\Delta k/2$ of k_h. Other spectra, such as the potential energy $E_P(k_h)$ and the transfer and buoyancy flux spectra discussed below, are computed similarly.

Over the range of parameters considered here, the kinetic energy spectra are extremely sensitive to Fr_h and Re. Figure 8.2a shows $E_K(k_h)$ from three simulations with the same stratification $Fr_h \approx 0.02$ and different Re. All the spectra are peaked at $k_h/\Delta k = 3$, where kinetic energy is injected by forcing. For the largest Reynolds number (which has $Re = 9400$ and $Re_b = 2.6$),

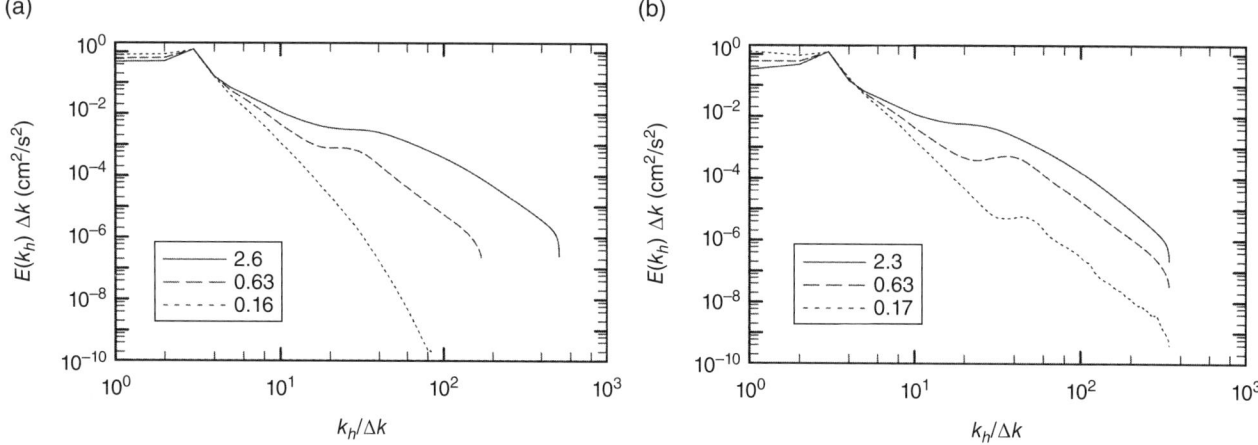

Figure 8.2. Horizontal wave number spectra of kinetic energy in simulations with (a) fixed $N = 0.566$ s^{-1} and different ν (and hence different Re and Re$_b$, i.e., runs A6, B4, and C2), and (b) fixed $\nu = 0.0625$ cm^2/s and different N (and hence different Fr$_h$ and Re$_b$, i.e., runs A5, B5, and C5). Dash patters denote different Re$_b$.

the spectrum exhibits a short power law range downscale of the forcing with a slope of around -3, followed by a broad spectral bump at larger k_h. As the Reynolds number (and hence Re$_b$) decreases, the spectrum steepens and the bump is reduced. Similar behavior is displayed in Figure 8.2b, which shows spectra from three simulations with the same viscosity and different stratifications. The spectra steepen as the Froude number (and hence Re$_b$) decreases, while the spectral bump shrinks and moves to higher k_h.

These changes to the kinetic energy spectrum with Fr$_h$ and Re can be partially accounted for by a dependence on Re$_b$ alone [as in *Brethouwer et al.*, 2007]. Spectra from all simulations are plotted Figure 8.3, where they are arranged into groups of approximately equal values of Re$_b \approx 2$, 0.6, 0.2. The forcing and power law portions of the spectra collapse fairly well at constant Re$_b$, at least for k_h not too large. The spectral slopes of the power law range are plotted in Figure 8.4 (slopes are computed with a least squares fit over $4 \leq k_h/\Delta k \leq 10$; note that for the lowest Re simulations in Figure 8.3a, there is no clear distinction between the power law range and spectral bump). The changes in slope obtained at constant Re$_b$ are for the most part much smaller than changes at constant Fr$_h$ or Re. All spectra are steeper than $-\frac{5}{3}$, which as discussed in Section 8.2.3 is to be expected at these values of Re$_b$. Overall, smaller Re$_b$ yield steeper spectra: Slopes are around -3, -4, and -5 for Re$_b \approx 2$, 0.6, 0.2. The collapse is very good for smaller Re$_b$, where the slopes vary by less than 10% at constant Re$_b$. A greater spread in slopes is found for Re$_b \approx 2$: In these simulations, there is a clear steepening from around -2 to -3 as Fr$_h$ decreases and Re increases, even though Re$_b$ is approximately constant. The slopes appear to have not quite converged in this case and may steepen below -3 for even smaller Fr$_h$ and larger Re.

At larger wave numbers the collapse of the spectra in Figure 8.3 is not particularly good. For each Re$_b$, the spectra obtained with the lowest Reynolds numbers have only a short power law range between the forcing and dissipation wave numbers. However, a spectral bump eventually emerges in each case as Re increases and Fr$_h$ decreases. As Re$_b$ decreases, higher values of Re (and hence smaller Fr$_h$) are required for a spectral bump to appear: For Re$_b \approx 0.2$, only the two highest-Re cases exhibit a bump downscale of the steep -5 spectrum. The position of the bump appears to move to larger k_h as Re increases and Fr$_h$ decreases, but its shape is quite variable; in general, it is broader for large Re$_b$, and it narrows (or disappears entirely) as Re$_b$ decreases. These findings are reminiscent of the buoyancy-scale bumps described by *Waite* [2011], which were located at k_h around the buoyancy wave number $k_b \equiv 2\pi/L_b \equiv N/U$. The possible relationship between the spectral bumps in Figure 8.3 and the buoyancy scale is investigated further below.

8.3.3. Energy Budget

The spectral budget of kinetic and potential energy is governed by the equations

$$\frac{\partial}{\partial t} E_K(k_h) = T_K(k_h) + B(k_h) - D_K(k_h) + F(k_h), \quad (8.26)$$

$$\frac{\partial}{\partial t} E_P(k_h) = T_P(k_h) - B(k_h) - D_P(k_h). \quad (8.27)$$

The terms $T_K(k_h)$ and $T_P(k_h)$ are the transfer spectra of kinetic and potential energy, which represent conservative exchanges of energy between different wave numbers by nonlinear interactions. The $B(k)$ term is the buoyancy flux, which is given by the cross spectrum of vertical velocity and buoyancy; it describes the wave number local conversion of potential to kinetic energy, and so it appears in both equations (8.26) and (8.27) with opposite signs. The $F(k_h)$ term denotes injection of kinetic energy by

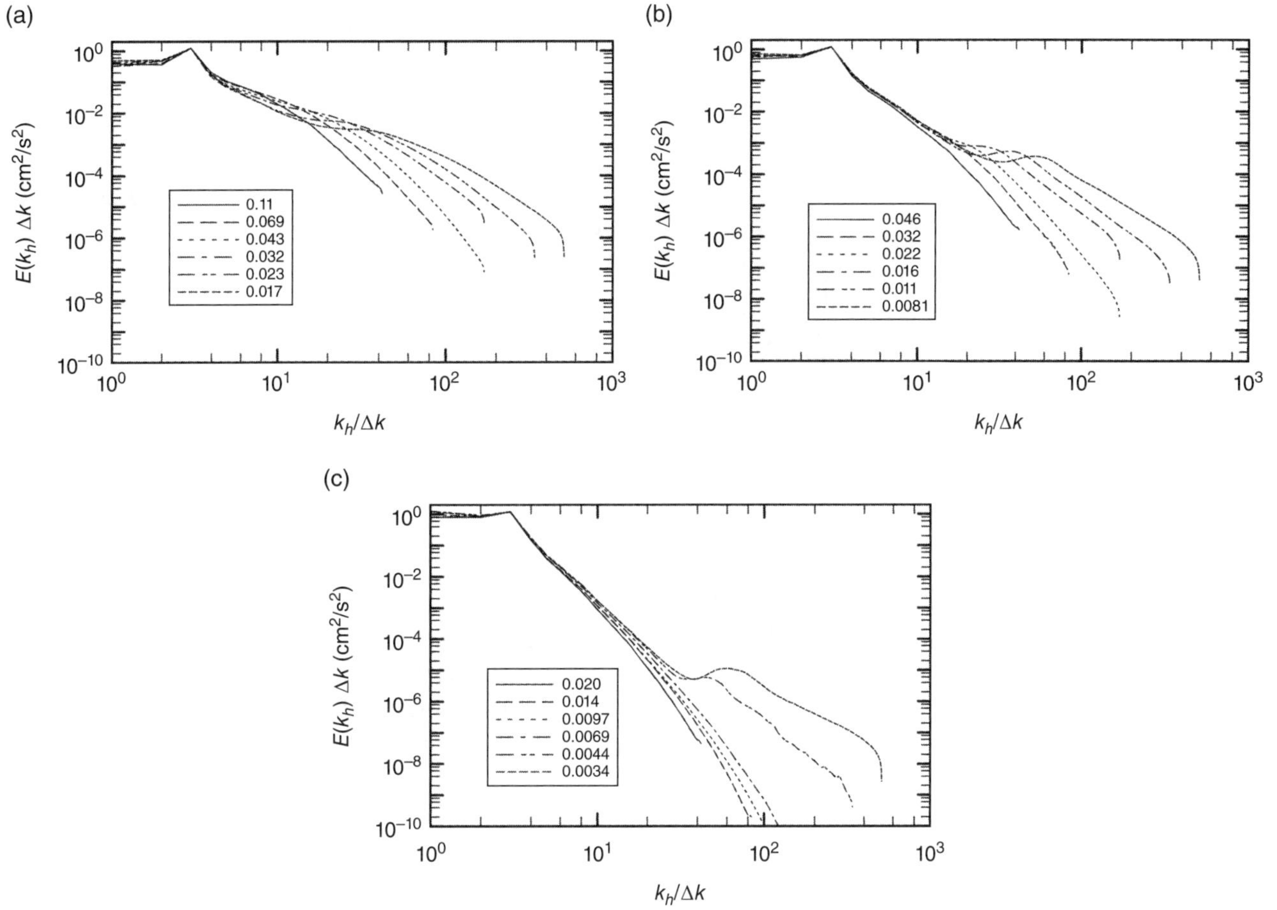

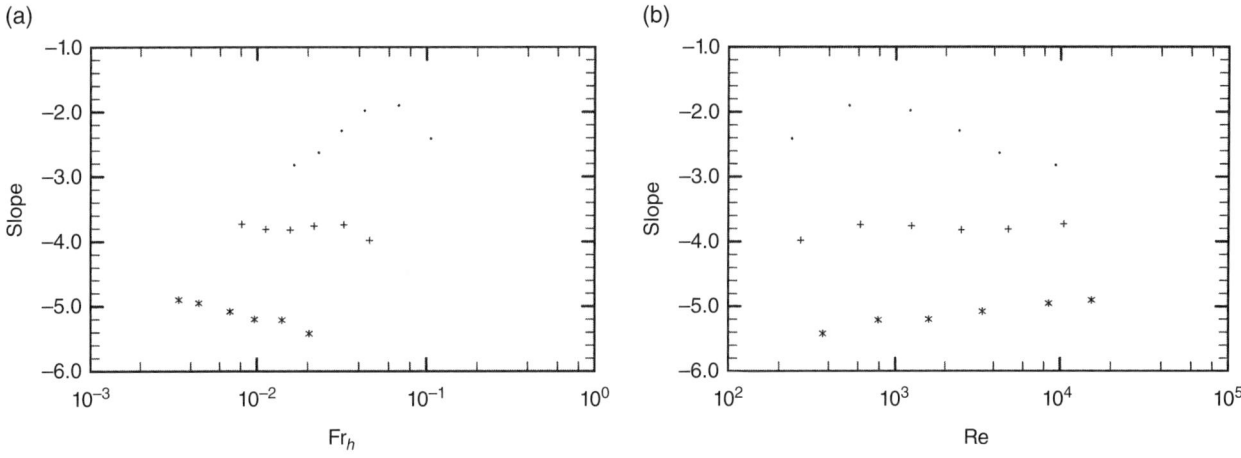

Figure 8.3. Horizontal wave number spectra of kinetic energy in simulations with $Re_b \approx$ (a) 2, (b) 0.6, and (c) 0.2 (simulation sets A, B, and C). Dash patterns denote different Fr_h.

Figure 8.4. Spectral slopes of the horizontal wave number spectra of kinetic energy, plotted vs. (a) Fr_h and (b) Re for $Re_b \approx 2$ (·), 0.6 (+), and 0.2 (∗) (simulation sets A, B, and C).

forcing (potential energy is not directly forced in these simulations), while $D_K(k_h)$ and $D_P(k_h)$ are the viscous and diffusive dissipation of kinetic and potential energy. These dissipation terms include the contributions from both horizontal and vertical gradients, and so their effects are not necessarily dominated by large k_h, as in a typical dissipation range at large Re.

Figure 8.5 shows the transfer and buoyancy flux spectra from three simulations with the same stratification $Fr_h \approx 0.02$ and different Re. Dissipation and forcing

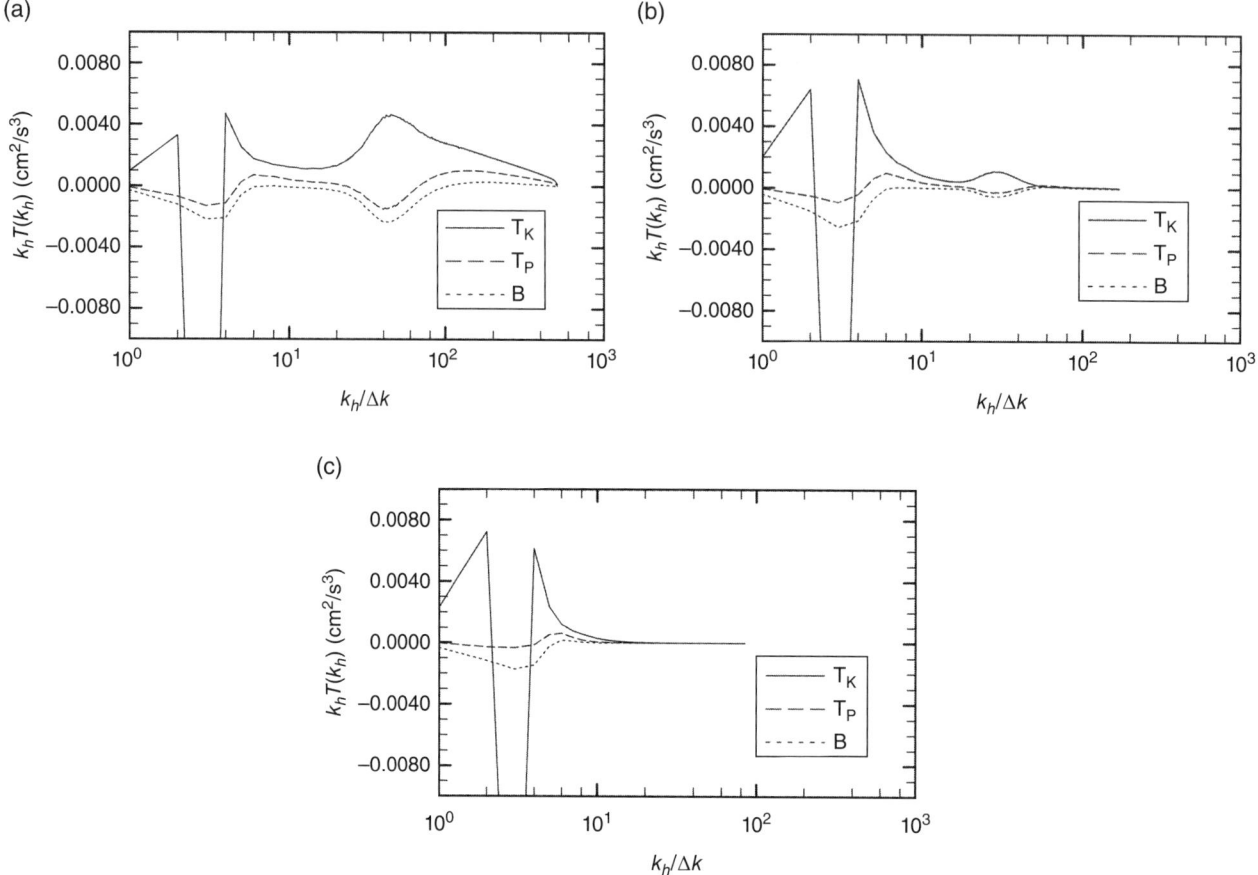

Figure 8.5. Horizontal wave number spectra of kinetic and potential energy transfer and buoyancy flux in simulations with fixed $N = 0.566$ s^{-1} and different ν (and hence different Re and Re$_b$), corresponding to Re$_b$ = (a) 2.6, (b) 0.63, and (c) 0.16 (runs A6, B4, and C2, as in Figure 8.2a). Spectra are multiplied by k_h to preserve area under the curves with loglinear axes.

terms are not shown but can be deduced as residuals since the spectra are approximately stationary. For the largest Re (Figure 8.5a), three distinct spectral ranges are apparent. Small wave numbers are dominated by the forcing: Kinetic energy is injected around $k_h/\Delta k \approx 3$, where it is primarily removed by nonlinear interactions and sent downscale (the rest is dissipated directly at k_f or converted to potential energy). For $k_h \gtrsim k_f$, there is a range over which $T_K(k_h)$ and (to a lesser extent) $T_P(k_h)$ are positive and $B(k_h) \approx 0$. If this range of k_h were a true inertial subrange, the transfer would be zero. The positive transfer implies that some of the kinetic energy injected into these scales is dissipated, which is to be expected since Re$_b$ = 2.6 is not very large.

Downscale of this plateau of positive transfer, there is a major transition in the spectral energy balance. A large peak in $T_K(k_h)$ around $k_h/\Delta k \approx 40$ points to a significant injection of kinetic energy. Some of this energy is converted to potential energy via the negative peak in buoyancy flux around the same scale, and the rest is dissipated. As the Reynolds number (and hence Re$_b$) decreases, strong dissipation is felt over a wider range of k_h, and the small-scale peaks in transfer and buoyancy flux are suppressed (Figures 8.5b and 8.5c). In the smallest Re$_b$ case, the energy injected by forcing is transferred only to nearby wave numbers, where it is dissipated directly.

Figure 8.6 shows transfer and buoyancy flux spectra from three simulations with approximately equal buoyancy Reynolds numbers around 2. These spectra are consistent with the picture described above: there is a short forcing range, followed by a plateau of positive transfer and negligible buoyancy flux, followed by large peaks of positive $T_K(k_h)$ and negative $T_P(k_h)$ and $B(k_h)$. As Fr$_h$ decreases and Re increases, the plateau gets longer as the peaks move downscale (in Figure 8.6a, the peaks are adjacent to the forcing range and no plateau is obtained). The position of the peak moves to larger k_h as Fr$_h$ decreases and Re increases, similar to the spectral bumps in Figure 8.3a. The magnitude of the bump does not diminish with decreasing Fr$_h$, suggesting that even smaller

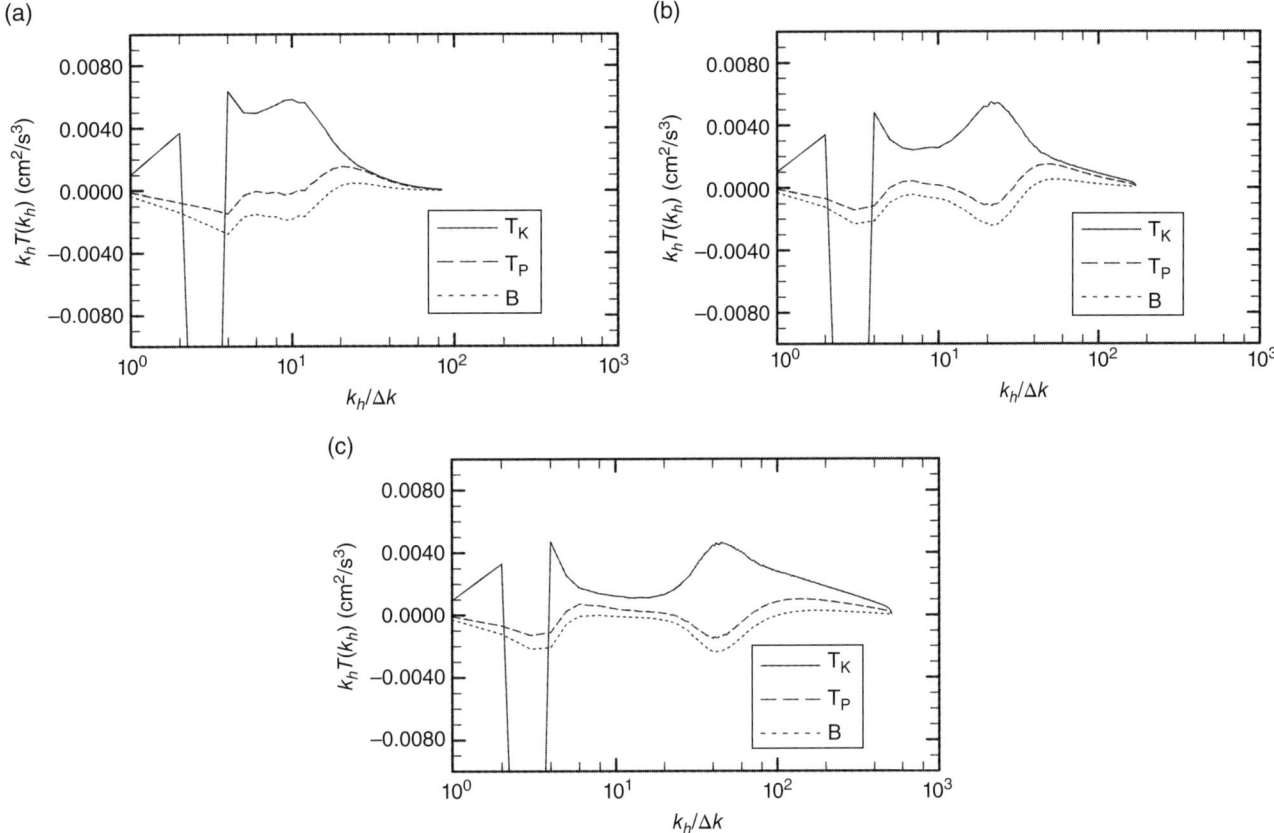

Figure 8.6. Horizontal wave number spectra of kinetic and potential energy transfer and buoyancy flux in simulations with fixed $Re_b \approx 2$ and different Fr_h and Re, corresponding to Fr_h = (a) 0.069, (b) 0.032, and (c) 0.017 (runs A2, A4, and A6). Spectra are multiplied by k_h to preserve area under the curves with loglinear axes.

Fr_h with $Re_b \approx 2$ would yield a similar injection of kinetic energy and negative buoyancy flux at even larger k_h.

At the smallest values of $Re_b \approx 0.2$, Figure 8.5c suggests that the transfer becomes very small just downscale of the forcing, where the energy spectra have steep slopes of around −5. This range is magnified with logarithmic axes in Figure 8.7, which shows transfer and buoyancy flux spectra for two simulations with $Re_b \approx 0.2$. The left panel is representative of the cases when $Fr_h \gtrsim 0.007$. For all $k_h \gtrsim k_f$, the total transfer dominates the buoyancy flux by at least an order of magnitude and is balanced by the vertical part of the dissipation. It has a spectral slope of around -4 at large scales but falls off faster than a power law at larger k_h. As observed above for larger Re_b, a small-scale transition emerges in the energy balance at sufficiently small Fr_h and large Re (Figure 8.7b), where a small bump of positive transfer and negative buoyancy flux emerges at large k_h. These bumps are quite intermittent, appearing and disappearing over the averaging interval $1200 \leq t \leq 2000$ s. In the averaged spectra in Figure 8.7b, they are centered around $k_h/\Delta k \approx 60$.

8.3.4. Length Scales

In physical space, these simulations exhibit the familiar structure seen in a number of recent studies of stratified turbulence: thin layers of predominantly horizontal velocity at large horizontal scales along with, in some cases, shear instabilities and small-scale turbulence [e.g., *Laval et al.*, 2003; *Riley and deBruynKops*, 2003; *Brethouwer et al.*, 2007; *Waite*, 2011; *Bartello and Tobias*, 2013]. Figure 8.8 shows representative snapshots of the y component of vorticity, which is dominated by the vertical shear $\partial u/\partial z$, for three different Re_b values (the simulation with the largest Re is shown in each case). For $Re_b = 2.6$, there are several patches resembling Kelvin–Helmholtz instabilities, some of which appear to have transitioned to smaller-scale turbulence. However, much of the domain remains quiet, with a smooth, layerwise structure. As Re_b decreases, the regions of instability become increasingly intermittent and without the associated breakdown into smaller-scale turbulence; at the time of the snapshots in Figure 8.8, only a few patches of instability are visible for $Re_b = 0.69$ and one for $Re_b = 0.18$.

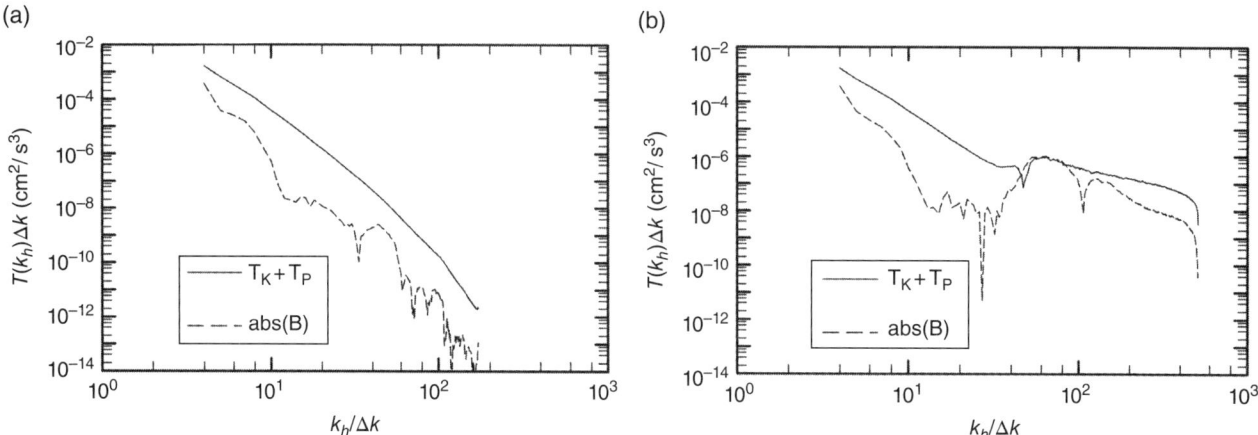

Figure 8.7. Horizontal wave number spectra of total energy transfer and buoyancy flux in simulations with fixed $Re_b \approx 0.2$ and different Fr_h and Re, corresponding to $Fr_h =$ (a) 0.0069 and (b) 0.0034 (runs C4 and C6). Log-log axes are used for clarity. As a result, only $k_h/\Delta k \geq 4$ are shown, and the absolute value of the buoyancy flux is taken. The buoyancy flux is positive over almost all k_h, but the peaks around $k_h/\Delta k \approx 40$ in (a) and 60 in (b) are negative.

Figure 8.8. Vertical (x,z) slices of the y component of vorticity in simulations with fixed $\nu = 0.03125$ cm^2/s and different N (and hence different Fr_h and Re_b), corresponding to $Re_b = 2.6$ (top), 0.69 (middle), and 0.18 (bottom) (runs A6, B6, and C6). Shading is saturated to white/black at ± 3 s^{-1}. Fields in (a) and (b) are plotted at the final time of the simulations $t = 2000$ s. Panel (c) is plotted at $t = 1900$ s, which was chosen to illustrate the lone patch of Kelvin-Helmholtz billows visible at this time.

The vertical scale of the layers seen in Figure 8.8 can be measured from the first moment of the vertical wave number spectrum of kinetic energy $E_K(k_v)$ (which is computed in an analogous fashion to the horizontal spectrum). We define the characteristic vertical scale to be

$$L_v \equiv 2\pi \frac{\int E_K(k_v)\,dk_v}{\int k_v E_K(k_v)\,dk_v} \qquad (8.28)$$

(a similar definition based on the $\frac{1}{2}$ moment was used by *Waite and Bartello* [2004]; *Brethouwer et al.* [2007] used (8.28)). As discussed in Section 8.2, the vertical scale of stratified turbulence is expected to scale like L_{visc} when the layer thickness is set by viscous coupling and L_b when the layers break down into shear instabilities and small-scale three-dimensional turbulence. In Figure 8.9, the vertical scale in all simulations is plotted against L_b and L_{visc}. The buoyancy and viscous scales are similar in these simulations, since (8.18) and (8.21) imply that

$$\frac{L_b}{L_{\text{visc}}} = \sqrt{\text{Re}_b}, \qquad (8.29)$$

and our $\sqrt{\text{Re}_b}$ values range from 0.39 to 1.6. Nevertheless, the L_{visc} scaling is more convincing than the L_b scaling. The layer thickness in these simulations therefore seems to be set by viscosity, which is consistent with our relatively small values of Re_b [as in *Brethouwer et al.*, 2007]. Note, however, that (8.29) can also be interpreted as a vertical Froude number for $L_v \sim L_{\text{visc}}$, which is therefore not much smaller than $O(1)$. Since U and L_v are based on mean quantities, we expect our simulations to have patches with locally larger U and smaller L_v, which would therefore be susceptible to instabilities like those visible in Figure 8.8.

In the horizontal, the spectral bumps described above suggest a definition of a transition length scale. Following *Waite* [2011], let k_m be the horizontal wave number of the local minimum of buoyancy flux seen in many of the transfer spectra in Figures 8.5–8.7, and define $L_m \equiv 2\pi/k_m$. This is the horizontal length scale where kinetic energy is injected by nonlinear interactions and partially converted to potential energy when Fr_h is sufficiently small and Re is sufficiently large. In Figures 8.10a and 8.10b, L_m is plotted against the buoyancy and viscous scales for the simulations in which a clear bump in $B(k_h)$ could be identified. For the range of parameters considered here, L_m scales fairly well with both L_b and L_{visc}; a wider range of Re_b would possibly be able to distinguish between these two scalings. By contrast, L_m does not scale particularly well with either the Ozmidov or Kolmogorov scales (Figures 8.10c and 8.10d), which are both much smaller than L_m. *Waite* [2011] found $L_m \sim L_b$ in simulations with viscous effects reduced by hyperviscosity, which models the regime of effectively large Re_b. The small-scale injection of kinetic energy and associated spectral bumps

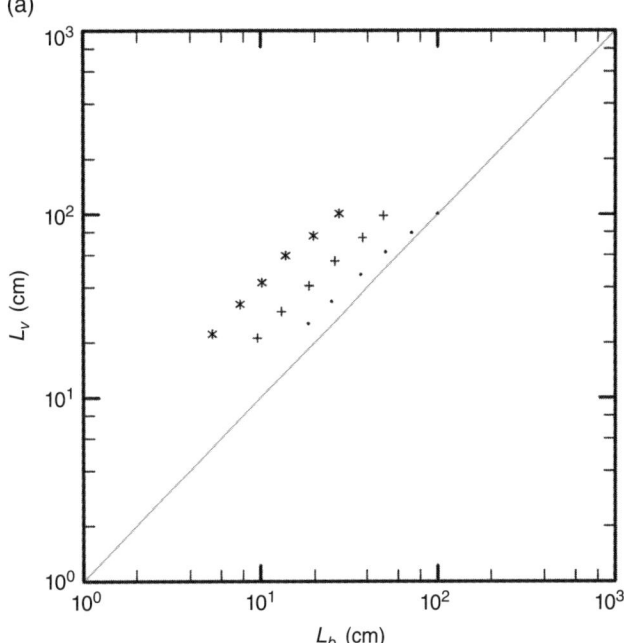

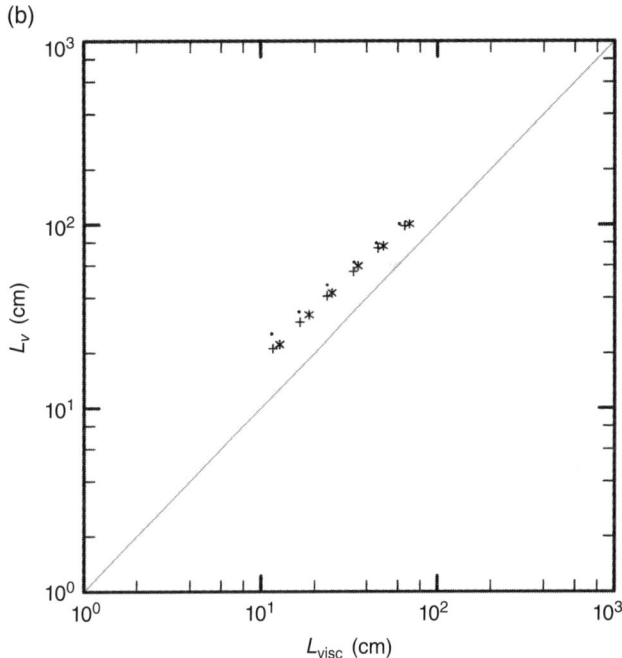

Figure 8.9. Characteristic vertical length scale L_v plotted against (a) the buoyancy scale L_b and (b) the viscous scale L_{visc}. Symbols denote $\text{Re}_b \approx 2$ (·), 0.6 (+), and 0.2 (∗) (simulation sets A, B, and C).

therefore appear to be signatures of the vertical layer thickness in the horizontal spectrum. This connection between vertical and horizontal is consistent with the roll-up of thin layers of horizontal vorticity into more isotropic Kelvin–Helmholtz billows, as seen in Figure 8.8.

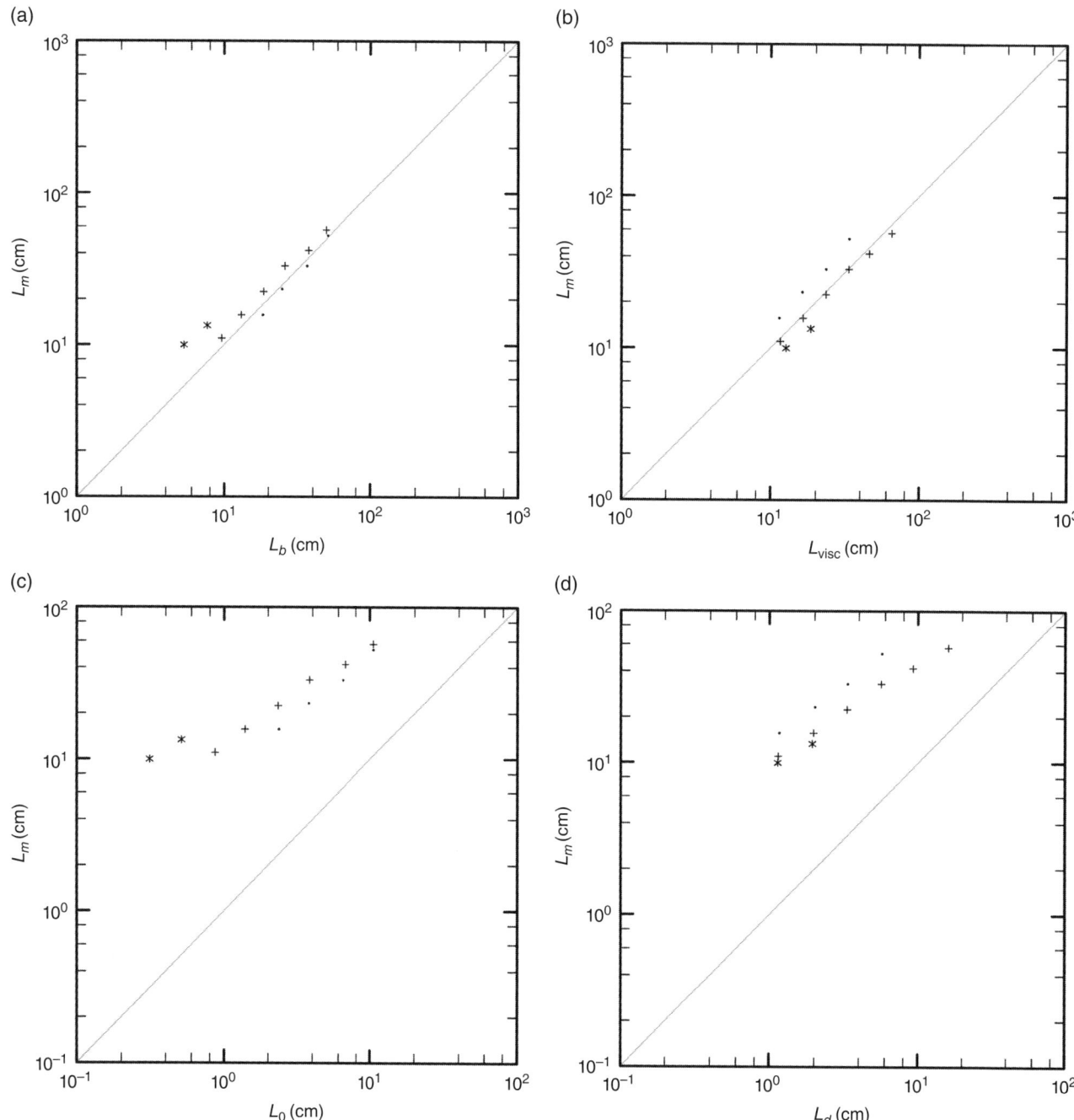

Figure 8.10. Horizontal scale L_m of minimum small-scale buoyancy flux plotted against (a) the buoyancy scale L_b, (b) the viscous scale L_{visc}, (c) the Ozmidov scale L_O, and (d) the Kolmogorov scale L_d. Symbols denote $\text{Re}_b \approx 2$ (·), 0.6 (+), and 0.2 (∗) (simulation sets A, B, and C).

8.4. DISCUSSION AND CONCLUSIONS

The stratified turbulence simulated in these numerical experiments is surprisingly reminiscent of the picture painted by *Lilly* [1983], despite the significant advances in understanding that have occurred over the last three decades. Of course, the energy transfer in these simulations is predominantly downscale, and there is no evidence of *Lilly's* [1983] hypothesized inverse cascade. Nevertheless, the layerwise structure that he anticipated is clearly visible. The layer thickness scales like the viscous scale L_{visc}, but for laboratory parameters with $\text{Re}_b \sim O(1)$, the viscous scale is very close to the buoyancy scale L_b. As a result, the layers in these simulations, despite being coupled by vertical viscosity, are thin enough to yield $Fr_v \sim O(1)$ and, at least intermittently, Kelvin-Helmholtz instabilities at small scales. For the larger values of $Re_b \approx 2$, these instabilities appear to break down into even smaller-scale three-dimensional turbulence, as foreseen by *Lilly* [1983].

The turbulent statistics in these simulations are quite sensitive to changes in Fr_h and Re, and as *Brethouwer et al.* [2007] have argued, some of this sensitivity appears as a dependence on Re_b. In particular, the kinetic energy spectrum has a power law range with a slope that steepens with decreasing Re_b. For the range of parameters considered here, the slope goes from around -3 to -5 as Re_b goes from 2 down to 0.2. These spectra are all steeper than *Lindborg's* [2006] proposed $-\frac{5}{3}$, which is to be expected at $O(1)$ values of Re_b, since vertical dissipation is non-negligible across all k_h. The very steep -5 spectrum at small Re_b has been observed in other simulations where the buoyancy scale is inside the vertical dissipation range (viscous or ad hoc) and appears to be the asymptotic behavior for $Re_b \to 0$ [*Laval et al.*, 2003; *Waite and Bartello*, 2004; *Brethouwer et al.*, 2007; *Kurien and Smith*, 2014]. At small Re_b, the entire range of k_h is essentially a dissipation range for the vertical part of the viscous term; energy injected at large scales is dissipated at large scales, and no energy cascade to small horizontal scales occurs. Furthermore, *Waite* [2013] recently showed that the potential enstrophy is approximately quadratic in this limit, by contrast with the larger Re_b regime where higher order terms are significant.

The dependence of the turbulent statistics on Re_b does not explain all of the sensitivity to Fr_h and Re observed in these simulations. In fact, there does seem to be some significant variability in the slope of the power law part of the spectrum, even at constant Re_b. For our largest Re_b values, the kinetic energy spectrum steepens as Fr_h decreases (with a corresponding increase in Re to keep $Re_b \approx 2$). The spectral slopes range from -2 to -3, and the steepening of these spectra has not quite converged, even for $Fr_h = 0.017$ and Re = 9400; further steepening may therefore occur as Fr_h is decreased more. The size of the vertical dissipation, as approximated by Re_b^{-1} in (8.22), is therefore not the only parameter that determines the shape of the spectra, even at small Fr_h and large Re. It would be interesting to see if this behavior occurs at higher values of Re_b and whether it yields spectra steeper than $-\frac{5}{3}$ for $Re_b \gg 1$.

Furthermore, many of these simulations exhibit spectral bumps at small horizontal scales, and the shape and position of these bumps are not solely determined by Re_b. These bumps appear to result from the injection of kinetic energy by nonlinear interactions, and their position scales like L_b and L_{visc} (it is not possible to distinguish between these scalings for $Re_b \sim O(1)$). Neither L_b nor L_{visc} is a simple function of Re_b alone, since

$$L_b \equiv 2\pi L_h \sqrt{\frac{Re_b}{Re}}, \quad L_{\text{visc}} \equiv 2\pi L_h \frac{1}{\sqrt{Re}}. \quad (8.30)$$

Both depend on Re in addition to Re_b, and both become smaller as Re is increased at fixed Re_b. *Waite* [2011] described spectral bumps with similar scaling in simulations that employed hyperviscosity to reduce viscous effects at large scales. In those simulations, a direct, non-local transfer of kinetic energy from large horizontal scales to L_b was found to be responsible for the bumps in the spectra. As in *Waite* [2011], the spectral bumps in the DNS in this chapter are consistent with Kelvin-Helmholtz instabilities transferring energy directly from large scales to the billow scale. For our largest $Re_b \approx 2$, the billows break down into smaller-scale turbulence and the peaks are broad; for smaller Re_b, the instabilities become more intermittent and seem to be directly damped by viscosity, and the peaks are narrower. It is interesting that these small-scale transitions occur even for small $Re_b \approx 0.2$, at least for sufficiently large Re and small Fr_h. Even though the vertical Froude number based on mean quantities is smaller than 1, it is likely that there are intermittent patches of locally small Richardson number and subsequent instability, as seen in Figure 8.8c. Presumably, there is a threshold Re_b below which no instabilities occur for any Re and Fr_h, but these simulations show that such a cutoff value is less than 0.2.

Given the similarity between the simulations presented in this chapter and those of *Waite* [2011], it seems reasonable to speculate that the location of the spectral bumps seen here will scale with the buoyancy scale L_b in DNS with $Re_b \gg 1$. This scaling would suggest the existence of a distinct spectral range between the buoyancy scale, where kinetic energy in injected by instability of the large-scale flow, and the Ozmidov scale, where these instabilities ultimately break down into three-dimensional isotropic turbulence. However, reproducing this range requires large Re_b along with small Fr_h and thus Reynolds numbers and numerical resolutions that are beyond the laboratory regime; *Bartello and Tobias* [2013] suggest $Re \gtrsim 10^5$, which is an order of magnitude larger than the highest Re values here. This range of scales between L_b and L_O is a key feature of geophysical stratified turbulence that is missing in laboratory-scale experiments and most numerical simulations because of the extremely high computational cost of resolving it. Indeed, typical values of L_b and L_O in the atmosphere are $O(1)$ km and $O(10)$ m, and so this scale range is not normally resolved in mesoscale simulations with $\Delta x \sim O(1)$ km. Interestingly, bumps in the atmospheric kinetic energy spectrum have been observed at this scale [e.g. *Vinnichenko*, 1970].

These DNS results underscore the care that must be taken when using laboratory-scale stratified turbulence as a proxy for the atmospheric mesoscale and oceanic sub-mesoscale. In particular, steep spectral slopes from experiments with $Re_b \sim O(1)$ should not be extrapolated too literally to geophysical scales. All else being equal, spectra do get shallower as Re_b increases, although it is still an open question whether the limiting slope for $Re_b \to \infty$ is $-\frac{5}{3}$ or something else. Nevertheless, it is quite encouraging that these simulations are able to capture some basic phenomena that are expected to hold at larger Re_b. In particular, the transfer of kinetic energy to small horizontal scales by Kelvin–Helmholtz instabilities appears to be quite robust, even for relatively modest values of Re_b, suggesting that the instabilities (if not the subsequent turbulent breakdown) are realizable in laboratory experiments [*Augier et al.*, 2014].

A more serious concern about the geophysical applicability of idealized laboratory experiments and numerical simulations such as these is the question of how energy gets into the large scales to begin with. Simulations and experiments necessarily employ ad hoc methods to inject energy at large scales, either through initial conditions and/or forcing. But in the real atmosphere and ocean, the large-scale motion is not stratified turbulence at all, but rather quasi-geostrophic turbulence forced by radiative heating, surface fluxes, and baroclinic instability. The cascade through the atmospheric mesoscale and oceanic sub-mesoscale that is envisioned by the stratified turbulence hypothesis must be forced by the breakdown of this larger-scale motion, for which the effects of rotation cannot be ignored [e.g., *Bartello*, 2010; *Molemaker et al.*, 2010; *Vallgren et al.*, 2011]. The extent to which the influence of this large-scale motion extends down into the mesoscale/sub-mesoscale and beyond is still an open question that requires further study.

Acknowledgments. This work was supported by a grant from the Natural Sciences and Engineering Research Council of Canada. Computations were performed on the GPC supercomputer at the SciNet HPC Consortium [*Loken et al.*, 2010]. SciNet is funded by the Canada Foundation for Innovation under the auspices of Compute Canada, the Government of Ontario, Ontario Research Fund, Research Excellence, and the University of Toronto.

REFERENCES

Almalkie, S., and S. M. de Bruyn Kops (2012), Kinetic energy dynamics in forced, homogeneous, and axisymmetric stably stratified turbulence, *J. Turbul.*, *13*, N29.

Augier, P., P. Billant, M. E. Negretti, and J.-M. Chomaz (2014), Experimental study of stratified turbulence forced with columnar dipoles, *Phys. Fluids*, *26*, 046603.

Babin, A., A. Mahalov, B. Nicolaenko, and Y. Zhou (1997), On the asymptotic regimes and the strongly stratified limit of rotating Boussinesq equations, *Theor. Comput. Fluid Dyn.*, *9*, 223–251.

Bartello, P. (2000), Potential vorticity, resonance and dissipation in rotating convective turbulence, in *Geophysical and Astrophysical Convection*, edited by R. H. Kerr, P. Fox, and C.-H. Moeng, pp. 309–321, Gordon and Breach, Amsterdam.

Bartello, P. (2010), Quasigeostrophic and stratified turbulence in the atmosphere, in *IUTAN Symposium on Turbulence in the Atmosphere and Oceans*, vol. IUTAM Bookseries 28, edited by D. G. Dritschel, pp. 117–130, Springer, New York.

Bartello, P., and S. Tobias (2013), Sensitivity of stratified turbulence to the buoyancy Reynolds number, *J. Fluid Mech.*, *725*, 1–22.

Billant, P., and J.-M. Chomaz (2000), Three-dimensional stability of a vertical columnar vortex pair in a stratified fluid, *J. Fluid Mech.*, *419*, 65–91.

Billant, P., and J.-M. Chomaz (2001), Self-similarity of strongly stratified inviscid flows, *Phys. Fluids*, *13*, 1645–1651.

Brethouwer, G., P. Billant, E. Lindborg, and J.-M. Chomaz (2007), Scaling analysis and simulation of strongly stratified turbulent flows, *J. Fluid Mech.*, *585*, 343–368.

Charney, J. G. (1971), Geostrophic turbulence, *J. Atmos. Sci.*, *28*, 1087–1095.

Cho, J. Y. N., Y. Zhu, R. E. Newell, B. E. Anderson, J. D. Barrick, G. L. Gregory, G. W. Sachse, M. A. Carroll, and G. M. Albercook (1999), Horizontal wavenumber spectra of winds, temperature, and trace gases during the Pacific Exploratory Missions: 1. Climatology, *J. Geophys. Res.*, *104*, 5697–5716.

Dewan, E. M. (1979), Stratospheric wave spectra resembling turbulence, *Science*, *204*, 832–835.

Drazin, P. G. (1961), On the steady flow of a fluid of variable density past an obstacle, *Tellus*, *13*, 239–251.

Durran, D. R. (2010), *Numerical Methods for Fluid Dynamics*: With Applications to Geophysics, Springer, New York.

Embid, P. F., and A. J. Majda (1998), Low Froude number limiting dynamics for stably stratified flow with small or finite Rossby numbers, *Geophys. Astrophys. Fluid Dyn.*, *87*, 1–50.

Gage, K. S. (1979), Evidence for a $k^{-5/3}$ law inertial range in mesoscale two-dimensional turbulence, *J. Atmos. Sci.*, *36*, 1950–1954.

Gage, K. S., and G. D. Nastrom (1986), Theoretical interpretation of atmospheric wavenumber spectra of wind and temperature observed by commercial aircraft during GASP, *J. Atmos. Sci.*, *43*, 729–740.

Godoy-Diana, R., J.-M. Chomaz, and P. Billant (2004), Vertical length scale selection for pancake vortices in strongly stratified viscous fluids. *J. Fluid Mech.*, *504*, 229–238.

Hebert, D. A., and S. M. de Bruyn Kops (2006a), Relationship between vertical shear rate and kinetic energy dissipation rate in stably stratified flows, *Geophys. Research Lett.*, *33*, L06602.

Hebert, D. A., and S. M. de Bruyn Kops (2006b), Predicting turbulence in flows with strong stable stratification, *Phys. Fluids*, *18*, 066602.

Herring, J. R., and O. Métais (1989), Numerical experiments in forced stably stratified turbulence, *J. Fluid Mech.*, *202*, 97–115.

Hopfinger, E. J. (1987), Turbulence in stratified fluids: A review, *J. Geophys. Res.*, *92*, 5287–5303.

Khani, S., and M. L. Waite (2014), Buoyancy scale effects in large-eddy simulations of stratified turbulence, *J. Fluid Mech.*, *754*, 75–97.

Kimura, Y., and J. R. Herring (2012), Energy spectra of stably stratified turbulence, *J. Fluid Mech.*, *698*, 19–50.

Kolmogorov, A. N. (1941), The local structure of turbulence in incompressible viscous fluid for very large Reynolds number, *Dok. Akad. Nauk. SSSR*, *30*, 301–305.

Kraichnan, R. H. (1967), Inertial ranges in two-dimensional turbulence, *Phys. Fluids*, *10*, 1417–1423.

Kurien, S., and L. M. Smith (2014), Effect of rotation and domain aspect-ratio on layer formation in strongly stratified Boussinesq flows, *J. Turbul.*, *15*, 241–271.

Laval, J.-P., J. C. McWilliams, and B. Dubrulle (2003), Forced stratified turbulence: Successive transitions with Reynolds number, *Phys. Rev. E*, *68*, 036,308.

Lilly, D. K. (1983), Stratified turbulence and the mesoscale variability of the atmosphere, *J. Atmos. Sci.*, *40*, 749–761.

Lilly, D. K., G. Bassett, K. Droegemeier, and P. Bartello (1998), Stratified turbulence in the atmospheric mesoscale, *Theor. Comput. Fluid Dyn.*, *11*, 139–153.

Lin, J. T., and Y. H. Pao (1979), Wakes in stratified fluids: A review, *Annu. Rev. Fluid Mech.*, *11*, 317–338.

Lindborg, E. (2006), The energy cascade in a strongly stratified fluid, *J. Fluid Mech.*, *550*, 207–242.

Lindborg, E., and J. Y. N. Cho (2001), Horizontal velocity structure functions in the upper troposphere and lower stratosphere. 2. Theoretical considerations, *J. Geophys. Res.*, *106*, 10,233–10,241.

Loken, C., et al. (2010), SciNet: Lessons learned from building a power-efficient top-20 system and data centre, *J. Phys. Conf. Ser.*, *256*, 012,026.

Lumley, J. L. (1964), The spectrum of nearly inertial turbulence in a stably stratified fluid, *J. Atmos. Sci.*, *21*, 99–102.

Moin, P., and K. Mahesh (1998), Direct numerical simulation: A tool in turbulence research, *Annu. Rev. Fluid Mech.*, *30*, 539–578.

Molemaker, M. J., J. C. McWilliams, and X. Capet (2010), Balanced and unbalanced routes to dissipation in an equilibrated eady flow, *J. Fluid Mech.*, *654*, 35–63.

Munk, W. (1981), Internal waves and small-scale processes, in *Evolution of Physical Oceanography*, edited by B. A. Warren and C. Wunsch, pp. 264–291. MIT Press, Cambridge, Mass.

Nastrom, G. D., and K. S. Gage (1985), A climatology of atmospheric wavenumber spectra observed by commercial aircraft, *J. Atmos. Sci.*, *42*, 950–960.

Ozmidov, R. V. (1965), On the turbulent exchange in a stably stratified ocean, *Izvestia Akad. Nauk. SSSR Atmos. Oceanic Phys. Ser.*, *1*, 853–860.

Praud, O., A. M. Fincham, and J. Sommeria (2005), Decaying grid turbulence in a strongly stratified fluid, *J. Fluid Mech.*, *522*, 1–33.

Riley, J. J., and S. M. deBruynKops (2003), Dynamics of turbulence strongly influenced by buoyancy, *Phys. Fluids*, *15*, 2047–2059.

Riley, J. J., and M.-P. Lelong (2000), Fluid motions in the presence of strong stable stratification, *Annu. Rev. Fluid Mech.*, *32*, 613–657.

Riley, J. J., and E. Lindborg (2008), Stratified turbulence: A possible interpretation of some geophysical turbulence measurements, *J. Atmos. Sci.*, *65*, 2416–2424.

Riley, J. J., R. W. Metcalfe, and M. A. Weissman (1981), Direct numerical simulations of homogeneous turbulence in density-stratified fluids, in *Nonlinear Properties of Internal Waves*, edited by B. J. West, pp. 79–112.

Smith, L. M., and F. Waleffe (2002), Generation of slow large scales in forced rotating stratified turbulence, *J. Fluid Mech.*, *451*, 145–168.

Smyth, W. D., and J. N. Moum (2000), Anisotropy of turbulence in stably stratified mixing layers, *Phys. Fluids*, *12*, 1343–1362.

Staquet, C., and J. Sommeria (2002), Internal gravity waves: From instabilities to turbulence, *Annu. Rev. Fluid Mech.*, *34*, 559–593.

Sullivan, N. P., S. Mahalingam, and R. M. Kerr (1994), Deterministic forcing of homogeneous, isotropic turbulence, *Phys. Fluids*, *6*, 1612–1614.

Tulloch, R., and K. S. Smith (2009), Quasigeostrophic turbulence with explicit surface dynamics: Application to the atmospheric energy spectrum, *J. Atmos. Sci.*, *66*, 450–467.

Tung, K. K., and W. W. Orlando (2003), The k^{-3} and $k^{-5/3}$ energy spectrum of atmospheric turbulence: Quasigeostrophic two-level model simulation, *J. Atmos. Sci.*, *60*, 824–835.

Vallgren, A., E. Deusebio, and E. Lindborg (2011), Possible explanations of the atmospheric kinetic and potential energy spectra, *Phys. Rev. Lett.*, *107*, 268501.

VanZandt, T. E. (1982), A universal spectrum of buoyancy waves in the atmosphere, *Geophys. Res. Lett.*, *9*, 575–578.

Vinnichenko, N. K. (1970), The kinetic energy spectrum in the free atmosphere – 1 second to 5 years, *Tellus*, *22*, 158–166.

Waite, M. L. (2011), Stratified turbulence at the buoyancy scale, *Phys. Fluids*, *23*, 066,602.

Waite, M. L. (2013), Potential enstrophy in stratified turbulence, *J. Fluid Mech.*, *722*, R4.

Waite, M. L., and P. Bartello (2004), Stratified turbulence dominated by vortical motion, *J. Fluid Mech.*, *517*, 281–308.

Waite, M. L., and P. Bartello (2006), Stratified turbulence generated by internal gravity waves, *J. Fluid Mech.*, *546*, 313–339.

Waite, M. L., and C. Snyder (2013), Mesoscale energy spectra of moist baroclinic waves, *J. Atmos. Sci.*, *70*, 1242–1256.

Section III: Atmospheric Flows

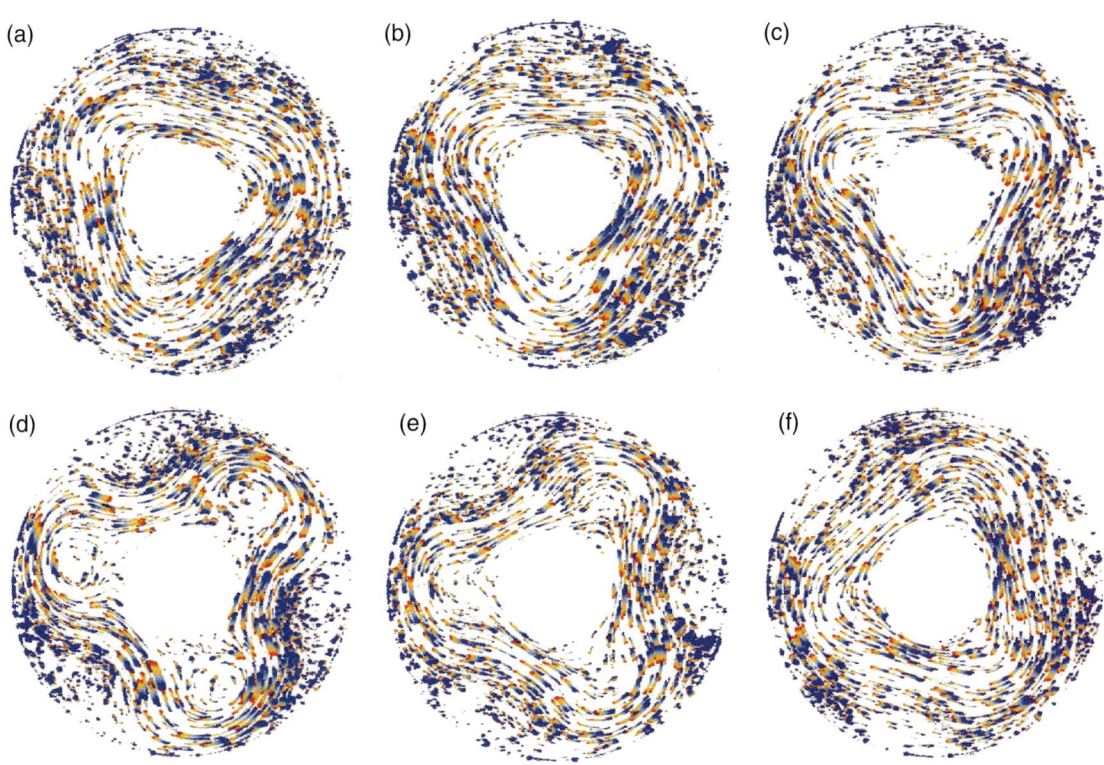

Figure 1.9. Typical horizontal flow fields (streak photographs) during an "amplitude vacillation" cycle of the rotating annulus in the same system as in Figures 1.8–1.13.

Modeling Atmospheric and Oceanic Flows: Insights from Laboratory Experiments and Numerical Simulations,
First Edition. Edited by Thomas von Larcher and Paul D. Williams.
© 2015 American Geophysical Union. Published 2015 by John Wiley & Sons, Inc.

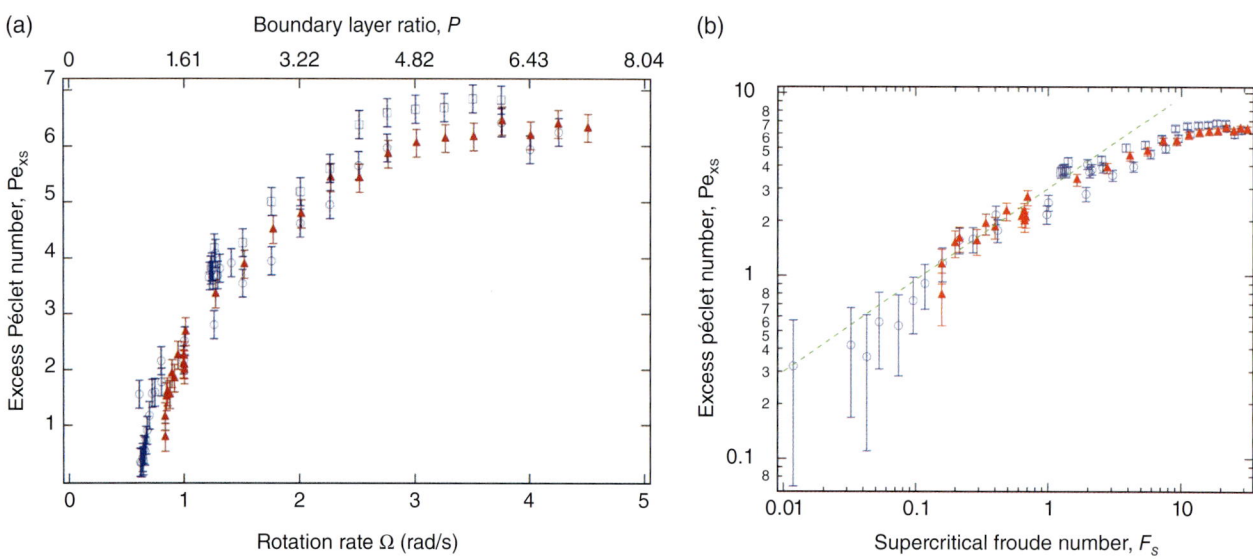

Figure 1.22. Experimental measurements of dimensionless integrated heat transport (excess Péclet number) attributable to baroclinic eddies in a rotating annulus experiment derived from data of *Read* [2003] and presented (a) as a function of Ω and $\mathcal{P}$ and (b) as a function of "supercritical" Froude number $\mathcal{F}_s$ (see text). Pe_{xs} may be compared with the nonrotating total Péclet number of around 10. Open circles are for $m = 2$ flows, filled triangles for $m = 3$, and open squares for $m = 4$ dominated flows.

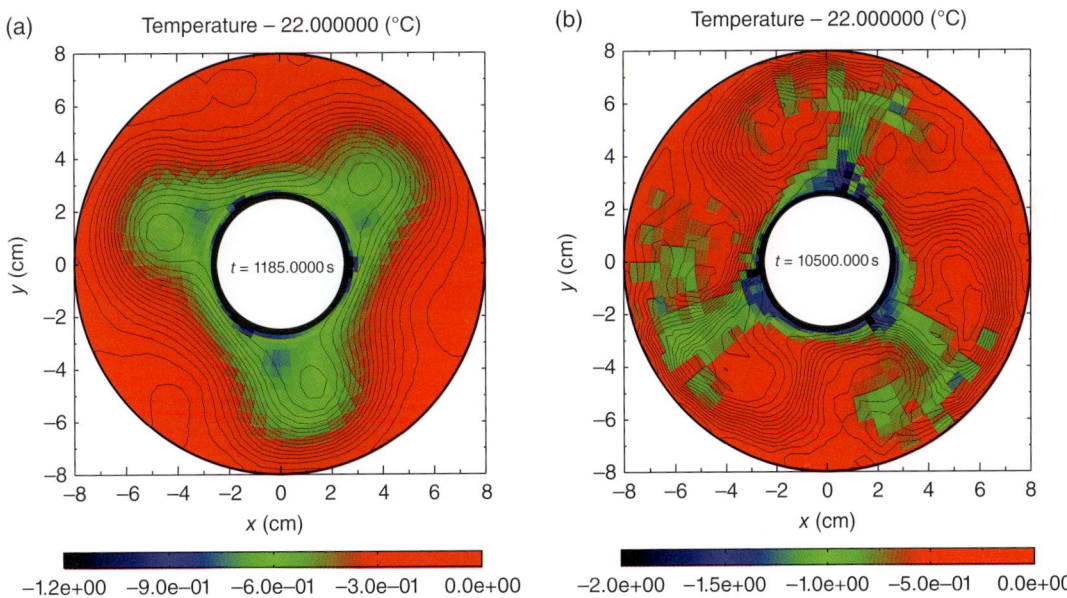

Figure 1.27. Representative temperature fields (colors) and horizontal stream function (contours) produced from assimilated horizontal velocity observations obtained in the same system as shown in Figures 1.8–1.13 and 1.21–1.26. Fields are plotted for regular (a) $\Omega = 0.875$ rad/s, $T_b - T_a \approx 4.07°C$ and chaotic flow (b) $\Omega = 3.1$ rad/s, $T_b - T_a \approx 4.02°C$ at $z = 9.7$ cm above the base of the annulus. Temperatures are relative to 22°C. Adapted from *Young and Read* [2013] with the permission of John Wiley & Sons, Inc.

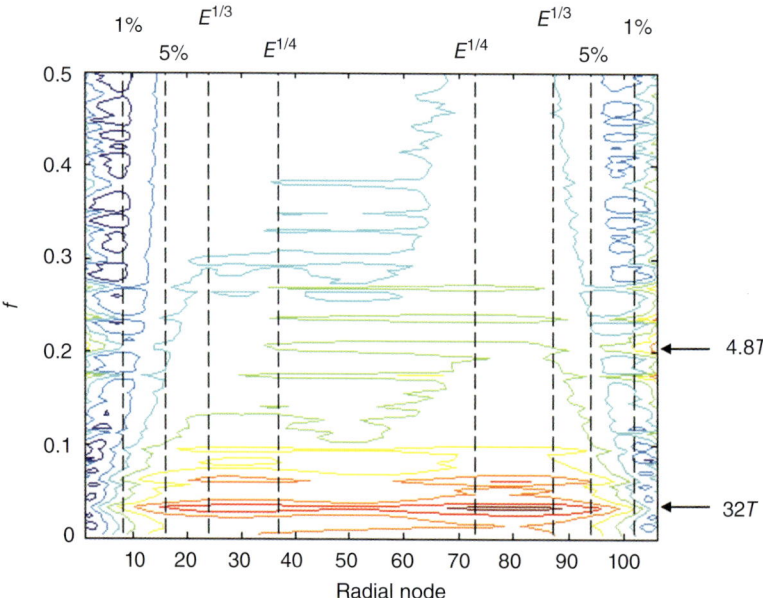

Figure 3.13. Radius-frequency contour plot of azimuthally averaged power spectral magnitude for structural vacillation in air-filled annulus.

Figure 4.7. Temperature distribution of a heated rotating disk shown by temperature-sensitive floating crystals [*Cederholm and Lundell*, 1998]. The photograph is obtained as a long-time exposure using stroboscopic light. Copyright © Cederholm & Lundell, 1998. Reprinted with permission.

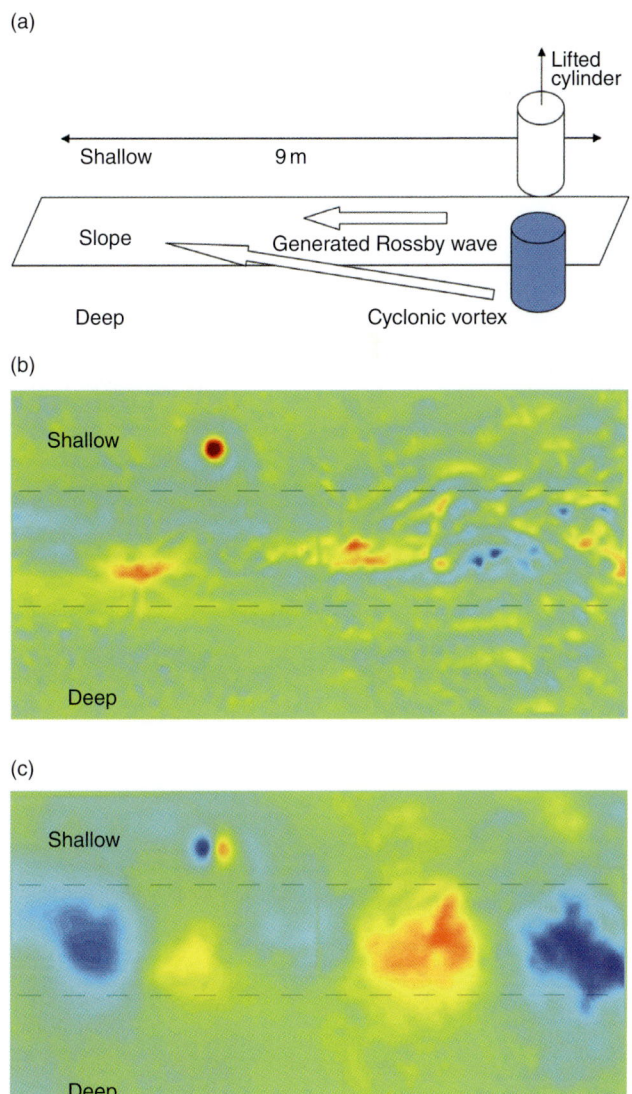

Figure 7.3. Preliminary experimental study of topographic waves excited by the passage of a cyclonic vortex (unpublished results by Zavala Sansón and van Heijst). (a) Schematic view (not to scale) of the experiments: A cyclonic vortex generated by the collapse technique drifts along the topographic slope and generates topographic Rossby waves traveling with shallow water to the right. (b) Relative vorticity surfaces measured in a horizontal plane at time $t = 14T$, with $T = 30$ s the rotation period of the Coriolis platform. Red (blue) colors indicate positive (negative) values. The domain shown is a 3.5 m × 1.8 m rectangle. The submerged slope is 0.5 m wide (approximated separation between dashed lines) and centerd in the figure. The vortex is generated near the lower right corner. The small, strong cyclonic vortex at the shallow side is a spurious vortex generated during the experiment and does not play any role on the evolution of the topographic waves over the slope. (c) Corresponding surfaces of the velocity component v, perpendicular to the topography contours.

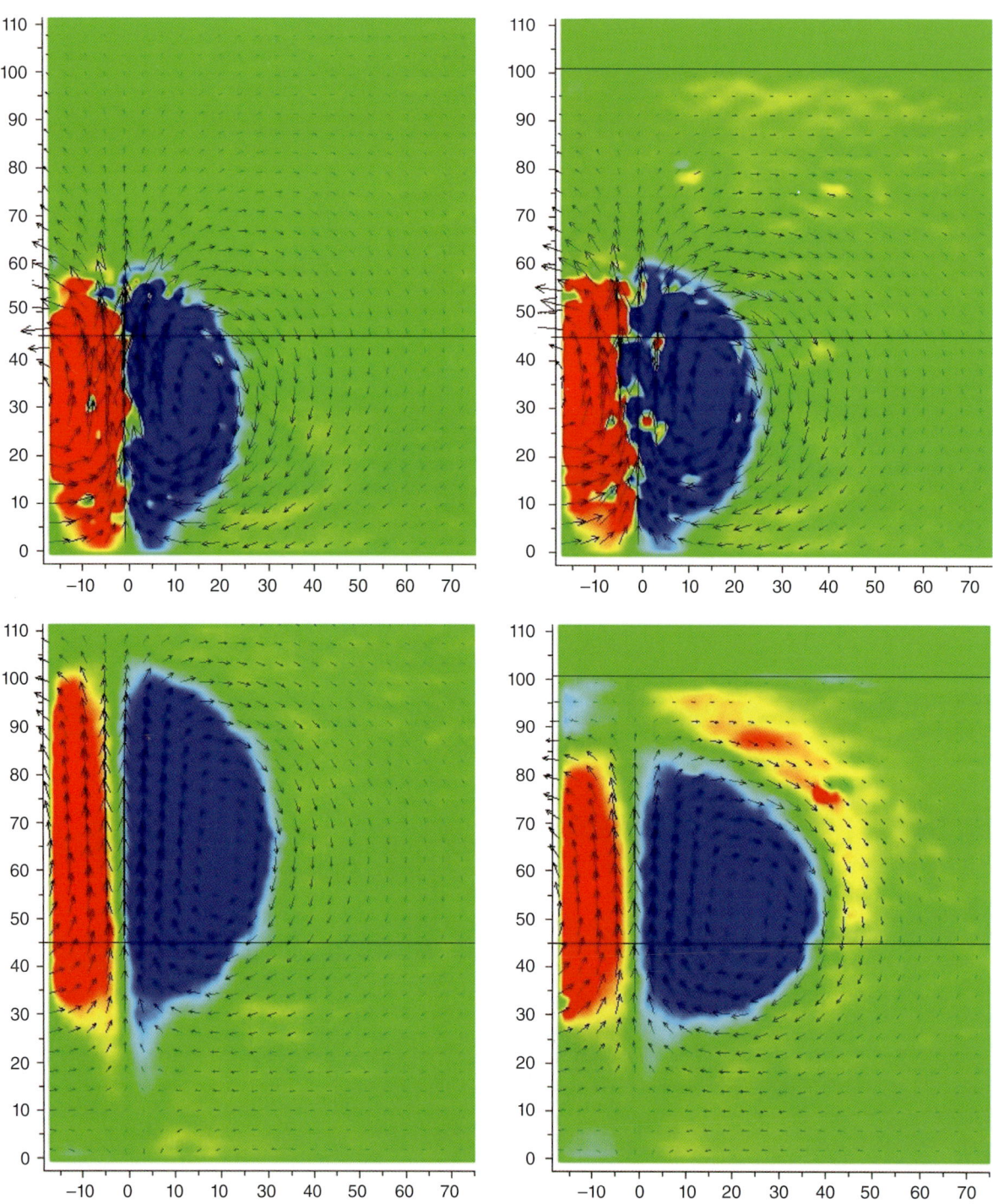

Figure 7.8. Experimentally measured flow evolution of a dipolar vortex encountering a weak slope topography with angle $\alpha = 3.4°$ (left panels) and a steeper slope with inclination $\alpha = 6.8°$ (right panels). Adapted from *van Heijst et al.* [2012]. The snapshots are taken at $t = 3$ s (top) and $t = 9$ s (bottom). The graphs show the horizontal flow field (vectors) and the ω_z vertical vorticity component (red/blue colors indicate positive/negative values) measured at the free surface. The horizontal black lines indicate the position of the edge of the slope.

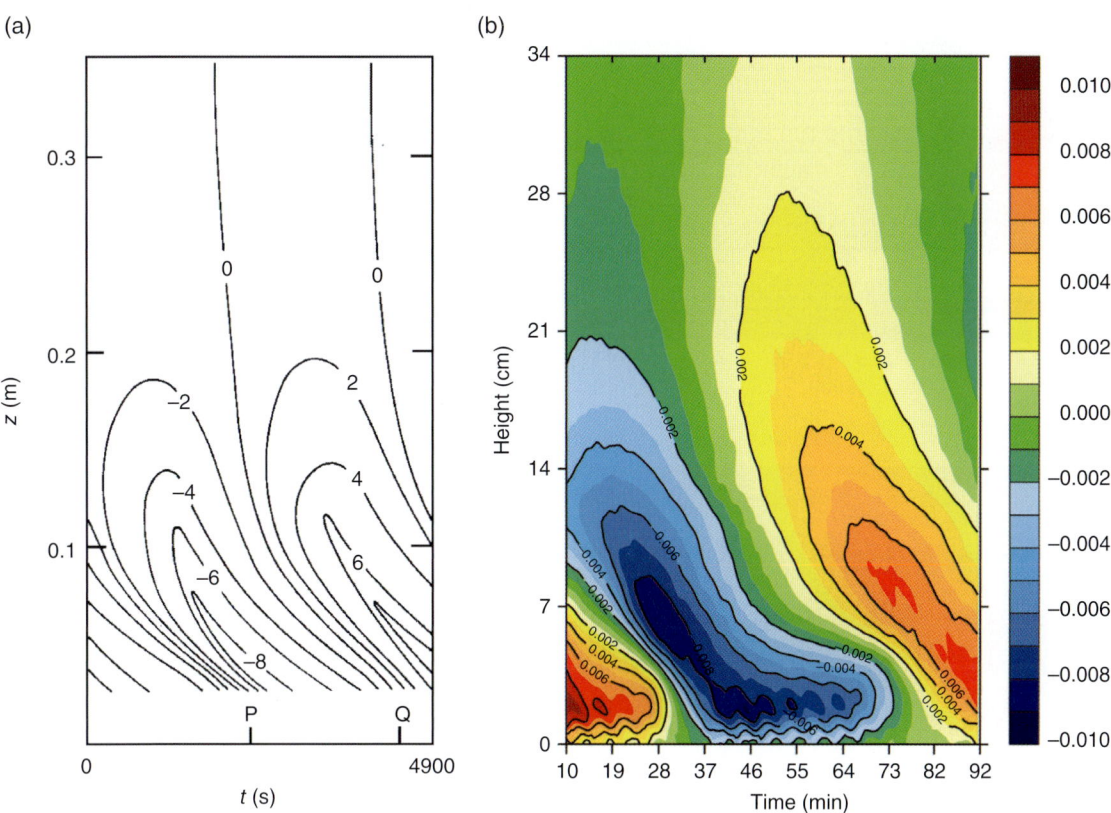

Figure 9.3. Time-height cross section of the observed zonal mean zonal flow velocity component (panel (a) adapted from Figure 10 in *Plumb and McEwan* [1978], contour lines in mm/s), compared to the result of the 3D numerical simulation at $y = L_y/2$ in panel (b), reproduced from Figure 2b in *Wedi and Smolarkiewicz* [2006], contour lines in m/s. © American Meteorological Society. Used with permission. According to *Plumb and McEwan* [1978], the lowest 2 cm in panel (a) could not be observed due to restrictions of the viewing window.

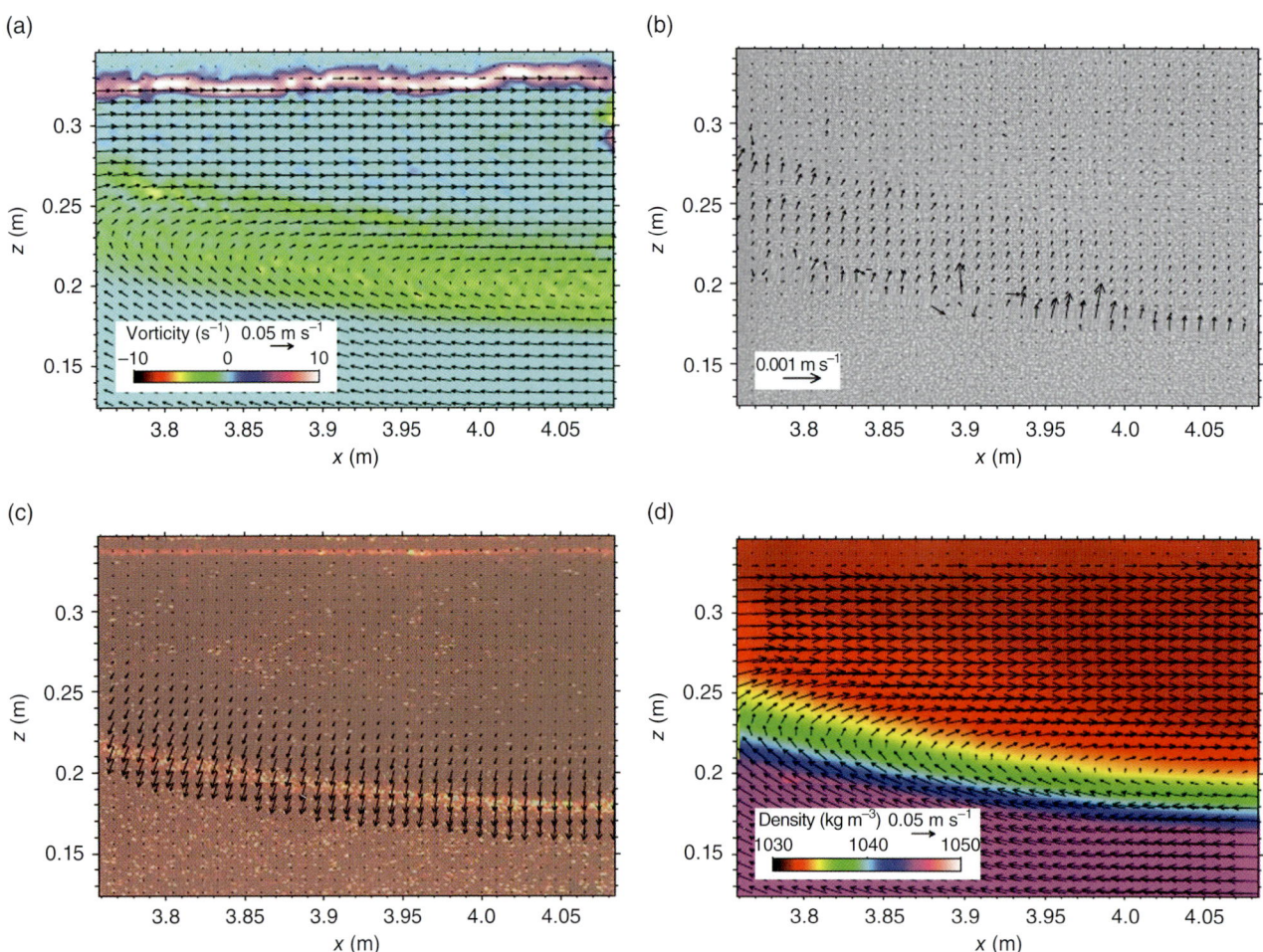

Figure 10.15. Combined PIV-schlieren examination of the passage of an internal solitary wave of depression in an approximately two-layer fluid. (a) Original velocity field computed using PIV superimposed on vorticity. (b) Apparent velocity resulting from distortions at density interface superimposed on background image of random dots. (c) Corrected displacements that account for strong density gradient at the interface superimposed on false color image of particles in the tank. (d) Corrected velocity field superimposed on contours of density. Reproduced from Figure 12 of *Dalziel et al.* [2007].

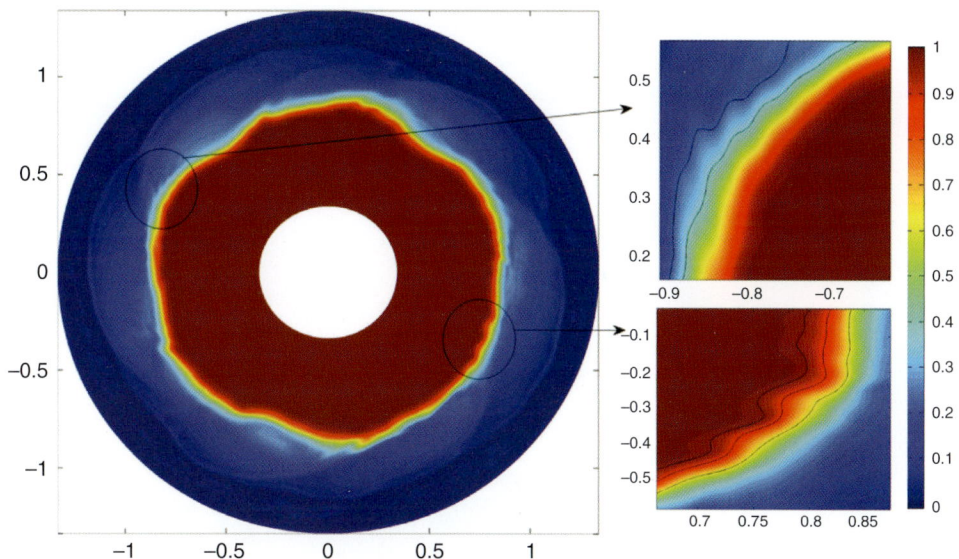

Figure 11.13. Density front for 3D simulation with Sc = 100 for $N_\theta \times N_r \times N_z = 1201 \times 301 \times 257$, Bu = 0.62, and Ro = 0.67. Right: zoom in on the encircled area showing the small-scale perturbations.

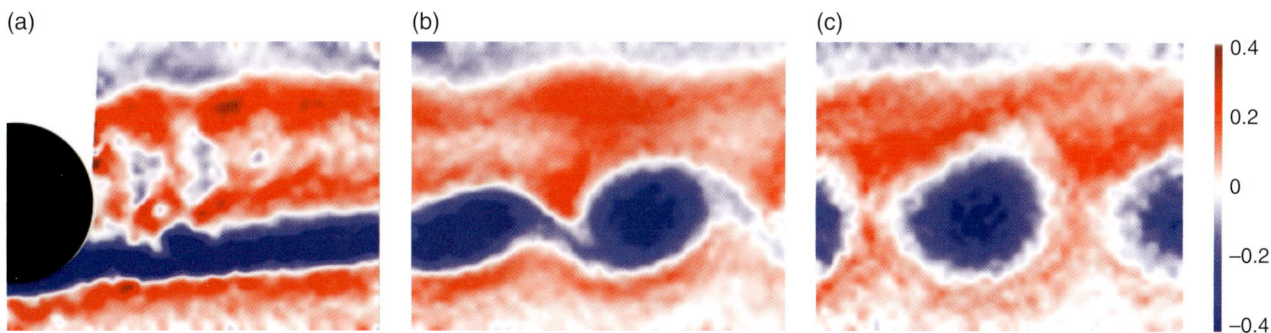

Figure 14.4. Surface vorticity field of a large-scale wake flow within an upper shallow-water layer at (a) $t = 0$, (b) $t \simeq 2T_0$, and (c) $t \simeq 5T_0$ when $Ro_I = 0.19$, Bu = 0.11, and Re = 800 [*Perret et al.*, 2006b]. Blue color corresponds to anticyclonic and red to cyclonic vorticity. The cylinder ($D = 7$ cm) is drawn in black and its shadow in the laser sheet is in white on image (a).

Figure 14.7. Dye visualization of anticyclonic destabilization within an upper shallow-water layer (h_c = 5 cm) corresponding to $Ro_I \simeq 1$, $Bu \simeq 0.5 - 1.5$, and $Re = 10000$ [*Teinturier et al., 2010*]. Different dye colors were released on each side of the cylinder (D = 50 cm): black (red) in the anticyclonic (cyclonic) boundary layer.

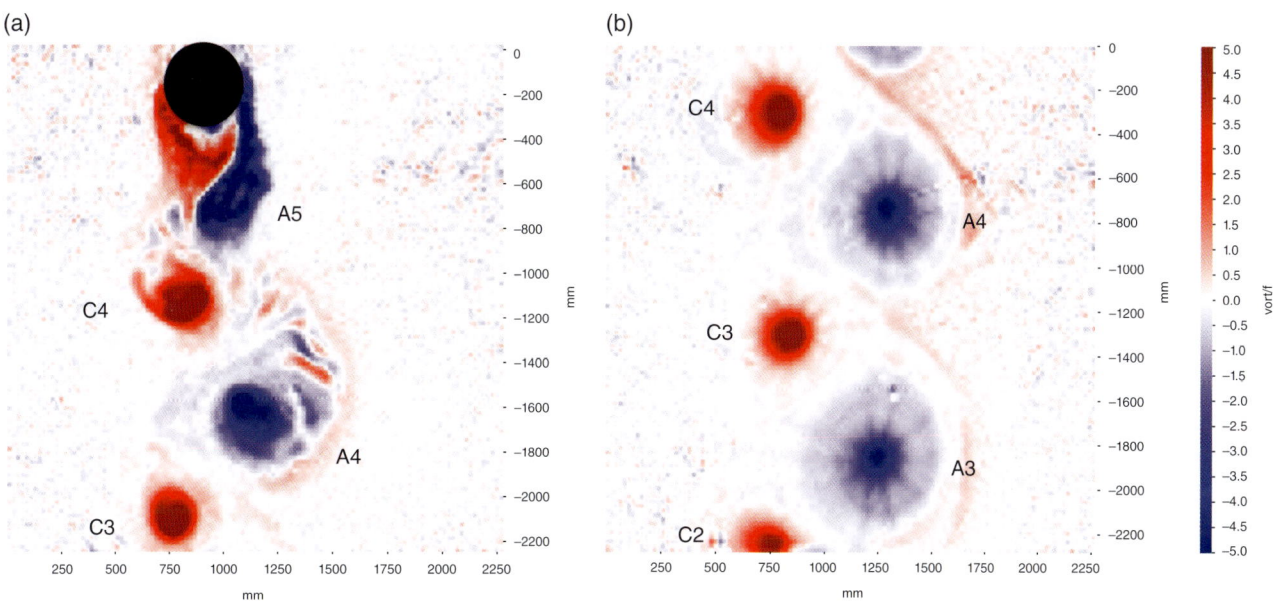

Figure 14.8. Surface vorticity field of an intense submesoscale wake within a thin (h_C = 6.7 cm) and stratified (N/f = 13) upper layer at (a) $t = 0$ and (b) $t \simeq 2T_0$ when Ro$_l$ = 2.8, Bu = 48, and Re = 12,500 [*Stegner et al.*, 2012]. The color panel quantifies the relative vorticity amplitude $-5 < \zeta/f < 5$ in blue (red) for anticyclonic (cyclonic) vorticity.

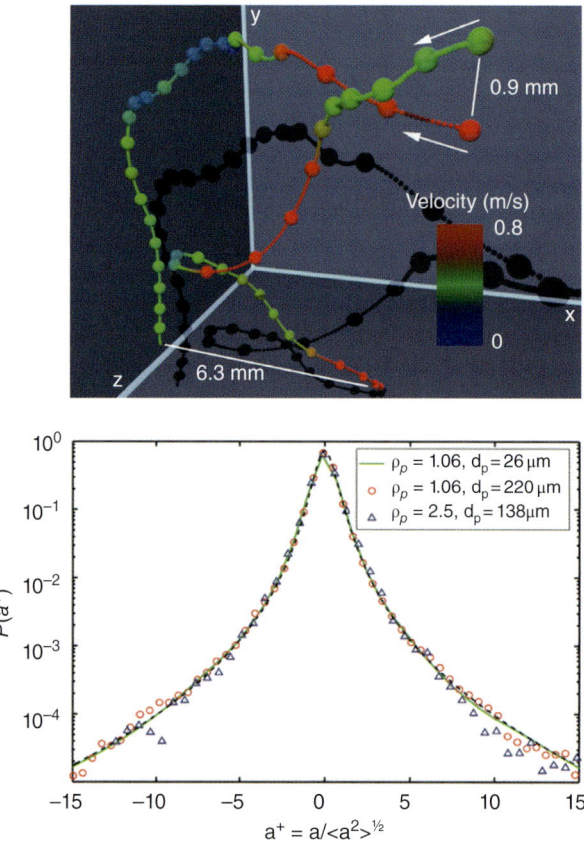

Figure 15.4. Top: Example of high-resolution reconstruction of particle trajectories in experiment shown in Figure 15.2. The small spheres mark every other measured position of the particles and are separated by 0.074 ms; the large spheres mark every 30th position. Color indicates the velocity of the particle along its trajectory. From *Bourgoin et al.* [2006]. Bottom: One component acceleration statistics measured in a von Kármán swirling flow seeded with particles of different size and density. From *Xu and Bodenschatz* [2008].

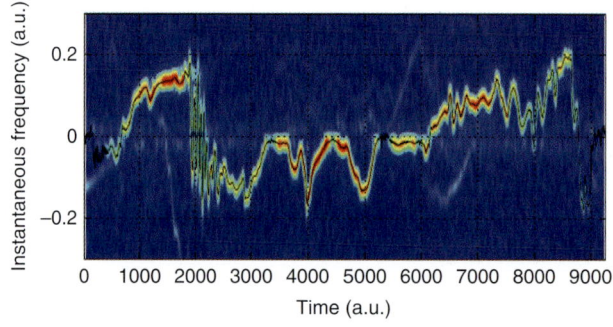

Figure 15.11. Example of reconstruction of the instantaneous frequency scattered by a moving sphere in a swirling flow of water. The color image represents the spectrogram computed with usual Fourier analysis time-frequency analysis (the width of the frequency trace illustrates the incertitude due to Heisenberg constrains) to which is superimposed the estimation given by the AML method (solid black line). From *Mordant* [2001].

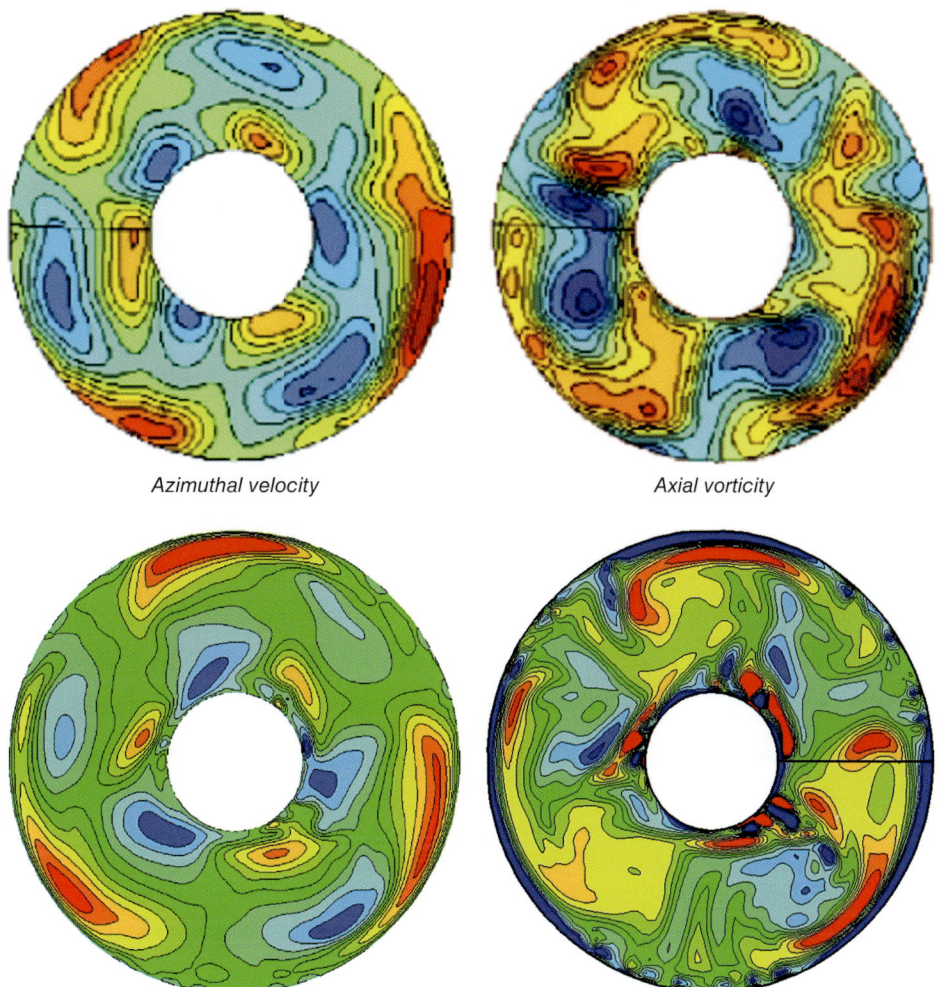

Figure 16.8. Comparison of flow characteristics at midheight between experimental measurements (top) and computed solutions (bottom) for the structural vacillation regime in the liquid-filled cavity, Pr = 16.

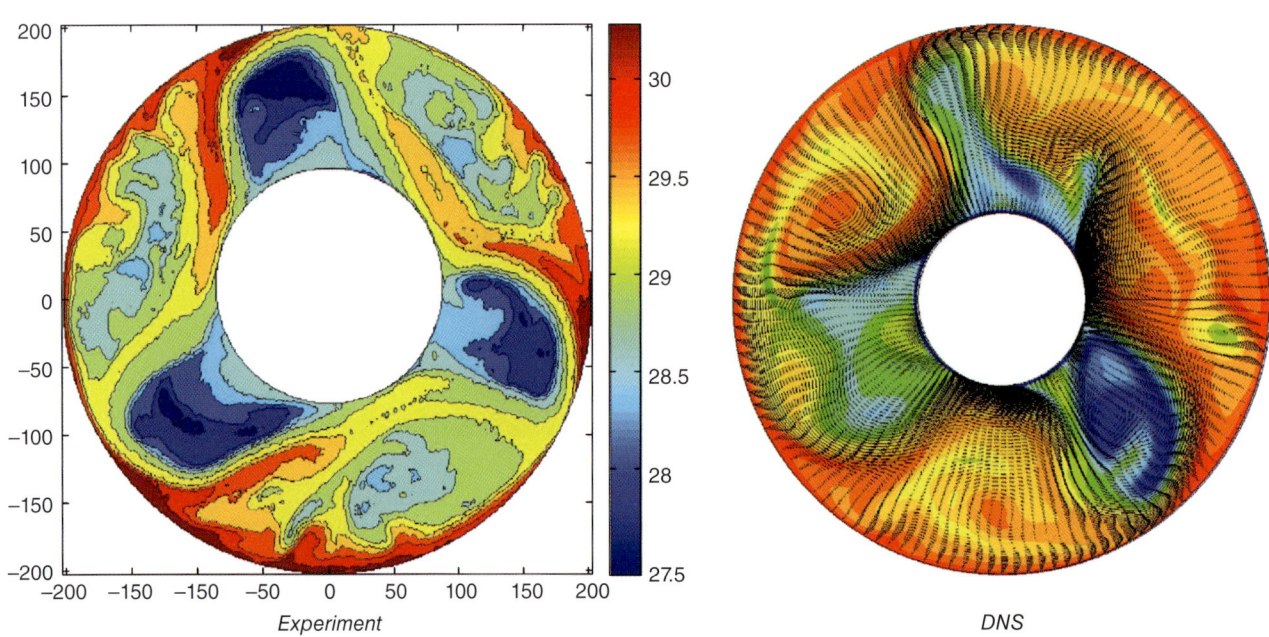

Figure 16.10. Instantaneous temperature field at the open upper free surface for the water-filled cavity.

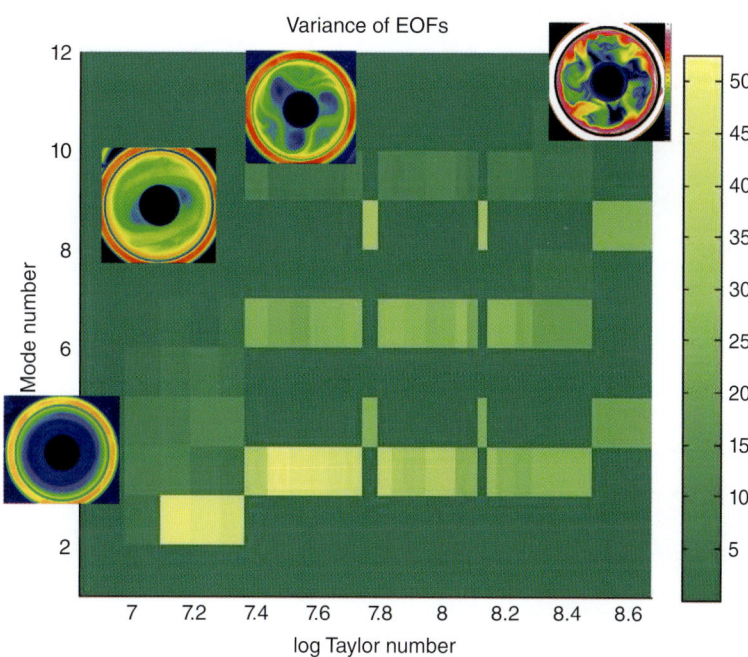

Figure 17.6. EOF variance spectra obtained along a transection through the wave regime. Left: Each circle in the Ta-Ro diagram corresponds with an experiment. Right: Distribution of the variance (in % of the total variance) for the first 12 wave modes as a function of Ta. See also Table 17.1.

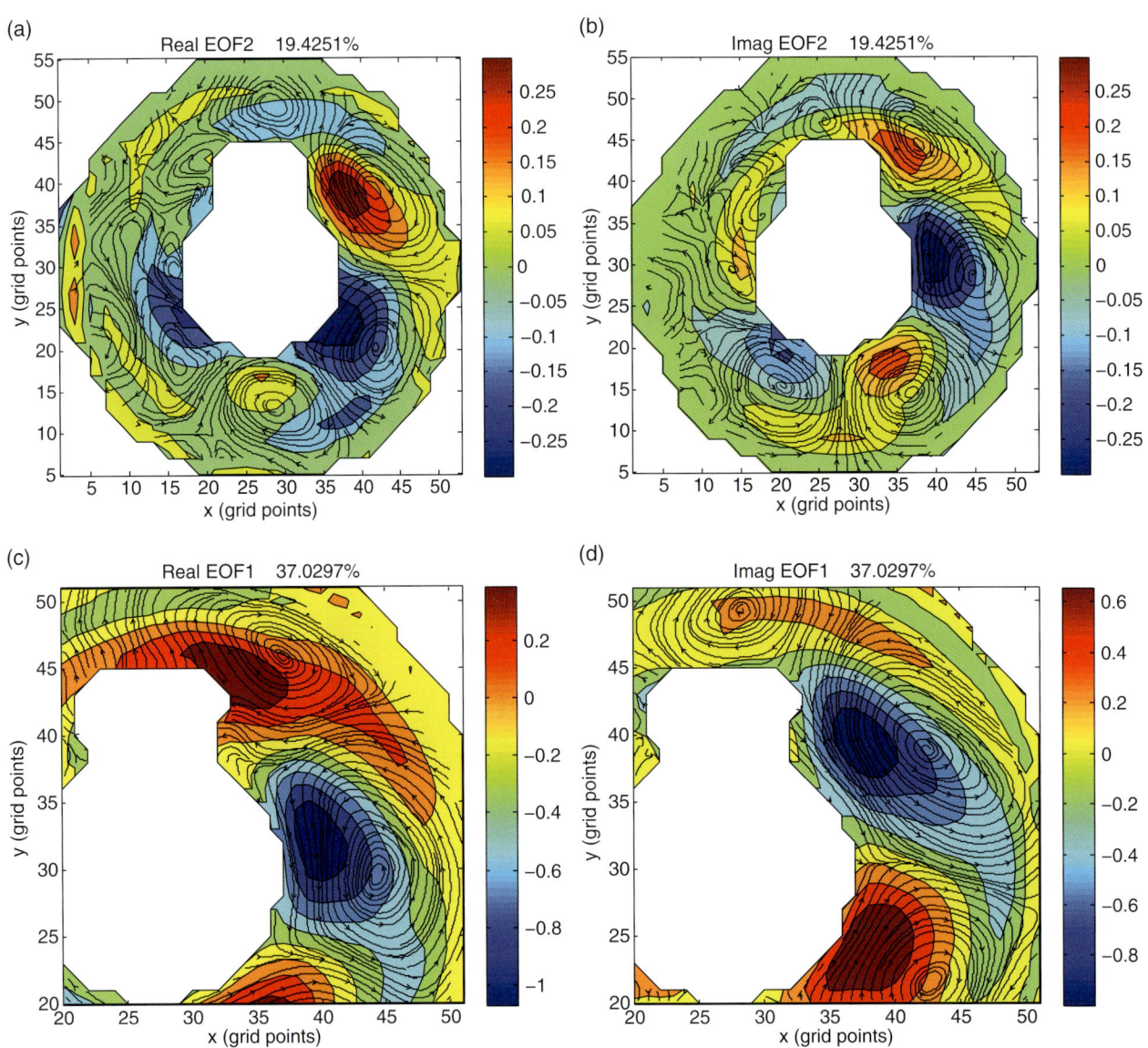

Figure 17.11. Real and imaginary parts of CEOFs: surface temperature and velocity: (a,b) CEOF2 for full annulus; (c,d) CEOF1 of upstream region. The CEOFs resolve coupled heat and velocity anomalies propagating anticlockwise toward the barrier.

9

Numerical Simulation (DNS, LES) of Geophysical Laboratory Experiments: Quasi-Biennial Oscillation (QBO) Analogue and Simulations Toward Madden–Julian Oscillation (MJO) Analogue

Nils P. Wedi

9.1. INTRODUCTION

It is human nature to seek and be fascinated by the self-organization of coherent spatiotemporal structures, common in many disciplines such as meteorology, astrophysics, biology, human sciences, and economy [*Kondepudi and Prigogine*, 1998]. Among many other interesting phenomena, the equatorial atmosphere offers two distinct intraseasonal (ir)regularities, the quasi-biennial oscillation (QBO) and the Madden-Julian oscillation (MJO). Both phenomena provide intriguing clues into some of the fundamental mechanisms underlying Earth's climate system and thus provide excellent topics for laboratory-scale experimentation as well as associated numerical simulations.

Previous chapters considered specific aspects of baroclinic and barotropic flows that can be reproduced in laboratory studies and that are ubiquitous in global atmospheric flows. In this chapter, we consider two conspicuous phenomena that are specific to the equatorial zone of rotating planets, where the quasi-horizontal "balance" model obtained from geostrophic theory does not generally apply [*Saujani and Shepherd*, 2006]. Instead, planetary scales $L \gtrsim 3000$ km in the tropics are commonly viewed as dominated by linear waves, described by the linear shallow-water theory with nonlinear advective terms neglected, with characteristic time scale significantly shorter than that of synoptic-scale systems [*Matsuno*, 1966; *Gill*, 1980; *Wheeler and Kiladis*, 1999; *Yano and Bonazzola*, 2009]. Consistently, a substantial part of the observed large-scale tropical variability is explained by linear equatorial wave motions. This is illustrated by the finding in *Zagar et al.* [2005] that convectively coupled equatorial waves [1] match a dominant portion of the horizontal structure of statistical deviations from a linear wave model that is employed for optimizing the multivariate use of observations in the ECMWF global data assimilation for numerical weather prediction (NWP). However, low-frequency planetary-scale waves in the tropics, with relatively small meridional-to-zonal wavelength aspect ratio, can be described by the same barotropic vorticity equation as obtained by *Charney* [1963] for synoptic-scale motions [*Maicun*, 1987].

The prevalent mix of Rossby, Kelvin, and gravity waves in the equatorial atmosphere propagate horizontally as well as vertically and substantially influence stratospheric motions and subsequently the global atmospheric circulation. Multiscale convective heating, extratropical wavedriving, and orographic forcings provide important sources of small-scale (yet influential) vertically propagating gravity waves [*Fritts and Alexander*, 2003; *Alexander*, 2010]. A recent review of the current theoretical understanding of stratospheric equatorial dynamics including the QBO can be found in the work of *Gray* [2010].

In this chapter the direct numerical simulation (DNS) of the laboratory analogue of the QBO and the large-eddy simulation (LES) of a proposed laboratory analogue of MJO-like tropical dynamics serve as examples of a deeper underlying concept. By using the same numerical apparatus, it is possible to show the formation of coherent spatiotemporal structures similar to the ones observed in the tropical atmosphere: Either by vertical oscillations

---

[1] After *Wheeler and Kiladis* [1999], equatorial wave patterns in the spectral analysis of satellite-observed data records have been termed "convectively coupled equatorial waves."

Numerical Aspects Section, European Centre for Medium-Range Weather Forecasts, Reading, United Kingdom.

Modeling Atmospheric and Oceanic Flows: Insights from Laboratory Experiments and Numerical Simulations, First Edition. Edited by Thomas von Larcher and Paul D. Williams.
© 2015 American Geophysical Union. Published 2015 by John Wiley & Sons, Inc.

of upper or lower boundaries in a zonally periodic channel, forcing gravity wave motions in a stratified fluid, or, in the other case, by meridional meandering of the lateral boundaries of a zonally periodic equatorial β-plane, forcing Rossby wave motions in a rotating stratified fluid. Hence, the differentiation between the two systems is on the basis of the forces that drive the motion.

9.2. QUASI-BIENNIAL OSCILLATION

The QBO represents the dominant variability in Earth's equatorial stratosphere, exhibiting quasi-regular zonal mean wind reversals with an average period of approximately 28 months. Figure 9.1 shows the unfiltered zonal mean zonal wind over a 25 year period as analyzed by the ERA40 reanalysis dataset [*Uppala et al.*, 2005], clearly showing the alternating quasi-regular wind regimes of easterly and westerly winds and the distinct downward propagation between 20 and 40 km height at a rate of approximately 1 km per month. The QBO fundamentally results from the interaction of vertically propagating waves with the horizontal background flow.

The principal interaction of waves with a mean flow [*Eliassen and Palm*, 1961] may be illustrated in a viscous, nonrotating Boussinesq fluid [*Plumb*, 1977] by the momentum equations averaged in a horizontally periodic domain as

$$\frac{\partial U}{\partial t} - \nu \frac{\partial^2 U}{\partial z^2} = \sum_i \frac{\partial F_i}{\partial z}, \quad (9.1)$$

where $U := \overline{u}^{xy}$ denotes the horizontally averaged (mean) flow, ν denotes the kinematic viscosity, and $F_i := \overline{u'w'}^{xy}$ expresses the ith contribution to the averaged nonlinear momentum flux from a spectrum of waves. Most atmospheric research of the QBO is now devoted to finding the precise physical origins of the right-hand side of equation (9.1) and their realizability in the context of numerical weather prediction and climate modeling. Long-term observations of temperature fluctuations of Jupiter's equatorial stratosphere suggest a similar mechanism for the quasi-quadrennial oscillation (QQO) [*Lie and Read*, 2000]. Furthermore, it has been suggested that dissipative wave–mean flow interactions are responsible for superrotation phenomena on Venus [*Leovy*, 1973; *Fels and Lindzen*, 1974; *Hou and Farrell*, 1987; *Yamamoto and Takahashi*, 2003], the redistribution of chemical constituents in solar-type stars [*Charbonnel and Talon*, 2005], and the recently observed equatorial oscillation on Saturn [*Schinder et al.*, 2011]. Moreover, gravity wave–driven shear flow oscillations in the equatorial plane of the solar radiative interior have been proposed to help explain the solar internal rotation, an outstanding problem in stellar physics [*Rogers et al.*, 2008].

The remarkable laboratory experiment of *Plumb and McEwan* [1978] demonstrates the principal mechanism of the periodically reversing winds of the QBO. The laboratory setup consists of a cylindrical annulus filled with density-stratified salty water forced at the lower boundary by an oscillating membrane. At sufficiently large forcing amplitude the wave motion generates an oscillation in the zonal mean zonal flow with relatively long periods compared to the period of the forcing oscillations. The laboratory experiment is often employed to explain the basic mechanism of the atmospheric QBO [*Baldwin et al.*, 2001].

In the laboratory, the averaged momentum flux has been attributed almost exclusively to viscous internal wave dissipation [*Plumb*, 1977; *Plumb and McEwan*, 1978; *McIntyre*, 2003], in analogy to equatorial Kelvin and Rossby gravity wave attenuation by infrared cooling and hence consistent with the theory of *Holton and Lindzen* [1972]. The latter theory motivated the laboratory experiment and its corresponding conceptual synthesis [*Plumb*, 1977; *Plumb and McEwan*, 1978; *Plumb and Bell*, 1982] that established the fundamental picture of the QBO as forced by large-scale upward propagating waves, with the amplitude and the rate of downward propagation determined by the waves' phase speeds and intensity, respectively [*Lindzen*, 1987].

Notwithstanding, the two-wave model with oppositely traveling Kelvin and Rossby gravity waves [*Holton and Lindzen*, 1972; *Plumb*, 1977] is incomplete. Observations showed that there must be an additional easterly gravity wave mode required to account for the easterly acceleration of the QBO [*Lindzen and Tsay*, 1975]. Furthermore, tropical upwelling, a climatic mean upward motion of the tropical atmosphere, was shown to necessitate contributions to the mean-flow momentum budget of other wave types. In particular, *Dunkerton* [1997] argued that shorter-scale gravity waves contribute up to 70% to the forcing, thus indicating a considerable uncertainty about the precise origin and the nature of the waves responsible for driving the QBO. Consistently, high-resolution 3D simulations demonstrated insufficient provision of wave momentum flux by the classical two-wave model, given realistic amplitudes of Kelvin and Rossby gravity waves [*Takahashi and Boville*, 1992]. By relaxing the simplifying assumptions characteristic of 1D and 2D mechanistic models, Takahashi and Boville found wave-wave interactions of Kelvin and Rossby gravity waves, and subsequent modifications of the background wind, to be important for the simulated QBO period.

Following *Andrews et al.* [1987] and *McIntyre* [2003], a "critical layer" is a not necessarily zonally continuous height band, in which a group of waves are attenuated by a variety of dissipative processes. A critical surface or level, first elucidated by *Bretherton* [1966] and *Booker and*

Bretherton [1967], represents an upper asymptotic limit to a critical layer. An earlier explanation of the QBO in fact suggested that the stratospheric mean-flow oscillation is driven by critical layer attenuation of a spectrum of gravity waves [*Lindzen and Holton*, 1968]. Similarly, *Dunkerton* [1981] and *McIntyre* [1994] considered wave transience and wave breaking as chronologically more important primary causes of the zonal mean-flow oscillation in the atmosphere. Hence, the laboratory experiment of Plumb and McEwan has also been criticized for its apparent fundamental difference to the QBO [*Dunkerton*, 1981]. However, the DNS of the laboratory experiment [*Wedi and Smolarkiewicz*, 2006] suggests in fact the opposite: The laboratory experiment with its distinct gravity wave sources, a chronology of wave interference, critical layer formation, and subsequent downward propagation of the critical layer, leaving behind a uniform zonal mean zonal flow, portray a picture of the laboratory experiment that is much closer to the real atmosphere than perhaps originally thought. DNS means here integrating the Navier-Stokes equations under the Boussinesq approximation for salty water without any parametrizations, resolving the fluid motion down to the Kolmogorov length scale $\eta = (\nu^3/\epsilon)^{1/4}$, where ν is the kinematic viscosity and ϵ denotes the kinetic energy dissipation rate. In this case grid sizes are $\Delta x = \mathcal{O}(\eta)$; see *Moin and Mahesh* [1998] for a review of DNS. With these grid sizes the dissipation scale of the density is not resolved but justified a posteriori (see Section 3e in *Wedi and Smolarkiewicz* [2006]).

In large-scale numerical simulations, *Horinouchi and Yoden* [1998] found a critical-layer mechanism and contributing waves consistent with *Lindzen and Holton* [1968] to be responsible for QBO-like oscillations. Notably, the latter study used an "aqua-planet" idealization with parametrized convection, exciting a spectrum of shorter-scale gravity waves. The intercomparison study of several global circulation models (GCMs) [*Horinouchi et al.*, 2003] corroborates a correlation between the simulated QBO and convective processes, while finding equatorial Kelvin and Rossby gravity waves unimportant relative to other types of wave forcing. Moreover, the reanalysis dataset ERA-15 [*Gibson et al.*, 1999] shows no significant spectral signals of equatorially trapped planetary waves [*Tindall*, 2003].

Consistently, successful numerical modeling of a realistic 3D atmospheric QBO as well as realistic representations of tropospheric and stratospheric global circulations require the parametrization of nonorographic, vertically propagating gravity waves [*Scaife et al.*, 2000; *Giorgetta et al.*, 2002; *Orr et al.*, 2010].

9.2.1. Laboratory Experiment

The laboratory experiment of *Plumb and McEwan* [1978] was conducted in a transparent cylindrical annulus (radii $a = 0.183$ m and $b = 0.3$ m) filled with density-stratified salty water to a height of $z_{ab} = 0.43$ m. The lower boundary consisted of a thin rubber membrane oscillating with a constant frequency $\omega_0 = 0.43$ s^{-1} and amplitude $\epsilon = 0.008$ m [*Plumb and McEwan*, 1978]. After some time, a zonal mean zonal flow (visualized by polystyrene pellets) can be observed for approximately 30–40 min in one direction and then 30–40 minutes in the other direction, driven exclusively by the gravity wave trains excited by the boundary oscillation.

The laboratory experiment was repeated by the GFD-Dennou Club at the University of Kyoto. A photograph of their cylindrical annulus can be seen in Figure 9.2. Two main differences to the original Plumb and McEwan setup exist: (i) the oscillating membrane was placed at the top of the annulus initiating an oscillation with an apparent *upward* propagation of the mean flow as opposed to a *downward* propagation in the original experiment and (ii) several water tank sizes were available (Shigeo Yoden, personal communication). For example, the setup described

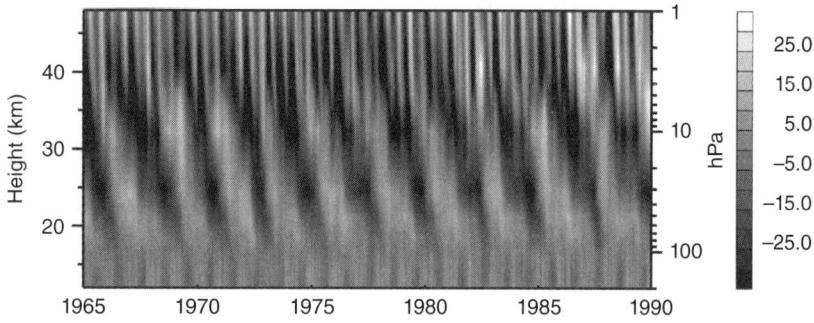

Figure 9.1. Quasi-biennial oscillation as analyzed in ERA40 in a time-height representation. The figure shows the equatorial analyzed (unfiltered) zonal mean zonal wind between 200 hPa and 1 hPa averaged between ±1° latitude for the period 1965–1990. The contour interval is 5 m/s.

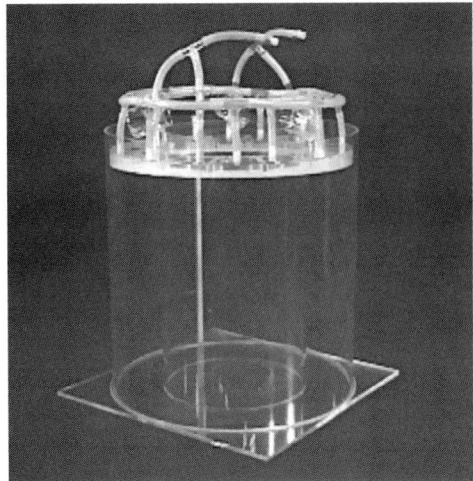

Figure 9.2. Cylindrical annulus used for the QBO experiments conducted at the University of Kyoto as part of the GFD-Dennou Club. Used with permission, *Sakai* [1997] and *Otobe et al.* [1998]. The oscillating membrane can be seen at the top of the annulus, while in the original experiment of Plumb and McEwan the membrane was mounted at the bottom.

by *Otobe et al.* [1998] is equivalent to the original Plumb and McEwan experiment, whereas the experimental setup described by *Sakai* [1997] used an approximately twice larger domain ($2L_x, 2L_y$). Their laboratory experiments and their detailed descriptions of the experiment [cf. *Sakai*, 1997] give interesting insights into the sensitivity and the difficulties encountered when trying to exactly reproduce the oscillation period obtained by Plumb and McEwan. For example, the stratification degraded in the near-membrane layers, influencing the mean-flow oscillation period. Equally, the tank size and the associated wavelengths of the forcing have a substantial influence on the oscillation period [*Wedi and Smolarkiewicz*, 2006]. The laboratory experiments at the University of Kyoto show a range of observed mean-flow oscillation periods of 45–120 min [*Otobe et al.*, 1998], with several at approximately 1 h. Nevertheless, all these results show that the emerging zonal mean zonal flow oscillation period is always much longer than the inverse of the frequency of the forcing waves.

9.2.2. DNS of the Laboratory Analogue of the QBO

Incorporating the rapidly undulating boundaries of the laboratory experiment into the numerical algorithm via time-dependent curvilinear coordinates allows to reproduce the experimental setup while minimizing numerical uncertainties. *Wedi and Smolarkiewicz* [2006] presented results of the first DNS of the phenomena that lead to the zonal mean-flow reversal in the laboratory analogue of the QBO. The detailed study of the parametric and numerical sensitivities reveal the dominant role of wave-wave and wave–mean flow interactions in the laboratory flow, with critical layer absorption and viscous dissipation chronologically secondary to instabilities from nonlinear flow interactions [*Wedi and Smolarkiewicz*, 2006]. These findings elevate the importance of the laboratory setup for its fundamental similarity to the atmosphere, where such instabilities are observed to occur.

9.2.2.1. Numerical Model.
The Boussinesq equations of motion, an accurate approximation for salty water [*Gill*, 1982], for a nonrotating, density-stratified, viscous fluid are cast in a time-dependent curvilinear framework [*Prusa and Smolarkiewicz*, 2003; *Smolarkiewicz and Prusa*, 2005]:

$$\frac{\partial \left(\rho^* \overline{v}^{sk}\right)}{\partial \overline{x}^k} = 0,$$
$$\frac{dv^j}{d\bar{t}} = -\widetilde{G}_j^k \frac{\partial \pi'}{\partial \overline{x}^k} - g\frac{\rho'}{\rho_0}\delta_3^j + \mathcal{C}^j + \mathcal{F}^j,$$
$$\frac{d\rho'}{d\bar{t}} = -\overline{v}^{sk}\frac{\partial \rho_e}{\partial \overline{x}^k} + \mathcal{F}_\rho. \quad (9.2)$$

Here, $\rho^* := \rho_0 \overline{G}$, with $\overline{G}$ denoting the Jacobian of the transformation between physical (t, x, y, z) and computational $(\bar{t}, \bar{x}, \bar{y}, \bar{z})$ space. Indices $j, k = 1, 2, 3$ correspond to the $\bar{x}, \bar{y}, \bar{z}$ components, respectively; summation is implied by repeated indices unless stated otherwise. The total derivative is $d/d\bar{t} = \partial/\partial \bar{t} + \overline{v}^{*j}(\partial/\partial \overline{x}^j)$, where $\overline{v}^{*j} := \dot{\overline{x}}^j$ denotes the contravariant velocity. The solenoidal velocity, satisfying the mass continuity equation in (9.2), is $\overline{v}^{sj} := \overline{v}^{*j} - \partial \overline{x}^j/\partial t$. The components of physical velocity v^j are related via $\overline{v}^{sk} = \widetilde{G}_j^k v^j$, where $\widetilde{G}_j^k = (\partial \overline{x}^k/\partial x^j)$ are transformation coefficients; ρ' and π' denote density and normalized pressure perturbations, respectively, with respect to the static ambient state characterized by the linearly stratified profile $\rho_e = \rho_0(1 - (N^2/g)z)$, g symbolizes the gravitational acceleration, ρ_0 is a constant reference density, and δ_3^j is the Kronecker delta. The change of density due to the diffusivity of salt in water, $\mathcal{F}_\rho \sim \nabla \cdot \kappa \nabla \rho'$, and the momentum dissipation, $\mathcal{F} \sim \nabla \cdot \nu(\nabla \mathbf{v} + \nabla \mathbf{v}^T)$, are detailed in the work of *Smolarkiewicz and Prusa* [2005]. Here, we specify a kinematic viscosity $\nu = 1.004 \times 10^{-6}$ m^2/s and a diffusivity of salt in water κ (= 1.5×10^{-9} m^2/s). The Coriolis force terms $\mathcal{C}^j \equiv 0$.

The generalized Gal-Chen coordinate transformation assumes identity transformations $\bar{t} = t$, $\bar{x} = x$, and $\bar{y} = y$, but

$$\bar{z} = H_0 \frac{z - z_s(x,y,t)}{H(x,y,t) - z_s(x,y,t)}, \quad (9.3)$$

whose theoretical development and efficient numerical implementation were discussed thoroughly by *Wedi and Smolarkiewicz* [2004]. The transformation (9.3) allows for time-dependent upper, $H(x, y, t)$, and lower, $z_s(x, y, t)$, boundary forcing without small-amplitude approximations.

The governing equations (9.2) are discretized in the transformed space using a second-order-accurate, optionally semi-Lagrangian or Eulerian, nonoscillatory forward-in-time (NFT) approach, broadly documented in the literature (cf. *Smolarkiewicz and Prusa* [2002] and *Smolarkiewicz* [2006] for a recent review). Here we employ the flux form Eulerian, semi-implicit version of the algorithm unless stated otherwise. All prognostic equations in (9.2) are integrated using the trapezoidal rule, treating all inviscid forcing on the right-hand side implicitly. The viscous and diffusive terms are computed explicitly to first-order accuracy, (see Section 3.5.4 in *Smolarkiewicz and Margolin* [1998]). Together with the curvilinearity of the coordinates, this leads to a complicated elliptic problem for pressure (see Appendix A in *Prusa and Smolarkiewicz* [2003] for the complete description) solved iteratively using a preconditioned nonsymmetric Krylov subspace method [*Smolarkiewicz and Margolin*, 2000].

9.2.2.2. DNS Simulation.
In the numerical simulation an initially stagnant fluid is forced by an oscillating lower boundary. The cylindrical laboratory tank is represented by a zonally periodic, rectangular computational domain of $639 \times 38 \times 188$ grid intervals with $L_x = 2\pi(a+b)/2$, $L_y = b - a$, $L_z = 0.43$ m. The impermeable lower and lateral boundaries are no slip. At the upper boundary a free-slip rigid lid is assumed. The lower surface is prescribed as a linear combination of elementary shape functions

$$z_s(x, y, t) = \sum_{\eta=0}^{\mathcal{N}+1} \epsilon_\eta \xi_\eta(x) \sin\left(\frac{\pi}{L_y} y\right) \sin(\omega_0 t + \phi_\eta), \quad (9.4)$$

with individual amplitudes ϵ_η and zonal profiles

$$\xi_\eta = \begin{cases} \cos^2(\pi r_\eta / 2L_\eta) & \text{if } \|r_\eta/L_\eta\| \leq 1, \\ 0 & \text{otherwise}, \end{cases} \quad (9.5)$$

where $L_\eta = L_x/\mathcal{N}$ and $r_\eta = x + (\eta - 0.5)L_\eta$. In general, equations (9.4) and (9.5) describe a subdivision of the periodic domain into $\mathcal{N}$ independently oscillating chambers. For $\mathcal{N} = 16$, $\epsilon_\eta = 0.008$ m $\forall \eta$, and $\phi_\eta = \pi$ for even η ($\phi_\eta = 0$ for odd η), the equation collapses into the symmetric solution of the 2D wave equation employed in the original Plumb and McEwan paper. In the original experiment, $H = H_0$. In the case of the Kyoto experiment, the equivalent numerical simulations are obtained by setting $z_s = 0$ in (9.3) and using $H(x, y, t) = H_0 - z_s(x, y, t)$ with z_s from (9.4).

The bottom boundary in the latter simulations was assumed rigid no slip. The length of the oscillation period has been found to be moderately sensitive to the choice of the bottom boundary condition [*Wedi and Smolarkiewicz*, 2006].

For the initial condition a static state was assumed with buoyancy frequency $N = 1.57$ s^{-1}. The integration time was several hours with a time step $dt = 0.05$ s. QBO-like oscillations are a canonical example of long-time behavior resulting from short-term fluctuations. Therefore, one may expect the initial conditions to be of minor relevance to the long-term solution. Nevertheless, the choice of initial conditions can be important for practical reasons. The zonal mean zonal flow oscillation was seeded with a zonal background flow u_e in the near-membrane layers using $u_e = u_0 \{1 - 0.5\delta[1 + \tanh(z - d_0)/\gamma]\}$ with $u_0 = 0.02$ m/s, $\delta = 0.9999999$, $d_0 = 0.06$ m, and $\gamma = \Delta z$. However, it was found later that this is not necessary, as all that is needed to start an oscillation is the addition of *zonally asymmetric* noise. If this is not done, the simulation may not develop an oscillation for many hours of simulation time, after which finally numerical truncation errors have sufficiently accumulated to provide the same effect (see also Section 3b in *Wedi and Smolarkiewicz* [2006]).

Figure 9.3 illustrates the overall agreement of the solution structure between the laboratory experiment (adapted from Figure 10 of *Plumb and McEwan* [1978]) and the numerical simulation of the laboratory analogue.

The detailed chronology of the zonal mean zonal flow reversal in the DNS of the laboratory analogue is sketched in Figure 9.4. It describes the turbulent breakdown of gravity wave trains with viscous dissipation and critical layer absorption chronologically secondary to instabilities from nonlinear flow interactions. The latter motivated *Fetecau and Muraki* [2011] to refine the modulation theory of the time evolution of slowly varying wave trains in a density-stratified fluid with coupling to the mean flow (as, e.g., applied in *Plumb* [1977]). The authors derived high-order corrections to the modulation theory and found self-regularizing properties of these corrections, whereby the growth of unstable modes is essentially tempered by nonlinearity. Taking into account these corrections resulting from nonlinearity, they found excellent agreement between solutions of the corrected modulation system and corresponding variables extracted from the numerical solutions of the nonlinear Boussinesq equations. The corrected modulation system thus provides a mathematical pathway commensurate with the original explanation of the laboratory experiment and the DNS results. In agreement with the original analysis of the laboratory experiment, the most dominant sensitivity on the length of the zonal mean zonal flow oscillation period is by the forcing wave number and thus the shape of the oscillating membrane [*Wedi and Smolarkiewicz*, 2006].

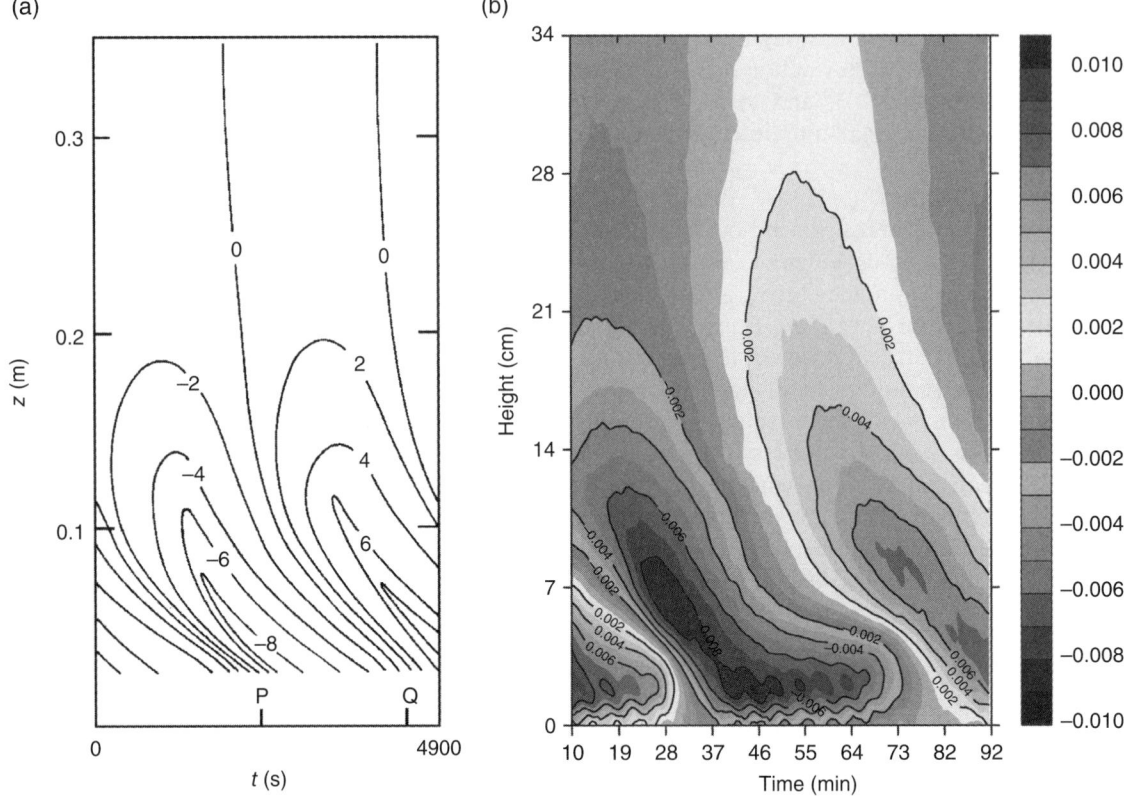

Figure 9.3. Time-height cross section of the observed zonal mean zonal flow velocity component (panel (a) adapted from Figure 10 in *Plumb and McEwan* [1978], contour lines in mm/s), compared to the result of the 3D numerical simulation at $y = L_y/2$ in panel (b), reproduced from Figure 2b in *Wedi and Smolarkiewicz* [2006], contour lines in m/s. © American Meteorological Society. Used with permission. According to *Plumb and McEwan* [1978], the lowest 2 cm in panel (a) could not be observed due to restrictions of the viewing window. For color detail, please see color plate section.

9.3. MADDEN-JULIAN OSCILLATION

The MJO [*Madden and Julian*, 1971], named after the scientists who first discovered the anomaly in time series of atmospheric datasets, is the main intraseasonal atmospheric fluctuation in the equatorial troposphere affecting weather in large parts of the world. Polynesian seamen are believed to have used the phenomenon to sail eastward in the trade wind–dominated equatorial Pacific 4500 years ago [*Zhang*, 2005]. Despite substantial efforts, a reliable forecasting of the MJO and understanding its underlying dynamical mechanisms remain key challenges in atmospheric science. It is not strictly an oscillation as its period varies and its appearance is episodic [*Hartmann and Hendon*, 2007]. Existing theories stress the importance of the feedback mechanisms between convection, large-scale wave dynamics and surface fluxes; see *Zhang* [2005] for a comprehensive review. However, while a synthesis of the theories and observations explains important aspects of the MJO life cycle, a unifying theory is still elusive for the basic mechanism that would also explain the persistent and ubiquitous modeling difficulties with state-of-the-art global NWP and climate models. The complexity of the processes involved, due to their multiscale nature ranging from micro to global scales, render the MJO an intriguing problem in fluid dynamics, not least because of its appearance as a solitary structure, with its episodic appearance and its slow (5 m/s), nearly dispersionless eastward motion.

What makes solitary wave theories with Korteweg-de Vries (KDV)-like solutions particularly attractive is that they extend the linear shallow-water theory [*Matsuno*, 1966], commonly used to explain different modes of equatorial wave motions, to the weakly nonlinear regime. Remarkably, most or all spectral signals of satellite-observed outgoing long-wave radiation (OLR), a proxy for cloudiness, can be explained via the linear theory of equatorially trapped waves, except for the dominant low-frequency spectral peak of the MJO [*Wheeler and Kiladis*, 1999].

In addition, moist physics and in particular the coupling of the large-scale flow to the deep convection parametrization [*Hirons et al.*, 2013], or in fact the absence of deep

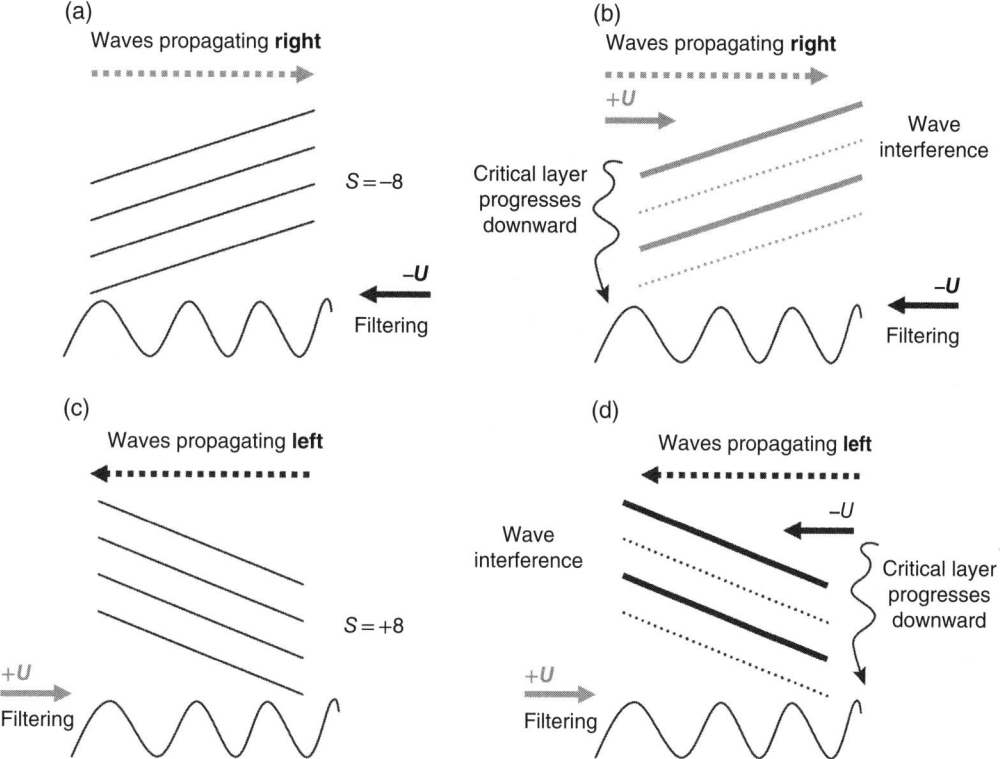

Figure 9.4. The principle mechanism of the oscillation in the laboratory experiment of *Plumb and McEwan* [1978] is shown in four representative phases of the flow (reproduced from Figure 19 in *Wedi and Smolarkiewicz* [2006]): (a) There is a zonal mean zonal flow $-U$ allowing gravity waves with a single dominant horizontal wave number $s = -8$; (b) wave-wave mean-flow interaction generates a critical layer that propagates downward generating a zonal mean zonal flow $+U$; (c) the reversed mean flow arriving at the oscillating membrane filters the waves of opposite direction to allow only waves with horizontal wave number $s = +8$; (d) wave-wave mean-flow interaction generates a critical layer for the opposite traveling waves generating a downward propagating zonal mean zonal flow $-U$.

convection parametrization [*Miura et al.*, 2007], appear to have a substantial influence on the predictive skill of MJO events in actual forecast simulations. The interaction of planetary-scale, lower-tropospheric moisture anomalies with convective processes have been suggested as a fundamental mechanism of the MJO [*Majda and Stechmann*, 2009]. The effects of moist physics are not least manifested in the fact that convectively coupled equatorial waves map to linear waves with a smaller effective equivalent depth than what would be assumed for dry conditions.

All this makes the MJO complex to study in a laboratory-scale environment and a laboratory analogue to the MJO does not (yet) exist. In fact, the search for an explanation of the MJO has been called the "search for the Holy Grail of tropical atmospheric dynamics" [*Raymond*, 2001], and the next section attempts to contribute to this search by reproducing the following prominent features of the MJO at laboratory scales: the slow eastward propagation of a near dispersionless solitary phenomenon and its anomalous horizontal quadrupole structure in the upper part of the domain.

9.3.1. LES of Laboratory Analogue for MJO-Like Tropical Dynamics

The same numerical apparatus with time-dependent curvilinear coordinates that has been successfully applied for the QBO analogue is employed here to construct/propose a virtual laboratory-scale experiment, where the generation of solitary structures is excited and maintained via zonally propagating meanders of the meridional boundaries of a zonally periodic β-plane. The proposed experimental setup of the laboratory analogue for MJO-like tropical dynamics and its scaled mapping onto the equatorial troposphere is given in Table 9.1. From the dimensions it becomes clear that DNS of the proposed experiment is still (just) out of reach when considering a minimum requirement of $4300 \times 4000 \times 110$ grid points

Table 9.1. Experimental parameters in the LES of a laboratory analogue for MJO-like tropical dynamics.

Symbol	Laboratory Scale	Atmospheric Scale	Description
L	2 m	4000 km	Length scale
U	0.05 m/s	50.0 m/s	Velocity scale
β'	0.093 m^{-1}s^{-1}	2.3×10^{-11} m^{-1} s^{-1}	Coriolis β
N	1.566 s^{-1}	0.01 s^{-1}	Brunt-Väisällä frequency
$u_e(v_e = 0, w_e = 0)$	0.05 m/s	50.0 m/s	Ambient flow
x_0	4.3 m	8600 km	Zonal domain length
y_0	4.0 m	8000 km	Meridional domain width
H_0	0.11 m	17000 m	Vertical domain height
T_1, T_2	120 s, 100 s	2.7 days, 2.3 days	Forcing periods
a	0.2 m	400 km	Forcing amplitude
$\lambda = x_0/s(s = 6)$	0.717 m	1430 km	Forcing wavelength
$L_D = NH_0/\beta L$	0.926 m	1850 km	Equatorial deformation radius
$T_D = L_D/U$	18.5 s	10 h	Equatorial deformation time

Reproduced from *Wedi and Smolarkiewicz* [2010].

for a simulation of upto 8 h, representing a laboratory-scale "climate" realization. Equally, the proposed laboratory analogue experiment presents immense technical challenges since it would require to emulate a zonally periodic equatorial β-plane while maintaining constant vertical stratification for several hours in the presence of bottom heating and zonally propagating meanders of the meridional sidewalls.

The laboratory-scale climate simulations described here thus broadly fit into the class of implicit large-eddy simulations (ILESs). ILESs dispense with explicit subgrid-scale models and exploit truncation properties of high-resolution (nonoscillatory) finite-volume methods to mimic the spectral viscosity of standard LESs [*Domaradzki et al.*, 2003; *Margolin et al.*, 2006; *Piotrowski et al.*, 2009].

The simulations capture details of the formation of solitary structures and of their impact on the convective organization [*Wedi and Smolarkiewicz*, 2008, 2010]. The horizontal structure and the propagation of anomalous stream function patterns, a diagnostic typically used in tracing the equatorial MJO, are similar to archetype solutions of the Korteweg-deVries equation, which extends the linear shallow-water theory, commonly used to explain equatorial wave motions, to a weakly nonlinear regime for small Rossby numbers. In this way, *Wedi and Smolarkiewicz* [2010] attributed the origin and evolution of periodically reoccuring anomalous flow patterns in the equatorial troposphere to resonant nonlinear wave dynamics. As a result, the MJO may be understood as a quasi-horizontal and quasi-nondivergent synoptic-scale motion that is driven, or rather preconditioned, by coupling with extratropical weather and that persists at planetary scale due to nonlinearity. Such motion is governed by a particular solution of the conservation law for absolute vorticity $\nabla^2 \psi + f$ [*Charney*, 1963]. Consequently, the process of convection is found to be important but chronologically secondary to the MJO evolution.

There are MJO case studies [*Hsu et al.*, 1990] that support this view. Moreover, *Vitart and Jung* [2010] recently confirmed a substantial influence of the Northern Hemisphere extratropics on the skill in predicting the MJO by relaxing the Northern Hemisphere (and in particular the North Pacific) to the "known" analyzed state. The predictive skill is even further improved when also the Southern Hemisphere is relaxed toward the analysis (Vitart, personal communication). The latter suggests that the MJO is a global phenomenon not only in appearance but also in its initiation and propagation properties.

9.3.1.1. Numerical Model. The numerical model is as described in Section 9.2.2.1. In addition, the Coriolis force terms on the equatorial β-plane[2] are now given as $\mathcal{C}^1 = +\beta y(v^2 - v_e^2)$, $\mathcal{C}^2 = -\beta y(v^1 - v_e^1)$, and $\mathcal{C}^3 = 0$. Diabatic and frictional terms, emulating boundary layers adjacent to $r = y, z$ boundaries, are of the form $\mathcal{F}_\rho(r) := \tau_\rho^{-1} e^{-r/h}(\rho - \rho_b)$ and $\mathcal{F}^j(r) := \tau_{vj}^{-1} e^{-r/h}(v^j - v_b^j)$, with subscript b denoting a prescribed boundary value; the attenuation time scales are $\tau_\rho = \Delta z^2/\kappa$ and $\tau_{vj} = 0.125\tau_\rho$ (assuming diffusivity of heat in water, $\kappa = 1.39 \times 10^{-7}$ m^2/s) and height scale $h = 2\Delta z$, where Δz is the vertical grid size. Given the model's formulation in density, heating is included indirectly via the gradient of density at the lower boundary,

---

[2]The shallow atmosphere (or "traditional" approximation) has been applied here for consistency with simulations using the integrated forecasting system (IFS), the operational global forecast model at ECMWF; see also *Wedi and Smolarkiewicz* [2009].

which induces convective vertical motions. Weak, moderate, and strong heating is defined by the boundary value of the density ratio $(\rho_0 - \rho_b)/\rho_b$=0.0016, 0.0048, 0.0096, respectively. The velocity values at the lower and lateral boundaries are set to zero ($v_b^j = 0$) unless specified otherwise. The upper boundary is free slip, with $\mathcal{F}^j(z = H_0) = 0$ and $\mathcal{F}_\rho(z = H_0) = 0$. The ambient flow prescribed as $u_e = 0.05$ m/s ($v_e = 0, w_e = 0$) is used together with the lateral boundary meander to mimic an extratropical anomalous flow regime.

The particular coordinate mapping employed for the simulation of lateral boundary meanders assumes the identity transformations $\bar{t} = t$, $\bar{x} = x$, and $\bar{z} = z$, and

$$\bar{y} = y_0 \frac{y - y_S(x, y, t)}{y_N(x, y, t) - y_S(x, y, t)}, \quad (9.6)$$

where y_S and y_N are the southern and northern domain boundaries, respectively, and y_0 denotes the domain size in meridional direction. Numerically, the elements of the Jacobi matrix are evaluated in $(\bar{t}, \bar{\mathbf{x}})$ leading to the required subset of (nontrivial) coefficients [*Wedi and Smolarkiewicz*, 2004] $\widetilde{G}_1^1 = \mathcal{G}^{-1}\partial y/\partial \bar{y}$, $\widetilde{G}_1^2 = \mathcal{G}^{-1}\partial y/\partial \bar{x}$, $\widetilde{G}_2^1 = \mathcal{G}^{-1}\partial x/\partial \bar{y}$, and $\widetilde{G}_2^2 = \mathcal{G}^{-1}\partial x/\partial \bar{x}$ with $\mathcal{G}^{-1} = (\partial y/\partial \bar{y}\, \partial x/\partial \bar{x} - \partial y/\partial \bar{x}\, \partial x/\partial \bar{y})^{-1}$.

The simulations on the zonally periodic β-plane are represented by 128 × 128 × 64 grid points and the experimental parameters and their corresponding (rescaled) atmospheric values are summarized in Table 9.1. The model uses a time step $\Delta t = 0.1$ s.

The meridional boundaries are specified by the superposition of two waves with frequencies $\omega_1 = 2\pi/T_1$ and $\omega_2 = 2\pi/T_2$,

$$y_S(x, y, t) = -\frac{y_0}{2} + a \cos\left(\frac{\omega_2 - \omega_1}{2}t\right)$$
$$\times \sin\left(k_x x - \frac{\omega_1 + \omega_2}{2}t\right), \quad (9.7)$$

and $y_N(x, y, t) = -y_S(x, y, t)$, where y_0 denotes the domain size in the meridional direction (and analogously x_0 specifies the domain size in the zonal direction) and $k_x = 2\pi/\lambda$ with zonal wavelength λ. Such a forcing prescribes a boundary meander (a similar forcing is provided in case d of Table 1 of *Malanotte-Rizzoli et al.* [1988]) propagating eastward with the mean phase velocity $(\omega_1 + \omega_2)/2k_x$ and pulsating with the frequency $(\omega_2 - \omega_1)/2$. Unless specified otherwise, the boundary forcing is activated only after some fixed initial period and the forcing amplitude a is one-tenth of the meridional extent of the domain. As in the case for the QBO analogue, equations (9.6) and (9.7) together with (9.2) allow for a time-dependent meridional boundary forcing free of small-amplitude approximations.

The 3D simulation, including stratification and bottom boundary thermal forcing, is illustrated in Figure 9.5. The 3D organization of the solitary structure that emerges is moving slowly to the right. A broader quadrupole structure spanning the whole meridional extent can be seen in the anomalous wind field. One can identify one area with predominantly upward motions and one area with predominantly downward motions that move to the right with the quadrupole structure. Overall the 3D organisation is reminiscent of the suggested archetypical MJO structure illustrated in Figure 9.6a in *Moncrieff* [2004].

9.4. DISCUSSION

DNSs or LESs of geophysical laboratory experiments represent a stringent challenge for model numerical cores. In order to minimize numerical uncertainties, the presented numerical formulation uses time-dependent, generalized curvilinear coordinates, pairing the mathematical apparatus of Riemannian geometry with modern CFD.

While a realistic simulation of a laboratory experiment does not necessarily guarantee that the same numerics perform well at global scales on the sphere, convergence in the the limit of grid sizes $\Delta x = \mathcal{O}(\eta)$ builds confidence in the overall numerical procedure. Moreover, with increasing resolution of weather and climate simulations, it is relevant how the numerics of a model deals with the influence of small-scale fluctuations on the larger scale flow when such a mechanism is only partially or not at all parametrized. In particular, "virtual" laboratory setups provide another diagnostic tool to study the dynamics of selected flow phenomena while examining the role of implicit or explicit dissipation. This has been demonstrated in the DNS of the laboratory experiment of Plumb and McEwan for the QBO [*Wedi and Smolarkiewicz*, 2006]. The basic mechanism in the laboratory experiment and in the numerical simulations is the sequence of gravity wave excitation by simple fluctuations of the upper or lower boundary, subsequent wave-wave mean-flow interactions, critical layer formation followed by wave breaking, and the emergence of a long-time zonal mean zonal flow oscillation entirely driven by the wave momentum flux changes. All of these gravity wave processes and subsequent zonal mean-flow changes are found in the atmosphere, and hence the accuracy of different numerical choices for the simulation of wave-driven flow phenomena is equally relevant to future weather and climate predictions.

For example, a comparison of the zonal mean zonal flow reversal in numerical simulations of the QBO analogue with a flux form Eulerian and a semi-Lagrangian advection algorithm [*Wedi*, 2006] showed the onset of critical layers in different spatial positions for the latter, creating different bifurcation points for the flow development. Since, in general, flux form schemes have higher-order truncation errors proportional to the differentials

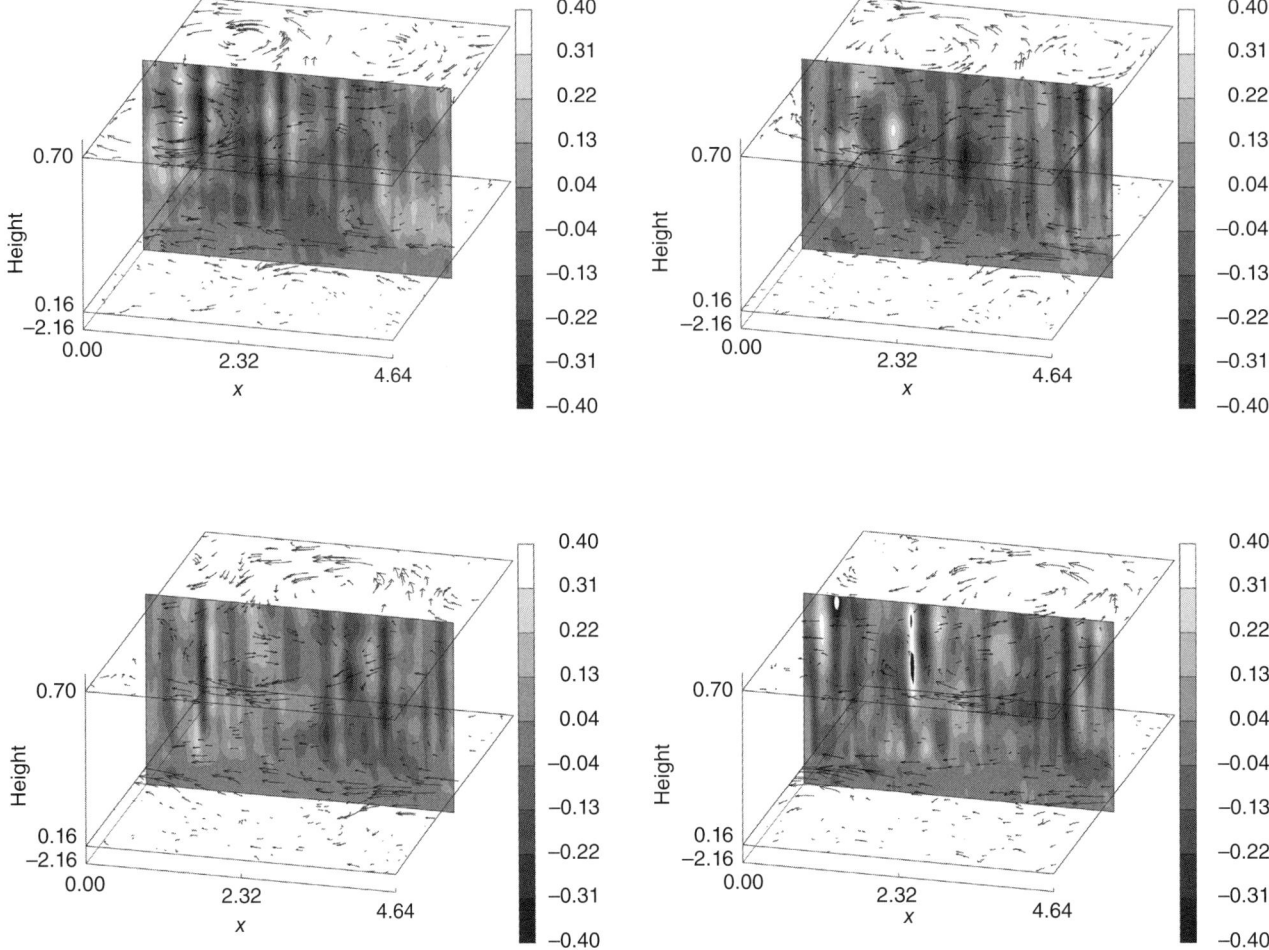

Figure 9.5. A sequence of four time instances are shown in a 3D view illustrating the instantaneous, meridionally averaged vertical motion field of the simulation described in the text. The colors with dark shading correspond to downward motions and lighter shading corresponds to upward motions (mm/s). The anomalous (time-mean subtracted) horizontal wind field is plotted as vectors illustrating the streamlines in two distinct vertical layers. The sequence is from top left to bottom right.

of fluxes of the primitive variables rather than to the differentials of the variables themselves (as characteristic of advective form schemes), *Smolarkiewicz and Margolin* [1997] concluded that the overall accuracy of the approximation increases when the fluxes of the variables exhibit a greater degree of homogeneity than the variables themselves. This may be the case in the QBO analogue simulations with fairly steady wave momentum fluxes below the critical layer. It is not clear if, in general, higher resolution simulations are characterized by less homogeneity of the prognostic variables compared to their associated fluxes. However, this may be relevant since emerging new dynamical core developments almost exclusively use flux form formulations in contrast to most existing operational NWP models today that employ the semi-Lagrangian technique [*Williamson*, 2007].

It appears that with increased resolution, and lacking suitable parametrizations, the role of higher-order truncation errors of the numerical core itself gains importance, in particular also for the organization of convection, as implied by the ILES simulations by *Piotrowski et al.* [2009] and the ones described above. For example, in the case of the MJO-like solitary structures, when the boundary meander is stopped after some time, the solitary structures exhibit an extraordinary persistence [*Wedi and Smolarkiewicz*, 2010]. Figure 9.6a shows the result of a flux form Eulerian simulation that has been restarted using the same flux form Eulerian scheme, whereas Figure 9.6b shows the result of the simulation restarted using a semi-Lagrangian scheme. Visibly the persistence and the solitary anomaly in velocity potential is lost if the simulation is continued with the semi-Lagrangian scheme. The results imply also a decisive influence of the numerical model core on the formation and the persistence of temporal flow anomalies, which is consistent with the large discrepancies found in the

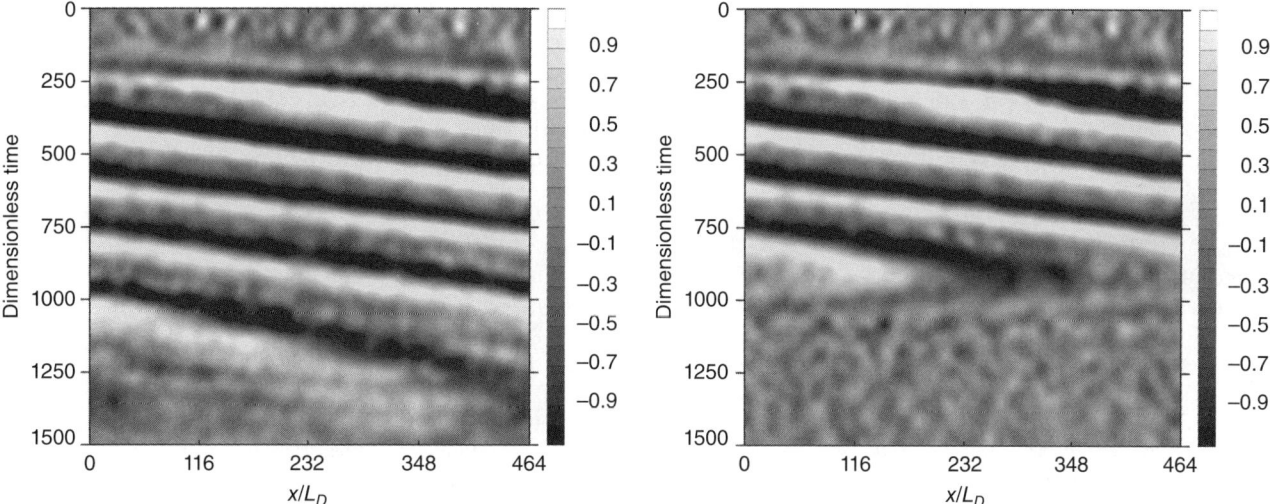

Figure 9.6. Hovmöller diagrams of the temporal anomaly of velocity potential ($\times 10^{-5}$ m^2/s) for the simulation of a stably stratified flow with uniform bottom heating and boundary layer friction at the bottom and the lateral walls of an equatorial beta plane. Left: Simulation result with a flux form Eulerian scheme, where the lateral boundary oscillation has been stopped at nondimensional time $t^* = 762$ and the model restarted. Right: Simulation result when the model is continued using a semi-Lagrangian scheme. Note that the data displayed in both panels are identical up to $t^* = 762$. The Hovmöller data have been averaged over the near-equatorial region $\pm 0.844 L_D$ from the mid-channel and lowpass filtered to attenuate all frequencies larger and equal to the beat frequency of the boundary oscillation. Both time and the zonal length are nondimensionalized using the internal Rossby radius of deformation L_D.

Available Potential Energy (APE) model intercomparison [*Blackburn et al.*, 2013; *Williamson et al.*, 2011] for temporal anomalies of precipitation.

In summary, the presented examples serve as a compelling reminder of the multiscale nature of atmospheres and oceans, the decisive role of small-scale fluctuations, and the numerical intricacies associated with the simulation of spatial or temporal anomalies. Certainly many questions remain, for example, regarding the interaction of convection and large-scale atmospheric waves and the role of atmospheric turbulence in general. But it has been shown that (I)LES and DNS of laboratory analogues can complement a hierarchy of reduced models, to narrow the gap between the theoretical understanding of Earth's climate system, and the growing complexity of comprehensive global-scale weather and climate simulations [*Held*, 2005].

Acknowledgments. I am extremely grateful to Dr. Piotr Smolarkiewicz for his supervision, collaboration, enthusiasm, and advice on many aspects of the simulations described in the text as well as for the continued development of the EULAG model.

REFERENCES

Alexander, M. J. (2010), Gravity waves in the stratosphere, in *The Stratosphere: Dynamics, Transport, and Chemistry*, Geophys. Monogr. Ser., vol. 190, editors by L. M. Polvani, A. H. Sobel D. W. Waugh, pp. 109–121, AGU, Washington, D. C.

Andrews, D. G., J. R. Holton, and C. B. Leovy (1987), *Middle Atmosphere Dynamics*, International Geophysics series, vol. 40, Academic, London.

Baldwin, M. P., et al. (2001), The quasi-biennial oscillation, *Rev. Geophys.*, 39(2), 179–229.

Blackburn, M., et al. (2013), The aqua planet experiment (APE): Control SST simulation, *J. Met. Soc. Japan*, special Issue, The Aqua Planet Experiment (APE), *J. Met. Soc. Japan*, 91A, pp. 17–56.

Booker, J. R., and F. P. Bretherton (1967), The critical layer for internal gravity waves in a shear flow, *J. Fluid Mech.*, 27, 513–539.

Bretherton, F. P. (1966), The propagation of groups of internal gravity waves in a shear flow, *Q. J. R. Meteorol. Soc.*, 92, 466–480.

Charbonnel, C., and S. Talon (2005), Influence of gravity waves on the internal rotation and Li abundance of solar-type stars, *Science*, 309, 2189–2191.

Charney, J. (1963), A note on large-scale motions in the tropics, *J. Atmos. Sci.*, 20, 607–609.

Domaradzki, J., Z. Xiao, and P. Smolarkiewicz (2003), Effective eddy viscosities in implicit large eddy simulations of turbulent flows, *Phys. Fluids*, 15, 3890–3893.

Dunkerton, T. J. (1981), Wave transience in a compressible atmosphere. Part II: Transient equatorial waves in the quasi-biennial oscillation., *J. Atmos. Sci.*, 38, 298–307.

Dunkerton, T. J. (1997), The role of gravity waves in the quasi-biennial oscillation, *J. Geophys. Res.*, 102(D22), 26,053–26,076.

Eliassen, A., and E. Palm (1961), On the transfer of energy in stationary mountain waves, *Geofys. Publ.*, *22*, 1–23.

Fels, S. B., and R. S. Lindzen (1974), The interaction of thermally excited gravity waves with mean flows, *Geophys. Fluid Dyn.*, *6*, 149–191.

Fetecau, R. C., and D. J. Muraki (2011), A dispersive regularization of the modulational instability of stratified gravity waves, *Wave Motion*, *48*(7), 667–679.

Fritts, D. C., and M. J. Alexander (2003), Gravity wave dynamics and effects in the middle atmosphere, *Rev. Geophys.*, *41*, 1–68.

Gibson, J. K., P. Kallberg, S. Uppala, A. Hernandez, A. Nomura, and E. Serrano (1999), ECMWF re-analysis project report series: 1. ERA-15 description, Tech. Rep., ECMWF, Shinfield Park, Reading, Mass.

Gill, A. (1980), Some simple solutions for heat-induced tropical circulation, *Q. J. R. Meteorol. Soc.*, *106*, 447–462.

Gill, A. (1982), *Atmosphere-Ocean Dynamics*, *International Geophysics Series*, vol. 30, Academic, London.

Giorgetta, M. A., E. Manzini, and E. Roeckner (2002), Forcing of the quasi-biennial oscillation from a broad spectrum of atmospheric waves, *Geophys. Res. Lett.*, *29*(8), 86-1–86-4.

Gray, L. J. (2010), Stratospheric equatorial dynamics, in *The Stratosphere: Dynamics, Transport, and Chemistry*, *Geophys. Monogr. Ser.*, vol. 190, editors by L. M. Polvani, A. H. Sobel D. W. Waugh, pp. 93–107, AGU, Washington, D. C.

Hartmann, D. L., and H. H. Hendon (2007), Resolving an atmospheric enigma, *Science*, *318*, 1731–1732.

Held, I. (2005), The gap between simulation and understanding in climate modelling, *Bull. Am. Meteor. Soc.*, *86*, 1609–1614.

Hirons, L. C., P. Inness, F. Vitart, and P. Bechtold (2013), Understanding advances in the simulation of intraseasonal variability in the ECMWF model. Part I: The representation of the MJO, *Q. J. R. Meteorol. Soc.*, doi:10.1002/qj.2060, avail.

Holton, J. R., and R. S. Lindzen (1972), An updated theory for the quasi-biennial cycle of the tropical stratosphere, *J. Atmos. Sci.*, *29*, 1076–1079.

Horinouchi, T., and S. Yoden (1998), Wave-mean flow interaction associated with a QBO-like oscillation simulated in a simplified GCM, *J. Atmos. Sci.*, *55*, 502–525.

Horinouchi, T., et al. (2003), Tropical cumulus convection and upward-propagating waves in middle-atmospheric GCMs, *J. Atmos. Sci.*, *60*, 2765–2782.

Hou, A. Y., and B. F. Farrell (1987), Superrotation induced by critical level absorption of gravity waves on Venus: An assessment, *J. Atmos. Sci.*, *44*, 1049–1061.

Hsu, H.-H., B. J. Hoskins, and F.-F. Jin (1990), The 1985/86 intraseasonal oscillation and the role of the extratropics, *J. Atmos. Sci.*, *47*, 823–839.

Kondepudi, D., and I. Prigogine (1998), *Modern Thermodynamics: From Heat Engines to Dissipative Structures*, Wiley, New York.

Leovy, C. B. (1973), Rotation of the upper atmosphere of Venus, *J. Atmos. Sci.*, *30*, 1218–1220.

Lie, X., and P. L. Read (2000), A mechanistic model of the quasi-quadrennial oscillation in Jupiter's stratosphere, *Planetary Space Sci.*, *48*, 637–669.

Lindzen, R. S. (1987), On the development of the theory of the QBO, *Bull. Am. Meterol. Soc.*, *68*(4), 329–337.

Lindzen, R. S., and J. R. Holton (1968), A theory of the quasi-biennial oscillation, *J. Atmos. Sci.*, *25*, 1095–1107.

Lindzen, R. S., and C.-Y. Tsay (1975), Wave structure of the tropical stratosphere over the Marshall Islands area during 1 April–1 July 1958, *J. Atmos. Sci.*, *32*, 2008–2021.

Madden, R. A., and P. R. Julian (1971), Detection of a 40–50 day oscillation in the zonal wind in the tropical Pacific, *J. Atmos. Sci.*, *28*, 702–708.

Maicun, L. (1987), On the low-frequency, planetary-scale motion in the tropical atmosphere and oceans, *Adv. Atmos. Sci.*, *4*, 249–263.

Majda, A. J., and S. N. Stechmann (2009), The skeleton of tropical intraseasonal oscillations, *Proc. Natl. Acad. Sci.*, *106*(21), 8417.

Malanotte-Rizzoli, P., R. E. Young, and D. B. Haidvogel (1988), Numerical simulation of transient boundary-forced radiation. Part II: The Modon regime, *J. Phys. Oceanogr.*, *18*, 1546–1569.

Margolin, L. G., W. J. Rider, and F. F. Grinstein (2006), Modeling turbulent flow with implicit les, *J. Turbul.*, *7*, 1–27.

Matsuno, T. (1966), Quasi-geostrophic motions in the equatorial area, *J. Meteor. Soc. Japan*, *44*, 25–43.

McIntyre, M. E. (1994), The quasi-biennial oscillation (QBO): Some points about the terrestrial QBO and the possibility of related phenomena in the solar interior, in *The Solar Engine and Its Influence on the Terrestrial Atmosphere and Climate*, *Nato ASI Subseries I*, vol. 25, edited by E. Nesme-Ribes, pp. 293–320, Springer-Verlag, Heidelberg.

McIntyre, M. E. (2003), On global-scale atmospheric circulations, in *Perspectives in Fluid Dynamics: A Collective Introduction to Current Research*, edited by G. Batchelor, H. Moffatt, and M. Worster, pp. 557–624, Cambridge Univ. Press, Cambridge, UK.

Miura, H., M. Satoh, T. Nasuno, A. T. Noda, and K. Oouchi (2007), A Madden-Julian oscillation event realistically simulated by a global cloud-resolving model, *Science*, *318*, 1763–1765.

Moin, P., and K. Mahesh (1998), Direct numerical simulation: A tool in turbulence research, *Annu. Rev. Fluid Mech.*, *30*, 539–578.

Moncrieff, M. W. (2004), Analytic representation of the large scale organization of tropical convection, *J. Atmos. Sci.*, *61*, 1521–1538.

Orr, A., P. Bechtold, J. Scinoccia, M. Ern, and M. Janiskova (2010), Improved middle atmosphere climate and analysis in the ecmwf forecasting system through a nonorographic gravity wave parametrization, *J. Climate*, *23*, 5905–5926.

Otobe, N., S. Sakai, S. Yoden, and M. Shiotani (1998), Visualization and WKB analysis of the internal gravity wave in the QBO experiment, *Nagare Japan Soc. Fluid Mech.*, *17*(3), http://www.nagare.or.jp/mm/98/index.htm.

Piotrowski, Z. P., P. K. Smolarkiewicz, S. P. Malinowski, and A. Wyszogrodzki (2009), On numerical realizability of thermal convection, *J. Comput. Phys.*, *228*, 6268–6290.

Plumb, R. A. (1977), The interaction of two internal waves with the mean flow: Implications for the theory of the quasi-biennial oscillation, *J. Atmos. Sci.*, *34*, 1847–1858.

Plumb, R. A., and R. C. Bell (1982), A model of the quasi-biennial oscillation on an equatorial beta plane, *Q. J. R. Meteorol. Soc.*, *108*, 335–352.

Plumb, R. A., and D. McEwan (1978), The instability of a forced standing wave in a viscous stratified fluid: A laboratory analogue of the quasi-biennial oscillation, *J. Atmos. Sci.*, *35*, 1827–1839.

Prusa, J. M., and P. K. Smolarkiewicz (2003), An all-scale anelastic model for geophysical flows: Dynamic grid deformation, *J. Comput. Phys.*, *190*, 601–622.

Raymond, D. J. (2001), A new model of the MaddenJulian Oscillation, *J. Atmos. Sci.*, *58*, 2807–2819.

Rogers, T. M., K. B. MacGregor, and G. A. Glatzmaier (2008), Non-linear dynamics of gravity wave driven flows in the solar radiative interior, *Mon. Not. R. Astron. Soc.*, *387*, 616, 630.

Sakai S. (1997), Atmosphere and ocean in a laboratory, GFD-Dennou Club, http://www.gfd-dennou.org/library/gfd_exp/.

Saujani, S., and T. G. Shepherd (2006), A unified theory of balance in the extratropics, *J. Fluid Mech.*, *569*, 447–464.

Scaife, A., N. Butchart, C. D. Warner, D. Stainforth, and W. Norton (2000), Realistic quasi-biennial oscillations in a simulation of the global climate, *Geophys. Res. Lett.*, *27*(9), 3481–3484.

Schinder, P. J., et al. (2011), Saturn's equatorial oscillation: Evidence of descending thermal structure from Cassini radio occultations, *Geophys. Res. Lett.*, *38*(L08205), 1–5.

Smolarkiewicz, P. K. (2006), Multidimensional positive definite advection transport algorithm: An overview, *Int. J. Numer. Methods Fluids*, *50*, 1123–1144.

Smolarkiewicz, P. K., and L. G. Margolin (1997), On forward-in-time differencing for fluids: An Eulerian/semi-Lagrangian non-hydrostatic model for stratified flows, *Atmos. Ocean Special*, *35*, 127–152.

Smolarkiewicz, P. K., and L. G. Margolin (1998), MPDATA: A finite difference solver for geophysical flows, *J. Comput. Phys.*, *140*, 459–480.

Smolarkiewicz, P. K., and L. G. Margolin (2000), Variational methods for elliptic problems in fluid models, in *Proceedings of 2000 ECMWF Workshop on Developments in Numerical Methods for Very High Resolution Global Models*, pp. 137–159, Eur. Cent. for Medium-Range Weather Forecasts, Reading, UK.

Smolarkiewicz, P. K., and J. M. Prusa (2002), Forward-in-time differencing for fluids: Simulation of geophysical turbulence, in *Turbulent Flow Computation*, edited by D. Drikakis and B. Guertz, pp. 279–312, Series: Fluid Mechanics and Its Applications, Vol. 66, Springer.com.

Smolarkiewicz, P. K., and J. M. Prusa (2005), Towards mesh adaptivity for geophysical turbulence: Continuous mapping approach, *Int. J. Numer. Methods Fluids*, *47*, 789–801.

Takahashi, M., and B. A. Boville (1992), A three-dimensional simulation of the equatorial quasi-biennial oscillation, *J. Atmos. Sci.*, *49*, 1020–1035.

Tindall, J. C. (2003), Dynamics of the tropical tropopause and lower stratosphere, Ph.D. thesis, University of Reading, Reading, UK.

Uppala, S. M., et al. (2005), The ERA-40 re-analysis, *Q. J. R. Meteorol. Soc.*, *131*, 2961–3012.

Vitart, F., and T. Jung (2010), Impact of the northern hemisphere extratropics on the skill in predicting the Madden-Julian Oscillation, *Geophys. Res. Lett.*, *37*(L23805), 1–6.

Wedi, N. P. (2006), The energetics of wave-driven mean flow oscillations, *Int. J. Numer. Methods Fluids*, *50*(10), 1175–1191, special issue: Multidimensional Positive Definite Advection Transport Algorithm Methods.

Wedi, N. P., and P. K. Smolarkiewicz (2004), Extending Gal-Chen & Somerville terrain-following coordinate transformation on time-dependent curvilinear boundaries, *J. Comput. Phys.*, *193*, 1–20.

Wedi, N. P., and P. K. Smolarkiewicz (2006), Direct numerical simulation of the Plumb-McEwan laboratory analog of the QBO, *J. Atmos. Sci.*, *63*(12), 3226–3252.

Wedi, N. P., and P. K. Smolarkiewicz (2008), A reduced model of the Madden-Julian oscillation, *Int. J. Numer. Methods Fluids*, *56*, 1583–1588.

Wedi, N. P., and P. K. Smolarkiewicz (2009), A framework for testing global nonhydrostatic models, *Q. J. R. Meteorol. Soc.*, *135*, 469–484.

Wedi, N. P., and P. K. Smolarkiewicz (2010), A nonlinear perspective on the dynamics of the MJO: Idealized large-eddy simulations, *J. Atmos. Sci.*, *67*, 1202–1217.

Wheeler, M., and G. N. Kiladis (1999), Convectively coupled equatorial waves: Analysis of clouds and temperature in the wave number-frequency domain, *J. Atmos. Sci.*, *56*, 374–399.

Williamson, D. L. (2007), The evolution of dynamical cores for global atmospheric models, *J. Meteor. Soc. Japan*, *85B*, 241–269.

Williamson, D. L., et al. (2011), APE-atlas, Tech. Rep. TN-484, NCAR, Boulder, Colo.

Yamamoto, M., and M. Takahashi (2003), The fully developed superrotation simulated by a general circulation model of a Venus-like atmosphere, *J. Atmos. Sci.*, *60*, 561–574.

Yano, J.-I., and M. Bonazzola (2009), Scale analysis for the large-scale tropical atmospheric dynamics, *J. Atmos. Sci.*, *66*, 159–172.

Zagar, N., E. Andersson, and M. Fisher (2005), Balanced tropical data assimilation based on a study of equatorial waves in ECMWF short-range forecast errors, *Q. J. R. Meteorol. Soc.*, *131*, 987–1011.

Zhang, C. (2005), Madden-Julian oscillation, *Rev. Geophys.*, *43*, doi:10.1029/2004RG000158.

10

Internal Waves in Laboratory Experiments

Bruce Sutherland[1], Thierry Dauxois[2], and Thomas Peacock[3]

10.1. INTRODUCTION

Since the realization by physical oceanographers that transport and mixing by internal waves are an important component of the thermohaline circulation, there has been a resurgence in interest in their dynamics [*Polzin et al.*, 1997; *Munk and Wunsch*, 1998; *Ledwell et al.*, 2000]. Consequent studies have been designed to examine mechanisms for wave generation and interaction with topography. Theoretical studies have examined the means by which energy from the moon forcing the barotropic tide might be converted into internal wave energy as a consequence of oscillatory stratified flow over topography. This began with the pioneering studies of *Zeilon* [1912] and *Baines* [1974, 1982] and have since been extended, though still in the realm of linear theory, to examine the influence of more complex topography and stratification [*Balmforth et al.*, 2002; *Llewellyn-Smith and Young*, 2002; *Bühler and Muller*, 2007]. Related to these is the examination of scattering of internal waves by topography in which incident low-mode internal waves generate an oscillatory flow over topography that launches higher-mode internal waves [*Larsen*, 1969; *Robinson*, 1969; *Sandstrom*, 1969].

Just as an oscillatory flow over a rigid body generates internal waves, so does an oscillating body generate internal waves in otherwise stationary fluid. The particular circumstance of internal waves generated by oscillating cylinders and spheres has garnered much attention [*Görtler*, 1943; *Mowbray and Rarity*, 1967; *Thomas and Stevenson*, 1972; *Voisin*, 1991, 1994; *Hurley*, 1997; *Hurley and Keady*, 1997]. Even with such simple geometries and despite neglecting Coriolis effects, this work has revealed the importance of including viscosity to resolve singularities that occur along tangents to the oscillating body in the along-beam direction.

More recently, in the study of tidally generated internal waves, attention has turned to faster time-scale processes in which large-amplitude internal wave packets are generated during one cycle of the tide. This work extends earlier studies of steady uniformly stratified flow over topography (e.g., see *Baines* [1995]) to include consideration of nonuniform stratification and large-amplitude topography. Oceanographers have focused primarily upon the generation, propagation, and dissipation of internal solitary waves at the thermocline [*Pinkel*, 2000; *Klymak and Gregg*, 2004; *Klymak et al.*, 2006; *Li and Farmer*, 2011; *Alford et al.*, 2011].

By exploring large-amplitude and viscous effects, the results of laboratory experiments have often challenged existing theory. For example, they have revealed the importance of nonlinear processes in the scattering of internal waves from large-amplitude topography [*Peacock et al.*, 2009], they have demonstrated the importance of the viscous boundary layer in the generation of internal waves from oscillating bodies [*Sutherland and Linden*, 2002; *Flynn et al.*, 2003], and they have shown that boundary layer separation in stratified flow over steep topography reduces the effective topographic height while generating turbulence and internal waves in the lee of localized topography [*Baines and Hoinka*, 1985; *Sutherland*, 2002; *Aguilar and Sutherland*, 2006].

Laboratory experiments have entered a renaissance due to digitization technology, advancement in lasers and computer-controlled equipment, and increases in computational memory and speed, which have created valuable new analysis tools such as particle image velocimetry (PIV) and laser-induced fluorescence (LIF). As a result,

---

[1]Departments of Physics and of Earth & Atmospheric Sciences, University of Alberta, Edmonton, Alberta, Canada.

[2]Laboratoire de Physique, École Normale Supérieure, Lyon, France.

[3]Department of Mechanical Engineering, Massachusetts Institute of Technology, Cambridge, Massachusetts, United States of America.

Modeling Atmospheric and Oceanic Flows: Insights from Laboratory Experiments and Numerical Simulations, First Edition. Edited by Thomas von Larcher and Paul D. Williams.
© 2015 American Geophysical Union. Published 2015 by John Wiley & Sons, Inc.

it is now possible to make nonintrusive measurements of velocity and concentration in two and even three dimensions. These tools have provided new insights into problems involving turbulence and mixing that remain a challenge in computational fluid dynamics.

However, the study of stratified fluids remains an experimental challenge because light typically refracts differently through fluids of varying density. This can distort and smear the apparent positions of particles used in PIV and so lead to spurious predictions of flow speeds. On the other hand, the very fact that density and refractive index are related has provided other means to examine nonintrusively the structure of stratified fluid flow.

One visualization tool used in laboratory experiments of salt-stratified fluids is the shadowgraph. In this, a light source placed far behind the test section shines through the stratified fluid landing upon a translucent surface such as Mylar. At interfaces where the density rapidly changes, the light focuses and defocuses as it bends relatively more or less while passing through fluid of varying salinity and, hence, varying refractive index. If density variations due to internal waves are gradual, focusing may not be evident. The shadowgraph proves particularly useful in the examination of approximately two-layer fluids, in which case light focusing at the interface can be used to track the motion of interfacial waves. For internal waves in uniformly stratified fluid, the shadowgraph is particularly effective in the examination of waves that are close to breaking, as shown in Figure 10.1. In this experiment [*Koop and McGee*, 1986], sinusoidal topography is towed leftward beneath a shear flow whose speed increases leftward with height. At middepth in the experiments the waves encounter a critical level, where the background flow speed is close to the towing speed of the hills.

Another method taking advantage of the relationship between refractive index and density is called "schlieren" [*Schardin*, 1942; *Settles*, 2001]. In the traditional approach, light reflected from a parabolic mirror passes through a test section before striking a second parabolic mirror that refocuses the light. A knife edge at the focus acts as a filter on spurious signals, thus revealing index-of-refraction-dependent structures within the test section.

Mowbray and Rarity [1967] were the first to use traditional schlieren methods to visualize internal waves generated by a cylinder oscillating at a fixed frequency, ω. Provided ω was sufficiently small, they observed that the waves emanated vertically and horizontally from the cylinder in a cross pattern, as shown in Figure 10.2.

In this experiment, the fluid was a uniformly stratified salt solution whose density decreased linearly with height. The stratification can be represented by the buoyancy frequency N, defined in the Boussinesq approximation by

$$N^2 = -\frac{g}{\rho_0}\frac{d\bar{\rho}}{dz}. \qquad (10.1)$$

Figure 10.1. Internal wave breaking near a critical layer as visualized by shadowgraph. Reprinted from Figure 7 of *Koop and McGee* [1986].

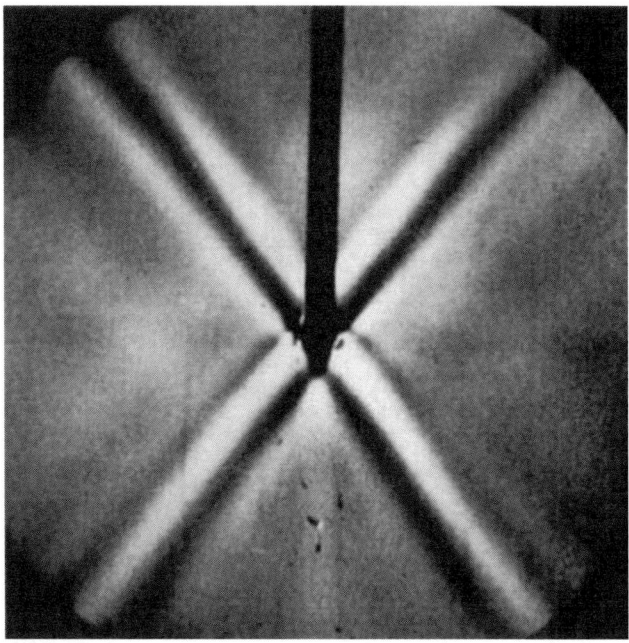

Figure 10.2. Pattern of internal waves generated by an oscillating cylinder as visualized by conventional schlieren methods. Reproduced from Plate 1(6) of *Mowbray and Rarity* [1967].

Here $\bar{\rho}(z)$ is the ambient density, ρ_0 is the characteristic density (e.g., that for fresh water at room temperature), and g is the acceleration of gravity. In agreement with the predicted dispersion relation of internal waves, *Mowbray* and Rarity [1967] found that the arms of the cross-pattern of waves formed a fixed angle Θ from the vertical that was related to the ratio of ω to N by $|\Theta| = \cos^{-1}(\omega/N)$.

Color filters have allowed schlieren to be more quantitative [Howes, 1984; Teoh et al., 1997; Chashechkin, 1999]. But the expense and physical constraints imposed by the need for well-aligned pairs of parabolic mirrors has limited the use of schlieren until recently.

Schlieren technology has advanced enormously since the mid-1990s. As a result of digitization technology and computers, "synthetic schlieren" was developed as an inexpensive, versatile and, most importantly, quantitative tool for sensitively measuring density perturbations in stratified fluids.

In what follows we examine how synthetic schlieren has been used to test theory and to develop new insights into the dynamics of internal waves. In the process we review an analysis method for separating out waves propagating in different directions and we describe a recently developed mechanism for generating waves that does not suffer some of the drawbacks of oscillating or towed rigid objects.

Section 10.2 briefly discusses how synthetic schlieren visualizes disturbances in a fluid through contrasting snapshots taken by a digital camera looking through the fluid at a black-and-white image of lines or dots. If the disturbances are small, the displacements of objects in the image can be computed and, from these, the magnitude of the disturbance calculated. This is described in Section 10.3 with the assumption that the disturbance in the tank is uniform across the line of sight. The treatment of axisymmetric and fully three-dimensional disturbances is described in Section 10.4. Other advances in generating internal waves and analyzing them using PIV are described in Section 10.5. Future directions are described in Section 10.6.

10.2. QUALITATIVE USE OF SYNTHETIC SCHLIEREN

Synthetic schlieren [Dalziel et al., 2000] makes away with the need for parabolic mirrors to straighten and refocus a localized light source. Instead, a camera is focused upon an image behind a tank filled with salt-stratified fluid.[1] Disturbances in the fluid displace isopycnal surfaces and so locally change the refractive index of the salt water through which light passes from a point on the image through the tank to the camera. The image apparently distorts as a result.

For example, Figure 10.3 shows how qualitative synthetic schlieren observes distortions of an image of horizontal black-and-white lines resulting when a set of model sinusoidal "hills" are towed from left to right over the surface of a tank filled with uniformly stratified fluid.

In the initial image, shown in Figure 10.3a, the hills immersed in the ambient are apparent near the top of the frame. The black-and-white lines are not in the tank, however. The image is situated approximately 10 cm behind the tank. After the hills are set in motion, various disturbances in the ambient can be seen as a result of the distortion of the image (Figure 10.3b). In the lee of each hill, boundary layer separation results in large perturbations that warp and blur the lines. Furthermore, below the hills the eye can barely make out smaller undulations of lines in the image.

These alterations can be enhanced through digital image processing. Each snapshot can be represented as an array of pixels with each pixel given a number corresponding to its intensity (e.g., 0 for black, 1 for white, and in between for gray). The image in Figure 10.3c is produced by taking the difference of the digitized snapshot in (b) from that in (a), then taking the absolute value and multiplying the result by an enhancement factor, typically 10. Thus even small changes to the image become obvious.

One advantage of synthetic schlieren is that its sensitivity can be increased by widening the distance between the test section and the image behind it. For example, it is easy to observe heat rising off one's hand if the image is several meters away.

10.3. SPANWISE-UNIFORM DISTURBANCES

10.3.1. Quantitative Synthetic Schlieren

When light passes through a medium whose refractive index changes in space, it is deflected in a manner well predicted by Snell's law. In the particular case of stably stratified fluid, the density ρ and hence refractive index n change with height z. The path of light passing in the y direction through the fluid at a small angle to the vertical from the y axis is given by [Sutherland et al., 1999]

$$\frac{d^2 z}{dy^2} \simeq \frac{1}{n_0} \frac{\partial n}{\partial z}. \quad (10.2)$$

Here n_0 is the characteristic refractive index of the fluid (e.g., $n_0 = 1.3330$ for pure water).

In uniformly stratified fluid, the vertical gradient of the refractive index can be related to the vertical density gradient:

$$\frac{\partial n}{\partial z} = \frac{dn}{d\rho} \frac{\partial \rho_T}{\partial z}. \quad (10.3)$$

Here ρ_T denotes the sum of the ambient density $\bar{\rho}(z)$ and the perturbation density $\rho(\vec{x}, t)$.

---

[1] Synthetic schlieren has also been called "background oriented schlieren" by Meier [2002], who used it to visualize and measure shock waves in air.

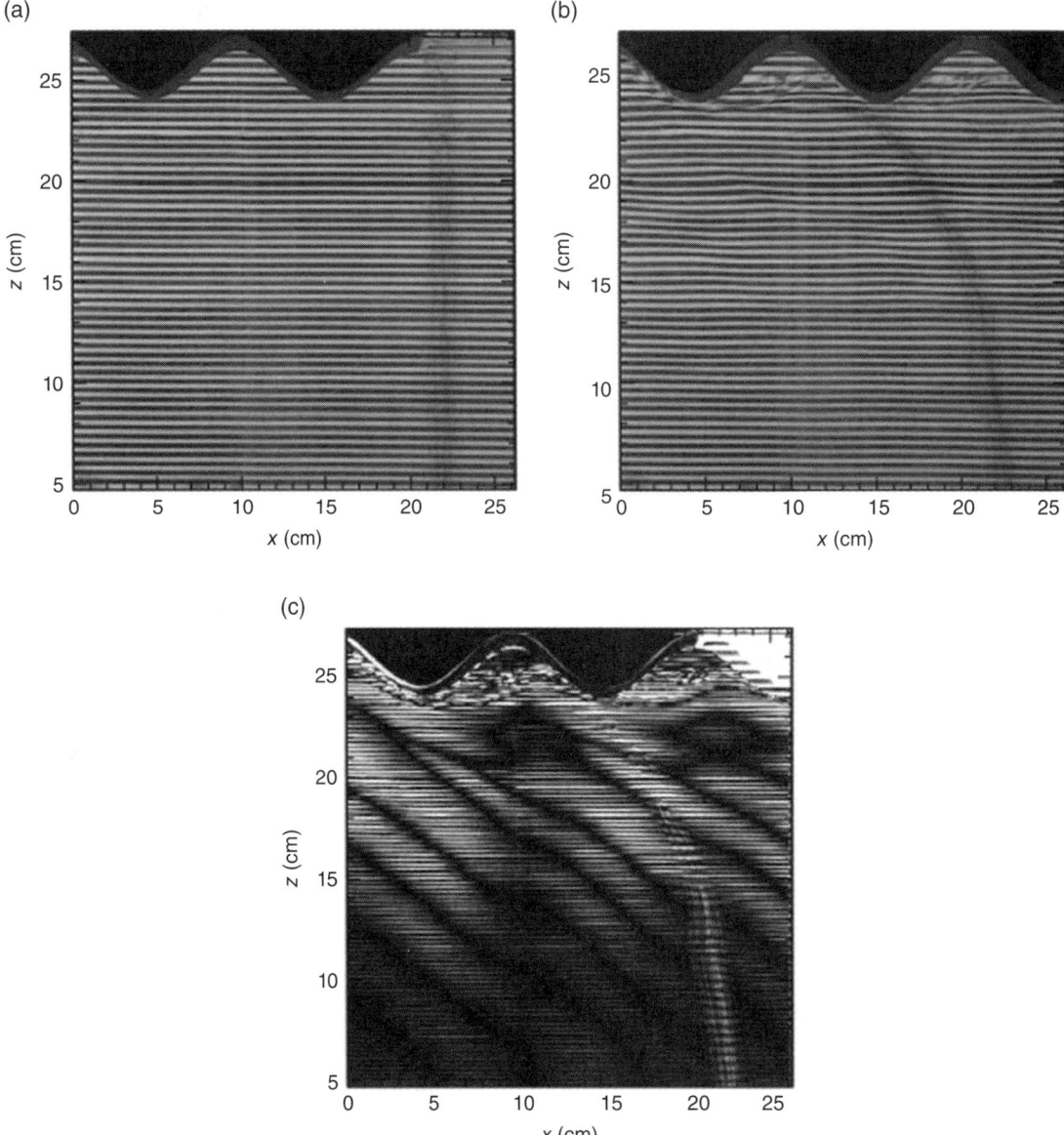

Figure 10.3. (a) Side view looking through tank filled with salt-stratified fluid with model inverted sinusoidal hills at the surface and an image of horizontal black-and-white lines placed behind the tank. (b) Side view after hills have been towed slowly a distance of one hill width. (c) Qualitative synthetic schlieren image produced by taking the absolute value of the difference of the digitized images shown in (a) and (b). Also evident in (a) and (b) is a dark vertical streak above $x \simeq 20$ cm. This is a vertical dye line suspended in the fluid itself. Its displacement can be used to determine mean horizontal flow, which is retrograde to the towing direction of the hill.

In computing image displacements, it is sometimes more intuitive to compute them in terms of the squared buoyancy frequency rather than density gradients. The local stratification resulting from both the background and perturbation density is expressed by the total squared buoyancy frequency:

$$N_T^2 = -\frac{g}{\rho_0}\frac{\partial \rho_T}{\partial z}. \tag{10.4}$$

Thus vertical variations of the refractive index can be written in terms of N_T by

$$\frac{\partial n}{\partial z} = -n_0 \gamma N_T^2, \tag{10.5}$$

in which the coefficient γ is defined so that

$$\gamma \equiv \frac{1}{g}\frac{\rho_0}{n_0}\frac{dn}{d\rho} \simeq 1.878 \times 10^{-4} \text{ s}^2/\text{cm}. \tag{10.6}$$

The rightmost empirical approximation assumes relatively weak concentrations of sodium chloride solutions [*Weast*, 1981].

Combining equations (10.2) and (10.5), we find that light follows a parabolic path when passing through a uniform stratification at a scant angle from the horizontal such that

$$z(y) = z_i + y \tan \phi_i - \frac{1}{2} \gamma N_T^2 y^2, \quad (10.7)$$

in which z_i is the height and ϕ_i is the angle at which the light ray enters the tank, In deriving (10.7), we have assumed that N_T is independent of y, which is the case for spanwise uniform disturbances in a tank. This assumption will be relaxed below in the consideration of axisymmetric and fully three-dimensional disturbances.

Synthetic schlieren is usually employed to measure perturbations to the ambient. Directly, it measures how the squared buoyancy frequency changes as a result of the compression and stretching of isopycnal surfaces. Over a distance $y = L_T$, light is deflected vertically by

$$\Delta z = -\frac{1}{2} \gamma \, \Delta N^2 \, L_T^2, \quad (10.8)$$

in which

$$\Delta N^2 \equiv N_T^2 - N^2 = -\frac{g}{\rho_0} \frac{\partial \rho}{\partial z} \quad (10.9)$$

is the change in the squared buoyancy due to the density perturbation ρ.

For example, if internal waves compress isopycnals so that the ambient N^2 locally increases by 10% from 1.0 to 1.1 s^{-2}, then the light deflects by 0.04 mm crossing a 20 cm wide tank. This is a small but discernible displacement that can be captured by a digital camera with sufficiently high resolution.

The apparent deflection is larger if the image is placed some distance behind the tank. Not only is the light deflected downward if the stratification increases, but also the angle of the light ray at the tank wall changes. So the apparent image displacement magnifies linearly as the image is moved further away.

Assuming the tank walls are negligibly thin, one can predict the total displacement of light from an object a distance L_o from one side of the tank to a camera on the other side of the tank to be

$$\Delta z(\Delta N^2) \simeq -\frac{1}{2} \gamma \, \Delta N^2 \, L_T^2 - \frac{n_0}{n_a} \gamma \, \Delta N^2 \, L_o L_T, \quad (10.10)$$

in which n_a is the refractive index of air and ϕ_0 is the angle from the horizontal at which light enters the camera from the object.

In the example above, if we now suppose the image is 20 cm behind the tank, then the displacement of the light path is 0.14 mm, which is much more easily discernible. If one pixel of the camera has a vertical resolution of 0.5 mm, then the disturbance in the tank will shift the image by about a third of a pixel, which can easily be observed by the change of intensity of light emanating from the edge of a line.

Through (10.10), we have found the forward equation in which known changes to the stratification enables us to predict the vertical displacement of a point in an object as seen by a camera looking through the stratified fluid. In the derivation of (10.7), because we have assumed the disturbance is spanwise uniform, it is a trivial matter to invert (10.10) so that an observed vertical displacement in an image can predict the change in the stratification:

$$\Delta N^2 \simeq -\Delta z \, \frac{1}{\gamma} \left[\frac{1}{2} L_T^2 + L_o L_T \frac{n_0}{n_a} \right]^{-1}. \quad (10.11)$$

If part of an image appears to deflect downward, it means that the stratification between it and the camera has locally become stronger.

This result is straightforwardly applied to the circumstance in which a camera looks through a tank at an image of horizontal black-and-white lines, as is the case in Figure 10.3. Even if density perturbations result in image distortions that shift the lines by a fraction of their width (which, in fact, is ideal if you wish easily to compute the line displacement), then (10.11) immediately predicts the change to the squared buoyancy frequency.

More processing is required if the line is displaced significantly. The calculation of displacements is even more difficult if the lines become magnified or contracted because second derivatives of the refractive index become significant (in which case a shadowgraph becomes the more useful, if qualitative, tool). The method breaks down entirely, as within the valleys of the model hills in Figure 10.3b, when the lines blur due to three-dimensional mixing.

The examination of an image of lines provides a relatively easy method to calculate small vertical displacements in an image and hence ΔN^2. But a far more informative, though computationally more intensive, application uses an image of dots [*Dalziel et al.*, 2000]. For example, Figure 10.4 shows a qualitative synthetic schlieren image produced with a beam of internal waves from an oscillating cylinder that passes in front of an image of regularly spaced black circles on a white background. The difference image is shown, analogous to that in Figure 10.3.

The cat-eye-like patterns have a black portion near the center of the dot surrounded on two sides by white, indicating how the dot shifted both horizontally and vertically as a result of internal waves passing between the camera and image. The angle and eccentricity of the

Figure 10.4. Pattern of internal waves generated by an oscillating cylinder in the upper left-hand corner as visualized by qualitative synthetic schlieren applied to a pattern of equally spaced dots behind the tank. Adapted from Figure 7c of *Dalziel et al.* [2000].

resulting elliptical disturbance give both components of the perturbation density gradient. Multiplying by $-g/\rho_0$, as in equation (10.9), the perturbation density gradient may be recast in terms of the perturbation squared buoyancy frequency: $(\Delta N^2_x, \Delta N^2_z) \equiv -(g/\rho_0)\nabla\rho$.

Alternately, if the image of dots is randomly distributed, then particle image velocimetry techniques can be used to measure horizontal and vertical displacements of portions of the image [*Dalziel et al.*, 2000]. Assuming the disturbance in the ambient is uniform along the line of sight, each displacement field can be used to find the perturbation density gradient. This can be integrated to compute the perturbation density field, as shown in Figure 10.5.

10.3.2. Separating Up from Down and Left from Right

A powerful analysis tool related to Hilbert transforms can be used to distinguish upward from downward propagating internal waves and simultaneously distinguish leftward from rightward propagating waves [*Mercier et al.*, 2008]. This can be applied to vertical or horizontal time series constructed from simulations, in situ observations, and laboratory experiments. Whether spanwise uniform, axisymmetric, or fully three dimensional, the method can distinguish propagation direction in the spatial component of the time series. Its application will be discussed here in the context of internal waves generated by an oscillating cylinder.

Consider Figure 10.6, which shows the ΔN^2 field computed using synthetic schlieren for an oscillating cylinder experiment. Imagining a Cartesian grid superimposed on the wave field with the origin at the center of the disturbance, the four arms of the cross consist of upward propagating waves emanating rightward and leftward in the first and second quadrants, respectively, and of downward propagating waves emanating leftward and rightward in the third and fourth quadrants, respectively. Hilbert transforms can extract each arm of the cross, as shown in Figure 10.7.

In application, the method does not formally compute the Cauchy principle value integral associated with a Hilbert transform. Instead, it employs filtering of Fourier transform images, specifically those of time series of the disturbances. For example, consider a synthetic schlieren employing an image of horizontal black-and-white lines behind a tank. During the evolution of a spanwise-uniform disturbance, one can compute $\Delta N^2(x,z,t)$, in which z is vertical, x is the along-tank coordinate, and t is time. Fixing an arbitrary horizontal location X, one can construct the vertical time series $\Delta N^2(z,t;x=X)$. Taking the double fast Fourier transform first in t and then in z gives the complex series coefficients $\widehat{\Delta N^2}(k_z,\omega)$, in which k_z is the vertical wave number and ω is the frequency.

To extract upward propagating disturbances, we use the fact that the group velocity of internal waves is positive if the vertical wave number is negative. So we set $\widehat{\Delta N^2}$ to zero if k_z is positive and leave the field untouched otherwise. An inverse Fourier transform then produces a filtered field $\Delta N^2_\uparrow(z,t;x=X)$ with only upward propagating disturbances. This process can be repeated at different horizontal locations until the entire evolution field of upward propagating disturbances is reproduced: $\Delta N^2_\uparrow(x,z,t)$.

We can similarly extract rightward propagating waves by Fourier transforming horizontal time series at successive $z=Z$, setting the coefficients of negative horizontal wave numbers to zero and then inverse transforming. The result of applying this to $\Delta N^2_\uparrow$, for example, gives the up and rightward wave beam shown in the top-left panel of Figure 10.7.

10.3.3. Partial Transmission and Reflection

The use of the Hilbert transform method described above has proven to be particularly useful in the study of internal wave propagation in nonuniformly stratified fluid. The intuitive understanding of their propagation is based upon ray theory, which assumes the small-amplitude wavepackets are quasi-monochromatic and that the background varies slowly compared to the wavelength [e.g., see *Lighthill*, 1978; *Sutherland*, 2010]. In particular, in a stationary fluid this predicts that waves reflect from a level

Figure 10.5. Quantitative synthetic schlieren applied to the circumstance shown in Figure 10.4, in which (a) the horizontal displacement of dots behind the tank is used to compute the (assumed spanwise-uniform) horizontal perturbation density gradient of fluid in the tank, as indicated by intensity of the gray scale. (b) Vertical dot displacements are used to compute the vertical density gradient. (c) The two components of the density gradient are integrated to find the perturbation density field. Reproduced from Figure 13 of *Dalziel et al.* [2000].

where the background buoyancy frequency is less than the wave frequency.

Following a theoretical approach analogous to that used in thin-film optics for light or in quantum mechanics for electrons, *Sutherland and Yewchuk* [2004] showed that internal waves can partially transmit (i.e., "tunnel") through a weakly stratified layer provided it is thin compared to the horizontal wavelength of the incident waves. For piecewise constant profiles of the background buoyancy frequency, they predicted the transmission coefficient T as a function of the relative frequency of the waves, ω/N_0, and their relative horizontal wave number $k_x L$ in which N_0 is the far-field buoyancy frequency and L is the depth of the thin stratified layer with buoyancy frequency N_1. The predictions are shown in Figure 10.8 for the cases of an unstratified layer, a weakly stratified layer, and a strongly stratified layer.

Counter to intuition based upon ray theory, one sees in particular that waves can partially reflect from a strongly stratified layer even though their frequency is always smaller than the background buoyancy frequency.

Of course, the phenomenon is well known in optics. Indeed, *Mathur and Peacock* [2010] made the analogy between internal waves and light showing that they behave like a Fabry-Perot multiple-beam interferometer. The resulting resonance of internal waves in a localized region of enhanced stratification was demonstrated in laboratory experiments, shown in Figure 10.9.

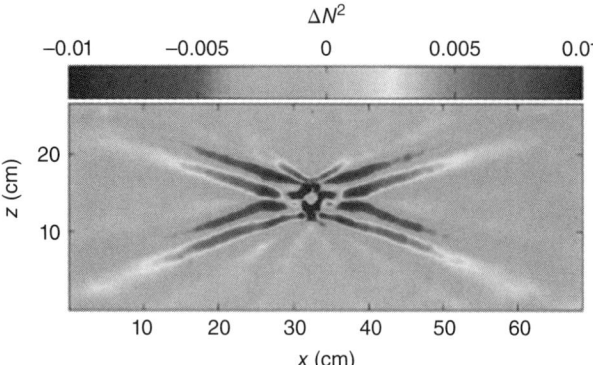

Figure 10.6. Change in the background squared buoyancy frequency (ΔN^2) due to internal waves generated by an oscillating cylinder, as measured by quantitative synthetic schlieren. The left image shows a snapshot of the ΔN^2 field; the right image shows the amplitude envelope of the wave beams. Reproduced from Figure 2 of *Mercier et al.* [2008].

The transmission of a small-amplitude wavepacket through arbitrary stratification and background flow can be computed through the solution of the Taylor-Goldstein equation [*Nault and Sutherland*, 2007, 2008]. These results were compared with laboratory experiments of internal waves incident upon a pycnocline [*Mathur and Peacock*, 2009] and of internal wave beams incident upon a weakly stratified layer [*Gregory and Sutherland*, 2010]. Using the Hilbert transform method, the incident, reflected, and transmitted waves could be distinguished and so transmission and reflection coefficients could be computed. Theory was found to be consistent with the experiments, but there was great sensitivity of the predicted transmission coefficient to the details of the stratification. For example, referring to Figure 10.8b with $\omega \simeq 0.8N$, one sees that the transmission coefficient increases rapidly from 0.4 to 1 as $k_x L$ increases from 0.7 to 1. And so, uncertainty in the measurement of k_x greatly increases the uncertainty in the predicted transmission. Likewise, with smooth N^2 profiles, the predicted transmission coefficient sensitively depends upon the smoothness of N, particularly if the incident wave frequency is close to the minimum value of N [*Gregory and Sutherland*, 2010]. For those intending to use theory to predict internal wave transmission, these experiments emphasize the importance of performing an error analysis for both incident internal wave properties and the structure of the ambient.

10.4. NON-SPANWISE-UNIFORM DISTURBANCES

The quantitative uses of synthetic schlieren described above assumed that any disturbances in the stratified fluid were uniform across the width of the tank. With this assumption, it was straightforward to relate displacements of images to changes in stratification through equation (10.10). It was likewise trivial to invert this equation and so infer changes in the stratification knowing the measured displacements, as in equation (10.11).

If disturbances are not spanwise uniform, one can still use Snell's law to write down expressions for the apparent displacement of an image due to light passing

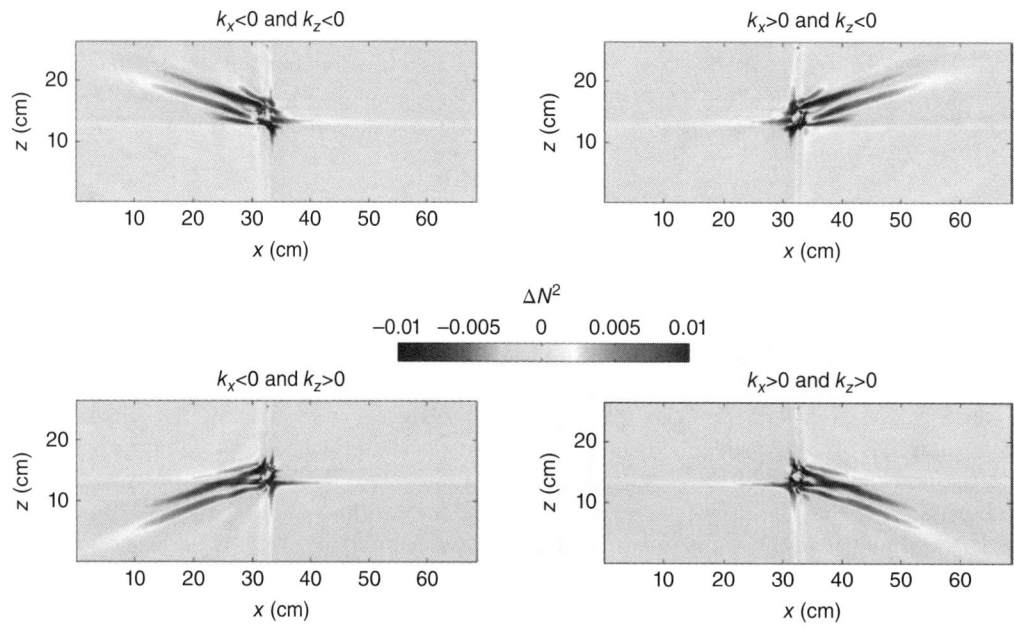

Figure 10.7. The four arms of the cross shown in Figure 10.6 determined by Fourier implementation of the Hilbert transform. Reproduced from Figure 3 of *Mercier et al.* [2008].

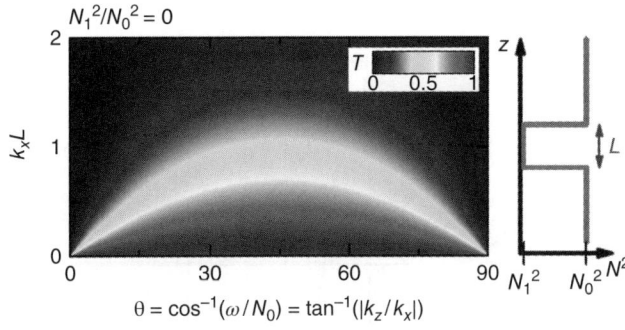
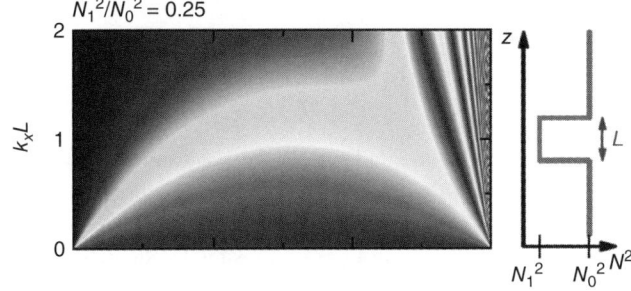
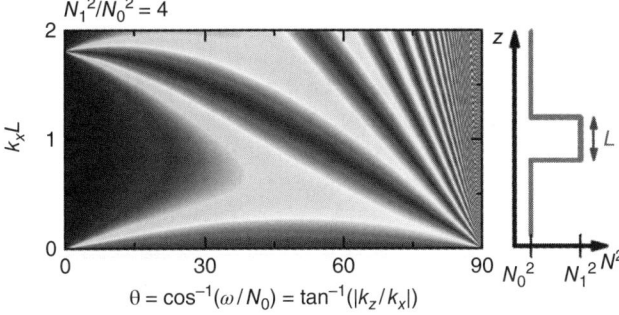

Figure 10.8. Predicted transmission coefficient for monochromatic internal waves of frequency ω and horizontal wave number k_x incident upon a region of depth L where the ambient buoyancy frequency is N_1 instead of N_0. Adapted from Figure 1 of *Sutherland and Yewchuk* [2004].

through fluid with known varying density (hence with known varying refractive index). The challenge is to invert this formula to find the change in stratification for given observed displacements.

For axisymmetric disturbances about a vertical axis, the procedure amounts to inverting a square matrix to determine $\nabla \rho$ from observed displacements of an image. For fully three-dimensional disturbances, tomographic inversion techniques are needed to reconstruct $\nabla \rho$ from displacements observed from multiple perspectives.

10.4.1. Axisymmetric Synthetic Schlieren

We consider the simplest case of reconstructing ΔN^2 from observed vertical displacements of an image, Δz [*Onu et al.*, 2003]. First we consider the vertical displacements at a fixed time and at a fixed height, so that $\Delta z(x)$ is taken to be a function only of the along-tank distance x. We seek the corresponding value of $\Delta N^2(r)$, which is assumed to be axisymmetric, varying with radius r.

The inversion problem begins with representing the along-tank direction by $n + 1$ discrete points $x_i = i\,dx$ for $i = 0, \ldots, n$ and by discretizing the radial disturbances by concentric rings of outer radius $r_j = (j + 1/2)\,dr$. So that the inversion problem is well posed, we take $dr \equiv dx$ and we set $j = 0, \ldots, n$, in which $j = 0$ signifies the innermost circle. The correspondence of the x and r coordinate systems is shown in Figure 10.10.

We assume that ΔN^2 is constant within each annulus in the central circle. And so we denote $(\Delta N^2)_0 = \Delta N^2$ for $0 \le r < dr/2$, $(\Delta N^2)_1 = \Delta N^2$ for $dr/2 \le r < 3\,dr/2$, $(\Delta N^2)_2 = \Delta N^2$ for $3\,dr/2 \le r < 5\,dr/2$, etc. Outside the outermost ring we assume the ambient is undisturbed so that $\Delta N^2 = 0$.

We now consider the path of light passing in the y direction from the far side of the tank through the disturbance field to the side of the tank nearest the camera (i.e., from top to bottom of the schematic in Figure 10.10).

Given values of ΔN^2 in each ring, we integrate equations (10.2) and (10.5), summing the discretized equations to determine the vertical position of light, $z(y)$, as it crosses each annulus. Doing so requires computing in advance the distance dy_{ij} that light from location x_i crosses the jth annulus (with the zeroth "annulus" being the central circle). Although they could be computed analytically, these geometric distances are straightforwardly determined by a numerical algorithm.

The result of the forward problem is a matrix set of equations,

$$\vec{\Delta z} = \mathbf{G}\,\vec{\Delta N^2}, \qquad (10.12)$$

in which $\vec{\Delta z}$ is the transpose of $(\Delta z(0), \Delta z(x_1), \ldots, \Delta z(x_n))$, $\vec{\Delta N^2}$ is the transpose of $((\Delta N^2)_0, (\Delta N^2)_1, \ldots, (\Delta N^2)_n)$, and $\mathbf{G}$ is a square matrix composed of the distances dy_{ij} and coefficient γ defined by (10.6).

Inverting $\mathbf{G}$, we can then determine the disturbance field knowing vertical displacements along a horizontal line:

$$\vec{\Delta N^2} = \mathbf{G}^{-1}\,\vec{\Delta z}. \qquad (10.13)$$

If the image is placed well behind the tank, the components of $\mathbf{G}$ are somewhat more complicated because one must consider the angle at which light enters the tank from the image as well as the vertical displacement of light. The extra terms may be added to components of $\mathbf{G}$, akin to the inclusion of the second term in equation (10.10) for the spanwise-uniform problem [*Onu et al.*, 2003].

Note that computing $\Delta N^2(r)$ need only be done using image displacements rightward of the center of the disturbance. Independently, one can compute $\Delta N^2(r)$ using

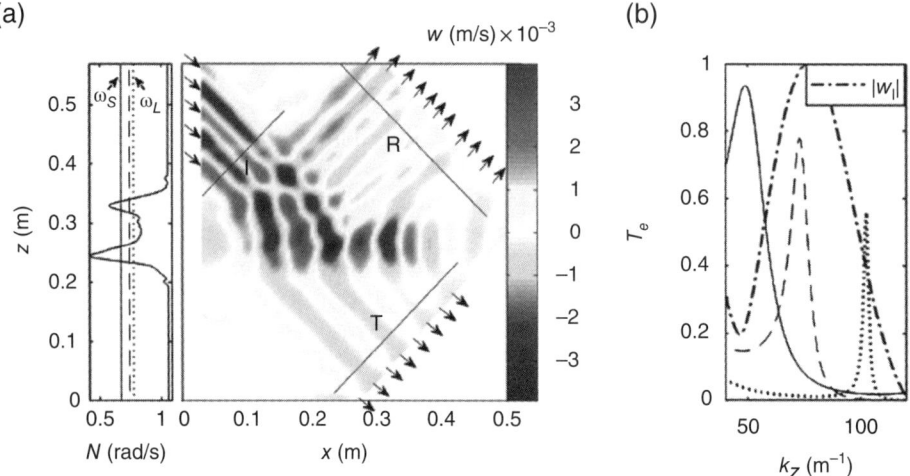

Figure 10.9. Schlieren image (color) showing the vertical velocity field associated with incident internal waves in nonuniformly stratified media partially transmitting (indicated by T) and reflecting (indicated by R) with partial trapping in a region of locally enhanced stratification for $0.27\,\text{m} \lesssim z \lesssim 0.33\,\text{m}$. The profile of the background stratification is shown on the left and transmission spectra are shown on the right for waves with frequencies indicated as the vertical lines through the buoyancy frequency profile shown (a). Adapted from Figure 3 of *Mathur and Peacock* [2010].

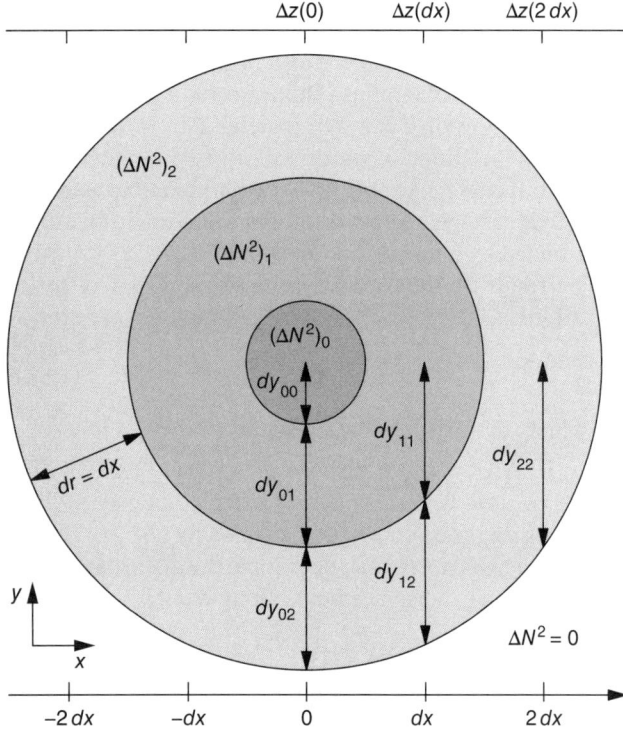

Figure 10.10. Discretization used to represent apparent displacements in an image behind a tank to axisymmetric disturbances within the tank. The disturbances are represented in terms of changes to the squared buoyancy frequency ΔN^2, which is assumed constant on annuli of width $dr = dx$, in which dx is the horizontal (e.g., pixel) resolution of the observed vertical displacements Δz of the object image.

image displacements leftward of center. Thus, comparing right and left gives a check on the accuracy of the assumption that the disturbance field was indeed axisymmetric.

Of course, the process of computing ΔN^2 at a particular height can be repeated at different heights so as to reconstruct a "snapshot" of $\Delta N^2(r, z)$. If the image is of dots instead of lines, one can compute horizontal as well as vertical components of the density gradient through this methodology.

An early application of axisymmetric schlieren examined the internal wave field surrounding a vertically oscillating sphere in uniformly stratified fluid [*Onu et al.*, 2003; *Flynn et al.*, 2003]. For example, Figure 10.11a shows the observed apparent vertical displacement of an image once the sphere had oscillated three times. No data were computed in the lower left-hand corner where the image was obscured by the sphere. The corresponding ΔN^2 field is shown in Figure 10.11b. As anticipated by theory, the along-beam amplitude decayed rapidly with distance from the center of the sphere as the conical wave beam expanded radially about the z axis. The theory predicted well the amplitude of the wave cones provided the sphere was sufficiently small (of radius 1.9 cm). But in experiments with a sphere of radius 3.2 cm, theory overpredicted the amplitude of the ΔN^2 field by as much as double, presumably because it neglected dynamics occurring within the viscous boundary layer surrounding the sphere [*Flynn et al.*, 2003].

This observation reveals a particularly useful aspect of the use of schlieren. Although the amplitude decays, the horizontal extent of the disturbance widens with distance from the origin. As a result, the vertical displacement

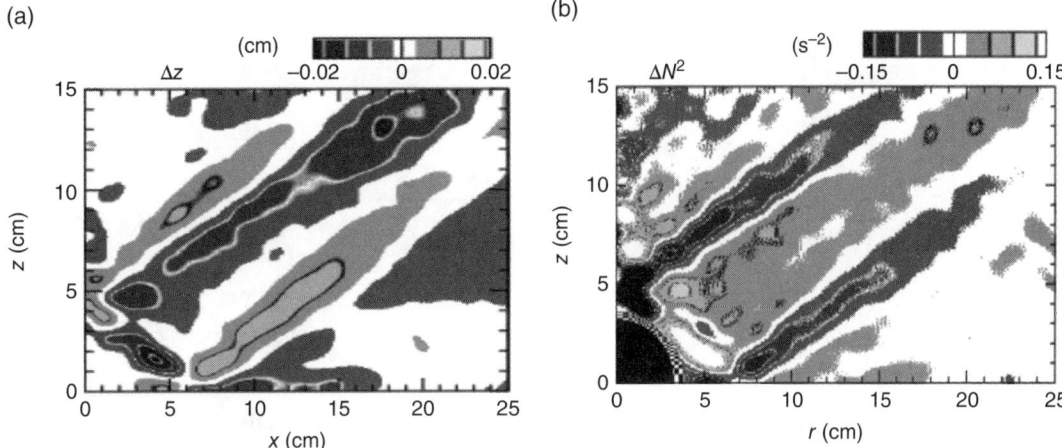

Figure 10.11. (a) Apparent vertical displacement $\Delta z(x, z)$ of horizontal lines in an image behind a tank in which a sphere (situated to the bottom left) oscillates in uniformly stratified fluid. (b) Corresponding change in squared buoyancy frequency $\Delta N^2(r, z)$ due to internal waves computed through axisymmetric synthetic schlieren. Adapted from Figures 2b and 4a of *Onu et al.* [2003].

signal does not weaken with distance away from the source. Indeed, the value of Δz in Figure 10.11a is largest near the top right-hand corner of the image. Hence, schlieren can extract signals over noise where in situ probe measurements or attempts to observe the motion of embedded particles might fail.

Since the development of the technique, it has been used to measure the laminar wake behind a falling sphere [*Yick et al.*, 2007] and internal waves above a plume in a stratified fluid [*Ansong and Sutherland*, 2010]. The latter case provided a model for internal wave generation by convective storms through the mechanical oscillator effect, in which the repeated rise and descent of cloud tops provide a forcing similar to that of an oscillating body.

10.4.2. Inverse Tomography

If the disturbance is fully three dimensional, then the problem of using synthetic schlieren to reconstruct the density gradient field from observed displacements of a single image is ill posed: Without invoking symmetry, it is impossible to reconstruct a three-dimensional object from its shadow. With multiple perspectives, however, it is possible to reconstruct and approximate the structure of the disturbance. In the medical use of magnetic resonance imaging (MRI), the method of tomographic reconstruction is well established. Making use of refractive index variations with air temperature, tomographic inversion has been used to measure the density of a supersonically expanding jet [*Faris and Byer*, 1988] and of two interacting jets [*Goldhahn and Seume*, 1988]. The latter was the first to employ the methodology of synthetic schlieren, recording the apparent displacement of an image of random dots, to determine the displacement of light rays.

Two approaches have since been taken to apply tomographic methods for the measurement of internal waves using synthetic schlieren. The Fourier-convolution approach of *Faris and Byer* [1988] and *Goldhahn and Seume* [1988] was used by *Hazewinkel et al.* [2011] in their study of internal wave attractors in a parabolic basin. The experiment itself was an extension of earlier studies into the formation of internal wave beams in spanwise-uniform, nonrectangular domains [*Maas et al.*, 1997; *Hazewinkel et al.*, 2008]. Because internal waves at a given frequency propagate at a fixed angle to the vertical, sloping sidewalls in the domain tend to focus the disturbances into a beam whose path effectively acts as an "attractor" for internal waves [*Maas and Lam*, 1995].

When a sphere was oscillated in stratified fluid within a paraboloidal basin, looking through the tank at different angles around the horizontal revealed attractor-like patterns in the observed displacement of images behind the tank. Four such images are shown in Figure 10.12. The information in these and several more images taken at different perspectives were combined through a convolution of their Fourier decompositions. The inverse transform of the result revealed the three-dimensional structure of the attractor, as shown in Figure 10.13.

A different approach follows that of the matrix inversion method used to measure axisymmetric disturbances [*Decamp et al.*, 2008]. At a fixed vertical level the observed image displacements could be represented by a vector with $2n$ entries, in which n is the number of pixels and the value is doubled to account for horizontal as well as vertical displacements. The perturbation density field (from which the density gradient is computed) could be discretized either in Cartesian or polar coordinates involving $N \equiv n_x \times n_y$ or $N \equiv n_r \times n_\theta$ points, respectively. For

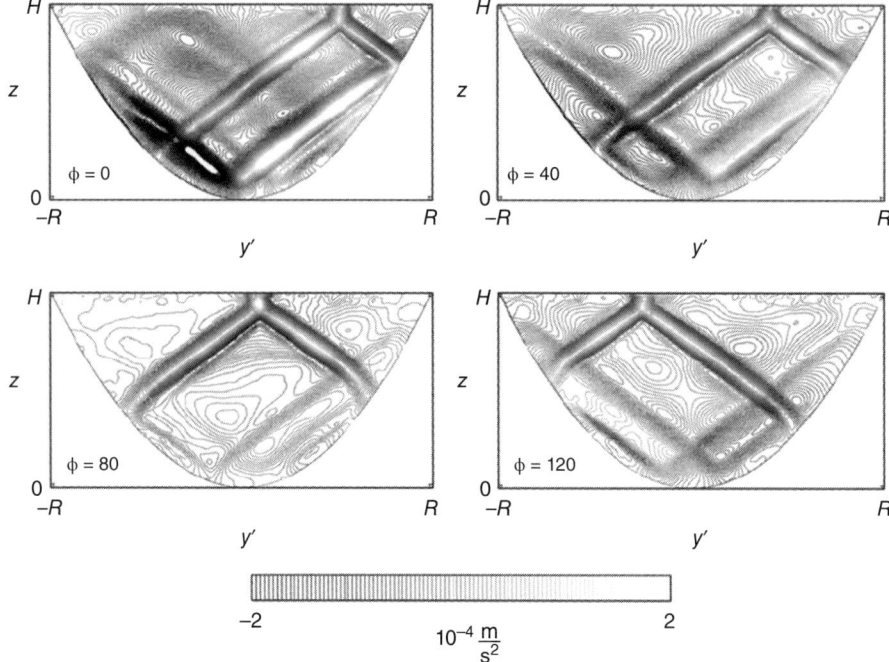

Figure 10.12. Image displacements recorded by different perspectives looking horizontally through a paraboloid filled with uniformly stratified fluid. The internal waves are generated by an oscillating sphere situated eccentrically near the surface. Reproduced from Figure 2 of *Hazewinkel et al.* [2011].

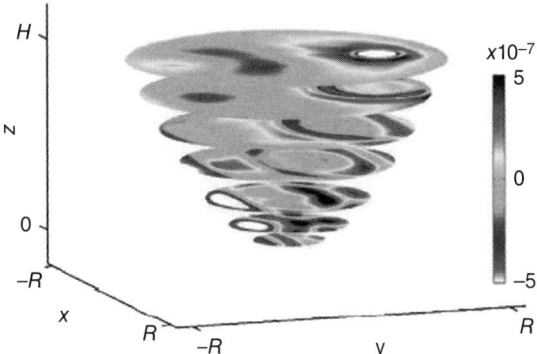

Figure 10.13. Tomographic reconstruction of internal wavefield inside a parabolic domain computed from many views of image displacements such as those in Figure 10.12. Reproduced from Figure 4a of *Hazewinkel et al.* [2011].

localized disturbances, the latter approach was found to be more effective.

The forward problem can thus be written as a coupled set of $2n$ equations in N unknowns. This is cast in matrix form analogous to equation (10.12):

$$\vec{\Delta} \equiv \left(\vec{\Delta x}^T, \vec{\Delta z}^T\right)^T = \mathbf{G}\,\vec{\rho}, \qquad (10.14)$$

in which the differentiation operators acting on elements of ρ to give $\nabla\rho$ are buried inside the components of $\mathbf{G}$, which is a $2n - N$ rectangular matrix. The resolution of the disturbance is chosen so that there are more unknowns than equations. The typical method to solve this system of equations is to multiply through by the transpose, $\mathbf{G}^T$, thus recasting the problem as N equations in N unknowns. Because the $N-N$ sparse matrix $\mathbf{G}^T\mathbf{G}$ is singular, it is typical to shift its eigenvalues by a so-called regularization parameter μ [*Zhdanov*, 2002]. Hence, the forward problem is written

$$\mathbf{G}^T\vec{\Delta} = \left(\mathbf{G}^T\mathbf{G} + \mu\mathbf{I}\right)\vec{\rho}, \qquad (10.15)$$

in which $\mathbf{I}$ is the identity matrix.

Rather than compute the inverse of the matrix multiplying $\vec{\rho}$ on the right-hand side of (10.15), it is efficient to solve iteratively using the biconjugate gradient method [*Golub and van Loan*, 1996].

This approach was tested against idealized disturbances by *Decamp et al.* [2008], who showed that a polar grid is best used provided the number of sectors is not a multiple of the number of perspectives. Even with just six perspectives, a cosine-times-Gaussian disturbance was well reproduced on a polar grid with 33 sectors and 40 rings.

Applying this method to internal wave fields generated in the laboratory has proved challenging in part because of the requirement to have multiple perspectives. In the work of *Hazewinkel et al.* [2011], the tank had curved sidewalls whose influence upon the path of light rays could be accounted for. In attempts to study non-axisymmetric waves generated (e.g., by a horizontally towed object) in a square tank, at most two perspectives at 90° might be recorded simultaneously, each with a camera on one side and the image on the other. To gain more perspectives, the experiment must be repeated but the generation mechanism reoriented within the tank to give the cameras a different perspective. This method requires perfect repeatability. Small changes can lead to large errors in the computation of $\nabla \rho$.

10.5. OTHER ADVANCES

Thus far we have focused upon the use of schlieren to examine internal waves in the laboratory. Here we mention other techniques used to generate and analyze internal waves.

10.5.1. Particle Image Velocimetry

Particle image velocimetry is now a well-established method used in the laboratory to measure flow fields nonintrusively. In this method, small particles are illuminated by a laser light sheet. Their displacements (or, more precisely, displacements of patches of particles in a window) are tracked between pulses of the laser. The technique has revolutionized laboratory experiments by providing a nonintrusive method that measures velocity at all points in the plane of the light sheet [*Fincham and Spedding*, 1997]. Using an oscillating mirror, one can also make multiple parallel light sheets that sequentially illuminate on a fast (typically microsecond) time scale [*Fincham*, 2006]. Thus the flow field can be reconstructed in three dimensions to within the resolution set by the separation between successive light sheets and the digital camera.

Using PIV in the study of internal waves poses additional challenges. Because light bends as it passes through stratified fluid, the position of particles in the flow can be misrepresented [*Dalziel et al.*, 2007]. One can try to eliminate particle distortions by adding another fluid to the ambient (e.g., alcohol) that cancels the refractive index change due to salinity, but this can also lead to problems with double diffusive behavior.

Without resorting to adding refractive index matching fluids, schlieren can be used to predict the distortion and so provide a correction to the digitized image of particles before they are processed to compute displacements.

For example, in the study of solitary waves by *Dalziel et al.* [2007], the direct application of PIV was hindered by distortions resulting from the sharp density gradient at the interface between the fresh and underlying salty water. Figure 10.14 shows the smearing and significant apparent particle displacement at a sharp density gradient. This is

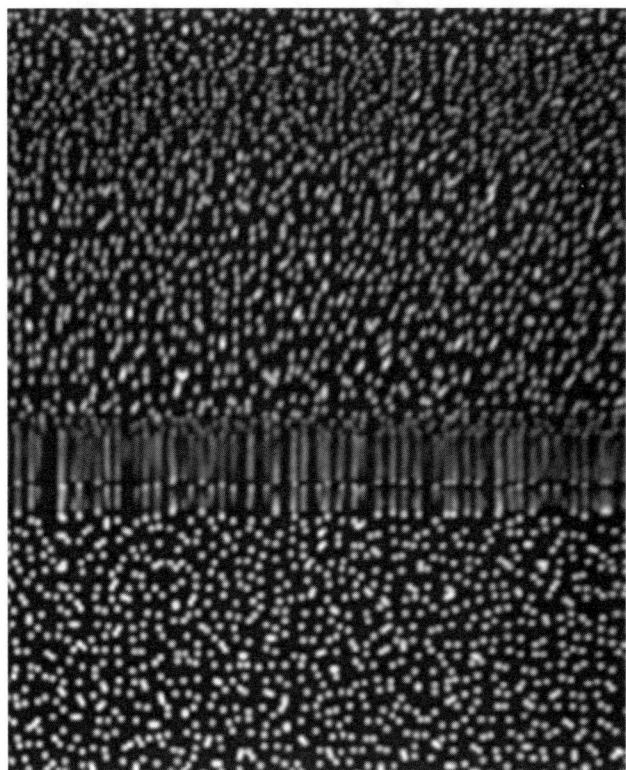

Figure 10.14. Image of random dots distorted by strong stratification at a density interface in an approximately two-layer fluid. Reproduced from Figure 6c of *Dalziel et al.* [2007].

not due to the vertical motion of the wave. It results from photons between the laser light sheet and observer being deflected as they pass through the interface.

Dalziel et al. [2007] addressed the issue by using schlieren to measure the density gradient and then using this information to correct for the apparent in situ particle displacement. The experimental configuration strobed between the camera recording the positions of particles in a laser light sheet in the fluid and it recording images of random dots on a screen behind the tank. This effectively rendered the schlieren and PIV measurements simultaneous. The result is shown in Figure 10.15. The corrected PIV image gives values of velocity and the schlieren measurements predicted the density. Importantly, the combined results measured the gradient Richardson number and so assessed the stability of internal solitary waves.

If the stratification is not too strong and disturbances in the fluid are not too large, then the distortion due to refractive index changes can be ignored and PIV can be applied directly. This method was used successfully in the measurement of internal waves generated by oscillatory flow over cylinders [*Zhang et al.*, 2007] spheres [*King et al.*, 2009] and a Gaussian-shaped hill [*Echeverri et al.*, 2009]. In these cases the distortions due to isopycnal

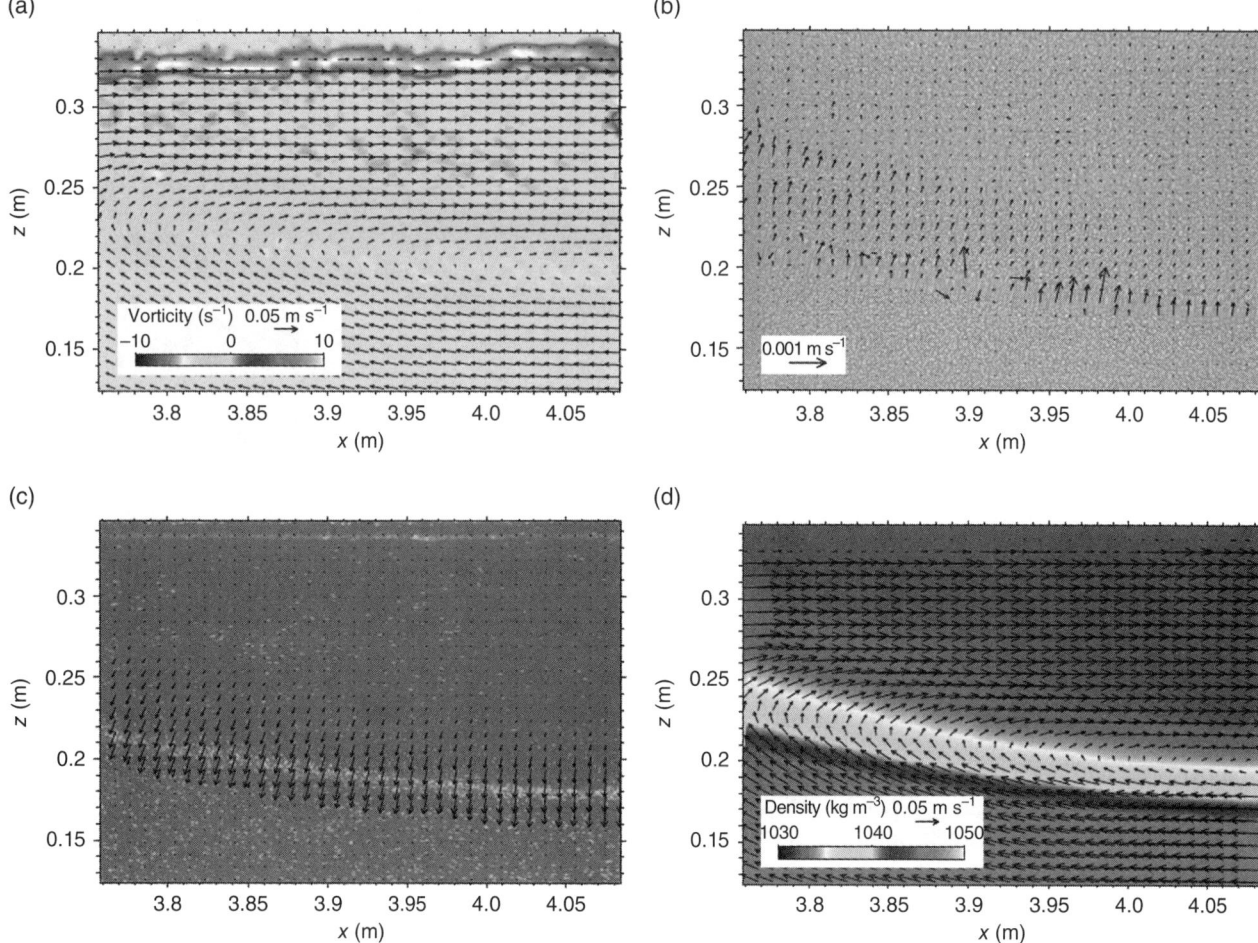

Figure 10.15. Combined PIV-schlieren examination of the passage of an internal solitary wave of depression in an approximately two-layer fluid. (a) Original velocity field computed using PIV superimposed on vorticity. (b) Apparent velocity resulting from distortions at density interface superimposed on background image of random dots. (c) Corrected displacements that account for strong density gradient at the interface superimposed on false color image of particles in the tank. (d) Corrected velocity field superimposed on contours of density. Reproduced from Figure 12 of *Dalziel et al.* [2007]. For color detail, please see color plate section.

displacements were not so large as to require corrections based upon schlieren.

An example of the use of PIV to measure internal wave amplitudes is shown in Figure 10.16. Here the color contours and arrows show the relative velocity field associated with internal waves generated when a Gaussian hill oscillated horizontally back and forth with maximum speed U. These are represented in a frame of reference moving with the hill, being equivalent to maximum flow rightward over the hill. In experiments (top images), the velocity could only be measured to the right of the hill. The structure of the beam is well reproduced by simulations (middle images) and theory, which predicts the far-field behavior (bottom images). This is true in subcritical cases (right), for which the slope of the wave beam is larger than the maximum slope of the hill, and in supercritical cases (left), for which the wave beam is tangent to the hill near its crest. The simulated amplitudes are smaller than what is observed, however. This can be attributed, in part, to the difficulty in capturing the viscous-dominated processes that occur in the generation region where the flow due to the waves moves along the hill slope. Coupling the nearly inviscid far-field dynamics with the viscous boundary layer dynamics remains an outstanding theoretical challenge.

10.5.2. "Fluo-Line" Technique

Laser-induced fluorescence is now frequently used in laboratory experiments to measure concentrations of fluorescent dye in the plane of a laser light sheet.

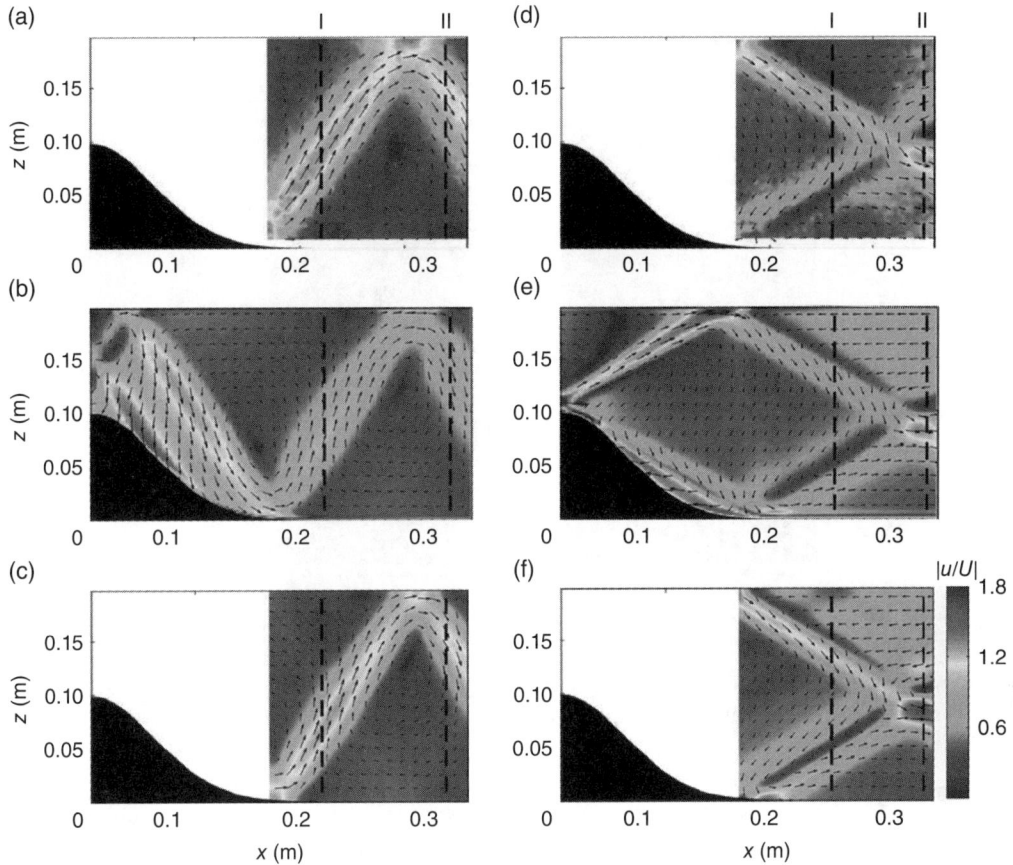

Figure 10.16. Experimental measurements (top), numerical results (middle), and theoretical predictions (bottom) of internal waves generated by oscillatory flow over a Gaussian hill in cases where the frequency of oscillation is subcritical (left) and supercritical (right). Reproduced from Figure 2 of *Echeverri et al.* [2009].

Consequently, this can be used to assess mixing and transport in fluids.

The technique has been used somewhat differently in the study of internal waves generated by a moving sphere in uniformly stratified fluid. Following the technique originally devised by of *Hopfinger et al.* [1991], *Voisin et al.* [2011] (see also *Ermanyuk et al.* [2011]) made thin, evenly spaced horizontal dye lines by soaking threads with fluorescein dye and slowly dragging them horizontally through a tank filled with uniformly stratified fluid. This created very thin markers of isopycnal surfaces that were clearly revealed as a sequence of lines in a vertical plane illuminated by a laser light sheet. The position of each line could be determined to subpixel accuracy by assuming a Gaussian vertical distribution of intensity.

An example of the displacement computed from successive dye lines in a plane passing through the center of a horizontally oscillating sphere is shown in Figure 10.17. Even where the displacement of lines is not obvious to the naked eye, they are clearly discerned by the digital analysis technique.

10.5.3. Novel Wave Generator

Typical methods for generating internal waves in the laboratory include oscillating a rigid body at constant frequency or towing a body horizontally at a constant speed. The former has the disadvantage that it creates four wave beams, as in Figure 10.2, or at least two if oscillated against a side boundary. Towed objects along a top or bottom boundary produce unidirectional waves, but towing piles up the stratified fluid ahead of the object forming what is called a "columnar mode." As a result, the endwall of the tank can influence the dynamics of flow over the obstacle [*Baines*, 1995].

A new mechanism for the generation of internal waves avoids these deficiencies [*Gostiaux et al.*, 2007; *Mercier et al.*, 2010]. In it vertically stacked flat plates periodically move back and forth providing forcing on the stratified fluid from the side. If the forcing is driven by a rotating spiral camshaft, as in Figure 10.18, the plates effectively move collectively as a vertically propagating wave whose vertical wavelength and amplitude are set by the

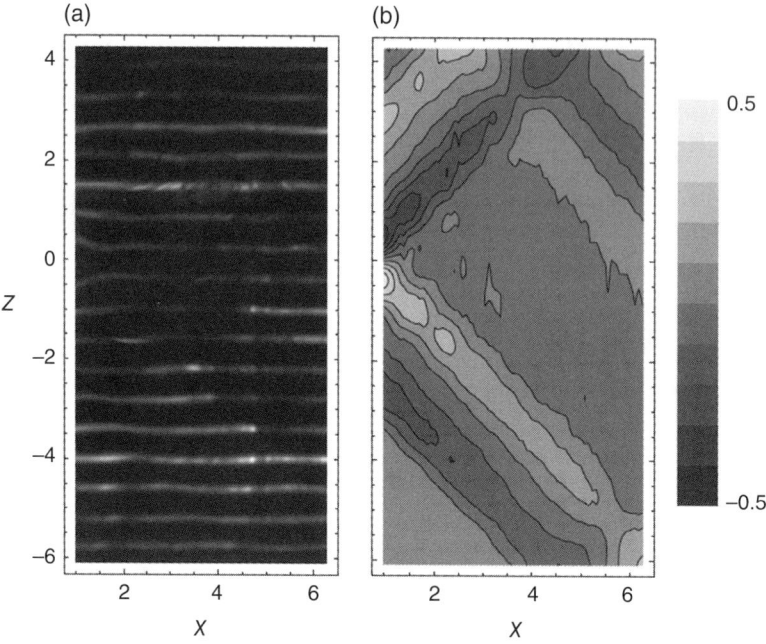

Figure 10.17. (a) Dye lines displaced to the right of a sphere oscillating horizontally about the origin in uniformly stratified fluid. (b) Computed displacement of the lines relative to the oscillating amplitude A. Reproduced from Figure 20 of *Voisin et al.* [2011].

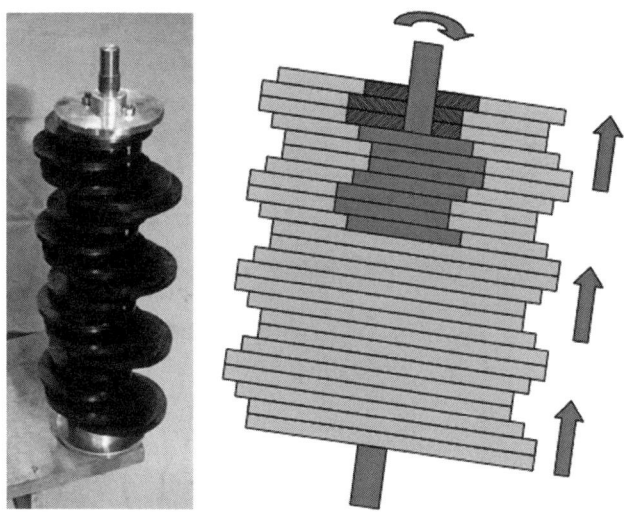

Figure 10.18. Camshaft and schematic cross section showing how the rotation of the shaft results in the back-and-forth oscillation of flat plates. Reproduced from Figure 2 of *Gostiaux et al.* [2007].

geometry of the camshaft and whose frequency is set by the rotation rate.

The mechanism thus acts like towed topography except that the translation of the periodic boundary is vertical. It does not generate columnar modes upstream. Nor is their boundary layer separation behind their crests. Because the boundary displacements are horizontal, large-amplitude forcing is less inclined to result in mixing of the stratified fluid.

This technique has been used in a variety of circumstances that have revealed important processes in the evolution of internal waves.

Satellite altimetry [*Egbert and Ray*, 2000] has recently been employed to observe the generation of oceanic internal waves by tidal flow over the continental margin and submarine sills. These have revealed the generation of low vertical-mode internal waves, associated with undulations of the thermocline. As well as theory and numerical simulations [*Balmforth et al.*, 2002; *Llewellyn-Smith and Young*, 2002; *Legg*, 2004], laboratory experiments have been performed to examine the generation of internal waves by Gaussian hill that oscillated horizontally back and forth with fixed frequency and amplitude [*Peacock et al.*, 2008; *Echeverri et al.*, 2009].

An outstanding question is how such low modes transfer their energy to smaller scales (high modes) so that they ultimately dissipate and mix the ocean.

The laboratory experiments by *Peacock et al.* [2009] examined one mechanism through which this may occur. When a low-mode internal wave is incident upon topography, the sloping sides of the hill refocus the energy into beams. This is shown in Figure 10.19.

The waves are created using the mechanism of oscillating stacked horizontal plates using the wave generator mechanism described above. Here, however, the rotating shaft is not spiral but is constructed with a vertically

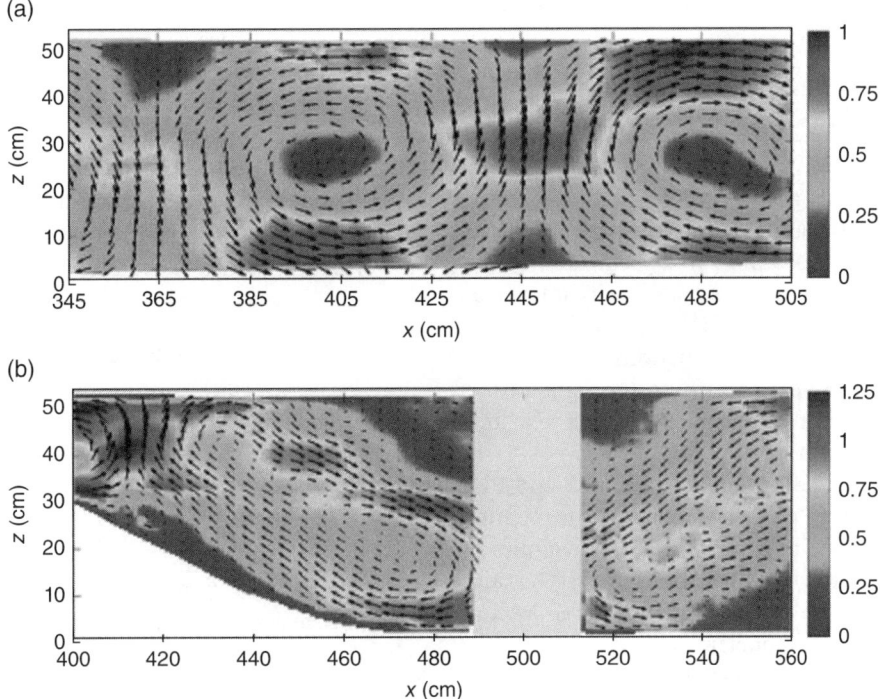

Figure 10.19. (a) Left-to-right propagating vertical mode 1 internal waves generated using horizontally oscillating flat plates, as in *Gostiaux et al.* [2007]. (b) Internal wave beam downstream of hill placed in front of the oncoming mode 1 waves. Adapted from Figures 2a and 3a of *Peacock et al.* [2009].

sinusoidal variation in a fixed plane so that rotating the shaft produces a mode 1 wave in a uniformly stratified medium [*Mercier et al.*, 2010]. In the absence of topography, PIV is used to reveal the mode 1 wave structure, which at one point in the phase exhibits forward motion at the surface and bottom and retrograde motion at mid depth. The flow directions reverse a half period later.

When this incident wavefield encounters a Gaussian hill, the structure of the wavefield changes significantly downstream. Just as an oscillating body creates internal wave beams in a stratified fluid, oscillations resulting from the incident low-mode internal waves create beamlike structures downstream. Thus energy from low modes is efficiently converted into higher modes. These ideas have recently been extended to the examination of internal waves incident upon the continental shelf [*Klymak et al.*, 2011].

10.6. DISCUSSION AND CONCLUSIONS

Several technological innovations have provided new tools for the study of internal waves in the laboratory. Here we have focused mostly upon the use of synthetic schlieren as a nonintrusive way to measure perturbation density gradients due to internal waves in continuously stratified media. When used to examine spanwise-uniform and axisymmetric disturbances, it has provided a useful check on the limitations of linear, inviscid theory, particularly with respect to the generation of internal waves from oscillating and steadily translating bodies.

Just as MRI revolutionized medicine, inverse tomography for schlieren has the potential to measure fully three-dimensional disturbances nonintrusively and continuously in time, provided the disturbances do not involve turbulent mixing and, hence, random scattering of light. However, several logistical obstacles remain to be overcome. In order to reconstruct relatively fine scale features, multiple perspectives from many angles must be recorded simultaneously or in rapid succession. But synthetic schlieren requires looking through a fluid at an object image on the opposite side. To have a large number of perspectives one must devise a method in which multiple cameras are not obstructed by multiple object images. Alternately, like MRI, one could construct a system in which the camera and object image rotate about a cylindrical tank on a fast time scale compared with that of internal waves. After image correction for the curvature of the tank there would remain the theoretical challenge to reconstruct the three-dimensional disturbance field from the images recorded continuously from changing perspectives.

PIV has provided another powerful tool for the nonintrusive examination of internal waves. It has the advantage of measuring in situ particle displacement, and hence velocities, in the plane of a laser light sheet. With multiple light sheets, the fully three-dimensional velocity field can be reconstructed within the spatial and temporal resolutions of the camera and laser.

Because light is significantly distorted where the refractive index changes due to rapid salinity changes, PIV is less effective at examining processes at density interfaces. However, synthetic schlieren and PIV can work in tandem, the former measuring density perturbations and using this information to predict how to correct the light distortion so that actual particle displacements can be measured more accurately through PIV.

To demonstrate the applications of schlieren and PIV, we have focused upon the phenomena of wave generation and propagation in nonuniform media. The dynamics of internal wave breaking with consequent mixing remains an outstanding challenge for experimentalists as well as theoretical and numerical modelers.

For example, in a process known as parametric subharmonic instability (PSI), internal waves resonantly transfer energy to subharmonic internal waves which may overturn and break or transfer energy to smaller-scale waves. PSI has been studied in laboratory experiments in which the displacement of horizontal dye lines were used to observe the evolution of resonantly excited mode 1 waves in a tank with square vertical cross section [*Benielli and Sommeria*, 1998]. The idealized numerical simulations of *MacKinnon and Winters* [2005] predicted that such resonance of internal tides might occur at 28.9°N latitude, which is the northern limit where subharmonic waves exist with frequencies lower than the inertial frequency f. Whether PSI actually occurs as catastrophically as they predicted in the ocean is presently under investigation. Laboratory experiments using the new technologies of schlieren and PIV may also provide new insights into the onset and energetics of PSI.

With increasing observations of internal solitary waves in the ocean, there is renewed interest in examining this phenomenon in the laboratory [*Grue et al.*, 2000; *Carr et al.*, 2008]. Although synthetic schlieren can work together with PIV to help correct apparent particle displacements within the flow [*Dalziel et al.*, 2007], it is not ideally suited to the study of interfacial waves. This is because the large curvature of the density field at the interface bends light to such a degree that an image behind the tank is distorted too much to compute apparent displacements.

Of course, one can track the motion of the interface by injecting dye there while the ambient is being established. The shadowgraph is also a useful tool. Other more innovative methods include the use of ultrasonic probes, which measure the interface displacement by recording the travel time of sound vertically through the ambient between a transmitter and receiver at a fixed depth straddling the interface [*Michallet and Barthélemy*, 1997, 1998].

Oceanographic observations continue to reveal the diversity and complexity of internal wave dynamics, sometimes inspiring and sometimes inspired by laboratory experiments. As digital cameras and image analyses continue to improve, the new techniques of schlieren and PIV are expected to continue stimulating new insights.

REFERENCES

Aguilar, D. A., and B. R. Sutherland (2006), Internal wave generation from rough topography, *Phys. Fluids*, *18*, Art. No. 066,603.

Alford, M. H., et al. (2011), Energy flux and dissipation in Luzon Strait: Two tales of two ridges, *J. Phys. Oceanogr.*, *41*, 2211–2222.

Ansong, J. K., and B. R. Sutherland (2010), Internal gravity waves generated by convective plumes, *J. Fluid Mech.*, *648*, 405–434.

Baines, P. G. (1974), The generation of internal tides over steep continental slopes, *Phil. Trans. R. Soc. Lond. A*, *277*, 27–58.

Baines, P. G. (1982), On internal tide generation models, *Deep-Sea Res.*, *29*, 307–338.

Baines, P. G. (1995), *Topographic Effects in Stratified Flows*, Cambridge Univ. Press, Cambridge, England.

Baines, P. G., and K. P. Hoinka (1985), Stratified flow over two-dimensional topography in fluid of infinite depth: A laboratory simulation, *J. Atmos. Sci.*, *42*, 1614–1630.

Balmforth, N. J., G. R. Ierley, and W. R. Young (2002), Tidal conversion by subcritical topography, *J. Phys. Oceanogr.*, *32*(10), 2900–2914.

Benielli, D., and J. Sommeria (1998), Excitation and breaking of internal gravity waves by parametric instability, *J. Fluid Mech.*, *374*, 117–144.

Bühler, O., and C. J. Muller (2007), Instability and focusing of internal tides in the deep ocean, *J. Fluid Mech.*, *588*, 1–28.

Carr, M., D. Fructus, J. Grue, A. Jensen, and P. A. Davies (2008), Convectively induced shear instability in large amplitude internal solitary waves, *Phys. Fluids*, *20*, 126,601, doi:10.1063/1.3030947.

Chashechkin, Y. D. (1999), Schlieren visualization of a stratified flow around a cylinder, *J. Vis.*, *1*, 345–354.

Dalziel, S. B., G. O. Hughes, and B. R. Sutherland (2000), Whole field density measurements, *Expt. Fluids*, *28*, 322–335.

Dalziel, S. B., M. Carr, J. K. Sveen, and P. A. Davies (2007), Simultaneous synthetic schlieren and PIV measurements for internal solitary waves, *Meas. Sci. Technol.*, *18*, 533–547.

Decamp, S., C. Kozack, and B. R. Sutherland (2008), Three-dimensional schlieren measurements using inverse tomography, *Expt. Fluids*, *20*, 747–758, doi:10.1007/s00348-007-0431-y.

Echeverri, P., M. R. Flynn, K. B. Winters, and T. Peacock (2009), Low-mode internal tide generation by topography: An experimental and numerical investigation, *J. Fluid Mech.*, *636*, 91–108.

Egbert, G. D., and R. D. Ray (2000), Significant dissipation of tidal energy in the deep ocean inferred from satellite altimeter data, *Nature, 405*, 775–778.

Ermanyuk, E. V., J.-B. Flor, and B. Voisin (2011), Spatial structure of first and higher harmonic internal waves from a horizontally oscillating sphere, *J. Fluid Mech., 671*, 364–383.

Faris, G. W., and R. L. Byer (1988), Three-dimensional beam-deflection optical tomography of a supersonic jet, *Appl. Opt., 27*, 5202–5212.

Fincham, A., and G. Spedding (1997), Low cost, high resolution DPIV for measurement of turbulent fluid flow, *Expt. Fluids, 23*, 449–462.

Fincham, A. M. (2006), Continuous scanning, laser imaging velocimetry, *J. Vis., 9*, 247–255.

Flynn, M. R., K. Onu, and B. R. Sutherland (2003), Internal wave generation by a vertically oscillating sphere, *J. Fluid Mech., 494*, 65–93.

Goldhahn, E., and J. Seume (1988), The background oriented schlieren technique: Sensitivity, accuracy, resolution and application to a three-dimensional density field, *Exp. Fluids, 43*, 241–249.

Golub, G. H., and C. F. van Loan (1996), *Matrix computations*, 3rd ed., Johns Hopkins Univ. Press, London.

Görtler, H. (1943), Über eine schwingungserscheinung in flüssigkeiten mit stabiler dichteschichtung, *Z. Angew. Math. Mech., 23*, 65–71.

Gostiaux, L., H. Didelle, S. Mercier, and T. Dauxois (2007), A novel internal waves generator, *Exp. Fluids, 42*, 123–130, doi:10.1007/s00348-006-0225-7.

Gregory, K., and B. R. Sutherland (2010), Transmission and reflection of internal wave beams, *Phys. Fluids, 22*, Art. No. 106,601, doi:10.1063/1.3486613.

Grue, J., A. Jensen, P.-O. Rusas, and J. K. Sveen (2000), Breaking and broadening of internal solitary waves, *J. Fluid Mech., 413*, 181–217.

Hazewinkel, J., P. van Breevoort, S. B. Dalziel, and L. R. M. Maas (2008), Observations on the wavenumber spectrum and evolution of an internal wave attractor, *J. Fluid Mech., 598*, 373–382.

Hazewinkel, J., L. R. M. Maas, and S. B. Dalziel (2011), Tomographic reconstruction of internal wave patterns in a paraboloid, *Exp. Fluids, 50*, 247–258.

Hopfinger, E. J., J.-B. Flor, J. M. Chomaz, and P. Bonneton (1991), Internal waves generated by a moving sphere and its wake in a stratified fluid, *Exp. Fluids, 11*, 255–261.

Howes, W. L. (1984), Rainbow schlieren and its application, *Appl. Opt., 23*, 2449–2460.

Hurley, D. G. (1997), The generation of internal waves by vibrating elliptic cylinders. Part 1: Inviscid solution, *J. Fluid Mech., 351*, 105–118.

Hurley, D. G., and G. Keady (1997), The generation of internal waves by vibrating elliptic cylinders. Part 2: Approximate viscous solution, *J. Fluid Mech., 351*, 119–138.

King, B., H. P. Zhang, and H. L. Swinney (2009), Tidal flow over three-dimensional topography in a stratified fluid, *Phys. Fluids, 21*, 116,601.

Klymak, J. M., and M. C. Gregg (2004), Tidally generated turbulence over the Knight Inlet sill, *J. Phys. Ocean., 34*(5), 1135–1151.

Klymak, J. M., R. Pinkel, C. T. Liu, A. K. Liu, and L. David (2006), Prototypical solitons in the South China Sea, *Geophys. Res. Lett., 33*(11), L11,607.

Klymak, J. M., M. H. Alford, R. Pinkel, R.-C. Lien, Y. J. Yang, and T.-Y. Tang (2011), The breaking and scattering of the internal tide on a continental slope, *J. Phys. Oceanogr., 41*, 926–945.

Koop, C. G., and B. McGee (1986), Measurements of internal gravity waves in a continuously stratified shear flow, *J. Fluid Mech., 172*, 453–480.

Larsen, L. H. (1969), Internal waves incident upon a knife edge barrier, *Deep Sea Res., 16*, 411–419.

Ledwell, J. R., E. Montgomery, K. Polzin, L. C. St. Laurent, R. Schmitt, and J. Toole (2000), Evidence for enhanced mixing over rough topography in the abyssal ocean, *Nature, 403*, 179–182.

Legg, S. (2004), Internal tides generated on a corrugated slope. Part I: Cross-slope barotropic forcing, *J. Phys. Oceanogr., 34*, 156–173.

Li, Q., and D. M. Farmer (2011), The generation and evolution of nonlinear internal waves in the deep basin of the South China Sea, *J. Phys. Oceanogr., 41*, 1345–1363.

Lighthill, M. J. (1978), *Waves in Fluids*, Cambridge Univ. Press, Cambridge, England.

Llewellyn-Smith, S. G., and W. R. Young (2002), Conversion of the barotropic tide, *J. Phys. Oceanogr., 32*, 1554–1566.

Maas, L. R. M., and F.-P. A. Lam (1995), Geometric focusing of internal waves, *J. Fluid Mech., 300*, 1–41.

Maas, L. R. M., D. Benielli, J. Sommeria, and F.-P. A. Lam (1997), Observation of an internal wave attractor in a confined stably stratified fluid, *Nature, 388*, 557–561.

MacKinnon, J. A., and K. B. Winters (2005), Subtropical catastrophe: Significant loss of low-mode tidal energy at 28.9°, *Geophys. Res. Lett., 32*, L15,605-1-5, doi:10.1029/2005GL023376.

Mathur, M., and T. Peacock (2009), Internal wave beam propagation in non-uniform stratifications, *J. Fluid Mech., 639*, 133–152.

Mathur, M., and T. Peacock (2010), Internal wave interferometry, *Phys. Rev. Lett., 104*, 118,501-1-4.

Meier, G. (2002), Computerized background oriented schlieren, *Exp. Fluids, 33*, 181–187.

Mercier, M. J., N. B. Garnier, and T. Dauxois (2008), Reflection and diffraction of internal waves analyzed with the Hilbert transform, *Phy. Fluids, 20*, Art. No. 086,601, doi:10.1063/1.2963136.

Mercier, M. J., D. Martinand, M. Mathur, L. Gostiaux, T. Peacock, and T. Dauxois (2010), New wave generation, *J. Fluid Mech., 657*, 308–334.

Michallet, H., and E. Barthélemy (1997), Ultrasonic probes and data processing to study interfacial solitary waves, *Exp. Fluids, 22*, 380–386.

Michallet, H., and E. Barthélemy (1998), Experimental study of interfacial solitary waves, *J. Fluid Mech., 366*, 159–177.

Mowbray, D. E., and B. S. H. Rarity (1967), A theoretical and experimental investigation of the phase configuration of internal waves of small amplitude in a density stratified liquid, *J. Fluid Mech., 28*, 1–16.

Munk, W. H., and C. Wunsch (1998), Abyssal recipes II: Energetics of tidal and wind mixing, *Deep-Sea Res.*, *45*, 1977–2010.

Nault, J. T., and B. R. Sutherland (2007), Internal wave tunnelling across a mixed region, *Phys. Fluids*, *19*, 016,601-1-8, doi:10.1063/1.2424791.

Nault, J. T., and B. R. Sutherland (2008), Beyond ray tracing for internal waves. Part I: Small-amplitude anelastic waves, *Phys. Fluids*, *20*, 106,601-1-10, doi:10.1063/1.2993167.

Onu, K., M. R. Flynn, and B. R. Sutherland (2003), Schlieren measurement of axisymmetric internal wave amplitudes, *Expt. Fluids*, *35*, 24–31.

Peacock, T., P. Echeverri, and N. J. Balmforth (2008), An experimental investigation of internal tide generation by two-dimensional topography, *J. Phys. Oceanogr.*, *38*, 235–242.

Peacock, T., M. J. Mercier, H. Didelle, S. Viboud, and T. Dauxois (2009), A laboratory study of low-mode internal tide scattering by finite-amplitude topography, *Phys. Fluids*, *21*, Art. No. 121,702, doi:10.1063/1.3267096.

Pinkel, R. (2000), Internal solitary waves in the warm pool of the western equatorial Pacific, *J. Phys. Oceanogr.*, *30*, 2906–2926.

Polzin, K. L., J. M. Toole, J. R. Ledwell, and R. W. Schmitt (1997), Spatial variability of turbulent mixing in the Abyssal Ocean, *Science*, *276*, 93–96.

Robinson, R. M. (1969), The effects of a vertical barrier on internal waves, *Deep-Sea Res.*, *16*, 421–429.

Sandstrom, H. (1969), Effect of topography on propagation of waves in stratified fluids, *Deep Sea Res.*, *16*, 405–410.

Schardin, H. (1942), Die schlierenverfahren und ihre anwendungen, *Ergebnisse der Exakten Naturwissenschaften*, *20*, 303–439.

Settles, G. S. (2001), *Schlieren and Shadowgraph Techniques: Visualizing Phenomena in Transparent Media*, Springer Verlag, Berlin and New York.

Sutherland, B. R. (2002), Large-amplitude internal wave generation in the lee of step-shaped topography, *Geophys. Res. Lett.*, *29*(16), Art. No 1769.

Sutherland, B. R. (2010), *Internal Gravity Waves*, Cambridge Univ. Press, Cambridge.

Sutherland, B. R., and P. F. Linden (2002), Internal wave excitation by a vertically oscillating elliptical cylinder, *Phys. Fluids*, *14*, 721–731.

Sutherland, B. R., and K. Yewchuk (2004), Internal wave tunnelling, *J. Fluid Mech.*, *511*, 125–134.

Sutherland, B. R., S. B. Dalziel, G. O. Hughes, and P. F. Linden (1999), Visualisation and measurement of internal waves by "synthetic schlieren." Part 1: Vertically oscillating cylinder, *J. Fluid Mech.*, *390*, 93–126.

Teoh, S. G., G. N. Ivey, and J. Imberger (1997), Laboratory study of the interaction between two internal wave rays, *J. Fluid Mech.*, *336*, 91–122.

Thomas, N. H., and T. N. Stevenson (1972), A similarity solution for viscous internal waves, *J. Fluid Mech.*, *54*, 495–506.

Voisin, B. (1991), Internal wave generation in uniformly stratified fluids. Part 1. Green's function and point sources, *J. Fluid Mech.*, *231*, 439–480.

Voisin, B. (1994), Internal wave generation in uniformly stratified fluids. Part 2. Moving point sources, *J. Fluid Mech.*, *261*, 333–374.

Voisin, B., E. V. Ermanyuk, and J.-B. Flor (2011), Internal wave generation by oscillation of a sphere, with application to internal tides, *J. Fluid Mech.*, *666*, 308–357.

Weast, R. C. (1981), *Handbook of Chemistry and Physics*, 62nd ed., CRC. Press, Boca Raton, Fla.

Yick, K.-Y., R. Stoker, and T. Peacock (2007), Microscale synthetic schlieren, *Exp. Fluids*, *42*, 41–48.

Zeilon, N. (1912), On tidal boundary-waves and related hydrodynamical problems, *Handl. K. svenska Vetens. Akad.*, *47*, 1–46.

Zhang, H. P., B. King, and H. L. Swinney (2007), Experimental study of internal gravity waves generated by supercritical topography, *Phys. Fluids*, *19*, 096,602.

Zhdanov, M. (2002), *Geophysical Inverse Theory and Regularization Problems*, Elsevier, Amsterdam, New York, and Tokyo.

11

Frontal Instabilities at Density–Shear Interfaces in Rotating Two-Layer Stratified Fluids

Hélène Scolan[1], Roberto Verzicco[2], and Jan-Bert Flór[3]

11.1. INTRODUCTION

Fronts in Earth oceans and atmosphere separate different temperature fluids or masses of air and therefore play a major role for the transport of heat and chemical or biological tracers in large-scale geophysical flows. Their dynamics and instability are crucial for weather forecasting, whereas their variability is a key element for understanding climate. First experiments that showed the occurrence of baroclinic modes were conducted in annular rotating tanks of which the exterior (or inner) cylinder was heated (cooled) [*Hide*, 1953, 1958; *Fultz et al.*, 1959; *Fowlis and Hide*, 1965].

Baroclinic instability arises at a density field that is inclined with respect to the horizontal. As mentioned, an inclined density gradient may be created in a rotating fluid by differential heating. In the quasi-geostrophic approximation, these flows are in thermal wind balance, i.e, the tilting of the vorticity due to shear in the fluid equals the baroclinic production of vorticity,

$$f\frac{\partial \vec{u}}{\partial z} = -\frac{g}{\rho_o}\vec{z} \times \vec{\nabla}_H \rho,$$

where u is the horizontal velocity, g the gravitational constant, and $\vec{\nabla}_H \rho$ the horizontal density gradient. Another manner to create a baroclinic front in thermal wind balance is achieved by applying a vertical shear across the density field. Such a shear can be obtained by using a rotating disk at the fluid surface [*Hart*, 1972]. For this type of flow the quasi-geostrophic two-layer model of *Phillips* [1954] and *Pedlosky* [1964, 1970] [see *Hart*, 1972] predicts well the observed wavelengths of the baroclinic instability. These flows depend on stratification, rotation, and the flow aspect ratio as expressed by the Burger number. This number indicates the baroclinicity of the flow. Strongly baroclinic flows have a small Burger number corresponding to a strongly inclined density field. This number serves as a control parameter for baroclinic instability (see Section 11.2.1). Various aspects of baroclinic instability have been considered. For small Burger numbers, periodic variations occur in the amplitude of the dominant baroclinic mode of the system, a nonlinear effect known as amplitude vacillation [*Hart*, 1980, 1985; *King*, 1979]. In considering the same forcing at the fluid surface for an annular flow, also *Bradford et al.* [1981], *Appleby* [1982], *Lovegrove et al.* [2000], and *Williams et al.* [2003, 2004a, 2004b, 2005] showed cases of amplitude vacillation, whereas for even smaller Burger numbers the flow is known to transit to chaos and turbulence [*Hide*, 2011; *Read et al.*, 1992; *Früh and Read*, 1997; *Eccles et al.*, 2009].

Other examples of fronts where the baroclinic instability is observed to develop are density intrusions in a rotating fluid, such as gravity currents, coastal flows, and vortex lenses created by the release of a fluid of a different density in a homogeneous (or stratified) rotating ambient [*Chia et al.*, 1982; *Bouruet-Aubertot and Linden*, 2002]. These systems generally consist of two layers, and the Phillips model for baroclinic instability shows good agreement with the observed instability wavelengths [*Griffiths and Linden*, 1981; *Cenedese and Linden*, 2002]. Interactions of the front with the Ekman boundary layer do not seem to influence the instability wavelength of these close to laminar fronts in small-scale laboratory experiments. Also the complex dynamics of critical layers at the interface do not significantly modify observations of the

---

[1] Atmospheric, Oceanic & Planetary Physics, University of Oxford, Oxford, United Kingdom.
[2] Department of Mechanical Engineering, Università di Roma Tor Vergata, Rome, Italy.
[3] Laboratoire des Ecoulements Géophysiques et Industriels (LEGI), Grenoble, France.

Modeling Atmospheric and Oceanic Flows: Insights from Laboratory Experiments and Numerical Simulations,
First Edition. Edited by Thomas von Larcher and Paul D. Williams.
© 2015 American Geophysical Union. Published 2015 by John Wiley & Sons, Inc.

Table 11.1. Dimensions and parameter regimes of experiments on the instability of baroclinic fronts and the parameter ranges considered.

Reference	Γ	γ	L (cm)	Bu	Ro	E_k	d
Hart [1972, 1973, 1976, 1979, 1980, 1981, 1985]	0.15, 0.039	0.36 or 0.67	12.5	0.02–0.1	0.02–0.25	$2.5 \times 10^{-7}, 3 \times 10^{-3}$ 3×10^{-5}	0.1–0.7
King [1979]	0.01-0.02	2.19	5.7	0.04–0.125	0.035–0.1	$1 \times 10^{-7} - 1 \times 10^{-5}$	0.1–0.7
Carrigan [1978]	0.027	0.67	7.62	0.0033–0.033	0.017–0.5	$1 \times 10^{-3} - 1 \times 10^{-2}$	—
Lovegrove et al. [1999, 2000]	0.006	2	6.25	0.06–0.25	0.1–10	$1 \times 10^{-6} - 1 \times 10^{-4}$	0.004–0.03
Williams et al. [2003, 2004a, 2004b, 2005, 2008, 2010]	0.006	2	6.25	0.03–333	0.1–100	$1 \times 10^{-6} - 3 \times 10^{-3}$	0.0003–0.03
Flór et al. [2011]	0.006	0.18	75	0.02–5	0.06–1.5	$7 \times 10^{-5} - 7 \times 10^{-4}$	0.01–0.1

Note: Flór et al. [2011] considered miscible salt-stratified fluids.

baroclinic instability. In this chapter, we focus on fronts that do not intersect with the top or bottom boundary.

We consider ageostrophic flows and discuss the recently investigated Rossby-Kelvin (RK) instability first discovered by *Sakai* [1989] and aspects of the frontal instability that are related to the thickness of the interface. *Sakai* [1989] considered ageostrophic modeling of baroclinic instability and, following *Hayashi and Young* [1987], investigated the resonance of different type of waves, such as Kelvin, Rossby, and Poincaré waves. The baroclinic instability was interpreted as a resonance (a coupling of the phase speeds) between two Rossby waves, one in the upper and one in the lower layer. The RK instability is a consequence of the resonance between a Rossby wave in one layer and Kelvin or Poincaré wave in the other layer. Though gravity waves move faster than Rossby waves, resonance occurs when the two wave frequencies match due to the Doppler shift. Recently, this approach has been continued for rectangular and annular geometries respectively by *Gula et al.* [2009a, 2009b], *Gula and Zeitlin* [2010], and *Gula et al.* [2010] showing especially significant growth rates for the resonance of a Rossby wave in the upper layer and a coastal Kelvin wave in the lower layer. *Sakai* [1989] used the term Rossby-Kelvin for all vorticity-gravity wave resonances, but here we will reduce the term "Rossby-Kelvin instability" to this particular resonance. These results were compared to experimental results by *Flór et al.* [2011].

The choice of a fluid interface between two miscible fluid layers is in between the limits of a very thin immiscible fluid interface [*Williams et al.*, 2005] and, at the other end, a linear density stratification. For two-layer fluids, it has been shown in numerical simulations [*Gula et al.*, 2009a] that the RK instability occurs only for thin interfaces. For interfaces with a thickness beyond a certain threshold, i.e. $\delta_\rho/(2H) > 0.16$, where $2H$ is the fluid depth and δ_ρ the interface thickness (see Section 11.3.2), the growth rate of ageostrophic modes was found to reduce to zero. Though we are not aware of simulations on the RK instability in linearly stratified fluids, in considering a very thick interface as an approximation of a linear stratification, one may expect also a zero growth rate there. In general, relatively thin density interfaces allow for a richer variety of shear instabilities, including Kelvin-Helmholtz and Hölmböe instability (see Section 11.3.3).

Small-scale waves at the immiscible fluid interface in the baroclinic unstable regime [*Lovegrove et al.*, 2000, *Williams et al.*, 2003, 2004a, 2004b, 2005] were interpreted as the spontaneous emission of inertia-gravity waves. In these experiments the optical activity of the two immiscible fluids gave access to very accurate measurements of interfacial perturbations. A constraint was that the reduced gravity was fixed by the limited type of optical fluids that is available (see Table 11.1). In a larger setup and filled with a salt-stratified two-layer fluid [*Flór et al.*, 2011], interfacial waves due to surface tension effects were eliminated, and, as mentioned, access to a larger range of scales of motion was achieved. Hölmböe instability was in part considered responsible for similar type of small waves, a mechanism that could also be efficient in the immiscible fluid experiments mentioned above [*Flór et al.*, 2011].

In this chapter, we report numerical results for a two-layer stratification with a smooth interface that have earlier been presented by *Scolan* [2011], and experimental results by *Flór et al.* [2011]. We discuss these interfacial waves next to the different instabilities observed in the parameter space set by Burger number, Rossby, and Ekman or the dissipation number, defined in the next section. In doing so, we consider experimental and numerical results of the setup depicted in Figure 11.1.

The governing equations and pertinent nondimensional numbers of this experimental system are presented in Section 11.2.1, followed by a brief description of the numerical approach in Section 11.2.2. The instability regime diagrams and the baroclinic instability including the recently reported observations of the RK instability are presented in Section 11.3.1. In Section 11.3.2 the secondary vertical circulation is investigated numerically.

FRONTAL INSTABILITIES AT DENSITY–SHEAR INTERFACES IN ROTATING TWO-LAYER STRATIFIED FLUIDS

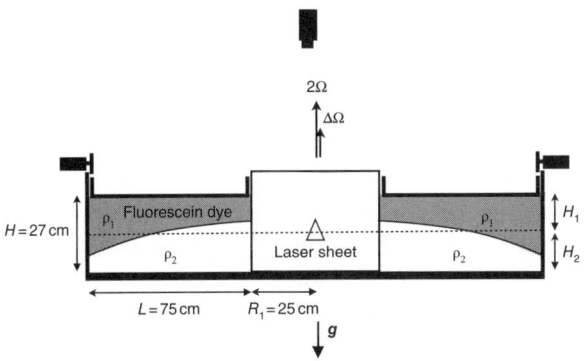

Figure 11.1. Sketch of experimental setup, with ρ_1 and ρ_2 the upper and lower layer fluid densities of depth H; the annular disk at the surface is driven by three motors with rotation $\Delta\Omega$.

We show the presence of interfacial Ekman layers and the influence of the Schmidt number. These properties are relevant for the small-scale instabilities discussed in Section 11.3.3. When the internal fluid interface is not discrete but continuous as in a salt-stratified two-layer fluid, questions about instability such as Kelvin-Helmholtz and Hölmböe instability, internal Ekman layers, as well as the effect of stratification on the RK instability come into play (Section 11.3.3). In the conclusions (Section 11.4) we discuss the different results in a geophysical context in considering realistic Burger and Rossby numbers in the atmosphere and oceans and a range of different experiments on baroclinic (Figure 11.2) instabilities, and small-scale instabilities.

11.2. EQUATIONS AND SCALES FOR LABORATORY EXPERIMENTS

The dynamics for a two-layer salt-stratified and rotating fluid with viscous boundary effects is described by the Navier-Stokes equations with boundary conditions. They are composed of the continuity equation, the equation of motion (including Coriolis force and centrifugal forces), and the conservation of salt. In the Boussinesq approximation, the general equations for a salt-stratified fluid in rotation are

$$\nabla \cdot \vec{u} = 0,$$

$$\frac{\partial \vec{u}}{\partial t} + (\vec{u}\cdot\nabla)\vec{u} = -\frac{\nabla p}{\rho_o} - 2\vec{\Omega}\times\vec{u} - g\frac{\rho - \rho_o}{\rho_o}\hat{z} + \nu\nabla^2\vec{u}$$
$$+ \frac{\rho - \rho_o}{\rho_o}\Omega^2 r\hat{r},$$

$$\frac{\partial S}{\partial t} + \nabla \cdot (S\vec{u}) = \kappa\nabla^2 S, \quad (11.1)$$

with $\vec{u}$ and p, the velocity and pressure field, respectively, ρ_o the mean density, i.e., $\rho_o = (\rho_1 + \rho_2)/2$, and the density field is assumed to vary as $\rho = \rho_o + \tilde{\rho}$ with the density perturbation $\tilde{\rho}$ assumed to be proportional to the salinity S, i.e. $\tilde{\rho}(\vec{r}) = \alpha S(\vec{r})$. The diffusivity of this scalar quantity S is represented by κ.

Following *Hart* [1972] the equations are nondimensionalized with length scale $L = R_2 - R_1$, velocity scale $U = \Delta\Omega L$, pressure $P = \rho_1 U^2$, time scale $T = 1/\Delta\Omega$, and for the density (salt concentration) $\Delta\rho = \rho_2 - \rho_1$. The experimental setup and notations are represented in Figure 11.1. The scaled equations then become

$$\frac{\partial \vec{u}}{\partial t} + (\vec{u}\cdot\nabla)\vec{u} = -\nabla p - \frac{1}{R_o}\hat{k}\times\vec{u} - \frac{B_u}{R_o^2 H/L}S\hat{z} + \frac{1}{R_e}\nabla^2\vec{u}$$
$$+ \frac{g'/g}{4R_o^2}Sr\hat{r},$$

$$\frac{\partial S}{\partial t} + \nabla \cdot (S\vec{u}) = \frac{1}{Pe}\nabla^2 S, \quad (11.2)$$

with the nondimensional parameters defined as the Burger number (or Froude number Fr = 1/Bu)

$$\mathrm{Bu} = \frac{g'H}{4\Omega^2 L^2},$$

the Rossby number

$$\mathrm{Ro} = \frac{\Delta\Omega}{2\Omega},$$

and the Ekman number

$$\mathrm{Ek} = \frac{\nu}{2\Omega H^2},$$

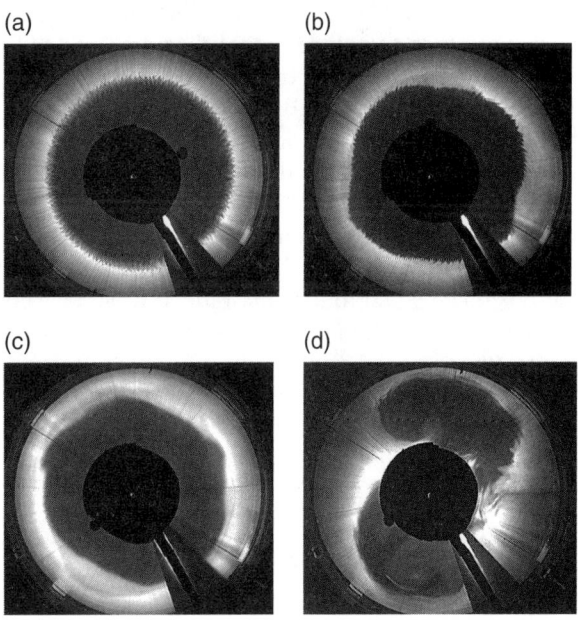

Figure 11.2. Typical observations of the instabilities reported in the diagram of Figure 11.3 with (a) the Hölmböe instability (H), (b) and (c) respectively, mode 4 and mode 6 of the RK instability, and (d) the baroclinic instability (BI) from *Flór et al.* [2011].

where H is the height for each layer, g' the reduced gravity $g' = 2g(\rho_2 - \rho_1)/(\rho_2 + \rho_1)$, Ω the background rotation and $\Delta\Omega$ the differential rotation of the disk. Often the dissipation number

$$d = \frac{\sqrt{\nu\Omega}}{H\Delta\Omega} = \frac{\sqrt{E_k}}{\sqrt{2}\,\text{Ro}}$$

is considered [*Hart*, 1972]. This number represents the dissipation of the spinning disk motion at the surface due to Ekman pumping. Alternatively, one can also consider the Reynolds number

$$R_e = \frac{R_o}{E_k}\left(\frac{L}{H}\right)^2$$

as dissipation parameter. For a given geometric configuration and stratification, the three other parameters are the ratio of accelerations $\Gamma = g'/g$, the aspect ratio $\gamma = H/L$, and for the diffusivity of momentum compared to salt, the Schmidt number $\text{Sc} = \nu/\kappa$. Instead of the Schmidt number, one can also consider the Peclet number $\text{Pe} = \Delta\Omega L^2/\kappa = \text{Re}\cdot\text{Sc}$.

Given an experimental setup, the flow can be determined by three parameters, d, Bu, and the Rossby number Ro. In experiments with immiscible fluids, g' is set by the available optically active fluids so that these experiments are associated with a Bu-Ro diagram (for a given range of rotation frequencies). A different reduced gravity g' would allow for the same Bu number but different Ro, Ek, or d numbers. The complete regime diagram is therefore set by Bu, Ro, and d. When viscous effects are negligible, this reduces to a two-dimensional diagram spanned by *Burger and Rossby* number. In order to compare different experiments, in *Flór et al.* [2011], the parameters were chosen equal to those in the work of *Williams et al.* [2005]. Because of the miscible fluid interface and larger setup, we expect that threshold critical values may vary in this regime diagram. Here, the nondimensional parameters are the same as before, but for a cylindrical tank, the length scale L is taken equal to the tank radius. Table 11.1 gives the experimental values of these systems so far tested in different experimental setups.

To examine the small-scale shear instabilities at the interface, we use the definition of *Alexakis* [2005] for the global Richardson number J_o, i.e. the gradient Richardson number Ri at the interface,

$$J_o = \text{Ri}(0) = JR = \frac{g'\delta_\nu}{(\Delta U)^2}R,$$

with the usual bulk Richardson number J, the total shear thickness δ_ν, and total density thickness δ_ρ of the interface and their ratio defined as $R = \delta_\nu/\delta_\rho$. In immiscible fluid layers, the total density thickness is given by twice the thickness of the Ekman layer.

From the general equations (11.1), the two-layer quasi-geostrophic model was derived to study the baroclinic instability of a geostrophic flow analytically. This simplified model can be obtained from an asymptotic development of the equations in the limit of small Rossby numbers [*Pedlosky*, 1987]. At order zero, the flow simply verifies the geostrophic and hydrostatic balances. At order one, the development gives the quasi-geostrophic equations. This two-layer model gives major properties for the baroclinic instability and allowed predictions for the threshold Burger number for the baroclinic instability in Hart's experiments [*Hart*, 1972]. For layers of the same depth and same viscosity, the potential vorticity equations for the two-layer quasi-geostrophic flow including Ekman boundary layers [*Hart*, 1972; *Pedlosky*, 1970],

$$\frac{d}{dt_1}\left[\nabla^2\psi_1 + \frac{1}{\text{Bu}}(\psi_2 - \psi_1)\right] = -d\left(\frac{3}{2}\nabla^2\psi_1 - \frac{1}{2}\nabla^2\psi_2 - 2\right),$$
$$\frac{d}{dt_2}\left[\nabla^2\psi_2 + \frac{1}{\text{Bu}}(\psi_1 - \psi_2)\right] = -d\left(\frac{3}{2}\nabla^2\psi_2 - \frac{1}{2}\nabla^2\psi_1\right),$$
(11.3)

with

$$\frac{d}{dt_i} = \left[\frac{\partial}{\partial t} + J(\psi_i,\ )\right],$$

where J is the Jacobian operator and ψ_i the stream function in cylindrical coordinates for the layer i. The same dimensional parameters Bu and d appear in the equations. The coupling between the layers occurs via the Burger number term and viscous effects in the form of Ekman pumping. Topographic effects are neglected since the bottom and disk are flat, and centrifugal effects on the shape of the interface are neglected [*Hart*, 1973]. Because of the quasi-geostrophic balance, inertia-gravity waves are filtered out, and the equations apply to length scales that are of the order of the deformation radius, for example, the scale of the Rossby waves.

By contrast, in the two-layer shallow-water model studied by *Sakai* [1989] and more recently by *Gula and Zeitlin* [2010] and *Gula et al.* [2010], gravity waves are not filtered out. Since this model contains no Rossby-number approximation or asymptotic development, it includes ageostrophic motions. It is an inviscid model and the effects due to Ekman pumping are neglected.

The quasi-geostrophic and the shallow-water model both rely on the shallow-water approximation, which assumes thin layers with a small aspect ratio and small vertical compared to horizontal gradients. The validity of this approximation in experimental configurations is achieved as long as $(H_j/L)^2\text{Ro}_j$ in layer j is small. For $H_1 = H_2$, it corresponds to the product of the square of the aspect ratio, γ^2, with the Rossby number. This number gives the ratio between the acceleration of the vertical velocity and the vertical pressure gradient. For aspect ratio 0.18 in each layer and Rossby number between 0.1 and 1, we have $\gamma^2 Ro$ between 0.003 and 0.03, which ensures that (a) the vertical acceleration can be neglected, (b) the hydrostatic balance is verified, and (c) the shallow-water approximation is valid. This condition is not always met in former experiments with larger aspect ratios (see Table 11.1).

Both models consider a jump in density with separated equations for each layer coupled via pressure and, if included, viscous effects. The dynamics of miscible interfaces are however not captured, and especially interactions between density and vorticity at the interface with shear instabilities are not included. Below we consider experiments and direct numerical simulations (DNSs) built on the full Boussinesq Navier-Stokes equations (11.2) in the annular configuration (see Section 11.2.2).

11.2.1. Experimental Modeling of Fronts

A baroclinic front in thermal wind balance is created in a two-layer rotating fluid with the shear across the interface driven by a rotating lid at the fluid surface. For the study of baroclinic instability, this forcing is preferable because the density difference is set from the beginning, and flow measurements of the velocity field and the density field are better accessible. The mixing of the miscible interface amounted to a very small percentage of the volume of the layers and therefore has a negligible influence on the buoyancy frequency at the interface. The flow dimensions of the experimental setup are indicated in Figure 11.1 and the parameter ranges are indicated in Table 11.1. In order to investigate the frontal evolution for the Bu number ranges indicated in Table 11.1, the disk rotation $\Delta\Omega$ was set constant for a very slowly increasing background rotation $\Omega(t)$ such that the values of the Ro, Ek, and Bu numbers changed correspondingly. This same method had initially been tested successfully by, e.g., *Williams et al.* [2005], and the comparison with the same experiment for fixed background rotation assured that Ekman pumping effects due to spin-up did not affect the flow dynamics and instability. The density difference across the interface, $\Delta\rho = \rho_1 - \rho_2$, and reduced gravity, $g' = g\Delta\rho/\bar{\rho}$ (with g the gravity constant), were kept constant and close or equal to the values by *Williams et al.* [2005] (see Table 11.1).

To visualize the flow, the upper fluid layer was dyed with fluorescein dye and illuminated by a horizontal laser sheet at middepth such that the inclined interface of the front was visible in the top-view images. The wavelengths, phase speeds, and growths of the instabilities are obtained from spatiotemporal sequences of the top-view frontal evolution, also known as Hövmüller diagrams. In addition, PIV (Particle Image Velocimetry) measurements from tracer particles gave access to typical velocity fields in each layer. The resolution in the experiments, especially the dye visualizations, was of the order of 3 mm for a tank of 1 m radius and allowed to measure the presence of small-scale waves.

The lack of information about the flow characteristics such as the local density gradient and three-dimensional velocity field motivated us to obtain additional flow information from DNSs for a series of almost identical, or at least comparable, initial values as in the experiments.

11.2.2. Numerical Approach

The numerical simulations were performed using the DNS code of *Verzicco et al.* [1997]. The governing equations are the Boussinesq equations, including Coriolis and centrifugal forces in the annular configuration. Equations are written in cylindrical coordinates (v_r, v_θ, v_z) and discretized on a staggered mesh by central second-order accurate finite difference approximation. Details of the numerical scheme are described by *Verzicco and Orlandi* [1996] and *Verzicco and Camussi* [1997]. The initial (nondimensional) density profiles are $\rho_2 = 1$ and $\rho_1 = 0$ to which a random value is added as a perturbation. At $t = 0$, the velocity field (v_r, v_θ, v_z) is at rest. At the solid walls, the no-slip condition is used, except for the top boundary, $z = 2H$, where the rotating disk imposes an azimuthal velocity $v_\theta = \Delta\Omega$.

This code has been tested for different initial conditions. The resolution of the grid amounted in most 3D simulations $N_\theta \times N_r \times N_z = 257 \times 97 \times 257$ or $97 \times 97 \times 97$. Tests with 2 times higher resolution grids as well as with 2 times lower resolution grids showed essentially the same results but with less detail.

In DNS, the full equations are solved, but care must be taken with the evolution at small scales. To assure stability, the grid size must be of the same order of magnitude as the Kolmogorov scale, i.e. the scale at which viscosity dissipates energy, $L_K = \text{Re}^{-3/4} \times l$, with l the integral scale and Re the Reynolds number. If the grid scale is larger than L_K, energy may accumulate artificially at this length scale without being dissipated, leading to a numerical instability. In the laboratory experiments, the Reynolds number based on the velocity of the rotated disk is of $O(50,000)$. Since this Reynolds number would require too long calculation times, a smaller Reynolds has been used that is based on a larger viscosity and a grid size that is adapted to the small-scale flow features of interest, as discussed below.

To reproduce the dynamics of the front in coherence with the laboratory experiments (Sc = 700), the Schmidt number is necessarily large and requires a high resolution. For small Schmidt numbers, the density interface was found to diffuse too rapidly, i.e., before the front becomes unstable. For high Sc numbers, scalar diffusion is much weaker than viscous diffusion so that its typical length scale is also much smaller than the Kolmogorov scale.

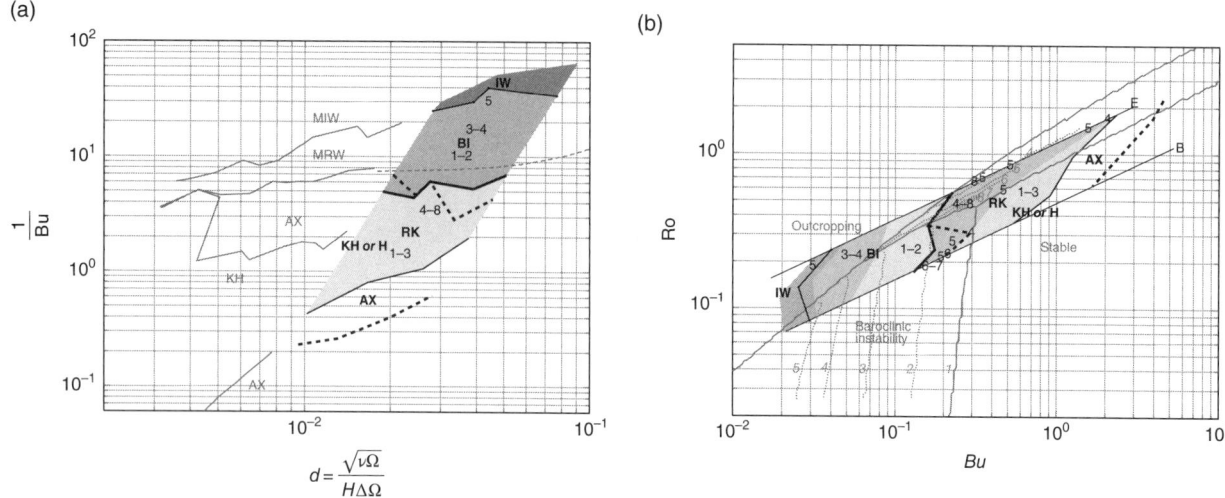

Figure 11.3. Regime diagrams of the different instabilities. (a) in (Bu, d)-space with the gray lines obtained for two-layer immiscible fluids [*Williams et al.*, 2005] separating axisymmetric flow (AX), mixed regular waves (MRWs), and mixed irregular waves (MIWs), and gray shaded areas representing regimes of Kelvin-Helmholtz or Hölmböe instability (KH or H), RK instability, and baroclinic instability (BI) (see Figure 11.2) found for approximately the same density difference g' in miscible fluids. The black dashed lines delimit the region of KH and Hölmböe instability, and the gray dashed line baroclinic instability according to quasi-geostrophic theory [*Hart*, 1972]; (b) in (Ro, Bu) space with the gray lines representing the theory of *Gula et al.* [2009a] for inviscid flows. Adapted from *Flór et al.* [2011].

The required resolution in this case should be estimated from the Batchelor scale for scalar diffusion, defined as $L_B = L_K \times \mathrm{Sc}^{-1/2}$ [*Batchelor*, 1959; *Buch and Dahm*, 1996; *Rahmani*, 2011]. But since singularities due to the relatively high Schmidt number and coarse grid in the present simulations occurred far from the relatively smooth interface, the interfacial dynamics were found to be well captured for a resolution rather estimated with L_K (grid step around 1–5 times L_K).

Given the numerical constraints and purpose of our study on rather small-scale phenomena, we thus found a compromise of a $N_\theta \times N_r \times N_z = 1 \times 700 \times 700$ grid with a Schmidt number of 700 and a Reynolds number of 3500 for the study of the secondary circulation. For the 3D simulation of the flow we choose a grid of $1201 \times 301 \times 257$ with a spatial resolution of 1–5 mm associated with Schmidt number of 100 and Reynolds number of 10,000.

Depending on the Reynolds number of the simulation and the associated value of the effective viscosity, the time corresponding to the spin-up took 2–40 dimensionless units.

11.3. FLOW EVOLUTION OF OBSERVED INSTABILITIES

Figure 11.3a summarizes the different instability regimes of the interface for a cyclonic forcing and constant density difference as a function of the inverse of the Burger number and dissipation number d. The evolution of the instability in parameter space changes from Kelvin-Helmholtz and RK instability to regular baroclinic instability and eventually the irregular change between different baroclinic modes known as the chaotic regime [*Flór et al.*, 2011]. These figures are obtained for increasing background rotation (i.e. for increasing inverse Burger number 1/Bu, and increasing d) and fixed disk rotation. With respect to results of *Williams et al.* [2005], *Flór et al.* [2011] showed RK and Hölmböe instability, both of which will be discussed in more detail in the sections below. When neglecting viscous effects, the full regime diagram reduces to a two-dimensional one as function of Ro and Bu, shown in Figure 11.3b.

Also amplitude vacillation has been observed [e.g., *Pedlosky*, 1987]. Typically, the oscillation occurs at the limit between RK and baroclinic instability. The flow oscillates in time between the axisymmetric state and the unstable baroclinic state. These periodic changes in wave amplitude are related to the transfer of energy between the wavefield and the zonal flow and have been observed before in two-layer immiscible fluid flows [*Hart*, 1976, 1979; *Read et al.*, 1992; *Früh and Read*, 1997] but for smaller Bu numbers, i.e, in the baroclinically unstable regime. In Figure 11.4a the very contrasted crests and troughs are separated by approximately 35 table rotation periods; the oscillation in mode 2 obtained from these data is represented in Figure 11.4b. Even though there is some noise and the amplitude of the mode gradually decreases,

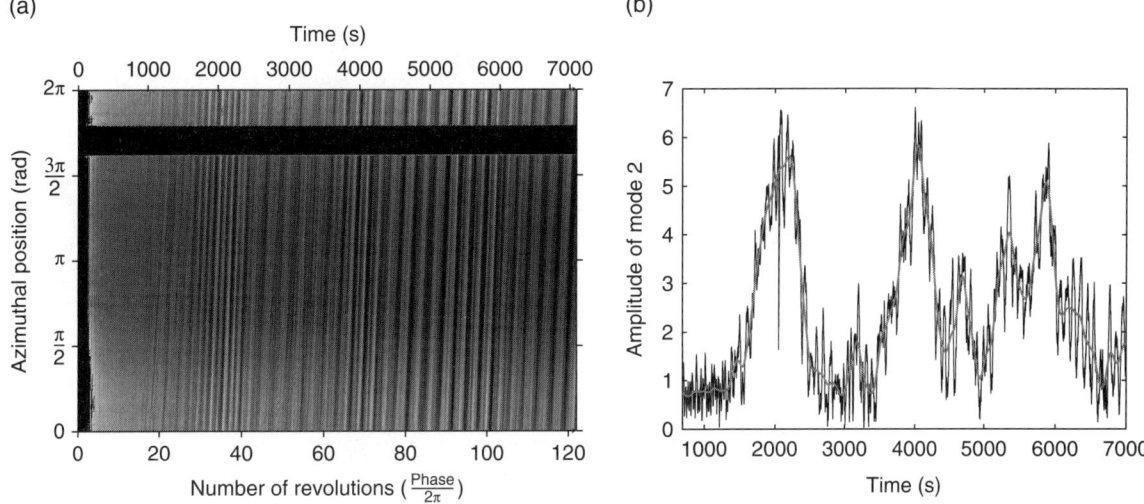

Figure 11.4. Amplitude vacillation of the baroclinic mode 2 shown in a Hovmöller diagram (a) with maxima at approximately 35, 70, and 100 rotation periods (x axis) and (b) the time evolution of the amplitude of the spatial Fourier mode 2. Experimental parameters: $d = 0.023$, $1/Bu = F = 3.27$.

this period can clearly be seen. This variation is most likely due to the gradual variation in stratification after more than 100 rotation periods. This confirms the occurrence of amplitude vacillation in two-layer miscible fluids with a smooth interface.

11.3.1. Rossby-Kelvin Instability

In order to further investigate the properties of the RK instability, a more specific simulation has been carried out for parameters in the area where both RK instability and Hölmböe instability are expected (region RK in Figure 11.3 with $Bu = 0.62$, $Ro = 0.67$). For this simulation, the grid resolution is chosen such that small-scale perturbations can be resolved. The Reynolds number and the Schmidt number are 10,000 and 100, respectively, compared to, respectively, 50,000 and 700 in the laboratory experiments (see Figure 11.5).

Figure 11.5 (right side) shows the top views of the interface height and the flow field represented by the stream function of each layer after having subtracted the mean velocity field. The structure in each layer appears clearly different. In the lower layer, linear stability theory predicts the flow to be separated into 12 segments (Figure 11.5b, left). After scrutinizing the image of Figure 11.5b (right), it is indeed possible to identify these 12 segments. Some of the separation lines do not extend to the inner cylinder but merge into a single line. This structure propagates along the outer boundary and can be identified as a Kelvin wave. In the upper layer, the structure of a Rossby wave is well reproduced with the presence of cyclones and anticyclones, and indeed 12 (or 13) cyclonic vortices can be counted. However, due to the interaction between modes, the structure appears not as regular as the linear prediction. A square lower mode structure can be observed on top of the Rossby mode that resonates with the Kelvin mode. Indeed, the temporal evolution of the mode amplitudes reveals that beyond the predominant modes 6 and 5, lower modes (2 and 3) are competing, thus confirming a nonlinear interaction between modes (see Figure 11.6). This may explain the observation of much higher growth rates than the growth rates predicted by the linear theory [*Gula et al.*, 2009a]. The general structure is nevertheless in good agreement with the RK instability (Figure 11.5 left side) (Gula, personal communication; see also *Gula et al.* [2009b]) and provides evidence of the resonance of a Rossby wave in the upper layer and a Kelvin wave propagating along the outer boundary in the lower layer, giving rise to the RK instability.

According to *Gula et al.* [2009a], the growth rate of the RK instability decreases when the density interface thickness increases. In almost all studies, interfaces of a negligible thickness are considered. Thick interfaces allow for critical layers with a much more complex dynamics. By contrast, the baroclinic instability does not seem to depend on the interface conditions and is not modified by the interface thickness. Nevertheless, Figures 11.3 and 11.6 show that the RK instability appears for higher Burger numbers than the threshold value for baroclinic instability. To further investigate the effect of the interface conditions on the growth rate of the RK instability, preliminary simulations were conducted for various Schmidt numbers and for parameters that are just below the (ageostrophic) threshold for baroclinic

220 MODELING ATMOSPHERIC AND OCEANIC FLOWS

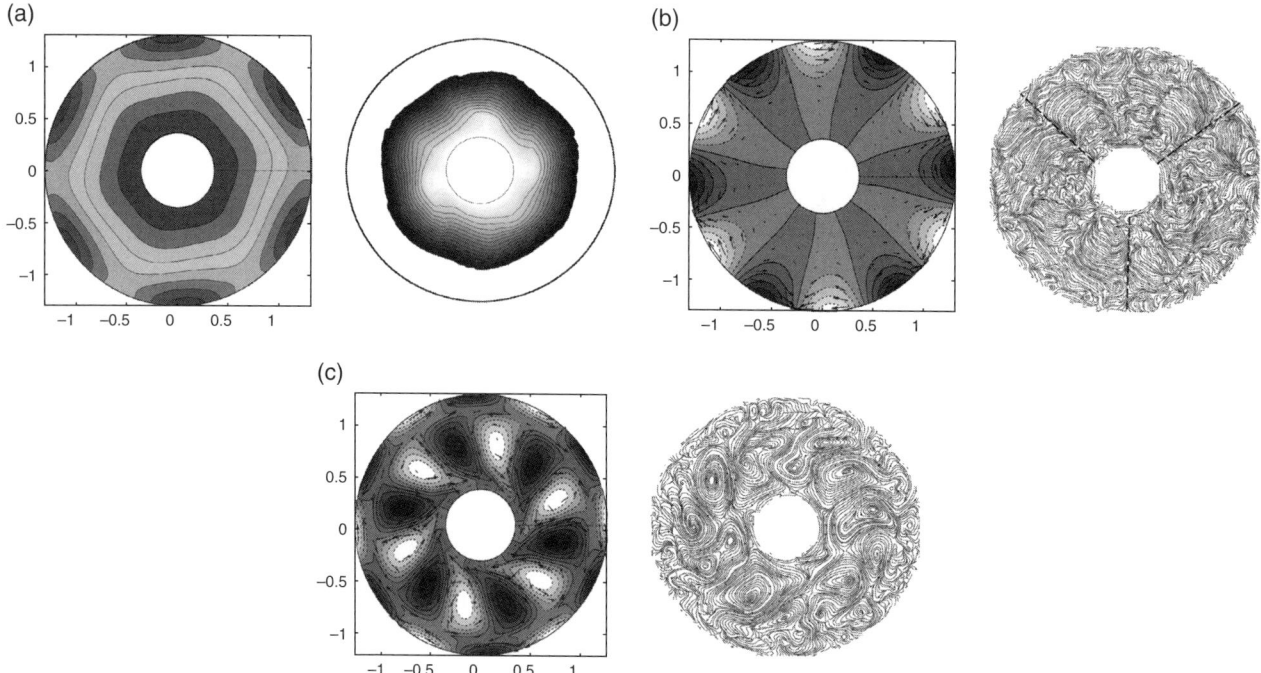

Figure 11.5. Comparison between a high-resolution 3D numerical simulation (right) and linear stability results (left) for the same parameters (Gula, personal communication): (a) interface height field; (b) perturbed velocity field in the lower layer showing a Kelvin wave; (c) perturbed velocity field in the upper layer showing a Rossby wave. The linear stability results also show the pressure field in shading. Numerical parameters: Bu = 0.62, Ro = 0.67, Ω = 0.077 rad/s, $\Delta\Omega$ = 0.104 rad/s, g' = 0.0598 m^2/s, $N_\theta \times N_r \times N_z = 1201 \times 301 \times 257$.

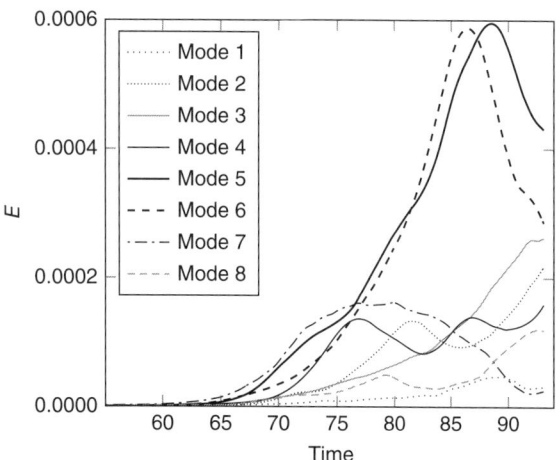

Figure 11.6. Evolution of the energy amplitude of the modes defined by $E(k) = \int_r \int_z \hat{u} \cdot \hat{u}^* \, dr \, dz$ with $\hat{u}(r, k, z)$ the Fourier transform of the velocity vector $u(r, \theta, z)$ in the azimuthal direction. Same simulation parameters as in Figure 11.5.

instability (see Figure 11.3). Different Schmidt numbers result in different interface thicknesses, especially for large times. The growth rate of the different amplitudes of the modes show a shift in growth rate with increasing Schmidt number. For low Schmidt numbers, the first growing RK modes are 5 and 6, whereas a baroclinic mode 2 started to grow at a later instant with a much higher growth rate. The type of instability was identified from the presence of a Rossby or a Kelvin wave in the top or bottom layer, respectively. With increasing Schmidt numbers, the mode number of both instabilities increased, and RK modes 7 and 8 were observed, followed by a faster growing baroclinic mode 4. These RK modes could well incline locally the interface by an increase in the local shear and thus decrease locally the Burger number. In this manner, the presence of RK modes could act as a finite amplitude perturbation and initiate the baroclinic instability for lower Bu numbers.

11.3.2. Interface Conditions and Ekman Circulation

To leading order, the flow generated by the rotating disk at the surface is horizontal. In the vertical, a second-order vertical Ekman circulation is generated by the fluid expelled at the boundary of the rotating disk. In the case of immiscible fluids, this circulation is associated with thin Ekman layers at the internal interface [e.g., *van Heijst*, 1984]. But when a tanh-like stratification profile is present, such as between two miscible fluids

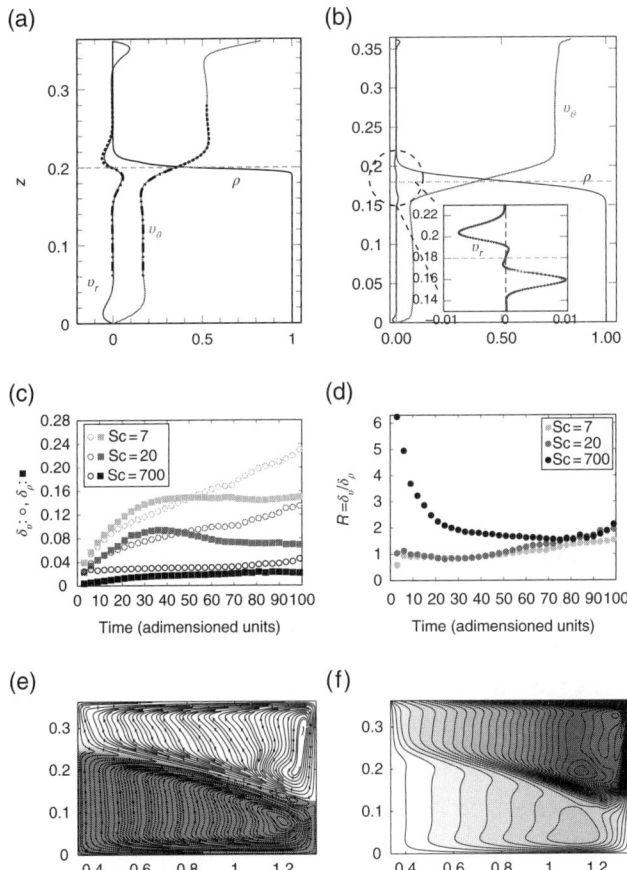

Figure 11.7. Properties of the Ekman layer with azimuthal and radial velocity and density profiles at mid radial position in the annulus after 18 rotation periods for (a) Ro = 0.4861 and Re = 3500 and (b) Ro = 0.0046 and Re = 547.6. The dot-dashed lines represent Ekman layers for a virtual boundary at the interface (the position of the interface is indicated in dashed grey). (c) Interface thicknesses δ_V (○) and δ_ρ (●) nondimensionalized by L for Schmidt numbers 7, 20, and 700 and (d) their ratio $R = \delta_V/\delta_\rho$. In (e) and (f) are shown, respectively, the streamlines and vortex lines along a r,z cross section. The contours of rv_θ are in the range [-1,1] from white to black and background shading for density. Numerical parameters for (a), (e), and (f): $N_\theta \times N_r \times N_z = 1 \times 700 \times 700$ and $g' = 0.06$, Bu = 0.3147, Sc = 700.

separated by a diffuse interface, the existence of such internal Ekman layers is not obvious. For the study of the instantaneous density and velocity profiles, we consider this problem numerically. Since the interface has no discontinuities and only axisymmetric stable flows are considered, the numerical modeling is adequate to capture the physics of the flow.

Figure 11.7a shows the profiles of azimuthal velocity, radial velocity, and density of a simulation with high Reynolds number and Sc = 700 for a resolution of $1 \times 700 \times 700$ supposing axisymmetric flow. Axisymmetric flows still allow for Ekman circulation. The Reynolds number is defined as $R_e = \Delta\Omega L^2/\nu$ and is approximately 3500. At $t = 0$ the interface is sharp in density and shear. After 18 rotation periods, there is a uniform azimuthal flow in the interior and an Ekman boundary near the interface, as can be deduced from the relative strength of the radial velocity and a variation in azimuthal velocity corresponding to an Ekman spiral. The dot-dashed lines in Figure 11.7a are the velocity profiles as obtained for a classical Ekman layer near a no-slip and zero vertical velocity boundary at the interface, such as for a virtual wall. The perfect agreement between the analytical solution and the numerical result confirms that indeed there are internal Ekman layers above and below the interface. The streamlines and vorticity lines in Figures 11.7e and 11.7f show the Ekman layers at each side of the inclined interface in space. In this simulation Ro is not very small (i.e. ≈ 0.48) allowing for some vertical motion and an inclination of the interface (see Figures 11.7e,f).

With increasing interface thickness, the radial velocity and therewith the Ekman pumping reduce in strength such that the flow tends to a unidirectional shear flow, for very thick interfaces. Especially for small Schmidt numbers, diffusion eventually weakens the density gradient so that the stratification and consequently the Burger number decrease. For such small Schmidt numbers the interface is a transient. It inclines more and more with its diffusion, until eventually the fluid is mixed and shear layers are represented by Taylor columns. The interaction of the front with top and bottom solid boundaries and the consequent transport of interfacial fluid via the horizontal and vertical boundary layers enhance the mixing. The evolution is then determined by the limited geometry of the annulus. To consider only the interface dynamics, the simulations are evolved for the time the density transport via the Ekman and Stewartson layers does not significantly affect the density in the bulk of the layers.

Figures 11.7c,d show the interface thickness in density and shear (Figure 11.7c) and the interface thickness ratio $R = \delta_V/\delta_\kappa$ as a function of time for different Schmidt numbers (Figure 11.7d). These thicknesses are obtained from fits with a tanh profile. For all Sc > 1, the initially very thin density interface increases more or less rapidly in thickness, thus reducing the value of R, and eventually reaches a value between 1 and 2. Due to entrainment, the density interface thickness stops growing while the shear thickness continuously increases. The interface thickness ratio R increases to values close to or higher than 2 if we follow the tendency of Figure 11.7d. This evolution shows that the diffusion of vorticity and density across an interface in a rotating fluid is not a linear diffusion problem but is governed rather by entrainment, leading to an eventual larger shear thickness than density thickness.

11.3.3. Small Interfacial Waves

Small-scale waves are observed in the different stages of the experiments reported in various papers on fronts in differentially rotating fluids [*Lovegrove et al.*, 2000; *Williams et al.*, 2005; *Flór et al.*, 2011] (see Figure 11.3). The nature of these waves has been a major question and has been debated in particular as a response to the paper of *Williams et al.* [2005]. The different candidates that have been considered are adjustment waves, spontaneous emitted inertia-gravity waves [*McIntyre*, 2009], waves due to Kelvin-Helmholtz or Hölmböe instability, and more recently also the instabilities of the boundary layer at the inner cylinder [*Jacoby et al.*, 2011] and the interfacial Ekman boundary layer [*Jacoby*, 2012]. Here we briefly review each of these wave types and subsequently focus on experiments in an annulus with a tanh density profile.

Adjustment waves occur when an initially imbalanced flow, during its transient process of adjustment toward a newly balanced geostrophic flow, radiates inertia-gravity waves [*Reznik et al.*, 2001]. Initially equilibrated flows may also radiate inertia-gravity waves, even though the flow is well described by the quasi-geostrophic equations. The emission of these ageostrophic waves are due to a "spontaneous imbalance," a process that may refer to different nonlinear mechanisms. First, these waves can indeed be emitted via a mechanism that is similar to the generation of noise in turbulent flow as described by Lighthill's theory for acoustics. Their radiation have been worked out for the shallow-water equations by Ford [1994]. Another mechanism of wave radiation that has been suggested [*Vanneste and Yavneh*, 2004] is due to advection: It induces a coupling between equilibrated motion and inertia-gravity waves by a Doppler shift effect.

When gravity waves have a smaller length scale than the balanced motion, the previous mechanisms cannot work since they describe waves with wavelengths of the same order as the large-scale balanced motion. The presence of waves in idealized simulations of baroclinic life cycles in continuously stratified flows [*O'Sullivan and Dunkerton*, 1995] is due not to the spontaneous emission of inertia-gravity waves mentioned above but to the nonlinear development of the baroclinic instability and the particular environment of the flow with strong horizontal deformation and large vertical shear [*Plougonven and Snyder*, 2005]. In particular, the vertical shear influences the propagation of wave packets. Long waves may either break or be absorbed. This wave capture phenomenon is relevant in shear flows and local wave sources such as, for instance, the radiation by a dipolar vortex in a continuously stratified fluid [*Snyder et al.*, 2007]. The amplitude of these waves is still small compared to the waves due to shear instabilities.

In order to know whether the observed waves are due to shear, we have to consider the Richardson number. Since the critical (local) Richardson number depends on the interface thickness in shear, δ_v, and density, δ_ρ, diffusive effects are expected to play an important role for the wavelength and growth of these instabilities. When the interface ratio $R = \delta_v/\delta_\rho$ exceeds a certain threshold, Hölmböe instability instead of Kelvin-Helmholtz instability is possible. Generally, $R > 2$ is needed for Hölmböe instability to occur. Diffusion largely affects the interface thicknesses, especially when the Reynolds number is low, and entrainment is not efficient enough to erode the diffused interface to a certain thickness. Stratified shear instabilities satisfy the general Miles criterion [*Miles*, 1961, 1963]. If we define the gradient Richardson number, $\mathrm{Ri}(z) = N^2/(dU/dz)^2$, where $N^2(z) = -(g/\rho_0)(\partial \rho/\partial z)$ is the Brunt-Väissälä frequency with a reference density ρ_0, this criterion states stability for $\mathrm{Ri}(z) \geq \tfrac{1}{4}$ everywhere in the profile and instability for $\mathrm{Ri}(z) < \tfrac{1}{4}$. Since gradients are difficult to measure, generally the bulk Richardson number J is used for experimental predictions. Following Hölmböe [*Hölmböe*, 1962; *Lawrence et al.*, 1991; *Ortiz et al.*, 2002; *Tedford et al.*, 2009; *Carpenter*, 2009], one can write the dimensionless numbers

$$J \equiv \frac{g\Delta\rho\delta_v}{\rho_0(\Delta U)^2}, \quad R \equiv \frac{\delta_v}{\delta_\rho}, \quad a \equiv \frac{2d}{\delta_v},$$

where $\rho_0 = (\rho_1+\rho_2)/2$, $\Delta\rho = \rho_2-\rho_1$, $\Delta U = U_2-U_1$, and a is a potential shift of the center of the velocity profile. For an inviscid Boussinesq fluid, the stability of a stratified shear flow is then given by the Taylor-Goldstein equation [e.g., *Hazel*, 1972; *Pouliquen et al.*, 1994; *Alexakis*, 2005, 2007, 2009].

For simplicity, it is generally assumed that $\delta_v \gg \delta_\rho$, allowing for an idealized model with piecewise constant velocity profile and a jump in density (as in *Ortiz et al.* [2002]) and illustrated in Figure 11.8a. In such a piecewise model, J keeps the same definition, whereas $a=0$ and

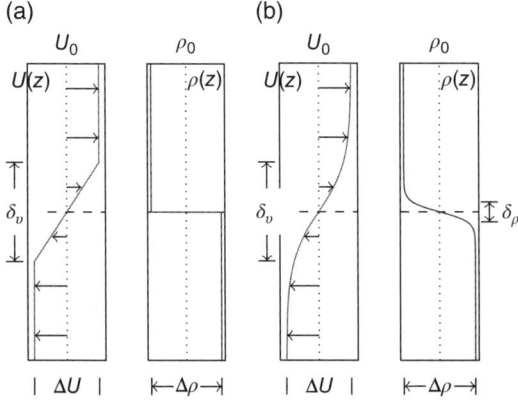

Figure 11.8. Density and velocity profiles in a stratified shear layer: (a) modelization with piecewise profiles; (b) hyperbolic tangent profiles.

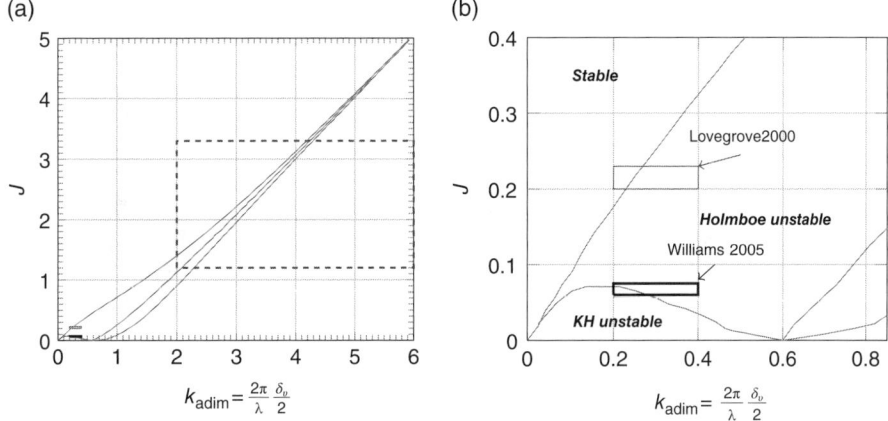

Figure 11.9. The $k - J$ instability diagram for piecewise profiles: (a) experimental areas for interfacial perturbations [*Flór et al.*, 2011] (dashed line) and (b) zoomed in on the regions corresponding to the experiments of *Lovegrove et al.* [2000] and *Williams et al.* [2005]. The dimensionless wave number $k = (2\pi/\lambda)(\delta_V/2)$ is based on half the shear thickness $\delta_V/2$.

δ_ρ is not relevant. For the small-scale perturbations in the experimental fronts under consideration, the wavenumber and the value of the bulk Richardson number are presented in the k-J diagram in Figure 11.9), where we use the dimensionless wave number $k = (2\pi/\lambda)(\delta_V/2)$ based on half the shear thickness $\delta_V/2$. Apparently, the experimental data fall in the unstable domain of the Hölmböe instability, especially in the case of experiments with immiscible layers [*Lovegrove et al.*, 2000; *Williams et al.*, 2005] for which the validity of the model is optimal.

However, for a two-layer stratified shear flow with miscible layers, a hyperbolic tangent profile with finite thicknesses in density and in velocity must be taken into account for an appropriate approach of a diffusing density and shear layer (see Figure 11.8b). The variation of the gradient Richardson number varies drastically with R [*Alexakis*, 2005]. As a consequence, the stability diagram depends also on the thickness ratio (see Figure 11.10). Moreover, the literature on Hölmböe instability has long considered the threshold limit of Hölmböe instability at the value $R = 2.2$ (or even 2.4) and the recent results from Alexakis show that the location limit is eventually $R = 2$. As illustrated in Figure 11.10 for cases of R just above $R = 2$, the unstable Hölmböe domain in the diagram for the global Richardson number $J_o = JR$ is very thin and becomes a line at $R = 2$. Since $1.84 \leq R \leq 2$, i.e. R is in a part of the regime being located under or equal to 2, it suggests that (in that part of the regime) these small waves are likely not Hölmböe waves. Indeed, the DNS simulations confirmed the presence of small-scale perturbations superimposed on the RK mode (Figure 11.12) that, because of the values of R and J, could not be due to stratified shear instability. Since the value of R increases above 2 for Time > 100 (see Figures 11.7d and also c,e,f), there will be also a temporal evolution of the stability.

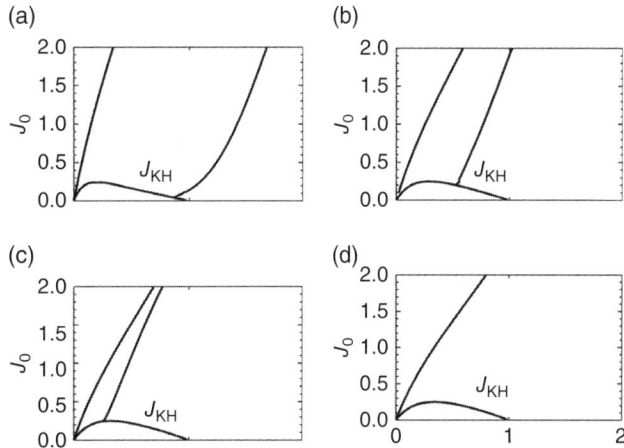

Figure 11.10. Variation of the $k - J_o$ diagram $(J_o = JR)$ as a function of the thickness ratio R (from *Alexakis* [2005]): (a) $R = 4$, (b) $R = 2.5$, (c) $R = 2.2$, (d) $R = 2$. In (d) the two stability boundaries $(J_1$ and $J_1 S)$ have collapsed together on a line and separate the nonsingular gravity wave modes on the left from singular neutral modes on the right.

For Sc = 700, initially the ratio $R > 2$ so that shear instabilities are likely to occur. Indeed, typical Hölmböe waves (being cusped shaped and with opposite phase direction below and above the interface) have been observed at the beginning of the experiments when the density interface was still relatively sharp (see Figure 11.11).

In the simulations, the ratio R is initially large and subsequently decreases to a critical value close to $R = 2$ (see Figure 11.12). Hölmböe instability is therefore possible but should eventually disappear. When the Schmidt number is less than 700, the asymptotic value tends to a value between 1 and 2, suggesting the possibility of

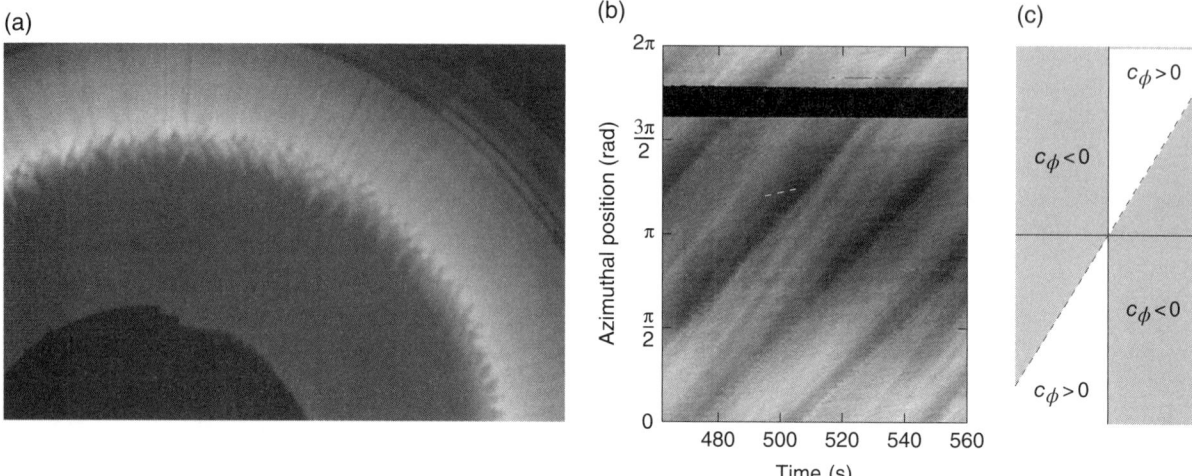

Figure 11.11. Observation of Hölmböe instability, with (a) cusped-shaped wave crests and (b) space-time diagram corresponding to the observation regime. The retrograde propagating Hölmböe waves are characterized by the oblique crests and troughs (indicated by the white dashed line). The diagram in (c) indicates the sign of the phase speed that corresponds to the inclination of the crests with respect to the mean flow (dashed line).

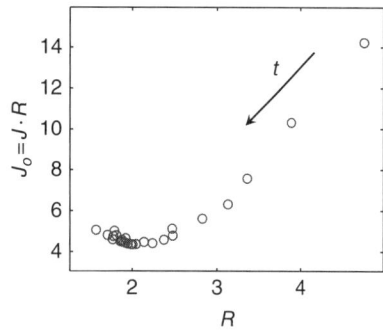

Figure 11.12. Numerical results for the evolution of the values of R and $J_O = JR$ as a function of time, with the initial time around $J_O \approx 14$ and $R > 4$. Note that R decreases below 2 and then increases again in time.

other waves than those due to Hölmböe instability (see Figure 11.7d). The DNS simulations indeed confirmed the presence of small-scale perturbations superimposed on the RK mode (Figure 11.13 and 11.14) that, because of the values of R and J, could not be due to stratified shear instability.

The origin of these waves is not understood. Next to the possibility of spontaneous emitted inertia-gravity waves mentioned above, a hypothesis is that the variations in interface slope as induced by the RK instability modes radiate inertia-gravity waves. The ambient presence of the large vertical shear and the presence of horizontal shear are in favor of the wave capture process mentioned above. Since the front is in an ageostrophic state, small-scale adjustment waves are not impossible. Further investigations are necessary to conjecture the type of wave radiation.

11.4. CONCLUSIONS AND DISCUSSION

We have focused our attention on experimental and numerical investigations of a baroclinic front in an annular tank. We summarize the results below and end with the consideration of some geophysical flows.

The numerical simulations confirm the presence of a lower layer Kelvin wave and upper layer Rossby wave for parameters in the domain where RK instability was seen experimentally. Preliminary simulations suggest that the change in interface thickness, such as due to different Schmidt numbers, modifies the appearance of instability modes. The growth rates of the higher RK modes increase, and the higher baroclinic modes appear for much lower Burger numbers. A possible mechanism is that the Bu number locally decreases due to the amplified inclination of the interface at positions where Kelvin waves and Rossby waves interact.

Quasi-geostrophic theory [*Hart*, 1972] predicts the baroclinic instability at a threshold Burger number of Bu < 0.12 (see, e.g., Figure 11.3). Calculations of the ageostrophic baroclinic instability [*Gula et al.*, 2009b] predict the first mode (mode 1) at Bu < 0.3 and mode 2 at Bu ≥ 0.2 being in excellent agreement with the experimental observations, noting that mode 2 is generally the first mode being observed.

Amplitude vacillation of the baroclinic instability seems a general feature in these flows. It appears also for the continuous stratifications here considered, but for larger Burger numbers than the threshold Bu-number value for baroclinic instability. The amplitude of the oscillation decreases in time most likely due to the mixing at the interface and corresponding change in Burger number.

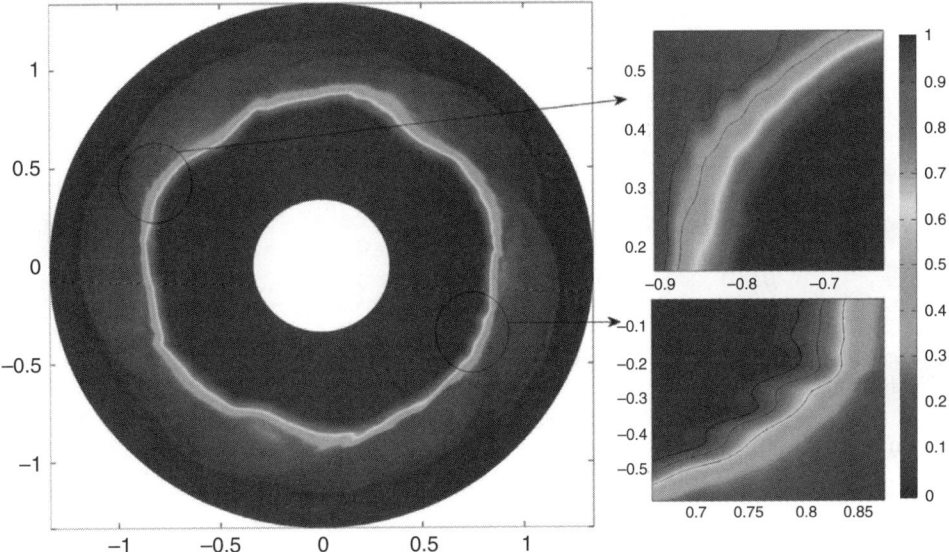

Figure 11.13. Density front for 3D simulation with Sc = 100 for $N_\theta \times N_r \times N_z = 1201 \times 301 \times 257$, Bu = 0.62, and Ro = 0.67. Right: zoom in on the encircled area showing the small-scale perturbations. For color detail, please see color plate section.

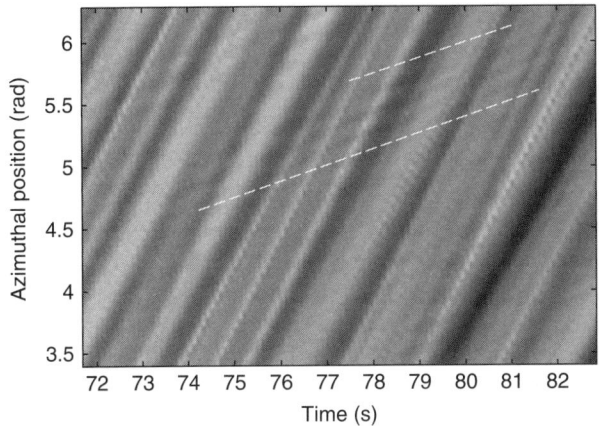

Figure 11.14. Hövmöller diagram, with small-scale perturbations located between the dashed white lines superimposed on the larger wave modes. The small-scale perturbations propagate much slower than the large wave mode. Numerical parameters are as in Figure 11.5.

Internal Ekman layers are considered in relation to interface conditions for different Schmidt numbers. In ageostrophic flows (i.e. for larger Ro numbers) there is an internal Ekman layer when the interface is sharp. But there is a transition to a shear layer (without Ekman pumping) for thicker interfaces and when the Ro number is really small, i.e., $O(0.005)$. In the latter case, the characteristic properties of an Ekman layer (Ekman pumping and vertical circulation) are lost. The effect of strong density diffusion (small Schmidt number) on the Ekman layer is that the Ekman layer disappears and turns into a unidirectional shear layer. For large Schmidt numbers the Ekman layer can be maintained until entrainment modifies the density and velocity distribution.

Small-scale waves have been observed in all experiments. In laboratory experiments, the precise interface conditions are generally not known, so that it is difficult to determine whether Hölmböe, Kevin-Helmhotlz, or other instabilities may occur. Their existence depends on the value of the global Richardson number and the value of the interface thickness ratio, R. The knowledge of the Ekman layer and density layer thicknesses in the work of *Williams et al.* [2005] allows for an accurate estimation of R, in contrast to the modeling with miscible fluids in continuously stratified fluids. Varying the Schmidt number modifies the eventual interface thickness. The numerical results show that, only at the start of the experiment when diffusion processes dominate because of the thin interface, the initial value of R may be smaller as well as larger than 2. Eventually, when entrainment processes dominate, R tends to a minimal value of 2 whereafter it increases again.

A linear diffusion of density and momentum would suggest diffusion rates and related layer thicknesses that are reflected directly by the Schmidt number value. The present results indicate that even for the low values of Sc values in the atmosphere (Sc = 1) and the ocean (Sc = 7), Hölmböe waves could occur, and the ratio R tends to a value of 2 and increases again afterward (see Figure 11.12) thus allowing for Hölmböe instability.

The RK instability in geophysical flows has so far not been considered. At difference with the baroclinic instability, the Kelvin waves that resonate with the Rossby wave for instability need a boundary along which

Table 11.2. Estimations of characteristics numbers of large scale oceanic and atmospheric cyclonic flows.

Geophysical Flows	N (rad/s)	$\gamma = H/(L/2)$	L (km)	Bu	Ro	ν_{eddy} (m^2/s)	d
Antarctic Polar Vortex	0.01–0.001	0.0023	6000	0.025–0.9	0.5	0.1–10	0.022–0.22
Cold Core Vortex rings	—	0.014–0.025	40–70	0.05–0.5	0.1–0.35	0.01	0.03–0.05
Gulf Stream	0.007	0.008–0.012	100	0.28	0.45	0.01	0.05
Antarctic Circumpolar Current	0.002	0.1–0.08	35–50	0.64–1.3	0.1–0.2	0.01	0.02
Agulhas Current	0.003	0.02	93	0.19	0.02–0.2	0.01	0.005–0.02

Note: For the ocean ν_{eddy} the vertical eddy viscosity is dominated by Ekman layers, i.e. $\nu = 0.01$ m^2/s [*Cushman-Roisin*, 1994]. For the case of circular currents, the radius or otherwise the half-width current, i.e. L/2 is taken for Ro and Bu number. The estimations for cold-core vortex rings are from *Olson* [1991] with for the maximum velocity a mean value of 1 m/s. Antarctic Circumpolar Current values are according to *Gille* [1994], who measured current widths of 35–50 km driven by surface winds, for which a wavelength of 150 km was found. The Rossby number is calculated from $U/(fL)$, taking $U = 20 - 40$ cm/s, and for d mean values are used. The values for the Agulhas current are estimated from *Webb* [1999] and *Beal et al.* [2006].

they propagate. Therefore most evident applications are oceanic, e.g. coastal fronts. To our knowledge, the possible existence of the RK instability in geophysical flows has not been considered yet. Below we further discuss some geophysical flows and their position in the Bu-d diagram presented in Figure 11.3.

To compare present results on frontal stability with fronts in nature, estimations of the nondimensional parameter values of some geophysical flows are presented in Table 11.2. The eddy viscosity is related to the friction in the Ekman boundary layer. Further it should be noted that the comparison with oceanic examples concern mainly outcropping flows (since they are visible whereas internal interfaces are not so often observed), in contrast to the present experimental results for internal interfaces. Supposing that the friction at the free surface is not so large to have an important effect on the instability, we compare these flows with the results presented in the diagram of Figure 11.3. The modeled parameter regime set by Ro, Bu numbers and the dissipation number d shown in Figures 11.3a and 11.3b almost covers entirely the parameter space of the real flows represented in Table 11.2. Further, the comparison in Figure 11.3 shows that we may well expect the occurrence of the RK instability in oceanic currents. Wavelengths are about 1 to a 1/3 of the current width, which corresponds to the size of the lenses observed in the ocean near the Gulf Stream, suggesting that the RK instability is a possible candidate that deserves further attention.

Acknowledgments. The authors acknowledge Jonathan Gula for providing the figures for linear stability in Figure 11.5, the contract CIBLE of the French region Rhone Alpes for financing this project, and the calculation centers IDRIS and CINES for the simulation time on their machines. Further, the authors acknowledge two anonymous referees for their corrections to the manuscript.

REFERENCES

Alexakis, A. (2005), On Holmboe's instability for smooth shear and density profiles, *Phys. Fluids*, 17, 084,103.

Alexakis, A. (2007), Marginally unstable Holmboe modes, *Phys. Fluids*, 19, 054,105.

Alexakis, A. (2009), Stratified shear flow instabilities at large Richardson numbers. *Phys. Fluids*, 21, 054,108.

Appleby, J. (1982), Comparative theoretical and experiment studies of baroclinic waves in a two-layer system, Ph.D. thesis, Leeds University, United Kingdom.

Batchelor, G. K (1959), Small-scale variation of convected quantities like temperature in turbulent fluid. Part 1. General discussion and the case of small conductivity, *J. Fluid Mech.*, 5, 113–133.

Beal, L.M., T.K Chereskin, Y.D Lenn, and S. Elipot (2006). The sources and mixing characteristics of the Agulhas current, *J. Phys. Oceanogr.*, 36, 2060–2074.

Bouruet-Aubertot, P., and P. F Linden (2002), The influence of the coast on the dynamics of upwelling fronts: Part i. Laboratory Experiments, *Dyn. Atmos. Oceans*, 36(1–3), 153–173.

Bradford, J., A.S. Berman and T.S. Lundgren (1981), Non-geostrophic baroclinic instability in a two-fluid layer rotating system, *J. Atmos. Sci.*, 38, 1376–1389.

Buch, Jr., K. A., and W. J. A. Dahm (1996), Experimental study of the fine-scale structure of conserved scalar mixing in turbulent shear flows. Part 1. Sc >> 1, *J. Fluid Mech.*, 317, 21–71.

Carpenter, J. (2009), Unstable waves on a sheared density interface, Ph.D. thesis, University of British Columbia, Vancouver.

Carrigan, C.R. (1978), Instability of a two-layer baroclinic flow in a channel, *Tellus*, 30, 468–471.

Cenedese, C., and P. F. Linden (2002), Stability of a buoyancy-driven coastal current at the shelf break, *J. Fluid Mech.*, 452, 97–121.

Chia, F., R. W. Griffiths, and P. F. Linden (1982), Laboratory experiments on fronts Part II: The formation of cyclonic eddies at upwelling fronts. *Geophys. Astrophys. Fluid Dyn.*, 19, 189–206.

Cushman-Roisin, B. (1986), Frontal geostrophic dynamics, *J. Phys. Oceanogra.*, *16*, 132–143.

Cushman-Roisin, B. (1994), *Introduction to Geophysical Fluid Dynamics*, Prentice-Hall, Englewood cliffs, NJ.

Eccles, F.J.R., P.L. Read, A.A. Castrejón-Pita, and T.W.N. Haine (2009), Synchronization of modulated traveling baroclinic waves in a periodically forced, rotating fluid annulus, *Phys. Rev. E*, *79*, 1:015202.

Faller, A. J. (1963), An experimental study of the instability of the laminar Ekman boundary layer, *J. Fluid Mech.*, *15*, 560–576.

Flór, J., H. Scolan, and J. Gula (2011). Frontal instabilities and waves in a differentially rotating fluid, *J. Fluid Mech.*, *685*, 532–542.

Ford, R. (1994), Gravity wave radiation from vortex trains in rotating shallow water, *J. Fluid Mech.*, *281*, 81–118.

Fowlis W. W., and R. Hide (1965), Thermal convection in a rotating annulus of liquid: Effect of viscosity on the transition between axisymmetric and non-axisymmetric flow regimes, *J. Atmos. Sci.*, *22*, 541–558.

Fritts, D.C. and M. J. Alexander (2003), Gravity wave dynamics and effects in the middle atmosphere, *Rev. Geophys. 41*(1), 1003.

Früh, W.G., and P.L. Read (1997), Wave interactions and the transition to chaos of baroclinic waves in a thermally driven rotating annulus, *R. Soc. Lond. Philos. Trans. Ser. A*, *355*, 101–153.

Fultz D., R.R. Long, G.V. Owens, W. Bohan, R. Kaylor and J. Weil (1959), Studies of thermal convection in a rotating cylinder with some implications for large-scale atmospheric motions, *Meteorol Monogr. 4*(21), 1–104.

Gille. S. T. (1994), Mean sea surface height of the Antarctic Circumpolar Current from GEOSAT data: Methods and application, *J. Geophys. Res.*, *99*, 18,255–18,273.

Griffiths, R., and P. Linden (1981), The stability of buoyancy-driven coastal currents. *Dyn. Atmos. Oceans*, *5*(4), 281–306.

Gula, J., and V. Zeitlin (2010), Instabilities of buoyancy-driven coastal currents and their nonlinear evolution in the two-layer rotating shallow-water model. Part 1. Passive lower layer, *J. Fluid Mech.*, *659*, 69–93.

Gula, J., R. Plougonven, and V. Zeitlin (2009a), Ageostrophic instabilities of fronts in a channel in a stratified rotating fluid, *J. Fluid Mech.*, *627*, 485.

Gula, J., V. Zeitlin, and R. Plougonven (2009b), Instabilities of two-layer shallow-water flows with vertical shear in the rotating annulus, *J. Fluid Mech.*, *638*, 27.

Gula, J., V. Zeitlin, and F. Bouchut (2010), Instabilities of buoyancy-driven coastal currents and their nonlinear evolution in the two-layer rotating shallow water model. Part 2. Active lower layer, *J. Fluid Mech.*, *665*, 209–237.

Hart, J.E. (1972), A laboratory study of baroclinic instability. *Geophys. Astrophys. Fluid Dyn.*, *3*, 181–209.

Hart, J.E. (1973), On the behavior of large-amplitude baroclinic waves, *J. Atmos. Sci.*, *30*, 1017–1034.

Hart J.E. (1976), The modulation of an unstable baroclinic wave field, *J. Atmos. Sci.*, *33*, 1874–1889.

Hart, J. (1979), Finite amplitude baroclinic instability, *Ann. Rev. Fluid Mech.*, *11*, 147–172.

Hart, J. (1980), An experimental study of nonlinear baroclinic instability and mode selection in a large basin, *Dyn. Atmos. Oceans*, *4*, 115–135.

Hart, J.E. (1981), Wavenumber selection in nonlinear baroclinic instability, *J. Atmos. Sci.*, *38*, 400–408.

Hart, J.E. (1985), A laboratory study of baroclinic chaos on the f-plane, *Tellus Ser. A*, *37*, 286.

Hart, J.E., and S. Kittelman (1986), A method for measuring interfacial wave fields in the laboratory, *Geophys. Astrophys. Fluid Dyn.*, *36*(2), 179–185.

Hayashi, Y., and W.R. Young (1987), Stable and unstable shear modes on rotating parallel flows in shallow water, *J. Fluid Mech.*, *184*, 477–504.

Hazel, P. (1972), Numerical studies of the stability of inviscid stratified shear flows, *J. Fluid Mech.*, *51*, 39–61.

Hide, R. (1953), Some experiments on thermal convection in a rotating liquid, *Q. J. R. Meteor. Soc.*, *79*(339), 161–161.

Hide, R. (1958), An experimental study of thermal convection in a rotating liquid, *R. Soc. Lond. Philos. Trans. Ser. A*, *250*, 441–478.

Hide, R. (2011) Regimes of sloping thermal convection in a rotating liquid "annulus," *Geophys. Astrophys. Fluid Dyn.*, *105*, 11–127.

Hölmböe, J. (1962), On the behavior of symmetric waves in stratified shear layers. *Geofys. Publ.*, *24*, 67–112.

Howard, L.N. (1961), Note on a paper of John W. Miles, *J. Fluid Mech.*, *10*, 509–512.

Jacoby, T.N.L. (2012), Inertia-gravity wave generation by boundary layer instabilities, Ph.D. thesis, University of Oxford, Oxford.

Jacoby, T. N. L., P. L. Read, P. D. Williams, and R. M. B. Young, (2011), Generation of inertia?gravity waves in the rotating thermal annulus by a localised boundary layer instability, *Geoph. Astr. Fluid Dyn.*, *105*(2–3): 161–181.

King, J.C. (1979), An experimental study of baroclinic wave interactions in a two-layer system, *Geophys. Astrophys. Fluid Dyn.*, *13*, 153–167.

Lawrence, G.A., F.K. Browand, and L.G. Redekopp (1991), The stability of a sheared density interface, *Phy. Fluids*, *3*, 2360–2370.

Lovegrove, A.F., P.L. Read, and C.J. Richards (1999), Generation of inertia-gravity waves by a time-dependent baroclinic wave in the laboratory, *Phys. Chem. Earth B*, *24*, 455–460.

Lovegrove, A.F., P.L. Read, and C.J. Richards (2000), Generation of inertia-gravity waves in a baroclinically unstable fluid, *Q. J. R. Met. Soc.*, *126*, 3233–3254.

McIntyre, M.,E (2009), Spontaneous imbalance and hybrid vortex–gravity structures, *J. Atmos. Sci.*, *66*, 1315–1325.

Miles, J.W. (1961), On the stability of heterogeneous shear flows, *J. Fluid Mech.*, *10*, 496–508.

Miles, J.W. (1963), On the stability of heterogeneous shear flows. Part 2. *J. Fluid Mech.*, *16*, 209–227.

Olson, D.B. (1991), Rings in the ocean, *Ann. Rev. Earth Planet. Sci.*, *19*, 283–311.

Ortiz, S., J. Chomaz, and T. Loiseleux, (2002), Spatial holmboe instability, *Phys. Fluids*, *14*, 2585–2597.

O'sullivan, D., and T.J. Dunkerton (1995), Generation of inertia–gravity waves in a simulated life cycle of baroclinic instability, *J. Atmos. Sci.*, *52*, 3695–3716.

Pedlosky, J. (1964), The stability of currents in the atmosphere and the ocean: Part I, *J. Atmos. Sci.*, *21*, 201–219.

Pedlosky, J. (1970), Finite-amplitude baroclinic waves, *J. Atmos. Sci.*, *27*, 15–30.

Pedlosky, J. (1987), *Geophysical Fluid Dynamics*, 2nd ed., Springer-Verlag, New York, 710 pp.

Phillips, N.A. (1954), Energy transformations and meridional circulations associated with simple baroclinic waves in a two-level, quasi-geostrophic model. *Tellus*, *6*, 273–286.

Plougonven, R. and C. Snyder (2005), Gravity waves excited by jet: Propagation versus generation, *Geophys. Res. Lett.*, *32*, L18802.

Plougonven, R., D.J. Muraki and C. Snyder (2005) A Baroclinic instability that couples balanced motions and gravity waves, *J. Atmos. Sci.*, *62*, 1545–1559.

Pouliquen, O., J.M. Chomaz, and P. Huerre (1994), Propagating Holmboe waves at the interface between two immiscible fluids, *J. Fluid Mech.*, *266*, 277–302.

Rahmani, M. (2011), Kelvin-Helmholtz instabilities in sheared density stratified flows, Ph.D. thesis, University of British Columbia, Vancouver.

Read, P.L., M.J., Bell, D.W. Johnson, and R.M., Small (1992), Quasi-periodic and chaotic flow regimes in a thermally driven, rotating fluid annulus, *J. Fluid Mech.*, *238*, 599–632.

Reznik, G.M., V. Zeitlin, and M. Ben Jelloul (2001), Nonlinear theory of geostrophic adjustment. Part 1. Rotating shallow-water model, *J. Fluid Mech.*, *445*, 93–120.

Sakai, S. (1989), Rossby-Kelvin instability: A new type of ageostrophic instability caused by a resonance between Rossby waves and gravity waves, *J. Fluid Mech.*, *202*, 149–176.

Scolan, H. (2011), Dynamique et stabilité de fronts: phénomènes agéostrophiques, Ph.D. thesis, Université de Grenoble, France.

Scolan, H., R. Verzicco, and J.B. Flór (2011), On density interfaces in rotating fluids, paper presented at the VII International Symposium on Stratified Flows, Rome, Italy.

Snyder, C., D.J. Muraki, R. Plougonven, and F. Zhang (2007), Inertia gravity waves generated within a dipole vortex, *J. Atmos. Sci.*, *64*, 4417.

Swarztrauber, P.N. (1974), A direct method for the discrete solution of separable elliptic equations, *SIAM J. Numer. Analy.*, *11*, 1136–1150.

Tatro, P. R., and E. L. Mollo-Christensen (1967), Experiments on Ekman layer instability, *J. Fluid Mech.*, *28*, 531–543.

Tedford, E.W., J.R. Carpenter, R. Pawlowicz, R. Pieters, and G.A. Lawrence (2009), Observation and analysis of shear instability in the Fraser River estuary, *J. Geophys. Res. (Oceans)*, *114*, C13, C11006.

van Heijst, G.J.F. (1984), Frontal upwelling in a rotating two-layer fluid, *Geophys. Astrophys. Fluid Dyn.*, *29*, 139–153.

Vanneste, J., and I. Yavneh (2004), Exponentially small inertia gravity waves and the breakdown of quasigeostrophic balance. *J. Atmos. Sci.*, *61*, 211–223.

Verzicco, R., and R. Camussi (1997), Transitional regimes of low-Prandtl thermal convection in a cylindrical cell. *Phys. Fluids*, *9*, 1287–1295.

Verzicco, R. and P. Orlandi (1996), A finite-difference scheme for three-dimensional incompressible flows in cylindrical coordinates, *J. Computat. Phys.*, *123*(2), 402–414.

Verzicco, R., F. Lalli, and E. Campana (1997), Dynamics of baroclinic vortices in a rotating, stratified fluid: A numerical study, *Phys. Fluids*, *9*, 419–432.

Webb, D.J. (1999), An analytic model of the Agulhas Current as a Western Boundary current with linearly varying viscosity *J. Phys. Oceanogr.*, *29*, 1517–1527.

Williams, P.D., P.L. Read, and T.W.N. Haine (2003),. Spontaneous generation and impact of inertia-gravity waves in a stratified, two-layer shear flow, *Geophys. Res. Lett.*, *30*, 24, 240000–1.

Williams, P.D., T.W.N. Haine, and P.L. Read (2004a). Stochastic resonance in a nonlinear model of a rotating, stratified shear flow, with a simple stochastic inertia-gravity wave parameterization. *Nonlin. Process. Geophys.*, *11*, 127–135.

Williams, P.D., P.L. Read, and T.W.N. Haine (2004b). A calibrated, non-invasive method for measuring the internal interface height field at high resolution in the rotating, two-layer annulus, *Geophys. Astrophys. Fluid Dyn.*, *98*, 453–471.

Williams, P.D., T.W.N. Haine, and P.L. Read (2005), On the generation mechanisms of short-scale unbalanced modes in rotating two-layer flows with vertical shear, *J. Fluid Mech.*, *528*:1–22.

Williams, P.D., T.W.N. Haine, and P.L. Read (2008), Inertia-gravity waves emitted from balanced flow: Observations, properties, and consequences, *J. Atmos. Sci.*, *65*, 3543.

Williams, P.D., P L. Read, and T.W.N. Haine (2010), Testing the limits of quasi-geostrophic theory: Application to observed laboratory flows outside the quasi-geostrophic regime, *J. Fluid Mech.*, *649*, 187–203.

Section IV: Oceanic Flows

12

Large-Amplitude Coastal Shelf Waves

Andrew L. Stewart[1], Paul J. Dellar[2], and Edward R. Johnson[3]

12.1. SHELF WAVES AND COASTAL CURRENTS IN THE LABORATORY

Coastal currents flowing along continental shelves are a complex dynamical feature of the global ocean. Such currents separate the coastal waters from the open ocean and so control the transport of both dynamical and passive tracers across and along the continental slope [*Nittrouer and Wright*, 1994]. Some of the most climatically important currents in the world ocean flow partly or entirely as shelf currents, so understanding their behavior represents an important oceanographic and dynamical problem. For example, the Antarctic Slope Front (ASF) [*Thompson and Heywood*, 2008] mediates transport of continental deep water onto the Antarctic continental shelf. This is responsible for preserving biological primary production around the Antarctic margins [*Prézelin et al.*, 2004] and controls the melting rate of ice shelves [*Martinson et al.*, 2008]. The Agulhas current, which flows over the continental shelf in the Mozambique Channel [*Bryden et al.*, 2005; *Beal et al.*, 2006, 2011], facilitates mass and heat exchange between the Indian and Atlantic oceans via thermocline water transported in Agulhas eddies [*Gordon*, 1985, 1986]. The Gulf Stream flows along the coastal shelf of North America until it separates at Cape Hatteras [*Stommel*, 1972; *Johns and Watts*, 1986; *Pickart*, 1995]. This current is associated with large meridional heat transport and closes the upper branch of the Atlantic Meridional Overturning Circulation [*Minobe et al.*, 2008].

The sharp increase in depth at the continental shelf break exerts a strong constraint on the flow via conservation of potential vorticity (PV) [see *Pedlosky*, 1987; *Vallis*, 2006]. This motivates a description of the flow in terms of topographic Rossby waves. *Longuet-Higgins* [1968] first showed that a continental shelf acts as a waveguide, approximating the bathymetry as a discontinuity in depth. Shelf wave theory was subsequently adapted to a wide range of bathymetric configurations and applied to regions of the real ocean [*Mysak*, 1980a, 1980b]. For example, *Mysak et al.* [1979] applied linear wave theory applied to double-exponential approximations of coastal depth profiles in the Pacific Ocean, while *Gill and Schumann* [1979] attempted to predict the path of Agulhas using an idealized representation of continental slope and deep ocean in a two-layer model. Linear shelf wave theory has since been extended to describe more realistic configurations such as continental shelves with along-shore depth variations [*Johnson*, 1985; *Johnson and Davey*, 1990], curved coastlines [*Kaoullas and Johnson*, 2010; *Johnson et al.*, 2012], and arbitrary isobath variations [*Rodney and Johnson*, 2012; *Kaoullas and Johnson*, 2012].

Coastal currents complicate this Rossby wave description of shelf dynamics, particularly when they flow over along-shore variations in the bathymetry and when they are driven by a time-dependent inflow or wind stress [*Allen*, 1980; *Brink*, 1991]. The development of numerical ocean models, particularly those adapted to large variations in bathymetry [e.g., *Haidvogel et al.*, 2008], has advanced our understanding of such situations considerably in recent decades. Yet laboratory studies of coastal dynamics have tended to focus either on topographic Rossby wave propagation or on coastal current evolution, so in this section we review separate selections of previous studies along each line of investigation. Both provide context for the experiments that serve as the focus of this

---

[1]Department of Atmospheric and Oceanic Sciences, University of California, Los Angeles, Los Angeles, California, United States of America.
[2]Oxford Centre for Industrial and Applied Mathematics, University of Oxford, Oxford, United Kingdom.
[3]Department of Mathematics, University College London, London, United Kingdom.

Modeling Atmospheric and Oceanic Flows: Insights from Laboratory Experiments and Numerical Simulations, First Edition. Edited by Thomas von Larcher and Paul D. Williams.
© 2015 American Geophysical Union. Published 2015 by John Wiley & Sons, Inc.

chapter, which describe the evolution of a coastal current in terms of large-amplitude shelf waves.

12.1.1. Shelf Waves

Ibbetson and Phillips [1967] performed some of the earliest laboratory experiments relevant to Rossby shelf wave dynamics. They constructed a rotating annulus between radii of 72.4 and 102.0 cm, similar to that described in Section 12.2. A background PV gradient was provided simply by the curvature of the free surface, required to balance the centrifugal force. Rossby waves were generated by the oscillatory rotation of a vertical paddle positioned across the breadth of the channel, with periods between 20 and 100 s. In an open annulus the damping rate of the resulting Rossby waves was found to be of the same order of magnitude as the theoretical prediction (see Section 12.3.1). When a 60° arc of the annulus adjacent to the paddle was enclosed, the waves intensified at the far boundary, equivalent to western boundary intensification in the real ocean.

Caldwell et al. [1972] performed the first experiments that directly represented the continental slope and deep ocean. They constructed an annular channel in which the fluid depth increased exponentially with radius over the inner half of its width and remained constant over the outer half, following the theoretical approximation of *Buchwald and Adams* [1968]. Rossby shelf waves were generated via one or two radially oscillating paddles and visualized via aluminium tracer particles and streak photography. The measured dispersion relation for the waves was found to agree closely with the linear theory of *Buchwald and Adams* [1968]. *Caldwell and Eide* [1976] used a similar setup to study Rossby waves over a shelf whose depth increased linearly with radius and Kelvin waves in a channel of uniform depth. Both were found to agree well with theoretical predictions once the assumption of a nondivergent horizontal velocity field was relaxed [*Buchwald and Melville*, 1977].

Colin de Verdiere [1979] examined Rossby wave-driven mean flows in a rotating cylinder of radius 31 cm. The curvature of the fluid's free surface supplied a background vorticity gradient, and an array of sources and sinks was used to excite a traveling wave with azimuthal wave number 12. The Rossby waves were found to generate a strong mean flow along planetary vorticity contours, equivalent to isobaths on a continental shelf. *Sommeria et al.* [1991] performed similar experiments in an annular channel between radii of 21.6 and 86.4 cm, with a depth that increased linearly with radius. Forcing was again supplied by an inner ring of sources and an outer ring of sinks, or vice versa, in the floor of the channel. The resulting net radial transport drove an azimuthal jet via the Coriolis force, whose direction was dictated by the relative radii of the sources and sinks. The jet supported unstable Rossby waves with azimuthal wave numbers ranging from 3 to 8.

Pierini et al. [2002] analyzed Rossby normal modes over a 2 m wide experimental slope connecting water depths of 30 and 60 cm. The slope was confined by walls to a finite length of 4.3 or 3.3 m, and the tank was rotated with periods ranging from 30 to 50 s. A paddle was used to drive deep fluid gradually and barotropically onto the slope, where it excited Rossby normal modes between the walls confining the slope. The phases of these modes agreed well with numerical solutions of a barotropic shallow-water model, but their amplitudes exhibited some substantial discrepancies.

Most recently, *Cohen et al.* [2010] generated topographic Rossby waves in an experimental slope configuration similar to that of *Caldwell et al.* [1972]. They constructed an annular channel between radii of 1 and 6.5 m, with a fluid whose depth was uniform in the innermost 1.5 m and decreased linearly by 0.4 m over the outermost 4 m. Shelf waves were excited by a radially oscillating paddle of length 2 or 0.4 m. The resulting wave velocities were measured using particle imaging velocimetry. The dispersion relation and radial velocity of the experimentally generated waves was found to agree closely with a theory derived from the linearized barotropic shallow-water equations. *Cohen et al.* [2012] performed similar experiments in an annular channel between radii of 0.25 and 1 m, with linear decrease in the depth from the center to zero at the outer edge. Topographic waves were generated via three radially aligned paddles and compared with linearized shallow-water theory. Approximating the continental boundary as a vertical wall, rather than a vanishing ocean depth, was found to distort the frequencies in the wave dispersion relation, resulting in disagreement with the experimental results.

12.1.2. Coastal Currents

Even the simplest representation of a coastal current flowing along a continental slope requires a more sophisticated treatment in the laboratory than topographic Rossby waves, and a much smaller body of analytical theory exists to predict their behavior. As a result, laboratory studies of coastal currents followed two decades after the first experiments with topographic Rossby waves. *Whitehead and Chapman* [1986] performed the first such study, generating a coastal current by attaching a rectilinear slope to the outside of a cylindrical tank. A buoyant gravity current was released at the outer edge of the tank, then propagated around the edge of the tank and then along the slope. The gravity current was found to widen and reduce its speed by a factor of up to 4 upon reaching the slope. If the current's speed fell below that of the first barotropic linear topographic wave, then it would radiate energy forward along the slope as a shelf wave.

Cenedese and Linden [2002] advanced this approach by constructing an annular channel between radii of 13 and 43 cm, with four axisymmetric topographic configurations: a flat bottom, a step, a raised shelf with a slope, and a raised step with a slope. Buoyant fluid was injected via an axisymmetric source at the inner wall of the annulus, forming a symmetric coastal current that was eventually subject to baroclinic instability. Introducing a shelf produced instabilities with slightly higher azimuthal wave numbers and resulted in secondary instability after axisymmetry was reestablished.

Cenedese et al. [2005] constructed a coastal geometry qualitatively similar to that discussed in Section 12.2, but in a rectilinear channel. An experimental slope of width 4 cm and height 6 cm separated a shelf from the deep ocean, each of width 40 cm. The channel was 84 cm long and bounded at each end by vertical walls, one of which was used to steer a baroclinic jet toward the slope. The experimental results agreed qualitatively with the quasi-geostrophic (QG) theory of *Carnevale et al.* [1999]: The jet either split between the tank edge and the slope or retroflected away from the slope.

Folkard and Davies [2001] were the first to investigate along-slope variations in the experimental topography, a crucial element of the experiments discussed in Section 12.2. They employed a rectangular channel with a linear slope that was broken by a gap of varying length filled with linearly stratified fluid. They then introduced a gravity current at one end of the channel, hugging either the slope or the "continental" channel wall. Breaking the slope was found to slow and destabilize the current, leading in some situations to the formation of persistent eddies at the upstream end of the gap. *Wolfe and Cenedese* [2006] employed a very similar experimental configuration and observed the same behavior in buoyant coastal currents with a gap in the slope. They also introduced slopes of various steepness in the gap and found that the current was stabilized whenever its width was smaller than that of the slope.

Sutherland and Cenedese [2009] extended this approach to a more realistically shaped trough in the experimental shelf. This configuration is conceptually opposite to the experiments described in Section 12.2, which represent a headland projecting out from the continent. Their experimental setup closely resembled that of *Whitehead and Chapman* [1986]: A buoyant coastal current was injected at one side of a cylindrical tank and then flowed around its edge until it encountered a rectilinear slope of height 20 cm and breadth ∼30 cm. The slope was interrupted for around 25 cm by a canyon of length ∼30 cm. The coastal current would separate from the bathymetry to cross the canyon if the width of the flow exceeded the radius of curvature of the isobaths, in some cases forming a counterrotating eddy in the canyon itself.

12.1.3. Large-Amplitude Shelf Waves

The theoretical and experimental studies of shelf waves described above focus almost exclusively on linear dynamics, for which a developed body of literature exists [e.g., *Mysak*, 1980b]. Yet in experiments featuring coastal currents, the evolution is characterized by strongly nonlinear features, such as the generation of eddies and large-amplitude deviations of the streamlines away from the isobaths [e.g., *Sutherland and Cenedese*, 2009].

In fact, a description of coastal currents in terms of nonlinear Rossby shelf waves is possible, but requires that the shelf break is approximated as a discontinuity in depth [*Longuet-Higgins*, 1968] and that the wave is long and evolves slowly in time. Under these assumptions, *Haynes et al.* [1993] derived a nonlinear wave equation based on a solution of the QG equations in the limit of long wavelength. *Clarke and Johnson* [1999] and *Johnson and Clarke* [1999] obtained a dispersive correction to this wave equation at next order in an asymptotic expansion based upon a small ratio of wave amplitude to wave length (see Section 12.4.2). *Johnson and Clarke* [2001] summarized the development and application of this theory.

This chapter focuses on experiments that are amenable to exactly this kind of theoretical description. We consider a channel similar to that of *Cenedese et al.* [2005], with a narrow slope separating much broader shallow and deep regions representing the continental shelf and open ocean, respectively. This is illustrated schematically in Figure 12.1. Our experiments elucidate the dynamics of a retrograde coastal current flowing past a headland, where the continental wall of the channel protrudes out, narrowing the continental shelf. Examples of such a configuration in the real ocean include the Agulhas current in the Mozambique Channel [*Bryden et al.*, 2005; *Beal et al.*, 2006, 2011] and the Gulf Stream at Cape Hatteras [*Stommel*, 1972; *Johns and Watts*, 1986; *Pickart*, 1995].

In Section 12.2 we describe our experimental setup and procedure and characterize the evolution of large-amplitude waves generated by retrograde flow past a headland. In Section 12.3 we briefly review the QG equations that underlie our nonlinear wave theory and numerical simulations. In Section 12.4 we adapt nonlinear shelf wave theory [*Johnson and Clarke*, 2001] to the annular channel, and in Section 12.5 we describe our numerical scheme for the QG equations. In Section 12.6 we compare the skill of our theory, numerical solutions, and experiments in predicting the characteristics of large-amplitude wave breaking. Finally, in Section 12.7 we summarize our findings and relate our results to previous studies of topographic Rossby waves and coastal currents.

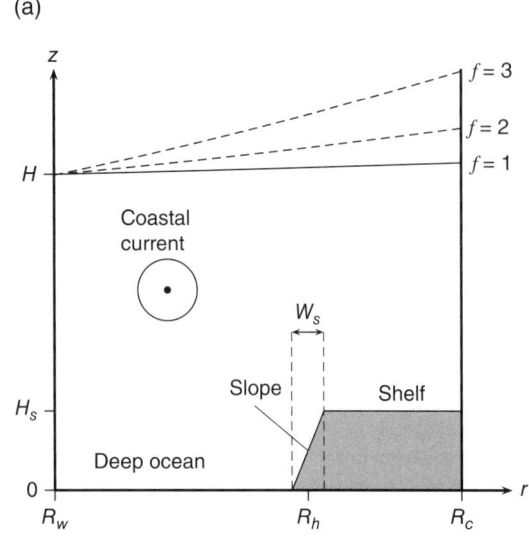

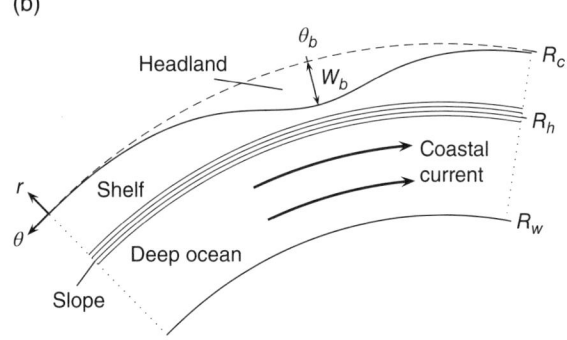

Figure 12.1. Idealized coastal ocean represented in our laboratory experiments and QG theory.

12.2. EXPERIMENTS WITH LARGE-AMPLITUDE SHELF WAVES

The purpose of our laboratory experiments is to capture the key features of continental shelves that are separated from the deep ocean by a relatively narrow slope. This configuration permits topographic Rossby waves whose amplitudes are large compared to the width of the slope; our goal is to elucidate the nonlinear dynamics of these waves. In this section we detail the setup of our experiments and illustrate their behavior via a reference experiment that characterizes our results.

In the laboratory, creating a straight, extended coastline with a flowing coastal current that is subject to a Coriolis force is logistically difficult. We therefore represent an idealized coastal ocean using a channel around the circumference of a rotating cylindrical tank, shown schematically in Figure 12.1. A retrograde coastal current of uniform density flows between rigid walls at $r = R_w$ and $r = R_c$, representing the oceanward extent of the current and the continental landmass, respectively. The flow is confined beneath a rigid lid at $z = H$ and above an ocean bed whose height ranges from 0 at $r = R_w$ to H_s at $r = R_c$ via a narrow slope of width W_s at $r = R_h$. This configuration also serves as the basis of the QG theory described in Section 12.3.5.

We conducted all of our experiments in a cylindrical tank of radius $R_c = 1.065$ m, shown in Figure 12.2. We constructed an inner wall of radius $R_w = 0.75$ m using a Plexiglas sheet weighted to the floor of the tank leaving an annular channel of width $L = 31.5$ cm. We built a flat shelf of width 10.75 cm and height $H_s = 5$ cm around the outer edge of the tank and attached a uniform slope of width $W_s = 2.5$ cm. The radius of the slope center is therefore $R_h = 0.945$ m. We painted the entire tank floor, shelf, and slope to ensure uniform bottom surface properties, as we found that bottom friction was by far the dominant source of dissipation in the experiments. The tank was filled with fresh water of uniform temperature to a depth $H = 20$ cm prior to each experiment.

This configuration differs from that of the real ocean in several respects:

1. The curvature of the channel influences the propagation of waves due to the radial dependence of any azimuthal volume element (see Section 12.4). We minimize this influence by using a channel whose width L is much smaller than the tank radius R_c.

2. In solid-body rotation, the upper surface of the fluid curves up toward the outer edge of the tank to balance the centrifugal force, as shown in Figure 12.1. The variation of the surface height across the channel reaches 6.6 cm when $f = 3$ rad/s, exceeding the 5 cm drop in depth across the experimental slope. However, even in this case the experimental slope is around 10 times steeper than the free surface and thus remains the dominant source of relative vorticity.

3. Real coastal currents are not bounded offshore by a wall but rather are bounded by the open ocean. However, the wall at $r = R_w$ is necessary to prevent the current interacting with itself across the tank, a situation that is more appropriate to a lake or small sea [e.g., *Johnson*, 1987; *Johnson and Kaoullas*, 2011].

4. The depths of continental shelves in the real ocean tend to be small compared to the full ocean depth. We used a relatively deep experimental shelf because we found that very shallow flows were damped too quickly by bottom friction. Furthermore, a small shelf is amenable to analysis using QG theory, for which a body of literature already exists [*Johnson and Clarke*, 2001]. These studies have demonstrated that the vortical dynamics captured by QG theory are most important for understanding shelf waves and that the mass constraints imposed by a shallow shelf are secondary. The relevance of two-dimensional experimental flows over topography to real geophysical flows is discussed at length in Chapter 10 of this volume.

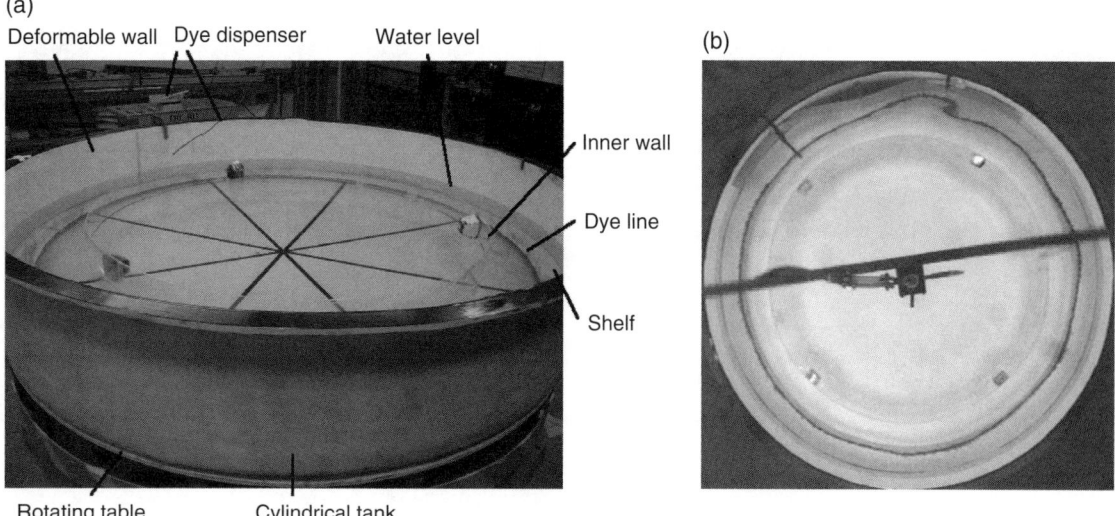

Figure 12.2. Photographs of (a) the rotating tank with elements of the experimental labeled and (b) our reference experiment from the overhead camera. The photograph in panel (a) was taken before attachment of the experimental slope in order to distinguish the shelf and before the base of the tank was repainted for the experiments.

5. Our experiments omit variations in density, which are dynamically important throughout the ocean. However, the barotropic dynamics are still informative, and strongly barotropic velocity profiles have been observed in the Agulhas current [*Bryden et al.*, 2005; *Beal et al.*, 2006].

All of the results described below concern shelf waves generated in the lee of a continental headland that protrudes out onto the continental shelf. In the laboratory we constructed a bump by attaching a sheet of plexiglass to the outer wall of the tank and then forcing its center out onto the shelf by placing a wedge behind it. This apparatus is visible in Figure 12.2. The outer wall of the channel is therefore given by $r = R_b(\theta)$. The bump in our annular channel, which was held fixed in all of the experiments below, is accurately described by

$$R_b = R_c - W_b(\theta), \quad W_b(\theta) = W_{b0} \operatorname{sech}^2\left(\frac{\theta - \theta_b}{\Theta_b}\right), \tag{12.1}$$

where $W_{b0} = 8.3$ cm, $\theta_b = 1.82$ rad, and $\Theta_b = 0.18$ rad. At the maximum extent of the protrusion, the width of the experimental shelf is reduced to 3.5 cm. This shape has been chosen to produce long waves with large amplitudes, which may be expected to exhibit nonlinear behavior without breaking.

In each experiment, we rotate the tank at a constant rate until the water within is brought to rest in the rotating frame. We then inject a line of dye into the fluid surface over the center of the slope at the PV front. This dye line serves as a passive tracer, intended to track the PV front as it is advected around the channel. We track the evolution of the dye line using an overhead camera that corotates with the tank, producing images like those shown in Figure 12.3. We generate a mean flow, representing along-shore transport by a coastal current, by rapidly changing the Coriolis parameter of the tank from $f - \Delta_f$ to f. This imparts a relative vorticity of $-\Delta_f$ to all of the water in the tank, inducing an azimuthal flow with sign opposite to that of Δ_f. For all of the experiments described herein, we use $\Delta_f > 0$, yielding a retrograde coastal current. We found that a retrograde current would develop large-amplitude shelf waves in the lee of the coastal protrusion, whereas a prograde current would not. This phenomenon may be understood via the nonlinear shelf wave theory described in Section 12.4. *Clarke and Johnson* [1999] showed that waves with the smallest amplitudes have the largest phase speeds unless the shelf line lies very close to one of the channel walls. Thus, disturbances formed by a prograde coastal current propagate rapidly downstream as small-amplitude waves, preventing a large wave envelope from forming on the continental shelf.

We have performed experiments over a range of tank rotation rates $f \in \{1, 1.5, 2, 2.5, 3\}$ rad/s and coastal current speeds corresponding to $\Delta_f \in \{0.01, 0.02, \ldots, 0.07\}$ rad/s. For the purpose of illustration, we define a reference experiment with $f = 1.5$ rad/s and $\Delta f = 0.03$ rad/s that characterizes the evolution of the shelf waves generated by our coastal current. Figure 12.3 shows images of the dye line close to the coastal protrusion at several times in this experiment. Between $t = 0$ s and $t = 17$ s, the coastal current advects fluid past the protrusion, developing a long wave with large amplitude. The rear portion of the wave then steepens continually

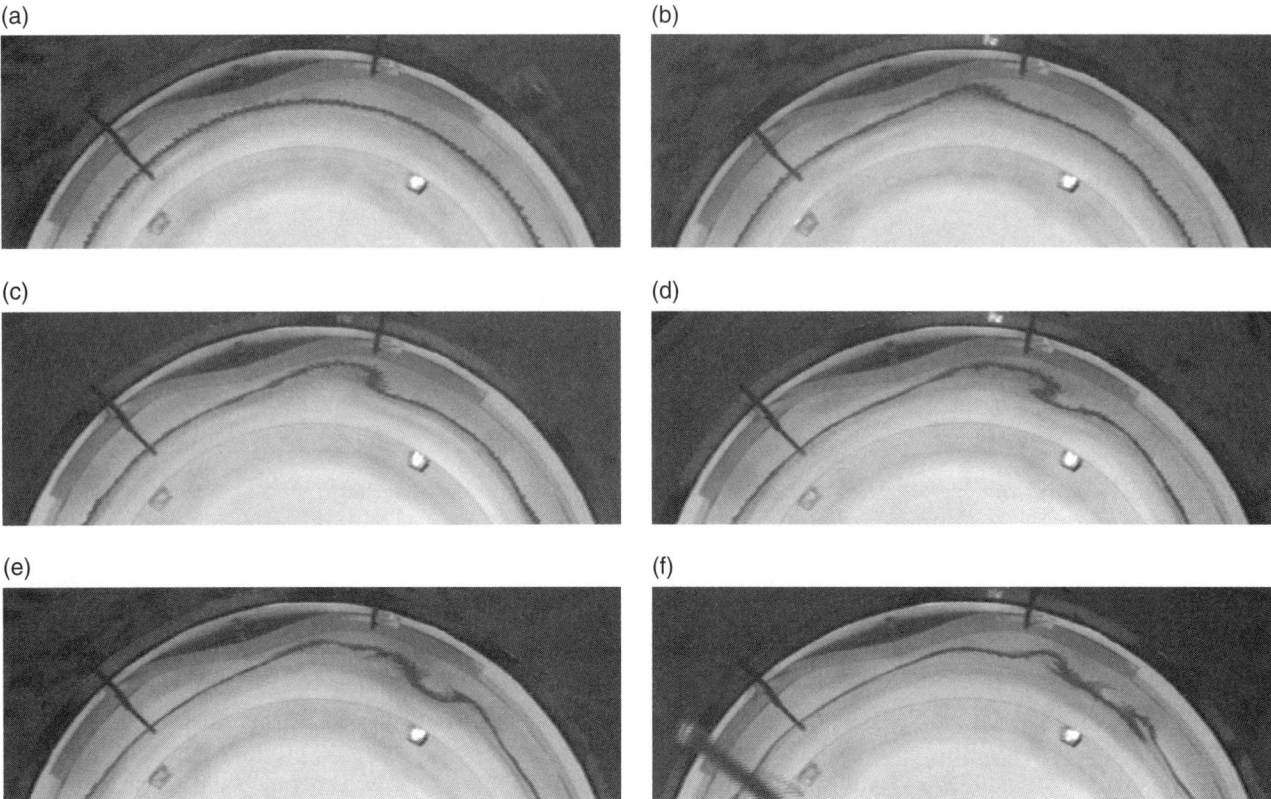

Figure 12.3. Snapshots from reference experiment at (a) $t = 0$ s, (b) $t = 17$ s, (c) $t = 29$ s, (d) $t = 45$ s, (e) $t = 66$ s, and (f) $t = 104$ s.

until $t = 29$ s, at which point the dye is aligned radially. By $t = 45$ s the wave has oversteepened and broken, and the position of the dye line is a double-valued function of azimuth. Thereafter, as the vorticity in the oversteepened wave envelope is removed by bottom friction, it gradually unwinds in the mean flow, forming an undulating wave train by $t = 66$ s. Eventually both sides of this wave train curl up on themselves, as shown at $t = 104$ s. In the following sections we will interpret this behavior using an adaptation of nonlinear shelf wave theory to the annulus and using numerical simulations of the barotropic QG shallow-water equations.

12.3. QUASI-GEOSTROPHIC MODEL EQUATIONS

We model the flow in our annular channel using the shallow-water QG equations [*Pedlosky*, 1987; *Vallis*, 2006]. The single-layer shallow-water equations describe the evolution of a homogeneous layer of fluid under the assumption that the flow is restricted to columnar motion and confined beneath a rigid lid at its upper surface. This assumption is consistent with a small ratio of vertical to horizontal length scales, $H/L \ll 1$, and appropriate for motions much slower than the surface gravity wave speed. The QG equations are derived thereafter by posing an asymptotic expansion for a small Rossby number, $\text{Ro} = U/fL \ll 1$, where U and L are characteristic horizontal velocity and length scales, respectively. Variations of the bottom topography h/H are also assumed to be $\mathcal{O}(\text{Ro})$.

For the configuration shown in Figure 12.1 we define the horizontal length scale as $L = R_c - R_w$ and the velocity scale as $U = \frac{1}{2}|\Delta_f| R_h$, so $L = 31.5$ cm and $U \approx 1.4$ cm/s. These scales yield an aspect ratio of $H/L \approx 0.6$ and a Rossby number of $\text{Ro} \approx 0.03$, while the variations of the bottom topography are characterized by $H_s/H = 0.25$. While these quantities are all smaller than 1, they are large enough to call QG theory into question. In practice, however, the rotation of the tank is sufficiently strong that the fluid adheres closely to columnar motion due to the Taylor-Proudman effect. *Williams et al.* [2010] showed that QG theory may provide qualitatively accurate results for rapidly rotating flows that lie far outside its formal regime of validity.

12.3.1. Interior Dynamics

Under the assumptions of shallow-water QG theory, the conservation of horizontal fluid momentum in our annular channel may be expressed in the form

$$\frac{\partial \boldsymbol{u}}{\partial t} + q\hat{\boldsymbol{e}}_z \times \boldsymbol{u} + \nabla \left[\Pi + \tfrac{1}{2}|\boldsymbol{u}|^2 - (f/H)\Psi \right] = -\kappa \boldsymbol{u}, \quad (12.2)$$

where $\boldsymbol{u} = -\nabla \times \psi \hat{\boldsymbol{e}}_z = u\hat{\boldsymbol{e}}_r + v\hat{\boldsymbol{e}}_\theta$ is the depth-independent horizontal velocity and Π is the density-normalized, ageostrophic pressure at the rigid lid, $z = H$. The PV is defined as

$$q = \zeta + \frac{fh(r)}{H}, \quad (12.3)$$

where $h(r)$ is the prescribed height of the bottom topography, given by

$$h(r) = H_s \begin{cases} 0, & r \leq R_h - \tfrac{1}{2}W_s, \\ \dfrac{r + \tfrac{1}{2}W_s - R_h}{W_s}, & |r - R_h| \leq \tfrac{1}{2}W_s, \\ 1, & R_h + \tfrac{1}{2}W_s \leq r, \end{cases} \quad (12.4)$$

as shown in Figure 12.1. Finally, Ψ is the first-order transport stream function

$$H\boldsymbol{u}_{\mathrm{ag}} - h(r)\boldsymbol{u} = -\nabla \times \Psi \hat{\boldsymbol{e}}_z, \quad (12.5)$$

where $\boldsymbol{u}_{\mathrm{ag}}$ is the (unknown) ageostrophic correction to the velocity. Taking the curl of (12.2) yields a material conservation law for PV, as modified by friction proportional to κ,

$$\frac{Dq}{Dt} = -\kappa \zeta, \quad \zeta = \nabla^2 \psi. \quad (12.6)$$

Here $D/Dt \equiv \partial/\partial t + \boldsymbol{u} \cdot \nabla$ is the advective derivative and ζ is the relative vorticity.

Motivated by our experimental observations, we have neglected any lateral viscous dissipation in (12.2) because bottom friction removes energy from the flow much more rapidly. The action of bottom friction is represented by a linear drag with the constant rate κ set by QG theory [*Pedlosky*, 1987],

$$\kappa = \frac{\sqrt{A_v f}}{H}. \quad (12.7)$$

The vertical viscosity A_v is here simply the molecular viscosity $\nu = 1 \times 10^{-6}\,\mathrm{m}^2/\mathrm{s}$. *Ibbetson and Phillips* [1967] found that the damping of Rossby waves in a similarly sized annulus was accurately described by (12.7).

We have separately conducted experiments without the bump in the outer wall, in which we released dye along a line of constant azimuth and then reduced f from $1.5\,\mathrm{r}$ to $1.3\,\mathrm{rad/s}$. This bestows upon the fluid a uniform angular velocity that may be expected to decay with an e-folding time of κ^{-1}. We estimate that, off the shelf, κ is around 1.8 times smaller than the theoretical value (12.7), while on the shelf it is around 1.5 times larger than (12.7). It is not clear why the variation in κ should be so large; simply replacing H by the actual depth $H - h$ in (12.7) cannot account for this. Nonetheless, (12.7) is a reasonable approximation to the bottom friction over the whole annulus. However, repeated measurements are not available to provide an accurate parameterization of κ, nor to explain the apparent discrepancy in κ between deeper and shallower waters.

12.3.2. Initial Conditions

As described in Section 12.2, we initiate our experiments by changing the tank's Coriolis parameter from $f - \Delta_f$ to f. This acceleration is relatively rapid, requiring only 1–2 s, so for the purposes of our model we treat it as an instantaneous modification of the fluid velocity in the rotating frame between $t = 0^-$ and $t = 0^+$. This avoids reformulating the QG equations for a frame rotating with variable velocity.

Relinquishing the QG approximation momentarily, we note that at $t = 0^-$ the fluid is in solid-body rotation with the tank and so adheres exactly to columnar motion. The absolute vertical vorticity ζ_a of any fluid column is then equal to that of the tank,

$$\zeta_a|_{t=0^-} = f - \Delta_f. \quad (12.8)$$

The acceleration of the tank modifies the absolute vorticity of its walls and base but leaves the absolute vorticity of the fluid instantaneously unchanged. Formally, we require that

$$\zeta_a|_{t=0^+} = f + \zeta|_{t=0^+} = \zeta_a|_{t=0^-}, \quad (12.9)$$

so our initial condition for the QG relative vorticity is

$$\zeta(r, \theta, 0^+) = -\Delta_f. \quad (12.10)$$

That is, the fluid acquires a relative vorticity that is everywhere equal to the change in the Coriolis parameter. One could follow a similar line of reasoning under the QG approximation, but the small-Rossby-number approximation of the PV (12.3) would introduce an $\mathcal{O}(H_s/H)$ error in (12.10).

Although (12.10) provides an initial condition for ζ, and therefore for q, the initial stream function $\psi(r, \theta, 0^+)$ is not uniquely determined by (12.10) alone. To invert (12.6) for ψ at $t = 0^+$, we require the stream function $\psi_0(0^+)$ on $r = R_b$, which corresponds to the along-channel transport. We obtain ψ_0 under the QG approximation by considering the total kinetic energy of the fluid in an inertial frame,

$$E(t) = \frac{1}{2} \iint_A \left[u^2 + (V + v)^2 \right] dA, \quad (12.11)$$

where V is the azimuthal velocity due to the rotation of the tank and A denotes the area of the annulus. In our model, only bottom friction can extract energy from the fluid, and we assume that this may be neglected during the rapid acceleration of the tank. We therefore require conservation of total energy across $t = 0$, i.e., $E^{(-)} = E^{(+)}$, where

$$E^{(-)} = \frac{1}{2} \iint_A \left(\frac{1}{2} fr\right)^2 dA, \quad (12.12a)$$

$$E^{(+)} = \frac{1}{2} \iint_A \left\{ u^2 + \left[\frac{1}{2}(f + \Delta_f)r + v\right]^2 \right\} dA. \quad (12.12b)$$

Thus, $\psi_0(0^+)$ must be chosen such that (12.12a) and (12.12b) are equal. For example, in a regular annulus ($W_b \equiv 0$), this yields $\psi_0(0^+) = \frac{1}{4}\Delta_f(R_c^2 - R_w^2)$, which corresponds to the intuitive result that the fluid acquires a uniform angular velocity opposite to the direction of the tank's acceleration. In general, $E^{(-)} = E^{(+)}$ must be solved numerically, but the long-wavelength asymptotic analysis described in Section 12.4 provides a very accurate approximate solution for the slowly varying channel shown in Figure 12.2.

12.3.3. Boundary Conditions

We apply no-flux boundary conditions at the inner and outer walls of the channel, requiring that both be streamlines of ψ. Without loss of generality we choose

$$\psi(R_w, \theta, t) = 0, \quad \psi(R_b, \theta, t) = \psi_0(t), \quad (12.13)$$

so ψ_0 corresponds to the counterclockwise along-channel transport. In a regular annulus ($W_b \equiv 0$) it is possible to derive an analytical evolution equation for $\psi_0(t)$ that excludes contributions from the unknown lid pressure Π. This is not possible when $W_b \neq 0$, because the bump in the outer wall may support an azimuthal pressure gradient that modifies the along-channel transport ψ_0. Instead, we determine additional boundary conditions by considering the circulations Γ_w and Γ_c around the inner and outer walls of the tank, respectively [*McWilliams*, 1977]. It follows from (12.2) that

$$\frac{d}{dt}\Gamma_c = -\kappa \Gamma_c, \quad \Gamma_c = \oint_{r=R_b} \mathbf{u} \cdot d\mathbf{r}, \quad (12.14a)$$

$$\frac{d}{dt}\Gamma_w = -\kappa \Gamma_w, \quad \Gamma_w = \oint_{r=R_w} \mathbf{u} \cdot d\mathbf{r}. \quad (12.14b)$$

In fact, we only need to ensure that either (12.14a) or (12.14b) is satisfied, because by Stokes's theorem

$$\Gamma_c - \Gamma_w = \iint_A \zeta \, dA \Rightarrow \frac{d}{dt}(\Gamma_c - \Gamma_w) = -\kappa(\Gamma_c - \Gamma_w). \quad (12.15)$$

The second equation in (12.15) follows from integrating (12.3) over the annulus. Condition (12.14b) determines the evolution of the outer wall stream function $\psi_0(t)$. We outline separate asymptotic and numerical strategies to solve this problem in Sections 12.4 and 12.5, respectively.

12.4. NONLINEAR SHELF WAVE THEORY

In interpreting the results of our laboratory experiments, it is instructive to compare with the predictions of the established nonlinear shelf wave theory [*Haynes et al.*, 1993; *Clarke and Johnson*, 1999; *Johnson and Clarke*, 1999, 2001]. This theory approximates the experimental/continental slope as a step, equivalent to $W_s \to 0$ in Figure 12.1, and describes the evolution of a fluid interface that lies initially over the shelf line, corresponding to the dye line in Figure 12.3. In this section we adapt this approach to an annular channel, improving upon the derivations of *Stewart* [2010] and *Stewart et al.* [2011] to utilize the initial and boundary conditions described in Sections 12.3.2 and 12.3.3, respectively.

12.4.1. Quasi-Geostrophic Dynamics Over a Step

In the limit of a vanishingly narrow slope ($W_s \to 0$), the height of the bottom topography in our annular channel becomes

$$h = H_s \mathcal{H}(r - R_h), \quad (12.16)$$

where $\mathcal{H}$ denotes the Heaviside step function. Neglecting the influence of bottom friction, the PV q is conserved exactly on fluid columns, and the dynamics may be described completely by the position of the interface that lies initially above the shelf line. With a view to describing waves whose length is much greater than their amplitude, we make the a priori assumption that this interface remains a single-valued function of azimuth and so may be denoted $r = R(\theta, t)$. At $t = 0$, we have $R(\theta, 0) = R_h$ by definition, and from (12.3) and (12.10) the initial PV is

$$q(r, \theta, 0) = -\Delta_f + Q\mathcal{H}(r - R_h), \quad (12.17)$$

where $Q = fH_s/H$. Thereafter, material conservation of q ensures that there is always a jump in PV at $r = R$,

$$q(r, \theta, t) = -\Delta_f + Q\mathcal{H}(r - R). \quad (12.18)$$

Using (12.3) and the definition of the relative vorticity (12.6), this may be rearranged as a Poisson equation for the stream function,

$$\nabla^2 \psi = Q[-\alpha + \mathcal{H}(r - R) - \mathcal{H}(r - R_h)], \quad (12.19)$$

where we define $\alpha = \Delta_f/Q$. Given the position of the interface $r = R$, inverting (12.19) yields a complete description of the flow in the annulus at any time t subject to the boundary equations described in Section 12.3.3. The evolution of the interface position $R(\theta, t)$ is determined by the requirement that particles on the interface remain on the interface, i.e., $(D/Dt)(r - R) = 0$. This condition may be rewritten as

$$\frac{\partial R}{\partial t} = -\frac{1}{R}\frac{\partial}{\partial \theta}\psi(R(\theta, t), \theta, t). \quad (12.20)$$

12.4.2. Asymptotic Solution

In general, (12.19) is not analytically tractable, so instead we obtain an asymptotic solution under the assumption of slow variations in azimuth and time. We first nondimensionalize (12.19) using the channel width $L = R_c - R_w$ as a length scale and Q^{-1} as a time scale,

$$r = L\hat{r}, \quad t = Q^{-1}\hat{t}, \quad q = Q\hat{q}, \quad \psi = QL^2\hat{\psi}. \quad (12.21)$$

Here hats denotes dimensionless variables. We then rescale under the assumption that azimuthal variations are characterized by $2\pi R_h$, the length of the channel at the shelf line. The parameter $\mu = (L/2\pi R_h)^2$, assumed to be asymptotically small, measures the ratio of radial to azimuthal length scales. We further assume that the flow evolves on a time scale that is $\mathcal{O}(\mu^{-1/2})$ longer than Q^{-1}, consistent with a velocity scale of QL and a length scale of $2\pi R_h$. This motivates rescaling θ and $\hat{t}$ as

$$\theta = \mu^{-1/2}\phi, \quad \hat{t} = \mu^{-1/2}\tau. \quad (12.22)$$

Under this scaling, (12.19) and (12.20) become

$$\hat{\psi}_{\hat{r}\hat{r}} + \frac{\hat{\psi}_{\hat{r}}}{\hat{r}} + \frac{\mu}{\hat{r}^2}\hat{\psi}_{\phi\phi} = \mathcal{H}(r - R) - \mathcal{H}(r - R_h) - \alpha, \quad (12.23a)$$

$$\frac{\partial \hat{R}}{\partial \tau} = -\frac{1}{\hat{R}}\frac{\partial}{\partial \phi}\psi\left(\hat{R}(\phi, \tau), \phi, \tau\right), \quad (12.23b)$$

The asymptotic parameter μ does not enter (12.23b) and only multiplies the azimuthal derivative in (12.23a).

We proceed by posing an asymptotic expansion of the stream function, $\hat{\psi} = \hat{\psi}^{(0)} + \mu\hat{\psi}^{(1)} + \cdots$. We solve (12.23a) at successive orders in μ subject to (12.13) in the form $\hat{\psi}^{(0)} = \hat{\psi}^{(1)} = 0$ on $r = R_w$ and $\hat{\psi}^{(0)} = \hat{\psi}_0^{(0)}(t), \hat{\psi}^{(1)} = \hat{\psi}_0^{(1)}(t)$ on $r = R_b$. For notational simplicity we present the solution in dimensional variables, writing $\psi = \psi^{(0)} + \psi^{(1)} + \cdots$. The leading-order stream function is

$$\psi^{(0)}/Q = -\frac{1}{4}\alpha(r^2 - R_w^2) + F\ln\left(\frac{r}{R_w}\right)$$
$$+ \frac{1}{4}(r^2 - R^2)\mathcal{H}(r - R) - \frac{1}{2}R^2\ln\left(\frac{r}{R}\right)\mathcal{H}(r - R)$$
$$- \frac{1}{4}(r^2 - R_h^2)\mathcal{H}(r - R_h)$$
$$+ \frac{1}{2}R_h^2\ln\left(\frac{r}{R_h}\right)\mathcal{H}(r - R_h), \quad (12.24)$$

where

$$F = \frac{1}{\ln(R_b/R_w)}\left[\frac{\psi_0^{(0)}}{Q} + \frac{1}{4}\alpha\left(R_b^2 - R_w^2\right)\right.$$
$$\left. + \frac{1}{4}\left(R^2 - R_h^2\right) + \frac{1}{2}R^2\ln\left(\frac{R_b}{R}\right) - \frac{1}{2}R_h^2\ln\left(\frac{R_b}{R_h}\right)\right]. \quad (12.25)$$

Substituting (12.24) and (12.25) into (12.20) yields a nondispersive nonlinear wave equation for $R(\theta, t)$, analogous to that studied by *Haynes et al.* [1993]. Solutions of this equation rapidly form shocks, so following *Clarke and Johnson* [1999], we continue the asymptotic solution to introduce dispersive terms,

$$\frac{\psi^{(1)}}{Q} = -\frac{1}{6}F_{\theta\theta}\ln^2\left(\frac{r}{R_w}\right) - \frac{1}{2}R_\theta^2\ln^2\left(\frac{r}{R}\right)\mathcal{H}(r - R)$$
$$+ \frac{1}{6}(RR_\theta)_\theta\ln^3\left(\frac{r}{R}\right)\mathcal{H}(r - R) + G\ln\left(\frac{r}{R_w}\right), \quad (12.26)$$

where

$$G = \frac{1}{\ln(R_b/R_w)}\left[\frac{\psi_0^{(1)}}{Q} + \frac{1}{6}F_{\theta\theta}\ln^2\left(\frac{R_b}{R_w}\right)\right.$$
$$\left. + \frac{1}{2}R_\theta^2\ln^2\left(\frac{R_b}{R}\right) - \frac{1}{6}(RR_\theta)_\theta\ln^3\left(\frac{R_b}{R}\right)\right]. \quad (12.27)$$

Higher-order corrections may be obtained by continuing the asymptotic solution, but the calculus becomes prohibitively complicated. Substituting (12.24)–(12.27) into (12.20) yields the nonlinear shelf wave equation,

$$\frac{\partial R}{\partial t} = -\frac{Q}{R}\frac{\partial}{\partial \theta}\left\{-\frac{1}{4}\alpha(R^2 - R_w^2) + F\ln\left(\frac{R}{R_w}\right)\right.$$
$$-\frac{1}{4}(R^2 - R_h^2)\mathcal{H}(R - R_h) + \frac{1}{2}R_h^2\ln\left(\frac{R}{R_h}\right)\mathcal{H}(R - R_h)$$
$$\left. + \gamma\left[-\frac{1}{6}F_{\theta\theta}\ln^2\left(\frac{R}{R_w}\right) + G\ln(R/R_w)\right]\right\}. \quad (12.28)$$

Here γ is simply a switch for the dispersive terms due to the first-order stream function (12.26). Setting $\gamma = 0$ recovers the leading-order nondispersive wave equation, while $\gamma = 1$ yields the full dispersive wave equation.

12.4.3. Azimuthal Transport

The nonlinear shelf wave equation (12.28) is closed except for the stream function $\psi_0 = \psi_0^{(0)} + \psi_0^{(1)}$ on the outer boundary, or equivalently the along-channel transport. We choose to constrain the transport using (12.14b) because (12.14a) is complicated by the bump in the outer wall. Under our asymptotic expansion, the stream function must satisfy (12.14b) at every order in μ.

In dimensionless variables, for the dissipation-free case $\kappa = 0$, this implies

$$\frac{d}{dt}\Gamma_w^{(n)} = 0, \quad \Gamma_w^{(n)} = \int_0^{2\pi} \left.\frac{\partial \psi^{(n)}}{\partial r}\right|_{r=R_w} R_w \, d\theta, \quad (12.29)$$

for all $n \in \{0, 1, \ldots\}$. Substituting (12.24) and (12.26) into (12.29) yields evolution equations for $\psi_0^{(0)}$ and $\psi_0^{(1)}$, respectively. For example, for $n = 0$ we obtain

$$\frac{d\psi_0^{(0)}}{dt} = -Q \frac{\int_0^{2\pi} \frac{\ln(R_b/R)}{\ln(R_b/R_w)} R R_t \, d\theta}{\int_0^{2\pi} \frac{d\theta}{\ln(R_b/R_w)}}. \quad (12.30)$$

We omit the corresponding expression for $\psi^{(1)}$ for the sake of brevity. Equation (12.30) describes the tendency of net radial vorticity fluxes to modify the along-channel transport. It is a consequence of our channel's finite length that the wave equation (12.28) acquires contributions that are dependent on the global behavior of the solution, whereas that of *Clarke and Johnson* [1999] in an infinite channel did not.

To complete the evolution of $\psi_0(t)$, we must also determine its initial condition following Section 12.3.2. Using $R(\theta, 0) = R_h$, equation (12.24) approximates the initial azimuthal velocity as

$$\left.\frac{\partial \psi^{(0)}}{\partial r}\right|_{t=0} = -\frac{1}{2}\Delta_f r + \frac{\psi_0^{(0)}(0) + (1/4)\Delta_f(R_b^2 - R_w^2)}{r \ln(R_b/R_w)}. \quad (12.31)$$

To a consistent order of approximation, we must omit the radial velocity (u^2) term in (12.12b), so the initial energy of the flow is

$$E^{(+)} = \frac{1}{2} \iint_A \left(\frac{1}{2}(f + \Delta_f)r + \left.\frac{\partial \psi^{(0)}}{\partial r}\right|_{t=0}\right)^2 dA. \quad (12.32)$$

Thus we may determine $\psi_0^{(0)}(0)$ by substituting (12.31) into (12.32) and solving $E^{(+)} = E^{(-)}$. For example, in a regular annulus with $R_b(\theta) \equiv R_c$, equation (12.31) gives the exact initial stream function ($\psi^{(0)}|_{t=0} \equiv \psi|_{t=0}$), and this procedure yields $\psi_0(0) = -\frac{1}{4}\Delta_f(R_c^2 - R_w^2)$. The corresponding azimuthal velocity $v|_{t=0} = -\frac{1}{2}\Delta_f r$ has uniform angular velocity, which is the expected response of the flow to a change in the tank's rotation rate.

12.4.4. Parameterizing Bottom Friction

Our neglect of bottom friction in (12.28) causes its solutions to diverge substantially from our experimental results. It is not possible to introduce κ exactly in our nonlinear wave theory because the vorticity of each fluid column depends sensitively on the times at which it has crossed the shelf line [*Stewart*, 2010]. Instead, we employ a crude representation of bottom friction to provide a source of dissipation in (12.28). We modify the PV equation (12.6) to an exact material conservation law,

$$\frac{D}{Dt}\left(\zeta e^{\kappa t} + \frac{fH_s}{H}\right) = 0. \quad (12.33)$$

This representation possesses identical conservation laws to (12.6) and (12.2) for the total vorticity and total energy in the annulus. In fact, its dynamics are identical to (12.6) away from the shelf line, where vorticity simply decays exponentially with rate κ. Fluid columns crossing the shelf line acquire a relative vorticity of constant magnitude $|Q|$ under (12.6), whereas under (12.33) they acquire a relative vorticity of magnitude $|Q|e^{-\kappa t}$.

All of the results discussed in this section may be rederived using (12.33). The nonlinear shelf wave equation acquires an additional factor of $e^{-\kappa t}$ multiplying the right-hand side of (12.28), while the azimuthal transport equation (12.30) is unchanged. The latter may seem contradictory, as the azimuthal transport must decay due to bottom friction. This is because ψ is a stream function for the modified vorticity $\zeta e^{\kappa t}$, and so ψ_0 differs from the true along-channel transport by a factor of $e^{-\kappa t}$.

12.4.5. Comparison with Experimental Flows

In Figure 12.4 we plot snapshots of a solution to (12.28) using parameters that correspond to the reference experiment shown in Figure 12.3. We solve (12.28) numerically by discretizing R on an azimuthal grid of 400 equally spaced points covering $\theta \in [0, 2\pi)$. We evaluate azimuthal derivatives spectrally via the fast Fourier transform, and we integrate (12.28) forward in time using third-order Adams–Bashforth time stepping. We employ an exponential Fourier filter [*Hou and Li*, 2007] to damp the mild instability that arises at high wave numbers due to aliasing error [*Boyd*, 2001].

Having assumed that the PV front remains a single-valued function of θ, it is perhaps unsurprising that the solution shown in Figure 12.4 is unable to capture the wave breaking shown in Figure 12.3. A large-amplitude wave develops in the lee of the bump in the outer wall, but instead of breaking it develops a dispersive wave train that spreads clockwise around the tank due to the strong retrograde mean flow. Although the mean flow decelerates rapidly due to bottom friction, the volume of water transported across the shelf line continues to increase even at $t = 104$ s, and the wave envelope does not collapse as in Figure 12.3.

The wave breaking shown in Figure 12.3c is actually captured more accurately by the nondispersive wave equation, corresponding to $\gamma = 0$ in (12.28), even though

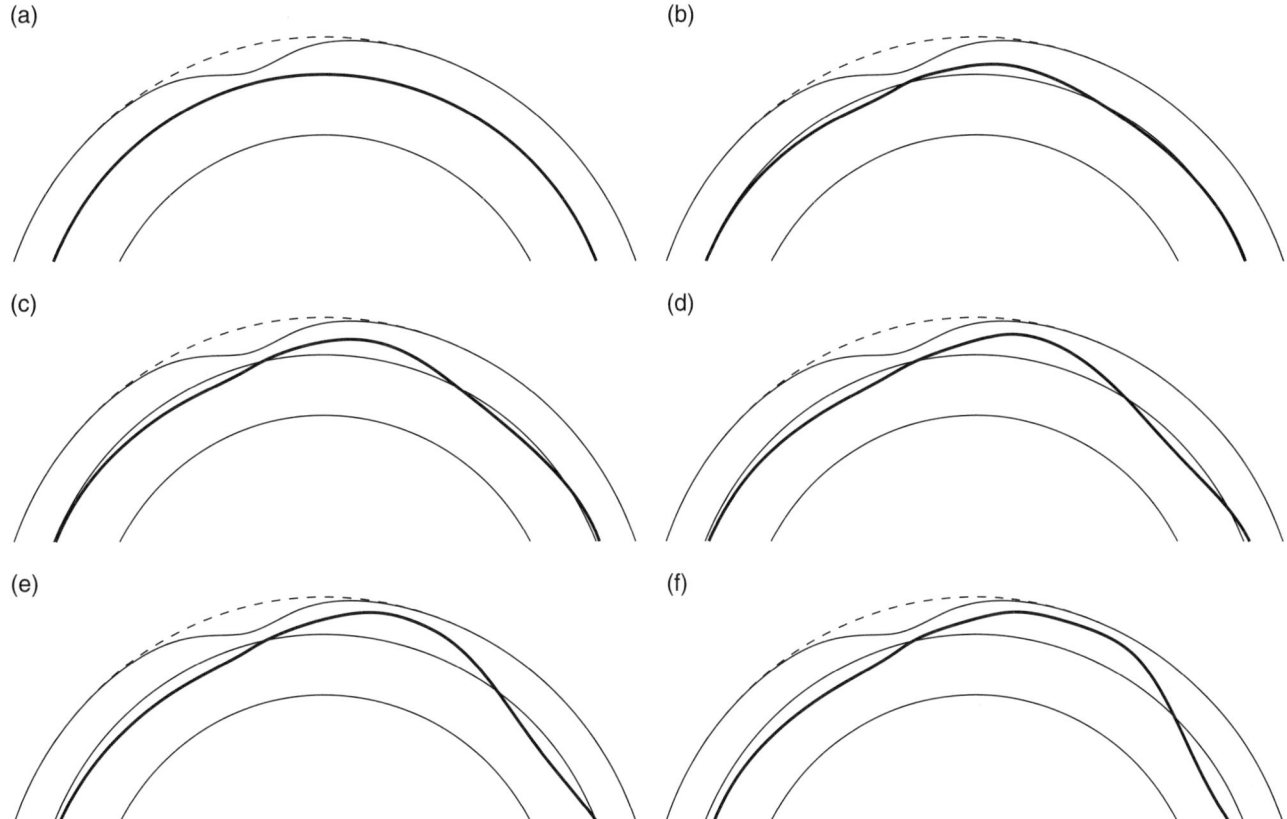

Figure 12.4. Solution of the nonlinear shelf wave equation (12.28) using parameters that correspond to the reference experiment in Figure 12.3. The thin solid lines show the positions of the inner and outer walls of the annulus, and the dashed line indicates the width of the protrusion. The thick solid line indicates the position of the PV front at (a) $t = 0$ s, (b) $t = 17$ s, (c) $t = 29$ s, (d) $t = 45$ s, (e) $t = 66$ s, and (f) $t = 104$ s.

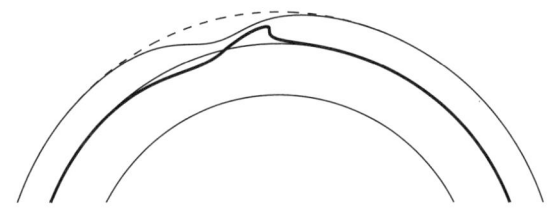

Figure 12.5. Solution at $t = 8.7$ s of the nondispersive wave equation corresponding to $\gamma = 0$ in (12.28) using parameters appropriate to the reference experiment.

this is formally a less accurate approximation. Figure 12.5 shows the computed solution to this equation on a grid of 7200 points in azimuth at $t = 8.7$ s, around which time the interface $R(\theta, t)$ forms a shock. This solution resembles the experimental wave shown in Figure 12.3c, but the wave breaking takes place much earlier in the evolution of the PV front. This may be due to the approximation of the slope as a step, which leads to a larger mean relative vorticity within the theoretical wave envelope.

12.5. NUMERICAL SOLUTION OF INVISCID QUASI-GEOSTROPHIC EQUATIONS

In this section we outline our algorithm for solving the QG model equations numerically. This approach is motivated in part by the apparent failure of the long-wave approximation employed in Section 12.4 to describe the behavior of our laboratory experiments. Additionally, these numerical solutions allow us to identify and explain deviations of the experimental results from QG theory and thus extrapolate to the behavior of coastal currents in the real ocean.

To ensure stability, we modify the QG model equations (12.6) to include a numerical viscosity term that smooths vorticity gradients at the scale of the numerical grid,

$$\frac{\partial q}{\partial t} = -J(\psi, q) - \kappa \zeta + A_n \nabla^2 \zeta, \quad (12.34a)$$

$$\nabla^2 \psi = q - \frac{fh}{H}. \quad (12.34b)$$

Here J is the two-dimensional Jacobian operator, q and ζ are related via (12.3), and A_n measures the numerical viscosity. We set $A_n = Qd^2$ so that numerical diffusion becomes comparable to advection at the scale of one grid box d over a dynamical time scale of Q^{-1}. Thus A_n is a purely numerical contribution to (12.34a), because $A_n \to 0$ as $d \to 0$. The numerical viscosity is not intended to represent the fluid's molecular viscosity, and (12.34a) and (12.34b) describe inviscid flow with numerical diffusion that acts only on the interior vorticity field. We therefore retain free-slip, rather than no-slip, boundary conditions at the channel walls.

12.5.1. Wall-Following Coordinates

The protrusion (12.1) in the outer wall of our annulus means that we cannot discretize the channel using a regularly spaced grid in r/θ coordinates without adaptation. We therefore transform the QG equations (12.34a) and (12.34b) into coordinates that follow the walls of the annulus,

$$\rho = R_w + \frac{(r - R_w)(R_c - R_w)}{R_b - R_w}, \quad \phi = \theta. \quad (12.35)$$

Here ρ is simply a rescaled radial coordinate that satisfies $\rho = R_w$ on $r = R_w$ and $\rho = R_c$ on $r = R_b$. Derivatives with respect to r and θ may be transformed to ρ, ϕ space by writing $q = q(\rho(r, \theta), \phi(\theta))$ and applying the chain rule,

$$\left.\frac{\partial q}{\partial r}\right|_\theta = \frac{R_c - R_w}{R_b - R_w} \left.\frac{\partial q}{\partial \rho}\right|_\phi, \quad (12.36a)$$

$$\left.\frac{\partial q}{\partial \theta}\right|_r = \left.\frac{\partial q}{\partial \phi}\right|_\rho + \frac{\rho - R_w}{R_b - R_w} \frac{dW}{d\phi} \left.\frac{\partial q}{\partial \rho}\right|_\phi. \quad (12.36b)$$

For example, the Jacobian operator in (12.34a) becomes

$$J(\psi, q) = \frac{1}{r(\rho, \phi)} \frac{R_c - R_w}{R_b - R_w} \left(\frac{\partial \psi}{\partial \rho} \frac{\partial q}{\partial \phi} - \frac{\partial \psi}{\partial \phi} \frac{\partial q}{\partial \rho} \right), \quad (12.37)$$

where $r(\rho, \phi)$ may be obtained by inverting (12.35). The Laplacian operator ∇^2 is considerably more complicated and includes a second-order cross-derivative in ρ and ϕ. We omit this expression for brevity.

12.5.2. Numerical Integration

We discretize our dependent variables ψ, ζ, and q on a grid of $N_\rho \times N_\phi$ points with regular spacings Δ_ρ and Δ_ϕ, respectively. We denote their positions as ρ_m for $m = 1, \ldots, N_\rho$ and ϕ_n for $n = 1, \ldots, N_\phi$, where $\rho_1 = R_w$ and $\rho_{N_\rho} = R_c$. We denote variables stored at (ρ_m, ϕ_n) as, for example, $\psi_{m,n}(t)$, and for now we retain a continuous dependence on time for notational convenience. We approximate the Jacobian (12.37) using a second-order energy-conserving discretization [*Arakawa*, 1966], equivalent to rewriting (12.37) as a flux form and replacing derivatives with second-order central differences. We approximate the Laplacian operator in (12.34a) and (12.34b) using straightforward second-order central differencing in ρ, ϕ coordinates.

Our numerical scheme is simplified by the fact that the initial vorticity (12.10) is uniform. As the numerical viscosity acts only in the interior, (12.34a) states that the vorticity remains constant along the boundary but decays exponentially with time scale κ^{-1}. This serves as a boundary condition for the relative vorticity, which may be written in ρ, ϕ coordinates as

$$\zeta(R_w, \phi, t) = \zeta(R_c, \phi, t) = \zeta_0(t), \quad \frac{d\zeta_0}{dt} = -\kappa \zeta_0. \quad (12.38)$$

with $\zeta_0(0) = -\Delta_f$. To simplify the presentation of our numerical scheme, we will assume for the moment that the stream function $\psi_0(t)$ on the outer wall is also a known function of time. We will explain how we evolve ψ_0 in Section 12.5.3.

We evolve the PV $q_{m,n}$ and stream function $\psi_{m,n}$ forward in time as follows. Given $q_{m,n}$ at all grid points at any time t, we first invert (12.34b) iteratively via successive overrelaxation to determine $\psi_{m,n}$ at all interior points subject to

$$\psi_{1,n}(t) = 0, \quad \psi_{N_\rho,n} = \psi_0(t), \quad n = 1, \ldots, N_\phi. \quad (12.39)$$

We then compute the right-hand side of (12.34a) at all interior grid points utilizing the vorticity boundary conditions

$$\zeta_{1,n}(t) = \zeta_{N_\rho,n}(t) = \zeta_0(t), \quad n = 1, \ldots, N_\phi, \quad (12.40)$$

to evaluate derivatives in the grid rows $m = 2$ and $m = N_\rho - 1$. This yields the time derivative of q at all interior grid points, which we use to step $q_{m,n}(t)$ forward in time using the third-order Adams-Bashforth scheme. We ensure that the fixed time step Δ_t satisfies the advective CFL condition throughout the integration.

For the purpose of comparison with our laboratory experiments, we track the position of the PV front that lies initially above the center of the slope. We accomplish this by advecting M passive tracer particles at (ρ_i, ϕ_i) for $i = 1, \ldots, M$ using the computed stream function $\psi_{m,n}(t)$. These particles are initially spread evenly around the shelf line, so

$$\rho_i(0) = R_w + \frac{(R_h - R_w)(R_c - R_w)}{R_b[\phi_i(0)] - R_w}, \quad \phi_i(0) = \frac{2\pi i}{M}. \quad (12.41)$$

Thereafter, at any time t the particle evolution is determined by

$$\frac{d\rho_i}{dt} = \frac{-1}{r(\rho_i,\phi_i)} \frac{R_c - R_w}{R_b(\phi_i) - R_w} \frac{\partial \psi}{\partial \phi}(\rho_i,\phi_i,t), \quad (12.42)$$

$$\frac{d\phi_i}{dt} = \frac{1}{r(\rho_i,\phi_i)} \frac{R_c - R_w}{R_b(\phi_i) - R_w} \frac{\partial \psi}{\partial \rho}(\rho_i,\phi_i,t). \quad (12.43)$$

At each time step, after inverting (12.34b) for $\psi_{m,n}$, we calculate $\partial\psi/\partial\rho$ and $\partial\psi/\partial\phi$ on each grid point via second-order central differencing. We then linearly interpolate these derivatives to compute the right-hand sides of (12.42) and (12.43). Finally, we integrate the particle positions forward in time using third-order Adams-Bashforth time stepping.

12.5.3. Azimuthal Transport

The above outline of our numerical algorithm assumes that the stream function $\psi_0(t)$ on the outer wall of the annulus is known. In fact, ψ_0 must evolve in such a way that either (12.14a) or (12.14b) is satisfied. We will focus on (12.14b) because it is more straightforward to evaluate numerically.

Following *Gresho* [1991a,1991b], we note that the solution ψ of (12.34b) may be decomposed as

$$\psi = \psi_P + \eta\psi_L, \quad (12.44)$$

where η is a constant to be determined and ψ_P and ψ_L solve Poisson's and Laplace's equations, respectively,

$$\nabla^2 \psi_P = \zeta, \qquad \nabla^2 \psi_L = 0. \quad (12.45)$$

We choose $\psi_P = \psi_L = 0$ on $\rho = R_w$ to ensure that ψ vanishes on the inner wall, and without loss of generality set $\psi_P = 0$ and $\psi_L = \psi_{L0}$ on $\rho = R_c$, where ψ_{L0} may be any nonzero constant. Given ζ, this completely defines ψ_P and ψ_L, and so if we can determine η, then ψ and ψ_0 follow trivially from (12.44).

We determine η by requiring that ψ_P and ψ_L satisfy (12.14b). We define Γ_P and Γ_L as the circulations due to ψ_P and ψ_L around the inner wall of the annulus,

$$\Gamma_j(t) = \int_0^{2\pi} \frac{R_w(R_c - R_w)}{R_b - R_w} \frac{\partial \psi_j}{\partial \rho} d\phi, \quad (12.46)$$

for $j \in \{P,L\}$. It follows from (12.44) that

$$\Gamma_P + \eta\Gamma_L = \Gamma_w, \quad (12.47)$$

where Γ_w is determined at all times by (12.14b) and both Γ_P and Γ_L may be computed directly via (12.46). Thus (12.47) defines η, so the stream function on the outer wall is given by

$$\psi_0 = \psi_{L0} \frac{\Gamma_w - \Gamma_P}{\Gamma_L}. \quad (12.48)$$

In practice, we define ψ_{L0} and compute ψ_L and Γ_L once at the start of the numerical integration, and then at each time step we compute ψ_P, Γ_P, and finally ψ.

The exception to the above procedure is at $t = 0$ when the initial circulation $\Gamma_w(0)$ is unknown. The initial transport $\psi_0(0)$ must then be chosen to ensure that (12.12a) and (12.12b) are equal. We compute $\psi_0(0)$ approximately using the asymptotic method described in Section 12.4.3, which defines $\psi_{m,n}(0)$ everywhere via (12.34b). We then calculate $\Gamma_w(0)$ directly from $\psi_{m,n}(0)$.

12.5.4. Convergence Under Grid Refinement

We verify our numerical scheme by defining a test case with parameters $f = 1.5$ rad/s and $\Delta_f = 0.02$ rad/s, whose behavior is qualitatively similar to that shown in Figure 12.7. We compute solutions using these parameters over a range of grid spacings Δ_ρ increasing from 1.25 mm to 1 cm by factors of $\sqrt{2}$. In each case we integrate (12.34a) and (12.34b) up to $t = 180$, the typical length of our laboratory experiments. We choose $\Delta_\phi = \Delta_\rho/R_c$ so that Δ_ρ is the largest absolute distance between any two adjacent grid cells. We measure convergence toward an "exact" reference solution obtained via Richardson extrapolation from the two smallest grid spacings, which we denote as $(\psi_e)_{m,n}$ and $(q_e)_{m,n}$. We define the error between each of our computed solutions and the "exact" solution at any time t using the discrete ℓ^2 norm

$$\epsilon_q(t) = \sqrt{\sum_{m,n} \left[q_{m,n}(t) - (q_e)_{m,n}(t)\right]^2}, \quad (12.49)$$

and similarly for $\epsilon_\psi(t)$.

In Figure 12.6 we plot ϵ_q and ϵ_ψ at four times over the course of our numerical integration. Both q and ψ exhibit convincing second-order convergence under grid refinement, owing to the smooth velocity field prescribed by our experimental conditions and the relatively short integration time. Our convergence study motivates the choice $\Delta_\rho = 2$ mm for the purpose of comparison with our experimental results. The numerical viscosity has an unduly strong influence on the solution if the grid spacing is much larger than this ($\gtrsim 5$ mm) and PV is poorly conserved in filaments that cross the slope. It so happens that at $\Delta_\rho = 2$ mm, our prescribed numerical viscosity $A_n \sim 1 \times 10^{-6}$ m^2/s is comparable to the molecular viscosity of water in the experiments.

12.5.5. Comparison with Experimental Flows

Figure 12.7 characterizes the behavior of solutions computed via the procedure outlined above. We use rotation parameters $f = 1.5$ rad/s and $\Delta_f = 0.03$ rad/s, as in our reference experiment. The grid spacing of $\Delta_\rho = 2$ mm results in a numerical grid of $N_\rho = 159$ by $N_\phi = 3347$ points. These parameters yield a numerical viscosity of $A_n = 1.5 \times 10^{-6}$ m^2 s^{-1} and a bottom friction κ

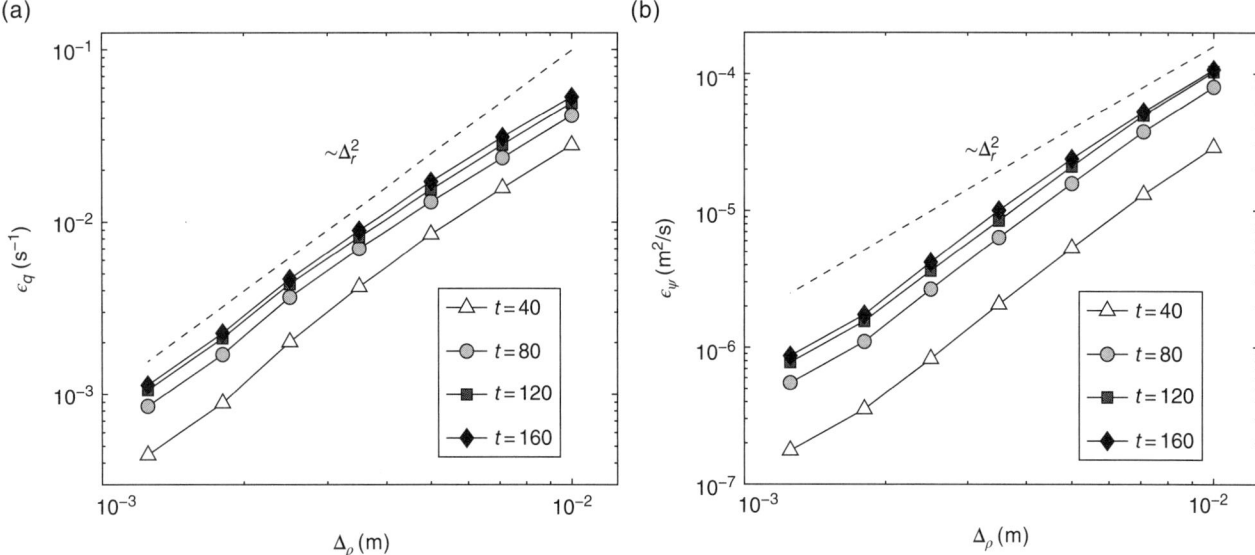

Figure 12.6. Verification of the numerical scheme described in Section 12.6 for the test case $f = 1.5$ rad/s, $\Delta_f = 0.02$ rad/s. The pointwise ℓ^2 error is plotted in (a) the PV and (b) the stream function between each solution and an "exact" solution is obtained via Richardson extrapolation.

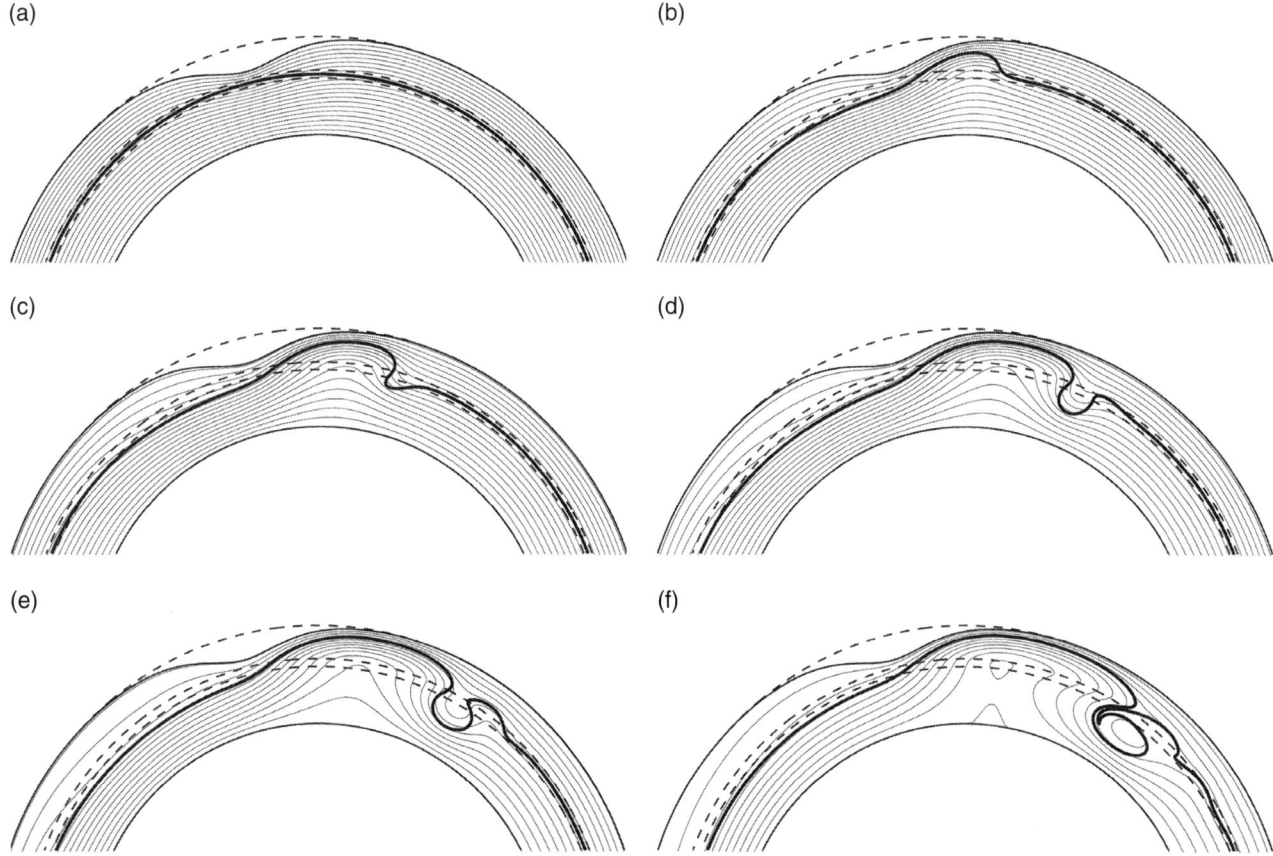

Figure 12.7. Evolution of the numerical solution of the QG model described in Section 12.5, at (a) $t = 0$ s, (b) $t = 17$ s, (c) $t = 29$ s, (d) $t = 45$ s, (e) $t = 66$ s, and (f) $t = 104$ s. The numerical parameters have been chosen to match the reference experiment shown in Figure 12.3. In each snapshot is plotted the position of the passively advected line of tracer (thick black line) and the instantaneous streamlines with a separation of 2 cm^2/s (thin gray lines). The thin black lines show the positions of the inner and outer walls of the annulus, and the dashed lines highlight the extent of the slope and the width of the protrusion.

of approximately 6.1×10^{-3} s^{-1}. The experimental dye is represented by $M = 3600$ tracer particles, visualized as a line in Figure 12.7.

The solution shown in Figure 12.7 is visibly similar to the evolution of the experimental dye line in Figure 12.3. In particular, the progression in Figure 12.7 of the solution from a long, smooth wave in panel (b) to a backward-breaking wave in panel (c) is strikingly similar to the behavior of the experiment. However, the steepening of the wave and in panel (d) the formation of the wave train appear to happen more rapidly than they do in Figure 12.3. By eye, Figure 12.3e corresponds better to Figure 12.7d than Figure 12.7e, despite being separated by 21 s. This may be due to imperfect conservation of PV on fluid columns crossing the experimental slope or our bottom friction coefficient κ may simply be too small. In Figure 12.7 the two sides of the wave train in panel (e) do eventually curl up on themselves, as shown in panel (f) and as observed in our reference experiment. In Section 12.6 we will quantify this comparison over a range of Coriolis parameters and coastal current speeds.

A much larger volume of deep water is retained on the shelf in Figure 12.7f than in Figure 12.3, associated with the growth and persistence of the large-amplitude shelf wave. The streamlines reveal that the growth of this wave leads to constriction of the along-shelf flow. This drives water off the shelf and across the slope far upstream of the protrusion, resulting in acceleration of the incoming flow in the deeper portion of the channel. In panels (d)–(f) the wave envelope drives an exchange across the slope, drawing inflowing deep water onto the shelf at the protrusion and then exporting it further downstream. At $t = 104$ s all inflowing deep water makes an excursion onto the shelf before continuing downstream. This is facilitated by patches of closed streamlines, corresponding to barotropic vortices/eddies in the flow that sustain the exchange of water across the slope.

12.6. SHELF WAVE BREAKING

The only consistent point of comparison between our numerical and experimental results is the formation of a breaking lee wave behind the bump in the outer wall of the channel. The evolution therafter varies widely across our parameter space in f and Δ_f. When the coastal current is weak and the background rotation is strong, the breaking wave rapidly rolls up into eddies, which are beginning to form in Figures 12.3f and 12.7f. A strong coastal current in the presence of weak background rotation will tend to inhibit wave breaking and lead to the formation of a wave train resembling that in Figures 12.3e and 12.7d. The wave breaking is therefore the most natural point of comparison between our theory, numerical solutions, and experiments.

12.6.1. Breaking Conditions

We obtain theoretical predictions of breaking wave properties using the nondispersive ($\gamma = 0$) form of the nonlinear shelf wave equation (12.28). The dispersive form ($\gamma = 1$) is formally more accurate but inhibits shock formation, which describes wave breaking in our theoretical framework. For each parameter combination (f, Δ_f) we solve (12.28) on an azimuthal grid of $N = 7200$ points using the method described in Section 12.4.5. We define the solution to have formed a shock, and thus the wave to have broken, when the gradient of the interface $r = R(\theta, t)$ satisfies

$$\max_{\theta \in [0, 2\pi)} \left| \frac{1}{R} \frac{\partial R}{\partial \theta} \right| > S_{\max}. \quad (12.50)$$

The gradient threshold $S_{\max}$ should be maximized to ensure that interface has come as close as possible to a discontinuity before (12.50) is satisfied. We use $S_{\max} = 20$ because the pseudospectral solution cannot reach much larger gradients on a grid of size $N = 7200$ without becoming subject to the Gibbs phenomenon.

In our numerical solutions we track the positions $(\rho_i(t), \phi_i(t))$ for $i = 1, \ldots, M$ of tracer particles that lie initially over the center of the slope. We deem the waveform comprised of these particles to have broken when the initial azimuthal ordering of the particles reverses, i.e.,

$$\phi_i(t) > \phi_{i+1}(t) \quad \text{for any} \quad i = 1, \ldots, M. \quad (12.51)$$

The addition $i + 1$ is taken modulo M because particle positions wrap around the annulus.

We identify wave breaking in our experimental results by processing the images of the tank to extract the positions of the dye line in each frame. Our focus on the initial development and breaking of the wave permits the use of a simplified algorithm, because the dyeline may be described as a single-valued function of azimuthal position. Each image is first filtered to remove shades dissimilar to that of the dye line. Then at each azimuthal position θ_p for $p = 1, \ldots, P$, we search radially to locate the midpoint of the dye line, which we denote as R_p. We thereby construct a series of points along the dye line in each frame, $(\theta_p, R_p(t))$, similar to our description of the interface $R(\theta, t)$ in our nonlinear wave theory. We omit azimuthal positions θ_p obscured by the clamps visible in Figure 12.2, and successive frames are compared to remove erroneous measurements due to the dye dispenser. We deem the wave to have broken by applying a condition analogous to (12.50) but using the mean gradient over a small range of θ to eliminate noise that arises from the image filtering. In practice, the choice of $S_{\max}$

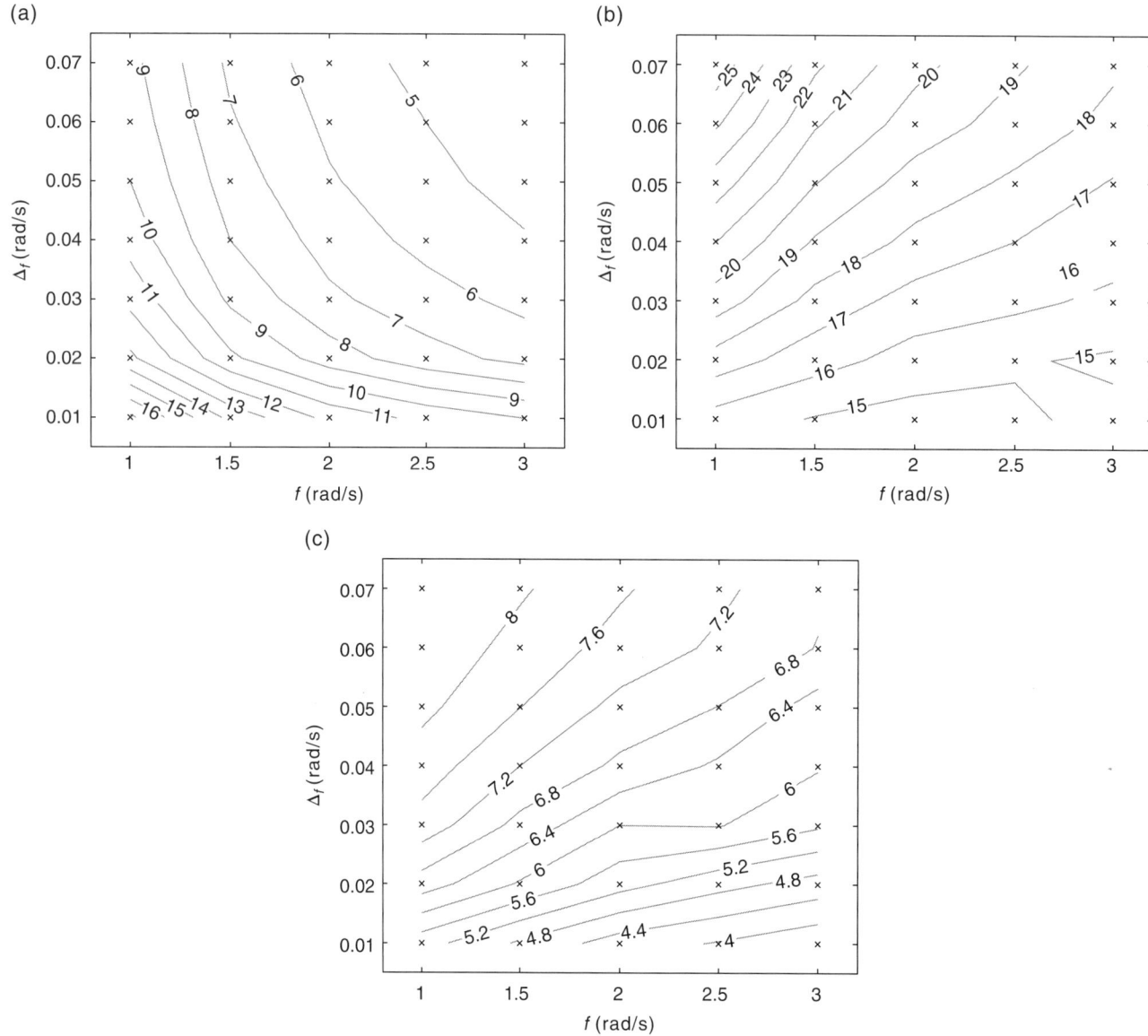

Figure 12.8. Contour plots of (a) the time of wave breaking in seconds, (b) the breaking wavelength in centimeters, and (c) the breaking wave amplitude in centimeters, calculated using the nondispersive ($\gamma = 0$) wave equation (12.28). The contours have been interpolated linearly between data points, which are indicated by crosses.

and the filtering must be adjusted to each experiment due to variations in the color of the dye line and the hues of the images.

12.6.2. Onset of Wave Breaking

The time at which the wave first breaks serves as a direct quantitative comparison of our theoretical, numerical, and experimental results. It is also an indication of how effectively PV is conserved as fluid crosses the shelf line, as waves of the same length will tend to steepen more quickly if they entrain a larger relative vorticity. In our theory, numerical solutions, and experiments, we define T_B as the smallest t for which the corresponding wave-breaking condition is satisfied. In Figures 12.8a, 12.9a, and 12.10a, we plot T_B over our experimental ranges of f and Δ_f, as calculated from our nonlinear wave theory, our numerical solutions, and our experiments, respectively.

The variation of T_B with f and Δ_f is qualitatively similar in our nonlinear wave theory and QG numerical solutions: A swift coastal current or strong rotation leads to the wave breaking much more rapidly than a slow current or weak rotation. However, solutions of the nonlinear wave equation break more than twice as

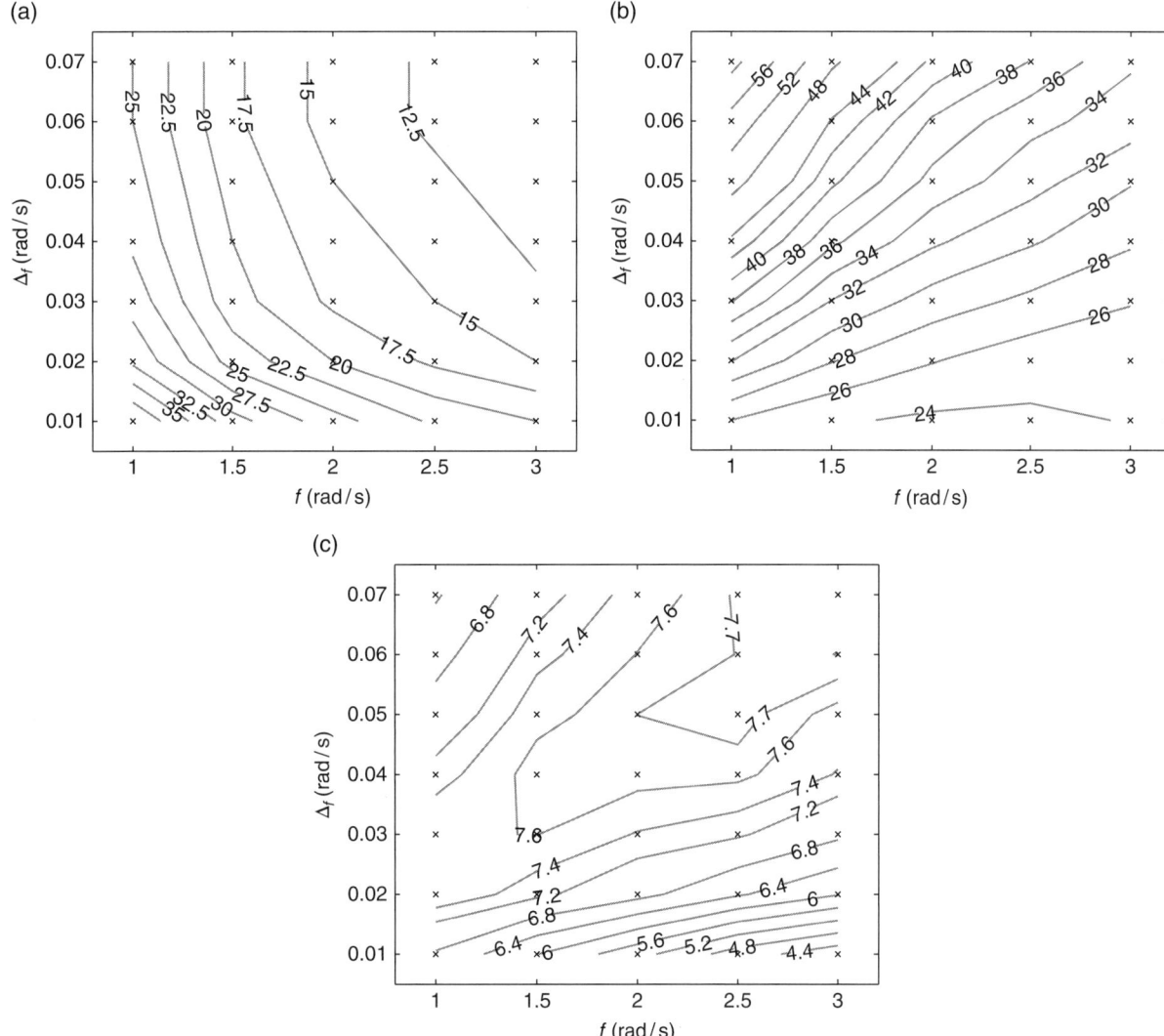

Figure 12.9. Contour plots of (a) the time of wave breaking in seconds, (b) the breaking wavelength in centimeters, and (c) the breaking wave amplitude in centimeters computed in numerical solutions of the QG model equations (12.34a) and (12.34b). The contours have been interpolated linearly between data points, which are indicated by crosses.

quickly. This reflects the fact that the continuous slope used in our numerical solutions results in a finite width PV front, whereas the theory describes a discontinuous front. The resulting wave envelops a smaller net relative vorticity, particularly during its initial formation, and therefore steepens less rapidly. The long-wave approximation used to derive (12.28) also fails when the gradient of the interface $R(\theta, t)$ becomes $\mathcal{O}(1)$, and this may exaggerate the rate at which the wave steepens and breaks.

Figure 12.10a shows that T_B is again qualitatively similar in our experiments and numerical solutions. In Figure 12.11a we plot the relative error in the numerically computed breaking times, $\Delta T_B = T_{B\text{numerical}}/T_{B\text{experiment}} - 1$. The QG model tends to underpredict T_B relative to the laboratory experiments, typically by a factor of $\sim \frac{2}{3}$. This indicates that PV is imperfectly conserved in our experimental channel, particularly given the similarities in the wavelength at breaking (see Section 12.6.3). This may be due to stronger bottom friction acting over the experimental shelf and slope, as discussed in Section 12.3.1.

12.6.3. Breaking Wavelength

We now turn our attention to the length of the lee wave at breaking. Apart from being a useful way to quantify the closeness of our QG solutions and laboratory experiments, the breaking wavelength determines the scale of the eddies that form behind the protrusion. We define L_B as the distance along the shelf line between the azmiuthal

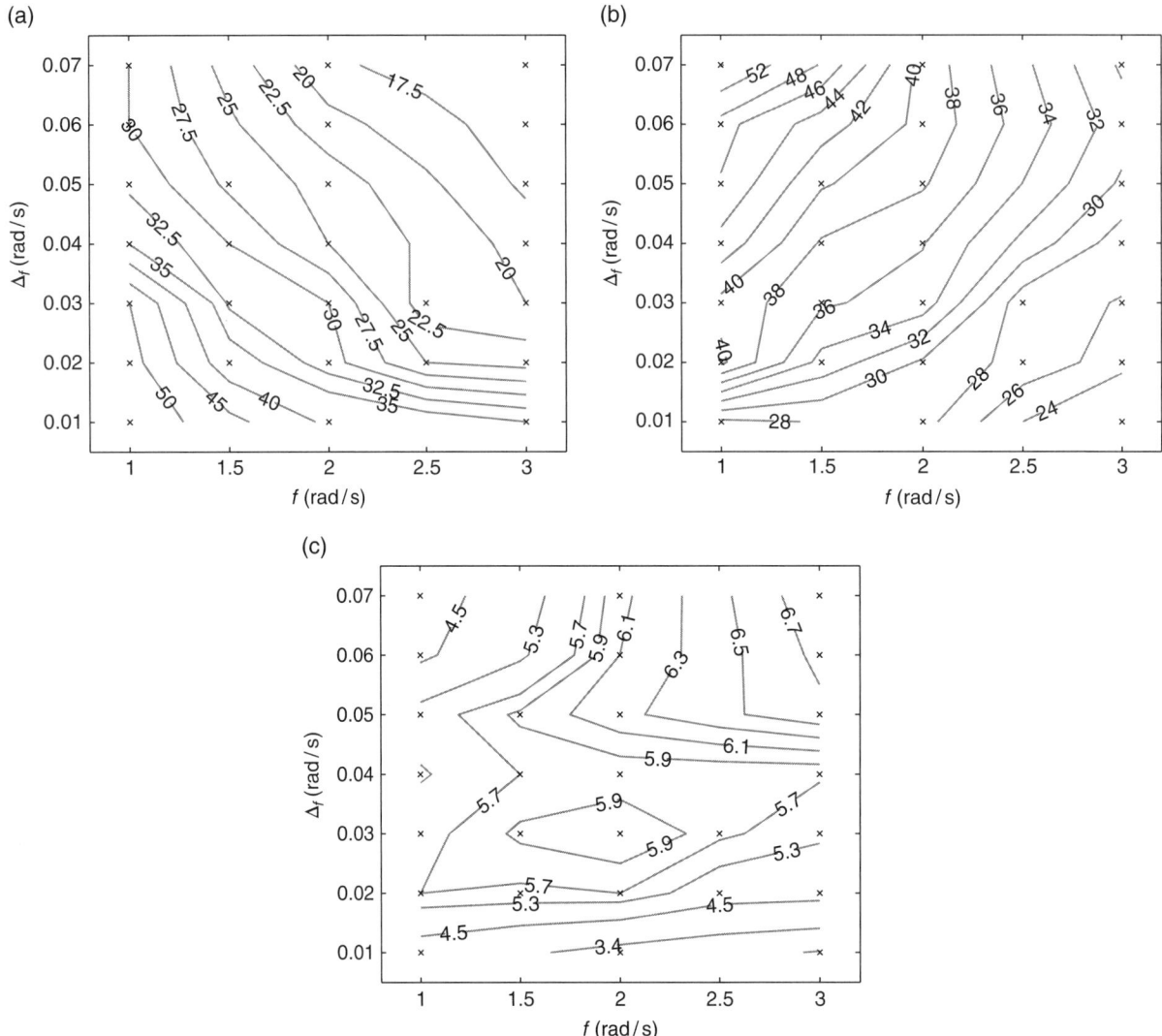

Figure 12.10. Contour plots of (a) the time of wave breaking in seconds, (b) the breaking wave length in centimeters, and (c) the breaking wave amplitude in centimeters, extracted from experimental results. The contours have been interpolated linearly between data points, which are indicated by crosses.

position (θ, ϕ_i, or θ_p) at which the wave has broken (θ_B) and the smallest subsequent $\theta > \theta_B$ at which the wave envelope lies below the shelf line $r = R_h$. In Figures 12.8b, 12.9b, and 12.10(b) we compare L_B between our nonlinear wave theory, numerical solutions, and experiments. We have converted L_B to distances in centimeters along the shelf line.

As with T_B in Section 12.6.2, we find that L_B has a qualitatively similar dependence on f and Δ_f in our theory and numerical solutions. A swift coastal current or weak background rotation yields a long wave at breaking, because water is drawn further onto the shelf by the mean flow before the relative vorticity it acquires can steepen and break the wave. Waves evolving under the nonlinear wave equation break at approximately half the length of the waves in our numerical solutions, consistent with their shorter breaking time. The nonlinear wave theory predicts an unusually large L_B for $f = 3$ rad/s and $\Delta_f = 0.01$ rad/s. In this corner of parameter space the wave speed given by (12.28) exceeds the mean-flow speed at the protrusion for sufficiently small perturbations of the interface R about the shelf line $r = R_h$. This allows the wave envelope to propagate further upstream, resulting in a larger wavelength at breaking.

The breaking wavelength shows good qualitative and quantitative agreement between our numerical solutions and laboratory experiments. Figure 12.11b shows that the relative error in the numerically computed breaking lengths, $\Delta L_B = L_{B\text{numerical}}/L_{B\text{experiment}} - 1$, is consistently below 20%. This is surprising because the experimental

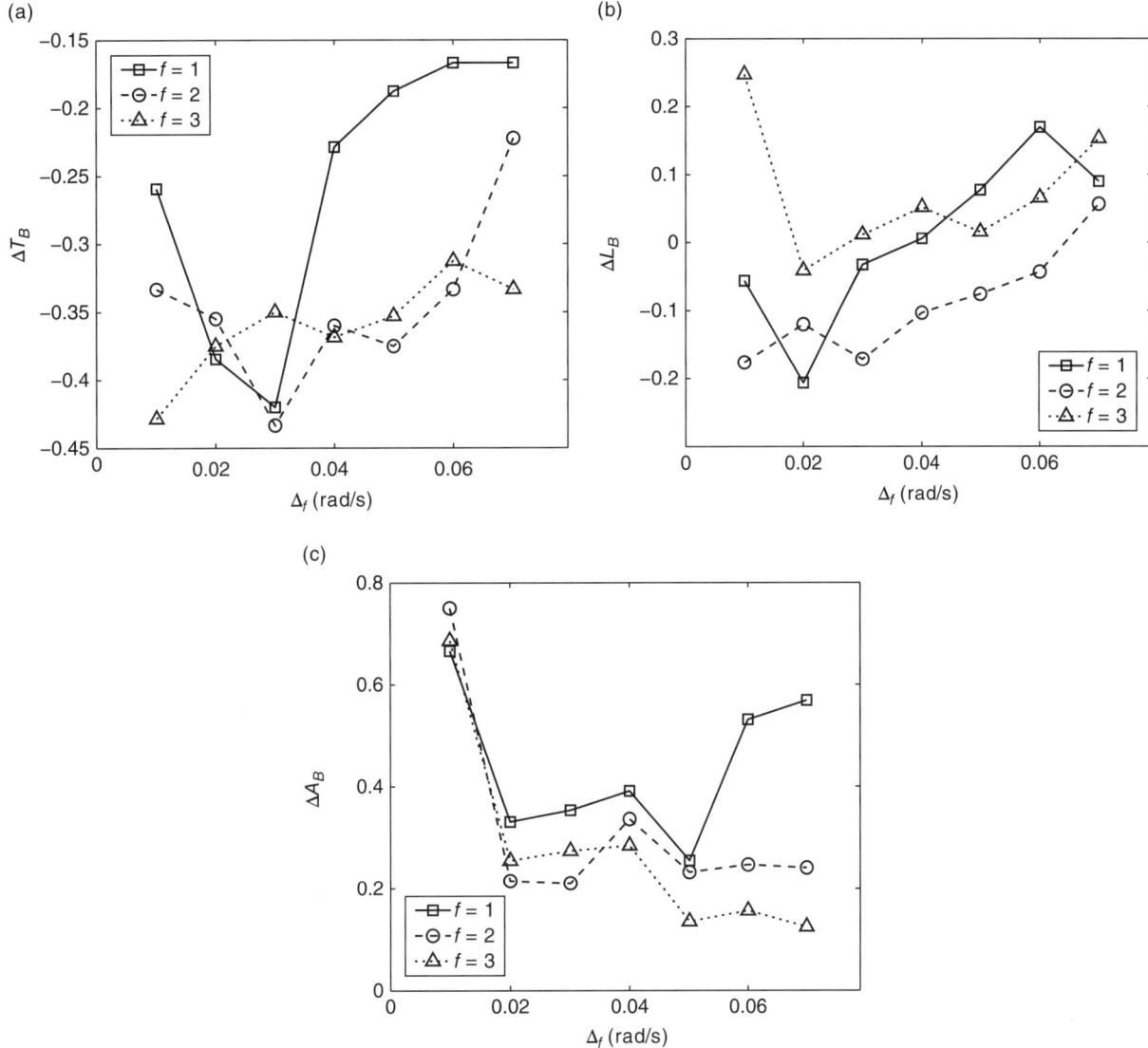

Figure 12.11. Relative error between numerical and experimental calculations for (a) the time of wave breaking, (b) the breaking wave length, and (c) the breaking wave amplitude.

waves take longer to break and so may be expected to be longer than numerically computed waves advected by the same mean flow. This may be due to a canceling effect with the smaller amplitudes of the experimental waves (see Section 12.6.4), which results in much shorter wavelengths due to the small aspect of the wave envelope.

12.6.4. Amplitude at Breaking

Lastly, we analyze the amplitudes of the waves at the point of breaking. This is directly relevant to the cross-slope exchange of ocean waters, as it serves as a measure of the volume transported onto the shelf due to advection past the protrusion. We define the breaking amplitude A_B as the largest radial extent of the wave envelope, measured from $r = R_h$, between θ_B and $\theta_B + L_B$. We plot A_B for our theory, numerical solutions, and experiments in Figures 12.8c, 12.9c, and 12.10c, respectively.

In this case there is a contrast between the amplitudes predicted by the nonlinear shelf wave theory and the numerical solutions. In Figure 12.8c, A_B is largest for swift coastal currents (large Δ_f) and weak background rotation (small f) and vice versa. Figure 12.9c exhibits a similar pattern when the coastal current is weak, but for sufficiently large Δ_f we find that A_B instead increases with f. This reflects the action of two competing effects on the amplitude of the wave envelope: Stronger relative vorticity within the wave envelope drives a stronger flow onto the shelf but also induces breaking more rapidly (see Section 12.6.2). The former is poorly represented in our long-wave theory, which implicitly assumes a small ratio of cross-slope to along-slope transport. We have recalculated Figue 12.8c using the full dispersive wave equation (12.28), with a much smaller gradient S_{max} in (12.50), but this still fails to capture the pattern shown in Figure 12.9c.

The pattern of A_B in Figure 12.10c is qualitatively similar to Figure 12.9c, albeit somewhat distorted because A_B is particularly sensitive to artifacts in the filtered experimental images. In Figure 12.11c we plot the relative error in the numerically computed amplitudes at breaking, $\Delta A_B = A_{B\text{numerical}}/A_{B\text{experiment}} - 1$. The wave amplitude is typically around 1–2 cm larger in our numerical solutions than in our experiments. We attribute this to our assumption of horizontally nondivergent flow under the QG approximation, which neglects depth changes in the mass conservation equation. Our numerical solutions therefore overestimate the volume of fluid drawn onto the shelf in the lee of the protrusion in the outer wall.

12.7. SUMMARY AND DISCUSSION

The large-scale flow of coastal currents is dominated by the interaction of the current with the strong topographic PV gradient at the continental shelf break, which results in complex behavior even in a strongly barotropic fluid. This has motivated previous experimental studies ranging from generation of isolated topographic Rossby waves [*Ibbetson and Phillips*, 1967; *Caldwell et al.*, 1972; *Cohen et al.*, 2010] to turbulent coastal currents [*Cenedese et al.*, 2005; *Sutherland and Cenedese*, 2009]. This chapter has focused on the flow of a retrograde coastal current past a headland that protrudes out onto the continental shelf. Examples of such a configuration include the flow of the Agulhas current through the Mozambique Channel [*Bryden et al.*, 2005; *Beal et al.*, 2006, 2011] and the Gulf Stream approach to Cape Hatteras [*Stommel*, 1972; *Johns and Watts*, 1986; *Pickart*, 1995].

In Section 12.2 we introduced our laboratory experiments whose setup is sketched in Figure 12.1 and illustrated in Figure 12.2. We constructed an annular channel in a rotating tank with a narrow slope leading to a raised shelf around the outer rim. The shelf was constricted over an azimuthal length $\sim \Theta_b$ by a protrusion in the outer wall, representing a continental headland. We generated a retrograde azimuthal mean flow, representing a coastal current, by changing the rate of the tank's rotation. This resulted in the formation of a large-amplitude Rossby shelf wave in the lee of the protrusion, which we visualized using a dye line that was positioned initially along the center of the PV front. Figure 12.3 characterizes the evolution of the flow, which develops a long, large-amplitude shelf wave that breaks and overturns. Thereafter the wave may unravel and form a wave train, as shown in Figure 12.3e, or may instead roll up into eddies.

To interpret the behavior of our experimental shelf waves, we introduced in Section 12.3 a QG shallow-water model. The model assumes nondivergent, geostrophically balanced barotropic flow beneath a rigid lid with small variations in depth. We applied no-flux boundary conditions for the channel walls and prescribed initial conditions that conserve the absolute vorticity of each fluid column and the total kinetic energy during the initial change in the tank's rate of rotation. Some discrepancy is to be expected between this model and our experiments, in which the surface height may vary across the channel and the shelf occupies a quarter of the water depth. However, QG theory has been shown to describe rapidly rotating laboratory flows far outside the formal range of validity [*Williams et al.*, 2010] and retains the vortical dynamics required to capture the evolution of shelf waves [*Johnson and Clarke*, 2001].

In Section 12.4 we derived a nonlinear wave theory to provide an intuitive description of the shelf wave evolution. Our theory follows that of *Haynes et al.* [1993] and *Clarke and Johnson* [1999] for a straight channel, approximating the slope as a discontinuity in depth and posing an asymptotic expansion in a small parameter that measures the ratio of the wave amplitude to the wavelength. The theory does not require the wave amplitude to be small, however: Making this additional assumption yields a Korteweg–de Vries-like equation [*Johnson and Clarke*, 1999]. Our nonlinear wave equation (12.28) improves upon the derivations of *Stewart* [2010] and *Stewart et al.* [2011] via the addition of an evolution equation (12.30) for the stream function on the outer wall. Figure 12.4 shows that the evolution of the first-order nonlinear wave equation initially resembles our experimentally generated waves, but dispersion prevents these solutions from capturing the wave breaking. By contrast, solutions of the nondispersive zero-order wave equation will always break but thereafter form a persistent shock that does not resemble the experimental flows.

To resolve the disparity between our experiments and our long-wave theory, in Section 12.5 we developed a numerical scheme that solves the QG model equations in a wall-following coordinate system. We advect an array of tracer particles that delineates the wave envelope, mimicking the dye line in our laboratory experiments. Figure 12.7 shows that the numerical shelf wave is qualitatively similar to our laboratory experiments, but the evolution occurs more rapidly. This may be due to imperfect conservation of PV or stronger bottom friction in the rotating tank. The numerical solution shows that regions of closed streamlines appear as the shelf wave curls up, forming eddies that transport water across the slope.

In Section 12.6 we compared shelf waves formed in our long-wave theory, numerical solutions, and laboratory experiments. The clearest point of comparison between the three is initial breaking of the wave; the evolution therafter varies widely, impeding quantitative comparison. In our long-wave theory a stronger coastal current (Δ_f) results in a larger PV flux across the slope, leading to longer shelf waves with larger amplitudes that break more rapidly. Stronger rotation f produces a larger relative vorticity in the wave envelope, resulting in shelf waves that break more rapidly at smaller wavelengths and amplitudes. Our numerical and experimental results exhibit similar patterns of breaking time (T_B) and wavelength at breaking (L_B), but the amplitude at breaking (A_B) may increase or decrease with f, depending on the size of Δ_f. We attribute this to competing tendencies for the wave to transport more water across the slope, but also to break more rapidly, when the relative vorticity in the wave envelope is larger. Figure 12.11 compares the properties of breaking in our numerical solutions and laboratory experiments. While L_B is consistently within around 20% error, our QG model underpredicts T_B and overpredicts A_B, in each case by a factor of $\frac{1}{3}$ on average. This is again consistent with imperfect conservation of PV or stronger bottom friction in the rotating tank.

The large amplitude of our coastal shelf waves distinguishes our experiments from previous studies of topographic Rossby wave generation, and the development of a nonlinear wave theory aids in interpreting our results (see Section 12.4). Previous investigations of Rossby waves over continental slopes, such as those of *Caldwell et al.* [1972], *Caldwell and Eide* [1976], and *Cohen et al.* [2010], have shown very good agreement with the corresponding theoretical dispersion relations. *Pierini et al.* [2002] found that numerical solutions of the barotropic shallow-water equations closely reproduced Rossby normal modes on an experimental slope between two walls. Our QG numerical solutions accurately describe the evolution of our experiments, as shown in Section 12.6. This indicates that flow in the rotating tank adheres closely to columnar motion, despite the formal requirements for the shallow-water QG approximation being poorly satisfied [see *Williams et al.*, 2010].

Solutions of the nonlinear shelf wave equation in Section 12.4 differ substantially from our experimental results. Our nonlinear wave theory cannot capture the overturning of the wave and the resulting break-up into eddies, due to the assumption of a single-valued interface $r = R(\theta, t)$. The theory should describe the flow accurately as long as the azimuthal gradients remain $\mathcal{O}(\mu^{1/2})$, which is certainly the case during the initial generation of the wave. The dispersive terms in (12.28) may be expected to inhibit wave steepening and thereby maintain a small amplitude-to-wavelength ratio, as in Figure 12.4. It is therefore unclear why the numerical and experimental shelf waves develop gradients that are $\mathcal{O}(1)$ within 10 − 50 s. The formation of a wave train in Figures 12.3 and 12.7 indicates that some dispersion takes place but that it is not sufficient to prevent breaking. This phenomenon is the subject of ongoing research.

Previous laboratory investigations of coastal currents have focused on buoyant gravity currents [*Whitehead and Chapman*, 1986; *Cenedese and Linden*, 2002; *Folkard and Davies*, 2001; *Wolfe and Cenedese*, 2006; *Sutherland and Cenedese*, 2009] or turbulent jets [*Cenedese et al.*, 2005]. Our experiments employ a barotropic coastal current generated via a rapid change in the tank's rotation rate. Such a configuration has received little attention in previous laboratory studies, though *Pierini et al.* [2002] generated a barotropic shoreward mean flow using a wide, slow-moving plunger. Our use of dye to visualize our experimental results implicitly describes the evolution of the coastal current as a wave. This may be insufficient for more realistic bathymetry like the coastal trough of *Sutherland and Cenedese* [2009], though their trough-crossing current resembles a trapped topographic Rossby wave [*Kaoullas and Johnson*, 2012]. The wavelike description of the current breaks down when eddies form, as in the experiments of *Folkard and Davies* [2001] and *Wolfe and Cenedese* [2006], and in the long-term evolution of our experimental and numerical coastal currents.

Our results show that retrograde coastal current flow past a continental headland shifts the PV front shoreward from the continental slope. The resulting large-amplitude lee wave breaks and subsequently tends to form eddies that exchange water between the coastal waters and the deep ocean. However, if the velocity of the current is sufficiently large, then the wave and any eddies are simply swept away downstream. Our parameter sweep in Section 12.6 yields insight into the eddies formed by the wave breaking and the resulting exchange of water across the shelf break. For the purpose of comparison with the real ocean, Δ_f measures the velocity of the coastal current, and f measures the PV jump between the continental shelf and the open-ocean. The lee wave is longest, resulting in the largest eddies, when the coastal current is strong and the PV jump is small. We measure the penetration of open-ocean water onto the experimental shelf using the amplitude of the lee wave at breaking. In general, water is transported further onto the shelf when the coastal current is strong and the PV jump is large, but the complete dependence on f and Δ_f is somewhat more complicated (see Section 12.6.4).

There is a developed body of theory that describes the evolution of coastal currents in terms of shelf waves [e.g., *Mysak*, 1980b; *Johnson and Clarke*, 2001]. Predictive models of coastal shelves require adaptation for the sharply varying bathymetry and must employ very high resolution to capture the shelf processes [e.g., *Haidvogel et al.*, 2008]. As a result, there remains scope for laboratory experiments to inform oceanographic research on the subject of continental shelf dynamics and to evaluate numerical ocean models. It would be particularly valuable to perform further experiments with strongly nonlinear flows, such as the evolution of turbulent boundary currents [*Sutherland and Cenedese*, 2009], generation of geostrophic eddies over the continental slope [*Pennel et al.*, 2012], or the nonlinear shelf waves considered here.

Acknowledgments. This research was supported by an EPSRC DTA award to A. L. S. and by the Summer Study Program in Geophysical Fluid Dynamics at Woods Hole Oceanographic Institution, funded by NSF grant OCE-0824636 and ONR grant N00014-09-10844. P. J. D.'s research was supported by an EPSRC Advanced Research Fellowship [grant number EP/E054625/1]. The authors thank Jack Whitehead for granting them use of the GFD laboratory at Woods Hole Oceanographic Institution and Anders Jensen for indispensible assistance and advice in constructing and conducting the experiments described herein.

REFERENCES

Allen, J. S. (1980), Models of wind-driven currents on the continental shelf, *Ann. Rev. Fluid Mech.*, *12*(1), 389–433.

Arakawa, A. (1966), Computational design for long-term numerical integration of the equations of fluid motion: Two-dimensional incompressible flow. Part I, *J. Comp. Phys.*, *1*(1), 119–143.

Beal, L. M., T. K. Chereskin, Y. D. Lenn, and S. Elipot (2006), The sources and mixing characteristics of the Agulhas Current, *J. Phys. Oceanogr.*, *36*(11), 2060–2074.

Beal, L. M., W. P. M. De Ruijter, A. Biastoch, and R. Zahn (2011), On the role of the Agulhas system in ocean circulation and climate, *Nature*, *472*(7344), 429–436.

Boyd, J. P. (2001), *Chebyshev and Fourier Spectral Methods*, Dover, Mineola, NY.

Brink, K. H. (1991), Coastal-trapped waves and wind-driven currents over the continental shelf, *Ann. Rev. Fluid Mech.*, *23*(1), 389–412.

Bryden, H. L., L. M. Beal, and L. M. Duncan (2005), Structure and transport of the Agulhas Current and its temporal variability, *J. Oceanogr.*, *61*, 479–492.

Buchwald, V., and W. Melville (1977), Resonance of shelf waves near islands, in *Waves on Water of Variable Depth*, edited by D. G. Provis and R. Radok, pp. 202–205, Springer, Berlin, Heidelberg.

Buchwald, V. T., and J. K. Adams (1968), The propagation of continental shelf waves, *Proc. Roy. Soc. A*, *305*(1481), 235–250.

Caldwell, D. R., and S. A. Eide (1976), Experiments on the resonance of long-period waves near islands, *Proc. R. Soc. A*, *348*(1654), 359–378.

Caldwell, D. R., D. L. Cutchin, and M. S. Longuet-Higgins (1972), Some model experiments on continental shelf waves, *J. Mar. Res.*, *30*(1), 39–55.

Carnevale, G. F., S. G. Llewellyn Smith, F. Crisciani, R. Purini, and R. Serravall (1999), Bifurcation of a coastal current at an escarpment, *J. Phys. Oceanogr.*, *29*(5), 969–985.

Cenedese, C., and P. F. Linden (2002), Stability of a buoyancy-driven coastal current at the shelf break, *J. Fluid Mech.*, *452*, 97–121.

Cenedese, A., C. Vigarie, and E. V. Di Modrone (2005), Effects of a topographic gradient on coastal current dynamics, *J. Geophys. Res.*, *110*(C9), C09,009.

Clarke, S. R., and E. R. Johnson (1999), Finite-amplitude topographic Rossby waves in a channel, *Phys. Fluids*, *11*, 107–120.

Cohen, Y., N. Paldor, and J. Sommeria (2010), Laboratory experiments and a non-harmonic theory for topographic Rossby waves over a linearly sloping bottom on the f-plane, *J. Fluid Mech.*, *645*(1), 479–496.

Cohen, Y., N. Paldor, and J. Sommeria (2012), Application of laboratory experiments to assess the error introduced by the imposition of "wall" boundary conditions in shelf models, *Ocean Modell.*, *41*, 35–41.

Colin de Verdiere, A. (1979), Mean flow generation by topographic Rossby waves, *J. Fluid Mech.*, *94*, 39–64.

Folkard, A. M., and P. A. Davies (2001), Laboratory studies of the effects of interrupted, sloping topography on intermediate depth boundary currents in linearly stratified fluids, *Dyn. Atmos. Ocean.*, *33*(4), 239–261.

Gill, A. E., and E. H. Schumann (1979), Topographically induced changes in the structure of an inertial coastal jet: Application to the Agulhas Current, *J. Phys. Oceanogr.*, *9*(5), 975–991.

Gordon, A. L. (1985), Indian-Atlantic transfer of thermocline water at the Agulhas Retroflection, *Science*, *227*(4690), 1030–1033.

Gordon, A. L. (1986), Interocean exchange of thermocline water, *J. Geophys. Res*, *91*(C4), 5037–5046.

Gresho, P. M. (1991a), Incompressible fluid dynamics: Some fundamental formulation issues, *Ann. Rev. Fluid Mech.*, *23*, 413–453.

Gresho, P. M. (1991b), Some current CFD issues relevant to the incompressible Navier-Stokes equations, *Comp. Meth. Appl. Mech. Eng.*, *87*, 201–252.

Haidvogel, D. B., et al. (2008), Ocean forecasting in terrain-following coordinates: Formulation and skill assessment of the Regional Ocean Modeling System, *J. Comp. Phys.*, *227*(7), 3595–3624.

Haynes, P. H., E. R. Johnson, and R. G. Hurst (1993), A simple model of Rossby-wave hydraulic behaviour, *J. Fluid Mech.*, *253*, 359–384.

Hou, T. Y., and R. Li (2007), Computing nearly singular solutions using pseudo-spectral methods, *J. Comput. Phys.*, *226*, 379–397.

Ibbetson, A., and N. Phillips (1967), Some laboratory experiments on Rossby waves in a rotating annulus, *Tellus*, *19*(1), 81–87.

Johns, W. E., and D. R. Watts (1986), Time scales and structure of topographic Rossby waves and meanders in the deep Gulf Stream, *J. Mar. Res.*, *44*(2), 267–290.

Johnson, E. (1987), Topographic waves in elliptical basins, *Geophy. Astrophys. Fluid*, *37*(4), 279–295.

Johnson, E. R. (1985), Topographic waves and the evolution of coastal currents, *J. Fluid Mech.*, *160*, 499–509.

Johnson, E. R., and S. R. Clarke (1999), Dispersive effects in Rossby-wave hydraulics, *J. Fluid Mech.*, *401*, 27–54.

Johnson, E. R., and S. R. Clarke (2001), Rossby wave hydraulics, *Ann. Rev. Fluid Mech.*, *33*(1), 207–230.

Johnson, E. R., and M. K. Davey (1990), Free-surface adjustment and topographic waves in coastal currents, *J. Fluid Mech.*, *219*, 273–289.

Johnson, E. R., and G. Kaoullas (2011), Bay-trapped low-frequency oscillations in lakes, *Geophys. Astrophys. Fluid*, *105*(1), 48–60.

Johnson, E. R., J. T. Rodney, and G. Kaoullas (2012), Trapped modes in coastal waveguides, *Wave Motion*, *49*(1), 212–216.

Kaoullas, G., and E. R. Johnson (2010), Geographically localised shelf waves on curved coasts, *Cont. Shelf Res.*, *30*(16), 1753–1760.

Kaoullas, G., and E. R. Johnson (2012), Isobath variation and trapping of continental shelf waves, *J. Fluid Mech.*, *700*(1), 283–303.

Longuet-Higgins, M. S. (1968), On the trapping of waves along a discontinuity of depth in a rotating ocean, *J. Fluid Mech.*, *31*, 417–434.

Martinson, D. G., S. E. Stammerjohn, R. A. Iannuzzi, R. C. Smith, and M. Vernet (2008), Western Antarctic Peninsula physical oceanography and spatio–temporal variability, *Deep Sea Res. Pt. II*, *55*(18), 1964–1987.

McWilliams, J. C. (1977), A note on a consistent quasi-geostrophic model in a multiply connected domain, *Dyn. Atmos. Oceans*, *1*, 427–441.

Minobe, S., A. Kuwano-Yoshida, N. Komori, S. P. Xie, and R. J. Small (2008), Influence of the gulf stream on the troposphere, *Nature*, *452*(7184), 206–209.

Mysak, L. A. (1980a), Topographically trapped waves, *Ann. Rev. Fluid Mech.*, *12*, 45–76.

Mysak, L. A. (1980b), Recent advances in shelf wave dynamics, *Rev. Geophys.*, *18*, 211–241.

Mysak, L. A., P. H. Leblond, and W. J. Emery (1979), Trench waves, *J. Phys. Oceanogr.*, *9*(5), 1001–1013.

Nittrouer, C. A., and L. D. Wright (1994), Transport of particles across continental shelves, *Rev. Geophys.*, *32*(1), 85–113.

Pedlosky, J. (1987), *Geophysical Fluid Dynamics*, Springer, New York and Berlin.

Pennel, R., A. Stegner, and K. Béranger (2012), Shelf impact on buoyant coastal current instabilities, *J. Phys. Oceanogr.*, *42*, 39–61.

Pickart, R. S. (1995), Gulf Stream-generated topographic Rossby waves, *J. Phys. Oceanogr.*, *25*(4), 574–586.

Pierini, S., A. M. Fincham, D. Renouard, M. R. D'Ambrosio, and H. Didelle (2002), Laboratory modeling of topographic Rossby normal modes, *Dyn. Atmos. Ocean.*, *35*(3), 205–225.

Prézelin, B. B., E. E. Hofmann, M. Moline, and J. M. Klinck (2004), Physical forcing of phytoplankton community structure and primary production in continental shelf waters of the Western Antarctic Peninsula, *J. Mar. Res.*, *62*(3), 419–460.

Rodney, J. T., and E. R. Johnson (2012), Localisation of coastal trapped waves by longshore variations in bottom topography, *Cont. Shelf Res.*, *32*, 130–137.

Sommeria, J., S. D. Meyers, and H. L. Swinney (1991), Experiments on vortices and Rossby waves in eastward and westward jets, *Nonlin. Top. Ocean Phys.*, *109*, 227–269.

Stewart, A. L. (2010), Nonlinear shelf waves in a rotating annulus, *Technical Report of the 2009 Geophysical Fluid Dynamics Program at Woods Hole Oceanographic Institution*, pp. 373–402.

Stewart, A. L., P. J. Dellar, and E. R. Johnson (2011), Numerical simulation of wave propagation along a discontinuity in depth in a rotating annulus, *Comput. Fluids*, *46*, 442–447.

Stommel, H. (1972), *The Gulf Stream: A Physical and Dynamical Description*, Univ. of California Press, Berkeley and Los Angeles, California, 1972.

Sutherland, D. A., and C. Cenedese (2009), Laboratory experiments on the interaction of a buoyant coastal current with a canyon: Application to the East Greenland Current, *J. Phys. Oceanogr.*, *39*(5), 1258–1271.

Thompson, A. F., and K. J. Heywood (2008), Frontal structure and transport in the northwestern Weddell Sea, *Deep-Sea Res. I*, *55*(10), 1229–1251.

Vallis, G. K. (2006), *Atmospheric and Oceanic Fluid Dynamics: Fundamentals and Large-Scale Circulation*, Cambridge Univ. Press, Cambridge, UK.

Whitehead, J., and D. C. Chapman (1986), Laboratory observations of a gravity current on a sloping bottom: The generation of shelf waves, *J. Fluid Mech.*, *172*, 373–99.

Williams, P. D., et al. (2010), Testing the limits of quasi-geostrophic theory: Application to observed laboratory flows outside the quasi-geostrophic regime, *J. Fluid Mech.*, *649*, 187–203.

Wolfe, C. L., and C. Cenedese (2006), Laboratory experiments on eddy generation by a buoyant coastal current flowing over variable bathymetry, *J. Phys. Oceanogr.*, *36*(3), 395–411.

13

Laboratory Experiments With Abrupt Thermohaline Transitions and Oscillations

John A. Whitehead

13.1. INTRODUCTION

Climate records indicate that ancient ocean temperatures occasionally change by many degrees within climatologically "fast" times (order of 50 years). Some of these are attributed to abrupt transitions of the thermohaline circulation regime [*Broecker et al.*, 1985; *Boyle*, 1990; *Keigwin and Jones*, 1994; *Keigwin et al.*, 1994; *Bard et al.*, 1996; *Broecker*, 1997; *Stocker and Wright*, 1991; *Stocker*, 2000; *Burns et al.*, 2003; *Weart*, 2003; and many others]. In addition, some numerical ocean circulation models produces abrupt transitions. The changes involve both salinity and temperature (henceforth always called thermohaline) changes [*Bryan*, 1986; *Cessi*, 1994; *Rahmstorf*, 1995; *Manabe and Stouffer*, 1995; *Whitehead*, 1998; *Rahmstorf and Ganopolski*, 1999, *Weaver et al.*, 1999]. The dynamics of such abrupt transitions is formulated in a pioneering mathematical box model study [*Stommel*, 1961]. This model has two well-mixed chambers of water connected side by side with one tube at the top and a second at the bottom. Both temperature and salinity diffuse through sidewalls to the chambers at different rates. Positive temperature and salinity diffuse into one chamber, and negative values diffuse in the other. The resulting flow has a range of the governing parameters in which there are two possible states, one with temperature dominance and the other with salinity dominance, each with a different flow rate and direction. Subsequent mathematical box models (reviewed by *Marotzke* [1994] and *Whitehead* [1995]) illuminate additional aspects that help us to understand how the abrupt thermohaline transitions arise and what their context might be. For example, ocean estuary mathematical box models have demonstrated abrupt thermohaline transitions [*Hearn and Sidhu*, 1999; *Bulgakov and Skiba*, 2003]. In actuality, no direct observations of abrupt thermohaline transitions in estuaries exist. Finally, abrupt transitions are also mathematically predicted for wind-forced convection [*Stommel and Rooth*, 1968] and in basins forced by surface stress alone [*Ierley and Sheremet*, 1995; *Jiang et al.*, 1995].

Mathematical box models and numerical simulations both have drawbacks. Although numerous box experiments readily produce abrupt thermohaline transitions, they are always subject to the criticism that they restrict the number of degrees of freedom that the flows can adopt. Mathematical box models cannot account for the large number of degrees of freedom that actual flows can include, making abrupt thermohaline transitions more prevalent than actually exist. Even though numerical models show that the abrupt thermohaline transitions do not vanish, a more convincing way to illustrate whether or not abrupt thermohaline transitions actually exist in a physical system is by developing controlled laboratory experiments.

Abrupt transitions in fluid mechanics are commonplace. For example, the sudden stall of an airplane wing along with all the dangers it produces to pilots and passengers, has been continually in the minds of aviators since the early part of the twentieth century. Generally, abrupt transitions occur within a finite range of the parameters that govern the flow. The flow in this range can have either one of two modes. Each mode is locally stable and the flow can be made to abruptly jump back and forth from one mode to another. Therefore, the transitions are said to have *hysteresis*, since the flow is determined by history, as well as by the governing parameters. Hysteresis is obviously a significant challenge to climate modeling if it exists (just as in airplane design). Depending on details of the

---

Department of Physical Oceanography, Woods Hole Oceanographic Institution, Woods Hole, Massachusetts, United States of America.

model's history, any given model might have one or the other climate.

The possibility of abrupt transitions is present in many areas of geophysical fluid dynamics. Virtually every natural body of fluid has two components that affect density and thereby the buoyancy flux of convection: temperature and water vapor affect air density, temperature and salinity affect ocean water density; temperature and different elements such as silicon, magnesium and hydrogen (water) affect the mantle density; temperature and sulfur affect the Earth's core, and heat and helium affect star density.

All aspects discussed above concerning abrupt thermohaline transitions have contributed to the motivation of producing the laboratory studies reviewed in this paper. In comparison to the many numerical and box models, so far, only a few laboratory experiments have produced these abrupt thermohaline transitions. The motivation to develop experiments with abrupt thermohaline transitions was also prompted by the fact that physical observations of thermohaline transitions in the ocean are unknown in historical or modern times. Therefore, before such experiments were completed, there were no direct scientific observations of the many theoretical ideas from simulations and models concerning abrupt thermohaline transitions. Devices to investigate abrupt thermohaline transitions require very precise controls of temperature, salinity, and in some cases, pumping rates. Generally, heat loss must be minimized too. Theory provides information on the parameter ranges required for hysteresis. This important information was used to design the devices. First, the flow must be driven by two components. Buoyancy force from temperature and salinity variations suffices for this. Furthermore, the boundary conditions must allow a flux of heat and salinity into the fluid at different rates. Then, the transitions are found when the buoyancy forces generated by temperature and salinity oppose each other.

Section 13.2 of this chapter reviews experimental observations of abrupt thermohaline transitions in the laboratory. This author and colleagues have performed all experiments, and much of the material is covered in *Whitehead* [2009]. Also, there is always a possibility that instead of an abrupt transition, the system oscillates back and forth between the two flow modes. Experiments that find such temperature and salinity oscillations are described in Section 13.3. Virtually no numerical simulations or ocean models produce similar oscillations, and their mechanism needs more investigation.

13.2. FOUR LABORATORY EXPERIMENTS SHOWING ABRUPT THERMOHALINE TRANSITIONS

Experimental apparatus used to find abrupt thermohaline transitions typically has one chamber containing fluid that is either heated or cooled. The chamber is connected by a passage consisting of some sort of opening to a reservoir kept at constant temperature and salinity. All known devices with enough precision to measure the range of hysteresis are sketched in Figure 13.1. The first one, called the *box experiment* (Figure 13.1a), has a well-mixed chamber connected laterally to a reservoir of fresh water at ambient room temperature (20° C) by two tubes one above the other [*Whitehead* 1996, 1998]. The chamber is heated below with a metal plate, and salt water is pumped into a sponge at the top surface. Heating causes water in the chamber to become lighter, but in contrast, the salt water influx causes it to become denser. The time constants for temperature and salinity are controlled by salt water pumping rate, chamber surface area, and depth.

The second, called the *slot experiment* (Figures 13.1b and 13.2), is like the first except that a tube replaces the sponge salinity source at the top of the chamber, and a vertical slot replaces the two tubes [*Whitehead et al.*, 2003]. Internal mixing from sinking of the salt water under the sponge is significantly reduced compared to the box experiment. Therefore, the possibility of one or more density layers within the chamber exists.

The third one, called the *layered experiment* (Figures 13.1c and 13.3), has the chamber connected to a reservoir with three tubes at the top, middle, and bottom [*Whitehead et al.*, 2005, *Whitehead and Bradley* 2006]. It is cooled from above instead of heated from below, and the reservoir contains a layer of fresh water with salt water below it, with both layers kept at 20° C by a heat transfer device labeled ILE in Figure 13.3. An early exploratory upside-down version of this 3-tube experiment had a chamber heated from below with a layer of salty water on the bottom below fresh water in the reservoir. This device is the first to document reproducible thermohaline oscillations [*teRaa* 2001]. Earlier runs by Bulgakov and his collaborators with the slot experiment also observed oscillations, but they were not reproducible.

The fourth, called the *cavity experiment* (Figure 13.1d), is geometrically the simplest. There is simply a cavity in the floor of a reservoir of fresh water at 20° C. The cavity is heated at the bottom, and salt water is pumped into it at a steady rate. The salt water is heated in the cavity; and it either spills out of the top lip of the cavity, which lies slightly above the floor of the reservoir, or it rises as a thermal to the top of the reservoir. The reservoir has a specially designed drain at both top and bottom so it can be flushed well enough to maintain fresh water at 20° C.

The experiments have progressively greater degrees of freedom. The box experiment has well mixed water inside the chamber so that no layers exist. The flows are limited to two modes in and out of the top tube, with corresponding flow out and in of the bottom tube. The slot

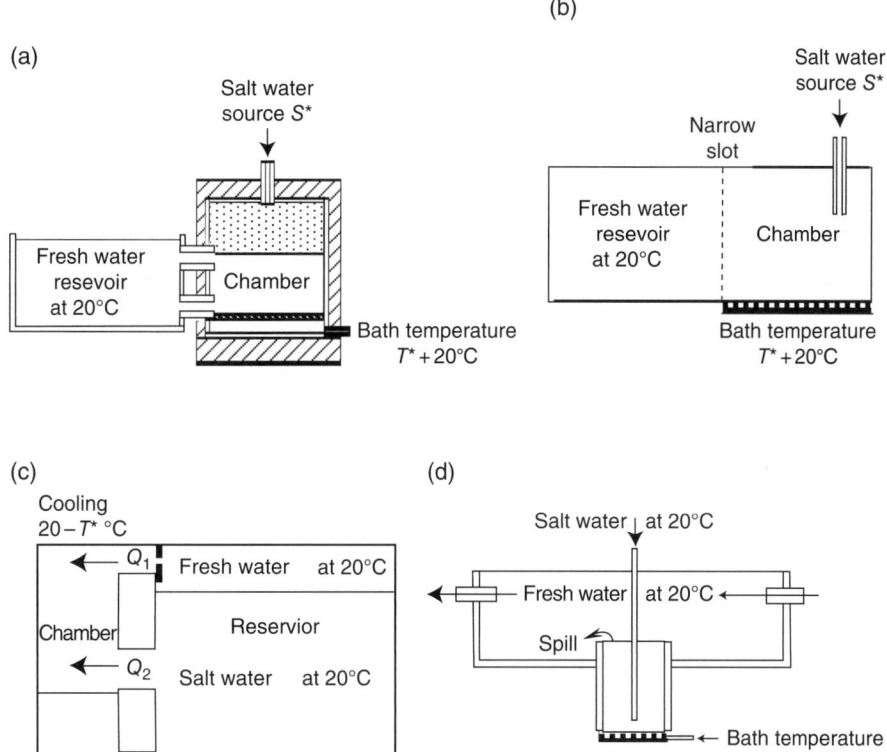

Figure 13.1. Four laboratory experimental configurations that have produced abrupt thermohaline transitions: (a) driving parameters T^*, S^*, and saltwater pumping rate (for a, b, and d) are indicated. (a) box experiment, (b) slot experiment, (c) layered experiment, and (d) cavity experiment. (Figure adapted from *Whitehead* 2009).

experiment is designed to limit the mixing so that layers in the box can exist. The layered experiment seeks to control flow in three layers, and the cavity experiment has fully three-dimensional flow.

As in theory and computer transitions, the abrupt thermohaline transitions that are produced in the experiments separate two flows that are distinctly different. This is most clearly seen in shadowgraphs of two flows on either side of an abrupt thermohaline transition in the slot experiment (Figure 13.2). Different names for the various modes are found in the literature, but for clarity in this review, we adopt a single set of names. The first such mode is flow driven primarily by temperature. The upper panel shows such a flow, and here it is called the temperature mode (called the T-mode in *Whitehead et al.* [2003]). This mode are found if the temperature of the bath above the reservoir temperature is greater in magnitude than a certain critical value T_T^*. Essentially, the flow is the same as the flow when salinity forcing is absent. In the top panel of Figure 13.2 this mode is shown by a shadowgraph. Water from the reservoir flows into the chamber at the bottom of the slot and hot salty water flows out

at the top as indicated by the arrows. The salty water is dyed, and the shadowgraph indicates that there is a lot of small-scale turbulent mixing. The mixing is provided by convection at relatively high Rayleigh numbers. The salt water injected by the tube is mixed and diluted to such an extent that the salinity makes negligible contribution to density. The hot plume rises to the top of the chamber where it exits through the top of the slot. The lower picture shows the salinity mode (called the S-mode by *Whitehead et al.* [2003]). This can be found if bath temperature is below a second critical value T_S^*. Three layers exist. The salty water sinks to the floor almost as though heating were absent and it forms a hot layer of water that flows out of the chamber at the bottom of the slot. Fresh water from the reservoir flows into the chamber along the top interface of the hot salty layer and forms the middle layer. It is heated by thermal conduction from the hot salty layer and rises to the top of the chamber accompanied by convection cells. The hot, fresh water then forms the top layer what exits the chamber through the top of the slot. Hysteresis happens because the experiments show that $T_T^* < T_S^*$, consequently *either* flow is found for

258 MODELING ATMOSPHERIC AND OCEANIC FLOWS

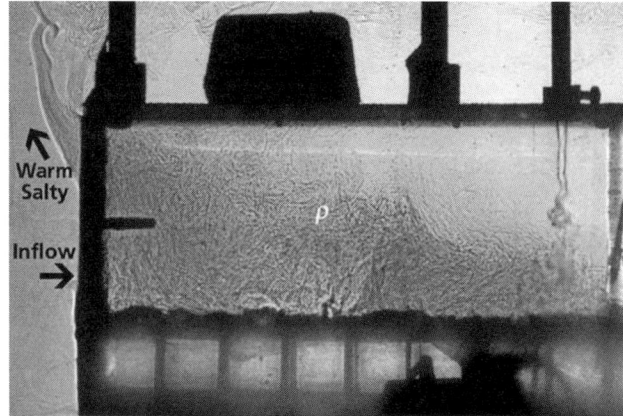

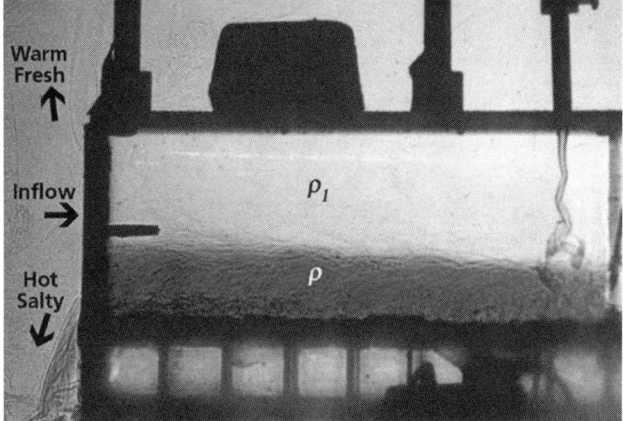

Figure 13.2. Side view of the slot experiment with the two different modes of flow at the same values of driving parameters. The top panel shows the temperature mode and the bottom panel shows the salt mode. BW rendition of a color figure [*Whitehead et al.*, 2003].

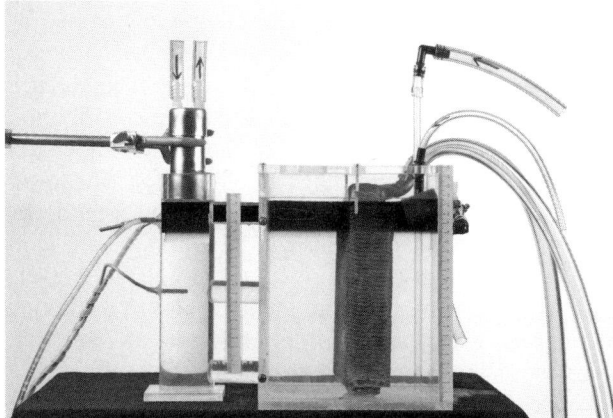

Figure 13.3. Layered experiment [*Whitehead* 2009].

bath temperatures between those two temperatures. The slot experiment is particularly appealing, because artificial mixing or suppression of mixing can trigger transition back and forth. If the salt mode exists, the flow can be converted to the temperature mode by inserting a small paddle into the chamber through the slot and mixing. If the temperature mode exists, the flow can be converted to the salt mode by inserting the paddle near the salinity source tube and suppressing the local turbulent mixing. Soon, a layer of salty water forms along the bottom and the salt mode forms.

The results are best summarized by plots of dimensionless temperature and salinity versus scaled bath temperature (Figure 13.4). For these runs, the value of salinity difference, and the pumping rates of salt water for the box, slot, and cavity experiments are each kept at one constant value. Experimental runs with different values of pumping rate were also studied in many of the papers, but results are inferior due to poorer coverage over parameter space, so they add little to the story. In all experiments, one complete run must always be conducted over a long enough time span to become convincingly steady. The necessary time is at least an hour and sometimes much longer. In regions of hysteresis, the two corresponding points with the same driving parameters are determined by how that particular run was initiated. One can start from an earlier run or start the apparatus with either fresh or salty water in the chamber. Transitions can be triggered by inserting temporary blocks in some of the tubes or by stirring, as mentioned above. Naturally, in points without hysteresis, the same point is found no matter how the flow is started.

Scaling the laboratory results is quite simple. The bath temperature, the temperature in the chamber, and the salinity of water in the chamber are plotted here using the density difference between salt and fresh water: $\bar{T}^* = \alpha T^*/\beta S_0$, $\bar{T} = \alpha T/\beta S_0$, and $\bar{S} = S/S_0$, where α is the coefficient of thermal expansion for water at 20° C, β is the density coefficient of salinity, and S_0 is salinity of the salt water pumped in at the source or, in the case of the layered experiment, the salinity of deep water. Since warm water has lower density than cold, the scaled temperature is inversely proportional to density but the scaled salinity is proportional to density. Thus, if scaled temperature and salinity are plotted together, (Panels a and c) the two opposing effects overlie each other and it is immediately observed as to whether temperature or salinity affects density more strongly. This is not true for Figure 13.4 panel b, where the scaled salinity is subtracted from unity to make the figure clear when compared to theory (shown by the curves).

Figure 13.4 shows a number of points. First density change from salinity always dominates over density change from temperature in the Salt Mode and density change from temperature always dominates over density change from salinity in the Temperature Mode, as expected. Second, the range of hysteresis is greatest for the box experiment and less so for the rest. A close relation between the experiments and box model theory similar to Stommel's are demonstrated in *Whitehead* (1996) and

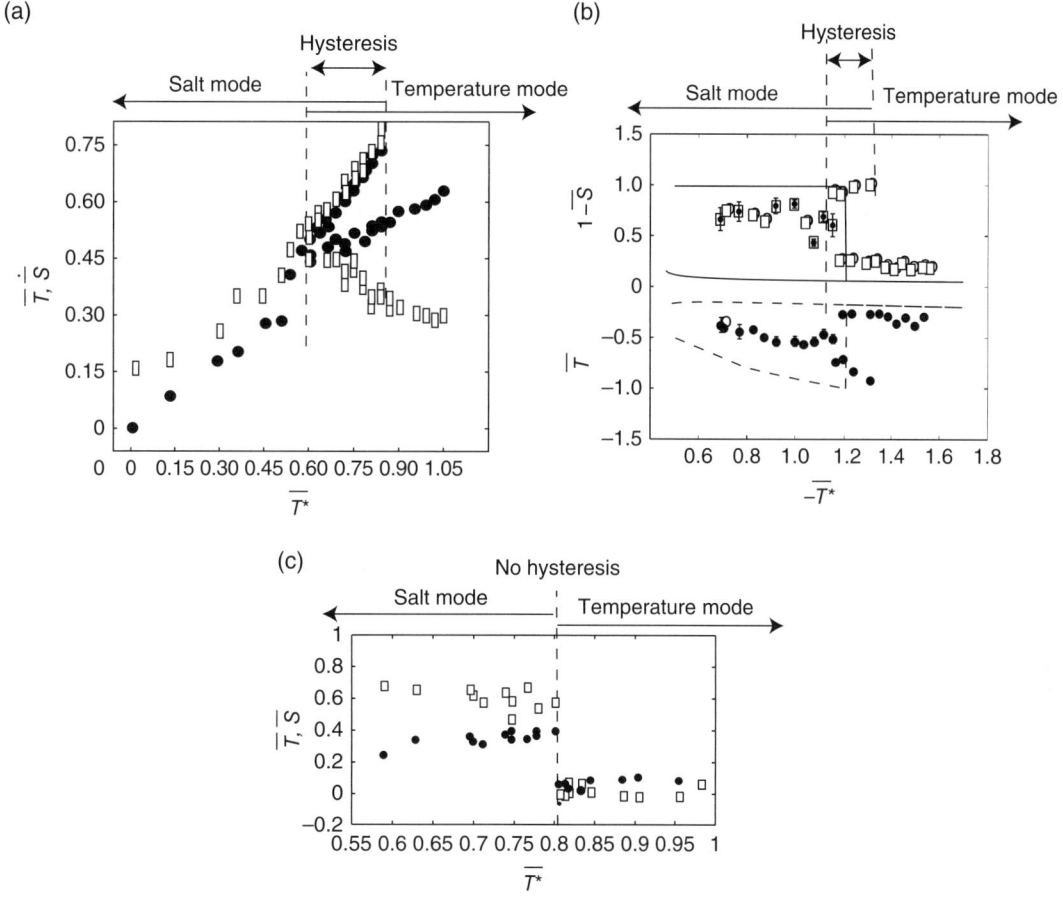

Figure 13.4. Measurements of dimensionless temperature (circles) and salinity (rectangles) versus driving temperature for (a) the box experiment, (b) the middle of the bottom layer of the layered experiment with the curves from the theory in *Whitehead*, [2000]. and (c) the middle of the cavity in the cavity experiment. The error bars in (b) are for the oscillation described in the next section and Adapted from *Whitehead* [2009].

Whitehead (1998), so it is not surprising to see sizeable hysteresis as the box model theory predicts. The theory for the layered experiment is approximately confirmed by the data in Figure 13.4b. The cavity experiment does not have a box model counterpart and hysteresis is too small to be resolved.

Figure 13.4 also indicates that density difference between the chamber and the reservoir changes sign upon an abrupt thermohaline transition from one mode to the other. This is more clearly seen with plots of dimensionless density from the box and cavity experiments (Figure 13.5). The greatest uncertainly arises because it is impossible to assign an exact value of temperature in the experiments since it varies within a chamber. Also, temperature at a point is time dependent, because active convection takes place. Only in the box experiment was the temperature in the chamber recorded over a long enough time to make good averages, and since the convection within the box is relatively even, the temperature is not a very strong function of the location of the temperature probe. Therefore, scatter in Figures 13.4a and 13.5a is relatively small and the data exhibit straight trends. The cavity experiment temperature data are not as linear (Figure 13.4c and 13.5b), because parcels of water of differing temperature and salinity are always present inside the cavity. Although conceivably one could average over long times to smooth out the time dependence, in practice, this was not possible with our existing equipment. Also, for the cavity experiment in the temperature mode, the average temperature is a function of elevation of the probe above the bottom boundary layer. Therefore, the scatter of the temperature and density data is larger than optimal in Figures 13.4d and 13.5b.

13.3. OSCILLATIONS

Oscillations were first discovered in the slot experiment. These did not prove to be reproducible in that

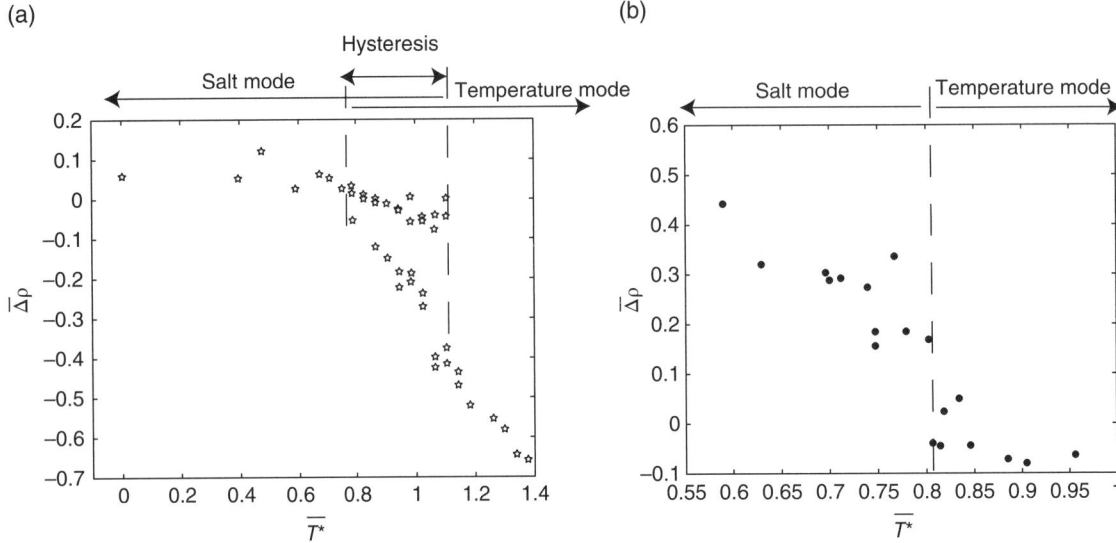

Figure 13.5. Density from (a) the box experiment and (b) the cavity experiment.

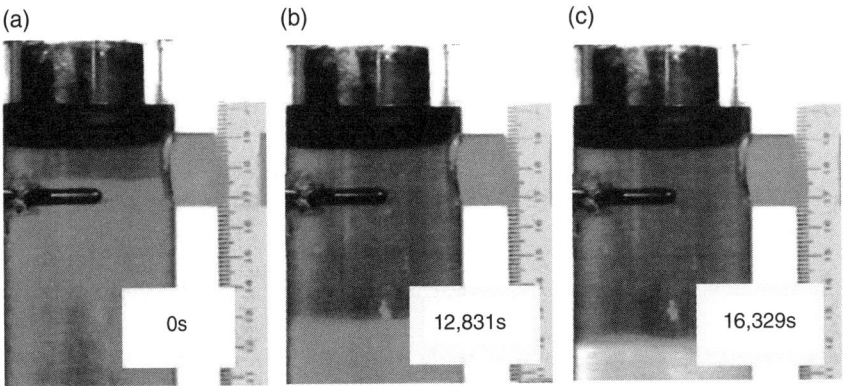

Figure 13.6. The oscillation cycle. (a) formation of a dyed freshwater layer over a very deep mixed layer of dyed salty water and clear salty water; (b) deepening of the top layer; (c) overturn and mixing of the deeper layer resulting in clear salt water lying below the top layer. Thereafter, a new fresh layer forms at the top as in (a). The previous top layer becomes the deep mixed layer, and the cycle repeats. [From *Whitehead et al.*, 2005].

experiment, so their existence is only mentioned briefly at the end of section 4.2 by *Whitehead et al.* [2003]. The next observations of oscillations were unexpectedly discovered in an early exploratory upside-down version of the layered experiment and were both robust and reproducible. The apparatus had a chamber heated from below with a layer of salty water flowing in along the bottom of a tank containing fresh water. The results are reported in a Geophysical Fluid Dynamics Fellow's report [*teRaa*, 2001]. This led to the development of the layered apparatus. The oscillations and their properties are documented more fully, and a supporting theory is developed in *Whitehead et al.* [2005]. This work was followed by oscillations reported by *Mullarney et al.* [2007]. They were unaware of the previous observations of oscillations. As in the te Raa apparatus, a layer of salt water flows over a hot plate on the bottom of a chamber of fresh water, but the apparatus is more than ten times larger.

The oscillation process in all of these experiments is the same, and it is very simple. In the layered apparatus, a layer of dyed freshwater forms at the top (Figure 13.6a). The interface at the base of this layer begins to move downward and the layer becomes much deeper (Figure 13.6b). Below this is a deeper older layer. The color of the deeper layer begins to fade dramatically as the freshwater source to this layer has been blocked by the new top layer. After some time, the interface between the old layer and the bottom salt water develops large waves that begin to break. Mixing between this deep layer and bottom salt water increases and the color difference between the lower layer

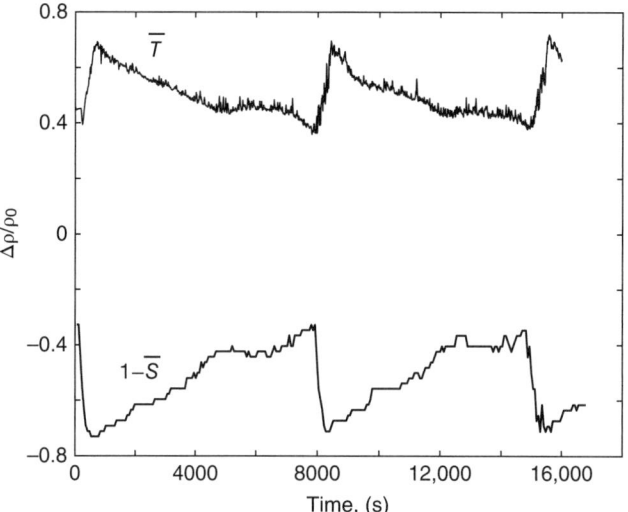

Figure 13.7. Density due to temperature and salinity versus time for a little over two experimental oscillation cycles. The top curve is from a temperature probe 2.54 cm below the top. The density (from salinity of the water) is withdrawn at a depth of 10 cm. The forcing strength is $-\alpha T^*/\beta S_0 = 0.85$.

and the salt water below it decreases. Suddenly the bottom layer's interface mixes away by direct overturning so that the top layer lies over purely salt water (Figure 13.6c). The cycle begins again as a new layer forms near the top. A time series of temperature at a depth of 2.54 cm and salinity at 10 cm depth is shown in Figure 13.7. The criterion for oscillations is $-\alpha T*/\beta S_0 < 1.2$. There is a small range where this overlaps the steady temperature mode, which is found for $1.15 < -\alpha T*/\beta S_0 < 1.35$.

The oscillation in the *Mullarney et al.* [2007] experiment is similar to the above oscillation although their apparatus was inverted compared to the layered experiment. It had an electric heating pad on one end of the bottom with ambient fresh water in the tank. A layer of salt water flowed along the bottom and over the heater. The oscillation occurred because the layer of salt water simply spread over the heater and after some time interval it mixed away. This was followed by another layer spreading that was followed by mixing, repeating the cycle. The criterion they found for oscillation is $0.038 < B_S/B_T < 0.067$, which differs from the criterion for the layered apparatus. The first parameter is the buoyancy flux of heat, $B_T = g\alpha F_T\,WL/\rho_0 c_p$. The second is the buoyancy flux of salinity, $B_S = g\beta S_0 q$. Here, g is the acceleration of gravity; heat flux per unit area is F_T; the volume flux of salt solution is q; the box length is L; the box width is W; the reference water density is ρ_0; the water's specific heat is c_p; and the other symbols are previously defined.

Neither experiment produces clear evidence of the complete criterion needed for oscillation. The balance is clearly between the formation of a layer and its destruction from mixing along the interface, separating the layer from ambient water. A simplified model of the oscillation [*Whitehead et al.*, 2005] includes a freshwater layer that deepens with time from interfacial mixing with salt water below the layer. This is followed by complete mixing and the start of a new layer at the top. It seems to produce the effects observed in the laboratory (Figure 13.8).

13.4. SUMMARY AND DISCUSSION

Temperature-salinity abrupt transitions and oscillations can be produced in the laboratory, as demonstrated by four different experiments. It is necessary to precisely control all driving temperatures and pumping rates to hold the experiment in the range of hysteresis. The findings generally confirm the presence of abrupt thermohaline transitions that were predicted by mathematical box models

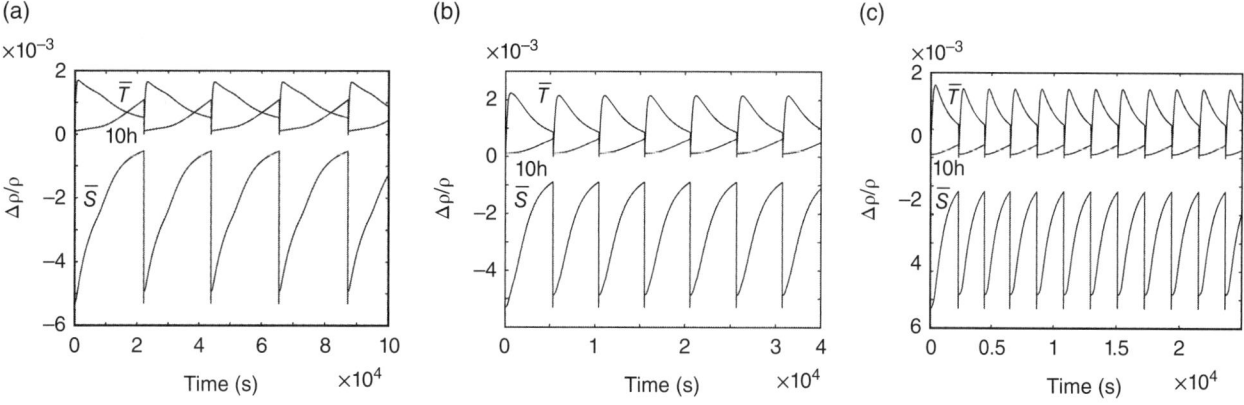

Figure 13.8. Time series from numerical calculations of density from temperature and salinity, (given in the same units as in Fig. 13.7) and layer depth h, (units multiplied by 10 and given in meters): (a) $-\alpha T^*/\beta S_0 = 0.5$ (b) $-\alpha T^*/\beta S_0 = 0.85$ (the value in Figure 13.7), (c) $-\alpha T^*/\beta S_0 \leq 1.15$. Adapted from *Whitehead et al.* [2005]).

and numerical simulations. Hysteresis appears to occupy a smaller range in experiments that allow more layers of stratified water in the interior [*Whitehead*, 2009]. In the cavity experiment the hysteresis range is less than the 1% range of resolution. The experiments uncovered nonlinear self-sustained oscillations that are not reported by box models and numerical simulations. These new oscillations exist in the hysteresis range of the layered experiment. The oscillations represent an excursion back and forth between layer formation and destruction by mixing so that the temperature, salinity, and flow patterns lie in two different "climates."

Double-diffusion processes include salt fingering, layer formation and flux between layers of two components. The two components affect buoyancy and have different diffusivities such as temperature and salinity [*Turner*, 1973, *Schmitt*, 1994]. The experiments described here most strongly indicate that double diffusive heat and salinity flux is clearly important during layer evolution in a number of the experiments. However, a close link between abrupt thermohaline changes and double-diffusion processes is almost completely unexplored, even though double diffusion can produce abrupt transitions (*Veronis*, 1965, *Whitehead* 2002). Experiments that are designed to clarify the role of double diffusion in conjunction with abrupt transitions or oscillations would be valuable. They would form a more complete understanding of two-component fluid mechanics.

We are not aware of any existing numerical experiments that possess no hysteresis with abrupt transitions as in the cavity experiment. Thus, this issue remains an open area of research. It is also clear that the manner in which thermohaline flows either oscillate or jump back and forth between different states as abrupt transitions is still poorly understood. All these are surprising since both nonlinear oscillations and abrupt transitions are understood mathematically.

REFERENCES

Bard E., B. Hamelin, M. Arnold, L. Montaggioni, G. Cabioch-parallel, G. Faure, and F. Rougerie (1996), Deglacial sea-level record from Tahiti corals and the timing of global meltwater discharge, *Nature, 382*, 241–244.

Boyle, E. A. (1990), Quaternary deepwater paleooceanography, *Science, 249*, 863–870.

Broecker W. S. (1997), Thermohaline circulation, the Achilles heel of our climate system: Will man-made CO_2 upset the current balance? *Science, 278*, 1582–1588.

Broecker, W. S., D. M. Peteet, and D. Rind (1985), Does the ocean-atmosphere system have more than one stable mode of operation? *Nature, 315*, 21–26.

Bryan, F. (1986), High-latitude salinity effects and interhemispheric thermohaline circulations, *Nature, 323*, 301–304.

Bulgakov, S. N., and Y. N. Skiba (2003), Are transitions abrupt in Stommel's thermohaline box model? *Atmosfera, 16*, 205–229.

Burns, S. J., D. Fleitmann, A. Matter, J. Kramers, and A. A. Al-Subbary (2003), Indian Ocean climate and an absolute chronology over Dansgaard/Oeschger events 9 and 13, *Science, 201*, 1365–1367.

Cessi, P. (1994), A simple box model of stochastically forced thermohaline flow, *J. Phys. Oceanogr., 24*, 1911–1920.

Hearn, C. J., and H. S. Sidhu (1999), The Stommel model of shallow coastal basins, *Proc. R. Soc. Lond., A455*, 3997–4011.

Ierley, G. R., and V. A. Sheremet (1995), Multiple solutions and advection-dominated flows in the wind-driven circulation Part 1: Slip, *J. Mar. Res., 53*, 703–733.

Jiang, S., F.-F. Jin, and M. Ghil (1995), Multiple equilibria, periodic, and aperiodic solutions in a wind-driven, double-gyre, shallow-water model, *J. Phys. Oceanogr., 25*, 764–786.

Keigwin, L. D., and G. A. Jones (1994), Western North-Atlantic evidence for millennial-scale changes in ocean circulation and climate, *J. Geophys. Res., 99* (C6), 12,397–12,410.

Keigwin, L. D., W. B. Curry, S. J. Lehman, and S. Johnsen (1994), The role of the deep-ocean in North-Atlantic climate-change between 70-Kyr and 130-Kyr ago, *Nature, 371*, 323–326.

Manabe, S., and R. J. Stouffer (1995), Simulation of abrupt climate change induced by freshwater input to the North Atlantic Ocean, *Nature, 378*, 165–167.

Marotzke J. (1994), Ocean models in climate problems, in *Ocean Processes in Climate Dynamics: Global and Mediterranean Examples*, edited by P. Malanotte-Rizzole and A.R. Robinson, pp. 79–109, Kluwer Academic, Dordrecht.

Mullarney, J., R. W. Griffiths, and G. O. Hughes (2007), The role of freshwater fluxes in the thermohaline circulation: Insights from a laboratory analogue, *Deep-Sea Res. I, 54*, 1–21.

Rahmstorf, S. (1995), Bifurcations of the Atlantic thermohaline circulation in response to changes in the hydrological cycle, *Nature, 378*, 145–149.

Rahmstorf, S., and A. Ganopolski, (1999), A simple theoretical model may explain apparent climate instability, *J. Climate, 12*, 1349–1352.

Schmitt, R. W. (1994), Double-diffusion in oceanography. *Annu. Rev. Fluid Mech., 26*, 255–285.

Stocker, T. F., (2000), Past and future reorganizations of the climate system, *Q. Sci. Rev., 19*, 301–319.

Stocker, T. F., and D. G. Wright (1991), Rapid transitions of the ocean's deep circulation induced by changes in surface water fluxes, *Nature, 351*, 729–732.

Stommel, H. (1961), Thermohaline convection with two stable regimes of flow, *Tellus, 3*, 224–230.

Stommel, H., and C. Rooth (1968), On the interaction of gravitational and dynamic forcing in simple circulation models, *Deep-Sea Res., 15*, 165–170.

te Raa, L. (2001), Convective oscillations in a laboratory model, GFD Fellow Report, http://gfd.whoi.edu/proceedings/2001/PDF/rep_teraa.pdf.

Turner, J. S. (1973), *Buoyancy Effects in Fluids*, Cambridge Univ. Press, New York.

Veronis, G. (1965), On finite amplitude instability in thermohaline convection. *J. Mar. Res., 23*, 1–17.

Weart, S. (2003), The discovery of rapid climate change, *Phys. Today*, *56*(8), 30–36.

Weaver, A. J., C. M. Bitz, A. F. Fanning, and M. M. Holland (1999), Thermohaline circulation: High-latitude phenomena and the difference between the Pacific and Atlantic, *Annu. Rev. Earth Planet Sci.*, *27*, 231–285.

Whitehead, J. A. (1995), Thermohaline ocean processes and models, *Annu. Rev. Fluid Mech.*, *27*, 89–114.

Whitehead, J. A. (1996), Multiple states in doubly-driven flow, *Phys. D*, *97*, 311–321.

Whitehead, J. A. (1998), Multiple T-S states for estuaries, shelves and marginal seas, *Estuaries*, *21*, 278–290.

Whitehead, J. A. (2000), Stratified convection with multiple states, *Ocean Modelling*, *2*, 109–121.

Whitehead, J. A. (2002), A boundary layer flow with multiple equilibria, *Phys. Fluids*, *14*, 2575–2577.

Whitehead, J. A., and K. Bradley (2006), Laboratory studies of stratified convection with multiple states, *Ocean Modelling*, *11*, 333–346.

Whitehead, J. A. (2009), Abrupt transitions and hysteresis in thermohaline laboratory models, *J. Phys. Oceanogr.*, *39*, 1231–1243.

Whitehead, J. A., M. L. E. Timmermans, W. G. Lawson, S. N. Bulgakov, A. M. Zatarian, J. F. A. Medina, and J. Salzig (2003), Laboratory studies of thermally and/or salinity-driven flows with partial mixing: Part 1 Stommel transitions and multiple flow states, *J. Geophys. Res.*, *108*, (C2), 3036, doi:10.1029/2001JC000902.

Whitehead, J. A., L. te Raa, T. Tozuka, J. B. Keller, and K. Bradley (2005), Laboratory observations and simple models of slow oscillations in cooled salt-stratified bodies, *Tellus*, 57A(5) 778-809, doi:10.1111/j.1600-0870.2005.00150.x.

14

Oceanic Island Wake Flows in the Laboratory

Alexandre Stegner

14.1. INTRODUCTION

One of the first rewards of meteorological satellites was the regular surveillance of cloud patterns over poorly observed oceanic areas. Many fascinating new patterns emerged, one of the most striking being the vortex streets downwind of small isolated islands [*DeFelice et al.*, 2001]. At first glance, such atmospheric vortex shedding looks similar to the incompressible two-dimensional von Karman street which forms behind a cylindrical obstacle. The classical two-dimensional wake is controlled by a single parameter, the Reynolds number $Re = U_0 D/\nu$ where U_0 is the free upstream velocity, D is the diameter of the cylinder, and ν is the horizontal molecular viscosity. If $1 < Re < 40$, a laminar separation is obtained and two steady vortices are formed immediately behind the cylinder where they remain attached. For larger Reynolds number a periodic vortex shedding of frequency f_s occurs. The dimensionless shedding frequency, in other words the Strouhal number $S_t = f_s D/U_0$, increases with the Reynolds number and approaches an asymptotic value of $S_t \simeq 0.21$ at high Reynolds number [*Wen and Lin*, 2001]. However, the direct application of these standard results to geophysical wakes is often misleading because the dissipation is highly turbulent and the molecular viscosity must be replaced by a turbulent eddy viscosity which is in general nonuniform and hard to estimate. Moreover, the atmosphere and the ocean are affected by Earth's rotation and the vertical stratification. We should then also take into account the relative size of the island (or the eddies) in comparison with the Rossby radius R_d, an intrinsic horizontal scale that controls the dynamics of rotating and stratified flows. For atmospheric vortex wakes, the characteristic horizontal scale of the coherent structures remains small (10–50 km) compared to the atmospheric Rossby radius (~ 1000 km). Hence, several experimental studies neglect the rotation and take into account only the atmospheric stratification [*Etling*, 1989; *Rotunno and Smolarkiewicz*, 1991] and the three-dimensional impact of the orography which may trigger both a wave [*Johnson et al.*, 2006] and a vortex wake. On the other hand, in the midlatitude oceans the Rossby radius associated to the first baroclinic mode [*Chelton et al.*, 1998] is much smaller (10–60 km) and becomes comparable or smaller than the typical island radius R. Hence, in the ocean the rotation cannot be neglected and we could encounter both mesoscale (R larger or equal to R_d) or submesoscale (R smaller than R_d) wakes. Surprisingly, the dynamical distinction between these two types of oceanic wakes was done only recently. The aim of this review is to present the most recent laboratory and theoretical results on the various dynamical regimes which control an idealized oceanic Karman wake at the mesoscale and submesoscale.

Considering oceanic flows, an important distinction should be made between shallow- and deep-water wakes, depending upon whether the dominant boundary stress is associated with the near-shore bottom or the coastal side of the island. If we consider shallow-water wakes, generated by small islands in shallow shelf sea or estuaries, the bottom drag is the primary source of vorticity generation [*Wolanski et al.*, 1984, 1996; *Alaee et al.*, 2004; *Neill and Elliott*, 2004]. For such configuration, where the bottom friction and the vertical mixing are strong, the stratification and the rotation are neglected. The equivalent Reynolds number is then the island wake parameter $P = (U_0 D/\kappa_z)(h/D)^2$, where h is the water depth and κ_z the turbulent diffusivity along the vertical. According to experimental and numerical studies [*Wolanski et al.*, 1996], for small P the bottom friction prevents the formation of eddies; when $P \gtrsim 1$ a stationary dipole emerges in the lee of the island and for larger values ($P \gtrsim 10$) a periodic shedding occurs. In this deep-water case, when the

Laboratoire de Météorologie Dynamique, CNRS and École Polytechnique, Palaiseau, France.

Modeling Atmospheric and Oceanic Flows: Insights from Laboratory Experiments and Numerical Simulations,
First Edition. Edited by Thomas von Larcher and Paul D. Williams.
© 2015 American Geophysical Union. Published 2015 by John Wiley & Sons, Inc.

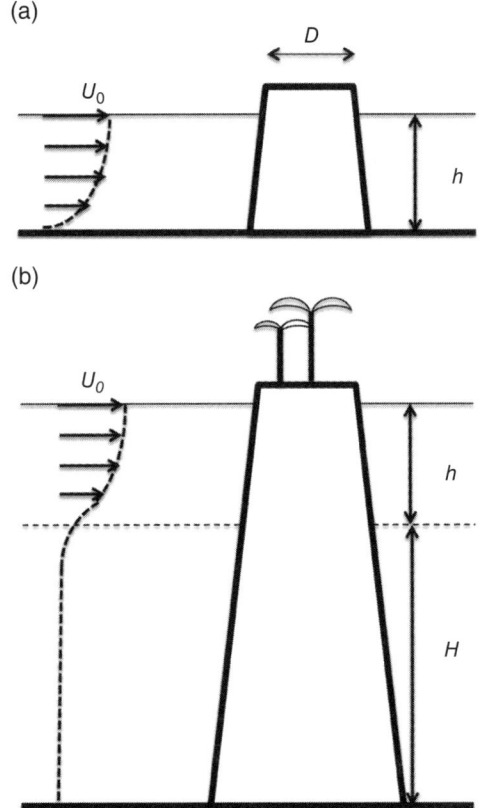

Figure 14.1. Oceanic current impinging on (a) shallow-water and (b) deep-water island configuration.

deep oceanic layer is much larger than the upper thermocline ($H \gg h$), the influence of the bottom drag could be neglected, as discussed by [*Tomczak*, 1988]. When the horizontal dissipation is weak, the (large) Reynolds number is not a crucial parameter anymore and the combined effects of rotation and stratification govern the wake pattern. We restrict this review to the cases of deep ocean wake (Figure 14.1b) where an upper surface current encounters an isolated island or archipelago.

14.2. DIMENSIONAL ANALYSIS AND DYNAMICAL PARAMETERS

In order to classify the various dynamical regimes of the idealized island wake, we first need to identify the main dimensionless parameters which control the flow. Assuming a symmetrical island shape with an effective diameter $D = 2R$ and an upstream surface current of thickness h (we consider here that the current has roughly the same height as the upper thermocline) above a deep oceanic layer of depth H, we get two geometric parameters:

$$\alpha = \frac{h}{R}, \qquad \delta = \frac{h}{h+H}.$$

The shallow-water parameter α is usually small in the ocean. For typical isolated islands such as Madeira [*Caldeira et al.*, 2002; *Caldeira and Sangra*, 2012], Gran canaria[*Aristegui et al.*, 1994; *Sangra et al.*, 2005], Hawaï [*Calil et al.*, 2008; *Chavanne et al.*, 2010], and even the small Aoga Shima [*Hasegawa et al.*, 2004], we get $\alpha = 0.05-0.005$. If the vortices generated downstream of the island keep such small values of α, they will satisfy the hydrostatic balance at the first order of approximation. The layer thickness ratio δ controls the dynamic interactions between the surface and the bottom layer. In the open ocean, this parameter is also small, $\delta = 0.1-0.02$, the deep bottom layer acts as a neutral layer with no motion, and the surface flow is not affected by the bottom dissipation induced by the seafloor roughness.

In order to quantify the impact of Earth rotation on the oceanic flow, we generally introduce a Rossby number. Taking into account the upstream current velocity U_0, the characteristic island radius R, and the local Coriolis parameter f, we can easily build the island Rossby number:

$$\text{Ro}_I = \frac{U_0}{fR}$$

However, a more accurate description of the vortex dynamics in the wake will be given by the vortex Rossby number:

$$\text{Ro} = \frac{V_{\max}}{f\, r_{\max}}$$

where $V_{\max}$ is the maximum velocity of the vortex and $r_{\max}$ the corresponding vortex radius. While the island and the vortex Rossby numbers have generally the same order of magnitude, the relative core eddy vorticity ζ_0/f could be much larger. Indeed, an intense anticyclone was detected in the lee of Oahu island [*Chavanne et al.*, 2010] with a finite vortex Rossby number $\text{Ro} \simeq 0.45$ and a higher relative core vorticity $\zeta_0/f \simeq -1.5$. For such large values the corresponding vortex will satisfy the cyclogeostrophic balance where the sum of the centrifugal and the Coriolis forces balance the pressure gradient. On the other hand, when a weak current encounters a large island, the corresponding Rossby number Ro_I will be small and both the wake flow and the eddies will satisfy the standard geostrophic balance.

The vertical density gradient is quantified by the Brunt-Vaisala frequency $N(z)$ (where $N^2 = -g\partial_z\rho/\rho$), and the relative strength of the oceanic stratification in comparison with the Earth rotation is given by the stratification parameter $S = N/f$. This parameter is large for the upper thermocline ($S \simeq 100$ according to Table 14.1) and induces a strong stratification regime for the deep-water wake. Another way to account for the vertical stratification is to introduce the Rossby deformation radius associated with the first baroclinic mode R_d. If we consider

Table 14.1. Characteristic scales and dynamical parameters for various oceanic wakes.

Islands	U_0 (cm/s)	R (km)	h (m)	$S = N/f$	Ro_I	Bu
Madeira	20–30	10–25	~100	~350	0.1–0.3	2–6
Gran Canaria	40	25	~100	—	~0.2	—
Hawaï (Oahu)	20–35	15–20	~100	150–200	0.2–0.4	1–3
Aldabra	40–60	30	~150	~150	0.6–0.8	~0.7
Aoga Shima	100–150	2	~100	~200	6–10	~100

a purely linear stratification in the upper layer, the latter is defined by $R_d = Nh/f = Sh$. We could then define a dimensionless Burger number from the relative island size:

$$\text{Bu} = \left(\frac{R_d}{R}\right)^2$$

For a geostrophic and circular vortex the Burger number is equal to the ratio of the kinetic over the potential energy. Hence, for mesoscale flows, where the typical horizontal scale R is equal to or larger than the Rossby radius R_d (Bu $\leq$ 1), there is a significant amount of potential energy available due to the isopycnal deviations within the thermocline. On the other hand, for the submesoscale flow, which corresponds to $R < R_d$ (Bu $\gg$ 1), the energy is mainly kinetic.

The Reynolds number is the most widely used parameter to quantify the dissipation. However, as far as we consider the nonisotropic turbulent ocean, it becomes a challenge to estimate an effective turbulent viscosity (i.e., the mean eddy diffusivity). This means that turbulent diffusivity depends on the spatial and temporal scales used for the averaging. Moreover, due to the vertical stratification, there is a net distinction between the horizontal diffusivity κ_h and vertical diffusivity κ_z. For an isolated island in deep water we can use a horizontal Reynolds number:

$$\text{Re} = \frac{U_0 D}{\nu_t}$$

where ν_t is the effective eddy (turbulent) viscosity. The distribution of tracers such as sulfur hexafluoride (SF6) was recently used to estimate the open-ocean eddy diffusivity [*Ledwell and Watson*, 1998], which varies from 1 to 10 m^2/s for horizontal scales of tens of kilometers. Hence, if we assume $\nu_t \simeq \kappa_h$, the horizontal Reynolds number will typically range from 10^2 to 10^4 for the above-mentionned islands (Table 14.1). Taking into account the rotation, we introduce the Ekman number

$$E_k = \frac{\kappa_z}{fh^2}$$

to quantify the vertical dissipation of shallow and coherent structures within the upper thermocline. The vertical diffusivity κ_z in the ocean is several orders of magnitude smaller than the horizontal one κ_h and the Ekman and Reynold numbers are therefore two independent parameters.

14.3. IDEALIZED LABORATORY SETUPS FOR OCEANIC ISLAND CONFIGURATION

We should first recognize that due to strong experimental constraints the laboratory experiments are always highly simplified in comparison with oceanic island flows. The turbulent coastal boundary layer strongly differs from the viscous boundary layer obtained in the laboratory and the interaction between an oceanic current and a complex island coast could hardly be reproduced on a rotating turntable. While the experimental forcing and the boundary layer detachment in the near wake may differ from the ocean, the global wake pattern (geometric structure, distance between eddies, cyclone-anticyclone asymmetries) and its dynamics (vortex interactions, secondary instabilities) will be close to the oceanic ones if the geometric and dynamic similarities are satisfied. In other words, the values of the main dimensionless numbers obtained in the laboratory should be comparable to the oceanic ones.

Among the geometric numbers, the similarity of the shallow-water parameter α is probably the most difficult to achieve. Indeed, to generate a Karman street which contains several eddies, the typical size of the rotating tank should be at least $10D$, where D is the typical island diameter. Hence, even on the world's largest turntable, the 13 m diameter Coriolis platform (http://coriolis.legi.grenoble-inp.fr/?lang=en)[1], the diameter $D = 2R$ should not exceed 1 m. For a layer thickness h of a few centimeters the shallow-water parameter $\alpha = h/R$ can hardly go below $\alpha \simeq 0.1$, which is still one order of magnitude larger than the oceanic values, $\alpha \simeq 10^{-2}$. In the pioneering experiments on rotating wakes [*Boyer and Kmetz*, 1983; *Boyer et al.*, 1984; *Tarbouriech and Renouard*, 1996] the shallow-water constraint was not strictly satisfied with $\alpha = 0.6$–6. Smaller values (see Table 14.2) were only reached with a two-layer configuration [*Perret et al.*, 2006b; *Teinturier et al.*, 2010; *Stegner et al.*, 2012] where the upper layer

---
[1] LEGI-Coriolis, Grenoble, France. http://coriolis.legi.grenoble-inp.fr/?lang=en

Table 14.2. Dynamical parameters of various laboratory experiments on rotating wakes.

Experiments	α	δ	Ro_I	N/f	Bu	Re
Boyer and Davies (1982)	1–6	1	0.02–0.4	0	—	50–200
Boyer and Kmetz (1983)	0.4–1.8	1	0.3–3.5	0	—	65–2300
Boyer et al. (1984)	0.6–3	1	1.4–4	0	—	2500–37,500
Tabouriech and Renouard (1996)	10	1	3–10	0	—	10,000
Stegner et al. (2005)	4–14	1	0.3–4	0	—	150–1400
Perret et al. (2006)	0.5	0.13	0.06–0.35	0	0.1–10	200–1000
Teinturier et al. (2010)	0.2	0.1–0.15	0.4–4	4–6	0.5–1.5	4000–30,000
Stegner et al. (2012)	0.15–0.6	0.1–0.15	0.3–3.5	7–14	4–70	2500–45,000

When the experiment is done in a single homogeneous fluid layer, the layer thickness ratio is set to unity. If this homogeneous layer has a free surface, we can use the layer thickness h to define a barotropic Rossby radius $R_d = \sqrt{gh}/f$ and the corresponding Burger number Bu = $(2R_d/D)^2$. When there is no stratification, we fix the stratification parameter to zero, $N/f = 0$.

could be thinner and not impacted by a strong bottom Ekman pumping. According to standard asymptotic expansion, the vertical acceleration will be negligible if $\alpha^2 \, \text{Ro} \ll 1$ [*Stegner*, 2007], hence for experimental configurations where $\alpha \simeq 0.1$ the hydrostatic balance will be satified even for finite Rossby number Ro $\simeq$ 1 as for the oceanic flows. Another geometric parameter, the vertical aspect ratio $\delta = h/(h + H)$, is also crucial. For the latter a close similarity could be achieved between the open ocean and the laboratory where $\delta \simeq 0.1$.

There are various ways to control the island Rossby number in rotating experiments. The relative velocity U_0, the size of the cylinder D, or the turntable rotation $\Omega_0 = f/2$ could be varied to satisfy the similarity with oceanic values from $Ro_I = 0.1$ to $Ro_I = 10$. Several experimental setups were used to generate a vortex street (Figure 14.2) and control the size of the eddies and their intensity. The first experimental investigations on rotating wakes [*Boyer and Davies*, 1982; *Boyer et al.*, 1984; *Tarbouriech and Renouard*, 1996] were made with a fixed cylinder while the mean speed U_0 of the upstream current was regulated by a hydraulic pump (Figure 14.2a). However, it could be easier to perform PIV (particle image velocimetry) measurements with a top-view camera if the fluid layer is at rest, on the rotating turntable, and the cylinder is translated at a constant speed (Figure 14.2b). Such configurations with high-resolution PIV measurements were used in recent experiments [*Stegner et al.*, 2005; *Perret et al.*, 2006b; *Teinturier et al.*, 2010; *Lazar et al.*, 2013b], especially when the rotating fluid layer is stratified (Figure 14.2c).

In order to mimic the oceanic density stratification of the thermocline, a salt stratification is preferentially used in the laboratory. A two-layer stratification which supports both baroclinic and barotropic motion could be used in a first step to mimic the upper themocline [*Perret et al.*, 2006b]. In order to satisfy a small vertical aspect

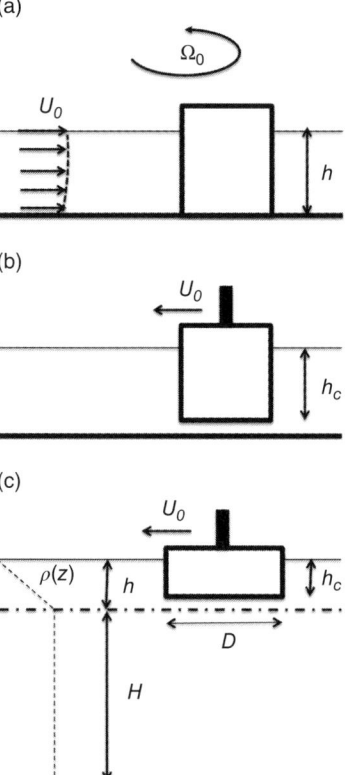

Figure 14.2. Various experimental setups used to generate a vortex wake in a rotating fluid layer: (a) homogeneous current impinging a fixed cylinder, (b) homogenous layer at rest with a drifting cylinder, and (c) stratified layers with a drifting cylinder confined in the upper layer.

ratio ($\delta \simeq 0.1$), the light upper layer is much thinner than the deep lower one. Thus, the baroclinic Rossby radius can be approximated by $R_d \simeq \sqrt{g^*h}/f$, where $g^* = g(\rho_2 - \rho_1)/\rho_2$ is the reduced gravity. The densities

ρ_1 and ρ_2 of the top and bottom layers are then adjusted to obtain a small or large Rossby radius R_d in comparison with the cylinder diameter. A more realistic stratification is obtained when the thin upper layer is linearly stratified (Figure 14.2c). In such case, we consider that the motion of the cylinder transfers momentum mainly over the heigth of the submerged cylinder h_c and the deformation radius associated with the first baroclinic mode is defined by $R_d = Nh_c/f$. Theses idealized salt stratifications are needed to study a wide range of Burger numbers from large mesoscale wake, Bu = 0.1, to small submesoscale wake, Bu = 70.

In the laboratory, the molecular viscosity controls both the horizontal and the vertical dissipation, and unlike the oceanic case, the Ekman and Reynolds numbers are not independent parameters. Sometimes it could be more convenient to use $1/E_k$ instead of the Ekman number because it evolves as the Reynolds number. Weak dissipation will correspond to a large value of $1/E_k$ according to

$$\frac{1}{E_k} = \frac{fh^2}{\nu} = \frac{\text{Re}}{\text{Ro}} \alpha^2,$$

where $\alpha = h/R$ is the shallow-water ratio parameter. If the later is small, $1/E_k$ could be much smaller than Re and the vertical dissipation will become more important than the horizontal one. In small facilities, typically 1 m diameter turntables, both Reynolds (Re = 10^2–10^3) and Ekman ($1/E_k = 10^2$–10^3) numbers correspond to viscous laminar dissipation. On a large-scale turntable, such as the 13 m diameter Coriolis platform, the Reynolds number may reach turbulent values up to 10^4–10^5 while the Ekman number will generally stay in the viscous range ($1/E_k = 10^2$–10^3) when the flow is confined in a shallow upper layer (Figure 14.2c).

14.4. MESOSCALE VORTEX WAKE

In this section, we consider large-scale wakes where the Burger number Bu is smaller than or equal to unity. In other words, the characteristic radius of the eddies, r_{max}, within the Karman street will be larger than or equal to the first baroclinic radius R_d.

14.4.1. Quasi-Geostrophic Wake

The quasi-geostrophic regime corresponds to small Rossby number $Ro_I \ll 1$ while the Burger number remains close to unity, Bu $\simeq$ 1. For this range of parameter, the cyclonic and anticyclonic eddies which are periodically shed behind a cylindrical obstacle are identical in size and intensity as for the classical Karman street. The distance L_e between two vortices having the same sign, positive for cyclonic and negative for anticyclonic, is roughly equal to five cylinder diameters. This is in agreement with the two-dimensional nonrotating flow in the limit of large Reynolds number, for which the Strouhal number, $S_t = f_s D/U_0 = D/L_e \simeq 0.2 - 0.24$, at large Reynolds numbers [*Wen and Lin*, 2001]. Both experimental [*Boyer and Kmetz*, 1983; *Stegner et al.*, 2005; *Teinturier et al.*, 2010; *Lazar et al.*, 2013b] and numerical studies [*Heywood et al*, 1996; *Perret et al.*, 2006a] confirm this result in a deepwater configuration when the bottom friction is negligible. Hence, a quasi-geostrophic wake will have the same pattern as the standard nonrotating Karman street even if the separation and the vortex shedding occur at higher critical Reynolds numbers [*Boyer and Kmetz*, 1983; *Boyer et al.*, 1984].

14.4.2. Large-Scale Geostrophic Wake

When the Rossby and the Burger numbers are both small, the flow is expected to follow the frontal geostrophic regime [*Cushman-Roisin*, 1986]. The isopycnal deviations scale as the ratio of the Rossby over the Burger number, $\lambda = Ro/Bu$, and when λ is finite, the evolution of cyclonic and anticyclonic eddies, confined in the upper surface layer, should differ strongly. Other studies on the stability of isolated eddies have shown that, beyond the quasigeostrophic regime, anticyclones tend to be more stable and coherent than their cyclonic counterparts [*Arai and Yamagata*, 1994; *Stegner and Dritschel*, 2000; *Baey and Carton*, 2002; *Graves et al.*, 2006]. Hence, when the mean island radius is two or three times larger than the local deformation radius, we could expect a significant cycloneanticyclone asymmetry of the vortex wake. Indeed, for some extreme cases, coherent cyclones do not emerge at all, and only an anticyclonic vortex street appears several diameters behind the circular island [*Perret et al.*, 2006b]. For these experiments, where the drifting cylinder was confined in a thin ($\delta \simeq 0.1$) and light (but non-stratified) surface layer, both dye visualization (Figure 14.3) and vorticity measurements (Figure 14.4) show this suprising behavior of large-scale geostrophic wake. This asymmetry was first explained by the linear stability analysis of parallel wake flows in the framework of rotating shallow-water equations [*Perret et al.*, 2006a; *Perret et al.*, 2011]. In the frontal regime, the most unstable mode is fully localized in the anticyclonic shear region. Hence, the anticyclonic perturbations, leading to large-scale anticyclones, have the fastest growth rates. Besides, a spatiotemporal analysis of the wake flow behind the cylinder reveals a change in the nature of the instability: For a large-scale cylinder the wake flow is convectively unstable [*Perret et al.*, 2006a] while the quasi-geostrophic wake and the classical Karman street is absolutely unstable [*Pier*, 2002; *Chomaz*, 2005]. Since the near wake flow is nearly parallel in the large-scale (frontal) regime, see Figures 14.3 and 14.4, this unstable wake would behave like a noise

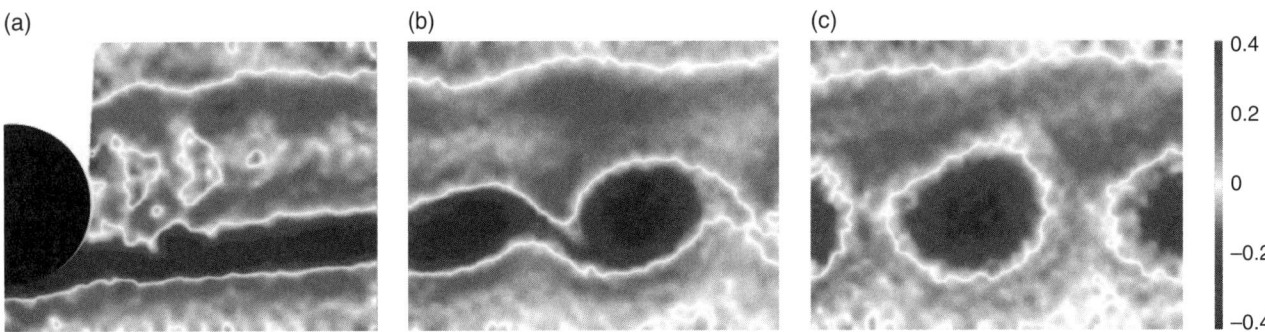

Figure 14.3. Dye visualization of an anticyclonic vortex street formed in the lee of a translating cylinder in an upper shallow-water layer at (a) $t = 0$ and (b) $t = 2T_0$ when $Ro_I = 0.36$, $Bu = 0.31$, and $Re = 1100$. The fluorescent dye was uniformly painted all around the cylinder ($D = 7$ cm) and released in the boundary layer during the translation. The turntable rotates clockwise and the cylinder starts to move at $t \simeq -6T_0$.

Figure 14.4. Surface vorticity field of a large-scale wake flow within an upper shallow-water layer at (a) $t = 0$, (b) $t \simeq 2T_0$, and (c) $t \simeq 5T_0$ when $Ro_I = 0.19$, $Bu = 0.11$, and $Re = 800$ [*Perret et al.*, 2006b]. Blue color corresponds to anticyclonic and red to cyclonic vorticity. The cylinder ($D = 7$ cm) is drawn in black and its shadow in the laser sheet is in white on image (a). For color detail, please see color plate section.

amplifier, the entire flow being convectively unstable. A small wave packet perturbation is amplified but propagates downstream in the flow and does not contaminate the upstream flow. Hence, large-scale anticyclones may form far downstream, unlike the classical Karman wake. In this frontal geostrophic regime, the typical radius of the anticyclone size is controlled by R_d, whereas for finite or large Bu the eddy size scales with the island radius R. The main consequence of the distortion or the disappearance of the cyclonic street is a strong increase in the Strouhal number. Indeed, the latter can be three times larger than the standard value reached in a classical Karman street when the surface deviation exceeds 10% of the unperturbed thickness. According to Figure 14.5, this strong variation of the Strouhal number is mainly governed by the relative interface deviation parameter $\lambda = Ro/Bu$ and is weakly affected by the Reynolds number. A similar increase of the Strouhal number and anticyclonic robustness was also found for the same range of parameters with the high-resolution numerical model (ROMS) of a surface intensified wake flow [*Dong et al.*, 2007].

14.5. SUBMESOSCALE VORTEX WAKE

When an oceanic current encounters a small island, smaller than the local Rossby radius, both the Burger and island Rossby numbers increase. When the island size is three or four times smaller than R_d, the Burger number becomes large and the island Rossby number is often finite. The relation between Ro_I and the local amplitude of the vorticity ζ/f is not well established and crucially depends on the dissipation, in other words the Reynolds number. Nevertheless, in both laboratory experiments and the ocean, for the most intense structures we generally get $|\zeta/f| \geq 2Ro_I$. It is well known that when the relative vorticity becomes finite, $|\zeta/f| \sim 1$, the rotation alters the stability of two-dimensional flow with respect to three-dimensional perturbations. The inertial [*Johnson*, 1963; *Yanase et al.*, 1993], centrifugal [*Kloosterziel and VanHeijst*, 1991; *Mutabazi et al.*, 1992], or symmetric [*Hoskins*, 1974; *Haine and Marshall*, 1998] instability may induce a selective destabilization of anticyclonic vorticity regions.

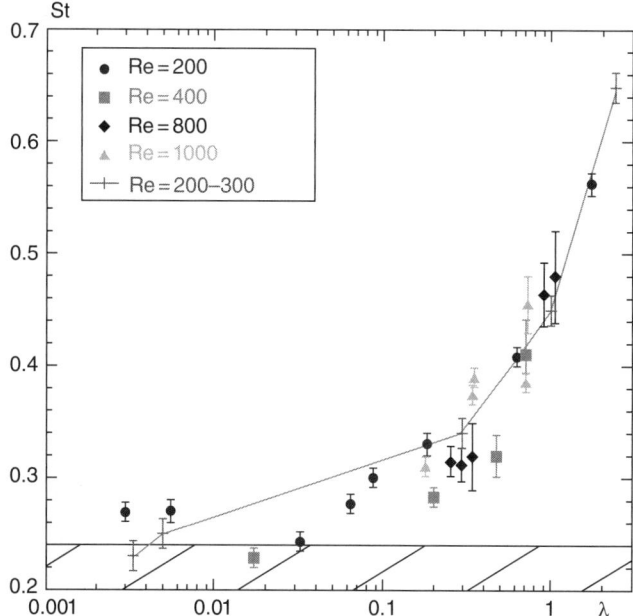

Figure 14.5. Strouhal number as a function of the relative interface deviations $\lambda = \text{Ro}_I/\text{Bu}$ for different Reynolds numbers. The dashed zone indicates the range of the Strouhal number for two and three-dimensional wakes at these Reynolds numbers. The solid line corresponds to numerical simulations of cylinder wakes in a rotating shallow-water layer.

14.5.1. Three-Dimensional Destabilization of Two-Dimensional Wake

The destabilization of intense anticyclones in a rotating vortex street was first studied in an homogeneous fluid layer [*Boyer and Davies*, 1982; *Boyer and Kmetz*, 1983; *Boyer et al.*, 1984; *Stegner et al.*, 2005]. Dye visualizations, such as Figure 14.6, reveal small-scale disturbances within anticyclonic eddies when the Reynolds Re or the inverse Ekman number $1/E_k$ was sufficiently large and the island Rossby number reached or exceed critical values close to unity: $\text{Ro}_I \simeq 0.6-0.8$ [*Boyer and Kmetz*, 1983; *Boyer et al.*, 1984; *Tarbouriech and Renouard*, 1996; *Stegner et al.*, 2005]. When quantitative velocity measurements were available, they showed a rapid decay of the negative core vorticity to the limit $\zeta/f \simeq -1$ while the cyclones could keep higher values for longer times [*Tarbouriech and Renouard*, 1996; *Stegner et al.*, 2005]. The symmetry between the cyclonic and the anticyclonic vortices in the lee of the obstacle is broken and only cyclonic eddies could survive with large core vorticity. Besides, the latest results [*Stegner et al.*, 2005] show that the strong ellipticity of detached vortices behind the cylinder governs the transient three-dimensional instability in the near wake. However, the application of these laboratory results to oceanic wakes could be misleading because these experiments were performed with a tall column cylinder, $\alpha = 3-10$, and without any stratification while island wakes in the deep ocean are confined within a shallow and stratified surface layer. Therefore, recent numerical and laboratory studies were performed with realistic stratification and a small shallow-water ratio, $\alpha \ll 1$.

High-resolution numerical models were needed to capture the three-dimensional unstable motions which emerged in the wake of an idealized island in deep water. The regional oceanic model system (ROMS) was used by *Dong et al.* [2007] to solve the rotating primitive equations with a high horizontal ($\Delta x = \Delta y = 250$ m) and vertical ($\Delta z = 25$ m) resolution. For large enough Reynolds number, when $\text{Ro}_I = 0.2-1$, $\alpha \simeq 0.01$, and $\text{Bu} \simeq 1$, the

(a)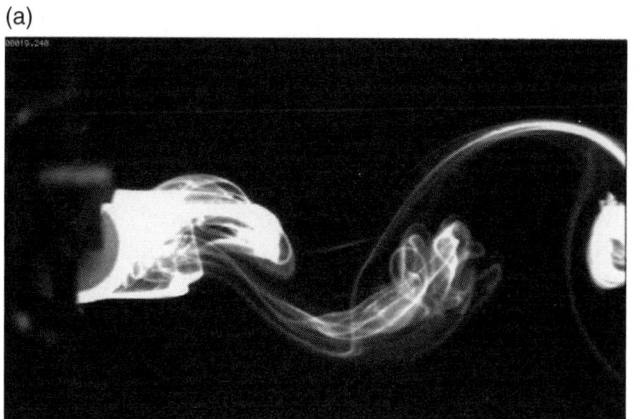

(b)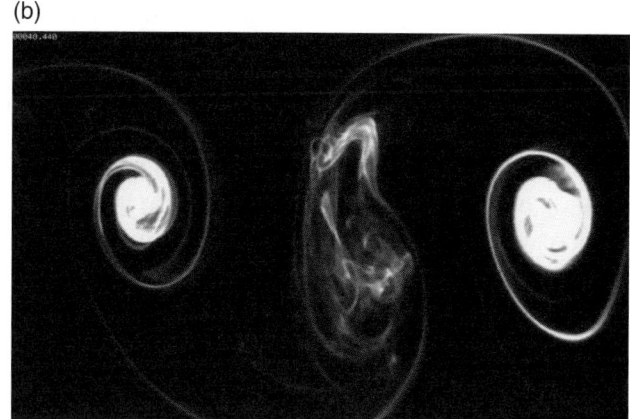

Figure 14.6. Dye visualization of anticyclonic destabilization within a vortex street at (a) $t = 0$ and (b) $t = 0.8T_0$ when $\text{Ro}_I = 2.5$, Re = 150, and $\alpha = 10$ [*Stegner et al.*, 2005]. The fluorescent dye was uniformly painted all around the cylinder ($D = 2$ cm) and released in the boundary layer during the translation. The turntable rotates clockwise and the cylinder starts to move at $t \simeq -3T_0$. The experimental configuration corresponds to figure 2 (b).

magnitude of the local vorticity exceeds unity and small-scale three-dimensional unstable perturbations tend to destroy completely the anticyclonic eddies. The horizontal and vertical scales of the perturbation decrease when the Reynolds number or the spatial resolution increases, in agreement with the linear stability analysis of the unstable inertial or centrifugal modes [*Potylitsin and Peltier*, 1998; *Kloosterziel et al.*, 2007; *Kloosterziel and Carnevale*, 2008; *Plougonwen and Zeitlin*, 2009]. The Princeton ocean model (POM) was used by *Hasegawa et al.* [2009] to simulate a uniform surface flow past a small square oceanic island ($D = 10$ km) with a high horizontal ($\Delta x = \Delta y = 1$ km) and vertical ($\Delta z = 10$ m) resolution. This study focuses on the dynamic impact of the wake flow (eddies and vertical mixing) on the biogeochemical cycles and the phytoplankton bloom. For this intense submesoscale wake, when $Ro_I = 2, \alpha \simeq 0.08$ and $Bu \simeq 2000$, the anticyclonic boundary layer does not roll up to form coherent anticyclones while cyclones are periodically shed and remain coherent even with a strong relative core vorticity ($\zeta/f = 4-5$ according to Figure 2a by *Hasegawa et al.*, [2009]). In the anticyclonic shear only small turbulent perturbations are visible on the vorticity field. Such small-scale and three-dimensional instability did not appear in previous numerical studies of oceanic wakes probably due to the limited vertical resolution.

A large rotating tank, such as the 13 m diameter Coriolis plateform, was needed to study the three-dimensional destabilization of rotating wakes confined in an upper shallow-water layer ($\alpha \ll 1$). Small-scale disturbances were first revealed by qualitative dye visualization in anticyclonic eddies (Figure 14.7) when the island Rossby number exceed a critical value around $Ro_I \simeq 0.8$ for a large Reynolds number $Re \simeq 5200$ and a weakly stratified layer $N/f \simeq 4-5$ [*Teinturier et al.*, 2010]. The typical horizontal scale of these unstable disturbances seems to be fixed by the upper layer thickness rather than the island radius. However, unlike experiments having large or finite shallow-water ratio [*Tarbouriech and Renouard*, 1996; *Stegner et al.*, 2005], the PIV measurements within the shallow-water layer ($\alpha \simeq 0.2$) show that the negative core vorticity of unstable anticyclones remains coherent for several rotation periods (i.e., several days). Hence, the shallow-water configuration seems to reduce the impact of inertial or centrifugal instabilities in the vortex street. For a higher stratification, this stabilization is even more pronounced. Indeed, several intense anticyclones, having a relative core vorticity down to $\zeta_0/f = -3$, were found to be stable in a shallow and strongly stratified ($N/f = 14$) wake where $Ro_I = 1.7$ and $Re = 15,000$ [*Lazar et al.*, 2013b]. Anticyclones become unstable for higher Rossby numbers ($Ro_I \geq 2$) but close to the marginal stability limit the signature of the inertial-centrifugal instability is not clearly visible on the vorticity field (Figure 14.8).

Figure 14.7. Dye visualization of anticyclonic destabilization within an upper shallow-water layer ($h_C = 5$ cm) corresponding to $Ro_I \simeq 1$, $Bu \simeq 0.5 - 1.5$, and $Re = 10000$ [*Teinturier et al.*, 2010]. Different dye colors were released on each side of the cylinder ($D = 50$ cm): black (red) in the anticyclonic (cyclonic) boundary layer. For color detail, please see color plate section.

Surprisingly, the core vorticity of both cyclonic and anticyclonic eddies keeps high values for a long time: $|\zeta_0/f| \simeq 4-5$. The cyclone-anticyclone asymmetry appears only on the velocity profiles: The maximal azimuthal velocity of the anticyclone decays much faster than its cyclonic counterpart. This anomalous decay of the mean azimuthal velocity profile is the experimental signature of the growth of inertial perturbations at the edge of an intense anticyclone. Similar signatures were found in numerical simulations [*Kloosterziel et al.*, 2007] and previous laboratory experiments [*Teinturier et al.*, 2010]. These experimental results show that the combined effect of a strong stratification and a moderate dissipation could strongly stabilize the intense submesoscale anticyclones. Hence, the threshold for three-dimensional destabilization of anticyclonic regions cannot be given by a single Rossby number, the critical Ro_I will crucially depend on

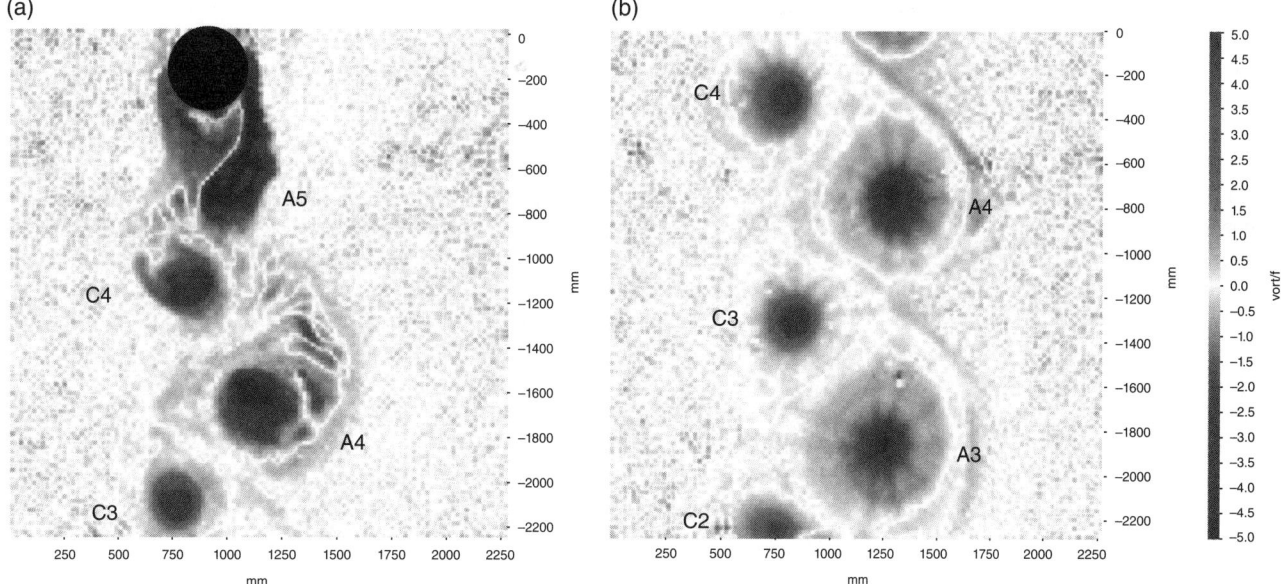

Figure 14.8. Surface vorticity field of an intense submesoscale wake within a thin ($h_c = 6.7$ cm) and stratified ($N/f = 13$) upper layer at (a) $t = 0$ and (b) $t \simeq 2T_0$ when $Ro_I = 2.8$, $Bu = 48$, and $Re = 12,500$ [Lazar et al., 2013b]. The color panel quantifies the relative vorticity amplitude $-5 < \zeta/f < 5$ in blue (red) for anticyclonic (cyclonic) vorticity. The experimental configuration corresponds to figure 2 (c). For color detail, please see color plate section.

the vertical dissipation (i.e., the Ekman number E_k), and the stratification is quantified at the submesoscale by the Burger number Bu.

14.5.2. Inertial-Centrifugal Instability of Shallow Stratified Anticyclones

A global stability analysis of the stratified and rotating wake flow is hard to achieve, especially due to the rapid evolution of the detached vortices in the near wake. However, the vortex street usually reaches a quasi-steady state in the far wake. If we neglect the vortex-vortex interactions, the stability of single and circular vortices could provide a first estimate of the threshold of inertial-centrifugal instability in the far wake.

For inviscid and circular vortices, the generalized Rayleigh criterion [*Kloosterziel and VanHeijst*, 1991; *Mutabazi et al.*, 1992] is a sufficient condition that all anticyclonic vortex columns are unstable to inertial-centrifugal perturbations if somewhere in the flow we get

$$\chi(r) = \left[\zeta(r) + f\right]\left[2\frac{V(r)}{r} + f\right] < 0, \quad (14.1)$$

where $V(r)$ is the azimuthal velocity profile and $\zeta(r) = \partial_r V(r) + V/r$ is the relative vorticity. The stratification or the shallow-water ratio will not change this criterion if the dissipation is negligible. According to this widely used criterion, the stability of circular anticyclones depends crucially on their velocity or vorticity profiles. For instance,

a Rankine vortex will become unstable when the vortex Rossby number $Ro = V_{max}/(f\, r_{max})$ is larger than $Ro \geq 0.5$ and a Gaussian vortex when $Ro \geq 0.31$. However, for a stratified and viscous vortex the Rayleigh criterion becomes less relevant. Close to the marginal stability limit, where the growth rates are strongly controlled by the vertical dissipation, the linear analysis of *Lazar et al.* [2013a] reveals that the instability is not sensitive to the velocity or the vorticity profiles. An analytical marginal stability limit, which depends only on three dimensionless parameters (Ro, Bu, E_k), was derived for anticyclonic Rankine vortex [*Lazar et al.*, 2013a]. The latter become linearly unstable to inertial modes if, in addition to the inviscid criterion (1), corresponding here to $Ro > 0.5$, the following equation is satisfied:

$$\frac{1}{E_k} \geq \left(\frac{8|a_0|}{3}\right)^3 Bu \frac{Ro^2}{\left(\sqrt{2Ro - 1}\right)^7}, \quad (14.2)$$

where $|a_0| = 2.3381$ is the first zero of the Airy function. The above marginal stability equation appears to be relevant for a wide variety of other vortices (parabolic, conical, Gaussian) and could then be used to build a "first guess" stability diagram for inertial-centrifugal destabilization of viscous and stratified anticyclones. Assuming that the vortex Rossby number Ro of the far wake vortices scales as the island Rossby number Ro_I in the turbulent ocean, we could extend such stability diagram to the wake flow.

14.6. CYCLONE-ANTICYCLONE ASYMMETRY

The above-mentioned experimental works show that a uniform upstream flow which encounters an idealized circular island may lead to a strong cyclone-anticyclone asymmetry in the downstream wake. Either cyclonic or anticyclonic eddies may become predominant in the oceanic wake. These asymmetries are due to the intrinsic dynamic properties of the flow and are mainly controlled by three dimensionless numbers: the island Rossby number Ro_I, the Burger number Bu, and the Ekman number E_k. In order to summarize these results and give a global picture, we plot in Figure 14.9 the approximate localization of the various dynamic regimes and the expected cyclonic or anticyclonic predominance in the (Bu, Ro_I) parameter range. The supercritical limit is drawn with a bold solid line; it corresponds to $Fd = U_0/C = Ro_I/\sqrt{Bu} = 1$, where C is the maximum phase speed of internal gravity waves. A supercritical incoming current is unrealistic in the ocean and therefore the region above this solid line ($Fd \geq 1$) is forbidden. Inside the gray area, corresponding to $Ro_I \leq 0.1$, the geostrophic balance should be fulfilled. In order to make a distinction between the mesoscale (Bu $\geq$ 1) and the submesoscale (Bu $\ll$ 1) dynamics, we draw an arbitrary line (thin dotted line) at Bu = 0.3. Of course, this vertical line is not an exact limit and does not correspond to any well-defined boundary between the mesoscales and the submesoscale. On the other hand, the marginal stability limit of three-dimensional destabilization could be calculated (equation (2)) or computed for idealized anticyclones. However, this limit depends crucially on the Ekman number. If the latter is well defined in viscous laboratory experiments, it is not the case for the turbulent oceanic thermocline. Hence, we plot three different stability curves which could be relevant for a viscous laboratory exepriment ($1/E_k = 1\,000$), an upper oceanic layer ($1/E_k = 20,000$) corresponding to a moderate diapycnal diffusivity $\kappa_z \simeq 10^{-4}$ m^2/s, or a fully inviscid layer ($E_k = 0$). The strongest anticyclonic destabilization and three-dimensional mixing will occur for submesocale wakes close to the supercritical limit.

14.7. FROM IDEALIZED LABORATORY FLOWS TO COMPLEX OCEANIC WAKE

These idealized laboratory experiments provide a general understanding of the dynamic process which impacts on deep oceanic island wake at various scales. However, various other processes not mentionned in this chapter may strongly impact the wake formation, the vortex street geometry, and its dynamic behavior. There is still a long way to reach realistic laboratory models of complex island wakes.

We learn from this review that various three-dimensional and ageostrophic processes are expected to impact oceanic island wakes. Indeed, to capture such small-scale processes, it is necessary to reach large Reynolds numbers and avoid an excessive dissipation at submesoscale. Therefore, realistic experiments should be conducted on sufficiently large turntables. The limited number of large rotating facilities, such as the 5 m Coriolis rotating basin in Trondheim or the 13 m Coriolis platform in Grenoble, may restrict the investigations.

Most of the experiments consider an idealized cylindrical or symmetrical island to generate the wake. However, even in the nonrotating case, an asymmetrical island coast may induce significant differences between the opposite side boundary layers. Hence, the symmetry of the detached eddies in the near-wake flow could obviously be broken by a realistic island coast. Few numerical works [*Caldeira and Sangra*, 2012; *Calil et al.*, 2008; *Dong and McWilliams*, 2007] take into account the complex island bathymetry and study its impact on the deep oceanic wake. A complex archipelago will induce multiple wake flows which may interact with each other [*Perret et al.*, 2011] and strongly influence the eddy formation in the near wake [*Caldeira and Sangra*, 2012; *Calil et al.*,

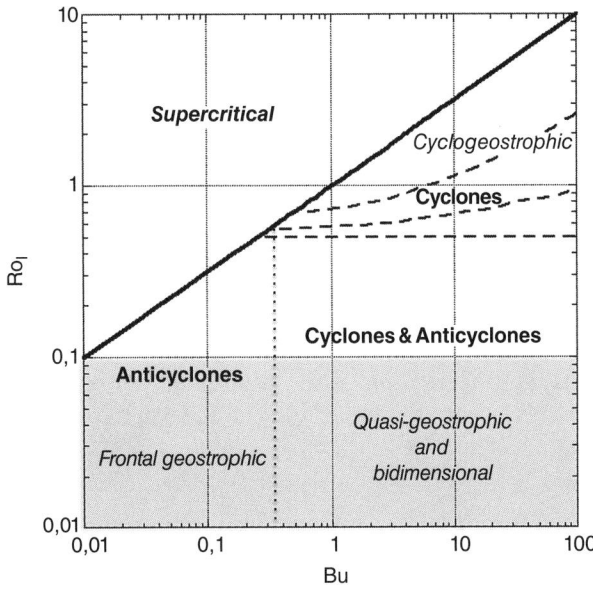

Figure 14.9. Dynamical regimes and cyclonic or anticyclonic predominance in the (Ro_I, Bu) parameter space. The bold solid line corresponds to the supercritical limit $F_d = 1$, while the horizontal dashed lines correspond to the inertial-centrifugal stability limit (i.e., equation (2)) for $E_k = 10^{-3}, 5 \times 10^{-5}, 0$. The dotted vertical line is an arbitrary delimitation between the mesoscale (Bu $\ll$ 1) and the submesoscale wake (Bu $\geq$ 1). The wake flow is expected to satisfy the geostrophic balance inside the gray area.

2008]. Exhaustive experimental investigations on rotating and stratified wakes which take into account asymmetrical island shapes or multiple islands interaction are missing. This is probably a new goal to achieve in the near future.

A common hypothesis for laboratory models of oceanic wake is to assume a steady and uniform upstream current. According to the standard Strouhal number, typical shedding periods of island wakes in the deep ocean should vary from one to several weeks. For such time intervals the variability of surface currents is not negligible and the upstream flow may change in direction and intensity. Hence, the transient response of the flow could become significant and affect the wake pattern. Indeed, when the upstream current is accelerated from rest, the first pattern to emerge in the lee of a cylindrical island is a symmetric dipole, also called the starting dipole [*Afanasyev and Korabel*, 2004], while regular and periodic shedding of opposite sign vortices will appear later on. The temporal variability of the incoming flow will disturb the quasi-steady vortex street and induce a wider variety of wake patterns. An unsteady wake forcing will then be a technical challenge for more realistic experiments on rotating wakes.

The role of local wind stress as a driver of oceanic vortices in the lee of mountainous island was recently highlighted in numerical studies of the Hawaiian [*Calil et al.*, 2008; *Yoshida et al.*, 2010; *Kersalé et al.*, 2011] or the Madeira [*Couvelard et al.*, 2012; *Caldeira et al.*, 2014] wake. For the Hawaiian case, the interaction between the North Equatorial Current and the archipelago is enough to generate eddies, but high-resolution winds stress curls are needed to get the correct intensities of the vortices as observed from the altimetry. For the Madeira case, the wind wake appears to be the main contributor to the generation of the oceanic eddies in the island wake [*Caldeira et al.*, 2014]. As far as experimental investigations are concerned, combining both an upstream flow and a local wind forcing on a rotating turntable seems hard to achieve and here we probably reach the limits of laboratory models.

Acknowledgments. The author gratefully acknowledges all the collaborators, G. Perret, S. Teinturier, A. Lazar, T. Dubos, J. M. Chomaz, R. Caldeira, and C. Dong, who took part in the experimental and numerical investigations on rotating shallow-water wakes. Warm thanks to the Coriolis team S. Vuiboud and H. Didelle for their irreplaceable expertise on rotating experiments. The author is also grateful to the French ANR, the 6th European Commission (EC) framework program Hydralab, and the sustainable development chair of the Ecole Polytechnique, who funded a large part of our studies these past years.

REFERENCES

Afanasyev, Y. D., and V. N. Korabel (2004), Starting vortex dipoles in a viscous fluid: Asymptotic theory, numerical simulations, and laboratory experiments, *Phys. Fluid*, *16* (11), 3850–3858.

Alaee, M. J., G. Ivey, and C. Pattiaratchi (2004), Secondary circulation induced by flow curvature and Coriolis effects around headlands and islands, *Ocean Dyn.*, *54*, 27–38.

Arai, M., and T. Yamagata (1994), Asymmetric evolution of eddies in rotating shallow water, *Chaos*, *4*, 163–175.

Aristegui, J., P. Sangra, S. Hernandez-Leon, M. Canton, A. Hernandez-Guerra, and J. L. Kerling (1994), Island-induced eddies in the canary islands, *Deep-Sea Res*, *41*(10) 1509–1525.

Baey, J. M., and X. Carton (2002), Vortex multipoles in two-layer rotating shallow-water flows, *J. Fluid Mech.*, *460*, 151–175.

Boyer, D. L., and P. A. Davies (1982), Flow past a circular cylinder on a β-plane, *philos. Trans. R. Soc. Lond. Seri. A*, *306*, (1496), 533–556.

Boyer, D. L., and M. L. Kmetz (1983), Vortex shedding in rotating flows, *Geophys. Astrophys. Fluid*, *26*, 51–83.

Boyer, D. L., M. L. Kmetz, L. Smathers, L. G. Chabert d'Hieres, and H. Didelle (1984), Rotating open channel flow past right circular cylinders, *Geophys. Astrophys. Fluid*, *30*, 271–304.

Caldeira, R. M. A., and P. Sangra (2012), Complex geophysical wake flows Madeira Archipelago case study, *Ocean Dyn.*, doi:10.1007/s10236-012-0528-6.

Caldeira, R. M. A. S. Groom, P. Miller, D. Pilgrim, N. Nezlin (2002), Sea-surface signatures of the island mass effect phenomena around Madeira island, northeast atlantic, *Remote Sens. Environ.*, *80*, 336–360.

Caldeira, R., A. Stegner, X. Couvelard, I. B. Araujo, P. Testor and A. Lorenzo "Evolution of an oceanic anticyclone in the lee of Madeira Island: In situ and remote sensing survey", *J. Geophys. Res. Oceans*, *119*, doi:10.1002/2013JC009493 (2014).

Calil, P. H. R., K. J. Richards, Y. Jia, and R. R. Bidigare (2008), Eddy activity in the lee of the Hawaiian Islands, *Deep Sea Res. P. II*, *55*, 1179–1194, doi:10.1016/j.dsr2.2008.01.008.

Chavanne, C., P. Flament, and K.-W. Gurgel (2010), Interactions between a submesoscale anticyclonic vortex and a front, *J. Phys. Oceanogr.*, *40*, 1802–1818, doi:10.1175/2010JPO4055.1.

Chelton, D. B., R. A. de Szoeke, M. G. Schlax, K. El Naggar, and N. Siwertz (1998), Geographical variability of the first-baroclinic Rossby radius of deformation, *J. Phys. Oceanogr.*, *28*, 433–460.

Chomaz, J. M. (2005), Global instabilities in spatially developing flows: Non-normality and nonlinearity, *Annu. Rev. Fluid Mech.*, *37*, 357.

Couvelard, X., R. M. A. Caldeira, I. B. Araujo, and R. Tome (2012), Wind mediated vorticity-generation and eddy-confinement, leeward of the Madeira Island: 2008 numerical case study, Dyn. Atmos. Oceans, 58, 128–149.

Cushman-Roisin, B. (1986), Frontal geostrophic dynamics, *J. Phys. Oceanogr.*, *16*, 132.

DeFelice, T. P. et al. (2001), Landsat-7 reveals more than just the surface feature in remote areas of the globe, *Bull. Am Meteorol. Soc.*, *81*, 1047.

Dong, C. M., and J. c. McWilliams (2007), A numerical study of island wakes in the Southern California Bight, *Cont. Shelf Res.*, *27*(9), 1233–1248.

Dong, C. M., J. C. McWilliams, and A. F. Shchepetkin (2007), Island wakes in deep water, *J. Phys. Oceanogr.*, *37* (4), 962–981.

Etling, D. (1989), On atmospheric vortex street in the wake of large islands, *Meteor. Atmos. Phys.*, *41*, 157.

Graves, L. P., J. C. McWilliams, and M. T. Montgomery (2006), Vortex evolution due to straining: A mechanism for dominance of strong, interior anticyclones, *Geophys. Astrophys. Fluid Dyn.*, *100*, 151–183.

Haine, T. W. N., and J. Marshall (1998), Gravitational, symmetric, and baroclinic instability of the ocean mixed layer, *J. Phys. Oceanogr.*, *28*, 634–658.

Hasegawa, D., H. Yamazaki, R. G. Lueck, and L. Seuront (2004), How islands stir and fertilize the upper ocean, *Geophys. Res. Lett.*, *16*, 31.

Hasegawa, D., M. R. Lewis, and A. Gangopadhyay (2009), How islands cause phytoplankton to bloom in their wakes? *Geophys. Res. Lett.*, *36*, L20605, doi:10.1029/2009GL039743.

Heywood, K.J., D. P. Stevens, and G. R. Bigg (1996), Eddy formation behind the tropical island of Aldabra, *Deep-Sea Res. Pt. I*, *43*(4), 555–578.

Hoskins, B. J. (1974), The role of potential vorticity in symmetric stability and instability, *Q. J. Roy. Meteor. Soc.*, *100*, 480–482.

Johnson, E. R., J. G. Esler, O. J. Rump, J. Sommeria, and G. G. Vilenski (2006), Orographically generated nonlinear waves in rotating and non-rotating two-layer flow, *P. R. Soc. A*, *462*, 3–20.

Johnson, J.A. (1963), The stability of shearing motion in a rotating fluid, *J. Fluid Mech.*, *17*(3), 337–352.

Kersalé, M., A. M. Doglioli, and A. A. Petrenko, (2011), Sensitivity study of the generation of mesoscale eddies in a numerical model of Hawaii islands, *Ocean Sci.*, *7*, 277–291.

Kloosterziel, R. C., and G. F. Carnevale (2008), Vertical scale selection in inertial instability, *J. Fluid Mech.*, *594*, 249–269.

Kloosterziel, R. C., and G. J. F. VanHeijst (1991), An experimental study of unstable barotropic vortices in a rotating fluid, *J. Fluid Mech.*, *223*, 1–24.

Kloosterziel, R. C., and G. F. Carnevale, P. Orlandi (2007), Inertial instability in rotating and stratified fluids: Barotropic vortices, *J. Fluid Mech.*, *583*, 379–412.

Lazar, A., A. Stegner, and E. Heifetz (2013a), Inertial instability of intense and stratified anticyclones. Part I: Linear analysis and marginal stability criterion, *J. Fluid Mech.*, *732*, 457–484.

Ledwell, J., and A. Watson (1998), Mixing of a tracer in the pycnocline, *J. Geophys. Res*, *103*, C10, 21499–21529.

Mutabazi, I., C. Normand, and J. E. Wesfreid (1992), Gap size effects on centrifugally and rotationally driven instabilities, *Phys. Fluids A-Fluid*, *4*(6), 1199–1205.

Neill, S. P., and A. J. Elliott (2004), Observations and simulations of an unsteady island wake in the Firth of Forth, Scotland, *Ocean Dyn.*, *54*, 324–332.

Perret, G., A. Stegner, T. Dubos, J. M. Chomaz, and M. Farge (2006a), Stability of parallel wake flows in quasigeostrophic and frontal regimes, *Phys. Fluids*, *18*, 126,602.

Perret, G., A. Stegner, M. Farge, and T. Pichon (2006b), Cyclone-anticyclone asymmetry of large scale wakes in laboratory, *Phys. Fluids*, *18*, 036,603.

Perret, G., T. Dubos, and A. Stegner (2011), How large scale and cyclogeostrophic barotropic instabilites favor the formation of anticyclonic vortices in the ocean, *J. Phys. Oceanogr.*, *41*, 303–328.

Pier, B. (2002), On the frequency selection of finite-amplitude vortex shedding in the cylinder wake, *J. Fluid Mech.*, *458*, 407.

Plougonven, R., and V. Zeitlin (2009), Nonlinear development of inertial instability in a barotropic shear, *Phys. Fluids*, *21*(10), 106, 601, doi:10.1063/1.3242283.

Potylitsin, P. G., and W. R. Peltier (1998), Stratification effects on the stability of columnar vortices on the f-plane, *J. Fluid Mech.*, *355*, 45–79.

Rotunno, R. and P. K. Smolarkiewicz (1991), Further results on lee vortices in low-Froude number flow, *J. Atmos. Sci.*, *48*, 2204.

Sangra, P., J. L. Pelegri, A. Hernandez-Guerra, I. Arregui, J. M. Martin, A. Marrero-Dıaz, A. Martınez, Ratsimandresy, A. W., and A. Rodrıguez-Santana (2005), Life history of an anticyclonic eddy, *J. Geophysi. Res.*, *110*, C03021.

Stegner, A. (2007), Experimental reality of geostrophic adjustment, in Nonlinear Dynamics of Rotating Shallow Water: Methods and Advances, edited by V. Zeitlin, pp. 323–377, Elsevier, Amsterdam.

Stegner, A. and D. G. Dritschel (2000), A numerical investigation of the stability of isolated shallow water vortices, *J. Phys. Oceanogr.*, *30*, 2562–2573.

Stegner, A., T. Pichon, and M. Beunier (2005), Elliptical-inertial instability of rotating Karman street, *Phys. Fluids*, *17*, 1–10.

Lazar, A., A. Stegner, R. Caldeira, C. Dong, H. Didelle, and S. Vuiboud (2013b), Inertial instability of intense and stratified anticyclones. Part II : Laboratory experiments, *J. Fluid Mech.*, *732*, 485–509.

Tarbouriech, L., and D. Renouard (1996), Stabilisation et destabilisation par la rotation d'un sillage turbulent, *C. R. Acad. Sci. III*, *323*, 323–391.

Teinturier, S., A. Stegner, S. Viboud, and H. Didelle (2010), Small-scale instabilities of an island wake flow in a rotating shallow-water layer, *Dyn. Atmos. Ocean*, *39*, 1–24, doi:10.1016/J.Dynatmoce.2008.10.006.

M. Tomczak (1988), Island wakes in deep and shallow water, *J. Geophys. Res.*, *93*, 5153–5154.

Wen, Y., and C. Y. Lin (2001), Two-dimensionnal vortex shedding of a circular cylinder, *Phys. Fluids*, *13*, 557.

Wolanski, E., J. Imberger, and M. L. Heron (1984), Island wakes in shallow coastal waters, *J. Geophys. Res.*, *89C*, 10,533–10,569.

Wolanski, E., T. Asaeda, A. Tanaka, and E. Deleersnijder (1996), Three-dimensional island wakes in the field, laboratory experiments and numerical models, *Cont. Shelf Res.*, *16*, 1437–1452.

Yanase, S., C. Flores, O. Metais, and J. J. Riley (1993), Rotating free-shear flows. 1. Linear-stability analysis, *Phys. Fluids A-Fluid*, *5*(11), 2725–2737.

Yoshida, S., B., Qiu, and P. Hacker (2010), Wind-generated eddy characteristics in the lee of the island of Hawaii, *J. Geophys. Res. Oceans*, *115*, C03019, doi:10.1029/2009JC005417.

Section V: Advances in Methodology

15

Lagrangian Methods in Experimental Fluid Mechanics

Mickael Bourgoin[1], Jean-François Pinton[2], and Romain Volk[2]

15.1. INTRODUCTION

Atmospheric and oceanographic flows are characterized by a strong complexity: Stratification, anisotropy, global rotation, inhomogeneity, turbulence, etc., are just a few of the main features involved. This chapter reviews some of the most recent advances in metrology relevant to investigate such complex flows in model laboratory experiments. We focus on high-resolution techniques which give access to the hierarchy of multiple spatial and temporal scales involved in this type of highly turbulent flows. As described below, among these techniques, Lagrangian approaches (where tracer particles are tracked in the flow) have undergone significant developments in the past decade. Lagrangian metrology has become one of the most versatile and accurate tools for the investigation of complex flows, particularly suited in regard to geophysical motivations, where transport issues are paramount.

In the last decades, the increase of performance in numerical simulations of fluid dynamics processes (CFD) has motivated an increasing demand in the accuracy of models and, naturally, experimental measurements. A particular challenge in the context of atmospheric and oceanographic research concerns the improvement of the multiscale description of flows and of energy cascade mechanisms. Geophysical flows are indeed characterized by a high turbulence intensity which results in an important hierarchy of relevant scales, with structures ranging from kilometers down to millimeters. In turbulent flows, the range of relevant scales (called the *inertial range*) between the energy injection scale L and the dissipative scale (also called Kolmogorov scale) η is directly related to the Reynolds number of the flow, Re: $L/\eta \propto \mathrm{Re}^{3/4}$ [*Tennekes and Lumley*, 1992]. Reynolds numbers of the order 10^6 are usual in geophysical flows, implying that at least four decades of spatial dynamics are typically involved. Similarly in the time domain, the ratio between the eddy turnover time T_L at injection scale and at dissipation scales τ_η goes as $T_L/\tau_\eta \propto \mathrm{Re}^{1/2}$, covering three decades of temporal dynamics. These dynamical ranges can be even further extended toward the largest scales due to inverse cascade mechanisms, which may become important in the atmosphere or the ocean, at scales where dynamics exhibits 2D properties, where flow structures can extend over hundreds of kilometers. To investigate related physics in laboratory experiments, with a typical dimension of the order of 1 m and typical correlation time scale of the order of 1 s, the investigation of a comparable hierarchy of scales pushes the smallest involved structures down to tens to hundreds of micrometers (or even smaller) in space with a typical time scale of fractions of milliseconds. This stimulates a permanent effort in the experimental community to develop measurements with increasing degrees of accuracy and resolution. Technological advances in high-speed digital imaging with the ability to record pictures with millions of pixels at rates exceeding thousands of frames per second have opened up a new era in experimental fluid dynamics and laboratory models of geophysical flows. In parallel, several innovative developments of scattering techniques (sound or light) have emerged. More recently, progress in embedded sensor and radio transmission technologies have led to the development of instrumented particles which probe the flow as they are entrained by the local motions.

Let us briefly recall that flow measurements are done using two approaches: the Eulerian description and the Lagrangian one (see Figure 15.1). In Eulerian techniques, the fluid velocity $\vec{v}_E(\vec{r}, t)$ is studied as a field varying in space $\vec{r}$ at time t. Although this field also experiences in

[1] *Laboratoire des Écoulements Géophysiques et Industriels, Université de Grenoble & CNRS, Grenoble, France.*
[2] *Laboratoire de Physique, École Normale Supérieure de Lyon, Lyon, France.*

280 MODELING ATMOSPHERIC AND OCEANIC FLOWS

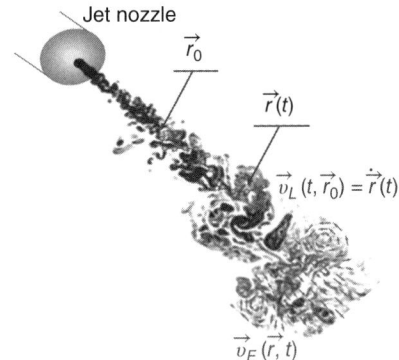

Figure 15.1. Eulerian versus Lagrangian description of a flow. In the Eulerian framework the flow is described in terms of the velocity field, while in the Lagrangian framework it is characterized from the trajectories of fluid tracers.

general instantaneous temporal fluctuations, in statistically stationary conditions, the time t is generally considered as a parameter which helps build ensemble averages by repeating the measurement (if the system is not statistically stationary, then time t becomes an actual variable of the problem which accounts for nonstationary effects). Eulerian measurements have for long been the most widely used in experimental fluid mechanics: Hot-wire measurements, particle image velocimetry (PIV), and laser Doppler velocimetry (LDV) are classical techniques. In the Lagrangian approach, instead of probing the flow at given fixed points $\vec{r}$ (where fluid particles constantly pass), and velocity is measured along the path of given fluid elements which are tagged and tracked individually. In this representation, the velocity $\vec{v}_L(t, \vec{r}_0)$ changes in the course of time t while space coordinates simply parametrize the initial position $\vec{r}_0$ of the fluid particle under investigation. In statistically homogeneous conditions, $\vec{r}_0$ is mainly a parameter which is considered either to improve statistical convergence by simultaneously tracking several particles with different initial separations or to address multiparticle problems such as mixing and dispersion.

Eulerian measurements have prevailed in experimental fluid mechanics for decades. Recent progress in Eulerian measurements mainly concern the improvement of PIV systems, which are now commonly available in 3D–3C configuration (where the three components of the velocity field are measured in a full 3D volume of the flow, for instance, using tomographic reconstruction), with an increased repetition rate (due to the newest high-speed camera technologies) giving access to time-resolved measurements.

On the other hand, Lagrangian measurements have now been developed to the point where Lagrangian particle tracking is among the most accurate fluid dynamics measurements. Several factors have contributed to what we can call the *Lagrangian revolution*: (i) The importance of Lagrangian approaches for modeling. The relevance of a Lagrangian description of the dynamics of a fluid in the context of mixing and transport phenomena has been realized since *Taylor* [1922] and *Richardson* [1926] (in particular for atmospheric and oceanic flows). It is now essential for the study of dispersion (pollutants, for instance). Modern stochastic models are best developed in the Lagrangian domain. (ii) Technological progress have given access to the ultrafast recording and processing performance required for Lagrangian tracking. This led for instance to the first measurements of acceleration of fluid particles in turbulent flows over a decade ago [*Voth et al.*, 2002; *Mordant et al.*, 2001]. We detail in the following sections some of the latest measurement techniques used in state-of-the-art laboratory experiments. We shall begin with optical techniques (Lagrangian tracking and extended laser Doppler velocimetry (ELDV)) before moving to acoustical ones (Lagrangian tracking and vorticity measurement). We close this chapter with an introduction to instrumented particles. As will be illustrated below, some of these techniques have already proven their relevance to investigate model flows with geophysical motivations (rotating fluids, stratified flows, turbulent transport phenomena, etc.).

15.2. OPTICAL TECHNIQUES

15.2.1. Particle Tracking

An important advance in Lagrangian measurements was done in 1997 by *Virant and Dracos* [1997], who developed 3D particle tracking velocimetry (PTV) based on the direct imaging of small particles seeding the flow. They used simultaneously four video cameras at a rate of 25 fps to access the 3D trajectories of several hundreds of particles at once. *Ott and Mann* [2000] developed a similar technique to study relative dispersion of fluid particles. In those two pioneering experiments, because of the low frame rate, particle dynamics could be resolved only for flows at moderate Reynolds numbers (Re < 4000 typically). *LaPorta et al.* [2001] used silicon strip detectors (initially developed for high-energy particle detection) at a frame rate up to 70 kHz, allowing the first fully resolved Lagrangian optical tracking measurements at Reynolds numbers approaching 10^5. However, only one particle at a time could be tracked with the silicon strip detectors. Subsequently, *Bourgoin et al.* [2006] developed a high-resolution 3DPTV facility similar to that of Virant and Dracos and Ott and Mann, but using ultrafast cameras at a repetition rate of 27 kHz, allowing the tracking of several hundred of particles in high-Reynolds-number regimes. In the context of atmospheric and oceanographic studies, similar 3D PTV systems have been recently implemented to investigate, for instance thermal convective flows [*Ni et al.*, 2012] as well as the influence of Coriolis force on the transport properties of rotating flows [*Del Castello and Clercx*, 2011]. We detail in the following how such multicamera tracking systems work.

15.2.1.1. Principle. The principle of optical particle tracking is conceptually very simple: It consists in filming the motion of particles and thereafter to reconstruct their trajectories. However, its practical implementation is a challenge, and several aspects that must be carefully considered will be discussed next.

Resolution Issues For high-resolution measurements, high-speed cameras with a large number of pixels are required. As already discussed, three decades of temporal resolution requires a repetition rate of at least 1 kHz (assuming large structures evolve with a typical time scale around 1 s), while four decades of spatial resolution would in principle require a sensor with at least $10^4 \times 10^4$ pixels. State-of-the-art high-speed cameras are typically capable of recording $10^3 \times 10^3$ pixel images at several thousands of frames per second, which yields over three decades in time and three decades in space. In practice, as discussed later, experimental noise generally requires to severely oversample the data, and this lowers the effective time resolution. On the other hand, several cameras are generally used simultaneously (as discussed below), which has the additional benefit of improving the effective resolution to about 1/10th of a pixel, hence recovering four decades of effective spatial resolution. The global resolution of a tracking system is generally a trade-off between temporal and spatial resolution, as higher repetition rates can be achieved by reducing the number of pixels and vice versa. However, due to the impressive progress in high-speed digital imaging technology, direct optical tracking has become one of the most accurate techniques in experimental fluid mechanics.

3D Issues Complex flows generally involve 3D structures which require tracking to be done in 3D. This has two consequences: (i) The flow has to be illuminated in volume (a laser sheet, as done for instance in PIV, is not sufficient) and (ii) particles must be tracked in 3D, hence requiring a stereoscopic configuration. In terms of illumination, as the tracers to be tracked are generally small (hence the diffused light is dim), the repetition rate is high (hence exposure time is short), and the light beam is enlarged (to illuminate a volume), high-power light sources are required. High power lasers

Figure 15.2. Sketch of a typical 3D particle tracking experiment. The central part of the bulk of the flow is illuminated using two expanded high-power laser beams. Three cameras record simultaneously the motion of small particle tracers in the bulk. From *Bourgoin et al.* [2006].

have been generally used [*Voth et al.*, 2002; *LaPorta et al.*, 2001; *Bourgoin et al.*, 2006], though alternative and less expensive solutions using high-power light-emitting diodes (LEDs) start to be developed [*Del Castello and Clercx*, 2011]. In terms of recording, the stereoscopic reconstruction requires at least two cameras with two different angles of view to be used simultaneously. In practice three or more cameras are used. Increasing the number of cameras has two main advantages: (i) It allows tracking more particles simultaneously, which is interesting to improve statistical convergence of the measurements, especially when multiparticle problems (for instance, related to dispersion issues) are investigated, and (ii) the redundancy for particles which are seen simultaneously by more than two cameras improves the accuracy of the 3D positioning of those particles, thus leading to an enhanced effective resolution. State-of-the-art optical Lagrangian systems using three or four high-speed cameras are capable of tracking several hundreds of particles with 1/10th of pixel of effective resolution. Figure 15.2 shows the three high-speed camera system implemented by *Bourgoin et al.* [2006].

Data Management Issues High-speed imaging experiments result in a huge data rate. For instance, 1 kHz acquisitions with three one megapixel sensors recording at a bit depth of 8 bits represent an effective data rate of a few gygabytes per second of recording. These usually require to couple the acquisition system to dedicated data storage and data processing servers.

The different steps of the data processing, essential for the optimization of the accuracy of the 3D tracking, are described next.

15.2.1.2. Reconstruction of 3D Trajectories.

Once the images of the tracers are recorded, the goal is to reconstruct the 3D trajectories of as many particles as possible. This operation requires three steps:

1. Particle Detection. Each image (at each time t) of each camera is analyzed to determine the position of the center of each visible particle. This step results in maps of the 2D position of the center of the particles on each frame of each camera.

2. 3D matching. The second step consists in combining at each given time t previous 2D maps of particle centers from the N cameras in order to reconstruct (by stereomatching) the 3D position of the center of the particles with the highest possible accuracy.

3. Lagrangian tracking. Finally, once the 3D positions of particles are found for all time steps, an appropriate tracking algorithm allows to reconnect the trajectories.

We describe briefly the key points of the previous steps in the following paragraphs. Further details and useful information can be found in the work of *Ouellette et al.* [2005].

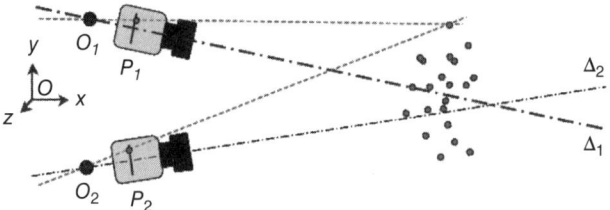

Figure 15.3. Optical tracking: 3D matching procedure.

Particle Detection *Ouellette et al.* [2005] have tested different algorithms for the detection of particle centers in 2D images. The choice of the best algorithm is a compromise between computation time and quality of the detection. The latter is quantified by both the accuracy with which the position of the center of the particles is determined and the number of particles correctly detected. The first step is to identify the local maxima of intensity on the image, indicating the presence of a particle. Then, the image around each maximum is analyzed to determine to the best accuracy the location of the center of the particle. For small particles (as generally used to seed the flow with tracers), the image of each individual particle does not exceed a few pixels. Under these conditions, simple algorithms based on the center of mass of intensity around the maximum are not sufficiently accurate. Algorithms based on neural networks can be very accurate, especially when images are very noisy, but relatively slow. A good compromise consists in fitting the local intensity profile by two Gaussians (one vertical and one horizontal), whose maxima define the center of the particle. The choice of two 1D Gaussian fits is preferred to that of one single 2D Gaussian because it is computationally significantly more efficient for almost the same accuracy. Ouellette et al. have shown that this method was typically capable of detecting 95% of particles and determining their position with subpixel accuracy.

3D Matching While the detection of particles can be made in the image space of each camera, 3D positioning and Lagrangian tracking must be made in real space (which is common to all cameras). The most widely used method to define the transformation for each camera between image space (in pixels) to real space (in real units) is based on a calibration method developed by *Tsai* [1987]. Each camera (let us say we consider camera i) is represented by a projection model defined by an optical axis Δ_i, an optical center O_i, and a projection plane P_i. The image of a particle X on the sensor of camera i is then simply given by the intersection of the line $O_i X$ with the plane P_i (see Figure 15.3). The model is generally defined by at least nine parameters for each camera: six external parameters for the absolute position of each camera (three coordinates for O_i and three angles for the orientation of the optical axis Δ_i) and three internal parameters (the distance $O_i P_i$, a coefficient for geometric aberrations and the aspect ratio of the pixels). Refinement of this basic model can be considered, for instance, by including several aberration coefficients (transverse and longitudinal). The parameters of the model are determined from the images of a calibration mask with known geometric properties. Once the parameters of the model for each camera are determined, the 3D matching is performed as follows (see Figure 15.3): Take the center of a particle x_i as previously determined in pixels on the projection of plane of one of the cameras; the real position X_i of the particle in real space then lies somewhere on the line of view $O_i x_i$. The intersection of such line of views from two (or more) cameras defines the absolute 3D position of the particle in real space. In theory two cameras are sufficient to determine this intersection. In practice, however, the lines of view rarely intersect due to slight imprecision in the calibration of the Tsai model. The 3D position is then defined as the point in real space which minimizes the distance to the different lines of view. Whenever a camera is added in the system, the redundancy of information provided by the additional line of view further restricts the possible 3D position of the particle. This greatly improves the effective spatial resolution of the 3D system. Ouellete et al. have shown that using three cameras instead of two gives an effective resolution of the order of one-tenth "equivalent pixel" (that is to say one-tenth of the spatial dimension whose image is the size of a pixel taking into account the magnification of the projective system). Thus the combination of three sensors of $10^3 \times 10^3$ pixels provides an effective spatial resolution of four decades in 3D. Adding a fourth camera is then essentially interesting to increase the number of particles actually followed. Indeed, ambiguous situations where a particle hides another one in the line of sight of a camera may occur. These ambiguities can be lifted by adding a fourth camera at a different angle in order to maximize the number of particles which are seen at any time by at least three cameras.

Trajectory Reconnections Lagrangian tracking consists in reconnecting particle trajectories between successive time steps. This requires to identify at time $t + 1$ particles already detected at time t. Lagrangian tracking algorithms are generally based on the minimization of a given cost function. The simplest algorithm, called *nearest neighbor*, simply consists in connecting a particle (let us say particle j) whose position at time $t + 1$ is $\vec{x}_j(t + 1)$ to the particle i whose position at time t minimizes the cost function $\phi_{ij} = \| \vec{x}_i(t) - \vec{x}_j(t + 1) \|$. This simple algorithm is accurate only if the interframe displacement is significantly less than the average interparticle separation. It is therefore generally limited to relatively diluted configurations. In higher seeding density situations, more sophisticated algorithms are required. As shown by Ouellette et al., one robust algorithm consists in defining a cost function ϕ_{ij} based on four consecutive images. Qualitatively, it is based on a smoothest acceleration criterion. Quantitatively it is implemented as follows: Assume trajectories have been reconnected up to time step t; the velocity of the particles is estimated from positions $\vec{x}_i(t)$ and $\vec{x}_i(t - 1)$, which allows one to estimate their probable position at time $t + 1$, $\vec{\tilde{x}}_i(t + 1)$; then the particle acceleration is estimated from $\vec{x}_i(t - 1)$, $\vec{x}_i(t)$ and $\vec{\tilde{x}}_i(t + 1)$ and this propagates at time $t + 2$ an estimation $\vec{\tilde{x}}_i(t + 2)$ of the position for each particle in the vicinity of $\vec{\tilde{x}}_i(t + 1)$; then the most probable trajectory is the one which minimizes the cost function $\phi_{ij} = \| \vec{\tilde{x}}_i(t + 2) - \vec{x}_j(t + 2) \|$.

15.2.1.3. Example of 3D Optical Tracking.
Figure 15.4a shows an example of tracking of pairs of particles by *Bourgoin et al.* [2006] in the high-Reynolds-number experiment previously shown in Figure 15.2. Figure 15.4a only shows two trajectories, but hundreds of such trajectories are simultaneously reconstructed. This allows a rapid statistical convergence of particle displacement, velocity, and acceleration statistics. Such data can be used to investigate different properties of the flow. In the study by Bourgoin et al. separation statistics are investigated in order to address the longstanding question of turbulent super diffusion. But time-resolved trajectories can be differentiated with time, once to obtain particle velocity and

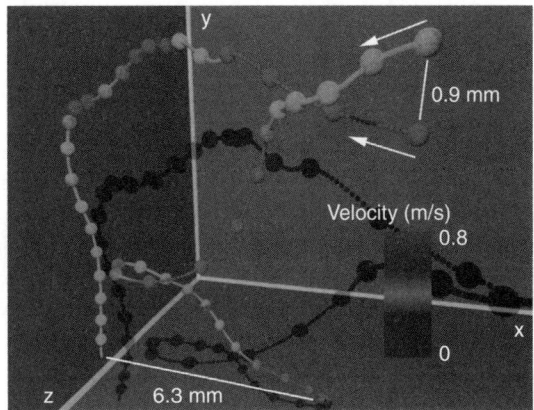

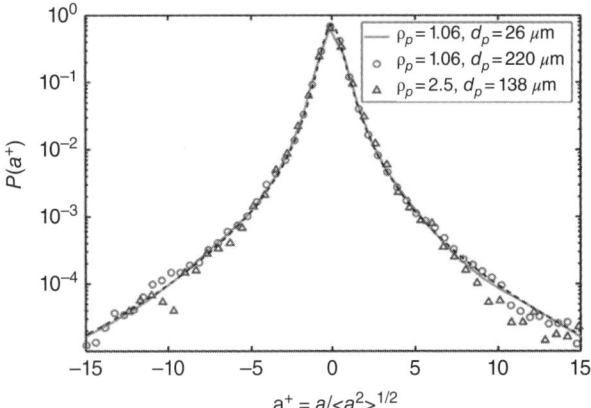

Figure 15.4. Top: Example of high-resolution reconstruction of particle trajectories in experiment shown in Figure 15.2. The small spheres mark every other measured position of the particles and are separated by 0.074 ms; the large spheres mark every 30th position. Color indicates the velocity of the particle along its trajectory. From *Bourgoin et al.* [2006]. Bottom: One component acceleration statistics measured in a von Kármán swirling flow seeded with particles of different size and density. From *Xu and Bodenschatz* [2008]. For color detail, please see color plate section.

twice to access particle acceleration. Optical tracking of small particles has shown the highly intermittent dynamics of such fluid tracers in turbulent flows. This is revealed for instance by measurements of acceleration statistics in von Kármán swirling flows (shown in Figure 15.4b), which exhibit highly non-Gaussian fluctuations corresponding to events of very high acceleration occurring with a probability order of magnitude larger than what would be expected for a normal random process with equivalent variance.

15.2.2. Extended Laser Doppler Velocimetry

As already mentioned, particle tracking is very demanding in terms of acquisition frequency, which needs to be much larger than the inverse of the Kolmogorov time scale $1/\tau_\eta = \sqrt{\epsilon/\nu}$ with a spatial resolution comparable with the Kolmogorov scale $\eta = (\nu^3/\epsilon)^{1/4}$ in order to access the very small scales of the particle motion. For typical water flows at the laboratory scale, one therefore needs to track particles with sizes in the range $10-100\,\mu$m with a sampling frequency larger than several kilohertz, which is a severe limitation in terms of camera specifications and cost of the experiment. To increase the temporal resolution at a modest cost, one possibility is to rely on scattering techniques using a reference wave (either using ultrasound or laser light) that will be scattered by the moving particles. This is the basis of the so-called laser Doppler velocimetry (an Eulerian measurement technique) and of extended laser Doppler velocimetry, which is its extension to Lagrangian measurements.

Principle of Laser Doppler Velocimetry The principle of laser Doppler velocimetry is quite simple: It uses two coherent laser beams (with wavelength λ_0) intersecting with an angle θ to create an interference pattern consisting of fringes perpendicular to the plane of the laser beams, separated by a distance $a = \lambda_0/(2\sin(\theta/2))$. When a particle crosses the fringes, it scatters light with an intensity $I(t)$ modulated with a frequency $f_p = u_\perp/a$, where $u_\perp$ is the component of velocity perpendicular to the fringes (Figure 15.5). If the measurement volume (region where the beams intersect) extends over a large region of space, a continuous detection of the instantaneous frequency $f_p(t)$ gives access to the evolution of the particle's velocity as a function of time. Such an extended laser Doppler velocimetry was developed by *Volk et al.* [2008, 2011] to perform velocity tracking of small particles in high-Reynolds-number flows.

Optical Arrangement In order to obtain interference fringes in a large region of space, only one laser beam is used: It is separated by a beam splitter into two coherent beams separately expanded using two pairs of lenses as telescopes. The sign of the velocity can be extracted if one creates a set of *traveling* interference fringes. This is achieved by shifting one of the optical beams at a frequency δf so that the actual modulation of the scattered signal is at frequency $f_p(t) = \delta f + u_\perp(t)/a$. Shifting is done by propagating the two laser beams through acousto-optic modulators (AOMs) with frequency shifts $f_1 = 40$ MHz and $f_2 = 40.1$ MHz so that the fringes are actually moving at constant velocity $v_f = a(f_2 - f_1) = a \cdot \delta f$.

Particle Detection In practice, the intensity needed for the measurement depends on the particles used as tracers

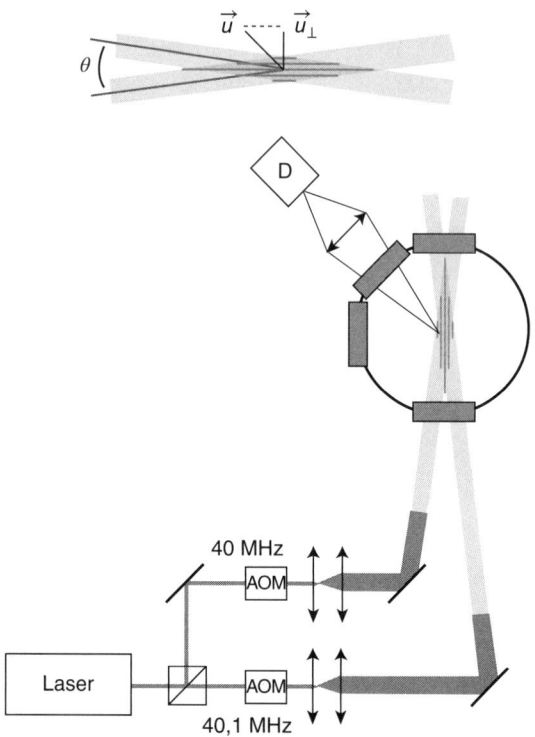

Figure 15.5. Optical arrangement of extended laser Doppler velocimetry detailed by *Volk et al.* [2008, 2011]. Two laser beams (with wavelength λ_0) forming an angle θ intersect to create an interference pattern with fringe spacing $a = \lambda_0/[2\sin(\theta/2)]$. Each beam is shifted in frequency with an acousto-optic modulator and expended using a telescope. The measurement volume is imaged onto a photodetector. A particle crossing the fringes scatters light modulated at frequency $f_p(t) = \delta f + u_\perp(t)/a$ with δf the frequency shift between the beams and $u_\perp(t)$ the component of velocity perpendicular to the fringes.

of the flow motion. Using a 1 W continuous argon laser with wavelength 514 nm and a small angle between the beams, one can obtain a fringe spacing $a = 41\,\mu$m. This is much larger than in classical LDV applications and allows to use as tracers small polystyrene fluorescent (with size $30\,\mu$m) or larger nonfluorescent particles, which are almost neutrally buoyant in water. For the fluorescent particles case, scattered light is weaker and the measurement volume has to be imaged on a low-noise photomultiplier with high gain. For particles larger than $100\,\mu$m, the scattered signal is stronger and the detection can be made using amplified photodiodes and using less than 0.5 W of laser power. As opposed to fluorescent particles, the optical contrast of the scattered signal strongly depends on the photodetector location and particle size; the detector is located in the plane of the beams and at 45° from the beams.

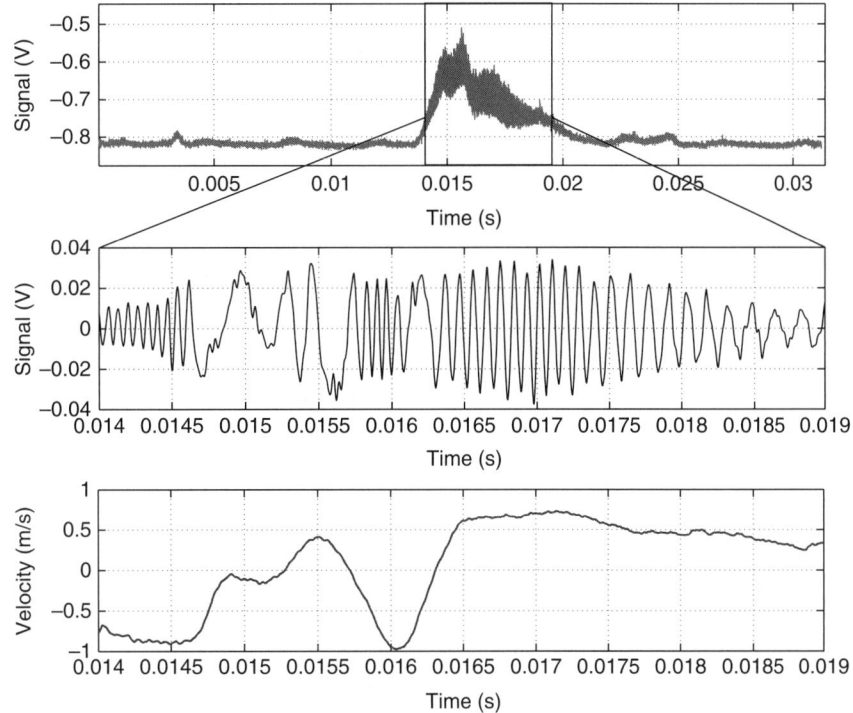

Figure 15.6. Typical optical signal measured with extended laser Doppler velocimetry. Top: Burst observed when a particle crosses the measurement volume. Middle: Real part of the complex signal obtained after filtering and demodulation from carrying frequency $\delta f = 100$ kHz. Bottom: Corresponding evolution of velocity as obtained from parametric estimation.

Signal Acquisition The use of two AOMs instead of one (for classical LDV) allows for a small frequency shift (100 kHz) so that raw data can be acquired using high-speed data acquisition board. Each time a particle crosses the measurement volume, it produces a burst of light with a signal of the following form (Figure 15.6a):

$$s(t) = \alpha(t) + \beta(t) \cos[2\pi \delta f \cdot t + \phi(t)], \quad (15.1)$$

with

$$\frac{d\phi(t)}{dt} = 2\pi \frac{u_\perp(t)}{a}, \quad (15.2)$$

where $\alpha(t)$ and $\beta(t)$ are slowly varying envelopes originating from the Gaussian radial profiles of the beams.

$$v(t) = \frac{1}{2\pi} \frac{d\phi(t)}{dt}.$$

The Doppler shift of the scattered signal due to the motion of the scatterer particle is represented as. In a typical situation, the diameter d of the beams is much larger than the fringe spacing a so that there is a scale separation between the fast modulation at frequency $v(t) = u_\perp(t)/a$ and the slow amplitude modulations $\alpha(t)$ and $\beta(t)$.

Signal Processing After running the experiment, the velocity is computed from the collection of light scattering signals $(s_i(t))_{[1,N]}$. This signal processing step is crucial as both time and frequency, i.e., velocity, resolutions rely on its performance. As the local frequency of the signal is varying in time, common time-frequency techniques based on Fourier analysis [Flandrin, 1998] (such as *short-time Fourier transform*) are usually too limited as the Heisenberg principle imposes that time resolution δ_t and frequency resolution δ_v must comply the inequality $\delta_t \delta_v > 1$, which means that one cannot have both high resolution in time (which is crucial to resolve the fastest dynamics of the particles) and frequency (which is crucial to have a good measurement of particle velocity, which according to relation (2) is directly given by $v(t)$). It is therefore necessary to overcome the Heisenberg principle limitation. Several methods exist, including Cohen class energetic estimators (such as Wigner-Ville and Choï Williams distributions) [Flandrin, 1998] which can be further refined using the *reallocation* technique [Flandrin, 1998; Kodera et al., 1976]. These methods are relatively time consuming in terms of computational processing and are generally adapted for situations where no information is a priori known on the form of the signal to be analyzed. In order to increase the frequency resolution with a small observation window, Mordant and coworkers introduced a fast demodulation algorithm with parametric estimation [Mordant et al., 2002, 2005]. It relies on a comparison between the measured signals $s_i(t)$ and the model given in equation (2). In practice, such a parametric estimate of amplitude and frequency modulations is very robust with respect to the experimental noise. The estimation is done in several steps:

1. As the time scale of $\alpha(t)$ (of order $d/u \sim 5$ ms) is very large as compared to $1/\delta f = 0.01$ ms, it is removed with high-pass filtering at several kilohertz.

2. To obtain an absolute definition of the local frequency through the evolution of the phase $\phi(t)$, the filtered signal $s'(t)$ is transformed into an analytical complex signal $\underline{x}(t)$: This is done using the Hilbert transform $\mathrm{HT}[s'](t)$ of the measured signal with the definition $\underline{x}(t) = s'(t) + i\mathrm{HT}[s'](t)$. The amplitude and frequency of the signal are then $\alpha(t) = \|\underline{x}(t)\|$ and $f_p(t) = \delta f + d\phi/dt$, respectively.

3. The complex signal $\underline{x}(t)$ is then demodulated from the carrying frequency by multiplication by $\exp(-2i\pi \delta f \cdot t)$. The real part of such a demodulated complex signal for a typical burst is displayed in Figure 15.6 (middle).

4. For a fast and precise measurement of the modulation frequency, an approximated maximum likelihood (AML) method is coupled with a Kalman filter so as to perform a parametric estimation of the instantaneous amplitude and frequency. In a moving window of duration δT, centered at time t, it assumes that the signal is made of a modulated complex exponential plus Gaussian noise $n(t)$ and compares the measured signal $\underline{x}(t)$ to the functional form

$$\underline{z}(t) = A(t) \exp\left[i2\pi \int_0^t \nu(t')\, dt' + i\psi\right] + n(t), \quad (15.3)$$

where $A(t)$ and $\nu(t)$ are the unknown amplitude and frequency to be estimated and ψ a constant phase originating from the initial particle position in the measurement volume. As an output, one obtains for each trajectory an estimate of the frequency $\nu(t) = u_\perp(t)/a$, amplitude $A(t)$, and a confidence criterion $h(t)$ which measures the quality of the estimation at each time step. This is done for each trajectory on a personal computer using Matlab to obtain a collection of trajectories to be further used to compute Lagrangian statistics of the flow.

Initially designed for acoustical Doppler measurements (described in the next section), the demodulation technique proved to be fast and accurate enough to perform Lagrangian ELDV measurements with typical time resolution of 10 μs and a sampling rate of 300 kHz. This represents the highest sampling rate ever used for Lagrangian measurements in high-Reynolds-number flows. It is particularly adapted to investigate flows with rapid and intense multiscale swirling structures. Therefore, although this technique has never been used (to our knowledge) in experiments with geophysical motivations, it is very likely to be a good candidate for the investigation of most extreme atmospheric events.

Particle Seeding Issues For practical applications, the particle seeding density has to be adjusted in order to be low enough so that one does not observe events with two particles at the same time in the measurement volume but high enough to observe several trajectories per second. For a fully turbulent flow with Reynolds number at the Taylor microscale, $\mathrm{Re}_\lambda \sim 600$, a collection of 15,000 trajectories with mean duration 20 Kolmogorov times (τ_η) is enough to converge velocity statistics, acceleration statistics, and acceleration autocorrelation function. In the case of acceleration autocorrelation $C_{aa}(\tau) = \langle a(t)a(t+\tau)\rangle/\langle a^2\rangle$, one gets access to an estimate of the local kinetic energy dissipation ϵ because the integral of the positive part of the curve is very close to the dissipation time $\tau_\eta = \sqrt{\nu/\epsilon}$ [*Volk et al.*, 2011]. As shown in Figure 15.7 (top), there is a good rescaling between the acceleration autocorrelation functions when time is measured in τ_η units for fully developed turbulent flows of water (with kinematic viscosity 10^{-6} m^2/s and dissipation scale 19 μm) and water-glycerol mixtures (kinematic viscosity 8×10^{-6} m^2/s, dissipation scale 90 μm). These two curves also show that particles with diameters $D = 5\,\eta$ (the dissipative scale) can still be considered tracers of the flow. As shown in 15.7 (bottom), this is no longer the case for larger particles for which one observes a decrease of particle acceleration variance following a power law $\langle a_D^2\rangle/\langle a_{\mathrm{tracer}}^2\rangle \propto (D/\eta)^{-2/3}$. This decrease of acceleration variance goes together with an increase of the particle acceleration autocorrelation time [*Qureshi et al.*, 2007; *Brown et al.*, 2009; *Volk et al.*, 2011].

15.3. ACOUSTIC TECHNIQUES

Whenever a sound wave encounters an obstacle or inhomogeneity along its propagation path, it is deflected from its original course, a phenomenon called acoustic scattering. The scatterer can be either a material obstacle or a physical inhomogeneity such as temperature or velocity gradient, which creates a contrast of acoustic impedance and influences the propagation of sound. The properties of the scattered acoustic wave depends upon the frequency of the incident wave, the shape and size of the obstacles, as well as their velocity. It is possible to take advantage of these scattering features to probe the dynamics of fluids. We detail here two recent techniques based on acoustic scattering: (i) Lagrangian acoustic tracking, which exploits the Doppler shift of the wave scattered by moving particles, and (ii) acoustical measurements of vorticity, which exploits the scattering properties of eddies in a fluid (with no need of seeding the flow). These techniques are intrinsically related to acoustic scattering properties and differ from more classical ones based on echo and time-of-flight measurements.

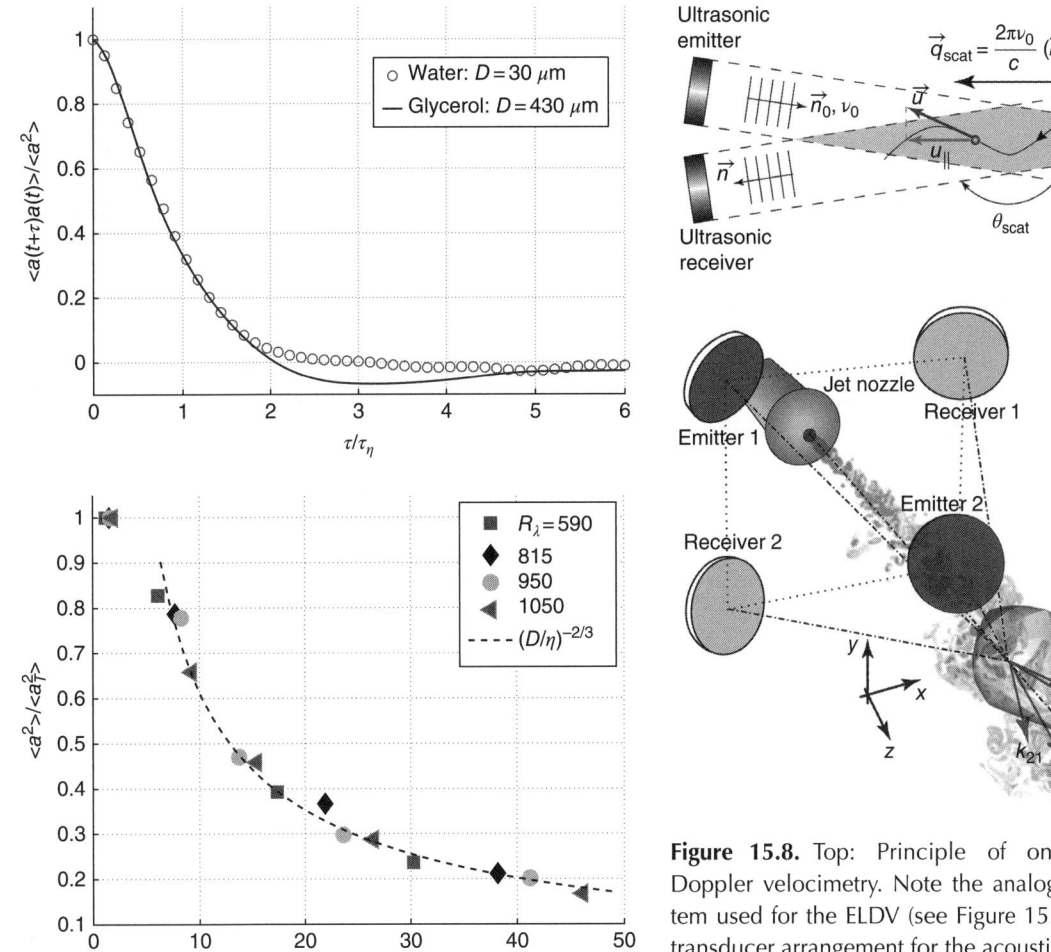

Figure 15.7. Top: Acceleration autocorrelation function $C_{aa}(\tau) = \langle a(t)a(t+\tau)\rangle/\langle a^2\rangle$ measured with ELDV in a fully developed turbulent flow. For both curves, time has been rescaled by the dissipative time $\tau_\eta = \sqrt{\nu/\epsilon}$. (o) The 30 μm fluorescent tracer particles for a water flow with dissipation scale $\eta = 19\,\mu m$. (−) Large 430 μm polystyrene particles behaving as tracers in the same (water-glycerol) flow with dissipation scale $\eta = 90\,\mu m$. For the two situations, the large-scale driving and dissipation $\epsilon = 20$ W/kg are the same. Bottom: Acceleration variance of particles with size D normalized by the one measured for tracers as a function of the ratio D/η. Particles with size $D/\eta > 5$ no longer behave as tracers of the flow motions.

15.3.1. Acoustic Doppler Lagrangian Tracking

15.3.1.1. Principle.
Acoustic Lagrangian tracking is based on the measurement of the Doppler shift of the acoustic wave scattered by a moving particle. Figure 15.8 (top) shows the principle of one-component ultrasonic Doppler velocimetry. An acoustic transducer emits a continuous ultrasonic wave at a given frequency ν_0 with a propagating direction $\vec{n}_0$. Whenever a particle crosses the acoustic beam of the emitter, it scatters the acoustic wave. An acoustic receiver then listens to the scattered wave in a specifc direction $\vec{n}_s$ ($\theta_s = (\vec{n}_0; \vec{n}_s)$ is the scatter angle). The intersection between the emitting and the receiving beams defines the measurement volume, where particles can be actually detected. Because the particles moves, the scattered wave is Doppler shifted and its frequency ν_s differs from ν_0 so that

$$\frac{\nu_s - \nu_0}{\nu_0} = \frac{\vec{V} \cdot (\vec{n}_0 - \vec{n}_s)}{c} = -2\frac{v_{\parallel}}{c}\sin\left(\frac{\theta_s}{2}\right), \quad (15.4)$$

Figure 15.8. Top: Principle of one-component acoustic Doppler velocimetry. Note the analogy with the optical system used for the ELDV (see Figure 15.5). Bottom: Example of transducer arrangement for the acoustic Lagrangian tracking of the three components of the velocity. Four transducers (two emitters and two receivers) are placed at the vertices of a square, tilted so that their axes cross at the same point on the jet axis, in a square-based pyramid configuration. The two emitters operate at two different working frequencies $\nu_{0,1}$ and $\nu_{0,2}$ and receivers listen to scattered waves in the vicinity of each of these frequencies. This arrangement is composed of four independent pairs of emitters-receivers capable of measuring four projections of the velocity, which gives a redundant 3D measurement, where redundancy improves the signal-to-noise ratio (SNR). Adding extra transducers would increase further the SNR.

where c is the speed of sound in the experimental conditions. For a given incident frequency ν_0 and a given scatter angle θ_s, the instantaneous frequency shift $\delta\nu(t) = \nu_s - \nu_0$ gives a direct measurement of the projection, $v_{//}$, of the particle velocity along $\vec{n}_0 - \vec{n}_s$ (note that this is an algebraic measurement: the sign of $v_{//}$ is given by the sign of the frequency shift). Hence, the continuous recording of the frequency shift $\delta\nu(t)$ gives a Lagrangian measurement of the velocity component $v_{//}(t)$ along the particles trajectory. It is interesting to note the analogy of this acoustic technique with the optical ELDV method previously described. The modulation of scattered light by a particle moving in the interference pattern in ELDV is indeed conceptually equivalent to the modulation of the Doppler-shifted acoustic wave scattered by a particle in the present configuration. Finally, we point out that the combination of several pairs of transducers and working frequencies allows to extend the measurement and access two or three components of the velocity (see Figure 15.8, bottom).

Such an acoustic Lagrangian tracking technique was first implemented by *Mordant et al.* [2001, 2002, 2005] in a pioneering study of Lagrangian turbulent statistics in a von Kármán swirling flow of water. In that case, piezoelectric elements were used as acoustic transducers (with typical emitting frequency operating in the megahertz range) and small polystyrene particles served as tracers. More recently, the same technique was ported to investigate opened air flows in a turbulent jet [*Poulain et al.*, 2004] and wind-tunnel experiments [*Qureshi et al.*, 2007, 2008]. It is interesting to note here that the same identical systems used by Poulain et al. and Qureshi et al. can also be easily ported to perform in situ measurements in real atmospheric flows. In these experiments Sell-type acoustic transducers were used, operated with typical frequencies around 100 kHz (ultrasounds at higher frequency are rapidly damped in air) and tracked particles were small millimeter soap bubbles either neutrally buoyant (bubbles are then inflated with helium) to have tracer behavior or intentionally heavier than air in order to address the question of the turbulent transport of inertial particles.

Pros and cons of scattering techniques compared to direct optical methods can be discussed. The main advantage of acoustics (shared with ELDV) is that Doppler shift measurements give a direct access to the tracer's velocity, while direct optical tracking requires to differentiate the position signal of the tracked particle to get velocity, an operation which is very sensitive to noise. Hence optical tracking usually requires severe oversampling if velocity and acceleration statistics are to be investigated. Other advantages of acoustic tracking are (i) the low cost of the required equipment, compared to expensive high-speed cameras; (ii) the possibility to probe flows in opaque fluids (as liquid metals for instance); (iii) the possibility to easily explore large volumes; and (iv) the ability to investigate open flows with large mean velocities. The last point is particularly important, for instance, in wind tunnel or jet experiments where efficient optical tracking generally requires to mount the camera on a platform moving at the mean wind speed [*Ayyalasomayajula et al.*, 2006] in order to track particles for sufficiently long times (typically comparable to the largest time scales of the flow) while acoustic tracking can be efficiently done using fixed transducers; moreover, a backscattering configuration ($\theta_{\text{scatt}} \lesssim 180°$) allows to significantly extend the streamwise dimension of the measurement volume [*Qureshi et al.*, 2007, 2008].

Among disadvantages, a strong limitation of acoustic tracking is its inability to accurately track several particles simultaneously. If more than one particle is present in the measurement volume, the signal recorded by the receiver superimposes the waves scattered by all the particles. Although signal processing strategies (discussed below) do exist to extract the contributions from each individual scatterer, the accuracy decreases with increasing number of particles.

15.3.1.2. Signal Processing and Doppler Shift Extraction. The acoustic signal recorded by the receiver combines a spectral component around the emitting frequency ν_0, which corresponds to echoes and reflections directly incoming in the receiver without being scattered by the particles and a Doppler-shifted component around ν_d resulting from the fraction of acoustic wave scattered by the moving particle. The information about particle velocities is entirely encoded in the Doppler shift $\delta\nu = \nu_s - \nu_0$. Therefore, a heterodyne downmixing is generally operated between the emitted and received signal which essentially results in shifting the emitting frequency ν_0 to zero. Figure 15.9 shows a typical spectrum of a downmixed

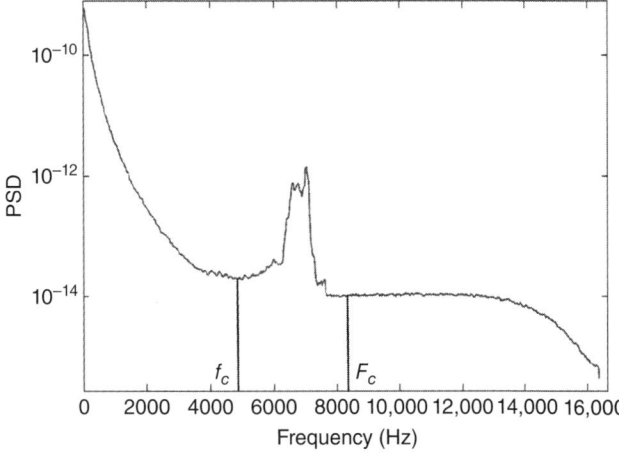

Figure 15.9. Typical spectrum of the donwmixed acoustic signal recorded by the receiver. The Doppler shift resulting from the scattering by moving tracers is visible around 7 kHz. From *Qureshi* [2009].

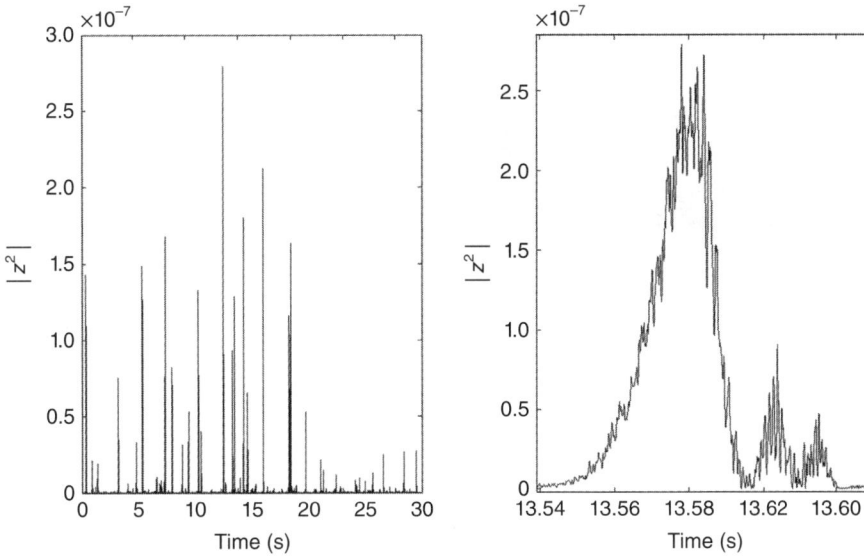

Figure 15.10. (a) Typical amplitude of the recorded down-mixed signal bandpass filtered around the Doppler peak (between frequencies f_c and F_c as shown in Figure 15.9). Peaks of high amplitude correspond to events where a particle travels into the measurement volume and scatters the incident acoustic wave toward the receiver. (b) Zoom on one such isolated event. From *Qureshi* [2009].

signal recorded in the wind tunnel experiment by *Qureshi et al.* [2007]. The peak at zero frequency corresponds to the emitting frequency v_0 (which was 80 kHz in this experiment) and the secondary peak (around 7 kHz) corresponds to the Doppler shift from the wave scattered by moving tracers (note that the central peak at the emitting frequency is enlarged by aerodynamic effects). The interest of heterodyne downmixing is that recording the original signal would require very high sampling rates, as the Doppler frequency would be here around $v_s = v_0 + \delta v \simeq 87$ kHZ), while it is only at 7 kHz after downmixing.

Figure 15.10a shows the amplitude $A(t)$ of the down-mixed signal versus time; each peak of amplitude corresponds to the passage of a particle in the measurement zone. The spectrum previously discussed was calculated from the entire times series and hence all time information has been lost: The observed Doppler peak corresponds to the superposition of spectral contributions from thousands of successive scattered traveling through the measurement volume. It is only interesting to extract global informations, such as the average velocity of the particles (given by central frequency of the Doppler peak) or the typical level of velocity fluctuation (given by the width of the Doppler peak). However, accessing the Lagrangian dynamics of the particles requires to extract the instantaneous Doppler shift $\delta v(t)$ for each individual particle.

This is achieved in two steps: First, each event corresponding to the passage of a particle is detected from the amplitude signal as shown in Figure 15.10a. (Figure 15.10b shows a zoom on such an individual particle event.) Second, the portion of signal corresponding to each individual event is analyzed with dedicated time-frequency tools to extract the instantaneous Doppler shift. For the same reasons previously discussed in the context of optical ELDV, the instantaneous Doppler shift can be efficiently extracted using an AML algorithm where after downmixing the signal scattered by one particle and recorded by the receiver is modeled in the form

$$z_i(t) = A_i(t) \exp[i2\pi \int_0^t \delta v_i(t')\, dt'] + n(t), \quad (15.5)$$

where $A_i(t)$ is the amplitude of the scattered acoustic wave, $\delta v_i(t)$ is the instantaneous frequency of the signal (subscript i indicates that we consider particle number i), and $n(t)$ represents an additive experimental noise. The AML algorithm determines for each particle i the best functions $A_i(t)$ and $\delta v_i(t)$ so that the model given by equation (15.5) matches as close as possible the actual recorded signal ($n(t)$ is assumed to be a random Gaussian noise whose amplitude is fixed according to the actual noise level of the experimental signal). Such a parametric algorithm is capable of overcoming the Heisenberg limitation (which would severely constrain the resolution of the measurement) due to the extra information given by the a priori imposed shape of the signal in equation (15.5) added in the signal processing. The gain in resolution offered by the AML method is illustrated by Figure 15.11. Also of interest is the fact that the AML algorithm yields a quantitative indicator of the relevance of expression (15.5) for the actual modeling of the scattering signal. This allows to discard spurious events from the statistical ensemble, for

290 MODELING ATMOSPHERIC AND OCEANIC FLOWS

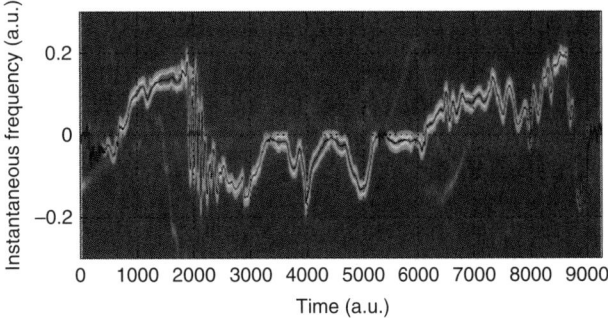

Figure 15.11. Example of reconstruction of the instantaneous frequency scattered by a moving sphere in a swirling flow of water. The color image represents the spectrogram computed with usual Fourier analysis time-frequency analysis (the width of the frequency trace illustrates the incertitude due to Heisenberg constrains) to which is superimposed the estimation given by the AML method (solid black line). From *Mordant* [2001]. For color detail, please see color plate section.

instance when two particles were simultaneously present in the measurement volume.

15.3.1.3. Example of Acoustical Particle Tracking.
Acoustical tracking has been used in different experimental facilities, including von Kármán swirling flows, turbulent jets, and wind tunnels to investigate both the characteristics of the flow and the dynamics of material particles transported by the flow. Figure 15.12 represents the probability density functions of the velocity and acceleration of material particles transported in a turbulent wind tunnel flow. Velocity is found to have Gaussian statistics while acceleration exhibits wide non-Gaussian tails, even for particles much denser than the fluid.

15.3.2. Vorticity Measurements

Measuring vorticity has always been a challenge in experimental fluid mechanics, especially when small scales are to be probed. We recall here that the vorticity $\vec{\Omega}$ of a velocity field $\vec{u}$ is given by the curl of $\vec{u}$ ($\vec{\Omega} = \vec{\nabla} \times \vec{u}$). A direct measurement of vorticity is usually done from spatial derivatives of the velocity field. This is typically the case with PIV or multiple hot-wire measurements. However, in either case, the spatial resolution is an issue when flows are highly turbulent as neither PIV nor multiple hot-wire probes are capable of resolving the smallest dissipative scales of the velocity fields, which are essential for an accurate estimation of a spatial derivative. We report in this section an elegant measurement of the vorticity of a flow based on the peculiar interaction between an acoustic wave and the vortical structures of a flow. In the context of atmospheric and oceanographic studies, this technique may be particularly suited to address questions related for

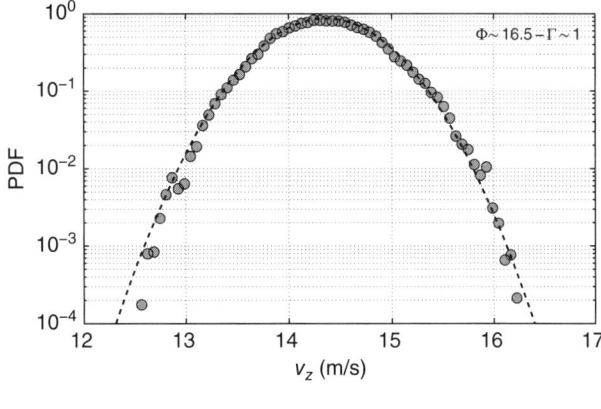

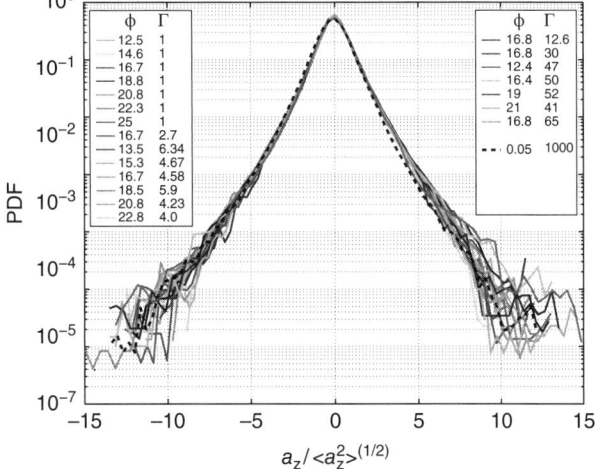

Figure 15.12. Top: One-component Lagrangian velocity PDF of finite-size part neutrally buoyant particles transported in a turbulent wind tunnel flow. From *Bourgoin et al.* [2011]. Bottom: Acceleration PDF of finite-size material particles with different sizes and density (parameters Φ and Γ) in the same wind tunnel flow. From *Qureshi et al.* [2008].

instance to the interaction between vortices and the influence of global rotation as well as the turbulent cascade of enstrophy at mesoscales for which the atmosphere and/or the ocean can be considered as 2D.

15.3.2.1. Principle.
The method relies on the interaction between an acoustic wave and the velocity gradients in the flow. The scattering properties from this acoustic-fluid interaction are non-trivial. Several theoretical and numerical studies can be found on the subject [*Obukhov*, 1953; *Kraichnan*, 1953; *Chu*, 1958; *Batchelor*, 1957; *Lund and Rojas*, 1989; *Llewellyn Smith and Ford*, 2001; *Colonius et al.*, 1994]. In particular, using a Born approximation, *Lund and Rojas* [1989] have shown that the scattered amplitude of a plane acoustic wave by a turbulent flow can be linearly related to the spatial Fourier transform of the vorticity field of the flow. This property can be

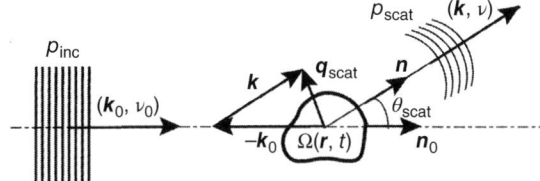

Figure 15.13. Typical implementation of acoustic scattering for probing the vorticity field of a flow. An ultrasonic emitter generates a plane acoustic wave in the direction $\vec{n}$ at the frequency v_0. A receiver listens to the acoustic wave scattered in a given direction $\vec{r}$. The relative amplitude of the scattered acoustic pressure to the incident acoustic pressure can be related to the spectral distribution of the vorticity field $\vec{\Omega}(\vec{x},t)$ according to relation 15.6. From *Poulain et al.* [2004].

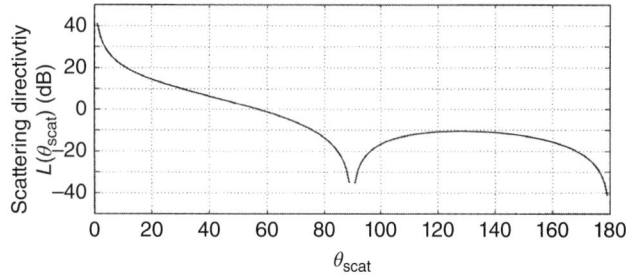

Figure 15.14. Angular factor $L(\theta_{scat})$. From *Poulain et al.* [2004].

qualitatively understood as the fact that each vortex in the flow acts as a scatterer which radiates a sound wave as it is perturbed by the incident impinging acoustic wave. The global scattered wave, results from the coherent average over the scatterer distribution. Figure 15.13 illustrates a typical acoustic scattering configuration which can be used to probe the vorticity of a flow. Lund et al. have shown that the acoustic pressure amplitude and the Fourier transform of the vorticity field are related as

$$p_{scat}\left(\vec{k},t\right) = L(\theta_{scat})\tilde{\Omega}_\perp\left(\vec{q}_{scat} = \vec{k}-\vec{k}_0,t\right) p_{inc}\left(\vec{k}_0,t\right), \tag{15.6}$$

where $\vec{k}_0$ and $\vec{k}$ are the vector wave numbers of the incoming and scattered acoustic waves, respectively, θ_{scat} is the scattering angle, p_{inc} and p_{scat} are the complex pressure amplitudes of the incoming and scattered acoustic wave, respectively, $L(\theta_{scat})$ is an angular factor which will be discussed further below, and $\tilde{\Omega}_\perp$ is the component of the space Fourier transform of the vorticity perpendicular to the scattering plane defined by the vector wave numbers of the incident and scattered acoustic waves (see Figure 15.13):

$$\tilde{\Omega}_\perp(\vec{q},t) = (\vec{n}_0 \times \vec{n}) \cdot \iiint \vec{\Omega}(\vec{r},t) e^{-i\vec{q}\cdot\vec{r}} d^3\vec{r}. \tag{15.7}$$

Equation (15.6) therefore shows that the amplitude of the scattered wave gives a direct measurement of one Fourier mode of the vorticity component $\Omega_\perp$. Interestingly, the Fourier mode at which the vorticity field is being probed is directly selected by the imposed scattering vector $\vec{q}_{scat} = \vec{k} - \vec{k}_0$. Hence, it is possible to reconstruct the complete vorticity spectra by spanning the explored scattering vector, and this can be done either by changing the angular position of the acoustic receiver or by changing the operating acoustic frequency (as $\|\vec{q}_{scat}\| \simeq 4\pi v_0/c \sin(\theta_{scat})$, assuming the Doppler shift $v - v_0$ remains small compared to v_0).

It is important to stress that this measurement is local in Fourier space, meaning that for a given scattering vector, only the mode of vorticity at wave number $\vec{q}_{scat}$ is actually measured. Hence the measurement is naturally global in space and the corresponding spectral mode is characterized across the entire measurement volume. Though the scattering structures of the flow are tracked as they move across the flow (this results for instance in a Doppler shift of the scattered acoustic wave), the recorded signal represents a coherent average of all structures at the probed scale simultaneously present in the measurement volume, and no information is extracted from individual scatterers. As a consequence this technique is not properly speaking of Lagrangian type, although the instrumentation is almost identical to that described in the previous section on Lagrangian acoustic tracking.

An important point to be considered is the angular factor $L(\theta_{scat})$ present in equation (15.6). Figure 15.14 shows the dependency of $L(\theta_{scat})$ with the scattering angle θ_{scat} as calculated by Lund et al. It shows a quadrupolar-like radiation pattern which diverges at zero angle (Born approximation fails in this limit) and vanishes for scattering angles $\theta_{scat} = 90°$ and $\theta_{scat} = 180°$ (back-scattering situation). Those two specific scattering angles are to be avoided for the vorticity measurement. On the contrary, they are optimal for particle tracking as no signal is then recorded from scattering effects of the fluid itself, and only the particles seeding the flow will be seen. This explains the back-scattering configuration chosen for the acoustic particle tracking described in the previous section.

15.3.2.2. Experimental Implementation and Typical Results.
Experimental evidence of ultrasonic scattering by vortical structures in a flow has first been given by *Baudet et al.* [1991] in the canonical configuration of the von Kármán vortex street behind a cylinder at low Reynolds number. Since then, the technique has been ported to turbulent flows at moderate Reynolds number (in a turbulent jet of air [*Poulain et al.*, 2004]) and at high Reynolds number (in a cryogenic turbulent jet of gaseous helium [*Bezaguet et al.*, 2002; *Pietropinto et al.*,

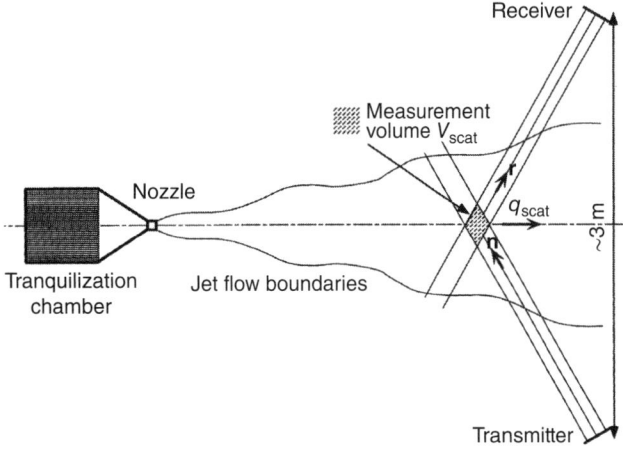

Figure 15.15. Example of implementation of vorticity measurement by acoustic scattering in a turbulent jet. From *Poulain et al.* [2004]. Note that the same configuration can be easily ported to open flows and in city measurements in the atmospheric boundary layer.

1999]). Figure 15.15 shows the schematic of the implementation of acoustic measurement of vorticity in a turbulent jet as done by *Poulain et al.* [2004]. In this experiment acoustic transducers are of Sell type consisting of a circular plane piston, with typical diameter of the order of 10 cm (larger and smaller transducers can be used depending on the extent of the flow to be probed), made of a thin mylar sheet (typically 15 μm thick). One important advantage of such transducers is their large bandwidth (typically between 1 and 200 kHz in air) which allows to span a wide range of scattering wave vectors $\vec{q}_{scat}$ with a fixed geometric arrangement (in particular with a fixed scattering angle). In the example, the scattering angle was kept fixed at constant value of the order of 60°. The choice of the working scattering angles responds to several criteria: (i) Given the angular factor dependence shown in Figure 15.14, angles close to 90° and 180° should be avoided. (ii) At small scattering angles the angular factor increases rapidly; however, unless thermal conditions in the experiment are very well controlled small scattering angles should be avoided as forward sound scattering is very sensitive to temperature gradients. (iii) Other practical criteria include, for instance, geometric constraints around the experiment, limitation of echoing effects and direct acoustic "blinding" from the emitter to the receiver (in particular via the secondary diffraction side lobes of the transducers). (iv) Beyond these practical considerations, the scattering angle should also be chosen in accordance with the physical properties of the flow to be probed. As already discussed, the amplitude of the scattering vector $q_{scat} = 4\pi \nu_0/c \sin(\theta_{scat}/2)$ defines the wave number at which the vorticity spectrum is being probed. It

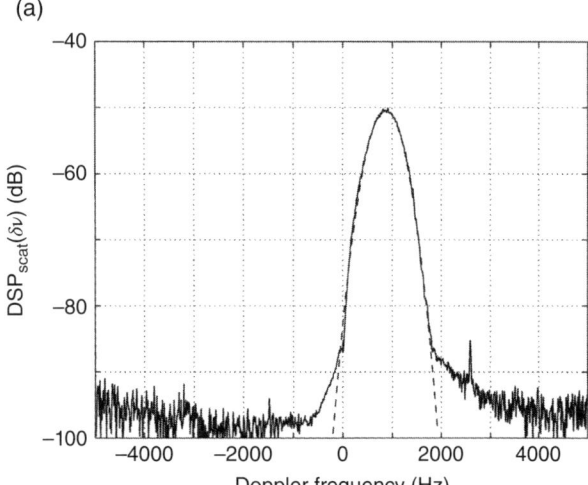

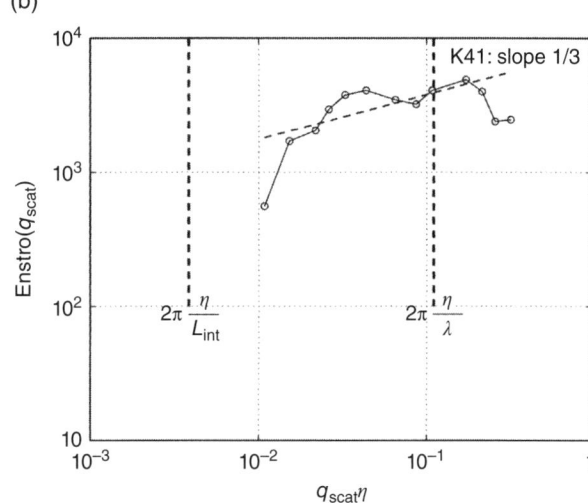

Figure 15.16. (a) Power spectral density of the signal scattered by the flow. (b) Discrete reconstruction of the spatial spectrum of the flow enstrophy. From *Poulain et al.* [2004].

can be selected by changing either the working frequency or the scattering angle. Hence, the scattering angle will be chosen so that wave numbers relevant to the investigated problem can be effectively spanned within the accessible range of operating frequencies of the acoustic transducers. In our case, an angle of 60° allowed the authors to probe a significant range of the inertial scales of the turbulent jet with a constant scattering angle by simply varying the working frequency ν_0 of the transducers.

As an example of results which can be obtained with this technique, we show in Figure 15.16a a typical power spectrum of the signal recorded by the acoustic receiver (note that the signal has been down-mixed exactly in the same way as explained for the acoustic Lagrangian measurement in the previous section). The maximum of the power spectrum occurs for a nonzero frequency which corresponds to the Doppler shift related to the average velocity of the jet at the location of the measurement volume. The range of detected frequencies around this maximum corresponds to the range of velocity statistically sampled by the vortices in the flow (within the range of spatial scales selected according to the wave number q_{scat}) which scattered the acoustic wave. As a consequence, the shape of the power spectrum directly reflects the statistical properties of the velocity field of the flow. It is a Gaussian, centered around a frequency which corresponds to the mean stream velocity of the jet flow at the location of the measurement volume and the width of the Gaussian corresponds to the standard deviation of the carrier velocity field. Poulain et al. have indeed shown that in their turbulent jet the power spectrum is well fitted by a Gaussian:

$$\text{PSD}_{\text{scat}}(\delta \nu) = \frac{A(\nu_0)}{\sqrt{2\pi}\delta\nu_{\text{rms}}} \exp\left(-\frac{(\delta\nu - \delta\nu_{\text{avg}})^2}{2\delta\nu_{\text{rms}}^2}\right) \quad (15.8)$$

and they have shown that the fitting Doppler shift frequencies $\delta\nu_{\text{avg}}$ and $\delta\nu_{\text{rms}}$ were in excellent agreement with hot-wire anemometry measurements of the mean and rms velocity of the jet flow. Perhaps more interestingly, they have shown that the quantity $q_{\text{scat}}^2 A(\nu_0)$ (where $A(\nu_0)$ is the maximum of the power spectral density) gives a direct estimate of the enstrophy spectrum of the flow at the given wave number q_{scat} once the transfer function $H(\nu_0)$ between the receiver and the emitter is applied. (An important aspect to be considered when using this method concerns the calibration of the acoustic transducers: While this is not crucial for the Lagrangian measurement previously described, which only relies on the frequency shift information of the scattered wave, a proper calibration of the transfer functions of the receiver and the emitter is required to extract the vorticity information, which is coded in the amplitude of the power spectral density.) By varying the working frequency ν_0 (and hence the scattering wave number q_{scat}), it is then possible to reconstruct the entire spectrum of enstrophy. Figure 15.16b shows the enstrophy spectrum of the jet flow investigated by Poulain et al. which is found to be in reasonable agreement with the Kolmogorov phenomenology of turbulence.

15.4. INSTRUMENTED PARTICLES

When performing Lagrangian measurement in a fluid flow, one is usually limited in track length because of the necessary finite size of the measurement volume. Using particle tracking velocimetry or Doppler velocimetry, one is also limited in the investigation of kinematic quantities (velocity, acceleration, vorticity, etc.). Tracking scalar quantities such as salinity or temperature along particle paths may also be of prime interest, in particular for stratified flows with geophysical motivations. One may therefore want to use instrumented particles (called *smart particles*) with embarked electronics able to measure scalar or kinematic quantities in the Lagrangian frame while continuously transmitting the information to the operator for data storage and postprocessing.

15.4.1. Lagrangian Temperature Measurement

A smart particle has been designed to measure continuously temperature along the particle trajectory using four thermistors, placed at the surface of the particle directly in contact with the fluid [*Gasteuil et al.*, 2007; *Shew et al.*, 2007]. It is made of a spherical capsule of diameter $D = 21$ mm containing temperature instrumentation, a radio frequency (RF) emitter, and a battery. It uses a resistance controlled oscillator LMC555 timer to create a square wave whose frequency depends on the temperature of the several thermistors. This square wave is used directly to modulate the frequency of the radio wave generated by the RF emitter in the range $22 - 26$ kHz about the carrying frequency $f_0 = 315$ MHz (see Figure 15.17

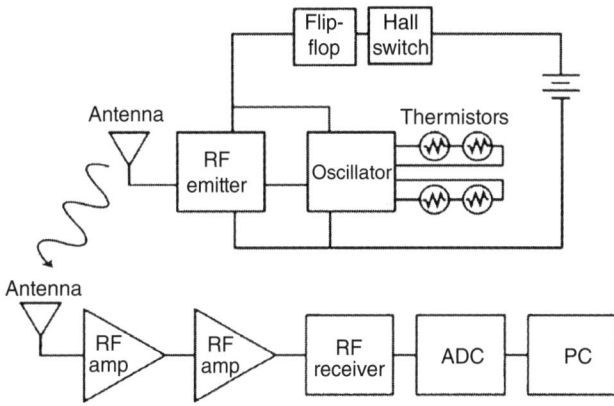

Figure 15.17. The smart particle measuring temperature uses four thermistors, placed directly in contact with the fluid, connected to a LMC555 timer to generate a square wave with frequency in the range $22 - 26$ kHz depending on the flow temperature [*Gasteuil et al.*, 2007; *Shew et al.*, 2007]. An RF emitter and receiver (MAX7044 and MAX1473 from Maxim Integrated Products) are used for frequency modulation and demodulation at 315 MHz. The emitter antenna is placed inside the capsule with emitting circuit tuned for emission at 315 MHz using a variable capacitor. The signal is received by a fixed antenna amplified and demodulated before acquisition at 10 MHz with high-speed DACQ. The slowly varying frequency of the square wave is then continuously measured using Labview standard library, then stored for further data analysis. A Hall switch and flip-flop are used for turning on and off the particle approaching a magnet close to the particle in order to save the battery when the experiment is not running.

and *Shew et al.* [2007] for more details). The entire mobile circuitry is powered with a coin cell battery which conditions the duration of the measurement, about 3 h. Using an antenna and RF receiver and amplifiers, it is possible to acquire directly the demodulated signal oversampling with a high-speed DACQ. The square-wave signal frequency is then directly measured using the standard Labview library, and time-varying frequency is converted to temperature using frequency-temperature calibration of the thermistors. The time resolution for this distant temperature measurement is about 10 Hz, which was suitable for instance, for studying turbulent thermal convection in laboratory experiments.

Measurement in Rayleigh-Bénard Convection. The particle was first used to investigate natural convection in a square tank with size 30 cm at high Rayleigh numbers, Ra $\sim 10^{10}$. Figure 15.18 (top) shows the time evolution of temperature along the particle trajectory with irregular oscillations caused by the motion of the particle crossing cold and hot regions in the vessel. Combining Lagrangian temperature measurement and particle

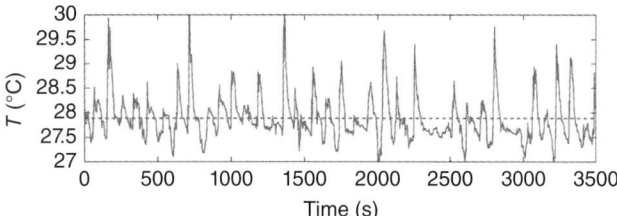

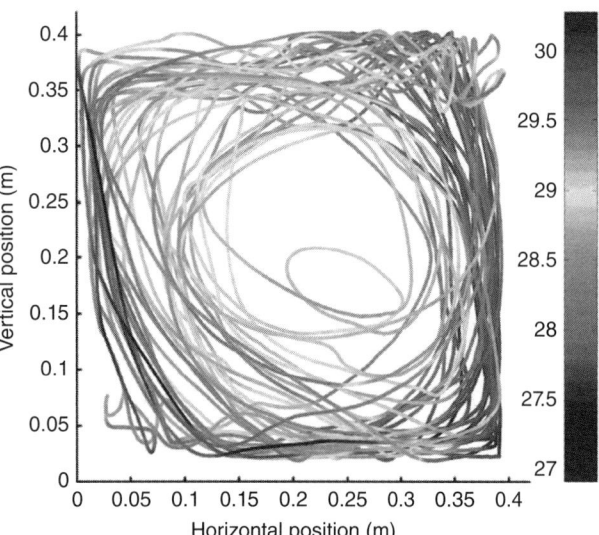

Figure 15.18. Top: Temporal evolution of temperature along a particle trajectory in turbulent Rayleigh-Bénard convection with square aspect ratio. Bottom: Combined PTV-instrumented particle measurement showing the $2d$ position of the particle $(X(t), Y(t))$ with local temperature $T(t)$ (see colorbar for values of temperature in Celsius). From *Shew et al.* [2007].

tracking velocimetry, it becomes possible to study the correlations between position and temperature as shown in Figure 15.18 (bottom). As all kinematic quantities (velocity or acceleration) can be obtained from PTV data, it is then possible to have information about the turbulent heat flux $q = \langle v'T' \rangle$ (with v' and T' fluctuating velocity and temperature) in the whole experiment volume with insight form the role of plumes in heat transport [*Gasteuil et al.*, 2007; *Shew et al.*, 2007].

15.4.2. Further Lagrangian Measurements

More recently the smart particle concept was extended to measurements of acceleration using numerical modulation and demodulation with suitable electronics [*Zimmermann et al.*, 2012]. The apparatus was designed from the earlier work by *Gasteuil* [2009] and built by smartINST S.A.S., an offspring company of CNRS and ENS de Lyon. It is a spherical particle with diameter 25 mm which embarks an autonomous circuit with a 3D acceleration sensor, a coin cell, and a wireless transmission system. It transfers the measured data to a data acquisition center (smartCENTER) which decodes, processes, and stores the signal delivered by the smart particle. The smart particle and smartCENTER measure, display, and store the three-dimensional acceleration vectors acting on the particle as it is advected in the flow. The acceleration sensor is an ADXL 330 (Analog Device). The three axes of the ADXL 330 are decoupled and form an orthogonal coordinate system attached to the chip package. This arrangement yields a 3D measurement of acceleration, including gravity, with a full scale of $\pm 3g$. The sensor has to be calibrated to compute the physical acceleration from the voltages of the accelerometer.

Other Sensors the smart particle technology allows the transmition of additional information originating from other sensors. Current developments aim at conductivity measurements which are of interest for salinity measurements and mixing issues in stratified flows.

15.5. CONCLUSIONS AND DISCUSSION

We have reported here some of the latest developments in Lagrangian characterization of flow dynamics in laboratory experiments. Although the Lagrangian approach has been already identified by Taylor and Richardson as a relevant description of geophysical flows, it is only recently that technological progress has allowed to develop platforms with sufficient accuracy and resolution. Some of these techniques, including 3D PTV and instrumented particles, have already been successfully used in model experiments with atmospheric and oceanographic motivations. But most of their advantages are still to be exploited. For instance, the versatility of 3D PTV, the ultimate temporal resolution of ELDV, the possibility to perform large-scale atmospheric in situ measurements with acoustic tracking, the capacity to directly access Fourier modes of vorticity using acoustic scattering, the capacity of instrumented particles to measure not only kinematic quantities but also heat transport, salinity, etc. show the many possible developments and advances that can still be expected from these techniques in addressing geophysical fluid dynamics issues.

Let us briefly summarize the advantages and drawbacks of the different techniques. High-resolution optical tracking has become one of the most accurate techniques in experimental fluid mechanics. It allows to track simultaneously hundreds of particles in 3D and thus the ability to address central questions related, for instance, to mixing and transport properties of flows. Its main drawback is the current cost of high-speed cameras. Acoustic tracking and ELDV, both based on Doppler velocimetry, are more affordable techniques, though they are limited to the tracking of essentially one particle at a time and are therefore not adapted to multiparticle studies. Their main advantage lies in the fact that they give a direct measurement of particle velocity (and not particle position as in optical tracking), hence limiting the increase of noise induced by the differentiation of position to access velocity. Similarly, measuring acceleration of the particles requires only one differentiation step while second derivatives must be estimated from optical tracking. These techniques are therefore very accurate to investigate the Lagrangian dynamics of individual particles. Instrumented particles have ported further the capacity of investigation of Lagrangian properties of flows by giving access not only to kinematic properties (as velocity or acceleration) but also to a Lagrangian description of almost any physical quantity for which a relevant sensor can be embedded in the particle. The main drawback for the moment concerns the size of the particle, which does not allow to probe scales smaller than about 1 cm. Finally, the acoustic vorticity measurement is unique as it gives a simple and accurate way to characterize the enstrophy spectrum of a flow, with intrinsic spectral resolution at a selected scale (including the smallest scales of the flow, which are hardly accessible with such classical techniques as PIV).

Note that we have not reported here all the possible extensions and add-ons of these methods, as, for instance, the use of digital holography [*Salazar et al.*, 2008; *Chareyron*, 2009], which allows to track particles in 3D with one single camera, or the tracking of particle rotational dynamics [*Zimmermann et al.*, 2011a, 2011b; *Klein et al.*, 2012], which allows to simultaneously investigate the translation and rotation of finite objects transported in a flow.

Experimental techniques in fluid mechanics are being constantly improved as new ideas combined with technological advances increase the resolution and the range of existing methods: For instance, cameras are ever faster and sensors better resolved; miniaturization and reduction of power consumption of electronic components will allow to reduce the size of instrumented particles; an important breakthrough in high resolution optical tracking is expected in the coming years due to FPGA (field programmable gate array) technology, which allows to process images on-board and hence to increase the effective data rate (for instance, particle detection could be done on-board and only the particle positions would be recorded). In this rapidly evolving context it is essential to promote an efficient interaction between fluid mechanics experimentalists and other communities, e.g., geophysicists, in order to develop the appropriate instrumentation for laboratory or field investigations.

REFERENCES

Ayyalasomayajula, S., A. Gylfason, L. R. Collins, E. Bodenschatz, and Z. Warhaft (2006), Lagrangian measurements of inertial particle accelerations in grid generated wind tunnel turbulence, *Phys. Rev. Lett.*, *97*, 144,507.

Batchelor, G. K. (1957), Wave scattering due to turbulence, in *Symposium on Naval-Hydrodynamics*, edited by F. S. Sherman, pp. 403–429, National Academy of Sciences, Washington.

Baudet, C., S. Ciliberto, and J.-F. Pinton (1991), Spectral analysis of the von Karman flow using ultrasound scattering, *Phys. Rev. Lett.*, *67*, 193–195.

Bezaguet, A., et al. (2002), A cryogenic high Reynolds turbulence experiment at CERN, *Adv. Cryo. Eng.*, *47*, 136–144.

Bourgoin, M., N. T. Ouellette, H. T. Xu, J. Berg, and E. Bodenschatz (2006), The role of pair dispersion in turbulent flow, *Science*, *311*, 835–838.

Bourgoin, M., N. M. Qureshi, C. Baudet, A. Cartellier, and C. Gagne (2011), Turbulent transport of finite sized material particles, *J. Phys. Conf. Ser.*, *318*, 012,005.

Brown, R. D., Z. Warhaft, and G. A. Voth (2009), Acceleration statistics of neutrally buoyant spherical particles in intense turbulence, *Phys. Rev. Lett.*, *103*, 194501.

Chareyron, D. (2009), Développement de méthodes instrumentales en vue de l'étude Lagrangienne de l'évaporation dans une turbulence homogène isotrope, Ph.D. thesis, Ecole Centrale de Lyon.

Chu, B. T. (1958), Non-linear interactions in a viscous heat-conducting compressible gas, *J. Fluid Mech.*, *3*, 494–514.

Colonius, T., S. K. Lele, and P. Moin (1994), The scattering of sound waves by a vortex: Numerical and analytical solutions, *J. Fluid Mech.*, *260*, 271–298.

Del Castello, L., and H. J. H. Clercx (2011), Lagrangian acceleration of passive tracers in statistically steady rotating turbulence, *Phys. Rev. Lett.*, *107*, 214,502.

Flandrin, P. (1998), *Time-Frequency/Time-Scale Analysis*, Academic, New York.

Gasteuil, Y. (2009), Instrumentation Lagrangienne en turbulence: Mise en oeuvre et Analyse, Ph.D. thesis, ENS Lyon.

Gasteuil, Y., W. L. Shew, M. Gibert, F. Chillà, B. Castaing, and J.-F. Pinton (2007), Lagrangian temperature, velocity, and local heat flux measurement in Rayleigh-B{é}nard convection, *Phys. Rev. Lett.*, *99*, 234,302.

Klein, S., M. Gibert, B. Antoine, and E. Bodenschatz (2013), Simultaneous 3D measurement of the translation and rotation of finite-size particles and the flow field in a fully developed turbulent water flow, *Meas. Sci. Technol.*, *24*, 024006.

Kodera, K., C. de Villedary, and R. Gendrin (1976), A new method for the numerical analysis of nonstationary signals, *Phys. Earth and Planet. Inter.*, *12*, 142–150.

Kraichnan, R. H. (1953), The scattering of sound in a turbulent medium, *J. Acoust. Soc. Am.*, *25*, 1096–1104.

LaPorta, A., G. A. Voth, A. M. Crawford, J. Alexander, and E. Bodenschatz (2001), Fluid particle accelerations in fuly developped turbulence, *Nature*, *409*, 1017.

Llewellyn Smith, S. G., and R. Fort (2001), Three dimensional acoustic scattering by vortical flows, *Phys. Fluids*, *13*, 2876–2889.

Lund, F., and C. Rojas (1989), Ultrasound as a probe of turbulence, *Phys. D*, *37*, 508–514.

Mordant, N. (2001), Mesure lagrangienne en turbulence: Mise en œuvre et analyse, Ph.D. thesis, Ecole Normale Supérieure de Lyon.

Mordant, N., P. Metz, O. Michel, and J.-F. Pinton (2001), Measurement of Lagrangian velocity in fully developed turbulence, *Phys. Rev. Lett.*, *87*, 214,501.

Mordant, N., J.-F. Pinton, and O. Michel (2002), Time-resolved tracking of a sound scatterer in a complex flow: Nonstationary signal analysis and applications, *J. Acoust. Soc. Amer.*, *112*, 108–118.

Mordant, N., P. Metz, J. F. Pinton, and O. Michel (2005), Acoustical technique for Lagrangian velocity measurement, *Rev. Sci. Instrum.*, *76*, 25,105.

Ni, R., S.-D. Huang, and K.-Q. Xia (2012), Lagrangian acceleration measurements in convective thermal turbulence, *J. Fluid Mech.*, *692*, 395–419.

Obukhov, A. M. (1953), Effect of weak inhomogeneities in the atmosphere on sound and light propagation, *Izv. Akad. Nauk. Seriya Geofiz.*, *2*, 155–165.

Ott, S., and J. Mann (2000), An experimental investigation of the relative diffusion of particle pairs in three-dimensional turbulent flow, *J. Fluid Mech.*, *422*, 207–223.

Ouellette, N. T., H. Xu, and E. Bodenschatz (2005), A quantitaive study of three-dimensional Lagrangian particle tracking algorithms, *Exper. Fluids*, *39*, 722–729.

Pietropinto, S., et al. (1999), Superconducting instrumentation for high Reynolds turbulence experiments with low temperature gaseous helium, *Phys. C*, *386*, 512–516.

Poulain, C., N. Mazellier, P. Gervais, Y. Gagne, and C. Baudet (2004), Spectral vorticity and Lagrangian velocity measurements in turbulent jets, *Flow Turbul. Combust.*, *72*, 245–271.

Qureshi, N. (2009), Experimental Investigation of finite sized inertial particles dynamics in wind tunnel grid generated turbulence, Ph.D. thesis, Université Joseph Fourier-Grenoble I.

Qureshi, N. M., M. Bourgoin, C. Baudet, A. Cartellier, and Y. Gagne (2007), Turbulent transport of material particles: An experimental study of finite size effects, *Phys. Rev. Lett.*, *99*, 184502.

Qureshi, N. M., U. Arrieta, C. Baudet, A. Cartellier, Y. Gagne, and M. Bourgoin (2008), Acceleration statistics of inertial particles in turbulent flow, *Eur. Phys. J. B*, *66*, 531–536.

Richardson, L. F. (1926), Atmospheric diffusion shown on a distance-neighbour graph, *Proc. R. Soc. Lond. Ser. A*, *110*, 709–737.

Salazar, J. P. L. C., J. de Jong, L. Cao, S. H. Woodward, H. Meng, and L. R. Collins (2008), Experimental and numerical investigation of inertial particle clustering in isotropic turbulence, *J. Fluid Mech.*, *600*, 245–256.

Shew, W. L., Y. Gasteuil, M. Gibert, P. Metz, and J.-F. Pinton (2007), Instrumented tracer for Lagrangian measurements in Rayleigh-B{é}nard convection, *Rev. Sci. Instrum.*, *78*, 65,105.

Taylor, G. I. (1922), Diffusion by continuous movements, *Proc. Lond. Math. Soc.*, *20*, 196–212.

Tennekes, H., and J. L. Lumley (1992), *A First Course in Turbulence*, MIT press, Cambridge, Mass.

Tsai, R. (1987), A versatile camera calibration technique for high accuracy 3d machine vision metrology using off-the-shelf tv cameras and lenses, *IEEE Trans. Robot. Autom.*, *RA-3*, 323.

Virant, M., and T. Dracos (1997), {3D} {PTV} and its application on Lagrangian motion, *Measur. Sci. Technol.*, *8*, 1539–1552.

Volk, R., N. Mordant, G. Verhille, and J.-F. Pinton (2008), Laser Doppler measurement of inertial particle and bubble accelerations in turbulence, *Eur. Phys. Lett.*, *81*, 34,002.

Volk, R., E. Calzavarini, E. Leveque, and J.-F. Pinton (2011), Dynamics of inertial particles in a turbulent von Karman flow, *J. Fluid Mech.*, *668*, 223–235.

Voth, G. A., A. LaPorta, A. M. Crawford, J. Alexander, and E. Bodenschatz (2002), Measurement of particle accelerations in fully developed turbulence, *J. Fluid Mech.*, *469*, 121–160.

Xu, H., and E. Bodenschatz (2008), Motion of inertial particles with size larger than Kolmogorov scale in turbulent flows, *Phys. D*, *237*, 2095–2100.

Zimmermann, R., Y. Gasteuil, M. Bourgoin, R. Volk, A. Pumir, and J.-F. Pinton (2011a), Rotational intermittency and turbulence induced lift experienced by large particles in a turbulent flow, *Phys. Rev. Lett.*, *106*, 154,501.

Zimmermann, R., Y. Gasteuil, M. Bourgoin, R. Volk, A. Pumir, and J.-F. Pinton (2011b), Tracking the dynamics of translation and absolute orientation of a sphere in a turbulent flow, *Rev. Sci. Instrum.*, *82*, 33,906.

Zimmermann, R., L. Fiabane, Y. Gasteuil, R. Volk and J.-F. Pinton (2013), Characterizing flows with an instrumented particle measuring Lagrangian accelerations, *New J. Phys.*, *15*, 015018.

16

A High-Resolution Method for Direct Numerical Simulation of Instabilities and Transitions in a Baroclinic Cavity

Anthony Randriamampianina[1] and Emilia Crespo del Arco[2]

16.1. INTRODUCTION

Baroclinic instability is recognized to be one of the dominant energetic processes in the large-scale atmospheres of terrestrial planets, such as Earth and Mars, e.g., *Pierrehumbert and Swanson* [1995], and in the oceans. Its fully developed form as sloping convection is strongly nonlinear and has a major role in the transport of heat and momentum in the atmospheric and oceanic motions. Its time-dependent behavior also exerts a dominant influence on the intrinsic predictability of the atmosphere and the degree of chaotic variability in its large-scale meteorology [e.g, *Pierrehumbert and Swanson*, 1995; *Read et al.*, 1998; *Read*, 2001]. On the other hand, the close analogy between the dynamics of the ocean and atmosphere has been reported by *Orlanski and Cox* [1973]: "Similar phenomena take place from the high frequency range characterized by internal gravity waves to the low range of frequencies dominated by quasi-geostrophic motion. Detailed temperature measurements indicate that the ocean has relatively large-scale density discontinuities that are very much like atmospheric fronts. The oceanic fronts have a characteristic slope which is determined by the density difference, rotation and vertical shear of the currents parallel to the front. Since atmospheric fronts are known to be baroclinically unstable, it appears to be appropriate to suspect the same mechanism may be present in the ocean."

Since the pioneering works of *Hide* [1958] in the 1950s, the differentially heated, rotating cylindrical annulus has been an archetypal means of studying the properties of fully developed baroclinic instability in the laboratory.

The system is well known to exhibit a rich variety of different flow regimes, depending upon the imposed conditions (primarily the temperature contrast ΔT and rotation rate Ω), ranging from steady, axisymmetric circulations through highly symmetric, regular wave flows to fully developed geostrophic turbulence [*Hide*, 1958; *Fowlis and Hide*, 1965].

With the exponential increase in computing power these last decades, direct numerical simulation has become an indispensable tool for investigating the complex spatiotemporal behaviors of baroclinic instability in the laboratory, complementarily with experiments. Even though it does not yet allow for a complete study of the fully developed turbulent regimes, it provides new insight into the mechanisms responsible for these disordered flows. Moreover, direct numerical simulation, free of uncertainties related to turbulence modelings and of imperfections of experimental setups, can supply more extensive data than measurements and thus facilitate the detailed analysis of the wave dynamics. In particular, it is useful to explore the different nonlinear flow regimes in the parameter space in order to accurately delineate a bifurcation diagram. Moreover, direct numerical simulation provides relevant information about the small-scale fluctuations that progressively destroy the regularity of the flow during the transition toward geostrophic turbulence. Thereby it can efficiently serve as a guide to experiments and also supplement measurements.

16.2. NUMERICAL MODEL

16.2.1. Background

The first numerical investigations devoted to baroclinic waves in the differentially heated rotating cylindrical annulus were reported by *Williams* [1969] based on

[1]*Laboratoire Mécanique, Modélisation & Procédés Propres, UMR 7340 CNRS, Aix Marseille Université, Marseille, France.*
[2]*Departamento de Física Fundamental, Universidad Nacional de Educación a Distancia (UNED), Madrid, Spain.*

Modeling Atmospheric and Oceanic Flows: Insights from Laboratory Experiments and Numerical Simulations,
First Edition. Edited by Thomas von Larcher and Paul D. Williams.
© 2015 American Geophysical Union. Published 2015 by John Wiley & Sons, Inc.

a second-order finite difference approximation in space and in time. The approach was implemented on staggered grids over a regular mesh for the pressure-temperature and the velocity components. An explicit second-order leap-frog scheme was employed to discretize both the space and time derivatives. However the integration domain was restricted to a sector, only admitting the dominant wave and its harmonics. The cavity was filled with water, assumed to satisfy the Boussinesq approximation, with the density variation applied to the gravitational acceleration [*Williams*, 1971].

Then *James et al.* [1981] carried out a combined laboratory and numerical study of the steady baroclinic waves. Similar to the model proposed by *Williams* [1969], staggered grids were used with a second-order finite difference formula, but without any arbitrary truncation of the full spectrum of the waves. A hyperbolic tangent transformation was introduced to stretch the mesh toward the boundaries in the radial and axial directions while keeping uniform distribution in the azimuthal direction with Fourier series. To avoid severe constraint on the time step stability condition associated with these small grid sizes within the boundary layers, a Dufort-Frankel scheme was used for the diffusion terms, taking into account eventual variations of the viscosity, unlike the formulation of *Williams* [1969]. Using a water-glycerol mixture as working fluid, properties were assumed variable, with quadratic and linear dependencies with the temperature respectively for the density and for the kinematic viscosity and the thermal diffusivity. However, the authors did not achieve direct comparison of results between experiment and numerical simulation under the same external conditions, partially inferred to the restricted coarse resolutions used, due to the existing computer capacity constraints.

Hignett et al. [1985] continued these studies and obtained the same wave flow structure under identical conditions for the laboratory experiment and numerical simulation by using a different combination of water and glycerol than *James et al.* [1981]. They introduced density variations in centrifugal acceleration, contrary to *James et al.* [1981]. They put forward the intransitivity of the flow for steady wave state, corresponding to the coexistence of different stable wave structures under the same external conditions. They also mentioned the occurrence of hysteresis cycles during the transition from axisymmetric to nonaxisymmetric solutions, as already observed by *Hide* [1958] with an open upper free-surface configuration.

A sophisticated version, *MORALS (Met Office/Oxford Rotating Annulus Laboratory Simulation)*, derived from the numerical tool proposed by *James et al.* [1981], was implemented at the University of Oxford, UK (AOPP), for the investigations of a wide spectrum of applications devoted to geophysical fluid dynamics.

The choice of high-resolution spectral technique in the present study stems from its ability to accurately predict the thresholds of the different bifurcations occurring during time-dependent flow regimes, resulting from its global character, in contrast with local finite difference discretization [*Gottlieb and Orszag*, 1977; *Canuto et al.*, 1987]. In particular, the accuracy of spectral techniques was discussed in detail by *Pulicani et al.* [1990] (see also *Randriamampianina et al.* [1990]) during the simulation of oscillatory convection at low Prandtl number. Moreover, the approach is well suited for the simulation of rotating flows in enclosures, where the boundary layer is three dimensional from its inception.

16.2.2. Governing Equations

The physical model, the so-called baroclinic cavity, consists of an annular domain of inner radius a, outer radius b, and height d rotating around its vertical axis of symmetry. The cavity is filled with a liquid defined by a Prandtl number Pr and is submitted to a temperature difference $\Delta T = T_b - T_a$ between the inner, cold, and outer, hot, cylinders closed by horizontal insulating rigid endplates. One specific configuration involving an open upper free surface is also considered.

In the meridional plane, the dimensional space variables $(r^*, z^*) \in [a, b] \times [0, d]$ have been normalized into the square $[-1, 1] \times [-1, 1]$, a prerequisite for the use of Chebyshev polynomials (where the asterisk denotes dimensional variables):

$$r = \frac{2r^*}{b-a} - \frac{b+a}{b-a}, \qquad z = \frac{2z^*}{d} - 1.$$

The fluid is assumed to satisfy the Boussinesq approximation [*Zeytounian*, 2003] with constant properties except for the density when applied to the Coriolis, centrifugal, and gravitational accelerations, where $\rho^* = \rho_0[1 - \alpha(T^* - T_0)]$, where α is the coefficient of thermal expansion and T_0 is a reference temperature $T_0 = (T_b + T_a)/2$. However, it was found that the contribution of density variation with the Coriolis term $\rho_0 \alpha T^* \Omega \mathbf{e}_z \times \mathbf{V}^*$ was negligible compared to the centrifugal and gravitational ones. Moreover, for the imposed external conditions, the variations with temperature of viscosity and thermal diffusivity remain small, keeping the value of Prandtl number almost constant (at least below the margins of error from measurements). The reference scales are the velocity $U^* = g\alpha \Delta T / 2\Omega$ and the time $t^* = (2\Omega)^{-1}$, and the nondimensional normalized temperature is $2(T^* - T_0)/\Delta T$ [*Randriamampianina et al.*, 2006].

Depending on the type of solution sought, axisymmetric or nonaxisymmetric, two different approaches are considered independently for the governing equations of the flow dynamics. In the first case, a vorticity stream function

Table 16.1. Summary of the dimensions of the system, the fluid properties, and the governing parameters for the liquid-filled cavity: Pr = 16.

Inner radius	a	4.5 cm
Outer radius	b	15. cm
Height	d	26. cm
Gap width	$L = b - a$	10.5 cm
Mean temperature	T_0	293 K
Temperature difference	$\Delta T = T_b - T_a$	2 K
Rotation rate	Ω	0.25–1.25 rad/s
Volume expansion coefficient	α	3.171×10^{-4} K^{-1}
Kinematic viscosity	ν	2.0397×10^{-2} cm^2/s
Thermal diffusivity	κ	1.2731×10^{-3} cm^2/s
Aspect ratio	$A = d/L$	2.47619
Curvature parameter	$R_c = (b + a)/L$	1.857
Prandtl number	$\text{Pr} = \nu/\kappa$	16.0215
Rayleigh number	$\text{Ra} = g\alpha \Delta T L^3/(\nu\kappa)$	2.7735×10^7
Froude number	$\text{Fr} = \Omega^2 L/g$	6.69×10^{-4}–1.67×10^{-2}
Taylor number	$\text{Ta} = 4\Omega^2 L^5/(\nu^2 d)$	$2.95 \times 10^6 - 7.37 \times 10^7$
Thermal Rossby number	$\Theta = g\, d\alpha\, \Delta T/(\Omega^2 L^2)$	2.3475–0.0939

with azimuthal velocity formulation is introduced. Not only does it reduce the number of equations to solve, in comparison with the primitive variable velocity-pressure formulation, but it also ensures a divergence-free velocity field irrespective of the mesh used. An influence matrix technique is implemented to treat the lack of boundary conditions for the vorticity coupled with the stream function [*Chaouche et al.*, 1990; *Randriamampianina et al.*, 2001, 2004]. For the three-dimensional solution, the primitive variables are directly solved. We present hereafter the details of the governing equations and the numerical method for the latter.

In a frame of reference rotating with the cavity, the resulting dimensionless system is written as [*Randriamampianina et al.*, 2006]

$$\frac{\partial \mathbf{V}}{\partial t} + \frac{2\,\text{Ra}}{A^2 \text{Pr Ta}} N(\mathbf{V}) + \mathbf{e}_z \times \mathbf{V}$$
$$= -\nabla \Pi + \frac{4}{A^{3/2} \text{Ta}^{1/2}} \nabla^2 \mathbf{V} + \mathbf{F}, \quad (16.1)$$

$$\nabla \cdot \mathbf{V} = 0, \quad (16.2)$$

$$\frac{\partial T}{\partial t} + \frac{2\,\text{Ra}}{A^2 \text{Pr Ta}} \nabla\cdot(\mathbf{V}T) = \frac{4}{A^{3/2} \text{Pr Ta}^{1/2}} \nabla^2 T, \quad (16.3)$$

with

$$\Pi = \frac{p + \rho_0 g z - \frac{1}{2}\rho_0 \Omega^2 r^2}{\rho_0 g \alpha\, \Delta T\, d/2},$$

$$\mathbf{F} = \frac{1}{2} T \mathbf{e}_z - \frac{\text{Fr}}{4A}(r + R_c)\, T \mathbf{e}_r,$$

where $\mathbf{e}_r$ and $\mathbf{e}_z$ are the unit vectors in the radial and axial directions, respectively, and $N(\mathbf{V})$ represents the nonlinear advection terms. The parameters governing the flow and the heat transfer are the aspect ratio A, the curvature parameter R_c, the Prandtl number Pr, the Rayleigh number Ra, the Froude number Fr, and the Taylor number Ta (see the definitions in Table 16.1). For a given fluid within a fixed geometry, the Taylor number is one of the two main control parameters traditionally used to analyze this system, following, e.g., *Fowlis and Hide* [1965] and *Hide and Mason* [1975]. The second parameter is the thermal Rossby number,

$$\Theta = \frac{g\, d\alpha\, \Delta T}{\Omega^2 (b-a)^2} \equiv \frac{4\,\text{Ra}}{\text{Pr Ta}},$$

introduced by *Hide* [1958] as a stability parameter, which gives a measure of the buoyancy strength over the Coriolis term and appears explicitly as coefficient of the advection terms in equations (16.1) and (16.3).

The "skew-symmetric" form proposed by *Zang* [1990] was chosen for the nonlinear advection term $N(\mathbf{V}) = [\mathbf{V}\cdot\nabla\mathbf{V} + \nabla\cdot(\mathbf{V}\mathbf{V})]/2$ in the momentum equation (16.1) to ensure the conservation of kinetic energy, a necessary condition for a simulation to be numerically stable in time.

16.2.3. Boundary Conditions

The boundary conditions are no-slip velocity conditions at all rigid surfaces,

$$\mathbf{V} = \mathbf{0} \quad \text{at } r = \pm 1 \quad \text{and at } z = \pm 1,$$

thermal insulation at horizontal rigid surfaces,

$$\frac{\partial T}{\partial z} = 0 \quad \text{at } z = \pm 1,$$

and constant-temperature conditions at the vertical sidewalls,

$$T = \pm 1 \quad \text{at } r = \pm 1.$$

In the case of an open upper wall, planar free-surface conditions are imposed, assuming the absence of vertical deformation along this boundary, in agreement with experimental observations for the control parameter values under consideration [*Harlander et al.*, 2011]:

$$\frac{\partial T}{\partial z} = \frac{\partial V_r}{\partial z} = \frac{\partial V_\phi}{\partial z} = V_z = 0 \quad \text{at } z = 1.$$

16.2.4. Solution Method

A pseudospectral collocation Chebyshev method is implemented in the meridional plane (r, z), in association with Galerkin-Fourier approximation in azimuth ϕ for the three-dimensional flow regimes to solve the primitive variable formulation described above. Each dependent variable is expanded in the approximation space $\mathcal{P}_{NM}$, composed of Chebyshev polynomials of degrees less than or equal to N and M in the r and z directions, respectively, while Fourier series are introduced in the azimuthal direction with K modes.

For each dependent variable f ($f \equiv V_r, V_\phi, V_z, T, p$), it reads

$$f_{NMK}(r, \phi, z, t)$$
$$= \sum_{n=0}^{N} \sum_{m=0}^{M} \sum_{k=-K/2}^{K/2-1} \hat{f}_{nmk}(t) T_n(r) T_m(z) \exp(ik\phi),$$

where T_n and T_m are Chebyshev polynomials of degrees n and m.

This approximation is applied at collocation points, where the differential equations are assumed to be satisfied exactly [*Gottlieb and Orszag*, 1977; *Canuto et al.*, 1987]. We have chosen the Chebyshev-Gauss-Lobatto distribution defined by a high concentration of points toward the boundaries, well suited to handle the thin dynamical and thermal boundary layers expected to develop at high values of the Taylor Ta and Rayleigh Ra numbers, which scale as $Ta^{-1/4}$ (Ekman layer, along horizontal walls) or $Ta^{-1/6}$ (Stewartson layer, along vertical cylinders) and $Ra^{-1/4}$, respectively:

$$r_i = \cos\left(\frac{i\pi}{N}\right) \quad \text{for } i \in [0, N],$$

$$z_j = \cos\left(\frac{j\pi}{M}\right) \quad \text{for } j \in [0, M],$$

and a uniform mesh in the azimuthal direction according to the Fourier series:

$$\phi_k = \frac{2k\pi}{K} \quad \text{for } k \in [0, K].$$

The time integration used is second-order accurate and is based on a combination of Adams-Bashforth (AB) and backward differentiation formula (BDF) schemes, chosen for its good stability properties [*Vanel et al.*, 1986]. The resulting AB/BDF scheme is semi-implicit, and for the transport equation of the velocity components in equation (16.1),

$$\frac{3f^{l+1} - 4f^l + f^{l-1}}{2\delta t} + 2\mathcal{N}(f^l) - \mathcal{N}(f^{l-1})$$
$$= -\nabla \Pi^{l+1} + \frac{4}{A^{3/2} Ta^{1/2}} \nabla^2 f^{l+1} + F_i^{l+1} \quad (16.4)$$

where $\mathcal{N}(f)$ is a "global term" including the advection terms $N(\mathbf{V})$ and the Coriolis term, F_i corresponds to the component of the forcing term $\mathbf{F}$, δt is the time step, and the superscript l refers to time level. The cross terms in the diffusion part in the (r, ϕ) plane resulting from the use of cylindrical coordinates in equation (16.1) are treated within $\mathcal{N}(f)$. The latter is discretized in time using a second-order explicit AB scheme in order to maintain an overall second-order time accuracy with the BDF scheme applied to the diffusion term, as shown in equation (16.4). An equivalent discretization applies for the transport equation (16.3) of the temperature. For the initial step, we have taken $f^{-1} = f^0$. At each time step, the problem then reduces to the solution of Helmholtz and Poisson equations. We recall that the same time scheme is also implemented for the integration of the axisymmetric system, even though steady state solutions are sought.

An efficient projection scheme is introduced to solve the coupling between the velocity and the pressure in equation (16.1). This algorithm ensures a divergence-free velocity field at each time step, maintains the order of accuracy of the time scheme for each dependent variable and does not require the use of staggered grids [*Hugues and Randriamampianina*, 1998; *Raspo et al.*, 2002]. At each time step, a preliminary Poisson equation for the pressure, directly derived from the Navier-Stokes equations, is first solved before integrating the governing system described above. It allows for a variation in time of the normal pressure gradient at boundaries [*Hugues and Randriamampianina*, 1998], which plays an important role for time-dependent flows, in particular in the presence of an open free surface. A complete diagonalization of operators yields simple matrix products for the solution of successive Helmholtz and Poisson equations at each time step [*Haldenwang et al.*, 1984]. The computations of eigenvalues, eigenvectors, and inversion of corresponding matrices from these Helmholtz and Poisson operators are performed once during a preprocessing step.

16.2.5. Computational Details

For the transition from the upper symmetric regime to the regular waves, the initial conditions corresponded to the steady axisymmetric solution at each azimuthal node to which a random perturbation was added to the temperature field in azimuth. Subsequently, the strategy consisted of progressively increasing the rotation rate without adding any further perturbations for the computation of the following successive three-dimensional solutions.

The length of time for each specific computed solution strongly depends on the fluid considered, e.g., on the Prandtl number, due to the very different temporal behaviors involved. Some values of the wave drift were reported for air by *Randriamampianina et al.* [2006]. Theoretically, only one drift period of the large-scale baroclinic waves is necessary to have a complete analysis of the flow. But close to a bifurcation, corresponding to significant changes on the flow structure and temporal behavior, several drift periods have been computed until the final state is reached. This was the case during the computations of the different bifurcations occurring at the transition between wave numbers $m = 2$ and $m = 3$ for air, in particular during the bifurcation between the quasi-periodic regimes QP2 and QP3 characterized by two and three incommensurate frequencies, respectively. On the other hand, the presence of small-scale inertia gravity waves in the cases of liquids required much longer drift periods of the baroclinic waves and higher resolutions to ensure grid independency of the solution than for air, therefore increasing the length of integration time to be simulated. The different meshes with the corresponding time step used are reported hereafter for each specific case treated.

For $Pr = 16$ with $\Delta T = 2\,K$ at $\Omega = 0.5125$ rad/s, about 7.62×10^{-5} CPU seconds per time step and per mode on the supercomputer *NEC–SX*5 (IDRIS, Orsay, France) were necessary to compute the different scales occurring simultaneously within the cavity once the transient was removed (see hereafter the corresponding time step and mesh used). The transient is assumed to be finished when a clear temporal behavior can be identified from the time evolution of one dependent variable taken at a fixed monitoring point and when its random behavior disappears. The transient is also associated with the flow structure observed.

16.2.6. Validation

The numerical tools have been completely developed by the team [*Chaouche et al.*, 1990; *Hugues and Randriamampianina*, 1998; *Raspo et al.*, 2002]. The three-dimensional solutions were previously validated by *Randriamampianina et al.* [1997] for a liquid-filled cavity (Pr = 13.07) with respect to the detailed results reported by *Hignett et al.* [1985] from a combined laboratory and numerical study. Comparisons have been carried out between our computations and their measurements for a regular steady three-wave flow (characterized by a dominant azimuthal wave number $m = 3$). Very close agreement has been obtained for the qualitative structure of the flow pattern and for the quantitative comparison of the radial variation of the azimuthal velocity at different heights. Particular attention has been paid to the grid effect on the solution, which has served as a basis for subsequent studies.

16.3. RESULTS

Hide [1958] and *Fowlis and Hide* [1965], from their pioneering experimental investigations of baroclinic instability using liquids as working fluids, have delineated three main classes of flow regimes: axisymmetric regimes, regular waves, and irregular waves or geostrophic turbulence (see also *Hide and Mason* [1975]). The regular wave regimes are composed of the steady waves, denoted S, and the vacillation regimes subdivided into amplitude vacillation, AV or MAV (modulated amplitude vacillation), and structural vacillation, SV. The steady waves are determined by a dominant azimuthal wave number m in space and characterized by periodic oscillations in time induced by the uniform angular drift of the waves with constant amplitude. The amplitude vacillation regimes are defined by periodic (AV), quasi-periodic, or chaotic (MAV) temporal behavior of the amplitude of the dominant wave number (for a detailed analysis of the amplitude vacillation phenomenon, see Chapter 3 in this book). The SV regime, an intermediate step before the transition toward geostrophic turbulence, is characterized by a spatiotemporal chaos but still with a well-defined dominant wave number, as shown by the experimental evidence of *Früh and Read* [1997] (see also *Read et al.* [2008]).

Three specific fluids have been considered in the present study, air, a water-glycerol mixture, and water, in order to get insight into the important role played by the Prandtl number on the spatiotemporal characteristics of the baroclinic instability. Indeed, the Prandtl number Pr is a parameter of particular interest, also in the context of other convection problems. *Fein and Pfeffer* [1976], who carried out a careful survey of the main flow regimes in a thermally driven annulus using mercury, water, or silicon oils, found significant differences in the onset of baroclinic instability in the region of the so-called lower symmetric transition at low Taylor number, where viscous diffusion and thermal diffusion are expected to play a major role. Some substantial differences in the onset of various types of regular waves were also noted at higher Taylor numbers. *Jonas* [1981] investigated the influence of Prandtl number on the incidence of various forms of vacillation using fluids with Pr ranging from 11 to 74. He reported that amplitude vacillation in particular was significantly more widespread at high Prandtl number, though the onset of "structural vacillation" close to the transition zone at high Taylor number was less sensitive to Pr. In most of the published studies so far, however, the range of Pr investigated has either been limited to relatively high values (using liquids based on water, silicon oils, or organic fluids such as diethyl ether) or very low Pr in liquid metals (mercury).

16.3.1. Transition Between Successive Wave Numbers in Air-Filled Cavity Pr = 0.7

The geometric configuration corresponds to the one used by *Fowlis and Hide* [1965] in their experimental studies of liquids, defined by an inner radius $a = 34.8$ mm, outer radius $b = 60.2$ mm, and height $d = 100$ mm. The cavity is filled with air, Pr = 0.7, and a temperature difference $\Delta T = 30$ K is imposed between the two cylinders. For the rotation rate values considered, a resolution of $N \times M \times K = 64 \times 96 \times 80$ was used in the radial, vertical, and azimuthal directions, respectively, with a dimensionless time step $\delta t = 0.1125$. The results are part of a previous work [*Randriamampianina et al.*, 2006].

Before our numerical investigations of the baroclinic instability using air as working fluid [*Randriamampianina et al.*, 2006], there was not yet any available experimental study devoted to this fluid with Pr = $\mathcal{O}(1)$. However, our findings have subsequently motivated the installation of a specific experimental rig at the university of Oxford, UK [*Castrejón-Pita and Read*, 2007]. Then the measurements confirmed a posteriori the computed results, especially the route to obtain the AV regime. Indeed, unlike in previous experimental works involving liquids, where the onset of the m AV regime was associated with a decrease of the rotation rate from the established steady wave regime mS (defined by a dominant azimuthal wave number m), the AV regime was observed when increasing the rotation rate in this study with air, during the transition between two successive steady waves regime, from an azimuthal wave number m to $m + 1$.

Hide [1958] reported that the transition from axisymmetric to regular wave regimes does not significantly depend on the value of the Prandtl number but rather depends on the thermal Rossby number through the empirical criterion $\Theta \leq \Theta_c = 1.58 \pm 0.05$. In the present simulation, the first regular steady wave was obtained for $\Theta = 1.488 < \Theta_c$ at $Ta = 1.8 \times 10^5$ [*Randriamampianina et al.*, 2006]. In Figure 16.1 we display the bifurcation

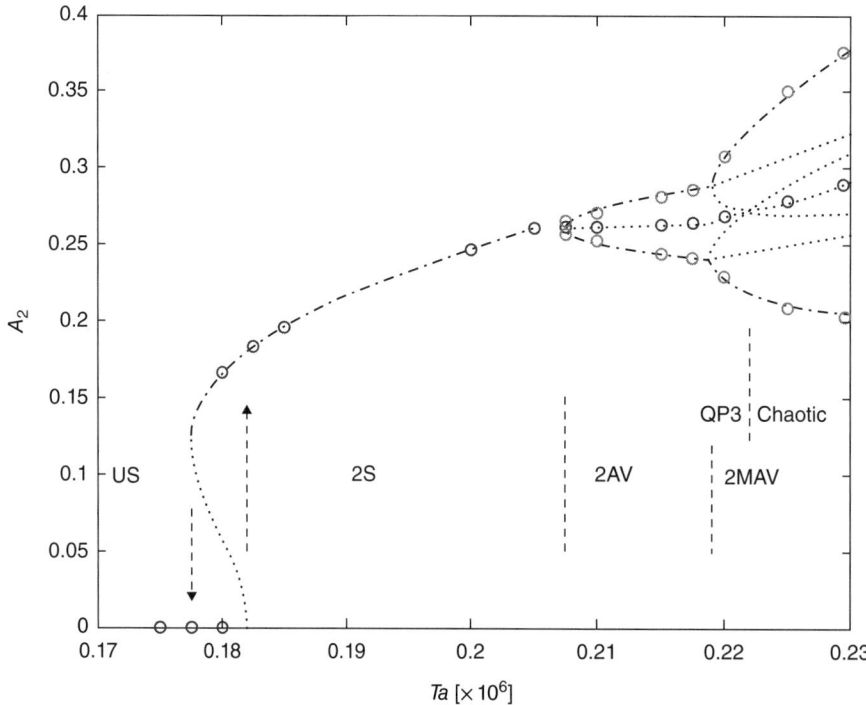

Figure 16.1. Amplitude of the dominant azimuthal wave number mode at midradius and midheight versus the Taylor number showing the bifurcation diagram for the transition from the upper symmetric regime to a steady wave and subsequent vacillations for the $m = 2$ flows in the air-filled cavity.

diagram showing the scenario for the transition from the upper symmetric regime to a steady wave and subsequent amplitude vacillations for the $m = 2$ flows. The, albeit narrow, hysteresis at the transition between the axisymmetric flow and the regular steady wave suggests a subcritical Hopf bifurcation, resulting from a zonal symmetry breaking. By using a weakly nonlinear stability analysis in the same air-filled configuration, *Lewis* [2010] confirmed the subcritical Hopf bifurcation observed during our numerical study. While *Hide and Mason* [1978] reported that such an hysteretic behavior was observed only if the upper boundary was a free surface, *Koschmieder and White* [1981] presented evidence for the possibility of small hysteresis in their experimental study of a water-filled cavity. On the other hand, *Castrejón-Pita and Read* [2007], using air as working fluid in their experiments, mentioned the occurrence of the so-called *weak waves*, characterized by $\Theta > \Theta_c$, prior to the onset of fully developed regular waves, but did not observe any hysteresis cycle. During this study, we did not find any hint of such weak waves.

By progressively increasing the rotation rate, the transition sequence from the upper symmetric US flow through all observed two-wave flows follows clear steps of increasing complexity before bifurcating to three-wave steady flow (the number before the letter denotes the dominant azimuthal wave number): US → 2S(P) → 2AV(QP2) → 2MAV(QP3) → 2MAV(NP) → 3S(P), as illustrated in Figure 16.1 showing the mean amplitude and the envelope of the vacillation. Here, P stands for periodic, QP for quasi-periodic (QP2 is characterized by two incommensurate frequencies, and QP3 by three frequencies) and NP for aperiodic regime. Steady wave solution 2S is obtained for $1.8 \times 10^5 \leq \text{Ta} \leq 2.05 \times 10^5$. The first 2AV regime, at $\text{Ta} = 2.1 \times 10^5$, likely occurs via a secondary Hopf bifurcation from an oscillatory flow, also known as a Neimark-Sacker bifurcation through a temporal symmetry breaking. It is characterized by a second frequency (QP2) resulting from the periodic oscillations of the amplitude, in addition to the wave drift observed during the regular steady flow. A further increase in rotation rate brings a third frequency coming from the modulation of the amplitude oscillations in the 2MAV regime. This corresponds to a continuation of "the quasi-periodic route to chaos" described by *Newhouse et al.* [1978], but the nature of the initial solution as a quasi-periodic 2MAV with three incommensurate frequencies was unusual. As shown by *Newhouse et al.* [1978], generic three-frequency flows are expected to be chaotic rather than periodic. To our knowledge, no previous example of such a flow has been reported from either numerical or experimental studies of baroclinic waves. The final type of flow dominated by $m = 2$ was a chaotic 2MAV regime that can be induced by a crisis as discussed in a similar baroclinic cavity by *Read et al.* [1998], in analogy with a noise-induced crisis in a multistable system (see also *von Larcher and Egbers* [2005]). Crisis is characterized by a sudden change in the flow dynamics and temporal behavior [see *Grebogi et al.*, 1983]. A further increase in rotation rate up to $\text{Ta} = 2.3 \times 10^5$ from this chaotic solution leads abruptly to the steady 3S regime, also due to a crisis. The temporal behaviors of all the AV and MAV solutions have been confirmed by the calculations of the corresponding largest Lyapunov exponent, reported in Figure 16.2. The quasi-periodic solutions have a largest Lyapunov exponent which cannot be distinguished from zero within the margin of error, while chaotic solutions are characterized by a positive Lyapunov exponent: $2.2 \times 10^5 < \text{Ta} < 2.3 \times 10^5$ [*Randriamampianina et al.*, 2006].

16.3.2. Liquid-Filled Cavity: Pr = 16

The details of the system, the fluid properties, and the governing parameters are summarized in Table 16.1. The configuration corresponds to one experimental rig used at the University of Oxford, UK [*Wordsworth*, 2009]. It consists of an annular domain of inner radius $a = 4.5$ cm, outer radius $b = 15$ cm, and height $d = 26$ cm. The cavity is filled with a liquid described by a Prandtl number Pr = 16 and is submitted to a temperature difference $\Delta T = 2$ K between the inner, cold, and outer, hot, cylinders closed by horizontal insulating rigid endplates. Four values of the rotation rate have been considered, covering different flow regimes of baroclinic waves. For $\Omega = 0.25, 0.35$, and 0.5125 rad/s, a mesh of $N \times M \times K = 128 \times 150 \times 256$ was used in the radial, axial, and azimuthal directions, respectively, with a dimensionless time step $\delta t = 0.0125$. For the rotation rate value $\Omega = 1.25$ rad/s, a refined resolution in the radial and azimuthal directions was necessary, $N \times M \times K = 150 \times 150 \times 320$, with a dimensionless time step $\delta t = 0.00625$.

The values of the control parameters used in the numerical simulation of the flow at these four rotation rates are represented in Figure 16.3 together with the experimental cases considered by *Wordsworth* [2009] in a (Ta, Θ) regime diagram. The slight difference between the measurements and the computations along the traverse corresponding to the temperature difference $\Delta T = 2$ K in Figure 16.3 results from the change operated on the outer radius of the experimental setup when drawing the diagram ($b_{\text{exp}} = 14.3$ cm instead of the value $b = 15$ cm used in the simulations; see Table 16.1). However, no significant differences were observed on the nature of the flow regime between measurements and computed solutions at identical control parameter values. Thus, in agreement with experimental investigations, the first value at $\Omega = 0.25$ rad/s, corresponding to $(\Theta, \text{Ta}) = (2.3475, 2.95 \times 10^6)$, yields a weak wave flow, while for the two others, at Ω values of 0.35

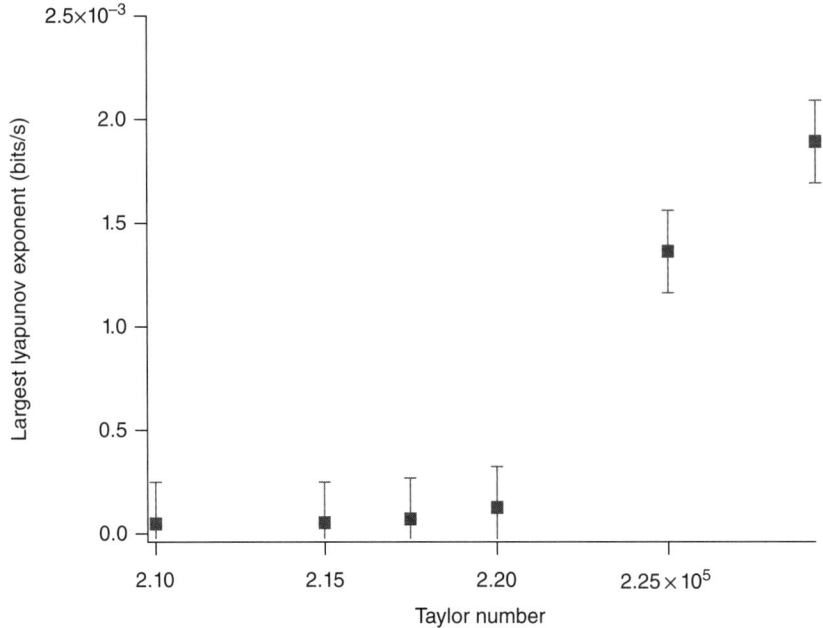

Figure 16.2. Largest Lyapunov exponent for the 2AV and 2MAV flows versus the Taylor number in the air-filled cavity.

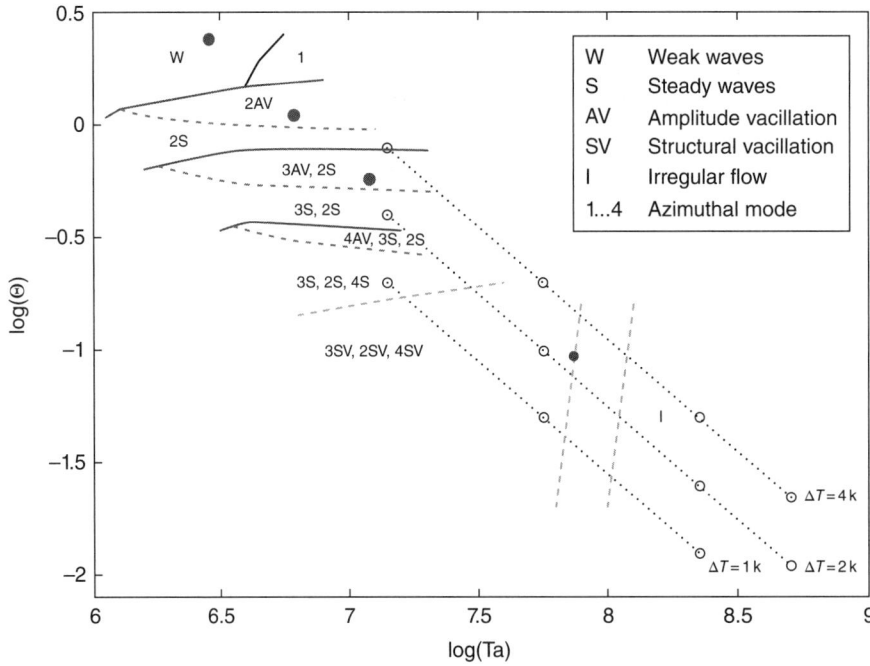

Figure 16.3. Regime diagram in the (Ta, Θ) plane for the liquid-filled cavity with Pr = 16 established from experimental investigations [*Wordsworth*, 2009]; the control parameter values from the measurements are represented by circles and from the computations by dots along the line for $\Delta T = 2$ K. The traverses in dotted lines correspond to values obtained when varying the rotation rate at fixed temperature difference ΔT (here 1 K, 2 K and 4 K). The continuous lines in blue indicate the transition between two flow regimes defined by different azimuthal dominant wave numbers, while the dashed lines in red delimit the transition between two types of temporal dependency.

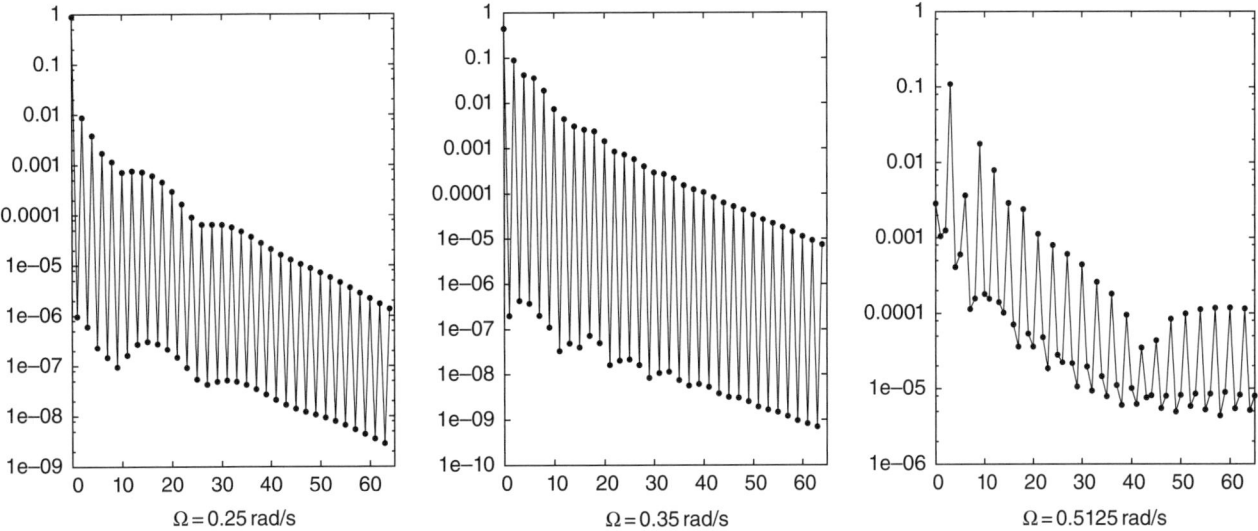

Figure 16.4. Spatial spectra of the time-averaged amplitudes of the azimuthal wave mode m of the temperature at midradius and midheight in the liquid-filled cavity, Pr = 16.

and 0.5125 rad/s, located at $(\Theta, Ta) = (1.1977, 5.78 \times 10^6)$ and $(0.5586, 1.24 \times 10^7)$ in the regime diagram, the flow evolves to a regular wave flow regime, characterized temporally by an amplitude vacillation but with different dominant azimuthal wave numbers: 2AV and 3AV, respectively. Even though experimental data were not available at these three specific values of the rotation rate, the regime diagram established from experiments pointed out the existence of these regimes at the corresponding control parameter values, as is cleary shown in Figure 16.3. The rotation rate $\Omega = 1.25$ rad/s with $(\Theta, Ta) = (0.0939, 7.37 \times 10^7)$ leads to a structural vacillation 3SV regime close to the transition zone as revealed by measurements. In the remainder of this section, comparisons of computed solutions with available experimental data [*Wordsworth*, 2009] are carried out for this rotation rate. Finally, preliminary results from a first attempt to obtain a turbulent flow using direct numerical simulation are discussed and compared with measurements [*Wordsworth*, 2009].

16.3.2.1. Weak Waves. At the lowest value of rotation rates considered, $\Omega = 0.25$ rad/s (Ta = 2.95×10^6), the simulation predicts a "weak wave" flow regime with a dominant azimuthal wave number $m = 2$, in agreement with experimental findings, as can be seen in Figure 16.3. Similar to the observations of *Castrejón-Pita and Read* [2007] during their experimental investigations in an air-filled cavity, the corresponding thermal Rossby number $\Theta = 2.3475$ is larger than the empirical critical value $\Theta_c = 1.58 \pm 0.05$ determined by *Hide* [1958] for the occurrence of a regular wave regime. This particular flow, developing prior to the onset of regular steady wave, is characterized by a small amplitude of the azimuthal variations of the temperature, $A_m/\Delta T < 0.01$ [*Hide and Mason*, 1978], where m refers to the dominant azimuthal wave number. Figure 16.4 shows the time-averaged spatial spectra of the amplitude of the azimuthal wave mode from Fourier analysis of the temperature at midradius and midheight $(r_{\text{mid}}, z_{\text{mid}})$. The fractional amplitude of the temperature $A_2/\Delta T \sim 0.005$ at $\Omega = 0.25$ rad/s is about 10 times smaller than that of the regular waves obtained at higher rotation rates, $\Omega = 0.35$ $(A_2/\Delta T)$ and 0.5125 rad/s $(A_3/\Delta T)$, for which $\Theta < \Theta_c$. Similar behaviors have been mentioned by *Castrejón-Pita and Read* [2007] from their experimental investigations of weak waves in an air-filled cavity. Moreover, it was found that the computed flow structure toward the upper half of the cavity remains broadly axisymmetric while the baroclinic waves are trapped toward the bottom of the cavity (Figure 16.5). This is consistent with the observations of *Hide and Mason* [1978] using liquids, but in contrast with the flow pattern reported by *Castrejón-Pita and Read* [2007] using air as working fluid with the weak waves visible at all heights of the cavity. This difference can be explained by the thermal stratification levels resulting from the different fluid properties, e.g., the Prandtl number. In the case of air, the Prandtl number is one order of magnitude lower than in liquids, leading to a more uniform density gradient along the vertical direction than in liquids, where a higher density gradient prevails in the lower part of the cavity. As a consequence, the baroclinic instability first develops near the bottom region in cavities filled with liquids.

Another feature of the weak waves comes from their angular drift velocity relative to the cavity. *Hide and Mason* [1978] found that weak waves drifted faster that the

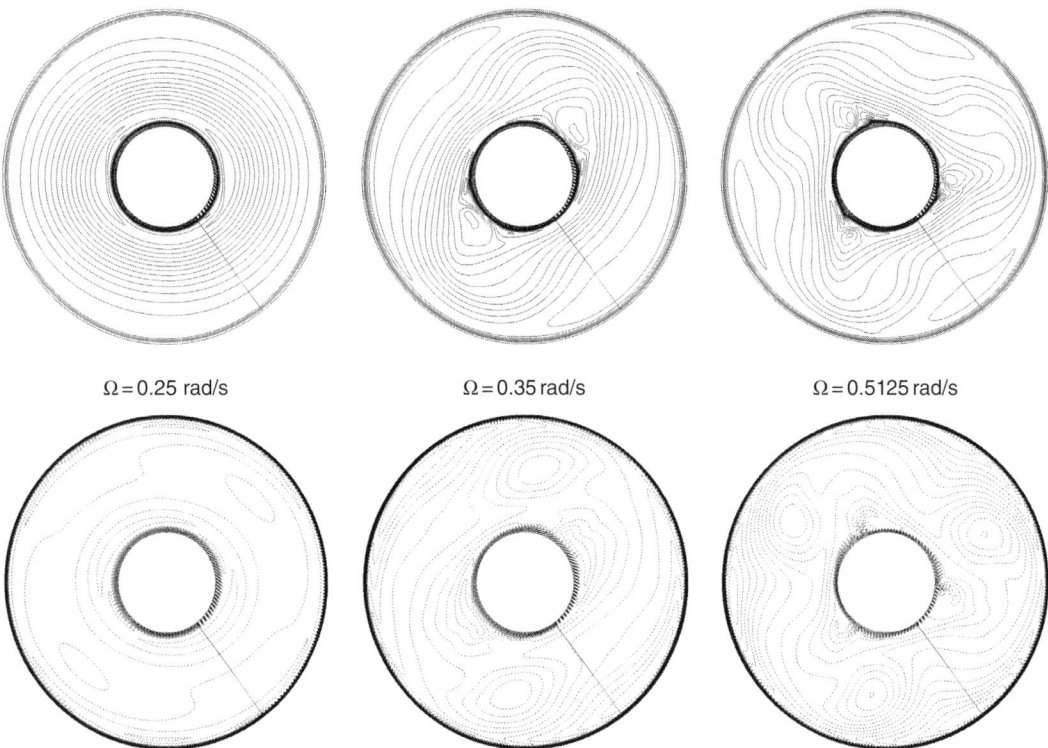

Figure 16.5. Instantaneous isotherms for the three rotation rate values at different heights of the liquid-filled cavity: top at midheight $z/d = 0.5$, bottom at $z/d = 0.19$.

strong waves, and *Castrejón-Pita and Read* [2007] reported a ratio up to 10 in their air-filled experiment. In our case, for the dimensionless drift frequency, we have obtained a ratio of 2 between the weak wave at $\Omega = 0.25$ rad/s and the next strong wave at $\Omega = 0.35$ rad/s, and a ratio of 1.22 between the two regular waves. These behaviors are consistent with the measurements of *Hide and Mason* [1978] in a liquid-filled cavity, keeping in mind that the computed solutions at $\Omega = 0.25$ rad/s and $\Omega = 0.35$ rad/s are characterized by the same dominant wave number $m = 2$, while at $\Omega = 0.5125$ rad/s, the dominant azimuthal wave number is $m = 3$. Both experimental investigations, using liquids [*Hide and Mason*, 1978] or air [*Castrejón-Pita and Read*, 2007], mentioned the marked transition between weak and strong waves. It is concluded that the presence of weak waves prevents the occurrence of a hysteresis cycle during the transition from the upper symmetric regime to regular waves, as these two phenomena were not observed simultaneously.

To our knowledge, the present numerical study represents the first simulation cleary showing such a weak wave flow occurring prior to the regular wave regime within a baroclinic cavity, in agreement with experimental findings.

16.3.2.2. Amplitude Vacillation Flow Regime.
At Ω values of 0.35 and 0.5125 rad/s, corresponding to $(\Theta, Ta) = (1.1977, 5.78 \times 10^6)$ and $(0.5586, 1.24 \times 10^7)$, respectively, in Figure 16.3, the simulation predicts two regular wave regimes characterized spatially by dominant azimuthal wave numbers $m = 2$ and $m = 3$, respectively, as revealed by instantaneous isotherms in Figure 16.5 and temporally by an amplitude vacillation. These solutions have been directly obtained by progressively increasing the rotation rate; e.g., the solution at the higher rotation rate $\Omega = 0.5125$ rad/s was computed using as initial conditions the one at lower rotation rate $\Omega = 0.35$ rad/s. The simulations were not able to capture any steady regular wave flow, although experiments reported 2S and 3S regimes in Figure 16.3, keeping in mind that 2AV (3AV) regimes resulted experimentally from steady waves 2S (3S) by decreasing the rotation rate (see Chapter 3). Moreover, as soon as the large-scale regular baroclinic structures arise, we have observed the spontaneous development of small-scale fluctuations particularly along the cold inner cylinder, even though experimental studies did not mention such a behavior. However, it can be explained by possible technical limitations to detect the very low level of these fluctuations at these rotation rates [*Read*, 1992]. Thus, in some particular regions, the computed flow exhibits locally a spatiotemporal chaotic behavior. As introduced previously by *Randriamampianina et al.* [2006] to describe the temporal characteristics, we use the

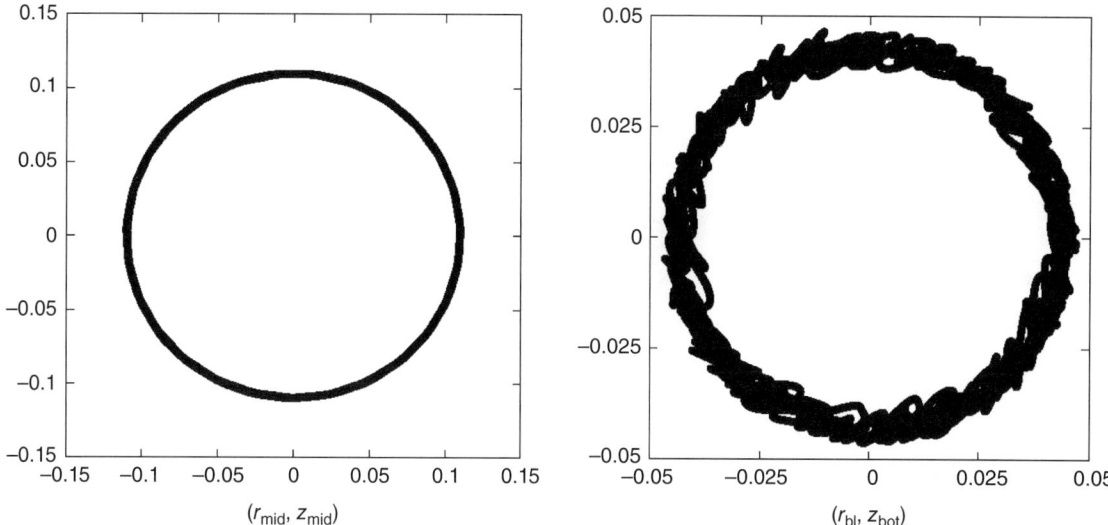

Figure 16.6. Phase space diagrams of the temperature field at two fixed (r, z) locations for $\Omega = 0.5125$ rad/s in the liquid-filled cavity, Pr = 16, showing the time series of the sine component versus that of the cosine component of the azimuthal dominant mode $m = 3$.

representation in the phase space based on time series of cosine and sine components of the dominant azimuthal wave number from a Fourier analysis of the temperature at two specific fixed (r, z) locations. We display these behaviors in Figure 16.6 at $\Omega = 0.5125$ rad/s, for which the intensity of the small-scale features was found highest between the computed AV solutions. In agreement with the experimental findings of *Wordsworth* [2009], the first plot at midradius and midheight (r_{mid}, z_{mid}) shows a "classical" 3AV profile defined by two frequencies: the wave drift represented by the large circle and the periodic oscillations of the wave amplitude, both related to the baroclinic instability [*Randriamampianina et al.*, 2006]. The second map, taken at a radius location r_{bl} inside the boundary layer along the inner cold cylinder and at a height near the bottom wall, z_{bot}, clearly exhibits a chaotic behavior corresponding to a 3MAV regime.

Since such different AV and MAV regimes were not observed simultaneously under fixed values of control parameters within the air-filled cavity [*Randriamampianina et al.*, 2006] but successively when increasing the values of rotation rate as shown in Figure 16.1, this localized 3MAV is directly ascribed to the presence of small-scale features. From their recent direct numerical simulation in a liquid-filled baroclinic cavity with Pr = 24.47, *Jacoby et al.* [2011] identified these small structures, occurring spontaneously and simultaneously with the large-scale baroclinic waves, as inertia-gravity waves (IGWs). The characteristics of the IGWs observed in the present configuration are discussed by *Randriamampianina* [2013]. The values of the sine and cosine components reported in Figure 16.6 are related to

IGW activity, more specifically to the balance between the two phenomena involved, with highest values associated with baroclinic instability. It reflects the variability of IGWs and the interaction between these two waves characterized by very different scales in time and in space. In particular, it puts forward the ability of the IGWs to induce locally a chaotic regime of the large-scale flow motion. Such a behaviour was not mentioned either by experiments [*Wordsworth*, 2009] or by previous simulations using liquids [*Hignett et al.*, 1985; *Jacoby et al.*, 2011]. From their numerical simulation based on finite difference approximation, *Hignett et al.* [1985] obtained also the AV regime, using a liquid defined by Pr = 13.07, but did not report the presence of these small-scale features. On the other hand, we did not observe the appearance of these fluctuations simultaneously with baroclinic waves when considering air in the present geometry, but rather we found the same flow structures reported by *Randriamampianina et al.* [2006] for air using different geometric dimensions. It clearly reveals the strong dependence on the Prandtl number of the baroclinic instability characteristics through the thermal stratification of the flow. Again we refer the reader to Chapter 3 in this book about the detailed analysis of the amplitude vacillation flow regime, particularly about the different mechanisms responsible for their occurrence.

16.3.2.3. Structural Vacillation Flow Regime. In spite of a well-defined dominant azimuthal wave number, this flow is characterized by the presence of small-scale fluctuations which progressively destroy the regularity of

Figure 16.7. Comparison of flow structures between experimental measurements at midheight $z/d = 0.50$ (top) and computed solutions at different heights (bottom) for the structural vacillation regime in the liquid-filled cavity, Pr = 16.

the large-scale baroclinic waves and therefore eventually lead to disordered flow [*Früh and Read*, 1997; *Read et al.*, 2008]. We have carried out comparisons of flow structures between our computed results and laboratory measurements obtained by *Wordsworth* [2009] (see also *Wordsworth et al.* [2008]). The rotation rate used in the experimental study was $\Omega = 1.3$ rad/s while in the simulation $\Omega = 1.25$ rad/s was used, corresponding to $(\Theta, \text{Ta}) = (0.0939, 7.37 \times 10^7)$, which is located close to the transition zone in the regime diagram (Figure 16.3), with a temperature difference $\Delta T = 2$ K. In both cases, a SV regime was obtained. The experiment reported 4SV, as revealed by instantaneous streaklines in Figure 16.7, but also 3SV as seen in Figure 16.8 from the azimuthal velocity and axial vorticity in a horizontal plane. This situation reflects the intransitivity phenomenon, inherent to rotating flows in cavities, with the coexistence of different stable flow structures, e.g., different dominant azimuthal wave numbers, under the same imposed external conditions (see also Figure 16.3). The simulation predicts a 3SV regime. The flow exhibits a chaotic behavior induced by the random presence of small-scale fluctuations over an almost regular arrangement of waves at midheight ($z/d = 0.5$), as can be seen in Figure 16.7. Such a flow structure was already observed in a rotating cavity under symmetrical boundary conditions to ensure mass conservation due to the antisymmetry of the flow with respect to the midheight [*Randriamampianina et al.*, 2001]. In particular, it is clearly visible in Figure 16.7 from the two instantaneous computed streaklines at different heights $z/d = 0.25$ and $z/d = 0.75$ located symmetrically with respect to the midheight, the onset and the growth of additional structures randomly induce the breakdown of the regularity of the large-scale baroclinic waves. This phenomenon, associated with the loss of symmetry of waves, known to be characteristic of the structural vacillation regime, ultimately leads to the fully disordered flow. The IGWs mentioned above at lower rotation rate values, $\Omega = 0.35$ rad/s and $\Omega = 0.5125$ rad/s, are found to be the small-scale fluctuations responsible for the chaotic behavior of this SV flow, in contrast with the air-filled cavity where the mechanism of transition resulted from radial buoyancy in a Rayleigh-Bénard-like rotating flow [*Read et al.*, 2008]. In the latter case the centrifugal acceleration was greater than the gravity everywhere inside the cavity, giving a local Froude number $\text{Fr}_r \equiv \Omega^2 r^*/g > 1$, $r^* \in [a, b]$, while in the present computations, $\text{Fr} = 1.67 \times 10^{-2}$ (see Table 16.1).

We have compared the instantaneous contours of the azimuthal velocity and of the axial component of the vorticity between the available experimental measurements [*Wordsworth*, 2009] and the computed solutions at midheight in Figure 16.8. We note the overall good agreement

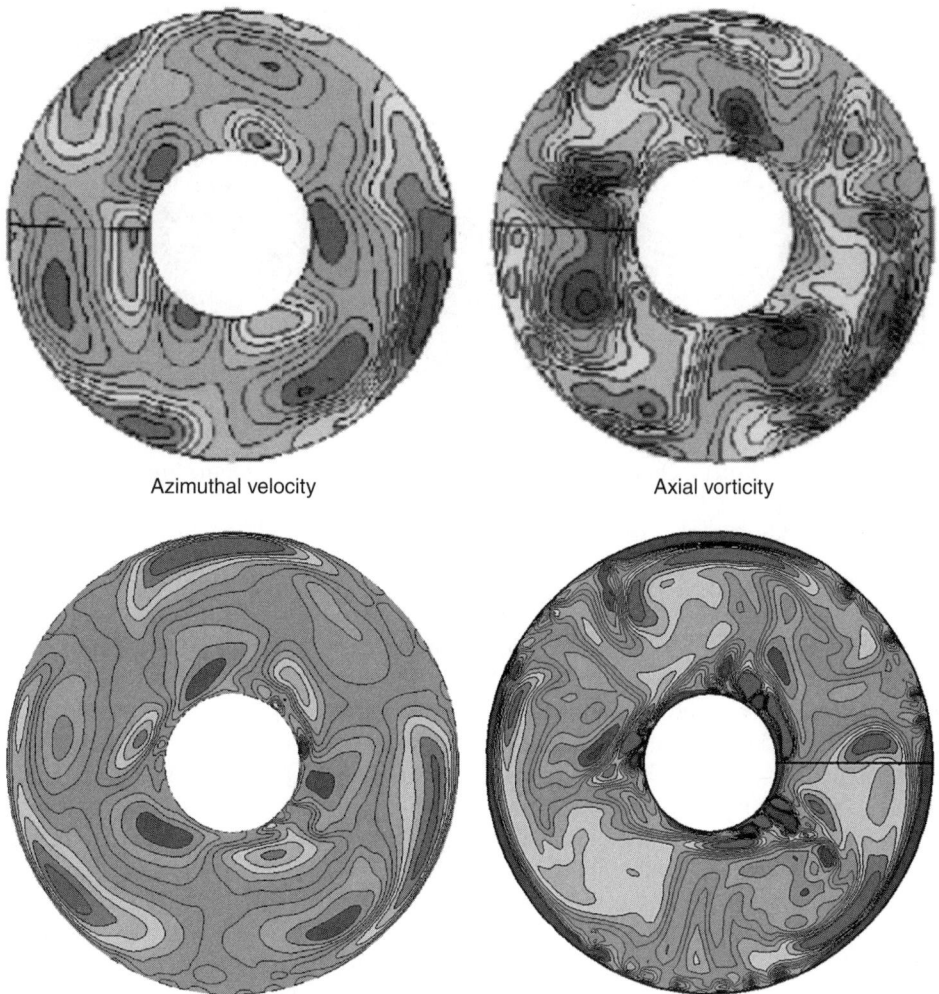

Figure 16.8. Comparison of flow characteristics at midheight between experimental measurements (top) and computed solutions (bottom) for the structural vacillation regime in the liquid-filled cavity, Pr = 16. For color detail, please see color plate section.

of the flow patterns obtained with the two approaches, in particular the similarity of large-scale structures, characterized by the same dominant azimuthal wave number $m = 3$. The slight discrepancy may result from the different isovalues chosen on contours by each approach. We note the presence of small-scale fluctuations initially developing along the inner cold wall, identified as IGWs and postulated to be the main mechanism responsible for the spatiotemporal chaotic behavior during the transition to turbulent flow regimes. Indeed, these small-scale features are expected to grow randomly in the whole cavity for the irregular waves. We refer to *Wordsworth et al.* [2008] for a detailed analysis of the experimental investigations.

16.3.2.4. Irregular Wave Regime. Direct numerical simulation was carried out to provide a first attempt to obtain the irregular wave regime in a baroclinic cavity and to compare the results with experimental data provided by *Wordsworth* [2009] (see also *Wordsworth et al.*

[2008]). The imposed external conditions represent a temperature difference between the two cylinders of $\Delta T = 2$ K with a rotation rate $\Omega = 3$ rad/s for the simulation, corresponding to $\Theta = 0.016$ and Ta $= 4.25 \times 10^8$, while for the experiment $\Omega = 3.9$ rad/s. The mesh used was $N \times M \times K = 256 \times 128 \times 512$ in the radial, axial, and azimuthal directions, respectively, with a dimensionless time step $\delta t = 0.000125$. The preliminary computed solution is compared with experimental measurements in Figure 16.9, showing the complex flow structure, where any dominant azimuthal wave number can be extracted as in the SV regime discussed above (see also the Figure 3 presented by *Wordsworth et al.* [2008]). The solution is obviously still far from its asymptotic state. However, this first result demonstrates the ability of the present numerical tool to compute the complex irregular waves in baroclinic cavities. The computed structures mimic very well the streak photographs illustrating irregular waves reported by *Hide and Mason* [1975] from their experimental studies in a

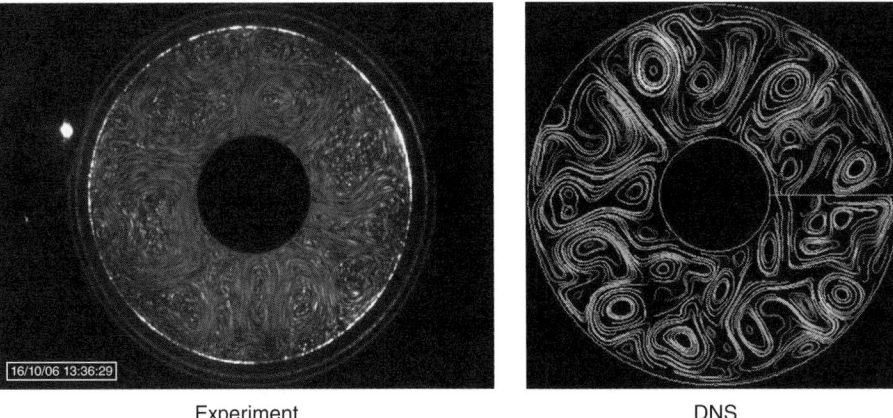

Experiment   DNS

Figure 16.9. Comparison between experimental visualization and computed solution for the irregular wave regime in the liquid-filled cavity, Pr = 16.

water-filled cavity. *Wordsworth et al.* [2008] reported a detailed analysis of the experimental investigations. The prohibitive CPU time cost resulting from an adequate mesh resolution required to resolve all the scales of the flow did not allow statistical convergence in a reasonable time. It also points out the actual limitations of computing power. An extension of the present numerical tool to parallel machines is in progress by implementing a domain decomposition technique, which should allow sufficient resolution to capture all the turbulence scales.

16.3.3. Open Free-Surface Water-Filled cavity: Pr = 7

Preliminary results are reported for an open upper free-surface cavity using water as working fluid, Pr = 7, and compared with available measurements from K. Alexandrov, Y. Wang, U. Harlander, and C. Egbers (*DFG-MetStroem reference experiment, BTU, Cottbus, Germany*). The geometric configuration used in the simulation is the one considered above in the liquid-filled cavity, with an inner radius $a = 4.5$ cm, outer radius $b = 15$ cm, and height $d = 26$ cm, while $a = 4.5$ cm, $b = 12$ cm, and $d = 13.5$ cm in the experimental setup [*Harlander et al.*, 2011]. Comparison concerns the temperature distribution at the upper free surface for a regular wave flow. It is recognized that the occurrence of baroclinic instability does not depend on the presence or absence of a top lid [*Hide and Mason*, 1978], even though significant differences can be observed. The open free surface, by inducing a shear flow, triggered by the use of Neumann conditions at this boundary in the numerical simulation, yields a motion at this top height with the associated temperature behavior. On the other hand, in the presence of a rigid lid, a part of the upper region remains at the same temperature as the outer hot wall, as was observed in the liquid-filled cavity with Pr = 16

[*Randriamampianina*, 2013]. Moreover, at low rotation rate for the later fluid in the weak wave regime, baroclinic instability was found to develop first near the bottom wall due to higher density gradient level toward this region (see Figure 16.5).

We display in Figure 16.10 the instantaneous temperature field at the open upper free surface obtained from simulation and infrared thermography. As different external conditions were used for the simulation ($\Delta T = 2$ K, $\Omega = 0.8$ rad/s) and for the experiment with $\Delta T = 8$ K, $\Omega = 0.47$ rad/s, the figure is presented to show qualitative behavior for a three-wave flow between the two approaches. It appears that the computed result did not yet reach the asymptotic state, as revealed by the asymmetry of the waves, in comparison with the measurements. *Harlander et al.* [2012] mentioned from their experimental investigations that a complete cycle can take about 27 revolutions. However, an overall agreement can be observed, in particular a pronounced incursion of "hot plumes" recirculating from the external hot cylinder toward the center of the cavity along the anticyclonic vortices, associated with a strong acceleration as seen from the superimposed velocity field. Such wave asymmetry can also be attributed to wave interaction during the establishment of baroclinic instability, as reported by *Harlander et al.* [2012].

16.4. CONCLUSIONS

Direct numerical simulations based on high-resolution pseudospectral methods were carried out for the investigation of the complex flow regimes occurring in a differentially heated rotating cylindrical annulus, the baroclinic cavity. The computed solutions have been compared with available laboratory measurements in configurations having an insulating top lid or an open free

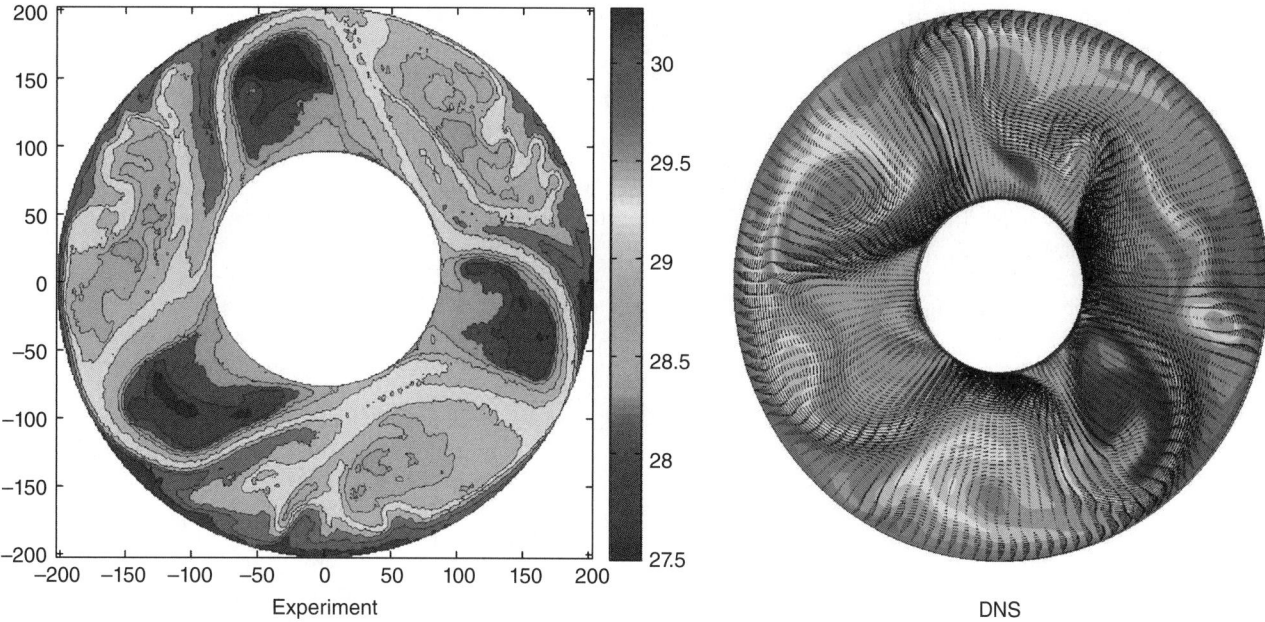

Figure 16.10. Instantaneous temperature field at the open upper free surface for the water-filled cavity. For color detail, please see color plate section.

surface. The approach was applied to describe the various spatiotemporal characteristics of baroclinic waves using three different fluids. It has allowed for a detailed analysis of the features observed during laboratory measurements. The results demonstrate the ability of the present numerical tool to reproduce the complex spatiotemporal behaviors and to capture the small-scale fluctuations responsible for the break of the regular waves to chaotic motion in baroclinic cavities. The simulations report the first realistic solutions of weak waves and structural vacillation regimes for liquids and amplitude vacillation for air, in agreement with experimental observations. Moreover, the computations point out the important role played by the Prandtl number on the baroclinic instability characteristics due to the differences in the thermal stratification levels. In particular, at high values of the Prandtl number, the spontaneous generation of IGWs was found simultaneously with the large-scale baroclinic waves. These small-scale features, initially developing along the inner cold cylinder, are postulated to be the mechanism responsible for the transition to irregular flows.

The extension of the approach to parallel computing, by implementing a domain decomposition technique, is in progress and is expected to allow for a direct numerical simulation of the fully developed turbulent flow regime in the baroclinic cavity.

Acknowledgments. The authors would like to acknowledge fruitful discussions with T. Jacoby and P. L. Read (AOPP, Oxford, UK), W.-G. Früh (SEPS, Heriot-Watt University, Edinburgh, UK), and R. Plougonven (LMD, Paris, France). The authors are indebted to R. Wordsworth (AOPP, Oxford, UK), U. Harlander, and C. Egbers (BTU, Cottbus, Germany) for their help in providing experimental measurements. The authors are grateful to the Spanish government for financial support, research project number FIS2011-24642. This work was granted access to the HPC resources NEC-SX8 of CCRT (CEA, France) and NEC-SX5 of IDRIS (CNRS, France) under the allocation 21444 made by GENCI (Grand Equipement National de Calcul Intensif). The authors are grateful to the anonymous referees for their constructive criticisms and suggestions.

REFERENCES

Alexandrov, K., Y. Wang, U. Harlander, and C. Egbers, DFG-MetStroem reference experiment, BTU, Cottbus, Germany.

Canuto, C., M. Hussaini, A. Quarteroni, and T. Zang (1987), *Spectral Methods in fluid Dynamics*, Springer-Verlag, Berlin.

Castrejón-Pita, A. A., and P. L. Read (2007), Baroclinic waves in an air-filled thermally driven rotating annulus, *Phys. Rev. E*, 75, 026,301, doi:10.1103/PhysRevE.75.026301.

Chaouche, A. M., A. Randriamampianina, and P. Bontoux (1990), A collocation method based on an influence matrix technique for axisymmetric flows in an annulus, *Comp. Meth. Appl. Mech. Eng.*, 80, 237–244.

Fein, J. S., and R. L. Pfeffer (1976), An experimental study of the effects of Prandtl number on thermal convection in a rotating, differentially heated cylindrical annulus of fluid, *J. Fluid Mech.*, 75, 81–112.

Fowlis, W. W., and R. Hide (1965), Thermal convection in a rotating annulus of liquid: Effect of viscosity on the transition between axisymmetric and non-axisymmetric flow regimes, *J. Atmos. Sci.*, *22*, 541–558.

Früh, W-G., and P. L. Read (1997), Wave interactions and the transition to chaos of baroclinic waves in a thermally driven rotating annulus, *Phil. Trans. R. Soc. Lond. A*, *355*, 101–153.

Gottlieb, D., and S. Orszag (1977), *Numerical Analysis of Spectral Methods: Theory and Applications*, CBMS-NSF Regional Conference Series in Applied Mathematics, SIAM, Philadelphia, Pennsylvania, USA.

Grebogi, C., E. Ott, and J. A. Yorke (1983), Crises, sudden changes in chaotic attractors, and transient chaos, *Phys. D*, *7*, 181–200.

Haldenwang, P., G. Labrosse, S. Abboudi, and M. Deville (1984), Chebyshev 3-d spectral and 2-d pseudospectral solvers for the Helmholtz equation, *J. Comput. Phys.*, *55*, 115–128.

Harlander, U., T. von Larcher, Y. Wang, and C. Egbers (2011), PIV- and LDV-measurements of baroclinic wave interactions in a thermally driven rotating annulus, *Exp. Fluids*, *51*, 37–49.

Harlander, U., J. Wenzel, K. Alexandrov, Y. Wang, and C. Egbers (2012), Simultaneous PIV and thermography measurements of partially blocked flow in a differentially heated rotating annulus, *Exp. Fluids*, *52*, 1077–1087.

Hide, R. (1958), An experimental study of thermal convection in a rotating fluid, *Philos. Trans. R. Soc. Lond.*, *A250*, 441–478.

Hide, R., and P. J. Mason (1975), Sloping convection in a rotating fluid, *Adv. Phys.*, *24*, 47–100.

Hide, R., and P. J. Mason (1978), On the transition between axisymmetric and non-axisymmetric flow in a rotating liquid annulus subject to a horizontal temperature gradient, *Geophys. Astrophys. Fluid Dyn.*, *10*, 121–156.

Hignett, P., A. A. White, R. D. Carter, W. D. N. Jackson, and R. M. Small (1985), A comparison of laboratory measurements and numerical simulations of baroclinic wave flows in a rotating cylindrical annulus, *Q. J. R. Met. Soc.*, *111*, 131–154.

Hugues, S., and A. Randriamampianina (1998), An improved projection scheme applied to pseudospectral methods for the incompressible Navier-Stokes equations, *Int. J. Numer. Meth. Fluids*, *28*, 501–521.

Jacoby, T. N. L., P. L. Read, P. D. Williams, and R. M. B. Young (2011), Generation of inertia-gravity waves in the rotating thermal annulus by a localised boundary layer instability, *Geophys. Astrophys. Fluid Dyn.*, *105*, 161–181.

James, I. N., P. R. Jonas, and L. A. Farnell (1981), A combined laboratory and numerical study of fully developed steady baroclinic waves in a cylindrical annulus, *Q. J. R. Met. Soc.*, *107*, 51–78.

Jonas, P. R. (1981), Some effects of boundary conditions and fluid properties on vacillation in thermally driven rotating flow in an annulus, *Geophys. Astrophys. Fluid Dyn.*, *18*, 1–23.

Koschmieder, E. L., and H. D. White (1981), Convection in a rotating, laterally heated annulus. the wave number transitions, *Geophy. Astrophys. Fluid Dyn.*, *18*, 279–299.

von Larcher, Th., and C. Egbers (2005), Experiments on transitions of baroclinic waves in a differentially heated rotating annulus, *Nonlin. Process. Geophys.*, *12*, 1033–1041.

Lewis, G. (2010), Mixed-mode solutions in an air-filled differentially heated rotating annulus, *Phys. D*, *239*, 1843–1854.

Newhouse, S. E., D. Ruelle, and F. Takens (1978), Occurrence of strange axiom a attractors near quasi-periodic flow on T^m, $m \geq 3$, *Commun. Math. Phys.*, *64*, 35–40.

Orlanski, I., and M. D. Cox (1973), Baroclinic instability in ocean currents, *Geophys. Fluid Dyn.*, *4*, 297–332.

Pierrehumbert, R. T., and K. L. Swanson (1995), Baroclinic instability, *Annu. Rev. Fluid Mech.*, *27*, 419–467.

Pulicani, J. P., E. Crespo Del Arco, A. Randriamampianina, P. Bontoux and R. Peyret (1990), Spectral simulations of oscillatory convection at low Prandtl number, *Int J. Numer. Meth. Fluids*, *10*, 481–517.

Randriamampianina, A. (2013), Inertia gravity waves characteristics within a baroclinic cavity, *Comptes Rendus Mécanique*. *341*, 547–552.

Randriamampianina, A., E. Crespo Del Arco, J. P. Fontaine, and P. Bontoux (1990), Spectral methods for two-dimensional time-dependent $Pr \to 0$ convection, *Notes Numer. Fluid Mech.*, *27*, 244–255.

Randriamampianina, A., E. Leonardi, and P. Bontoux (1997), A numerical study of the effects of Coriolis and centrifugal forces on buoyancy driven flows in a vertical rotating annulus, in *Advances in Computational Heat Transfer*, edited by G. De Vahl Davis and E. Leonardi, Begell House, Cesme, Turkey.

Randriamampianina, A., R. Schiestel, and M. Wilson (2001), Spatio-temporal behaviour in an enclosed corotating disk pair, *J. Fluid Mech.*, *434*, 39–64.

Randriamampianina, A., R. Schiestel, and M. Wilson (2004), The turbulent flow in an enclosed corotating disk pair: Axisymmetric numerical simulation and Reynolds stress modelling, *Int. J. Heat Fluid Flow*, *25*, 897–914.

Randriamampianina, A., W.-G. Früh, P. Maubert, and P. L. Read (2006), DNS of bifurcations in an air-filled rotating baroclinic annulus, *J. Fluid Mech.*, *561*, 359–389.

Raspo, I., S. Hugues, E. Serre, A. Randriamampianina, and P. Bontoux (2002), A spectral projection method for the simulation of complex three-dimensional rotating flows, *Comput. Fluids*, *31*, 745–767.

Read, P. L. (1992), Applications of singular systems analysis to baroclinic chaos, *Phys. D*, *58*, 455–468.

Read, P. L. (2001), Transition to geostrophic turbulence in the laboratory, and as a paradigm in atmospheres and oceans, *Surv. Geophys.*, *22*, 265–317.

Read, P. L., M. Collins, W.-G. Früh, S. R. Lewis, and A. F. Lovegrove (1998), Wave interactions and baroclinic chaos: A paradigm for long timescale variability in planetary atmospheres, *Chaos Solitons Fractals*, *9*, 231–249.

Read, P. L., P. Maubert, A. Randriamampianina, and W.-G. Früh (2008), DNS of transitions towards Structural Vacillation in an air-filled, rotating, baroclinic annulus, *Phys. Fluid*, *20*, 044,107.

Vanel, J. M., R. Peyret, and P. Bontoux (1986), A pseudospectral solution of vorticity-streamfunction equations using the influence matrix technique, in *Numerical Methods in Fluid Dynamics II*, edited by K. W. Morton and M. J. Baines, pp. 463–475, Clarendon Press, Oxford.

Williams, G. P. (1969), Numerical integration of the three-dimensional Navier-Stokes equations for incompressible flow, *J. Fluid Mech.*, *37*, 727–750.

Williams, G. P. (1971), Baroclinic annulus waves, *J. Fluid Mech.*, *49*, 417–449.

Wordsworth, R. D. (2009), Theoretical and experimental investigations of turbulent jet formation in planetary fluid dynamics, Ph.D. thesis, Linacre College, Oxford United Kingdom.

Wordswoth, R. D., P. L., Read, and Y. H. Yamazaki (2008), Turbulence, waves, and jets in a differentially heated rotating annulus experiment, *Phys. Fluid.*, *20*, 126,602.

Zang, T. A. (1990), Spectral methods for simulations of transition and turbulence, *Comp. Meth. Appl. Mech. Eng.*, *80*, 209–221.

Zeytounian, R. Kh. (2003), Joseph Boussinesq and his approximation: A contemporary review, *Comptes Rendus Mécanique*, *331*, 575–586.

17

Orthogonal Decomposition Methods to Analyze PIV, LDV, and Thermography Data of Thermally Driven Rotating Annulus Laboratory Experiments

Uwe Harlander[1], Thomas von Larcher[2], Grady B. Wright[3], Michael Hoff[4], Kiril Alexandrov[1], and Christoph Egbers[1]

17.1. INTRODUCTION

Already in the 1950s, an elegant laboratory experiment had been designed to understand how the atmospheric circulation transports heat from equatorial to polar latitudes (cf. the pioneering studies described by *Hide* [1958, 2010]). It consists of a cooled inner and heated outer cylinder mounted on a rotating platform, mimicking the heated tropical and cooled polar regions of Earth's atmosphere. Depending on the strength of the heating and the rate of rotation, different flow regimes had been identified in the gap: the zonal flow regime, wave regimes that can be classified by propagating waves of different wave numbers, and quasi-chaotic regimes where waves and small-scale vortices coexist.

The baroclinic annulus experiment, often called *the differentially heated rotating annulus of fluid*, has been accepted as a suitable laboratory model for the midlatitude large-scale flow in Earth's atmosphere. For example, *Fultz* [1961] and *Lorenz* [1964] used the heated rotating annulus as an analogy to the complex dynamics of the large-scale weather when they discussed problems related to climate variability.

Obviously, large-scale environmental flows and the flows observed in the rotating annulus show agreement on fundamental features. A large part of this agreement is owed to the baroclinic instability mechanisms that govern atmospheric and laboratory flows [*Pierrehumbert and Swanson*, 1995]. This fact makes the heated rotating annulus an inspiring experiment for the community of geophysical fluid dynamics, even in the computer age.

Baroclinic instability has been investigated in the annulus not only in numerous experimental studies but also theoretically [*Lorenz*, 1962] and numerically [*Miller and Gall*, 1983; *Lewis and Nagata*, 2004; *Randriamampianina et al.*, 2006; *von Larcher et al.*, 2013; *von Larcher and Dörnbrack*, 2014]. The two references mentioned last are discussed in more detail in chapters 2 and 16 of the present book.

Due to its relative simple geometry as well as to the well-definable forcing parameters, the rotating annulus is still of particular interest not only for research with respect to atmospheric sciences [*Gyüre et al.*, 2007; *Ravela et al.*, 2010] but also in the development of computational fluid dynamics (CFD) models where the annulus data can be used as reference for the validation of new numerical concepts. In this context it is worth noting that the rotating annulus experiment described here is a reference experiment within the German priority program Multiple Scales in Fluid Mechanics and Meteorology (MetStröm) that focuses on the development of model- as well as grid-adaptive numerical simulation concepts in multidisciplinary projects (see http://metstroem.mi.fu-berlin.de).

The flow regime that develops in the cylindrical gap of the annulus depends on the radial temperature gradient between the inner and the outer cylinder, ΔT, and on the rotation rate of the apparatus, Ω. Thus, a 2D parameter space (called regime diagram) spanned by the Taylor number ($\propto \Omega^2$) and by the thermal Rossby number

---

[1] Department of Aerodynamics and Fluid Mechanics, Brandenburg University of Technology (BTU) Cottbus-Senftenberg, Germany.
[2] Institute for Mathematics, Freie Universität Berlin, Berlin, Germany.
[3] Department of Mathematics, Boise State University, Boise, Idaho, United States of America.
[4] Leipzig Institute for Meteorology, University of Leipzig, Leipzig, Germany.

Modeling Atmospheric and Oceanic Flows: Insights from Laboratory Experiments and Numerical Simulations, First Edition. Edited by Thomas von Larcher and Paul D. Williams.
© 2015 American Geophysical Union. Published 2015 by John Wiley & Sons, Inc.

($\propto (\Delta T \ \Omega^{-2})$) includes all flow regimes. Such regime diagrams are the basis of many studies and they have been experimentally derived already by *Fowlis and Hide* [1965] and have been refined later by other authors e.g. [*Früh and Read*, 1997; *von Larcher and Egbers*, 2005a]. The range of azimuthal wave numbers m is restricted by the dimensions of the gap. *Hide and Mason* [1970] found an empirical law for the minimum and maximum wave number, $m_{min} \leq m \leq m_{max}$, known as the Hide criterion reading,

$$\frac{\pi}{4}\frac{b+a}{b-a} \leq m \leq \frac{3\pi}{4}\frac{b+a}{b-a}, \quad (17.1)$$

with a (b) as the inner (outer) radius of the gap.

One of the most fascinating aspects of the differentially heated rotating annulus is its rich time-dependent flow behavior. It is therefore not surprising that many studies have focused on this aspect. A phenomenon that has attracted much attention over many years is the so-called amplitude and structural vacillation, which is a modulation of the amplitude and the wave shape in distinct subregions of the regime diagram mentioned above. Wave dispersion and structural vacillation have been observed by e.g. *Pfeffer* and *Fowlis* [1968] using streak photographs and by *Harlander et al.* [2011] by particle image velocimetry (PIV). They showed the simultaneous presence of two subsequent wave modes and argued that some part of the vacillation might result from the different phase speeds of the two modes (see also *Yang* [1990]).

However, wave dispersion cannot explain the existence of multiple wave modes during a traverse of the regular wave regime. Therefore, *Lindzen et al.* [1982] numerically investigated a nonlinear version of Eady's baroclinic instability problem for the annulus, and *Barcilon and Drazin* [1984] investigated the problem by asymptotic techniques. In both studies, regions in the regime diagram could be identified where two modes with the same wave number may grow. Later, *Früh* [1996] and *Früh and Read* [1997] suggested that resonant wave triads are responsible for certain amplitude vacillations (see also the review on amplitude vacillations in this book in chapter 3). Such triads, besides the dominant mode, involve two other, weaker modes. Energy is redistributed between the members of a triad, and the dominant pattern vacillates with a characteristic time. Geostrophic turbulence, i.e., the irregular flow regime, is generally found at high rotation rates [*Morita and Uryu*, 1989; *Read et al.*, 1992; *Pfeffer et al.*, 1997].

The nonlinear behavior of the annulus flow motivated a number of contributions using nonlinear time series analysis to better understand the physical mechanisms. *Read et al.* [1992] and *Früh and Read* [1997], for example, used time series of temperature from probes in the fluid interior. In contrast, *Sitte and Egbers* [2000] and *von Larcher and Egbers* [2005a] used velocity time series that have been acquired by the optical laser Doppler velocimetry (LDV).

On the other hand, also linear, multivariate statistical techniques have been successfully applied to highlight certain aspects of the motion in the rotating annulus. [*Read*, 1993], e.g., used multivariate singular system analysis (MSSA) for phase portrait reconstructions of the annulus flow. Complex empirical orthogonal function (EOF) analysis have been applied to data from a rotating annulus with bottom topography [*Pfeffer et al.*, 1990]. The focus of this work was to identify features of the wave propagation as a function of the Taylor number. *Mundt and Hart* [1994] constructed a reduced low-dimensional model of two-layer baroclinic instability by projecting the governing equations onto the EOFs of numerical flow simulations. The same should be possible by using EOFs from annulus laboratory data [*Stephen et al.*, 1997, 1999]. Finally, *Read et al.* [2008] used EOFs deduced from numerical simulations to identify structural changes of the dominant modes in the annulus when the Taylor number is increased.

The differentially heated rotating annulus has also been used as a test bed for studies on weather predictability [*Young and Read*, 2008; *Ravela et al.*, 2010]. Young and Read studied the breakdown of predictability for numerically deduced irregular flow regimes. In this context breeding vectors (close relatives to singular vectors) play an important role. Such vectors are an orthogonal decomposition for flows with *nonorthogonal* eigenmodes.

The present chapter is organized as follows. In Section 17.2 we will give details on the experimental apparatus we use and the governing nondimensional parameters. Then, in section 17.3 we will present a summary of laboratory studies on annulus flows we performed over the previous few years. In particular, we describe the multivariate orthogonal decomposition techniques we applied to the laboratory data. In Section 17.3.1 we analyze PIV and LDV data at the transition between two different wave regimes by applying the complex EOF analysis and MSSA. Subsequently, in Section 17.3.2 we analyze data from an annulus with a broken azimuthal symmetry. Similar to *Pfeffer et al.* [1990], we are interested in the wave propagation characteristics in a rotating annulus with "topography". The data have been retrieved simultaneously by thermography and PIV measurements. Complex EOF analysis is able to decompose the flow into features typical for the flow up- and downstream of the annulus constriction. This study was motivated by specific large-scale ocean currents like the Antarctic Circumpolar Current where the "gap width" of the flow depends on longitude. In Section 17.3.3 we decompose surface temperature data of the annulus flow in principal oscillation patterns (POPs), that is, the linear eigenmodes, and in modes of maximal growth, called singular vectors (SVs). In contrast to the traditional approach, we deduce these modes from the data alone without using a linear model operator. Finally, in Section 17.3.4 we decompose the annulus flow in a purely rotational and a purely divergent part. This decomposition is based on radial basis functions (RBFs) and it might prove useful in discriminating different wave types in the flow. We close the chapter with Section 17.4, where we summarize our results and provide an outlook on future work.

17.2. EXPERIMENTAL SETUP, PARAMETERS, AND FLOW REGIMES

17.2.1. Setup

Our setup (Figure 17.1), described in more detail by *von Larcher and Egbers* [2005b], consists of a tank with three concentric cylinders mounted on a turntable that rotates around its vertical axis of symmetry. The inner cylinder is made of anodized aluminum; the middle and outer ones are made of borosilicate glass. The temperature of the inner and outer cylinders and the rotation rate of the apparatus are controlled by the experiment software which is programmed in LabVIEW®. Temperature sensors are part of the inner and outer side walls: at the inner wall at one azimuthal position in two different heights, at the outer wall at four equidistant positions in midheight of the annulus. The radial temperature difference between the outer and inner walls is realized by heating the fluid in the outer annulus-shaped chamber by a heating coil and by cooling the fluid in the inner cylinder by using a thermostat. Deionized water is used as working fluid in all experiments.

All experiments were conducted in the classical f-plane configuration though a sloping bottom could easily be inserted and then β-plane experiments could also be performed. Furthermore, the surface is free rather than a rigid lid. We refer the reader to *Fein* [1973] for a comparison of experiments with a free surface and a rigid lid and to *Mason* [1975] for details of the influence of a sloping bottom on the flow regimes. In cases of experiments on baroclinic channel flows with narrows, a barrier was inserted in the gap, as described in Section 17.3.2.

By keeping the temperature gradient fixed but varying the rotation rate of the apparatus, sequences of regime transitions can be observed. Once the radial temperature gradient is settled, the spin-up time is found to be less than 30 min for steady waves and up to 40 min for complex flows. Observations were usually done up to several hours per parameter point, extended partly in complex flow regimes.

17.2.2. Parameters

The shape of the annulus is defined by the radius ratio η and the aspect ratio Γ with

$$\eta = \frac{a}{b}, \qquad \Gamma = \frac{d}{b-a}, \qquad (17.2)$$

where the inner radius $a = 45$ mm, the outer radius $b = 120$ mm, and the fluid depth $d = 135$ mm (implying $\eta = 0.38$ and $\Gamma = 1.8$ for our apparatus).

Beyond these geometric parameters, the fluid motion is governed by the two dynamic control parameters, the rotation rate of the annulus, Ω, and the radial temperature difference in the cylindrical gap, ΔT. These parameters determine the nondimensional Taylor number Ta and thermal Rossby number Ro, as already mentioned above. The two numbers read

$$\text{Ta} = \frac{4\Omega^2 (b-a)^5}{v^2 d}, \qquad \text{Ro} = \frac{g\, d\alpha\, \Delta T}{\Omega^2 (b-a)^2}, \qquad (17.3)$$

where v is the kinematic viscosity, g the acceleration due to gravity, and α the volumetric expansion coefficient. The Taylor number measures the rotation rate with respect to the viscous effect, and the thermal Rossby number corresponds to the ratio of buoyancy and Coriolis terms and therefore indicates a thermal stratification of the flow.

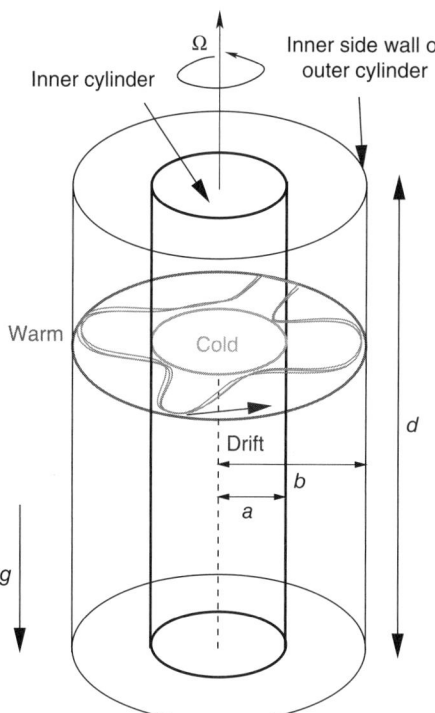

Figure 17.1. Sketch of the rotating annulus with illustration of a typical large-scale jet stream of wave number $m = 4$ that has a drift relative to the rotating reference system.

Another important parameter is the Prandtl number, defined by

$$\Pr = \frac{\nu}{\kappa}, \quad (17.4)$$

where κ is thermal conductivity and Pr describes the physical properties of the fluid, with Pr = 7.16 as the appropriate value for water at 20°C (see chapter 2 this volume and *Fowlis and Hide* [1965] for more details on the effect of the Prandtl number).

17.2.3. Flow regimes

By keeping $\Delta T = 7.5$ K constant and varying Ω in the range $3 \times 10^6 < \text{Ta} < 3 \times 10^8$, the regime diagram typically shows specific flow regimes characterized by the azimuthal wave number $0 \leq m \leq 4$, i.e., the axisymmetric $m = 0$ basic flow at $3 \times 10^6 < \text{Ta} < 6 \times 10^6$, the wave flow regime with $2 \leq m \leq 3$ at about $6 \times 10^6 < \text{Ta} < 1.5 \times 10^8$, and the regime with structural vacillations (called irregular wave regime when the wave structure is deformed in an irregular way) with dominant wave $m = 4$ at higher Taylor numbers (see Figure 17.2). Note, that $m_{max} = 4$ is in good agreement with the Hide criterion (eq. 17.1). Hysteresis occurs when one can find different regimes for the same control parameters where the parameters have been approached from different directions that is an increasing or decreasing Taylor number. Hysteresis is a well-known phenomenon for annulus experiments and can generically be observed in free-surface [*Sitte and Egbers*, 2000; *von Larcher and Egbers*, 2005b] and rigid lid experiments [*Cole*, 1971; *Hignett et al.*, 1985]. For our experimental setup a rather broad region of hysteresis exists at the transition between $m = 3$ and $m = 4$. The existence of hysteresis depends not only on Ta, Ro, and Pr but also on η and Γ [*von Larcher and Egbers*, 2005b].

Moreover, complex vacillating flow patterns are found in the transition from axisymmetric flow to the wavy flow regime (see Figure 17.3, upper row), where mode interaction between mode $m = 2$ and $m = 3$ occurs. Furthermore, structural (or shape) vacillations are found at high Taylor numbers (Figure 17.3, lower row), where cold cells separate from the inner wall, move radially outward, and then return to the inner cylinder while their outer boundary largely remains unaffected.

While the complex flow patterns in the first transition zone might be identified as a superposition of two coexisting waves with different zonal wave numbers and phase speeds, denoted as *interference vacillation* (IV), which was found to occur in experiments with a free surface [*Pfeffer and Fowlis*, 1968; *Kaiser*, 1970; *Harlander et al.*, 2011] as in rigid-lid experiments [*Früh and Read*, 1997], the structural vacillations observed at higher Taylor numbers could be consistent with an oscillation of a higher radial mode of the same azimuthal wave as observed in experiments by *Pfeffer et al.* [1980] and described theoretically by *Weng et al.* [1986].

17.3. RESULTS

Multivariate statistical techniques are suitable to understand better the variability of the heated rotating flow. We consider here four different methods: EOF analysis, MSSA, POP analysis, and SV analysis. It is worth to briefly mention the field of application of the different techniques. This enables the reader to assess which technique is good for what purpose. For many applications it is useful to decompose noisy multivariate data sets into subsets or subspaces. A few patterns might span the "signal subspace" where the noise is captured in the "noise subspace." The specification of relevant patterns can be done in many different ways, ranging from eigenmodes (e.g., in terms of Bessel and trigonometric functions for cylindrical geometry) to patterns that optimize certain statistical moments [*H. von Storch*, 1995].

EOFs are defined as those patterns that are powerful in explaining variance and thus the EOF method is the method of choice for analyzing the variability of fields. It is therefore widely used in the geosciences [*Lorenz*, 1956; *von Storch and Zwiers*, 1999]. This method also goes by different names, e.g., principal component analysis or proper orthogonal decomposition. The EOF method is able to find the spatial patterns of variability and their time variation and provides a measure for the relevance of each pattern. Simply speaking, the EOF method breaks the data into modes of variability that might (as is the case for our data) be interpreted as physical modes of the system. Strikingly, it can be shown that the EOF method provides the most efficient way of capturing the dominant components of a high-dimensional process with often surprisingly few modes [*Holmes et al.*, 1996]. However, more appropriate in our context is the use of the complex EOF (CEOF) analysis [*Pfeffer et al.*, 1990; *von Storch*

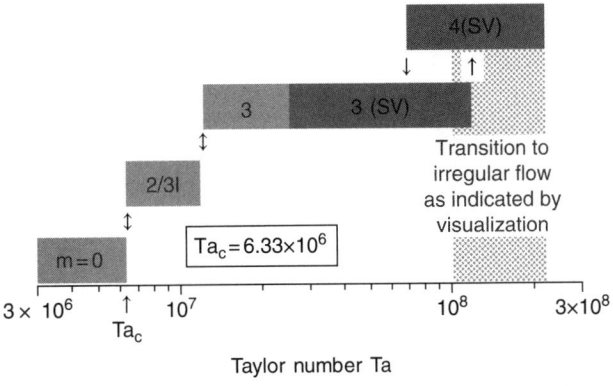

Figure 17.2. Regime diagram observed for $\eta = 0.38$, $\Gamma = 1.8$, and $\Delta T = 7.5$ K [*von Larcher and Egbers*, 2005b].

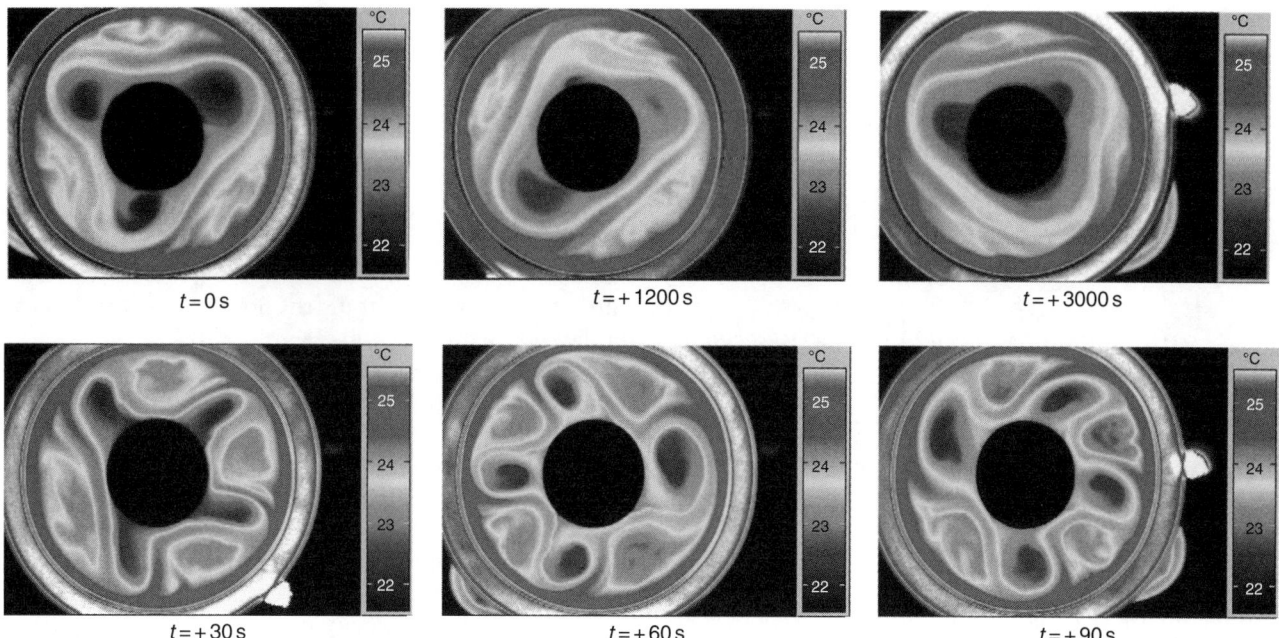

Figure 17.3. Sequence of thermographic measurements describing vacillating flows [*von Larcher and Egbers*, 2005b]. Top: Wave mode competition of wave number $m = 2$ and $m = 3$ at Ta $= 1.08 \times 10^7$, Ro $= 3.00$. Bottom: Structural vacillation flow, radial oscillation of a wavy flow of wave number $m = 4$ at Ta $= 7.65 \times 10^7$, Ro $= 0.41$, with t as relative time.

and Zwiers, 1999] that reveals propagating patterns of variability by single CEOFs, whereas pure EOFs capture only standing modes of variability.

The MSSA is a generalization of the single-time-series SSA method to multiple time series [*Broomhead and King*, 1986; *Read*, 1992; *Vautard*, 1995]. These time series may contain observations of a certain variable at different locations (as in our case) or even observations of different variables. In classical EOF analysis the dominant spatial patterns are captured by diagonalizing the covariance matrix. As discussed *Vautard* [1995], the coordinates of the state vector in the EOF analysis represent different locations in space at the same time. In an SSA, the state vector contains values at the same locations but at different time lags. CEOF analysis is a special case of the MSSA method; however, MSSA deals with more temporal degrees of freedom than spatial ones, allowing the investigation of spectral properties of the data. In contrast, CEOF contains only a single lag but a large number of spatial points. The MSSA method allows one to detect oscillating features in noisy time series where oscillations occur frequently only during certain time periods. The larger generality of the MSSA is purchased by a larger amount of computing time. A detailed description of the MSSA method is beyond the scope of the present chapter but is given by *Dettinger et al.* [1995], *Vautard* [1995], and *Elsner and Tsonis* [1996].

POPs are empirical, that is, data estimated normal modes [*Hasselmann*, 1988]. POPs are another way of decomposing a data set into a signal and noise subspace. To evaluate POPs, the system matrix corresponding to a linear model has to be found as described, e.g., by *von Storch and Zwiers* [1999]. Frequently, POPs correspond with EOFs, though this correspondence is not guaranteed from a mathematical point of view. Also the correspondence between true normal modes and POPs is not always obvious [*J.-S. von Storch*, 1995]. In real data, linearly unstable modes occur only in a nonlinearly saturated state. Thus POP modes are either neutral or damped. In contrast, a linear operator might allow for unstable modes that cannot be covered by any data based method. Nevertheless, POP analysis has proven to be useful in a broad range of applications and can be considered as one of the routine tools in climate research. From the empirically estimated system matrix, not only POPs can be computed. A further useful step is to estimate SV, from the system matrix. SVs correspond to those initial perturbations that grow in an optimal sense with respect to a chosen norm within a predefined time interval, the so-called optimization time. For large-scale baroclinic systems, SVs might play an important role and they might be even more relevant for real flows than unstable normal modes [*Badger and Hoskins*, 2001].

All the four mentioned orthogonal decompositions (EOF analysis, MSSA, POP analysis, SV analysis) are related via the data matrix. Let us briefly describe how. A variable X_i is observed at M different arbitrarily spaced points, $i = 1, 2, \ldots, M$, and at P different instances of time. Note that $t_{j+1} = t_j + j\Delta$, where Δ is the sampling interval, and that the first measurement was done at time t_1 and the last at time $t_P = t_N + n\Delta$, where n is an integer number. Subtracting the mean from each time series $x_i = X_i - \bar{X}$, we can put the data into a generalized $N \times M$ data matrix

$$\mathbf{F}^{n\Delta} = \begin{pmatrix} x_1(t_1 + n\Delta) & x_2(t_1 + n\Delta) & \cdots & x_M(t_1 + n\Delta) \\ x_1(t_2 + n\Delta) & x_2(t_2 + n\Delta) & \cdots & x_M(t_2 + n\Delta) \\ \vdots & \vdots & \ddots & \vdots \\ x_1(t_N + n\Delta) & x_2(t_N + n\Delta) & \cdots & x_M(t_N + n\Delta) \end{pmatrix},$$

where $n\Delta$, $n = 0, 1, 2, \ldots$, defines a time delay. An extended matrix F_m can be defined as

$$\mathbf{F}_m = \left(\mathbf{F}^{\Delta}, \ \mathbf{F}^{2\Delta}, \ \mathbf{F}^{3\Delta}, \ \ldots, \ \mathbf{F}^{m\Delta} \right). \quad (17.5)$$

Note that $\mathbf{F}_m$ is an $N \times mM$ matrix formed by the m submatrices $\mathbf{F}^{\Delta}, \ldots, \mathbf{F}^{m\Delta}$. EOFs are the eigenvectors of $(\mathbf{F}^{0\Delta})^T \mathbf{F}^{0\Delta}$, space-time EOFs (MSSA modes) with time window m are the eigenvectors of $\mathbf{F}_m^T \mathbf{F}_m$, POPs are the eigenvectors of $\mathbf{P} = ((\mathbf{F}^{1\Delta})^T \mathbf{F}^{0\Delta})((\mathbf{F}^{0\Delta})^T \mathbf{F}^{0\Delta})^{-1}$, and singular vectors with optimization time Δ and based on the Euclidean norm are the eigenvectors of $\mathbf{P}^T \mathbf{P}$. All these eigenvectors form an orthogonal basis. Therefore, the data vector at time i can be written as

$$\vec{F} = \sum_j a(t_i)_j \vec{\Phi}_j,$$

where $\vec{\Phi}_j$ is the jth eigenvector from an EOF, MSSA, POP, or SV analysis, and the coefficients $a(t_i)_j$ are found by a suitable projection of the eigenvectors on the data vector.

In Section 17.3.4 we discuss a novel orthogonal decomposition that is not premised on a statistical basis. Here we use a mesh-free data reconstruction method that is based on RBFs. Using these basis functions, we can decompose the horizontal velocity data into a sum of divergence-free and curl-free parts. Such a decomposition can be very useful in discriminating different wavefields in the annulus.

17.3.1. EOF and MSSA Analysis of Wave Interactions With PIV and LDV Measurements

In this section, we present the analysis of velocity data from classical f-plane thermally driven rotating annulus experiments with $\eta = 0.38$, $\Gamma = 1.8$, and Pr = 7.16 (see (17.2) and (17.4)), recovered by PIV and LDV (see *Harlander et al.* [2011] for details). The following questions will be addressed: (i) Can the statistical analysis detect coexisting wave modes during a traverse between two regular wave regimes? (ii) Can we find coexisting modes in the transition region to the quasi-chaotic regime or is the flow dominated by random fluctuations?

The data we used are sampled twice (PIV) or 20 times (LDV) per revolution of the annulus and cover time periods sufficiently long to allow for the application of multivariate statistical techniques. While the PIV data consist of the horizontal flow field, the LDV data consist of 20 time series regularly distributed along a circle in the annulus at mid-radius, i.e. $(a + b)/2$ (see Figure 17.1). Note that the sampling rate of the LDV measurement is 10 times larger than the one of the PIV measurements. However, with respect to a fixed spatial point, the LDV samples the data just once per revolution. Still, short-lived structures typical for more transient flows are better resolved in the LDV data.

The PIV system is used to measure the horizontal velocity components 15 mm below the fluid surface. Each experiment lasts typically 50τ, where τ is the revolution period of the annulus. We sample the PIV data with $\tau/2$, i.e. two observations per revolution. The PIV camera is mounted in an inertial frame above the cylinder, that is, the camera does not corotate with the cylinder. To obtain the velocity components in the corotating frame, we subtract the solid-body velocity $\vec{v} = \vec{\Omega} \times \vec{r}$ from each observed PIV velocity rigid-field. A preprocessing of the data is needed to eliminate erroneous vectors and to homogenize the data.

The radial velocity component is measured with the LDV that was fixed in the inertial frame, too. The measurements take place 2 mm below the fluid surface at midradius of the annulus. The large data set is reduced by an appropriate averaging. Furthermore, linear interpolation is applied to obtain a homogeneous data set with regular grid distance $\Delta\Phi = 18°$ and $\Delta t = \tau/20$. These preprocessed LDV data are then analyzed using the MSSA software toolkit by *Dettinger et al.* [1995].

The MSSA software package is particularly suited to detect (intermittent) oscillations in noisy time series as well as in multivariate data. With regard to the aims of our study, this makes the method particularly suitable to find structures in the transition region to the quasi-chaotic regime where the waves become more and more irregular. In the specific experimental setup used here, this regime occurs when the Taylor number is larger than 10^8 and the thermal Rossby number is smaller than 0.5 (cf. Figure 17.2).

Owing to the fixed PIV camera, the errors of PIV observations grow with growing angular velocity of the cylinder. In contrast, using LDV, the radial velocity component can be observed even for large angular velocities of the annulus with high accuracy. Thus, LDV data from the irregular wave regime will be analyzed by using the MSSA method (question (ii)). Instead, the preprocessed PIV data were analyzed by using the CEOF method with the focus on question (i).

A detailed description of our approach to find coupled propagating patterns with the CEOF is given by *Harlander et al.* [2011]. Briefly speaking, we use the Hilbert transform method to make the (horizontal) velocity components (u, v) complex and then formed extended time series by combining the two complex time series. With this new time series we built the covariance matrix and computed its eigenvectors (i.e., the CEOFs) and the corresponding time-dependent coefficients.

17.3.1.1. Analysis of PIV Measurements.

As described, PIV measurements were performed to detect complex flows during the traverse between two regular wave regimes. The measurements presented here have been conducted at Ta = 1.74×10^7, Ro = 1.30 (i.e., $\Omega = 0.50$ rad/s, $\Delta T = 6.7$ K). That parameter point is close to the transition from the steady wave regime of wave number $m = 3$ to the structural vacillation (SV) regime, i.e., $m = 3$ (SV) (cf. Figure 17.2, but note that here $\Delta T = 7.5$ K and that for larger ΔT transitions occur at larger Ta).

The eigenvalue spectrum (not shown) is dominated by the first eigenvalue that contains more than 40% of the total variance of the flow, and the second eigenvalue includes about 10% of the total variance. It should be noted that the variance distribution depends on the data quality and on Ro and Ta. The lower row of Figure 17.4 shows the corresponding real part of the CEOFs and the upper row their time evolution. Note that, in general, the real and imaginary parts of the CEOFs and their time series show a 90° phase difference [*von Storch and Zwiers*, 1999].

The velocity field can be reconstructed via (17.5). CEOF1 together with the time-dependent coefficient determines a prograde propagating wave (i.e., a wave propagating in the direction of the annulus revolution) with wave number $m = 3$; CEOF2, in contrast, determines a rather regular and slowly retrograde wave (propagating in the opposite direction of the annulus revolution) with wave number $m = 4$. By combining the first two patterns (which then contain about 50% of the total variance), a wavy jet flow with dominant wave number $m = 3$ is found which shows slow vacillations due to an interference with the (weak) mode pattern with $m = 4$. Further details and also patterns for different Ta and Ro are discussed in *Harlander et al.* [2011].

From the time-dependent coefficients (Figure 17.4, top panel), the drift rates of the dominant mode $m = 3$ and of the weak mode $m = 4$ are found to be 0.021 and −0.007 rad/s, respectively. Slow retrograde propagating modes are rather exceptional but have been reported earlier [*Früh and Read*, 1997]. It appears that the propagation of the weak mode is strongly affected by the dominant mode of the system and that linear wave theory fails to describe its anomalous retrograde propagation, as all unstable baroclinic modes should propagate with the volume-averaged mean flow (which was estimated to be approximately 0.009 rad/s), according to the linear Eady model.

To summarize the main results of the CEOF analysis, we find that in the rather stable $m = 3$ regime the presence of higher modes and their linear interaction with the leading wave mode can give rise to slow modulations. However, the existence of the $m = 4$ mode in the $m = 3$ wave regime cannot be explained by linear theory.

17.3.1.2. Analysis of LDV Measurements.

Next we discuss the results from the MSSA of the LDV data. As mentioned above, a strong feature of the MSSA is its ability to detect oscillating/propagating features in noisy data. Thus, the MSSA seems to be more suitable than EOF analysis to find excited propagating modes in flow with structural vacillations and "irregular wave regimes." Note that we call a wave irregular when it shows significant transient features. In contrast, for a turbulent flow regime a dominant wave can no longer be observed. Here we apply the MSSA to a parameter point in the irregular flow regime, i.e., at Ta = 3.76×10^8, Ro = 0.14 ($\Omega = 2.32$ rad/s, $\Delta T = 6.9$ K).

Figure 17.5 (top panel) shows the preprocessed LDV data. The data are presented in the form of a space-time diagram, where the abscissa runs from 0 to 2π, covering the spatial structure of the radial velocity at midradius of the annulus. Although the flow is much more noisy than for the PIV experiment, we clearly can identify a wave pattern with $m = 4$ that propagates prograde with a phase speed of 0.011 rad/s.

The eigenvalue spectrum is broad (Figure 17.5, upper right part). Noise is usually part of the flat tail in the eigenvalue spectrum. Here, we define the noise level to be at 1% of the total variance. Note that this is a qualitative measure, and it is not the exact signal-to-noise level of our experiments. The first two eigenvalues explain 36% of the total variance and the next two eigenvalues are also clearly above the defined noise level.

Similar to the description of flow field reconstruction using the CEOF method, reconstructing the data by using just the first two space-time EOFs (ST-EOFs) gives a filtered version of the original data (Figure 17.5, bottom left). Instructive is the reconstruction by ST-EOF 3 and 4, explaining at least 5% of the total variance (Figure 17.5, bottom right). The reconstruction reveals a wave pattern with wave number $m = 5$ that propagates essentially with the same phase speed as the dominant $m = 4$ wave. Finally, note that the $m = 5$ wave pattern shows slight amplitude vacillations. Moreover, its phase speed is less constant than the phase speed of the dominant wave.

Roughly speaking, for the irregular flow regime, the first wave modes seem to be less dispersive than for the

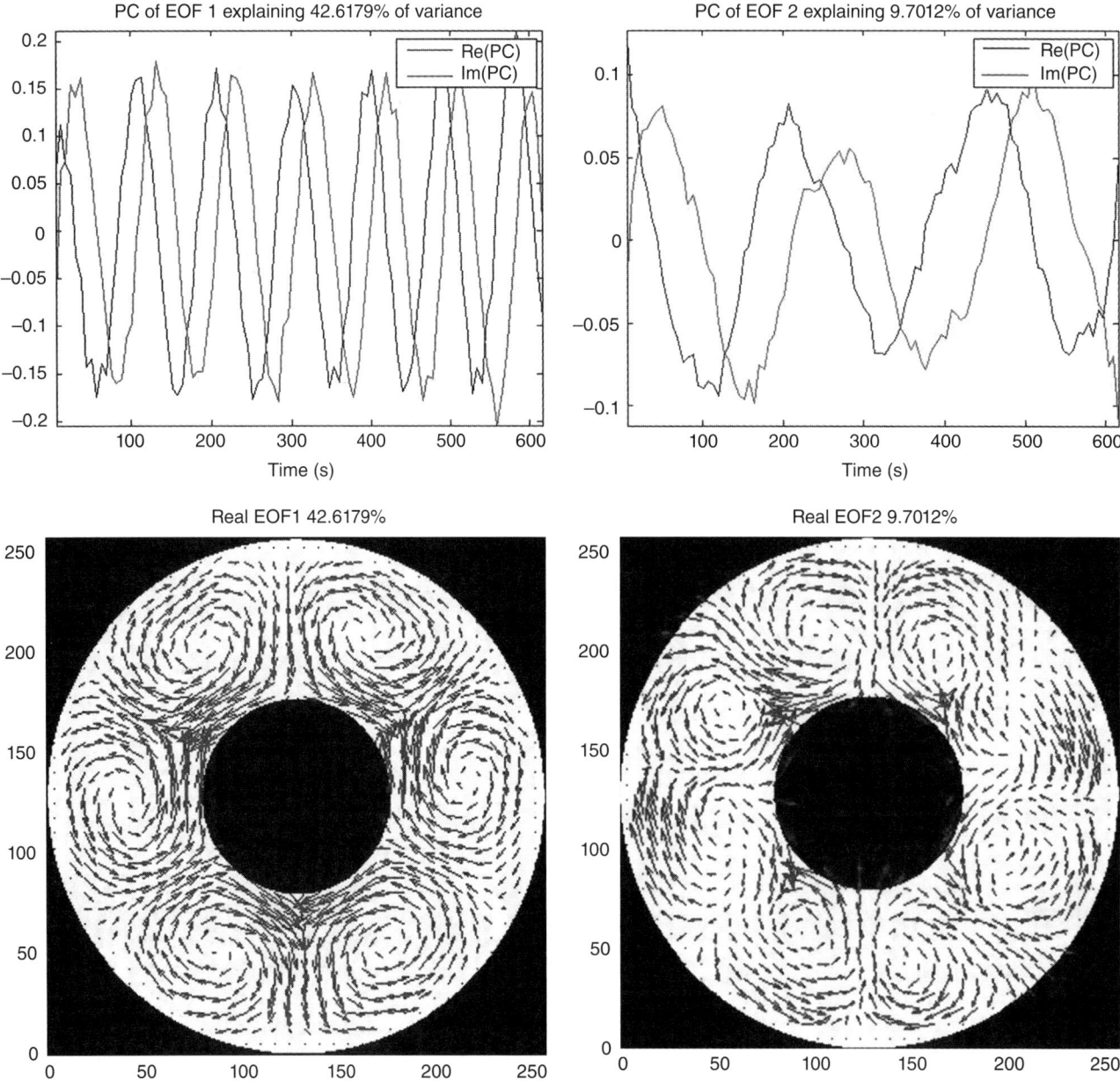

Figure 17.4. CEOF analysis of PIV measurements (given in corotating frame). Top: The principal component (PC) of CEOF1 (real part black, imaginary part red) (left) and of CEOF2 (right). Bottom: Real part of CEOF1 (left) and of CEOF2 (right). The imaginary parts of the two CEOFs are phase-shifted versions of the real parts.

vacillating wave regime discussed above. Classical Eady modes are nondispersive, and we might conclude that the first four EOFs do not resolve the nonlinear features of the wave modes in the irregular flow. Hence, EOFs from the noisy part of the spectrum need to be considered to address the irregularity of large Taylor/small Rossby number regimes.

Finally, in Figure 17.6 and Table 17.1 we show the variance distribution (in percent of the total variance) for the first 12 wave modes $m = 1, 2, \ldots, 12$ along a transection through the regime diagram, from the azimuthal flow regime ($m = 0$) to the slightly irregular $m = 4$ wave regime ($6.79 \times 10^6 \leq \mathrm{Ta} \leq 4.77 \times 10^8$, $5.73 \times 10^{-2} \leq \mathrm{Ro} \leq 4.0$, $\Delta T = 8\,\mathrm{K}$). The transition from $m = 0$ to $m = 2$ occurs at $(\mathrm{Ta}, \mathrm{Ro}) = (9.49 \times 10^6, 2.82)$. A dominant wave with $m = 2$ establishes, but other waves ($m = 3, \ldots, 6$) are also present in the variance spectrum. This indicates that the flow with zonal wave number 2 exhibits some vacillations. At $(\mathrm{Ta}, \mathrm{Ro}) = (2.3 \times 10^7, 1.19)$, the flow regime changes to $m = 3$ and we can find this wave and its harmonics

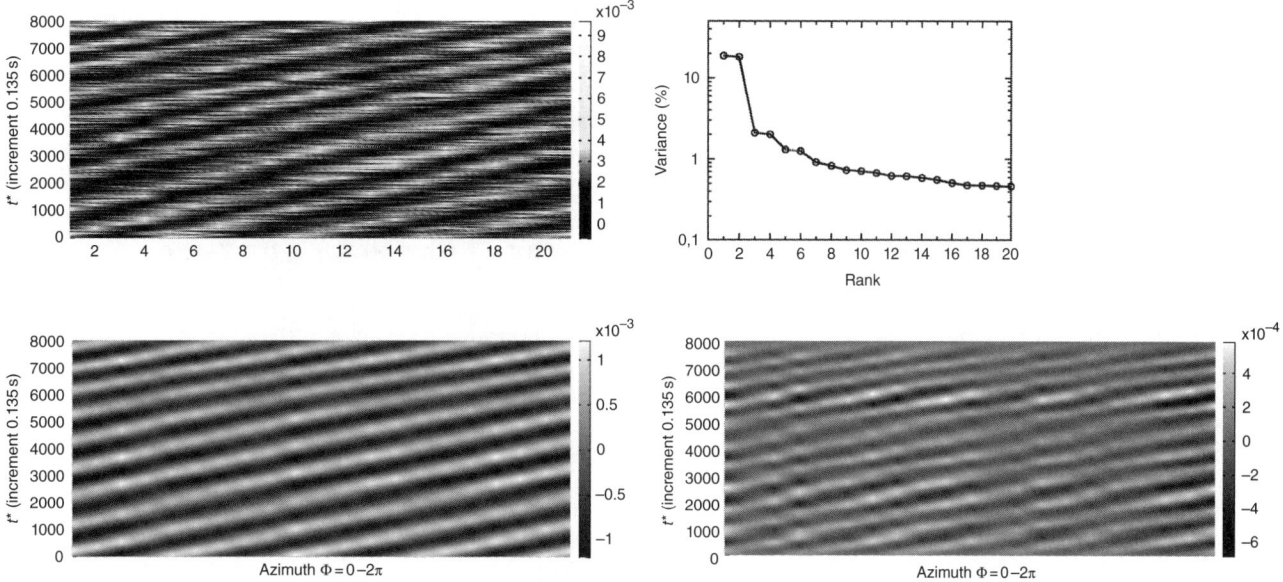

Figure 17.5. MSSA analysis of a wave mode of dominant wave number $m = 4$. Upper left: Space-time plot of the homogenized LDV data set, radial velocity [m/s] over azimuth $0 - 2\pi$. Upper right: The first 20 eigenvalues of the MSSA covariance spectrum. Note that a logarithmic scale has been used. Lower left: Reconstructed space-time plot using the first two MSSA eigenvectors. Lower right: Reconstructed space-time plot using the third and the fourth MSSA eigenvectors.

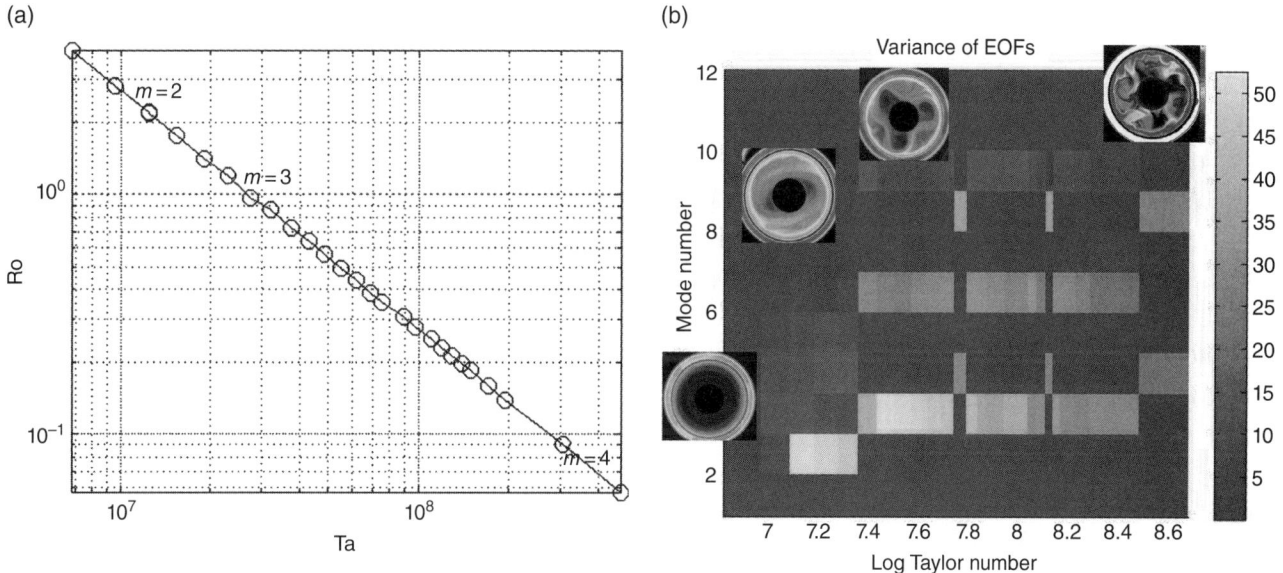

Figure 17.6. EOF variance spectra obtained along a transection through the wave regime. Left: Each circle in the Ta Rodiagram corresponds with an experiment. Right: Distribution of the variance (in % of the total variance) for the first 12 wave modes as a function of Ta. See also Table 17.1. For color detail, please see color plate section.

($m = 6, 9$) in the EOF variance spectrum. When we increase Ta, the dominant wave becomes weaker and the first harmonics stronger. At the transition to $m = 4$ (at (Ta, Ro) = $(3.04 \times 10^8, 9.01 \times 10^{-2})$) the first harmonic comprises as much variance as the $m = 4$ mode. The flow is rather irregular and it becomes more and more difficult to identify a dominant wave. The EOF variance spectrum starts to become broader. Within the $m = 3$ regime we find transitions to the $m = 4$ flow indicating regions of multiple equilibria. Note that the irregular wave flow must not be confused with turbulent flow. For the first a wave pattern is still present whereas for the latter it is not.

Table 17.1. Variance in % explained by the zonal wave modes $m = 1, 2, \ldots, 12$.

Ta	Ro	1	2	3	4	5	6	7	8	9	10	11	12
6.79E+06	4.00E+00	0.0	0.0	0.0	0.0	0.0	0.0	0.0	0.0	0.0	0.0	0.0	0.0
9.49E+06	2.82E+00	0.0	4.8	8.7	7.1	3.1	0.0	0.0	0.0	0.0	0.0	0.0	0.0
1.24E+07	2.21E+00	0.0	37.0	6.8	5.8	4.1	0.7	0.0	0.0	0.0	0.0	0.0	0.0
1.24E+07	2.19E+00	0.0	51.1	9.1	8.2	5.7	1.9	0.0	0.0	0.0	0.0	0.0	0.0
1.54E+07	1.76E+00	0.0	48.9	6.3	11.5	6.8	0.0	0.0	0.0	0.0	0.0	0.0	0.0
1.91E+07	1.40E+00	0.0	46.0	4.3	10.7	4.9	2.9	0.0	0.0	0.0	0.0	0.0	0.0
2.30E+07	1.19E+00	0.0	0.0	36.5	0.0	0.0	24.7	0.0	0.0	9.6	0.0	0.0	0.0
2.74E+07	9.67E-01	0.0	0.0	52.6	0.0	0.0	23.0	0.0	0.0	3.4	0.0	0.0	0.0
3.20E+07	8.64E-01	0.0	0.0	50.0	0.0	0.0	21.3	0.0	0.0	4.5	0.0	0.0	0.0
3.75E+07	7.24E-01	0.0	0.0	48.1	0.0	0.0	24.0	0.0	0.0	5.3	0.0	0.0	0.0
4.29E+07	6.40E-01	0.0	0.0	45.6	0.0	0.0	24.9	0.0	0.0	4.5	0.0	0.0	0.0
4.85E+07	5.62E-01	0.0	0.0	44.1	0.0	0.0	26.8	0.0	0.0	7.3	0.0	0.0	0.0
5.50E+07	4.89E-01	0.0	0.0	0.0	25.7	0.0	0.0	0.0	36.1	0.0	0.0	0.0	6.1
6.17E+07	4.40E-01	0.0	0.0	41.4	0.0	0.0	28.4	0.0	0.0	8.4	0.0	0.0	0.0
6.88E+07	3.85E-01	0.0	0.0	34.6	0.0	0.0	28.4	0.0	0.0	7.7	0.0	0.0	0.0
7.56E+07	3.54E-01	0.0	0.0	33.1	0.0	0.0	27.1	0.0	0.0	8.5	0.0	0.0	0.0
8.89E+07	3.08E-01	0.0	0.0	37.4	0.0	0.0	24.1	0.0	0.0	10.3	0.0	0.0	0.0
9.73E+07	2.80E-01	0.0	0.0	37.5	6 0.0	0.0	25.3	0.0	0.0	9.9	0.0	0.0	0.0
1.10E+08	2.51E-01	0.0	0.0	26.9	0.0	0.0	29.1	0.0	0.0	7.9	0.0	0.0	0.0
1.19E+08	2.30E-01	0.0	0.0	27.2	0.0	0.0	21.6	0.0	0.0	5.0	0.0	0.0	0.0
1.29E+08	2.11E-01	0.0	0.0	0.0	24.5	0.0	0.0	0.0	35.4	0.0	0.0	0.0	7.7
1.39E+08	1.97E-01	0.0	0.0	30.8	0.0	0.0	27.1	0.0	0.0	7.1	0.0	0.0	0.0
1.49E+08	1.84E-01	0.0	0.0	25.7	0.0	0.0	25.7	0.0	0.0	7.2	0.0	0.0	0.0
1.72E+08	1.59E-01	0.0	0.0	24.3	0.0	0.0	22.9	0.0	0.0	6.3	0.0	0.0	0.0
1.95E+08	1.39E-01	0.0	0.0	27.0	0.0	0.0	18.2	2.9	0.0	3.7	2.5	0.0	0.0
3.04E+08	9.01E-02	0.0	0.0	0.0	20.1	0.0	0.0	0.0	24.9	0.0	0.0	0.0	3.9
4.77E+08	5.73E-02	0.0	0.0	3.00	19.51	0.0	0.0	0.0	9.29	3.37	0.0	0.0	0.0

Note: Left two columns give the Taylor and Rossby number. The table corresponds with Figure 17.6, right.

Figure 17.6 illustrates that EOF decomposition is a powerful tool to classify regime transitions. Such an EOF-based classification is not restricted to laboratory data but should work as well for data from the real atmosphere.

In this section, we have presented the application of multivariate statistical methods to velocity data, and we have highlighted the particular abilities of the CEOF analysis and the MSSA to detect interactions. In the next section, we will apply the CEOF analysis to a more complex experiment with broken azimuthal symmetry and time-dependent boundary conditions. Moreover, in addition to the velocity, surface temperature has been measured as well.

17.3.2. Baroclinic Waves in Rotating Annulus with Barrier: CEOF Analysis

Several experiments with the differentially heated rotating annulus have been performed with modifications to the standard geometry (flat bottom, constant gap width). These geometric modifications were introduced to better represent certain aspects of natural flows, e.g., to further understand the dynamics of zonal flow over topography [*Weeks et al.*, 1997]. In ocean basins, the zonal flow is blocked by continents. Thus, several authors studied the case when a radial barrier is mounted in the annulus to understand better the western intensification of flows in

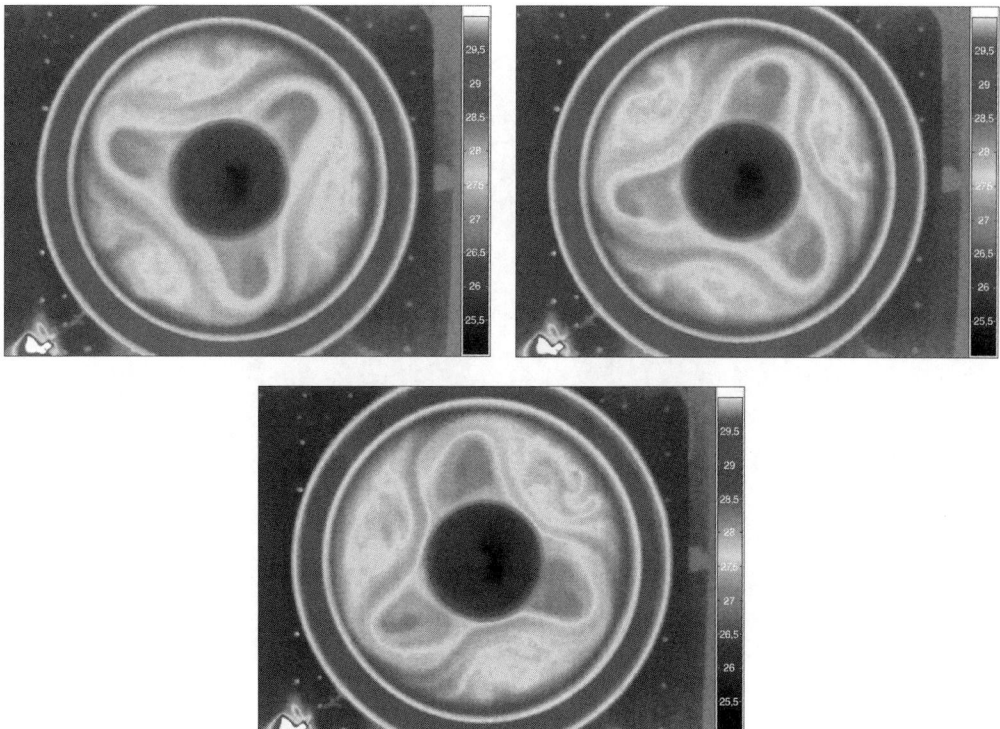

Figure 17.7. Sequence of surface temperature in the standard geometry at 69, 73, and 77 revolutions.

ocean basins [*Bowden and Eden*, 1968; *Maxworthy and Browand*, 1974; *Rayer et al.*, 1998].

Not all oceanic flows are completely blocked in the zonal direction. The Antarctic Circumpolar Current (ACC), for example, is not blocked, but the flow has to weave through the Drake Passage between the southern tip of South America and the Antarctic continent. The Drake Passage *partially* blocks the zonal flow. It is therefore straightforward to perform experiments with partial barriers to understand better the intermediate case between a free and fully blocked annulus flow.

In the experiment discussed here, the azimuthal flow is blocked at the inner cylinder and at the bottom (see Figure 17.9). At the barrier, the gap width is reduced from 75 to 43 mm and the fluid depth is reduced from 135 to 95 mm. More details can be found in the work of *Harlander et al.* [2012a].

Figure 17.7 shows a sequence of surface temperature images taken from an experiment in standard geometry with Ω = 4.6 rpm and ΔT = 2.8 K. We see the baroclinic wave after 69, 73, and 77 revolutions. Obviously, the wave is rather stable and rotates counterclockwise (that is, prograde) within a prograde mean flow. In contrast, in Figure 17.8, we see a comparable experiment (Ω = 4.8 rpm and ΔT = 4.0 K) but then with the barrier mounted. The most obvious new feature is wave breaking at the barrier and wave recovery downstream of the barrier. That is, the baroclinic wave never saturates but is invariably in a transient state. We can say that the barrier leads to a mechanically induced baroclinic life cycle. Life cycles play an important role for midlatitude atmospheric and oceanic flows where they occur due to linear growth, nonlinear saturation, and dissipative decay of large-scale waves. The experiment with the barrier opens the possibility to study baroclinic life cycles in a controlled way.

The experiment we discuss shows a slow periodic variation of the radial temperature difference with an amplitude of 1 K and a period of 26 min. The purpose of this variation of boundary conditions is to make sure that the wave breaking is due to the barrier and not due to regime transitions that occur for certain Taylor and Rossby numbers. We can exclude the latter from the fact that the flow looks very similar for maximum and minimum ΔT. It is important to note that during the experiment we observed the surface velocity and surface temperature simultaneously. This allows for analyzing coupled temperature and velocity anomalies and the evaluation of the mean and turbulent surface heat flux.

The CEOF analysis already applied in Section 17.3.1 is eminently suited for our data [*Pfeffer et al.*, 1990]. The CEOF analysis decomposes a propagating mode into a

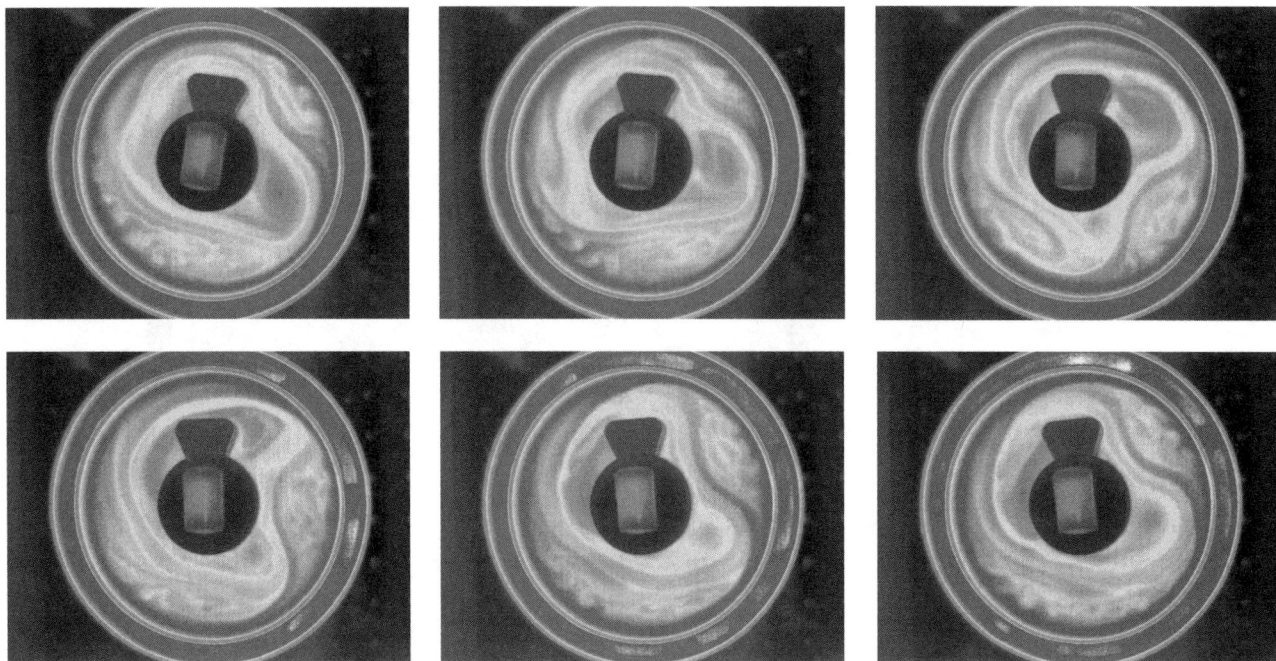

Figure 17.8. Sequence of surface temperature in the geometry with the barrier mounted. Top row from left to right: at 37, 39, and 43 rotations. Bottom row from left to right: at 45, 47, and 48 revolutions.

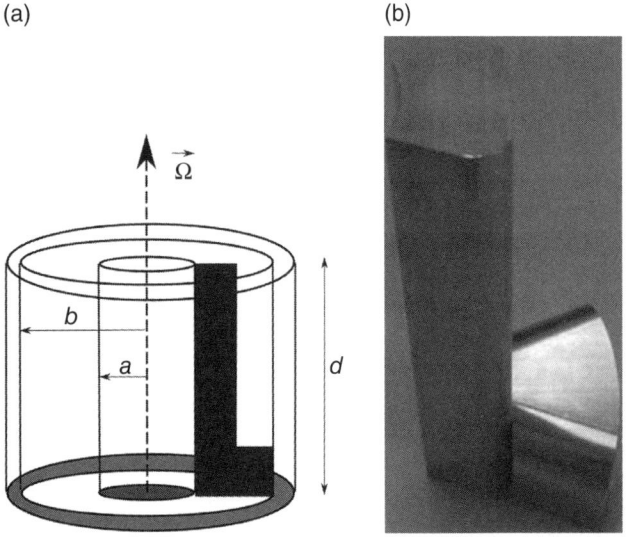

Figure 17.9. (a) Annulus with the barrier. (b) Photograph of the barrier.

single complex orthogonal function and not in two real functions. Therefore, amplitude and phase information is contained in a single CEOF. More specifically, we will apply a *combined* CEOF Analysis to find connected propagating anomalies of surface temperature and velocity [*Harlander et al.*, 2012a]. Frequently, the dominant components of a high-dimensional process are captured with often surprisingly few "modes" and thus the CEOF analysis is an efficient technique for dimension reduction.

We computed CEOFs over the full annulus and over subregions. The first 10 eigenvalues denoting the variance comprised in the corresponding CEOFs of the annulus region upstream of the barrier are shown in Figure 17.10a. We find that the first two local CEOFs explain more than 50% of the total variance in the region upstream of the barrier. For the CEOFs of the full annulus we find a very similar spectrum with contributions of 28.8% and 19.4% for CEOF1 and CEOF2, respectively. PC1 and PC2, that is, the temporal behavior of CEOF1 and CEOF2, are shown in Figures 17.10b and 17.10c for the upstream region. Note that real and imaginary parts are in quadrature, indicating propagating anomalies. Obviously, PC2 comprises for a significant part the slow periodic variation of the radial temperature difference mentioned above. In contrast, PC1 shows a strong wave oscillation and only a weak modulation due to the slow variation of the temperature difference. We find about 9 wave oscillations per 100 rotations. The PCs of the full annulus look very similar to the one of the upstream part; however, the order of PC1 and PC2 is reversed. That is, for the full annulus the first PC mainly contains the low-frequency forcing.

We now focus on the dominant spatial structure of the anomalies. CEOF2 of the full annulus and CEOF1 of the local domain are shown in Figures 17.11a–d. The CEOFs shown correspond to the PCs that mainly contain the waves and not the low-frequency forcing. Real parts

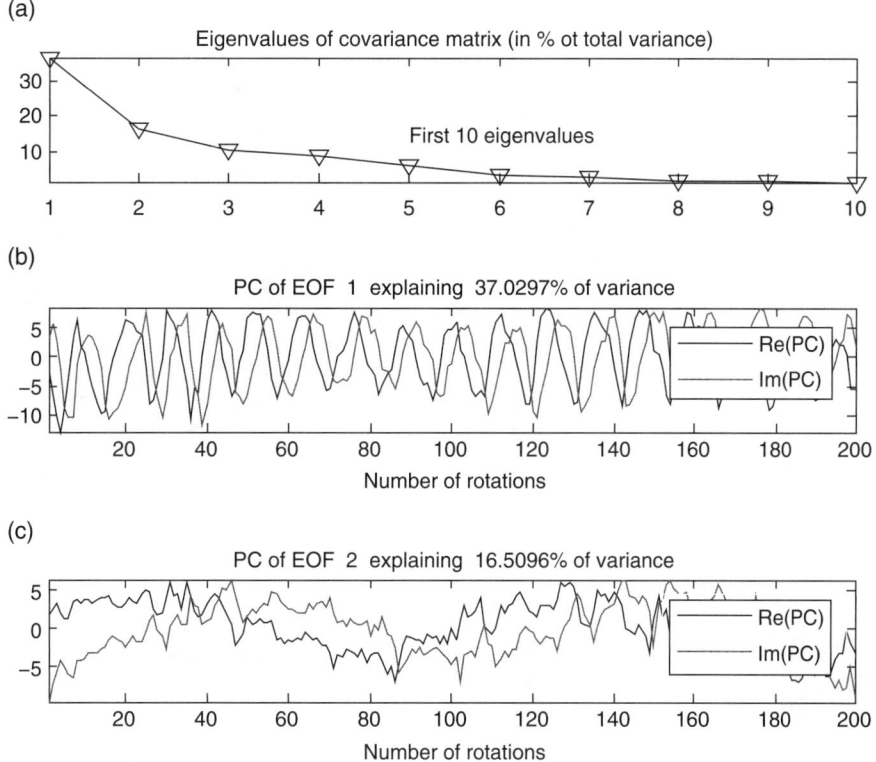

Figure 17.10. (a) Variance spectrum of a CEOF analysis. (b) Real and imaginary parts of PC1. (c) Real and imaginary parts of PC2. PC1 mainly captures the propagating wave whereas PC2 comprises a significant part of the low-frequency temperature forcing.

(a,c) and imaginary parts (b,d) of the CEOF show the propagating anomaly at two different phases with a phase shift of $\pi/2$. The patterns have to be read in the following way: Figure 17.11a,c represent the flow at the beginning of a cycle, and Figures 17.11b,d represent a quarter of a cycle later. The fields displayed in Figures 17.11a,c but multiplied by -1 show the flow at half of the cycle; Figure 17.11b,d multiplied by -1 give the flow at three quarters of the cycle. A quarters of the cycle later the cycle is complete and the starting point is reached again (that is, Figures 17.11a,c).

CEOF1 computed for the full annulus (not shown) contains a mix of waves and the forcing signal and is more difficult to interpret than CEOF2. In the temperature field we find the modulation pattern as an increase/decrease of the temperature along the outer boundary. We also find propagating wave structures and vortices in the upstream region and a locally fixed but pulsating vortex in the downstream region. This vortex can be seen even clearer in a local EOF analysis of the downstream region and is likely directly connected to the barrier and is not generated by baroclinic instability. The strength of the vortex changes due to the oscillating meridional temperature contrast.

CEOF2 (Figures 17.11a,b) clearly reveals that the wave structure is prominent in the upstream and weak in the downstream region. The pattern of the local CEOF1 (Figures 17.11c,d) is rather easy to interpret. It is clearly dominated by a regular train of vortices slowly traveling prograde towards the barrier. Note that the local CEOFs have been computed independently from the full annulus just for the local domain. Still, Figures 17.11c,d resemble very closely the upstream part of the full annulus CEOFs (Figures 17.11a,b). From the local CEOFs we find that temperature anomalies are not circular but show a prominent bulge that is opposed to the direction of the mean flow. We also find that the temperature maxima and minima do not correspond with the centers of the vortices. With respect to these centers, the temperature anomalies are shifted toward the inner cylinder and slightly downstream. We see that due to the bulges, a significant positive heat flux can be observed (positive anomalies are transported inward, negative anomalies outwards). This flux reduces the radial temperature gradient.

From Figures 17.11c,d we see further that the strength of a vortex continuously increases until the anomaly reaches the barrier. We further note a small cyclone within the constriction of the annulus (Figure 17.11d). This

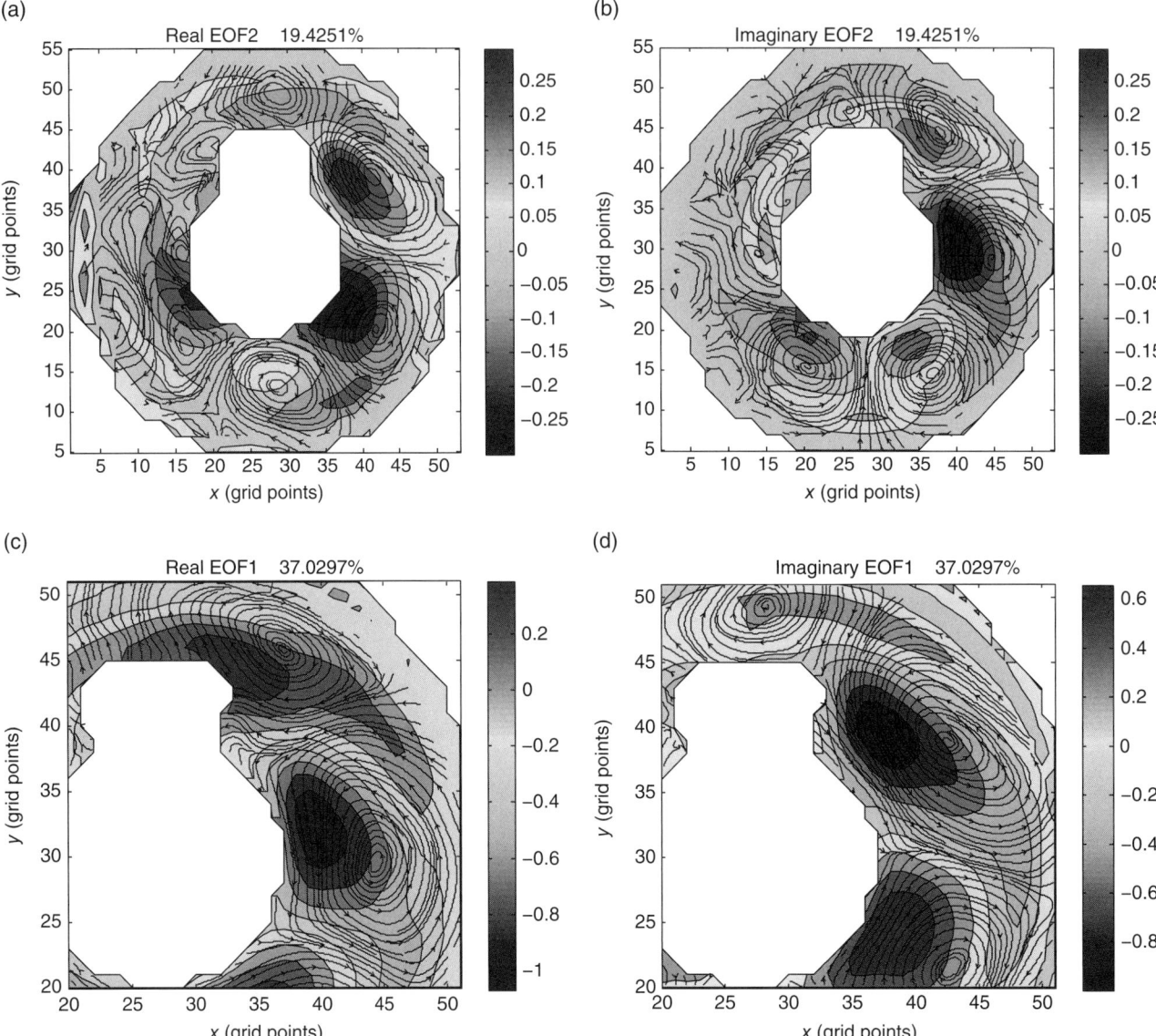

Figure 17.11. Real and imaginary parts of CEOFs: surface temperature and velocity: (a,b) CEOF2 for full annulus; (c,d) CEOF1 of upstream region. The CEOFs resolve coupled heat and velocity anomalies propagating anticlockwise toward the barrier. For color detail, please see color plate section.

demonstrates that vortices are important in the constriction even if this is not directly visible in Figure 17.8. However, the magnitude of the vorticity in the gap is weak compared to the vorticity in the region upstream of the barrier. Note finally that downstream of the barrier the anomalies become more turbulent and less coherent features can be observed (Figures 17.11a,b).

The present experiment of a differentially heated rotating annulus with a radial barrier that partially blocks the azimuthal component of the flow along the inner cylinder and the bottom enables us to present a rather general picture of the transient flow. This picture should hold for a certain range of Taylor and thermal Rossby numbers in the wave regime of the annulus. Roughly speaking, the annulus can be subdivided into the half upstream of the barrier, where waves amplify, and the half downstream of the barrier, where waves decay (see Figures 17.8 and 17.11). The dominant linear wave mode (in our case with azimuthal wave number 3) is unstable upstream but stable downstream of the barrier. In the upstream half, the azimuthal mean flow is moderate but with a significant positive radial eddy heat flux. In contrast, in the downstream half, we find an increased radial mean temperature gradient. The latter points to a weakened or even reversed radial eddy heat flux in the lee side of the barrier.

The experiment described has not been designed to resemble some particular large-scale flow that can be found in nature. However, it is tempting to point to similarities between experimental and real flows. The ACC is an annulus-like large-scale flow that is partially blocked at the Drake Passage, connecting the Pacific and the Atlantic oceans. For the ACC, eddy fluxes play a more central role in the dynamical and thermodynamical balances than in other oceans [*Rintoul et al.*, 2001]. Of course, many processes important for the ACC dynamics are not captured by the experiment, e.g., wind stress, coastal as well as bottom topography, and the high-latitude β-effect [*Harlander*, 2005; *Afanasyev et al.*, 2009, chapter 5 this issue]. Nevertheless, a meridional overturning circulation is present in the ACC and the experiment and baroclinic instability plays an important role for the ACC dynamics due to the meridional transport of heat toward the pole.

17.3.3. Principal Oscillation Patterns and Singular Vectors

In many cases, EOFs can be interpreted as the patterns of natural oscillations, i.e., as the eigenmodes of the system under consideration. From a mathematical point of view such an interpretation is not justified. However, empirically estimating the eigenmodes and then comparing them with the EOFs can reveal the connection between modes and patterns of variability [*Harlander et al.*, 2009b]. Empirically estimated normal modes are called principal oscillation patterns (POPs) in the meteorological literature [*Hasselmann*, 1988]. In the following we will compare EOFs and POPs computed from our annulus data. In a subsequent step we will then estimate optimal growing initial perturbations, called singular vectors from the POPs. Theoretically, SVs converge to POPs in the limit $t \to \infty$. As we will see, this holds for the empirically estimated SVs as well. SV growth might explain the increased irregularity that occurs for annulus flows at Taylor numbers larger that 10^8. We will briefly discuss this by employing laboratory data and model results.

17.3.3.1. Principal Oscillations Patterns. Generally speaking, the POP method is promising when the dynamical process under consideration is linear to first approximation. This sounds very restrictive in particular when geophysical applications are the target. However, the method has successfully been applied to a wide class of geophysical data [*von Storch and Zwiers*, 1999].

The linear system considered reads

$$\mathbf{x}(t + \Delta) = \mathbf{G}(\Delta)\mathbf{x}(t), \quad (17.6)$$

where $\mathbf{x}(t+\Delta)$ is the state vector at time $t+\Delta$ and $\mathbf{G}(\Delta)$ the propagator matrix that maps the state at time t to time $t + \Delta$. $\mathbf{G}(\Delta)$ bears the argument Δ since it can be estimated by

$$\mathbf{G}(\Delta) = \mathbf{C}(\Delta)\mathbf{C}(0)^{-1}, \quad (17.7)$$

where

$$\mathbf{C}(\Delta) = (\mathbf{F}^{1\Delta})^T \mathbf{F}^{0\Delta}, \quad (17.8)$$

$$\mathbf{C}(0) = (\mathbf{F}^{0\Delta})^T \mathbf{F}^{0\Delta} \quad (17.9)$$

are the covariance matrices with lag Δ and lag zero. The eigenvectors of $\mathbf{G}(\Delta)$ are the POPs, whereas the eigenvectors of $\mathbf{C}(0)$ are the EOFs. Usually, the POPs are sorted (in decreasing order) with respect to the e-folding times $\tau_i = -1/\ln(|\lambda_i|)$, where the λ_i are the eigenvalues of $\mathbf{G}(\Delta)$. The period of the POP is given by $T = 2\pi/\arg \lambda$.

We used surface temperature data to compute EOFs and POPs. The data have been recorded by an infrared camera that has a noncooled microbolometer detector with a spectral range of $7.5 - 14.0\,\mu$m and a temperature resolution smaller than 0.08 K with an accuracy of ± 1.5 K at 30°C. The spatial resolution of the infrared sensor is 640×480 pixels. To reduce the size of the covariance matrices we smoothed the data by using a running average over areas of 2×2 pixels. In Figure 17.12a we display the EOF1 that explains 27% of the total variance. Further, in Figure 17.12b, the real part of the least damped POP1 with a damping time of 1827.6 s and a period of 62.5 s for an experiment with $\Omega = 6$ rpm and $\Delta T = 8$ K is shown. It should be noted that EOF2 is a phase shifted version of EOF1 with nearly the same explained variance. Obviously, EOF1 and POP1 agree very well. Taking EOF1 and EOF2, as well as the real and imaginary parts of POP1 together, both patterns propagate with the same phase speed. This suggests that for the data considered the EOFs represent the eigenmodes of the system.

Let us next use the POPs to estimate the dominant patterns of nonmodal instability. This procedure needs an appropriate filtering and the EOF analysis is a suitable method for this purpose.

17.3.3.2. Empirical Singular Vectors. Instability is related to exponentially growing eigenmodes and thus to POPs with $|\lambda_i| > 1$. Interestingly, when finite time intervals are considered, growth rates of certain initial perturbations can exceed the growth rates of the most unstable modes. Moreover, even when all modes are damped ($|\lambda_i| < 1$), such particular initial perturbations can grow dramatically during finite time intervals. The perturbations with the largest growth rates are called singular vectors or optimal perturbations. They play an important role not only in atmospheric ensemble predictions [*Kalnay*, 2002] but also for the theory of instability and turbulence [*Trefethen et al.*, 1993; *DelSole*, 2007].

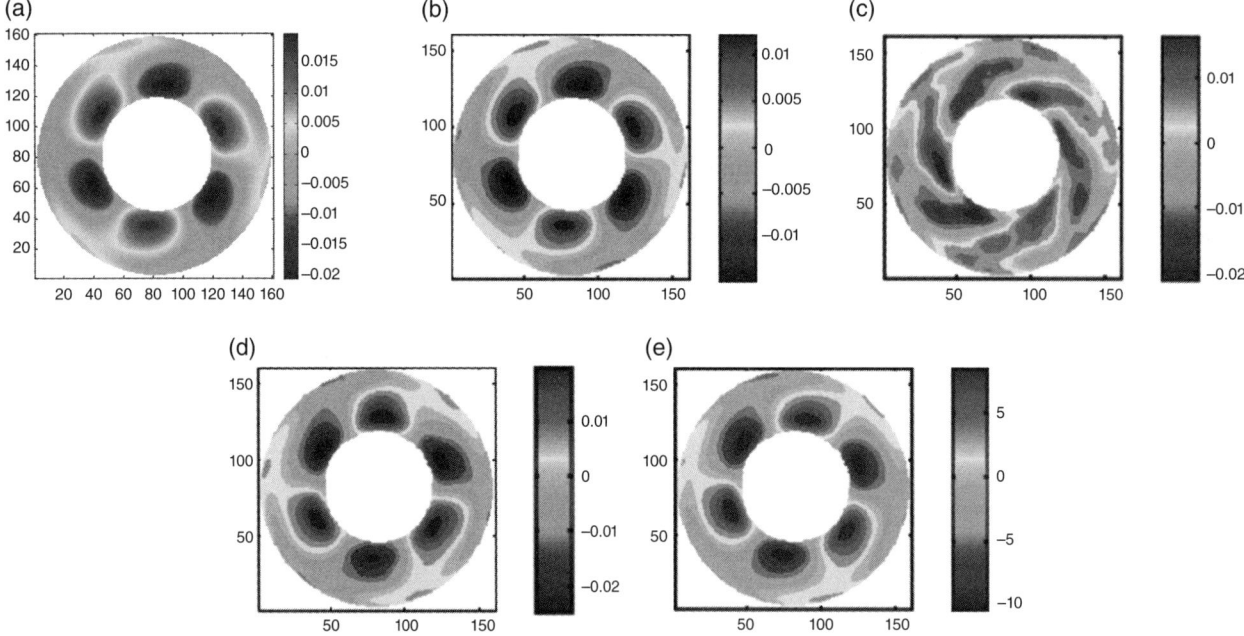

Figure 17.12. Experiment with parameters $\Omega = 6$ rpm, $\Delta T = 8$ K. (a) EOF1, 27% explained variance. (b) POP1, damping time 1827.6 s, period 62.5 s. (c) First singular vector at $t = 0$, $t_{opt} = 20$ s, $\sigma = 1.182$. (d) First singular vector at $t = t_{opt}$. (e) First singular vector at $t = 2500$ s.

The starting point for the SV analysis is the linear dynamical system

$$\frac{d\mathbf{x}}{dt} = \mathbf{Bx} \qquad (17.10)$$

with the system matrix $\mathbf{B}$. In atmospheric ensemble prediction linearization is generally done around a nonlinear solution and then $\mathbf{B}$ is time dependent. In contrast, in turbulence research linearization is done around a mean state. In that case $\mathbf{B}$ is time independent [*DelSole*, 2007]. We consider the mathematically simpler latter case.

The matrices $\mathbf{G}$ in (17.7) and $\mathbf{B}$ in (17.10) are connected via the equation

$$\mathbf{G}(\Delta) = \exp(\mathbf{B}\Delta) = \sum_{k=0}^{\infty} \frac{1}{k!} (\mathbf{B}\Delta)^k \qquad (17.11)$$

or

$$\mathbf{B} = \frac{1}{\Delta} \ln[\mathbf{G}(\Delta)] = \frac{1}{\Delta} \sum_{k=1}^{\infty} \frac{(-1)^{k+1}}{k} [\mathbf{G}(\Delta) - \mathbf{E}]^k, \qquad (17.12)$$

where $\mathbf{E}$ is the identity matrix. It is instructive to note that by discretizing the first term in (17.10) by $[\mathbf{x}(t + \Delta) - \mathbf{x}(t)]/\Delta$ and using (17.6), we find

$$\mathbf{B} = \frac{1}{\Delta}[\mathbf{G}(\Delta) - \mathbf{E}]. \qquad (17.13)$$

This corresponds to (17.12) when just the first term is kept. The simplification gives still good results for low-dimensional systems [*Harlander et al.*, 2009a] but it fails in general.

Having estimated the system matrix $\mathbf{B}$ from (17.12), we can compute the propagator $\mathbf{G}$ for any time interval Δ_0 by

$$\mathbf{G}(\Delta_0) = \exp(\mathbf{B}\Delta_0). \qquad (17.14)$$

When the L_2-norm is used to measures the growth of a perturbation in the time interval $0 \leq t \leq \Delta_0$, the SVs with optimazation time Δ_0 are given by the eigenvectors of the matrix $\mathbf{G}^T(\Delta_0)\mathbf{G}(\Delta_0)$. The eigenvalues of the matrix define the square of the growth rates.

Figure 17.12c shows the first singular vector at $t = 0$ estimated from surface temperature data of an experiment with $\Omega = 6$ rpm and $\Delta T = 8$ K. The data sampling rate was $\Delta = 5$ s, and the optimization time was set to be $\Delta_0 = 20$ s. For the time interval $0 \leq t \leq \Delta_0$ we obtain a growth rate of $\sigma = 1.182$. To reduce the noise in the data, we used the EOF filtering technique. Just 33 EOFs explain a total variance very close to 90%. Thus we used $j = 33$ in (17.5) to reconstruct the data from the EOFs. It should be noted that the growth rate increases by increasing the number of EOFs used for the reconstruction. It appears that the more EOFs that are available to support the SVs, the larger is the maximum growth rate. However, reconstructing the data by a large number of EOFs increases the noise, and at a certain number, the SVs seem to be dominated by noise and lose their physical meaning.

Figure 17.12d shows SV1 at the optimization time $t = \Delta_0$ computed by $\mathbf{x}(\Delta_0) = \mathbf{G}(\Delta_0)\mathbf{x}(0)$, where $\mathbf{x}(0)$ is the SV at $t = 0$ shown in Figure 17.12c. Obviously, the tilted troughs and ridges have turned up as expected [*Will et al.*, 2006]. The structure does not change much for longer times. Figure 17.12e shows the SV after $t = 125\Delta_0 = 2500$ s computed by $\mathbf{x}(N\Delta_0) = \mathbf{G}^N(\Delta_0)\mathbf{x}(0)$ with $N = 125$. As expected, this pattern agrees very well with POP1 shown in Figure 17.12b. For $t \to \infty$ the SVs converge to the corresponding normal modes [*Kalnay*, 2002].

Seelig et al. [2012] discussed SVs of the simple Lorenz annulus model [*Lorenz*, 1984] in the context of transitions to irregular flow and compared some numerically deduced SVs with data-based ones. The numerical model enabled the construction of a regime diagram in terms of singular vector growth rates, where the abscissa was the Taylor number and the ordinate the thermal Rossby number. Strikingly, the diagram based on singular vector growth strongly resembles the traditional bifurcation diagram for annulus flows [*Hide and Mason*, 1970; *Lorenz*, 1984]. The largest growth rates could be found in the irregular flow regime of the Lorenz model.

The findings from the simple numerical model suggest that the gradual increase of irregularity in the rather broad transition region to quasi-geostrophic turbulence might partly be addressed to singular vector growth. For laboratory experiments as well as for natural flows there is always a certain background noise level. Irregularities in the transition region might be seen as *extreme events* that arise from random excitation of singular vectors with unusual large growth rates [*DelSole*, 2007]. This process, together with nonlinear wave-wave interaction, could explain the gradual broadening of the spectrum when the rotating annulus flow transits to geostrophic turbulence [*Pfeffer et al.*, 1997]. Whether these ideas, derived from the low-order Lorenz model, can be transferred to real annulus flows is not clear yet. More data sets have to be analyzed by the techniques described above to address this question. However, it can be expected that the growth rates increase for irregular flows since more EOFs have to be considered to cover, say, 90% of the total variance for irregular flows.

17.3.4. Helmholtz-Hodge Decomposition of Annulus Flows

According to the Helmholtz-Hodge decomposition theorem, any suitably smooth vector field can be decomposed into the sum of a divergence-free field and a curl-free field [*Foias et al.*, 2008]. These two fields can be used to discriminate different wave types occurring in the annulus. Baroclinic waves and Rossby waves are divergence free, whereas inertia-gravity waves comprise a significant part of horizontal divergence. Presently, the process of spontaneous gravity wave emission is a major issue in atmospheric research. The differentially heated rotating annulus is a lab experiment suitable to systematically study spontaneous gravity wave emission in analogy to the atmosphere [*Williams et al.*, 2008]. To detect inertial-gravity waves in the experimental data, it is favorable to not use the full flow field but instead make use of the decomposition and analyze just the curl-free part of the flow.

The primary difficulties with computing the decomposition of the measured horizontal velocity at a given level of the cylindrical tank is that the PIV data do not line up on a nice grid, and the data may contain noise. To handle these two issues, we use a mesh-free reconstruction method based on radial basis functions (RBFs). The method employs *matrix-valued* kernels [*Narcowich and Ward*, 1994] and mimics the Helmholtz-Hodge decomposition of a 2D velocity field. It is similar to the method described by *Fuselier and Wright* [2009] for the surface of the sphere but is instead adapted for a 2D annular domain, for which dealing with boundaries becomes important. The method also provides a means of filtering the noise in the measured velocity fields and can be used to reconstruct the full 3D field in the rotating annulus. The key ingredients to the mesh-free reconstruction and decomposition technique are *divergence-free* and *curl-free* matrix-valued kernels. In this study, we construct these kernels from the scalar-valued Matérn radial kernels, which are popular for spatial statistics [*Stein*, 1999] and are given by

$$\phi_\nu(r) = \frac{1}{2^{\nu+1}\Gamma(\nu+1)}(\alpha r)^\nu K_\nu(\alpha r),$$
$$r \geq 0, \quad \nu > \frac{5}{2}, \alpha > 0, \quad (17.15)$$

where K_ν is the modified Bessel function of the second kind of order ν. Increasing ν in (17.15) increases the smoothness of the kernel, while increasing α increases its peakedness. Letting $\mathbf{x} = (x, y)$ and $\mathbf{x}_j = (x_j, y_j)$, the respective divergence-free and curl-free matrix-valued kernels are then defined as [*Narcowich and Ward*, 1994]

$$\Phi_\nu^{\text{div}}(\mathbf{x}, \mathbf{x}_j) = (-\nabla^2 \mathbf{I} + \nabla\nabla^T)\phi_\nu(\|\mathbf{x} - \mathbf{x}_j\|_2), \quad (17.16)$$

$$\Phi_\nu^{\text{curl}}(\mathbf{x}, \mathbf{x}_j) = -\nabla\nabla^T \phi_\nu(\|\mathbf{x} - \mathbf{x}_j\|_2), \quad (17.17)$$

where $\mathbf{I}$ is the 2×2 identity matrix and $\nabla\nabla^T$ is the Hessian matrix. By construction, the columns of Φ_ν^{div} are *divergence free*, while the columns of Φ_ν^{curl} are *curl free*.

Before discussing the exact details on the reconstruction and decomposition method, we note that since the present application involves boundaries, it is necessary to supplement the given data with boundary conditions to make the decomposition of 2D velocity field unique [*Foias et al.*, 2008]. We assume that both the divergence-free and curl-free parts of the field are parallel to the boundaries.

We enforce this condition on the reconstructed field at discrete locations on the boundary of the annulus.

Let $\mathbf{v}_j = (u_j, v_j)$, $j = 1, \ldots, N$, denote the normalized PIV measurements of the horizontal velocity field at any horizontal level ℓ of the cylindrical tank and let $\mathbf{x}_j = (x_j, y_j)$ denote the corresponding normalized locations of the measured field. Here we have normalized so that the outer radius of the tank is unity. Since the PIV data do not include measurements on the boundary, we must define these points. We choose the boundary points to be equally spaced on the inner and outer circles of the annulus with a density that is comparable to that of the interior points. We denote these boundary points by $\boldsymbol{\xi}_k$, $k = 1, \ldots, M$, and we let $\mathbf{n}_k$ denote the corresponding unit outward normal vector at $\boldsymbol{\xi}_k$. The matrix-valued kernel approximation of the field then takes the form

$$\tilde{\mathbf{v}}(\mathbf{x}; \nu_n, \nu_f) = \underbrace{\sum_{j=1}^{N} \Phi_{\nu_n}^{\text{div}}(\mathbf{x}, \mathbf{x}_j)\mathbf{a}_j + \sum_{k=1}^{M} (\Phi_{\nu_f}^{\text{div}}(\mathbf{x}, \boldsymbol{\xi}_k)\mathbf{n}_k)d_j}_{\tilde{\mathbf{v}}^{\text{div}}(\mathbf{x}; \nu_n, \nu_f)}$$

$$+ \underbrace{\sum_{j=1}^{N} \Phi_{\nu_n}^{\text{curl}}(\mathbf{x}, \mathbf{x}_j)\mathbf{a}_j + \sum_{k=1}^{M} (\Phi_{\nu_f}^{\text{curl}}(\mathbf{x}, \boldsymbol{\xi}_k)\mathbf{n}_k)c_j,}_{\tilde{\mathbf{v}}^{\text{curl}}(\mathbf{x}; \nu_n, \nu_f)}$$

(17.18)

where $\mathbf{a}_j = [a_j \ b_j]^T$, c_k, and d_k are determined by the following constraints:

$$\tilde{\mathbf{v}}(\mathbf{x}_i; \nu_n, \nu_f) = \mathbf{v}_i, \quad i = 1, \ldots, N,$$
$$\tilde{\mathbf{v}}^{\text{div}}(\boldsymbol{\xi}_i; \nu_f, \nu_f) \cdot \mathbf{n}_i = 0, \quad i = 1, \ldots, M, \quad (17.19)$$
$$\tilde{\mathbf{v}}^{\text{curl}}(\boldsymbol{\xi}_i; \nu_f, \nu_f) \cdot \mathbf{n}_i = 0, \quad i = 1, \ldots, M.$$

These constraints can be arranged into a $(2N + 2M) \times (2N + 2M)$ symmetric linear system of equations for determining the unknown coefficients.

We have introduced two smoothness parameters ν_n and ν_f in (17.18) to provide a mechanism for filtering the reconstructed field. The method we use for filtering is adapted from a technique first proposed by *Beatson and Bui* [2007] for scalar-valued RBF approximations. It involves fitting the noisy data with one smoothness parameter ν_n and then evaluating the resulting approximation with a larger smoothness parameter ν_f. This means the data is fit with one kernel but evaluated with a smoother yet similar kernel. As discussed by *Beatson and Bui* [2007], this kernel replacement technique corresponds to applying a low-pass filter to the approximations. Since the measurements are noisy and the boundary conditions are not, we only use ν_n in (17.19) when fitting the measurements. All evaluations of $\tilde{\mathbf{v}}$ are done with $\nu_n = \nu_f$ to filter out the noise. The resulting filtered approximation then satisfies the boundary conditions. Presently, there is no theory for selecting ν_n and ν_f in an "optimal" manner. Instead, the choice is somewhat by trial and error. In the experiments that follow, we found that $\nu_n = 3.5$ and $\nu_f = 5.5$ gave good results for several different flow parameter regimes and vertical measurement levels. Half-integer choices for the smoothness parameter also lead to significant simplifications in computing (17.15) [*Fasshauer*, 2007].

Because of the properties of Φ_ν^{div} and Φ_ν^{curl}, the expansions $\tilde{\mathbf{v}}^{\text{div}}$ and $\tilde{\mathbf{v}}^{\text{curl}}$ in (17.18) are divergence and curl free, respectively. Thus, the $\tilde{\mathbf{v}}$ mimics the Helmholtz-Hodge decomposition theorem. Furthermore, an approximation to the divergence-free or curl-free parts of the field can be obtained from these respective expansions.

In Figure 17.13, we show the reconstruction and decomposition of the velocity field for two sets of parameters measured with PIV close to the surface at $z = 120$ mm. For these data we set the shape parameter to $\alpha = 20.91$, which corresponds to the inverse of the minimum of the pairwise distances between the normalized sample locations.

Figures 17.13a,c show contour plots of the streamfunction for the divergence-free part $\tilde{\mathbf{v}}^{\text{div}}$ of the 120 mm fields, while Figures 17.13b,d show contours of the velocity potentials for the curl-free part of the fields $\tilde{\mathbf{v}}^{\text{curl}}$. As can be seen, the main pattern of the flow is quasi-geostrophic dynamics which is divergence free. The curl-free patterns shown in 17.13b,d can be interpreted as a deviation from pure quasi-geostrophic flow. We see that these deviations are strongest at the inner and outer boundaries of the annulus. There prominent axial flows can be expected due to the heating and cooling of the boundaries. The axial gradients of this flow component induce a horizontal divergence. While the divergence-free part shown in Figures 17.13a,c is rather robust, the curl-free part is more delicate and already small effects can perturb the symmetry of the patterns. Still, the curl-free part of both experiments is rather smooth and no small-scale wavelike features can be seen. The reason for this might be that the spatial resolution of the PIV observations is not high enough to resolve the transient, nongeostrophically balanced part of the flow.

We conclude this section by noting that we can also use (17.18) to compute the divergence of the reconstructed velocity field at any location in the 2D slice of the cylindrical tank. These approximations can be combined with the incompressibility assumption of the full 3D fluid in the rotating annulus to reconstruct the full velocity field of the fluid (see *Harlander et al.* [2012b] and the extended abstract on http://ltces.dem.ist.utl.pt/lxlaser/lxlaser2012/upload/92_paper_ecvgbw.pdf for details).

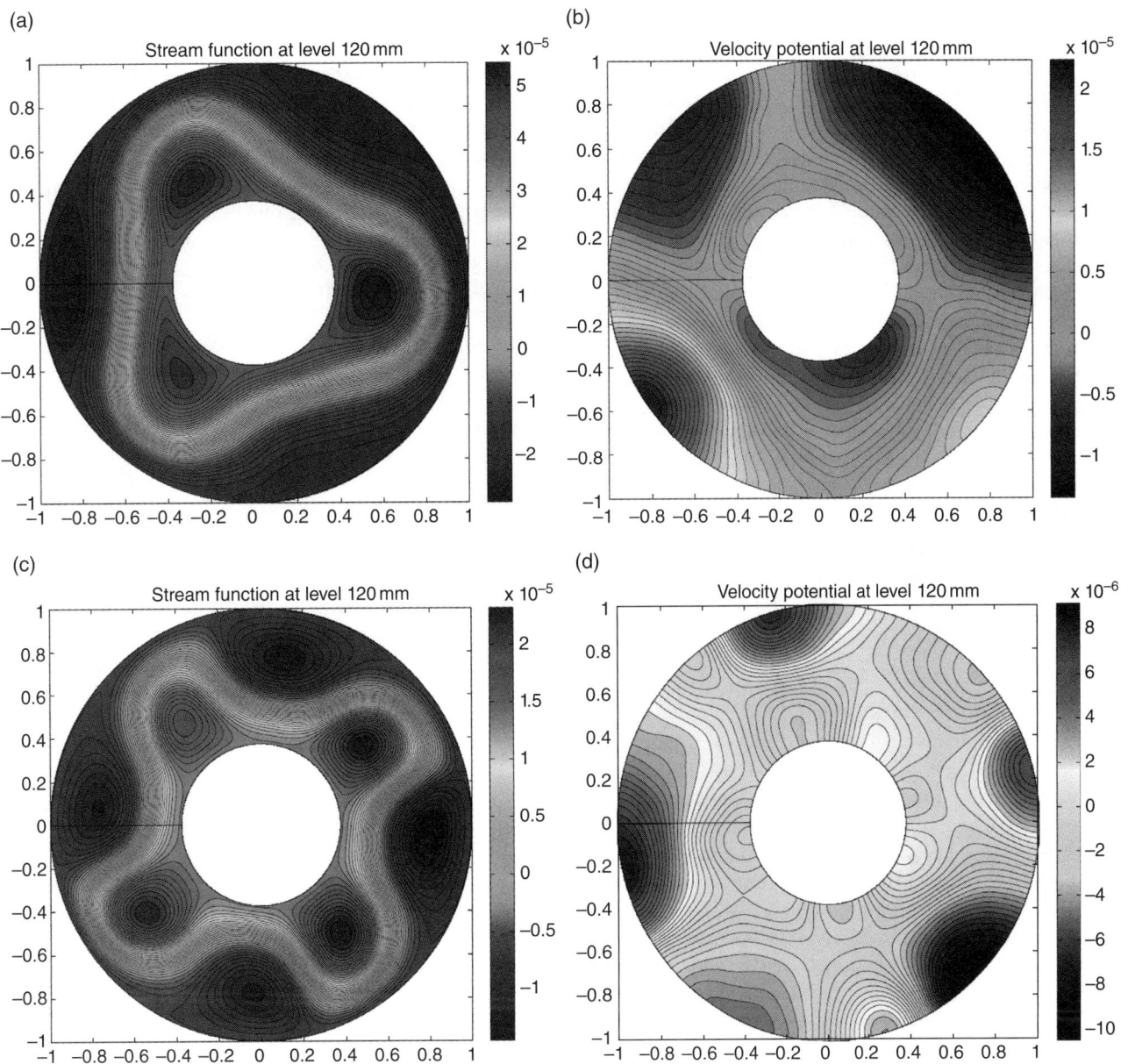

Figure 17.13. Helmholtz-Hodge decomposition for two sets of parameters $(\Omega, \Delta T, \text{Ta}, \text{Ro}, m) = (5.02\,\text{rpm}, 6.7\,\text{K}, 1.93 \times 10^7, 1.189, 3)$ for (a) and (b) and $(\Omega, \Delta T, \text{Ta}, \text{Ro}, m) = (15.00\,\text{rpm}, 8.0\,\text{K}, 1.72 \times 10^8, 1.150, 4)$ for (c) and (d): (a) stream function $m = 3$, (b) velocity potential $m = 3$, (c) stream function $m = 4$, (d) velocity potential $m = 4$.

17.4. CONCLUSION

This chapter summarized recent experimental work on the differentially heated rotating annulus. Some experimental techniques used are unique and the work on singular vectors and the Helmholtz-Hodge decomposition is novel. All the statistical and mathematical techniques discussed are versatile and can be applied to any kind of multivariate data, no matter whether they result from numerical models or field or laboratory observations.

Here we have applied the techniques solely on data from the heated rotating annulus that is operated in the lab of the Brandenburg University of Technology in Cottbus as part of the Multiple Scales in Fluid Mechanics and Meteorology initiative.

With the focus on geophysical fluid dynamics, the differentially heated and rotating annulus is an ideal test bed that allows for a rigorous and replicable testing of theories and computational tools. The atmosphere and oceans as well as many astrophysical fluids are stratified and rotate. The complex wave interactions and instabilities that shape their natural systems all have their equivalent in the annulus experiment.

Our understanding of the annulus flow is by far not complete and the work discussed in the present chapter can be seen as a starting point for further studies on the many facets of the experiment.

In Section 17.3.1 we studied transient features of baroclinic waves by using PIV and LDV measurements. The data sets involved could be expanded by *simultaneous* PIV and LDV observations. LDV is well suited to resolve fast local features like gravity waves excited close to the inner cylinder. At the same time, PIV can capture the state of the large-scale baroclinic wave that is particularly favorable for gravity wave excitation.

In Section 17.3.2 we discussed simultaneous surface temperature and PIV measurements. From these data the surface eddy heat flux can be derived [*Harlander et al.*, 2012a]. A future systematic study of the surface eddy heat flux for different flow regimes and annulus geometries would help to understand better the transient eddies and their feedback on the mean flow [*Wilson and Williams*, 2006].

In Section 17.3.3 patterns with large growth rates, so-called singular vectors, have been estimated from data [*Penland and Sardeshmukh*, 1995]. To our knowledge, no method is available yet that can do this for systems with time-dependent system matrices. In that case, linearization can no longer be done about a time mean state. Instead, linearization about a full nonlinear realization of the flow is necessary. Due to its reproducibility, the annulus experiment is well suited to provide data to test future statistical methods that can handle problems with time-dependent system matrices.

Finally, as already mentioned at the end of Section 17.3.4, from simultaneous observed surface temperature and 2D flow measurements, the 3D flow can be reconstructed by using radial basis functions [*Harlander et al.*, 2012b]. To further increase the reliability of such a novel reconstruction, the technique should be tested against stereo PIV observations that give 3D velocity fields on 2D laser slices. Combining such 3D flow observations with numerical simulations is a promising strategy to detect gravity waves and the excitation mechanisms in the stratified annulus [*Williams et al.*, 2008; *Scolan et al.*, 2013].

Acknowledgments. The work by Harlander, von Larcher, Alexandrov, and Egbers was partly funded by the German Science Foundation (DFG) under the DFG priority program MetStröm (SPP 1276/1-3) (grant EG 100/13-1-3). They thank Ulrich Achatz, Martin Baumann, Andreas Dörnbrack, Jochen Fröhlich, Vincent Heuveline, Stefan Hickel, Illia Horenko, Rupert Klein, Oswald Knoth, and Eric Severac from the MetStröm rotating annulus group for many helpful discussions and Yongtai Wang for help regarding measurements and data handling. In particular the authors thank Emilia Crespo del Arco and Mani Mathur for helpful comments that improved the chapter. The work of Wright was funded in part by the U.S. National Science Foundation (NSF) under grant DMS-0934581. Wright also wishes to thank Edward Fuselier of High Point University for discussing various ideas related to Helmholtz-Hodge decompositions on bounded domains.

REFERENCES

Afanasyev, Y. D., P. B. Rhines, and E. G. Lindahl (2009), Velocity and potential vorticity fields measured by altimetric imaging velocimetry in the rotating fluid waves, *Exp. Fluids*, *47*, 913–926.

Badger, J., and B. J. Hoskins (2001), Simple initial value problems and mechanisms for baroclinic growth, *J. Atmos. Sci.*, *58*, 38–49.

Barcilon, A., and P. G. Drazin (1984), A weakly nonlinear theory of amplitude vacillation and baroclinic waves, *J. Atmos. Sci.*, *22*, 3314–3330.

Beatson, R. K., and H. Q. Bui (2007), Mollification formulas and implicit smoothing, *Adv. Computat. Math.*, *27*, 125–149.

Bowden M., and H. F. Eden, (1968) Effect of a radial barrier on the convective flow in a rotating fluid annulus, *J. Geophys. Res.*, *73*, 6887–6895.

Broomhead, D. S., and G. P. King (1986), Extracting qualitative dynamics from experimental data, *Phys. D*, *20*, 217–236.

Cole, R. J. (1971), Hysteresis effects in a differentially heated rotating fluid annulus, *Quart. J. R. Meteor. Soc.*, *97*, 506–518.

DelSole, T. (2007), Optimal pertubations in quasigeostrophic turbulence, *J. Atmos. Sci.*, *64*, 1350–1364.

Dettinger, M. D., M. Ghil, C. M. Strong, W. Weibel, and P. Yiou (1995), Software expedites singular-spectrum analysis of noisy time series, *Eos Trans. AGU*, *76*(2), 12, 14, 21.

Fasshauer, G. E. (2007), *Meshless Approximation Methods with MATLAB*, Interdisciplinary Mathematical Sciences, Vol. 6, world Scientific Publishers, Singapore.

Fein, J. S. (1973), An experimental study of the effects of the upper boundary condition on the thermal convection in a rotating, differentially heated cylindrical annulus of water. *Geophys. Fluid Dyn.*, *5*, 213–243.

Foias, C., O. Manley, R. Rosa, and R. Temam (2008), *Navier-Stokes Equations and Turbulence*, Encyclopedia of Mathematics and Its Applications, Cambridge University Press, Cambridge.

Fowlis, W. W. and R. Hide (1965), Thermal convection in a rotating annulus of liquid: Effect of viscosity on the transition between axisymmetric and non-axisymmetric flow regimes, *J. Atmos. Sci.*, *22*, 541–558.

Früh, W.-G. (1996), Low-order models of wave interactions in the transition to baroclinic chaos. *Nonlin. Process. Geophys.*, *3*, 150–165.

Früh, W.-G. and P. L. Read (1997), Wave interactions and the transition to chaos of baroclinic waves in a thermally driven rotating annulus, *Phil. Trans. R. Soc. Lond. A*, *355*, 101–153.

Fultz, D. (1961), Development in controlled experiments on larger scale geophysical problems, *Adv. Geophys.*, 1–104.

Fuselier, E. J. and G. B. Wright (2009), Stability and error estimates for vector field interpolation and decomposition on the sphere with RBFs, *SIAM J. Num. Anal.*, *47*, 3213–3239.

Gyüre, B., I. Bartos, and I. M. Jánosi (2007), Nonlinear statistics of daily temperature fluctuations reproduced in a laboratory experiment, *Phys. Rev. E*, *76*, 037,301.

Harlander, U. (2005), A high latitude quasigeostrophic delta plane model derived from spherical geometry, *Tellus*, *57A*, 43–54.

Harlander, U., R. Faulwetter, K. Alexandrov, and C. Egbers (2009a), Estimating local instabilities from data with application to geophysical flows, *Advances in Turbulence XII, Springer Proceedings in Physics*, *132*, 163–167.

Harlander, U., H. Ridderinkhof, M. W. Schouten, and W. P. M. De Ruijter (2009b), Long term observations of transport, eddies, and Rossby waves in the Mozambique Channel, *J. Geophys. Res.*, *114*, C02003, doi:10.1029/2008JC004846.

Harlander, U., Th. von Larcher, Y. Wang, and C. Egbers (2011), PIV- and LDV-measurements of baroclinic wave interactions in a thermally driven rotating annulus, *Exp. Fluids*, *51*, 37–49, doi:10.1007/s00348–009–0792–5.

Harlander, U., J. Wenzel, K. Alexandrov, Y. Wang, and C. Egbers (2012a), Simultaneous PIV- and thermography-measurements of partially blocked flow in a heated rotating annulus. *Exp. Fluids*, *52*, 1077–1085, doi:10.1007/s00348-011-1195-y.

Harlander, U., G. B. Wright, and C. Egbers (2012b), Reconstruction of the 3D flow field in a differentially heated rotating annulus by synchronized particle image velocimetry and infrared thermography measurements, *16th Int. Symp. on Appl. of Laser Techniques to Fluid Mech., Lisbon, Portugal*, 9–12 July, 16:4.5.1.

Hasselmann, K. F. (1988), PIPS and POPS: The reduction of complex dynamical systems using principal interaction and oscillation patterns, *J. Geophys. Res.*, *93*, 11015–11021.

Hide, R. (1958), An experimental study of thermal convection in a rotating fluid, *Phil. Trans. R. Soc. Lond. A*, *250*, 441–478.

Hide, R. and P. J. Mason (1970), Baroclinic waves in a rotating fluid subject to internal heating, *Phil. Trans. Roy. Soc. Lond. A*, *268*, 201–232.

Hide, R. (2010), A path of discovery in geophysical fluid dynamics, *Astron. Geophys.*, *51*, 4.16–4.23.

Hignett, P., A. A. White, R. D. Carter, W. D. Jackson, and R. M. Small (1985), A comparison of laboratory measurements and numerical simulations of baroclinic wave flows in a rotating cylindrical annulus, *Q. J. Roy. Meteor. Soc.*, *111*, 131–154.

Holmes, P., J. L. Lumley, and G. Berkooz (1996), *Turbulence, Coherent Structures, Dynamical Systems and Symmetry*, Cambridge University Press, Cambridge.

Elsner J. B., and A. A. Tsonis (1996), *Singular Spectrum Analysis: A New Tool in Time Series Analysis*, Plenum, New York.

Kaiser, J. A. C. (1970), Rotating deep annulus convection 2. wave instabilities, vertical stratification and associated theories. *Tellus*, *22*, 275–287.

Kalnay, E. (2002), *Data Assimilation and Predictability*. Cambridge University Press, Cambridge.

Lewis, G. M. and W. Nagata (2004), Linear stability analysis for the differentially heated rotating annulus, *Geophys. Astrophys. Fluid Dyn.*, *98*, 279–299.

Lindzen, R. S., B. Farrell, and D. Jacqmin (1982), Vacillation due to wave interference: Applications to the atmosphere and to annulus experiments, *J. Atmos. Sci.*, *39*, 14–23.

Lorenz, E. N. (1956), Empirical orthogonal functions and statistical weather prediction. Scientific Report No. 1, Statistical Forecasting Project. Air Force Research Laboratories, Office of Aerospace Research, USAF, Bedford, MA.

Lorenz, E. N. (1962), Simplified dynamic equations applied to the rotating-basin experiments. *J. Atmos. Sci.*, *19*, 39–51.

Lorenz, E. N. (1964), The problem of deducing the climate from the governing equations, *Tellus*, *16*(1), 1–11.

Lorenz. E. N. (1984), Irregularity: A fundamental property of the atmosphere, *Tellus*, *36A*, 98–101.

Mason, P. J. (1975), Baroclinic waves in a container with sloping end walls. *Phil. Trans. R. Soc. Lond. A*, *278*, 397–445.

Maxworthy, T., and F. K. Browand (1974), Experiments in rotating and stratified flows: Oceanographic application, *Ann. Rev. Fluid Mech.*, *7*, 273–305.

Miller, T. L., and R. L. Gall (1983), A linear analysis of the transition curve for the baroclinic annulus, *J. Atmos. Sci.*, *40*, 2293–2303.

Morita O., and M. Uryu (1989), Geostrophic turbulence in a rotating annulus of fluid, *J. Atmos. Sci.*, *46*, 2349–2355.

Mundt, M. D., and J. E. Hart (1994), Secondary instability, EOF reduction, and the transition to baroclinic chaos, *Phys. D*, *78*, 65–92.

Narcowich F. J., and J. D. Ward (1994), Generalized Hermite interpolation via matrix-valued conditionally positive definite functions. *Math. Comp.*, *63*, 661–687.

Penland, C., and P. D. Sardeshmukh (1995), The optimal growth of sea surface anomalies, *J. Clim.*, *8*, 1999–2024.

Pfeffer, R. L., and W. W. Fowlis. (1968), Wave dispersion in a rotating differentially heated cylindrical annulus of fluids, *J. Atmos. Sci.*, *25*, 361–371.

Pfeffer, R. L., G. Buzyna, and R. Kung (1980), Time-dependent modes of thermally driven rotating fluids, *J. Atmos. Sci.*, *37*, 2129–2149.

Pfeffer, R. L., J. Ahlquist, R. J. Kung, Y. Chang, and G. Q. Li (1990), Study of baroclinic wave behavior over bottom topography using complex principal component analysis of experimental data. *J. Atmos. Sci.*, *47*, 67–81.

Pfeffer, R. L., S. R. Applequist, R. Kung, C. Long, and G. Buzyna (1997), Progress in characterizing the route to geostrophic turbulence and redesigning thermally driven rotating annulus, *Theor. Comput. Fluid Dyn.*, *9*, 253–267.

Pierrehumbert, R. T., and K. L., Swanson (1995), Baroclinic instability. *Ann. Rev. Fluid Mech.*, *27*, 419–467.

Randriamampianina, A., W.-G. Früh, P. L. Read, and P. Maubert (2006), Direct numerical simulations of bifurcations in an air-filled rotating baroclinic annulus, *J. Fluid Mech.*, *561*, 359–389.

Ravela, S., J. Marshall, C. Hill, A. Wong, and S. Stransky (2010), A realtime observatory for laboratory simulation of planetary flows, *Exp. Fluids*, *48*, 915–925.

Rayer, Q. G., D. W. Johnson, and R. Hide (1998), Thermal convection in a rotating fluid annulus blocked by a radial barrier. *Geophys. Astrophys. Fluid Dyn.*, *87*, 215–252.

Read, P. L. (1992), Applications of singular systems analysis to "baroclinic chaos". *Phys. D*, *58*, 455–468.

Read, P. L., (1993), Phase portrait reconstruction using multivariate singular systems analysis, *Phys. D*, *69*, 353–365.

Read, P. L., M. J. Bell, D. W. Johnson, and R.M. Small (1997), Quasi-periodic and chaotic flow regimes in a thermaly-driven, rotating fluid annulus, *J. Fluid Mech.*, *238*, 599–632.

Read, P. L., P. Maubert, A. Randriamampianina, and W.-G. Früh (2008), Direct numerical simulation of transitions towards structural vacillation in an air-filled, rotating, baroclinic annulus. *Phys. Fluids.*, *20*, 044107:1–17.

Rintoul, S. R., C. Hughes, and D. Olbers (2001), The Antarctic Circumpolar Current System, in *Ocean Circulation and Climate*, edited by G. Siedler, J. Gould, and J. Church, Elsevier Science, New York.

Scolan, H., J.-B. Flor, R. Verzicco. Frontal instabilities at a density-shear interface in rotating two-layer stratified fluid, chapter 11, this issue.

Seelig, T., U. Harlander, R. Faulwetter, and C. Egbers (2012), Irregularity and singular vector growth in the differentially heated rotating annulus, *Theor. Comp. Fluid Dyn.*, doi:10.1007/s00162-011-0255-5.

Sitte, B., and C. Egbers (2000), Higher order dynamics of baroclinc waves. in *Physics of Rotating Fluids*, edited by G. Pfister and, C. Egbers, pp. 355–375, Springer, Berlin.

Stein, M. L. (1999), *Interpolation of Spatial Data: Some Theory for Kriging*, *Springer Series in Statistics*, Springer-Verlag, New York.

Stephen, A. V., I. M. Moroz, P. L. Read, and W.-G. Früh (1997), A comparison of empirical orthogonal decomposition methods in baroclinic flows, *Dyn. Atmos. Oceans*, *27*, 649–660.

Stephen, A. V., I. M. Moroz, and P. L. Read (1999), POD analysis of baroclinic wave flows in the thermally-driven, rotating annulus experiment, *Phys. Chem. Earth (B)*, *24*, 449–453.

Trefethen, L. N., A. E. Trefethen, S. C. Reddy, and T. A. Driscoll (1993), On almost rigid rotations. part 2, *Science*, *261*, 578–584.

Vautard, R. (1995), Patterns in time: SSA and MSSA, in *Analysis of Climate Variability*, edited by H. Von Storch, and A. Navarra, Springer, Berlin.

von Larcher, Th., and C. Egbers (2005a), Dynamics of baroclinic instabilities using methods of nonlinear time series analysis, in J. Peinke, A. Kittel, S. Barth, and M. Oberlack, edited by *Progress in Turbulence*, edited by Springer, Berlin.

Von Larcher, Th., and C. Egbers (2005b), Experiments on transitions of baroclinic waves in a differentially heated rotating annulus, *Nonlin. Proc. Geophys.*, *12*, 1033–1041.

von Larcher, Th., A. Fournier, and R. Hollerbach (2013), *The influence of a sloping bottom endwall on the linear stability in the thermally driven rotating annulus with a free surface*, Theoretical and Computational Fluid Dynamics, *27*(3), 433–451.

von Larcher, Th. and A. Dörnbrack, *Numerical simulations of baroclinic driven flows in a thermally driven rotating annulus using the immersed boundary method*, Meteorologische Zeitschrift, to appear.

Von Storch, H., (1995), Statistical patterns: EOFs and CCA, in *Analysis of climate variability*, edited by H. Von Storch, and A. Navarra, Springer, Berlin.

Von Storch, H., and F. W. Zwiers (1999), *Statistical analysis in climate research*, by Storch, H. von and Zwiers, F.W. Cambridge University Press, Cambridge, 1999, No. of Pages x +484pp. Price US$65-00, ISBN 0 521 45071 3 (Hardback).

J.-S. von Storch (1995), Multivariate statistical modeling: POP-model as a first order approximation. in *Analysis of climate variability*, edited by H. von Storch and A. Navarra, Springer, Berlin.

Weeks, E. R., Y. Tian, J. S. Urbach, K. Ide, H.L. Swinney, and M. Ghil (1997), Transitions between blocked and zonal flows in a rotating annulus with topography. *Science*, *278*, 1598–1601.

Weng, H.-Y., A. Barcilon, and J. Magnan (1986), Transitions between baroclinic flow regimes, *J. Atmos. Sci.*, *43*, 1760–1777.

Wilson, C., and R. G. Williams (2006), When are eddy tracer fluxes directed downgradient? *J. Phys. Oceanogr.*, *36*, 189—201.

Will, A., U. Harlander, and W. Metz (2006), Climatological relevance of leading seasonal singular vectors. Part I: Energy, enstrophy and spatiotemporal variability, *Meteorologische Zeitschrift*, *15*, 463–472.

Williams, P. D., T. W. N. Haine, and P. L. Read (2008), Inertia-gravity waves emitted from balanced flow: Observations, properties and consequences. *J. Atmos. Sci.*, *65*, 3543–3556.

Yang, H., (1990), *Wave Packets and Their Bifurcations in Geophysical Fluid Dynamics*. Springer Verlag, New York.

Young, R. M. B., and P.L. Read (2008), Breeding and predictability in the baroclinic rotating annulus using a perfect model, *Nonlin. Processes Geophys.*, *15*, 469–487.

INDEX

2D boundary layer, 96
2D inverse cascade, 162
2D turbulence, 113, 161
2D wakes, three-dimensional destabilization of, 271–273
3D atmospheric QBO, numerical modeling of, 181. See also Quasi-biennial oscillation (QBO)
3D boundary layer, 96
3D destabilization, of two-dimensional wakes, 271–273
3D eddy-resolving simulation, 35
3D effects, 153, 155, 156
3D flow measurements, 155, 334
3D issues, in particle tracking, 281
3D matching, 282–283
3D nonlinear steady states, 88–89
3D numerical simulations, 155
3D optical tracking, example of, 283
3D particle tracking velocimetry (PTV), 280–283, 294, 295
3D simulated flows, 35
3D simulation, 187
3D trajectories, reconstruction of, 282–283
3-tube experiment, 256

Abrupt thermohaline transitions/oscillations, laboratory experiments with, 255–263
Abrupt transitions, 255–256
Absolute instability, 90
Antarctic Circumpolar Current (ACC) dynamics, 329
Acceleration autocorrelation functions, 286
Acceleration measurements, 294, 295
Acoustical particle tracking, example of, 290
Acoustical vorticity measurements, 286
Acoustic Doppler Lagrangian tracking, 287–290
Acoustic–fluid interaction, 290
Acoustic scattering, 286, 290–291
Acoustic techniques, for modeling atmospheric and oceanic fluid flow, 286–293
Acoustic tracking, 295
    advantages of, 288
    Lagrangian, 286, 287–290
Acoustic transducer calibration, 293
Acoustic vorticity measurement, 295
Acoustic waves, plane scattered amplitude of, 290–291
Acousto-optic modulators (AOMs), 284, 285
Adams–Bashforth scheme, 242, 243
AB/BDF scheme, 300
Adjustment waves, 222
Advection, vertical, 139
Advective heat transport/transfer, 25–26, 27
Ageostrophic model, 120, 124

Agulhas current, 231
Algorithms. *See also* Approximated maximum likelihood (AML) method; Computer simulations; GMRES iteration; Orthogonal decomposition methods
    fast demodulation, 285–286
    nearest neighbor, 283
Altimetric imaging velocimetry (AIV), 103, 105, 113
    observing flows using, 107
Altimetry
    in a GFD laboratory, 101, 102–105
    laboratory, 102–103
    optical, 103, 116
    satellite, 208
Altimetry technique, 101
Amplitude equations, 50
Amplitude modulations, 19
Amplitude oscillation, 16
Amplitude vacillation (AV), 15–16, 19, 71, 213, 218–219, 224, 302, 316
    baroclinic, 19, 20
    in baroclinic flows, 61–81
    toward chaos and turbulence, 76–77
    cycle, 19, 20
    defined, 62
    flow regime, 301, 302, 306–307
    mechanics of, 64–65
    modeling approaches for, 65–68
    modulated, 62
    periodic, 62
    phenomenology of, 61–64
    transition to, 64
Analysis correction method, 38
Angular drift velocity, 305–306
Angular factor, 291, 292
Angular velocity, 120, 140, 320. *See also* Rotation rates
Annular channels, 232, 233, 250
Annulus (annuli). *See also* Rotating annuli
    baroclinic, 77
    subdivided, 328–329
    thermally driven, 61
    two-layer flows in, 120
Annulus construction, 12–13
Annulus experiments
    boundary conditions of, 47
    differentially heated rotating fluid, 45
    fluid-flow governing equations of, 46–47
    role of, 12
Annulus flows, 334
    Helmholtz–Hodge decomposition of, 331–333

Modeling Atmospheric and Oceanic Flows: Insights from Laboratory Experiments and Numerical Simulations,
First Edition. Edited by Thomas von Larcher and Paul D. Williams.
© 2015 American Geophysical Union. Published 2015 by John Wiley & Sons, Inc.

INDEX

Annulus geometry, parameters for, 51
Annulus models
    eddy parameterizations in, 34–36
    Lorentz, 331
Annulus rotation rate, 317
Annulus systems, rotating, 12. *See also* Baroclinic annulus
Antarctic Slope Front (ASF), 231
Anticyclones, 269–270, 271. *See also* Cyclone–anticyclone asymmetry
    inertial–centrifugal instability of stratified, 273
    unstable, 272
Anticyclonic cyclogenesis, 18
Anticyclonic eddies, 272
Anticyclonic relative vorticity, 108
Anticyclonic rotation, 88, 95
Anticyclonic vorticity/vortices, 111, 147
Approximated maximum likelihood (AML) method, 286, 289
Aqua-planet idealization, 181
Archimedes's principle, 1
Asymmetry, cyclone–anticyclone, 274
Asymptotic solutions, 239
Atmospheres
    controlled experiments on, 38–39
    eddy heat transfer in, 27–28
    global circulation of, 10–12
Atmospheric circulation, 10–12, 315
Atmospheric dynamics, ocean dynamics and, 297
Atmospheric fluctuations, intraseasonal, 184
Atmospheric fronts, 297
Atmospheric mesoscale, cascade theories and, 162–163
Atmospheric vortex shedding, 265
Available potential energy (APE), 17
Axially rotating pipe flow experiments, 94–95
    results in, 97
Axially rotating pipe flows, 87, 90–91
Axisymmetric disturbances, 201
Axisymmetric flow, 13, 26, 75, 152
Axisymmetric flow regimes, 28–31, 302
    experimental verification of, 31
Axisymmetric noise, 183
Axisymmetric shallow vortex flow, 155
Axisymmetric sink vortex, 152
Axisymmetric solutions, 47–48, 50. *See also* Steady axisymmetric solution
Axisymmetric synthetic schlieren methods, 201–203
Axisymmetric/wave transition, 14–15
Azimuthal
    flow, 26, 325
    length scales, 239
    retrograde, 250
    mean velocity, 97
    periodicity, 96
    reflection symmetry, 91
    spatial spectrum, 76
    transport, 239–240, 243
    velocity, 91, 93, 221, 240
    velocity profile, 96
    vorticity equation, 29
    wave numbers, 19, 23, 24, 316

Background buoyancy frequency, 199
Background rotation, 155, 217
Balance, in rotating systems, 140
Balanced flow, 112
Balanced states, 127
Baroclinic amplitude vacillation, 19, 20, 61–81
Baroclinic annulus, 77
    primary flow transition in, 45–59
Baroclinic annulus experiment, 315
Baroclinic cavity instability/transition, direct numerical simulation of, 297–313
Baroclinic eddy transport
    quantifying, 31–32
    testing local closures for, 32–34
Baroclinic flows, 1, 73, 105
    amplitude vacillation in, 61–81
    strongly supercritical, 34
Baroclinic fronts, 217
Baroclinic instability (BI), 11, 14, 16, 18, 27, 39, 45, 65, 121, 124, 129, 213–214, 222, 310, 315
Baroclinic instability experiments, in rotating annuli, 119
Baroclinic instability theory, weakly nonlinear, 19, 21
Baroclinic life cycle, 325
Baroclinic two-layer model, 78–79
Baroclinic wave regime, 39
Baroclinic waves, 25–36, 45, 73, 74, 334
    large-scale, 308
    numerical investigations of, 297–301
    in rotating annulus with barrier, 324–329
    steady, 298
    traveling, 67
Baroclinic wave simulations, 33
Barotropically unstable flow, 39
Barotropic coastal currents, 251
Barotropic dynamics, 235
Barotropic interaction, 130–131
Barotropic stability, 17
Bathymetry, impact on instabilities, 134–136, 137
BEK boundary layer, 86, 87
BEK flows, 96
Benjamin–Feir instability, 70
β-drift, of monopolar vortices, 144–145
β-effect, topographic, 143–144, 145, 147, 149
β-plane approximation, 139
β-planes
    flows on polar, 101, 102
    planetary, 144
    topographic, 145, 149
β-plumes, 108
Biconjugate gradient method, 204–205
Bifurcating waves, 49
Bifurcation, 301. *See also* Double-Hopf bifurcation; Hopf bifurcation
Bifurcation diagrams
    one-parameter, 53
    two-parameter, 53, 54, 55, 56
Bifurcation sequences, 19, 21, 22
Bödewadt layer, 87
Body forces, unstably stratified, 85

Bottom friction, 237, 245, 247
  parameterizing, 240
  with rotation, 151–152
Bottom-induced vorticity, 153
Bottom plate, parabolic, 144
Bottom topography, experiments on flows over, 139–158
Boundary conditions, 127, 238, 256, 299–300
  of annulus experiments, 47
Boundary currents, stability of, 125–134
Boundary forcing, 187
Boundary layer feedback, low-order model of, 78–79
Boundary layer flow, 87
  laminar, 89–90
  rotating disk, 89
  turbulent rotating disk, 96–97
Boundary layer flow simulations, 97
Boundary layer instability, 95–96
Boundary layers
  BEK, 86, 87
  diffusive, 37
  laminar, 89
  role of, 71–72
  thermal, 72, 73–74
  three-dimensional, 96
  turbulent, 87, 97
  two-dimensional, 96
  velocity, 73
  viscous, 15, 193
Boundary layer separation, 193, 195
Boundary points, 332
Boussinesq approximation, 46–47, 215
Boussinesq equations, 182, 183
Boussinesq fluid, 15, 18, 35, 38, 180
Box experiment, 256
Box model theory, 258–259
Breaking conditions, 245–246
Breaking wave amplitude, 249–250
Breaking wavelength, 247–249
Breeding vectors, 316
Brunt–Väisälä frequency, 34, 63, 161
Buoyancy, 28
  centrifugal, 15
Buoyancy-driven boundary currents, stability of, 125–134
Buoyancy-driven flows, 1
Buoyancy-driven overturning circulation, 25
Buoyancy effects, centrifugal, 47
Buoyancy flux, 166, 167, 168, 169, 261
Buoyancy flux spectra, 168, 169
Buoyancy forces, 4, 159, 256
Buoyancy frequency, 63. *See also* Squared buoyancy frequency
Buoyancy Reynolds number (Re_b), 160, 163–164, 169
Buoyancy scale, 161, 162, 171, 173
Buoyancy time scale, 161
Buoyant coastal currents, 233
Buoyant gravity currents, 251
Burger number (B, Bu), 2, 15, 120, 128, 129, 213, 216, 219, 224, 267, 269–270, 271, 273
Bu–Ro diagram, 216, 218. *See also* Burger number (B, Bu); Rossby number (Ro)

Calibration methods, 282. *See also* Acoustic transducer calibration
Cartesian system, 102
Cascade theories, atmospheric mesoscale and, 162–163
Cavities
  liquid-filled, 303–310
  open free-surface water-filled, 310
Cavity experiment, 256, 257, 262
Center manifold theorem, 49–50
Centrifugal acceleration, density variations in, 298
Centrifugal buoyancy, 15
Centrifugal buoyancy effects, 47
Centrifugal force, 85
Chain rule, 242
Channel flows, 87
Chaos, amplitude vacillation toward, 76–77
Chaotic modulated amplitude vacillation, 77
Chaotic regime, 218
Chaotic state, 22
Chebyshev–Gauss–Lobatto distribution, 300
Chebyshev method, 300
Chebyshev polynomials, 298, 300
Circulation, of planetary atmospheres, 9–44
Circulation regimes, key classes of, 11–12
Climate simulations, laboratory-scale, 186
Cloud patterns, 265
Coastal currents, 112–113, 232–233, 235–236
  barotropic, 251
  buoyant, 233
  evolution of, 251–252
  in the laboratory, 231–234
  large-scale flow of, 250
  retrograde, 233, 234, 235, 251
Coastal current velocity profiles, 113
Coastal geometry, 233
Coastal shelf waves, large-amplitude, 231–253
Collapse technique, 147
Collocation method, 121
Columnar mode, 207
Complex EOF (CEOF) analysis, 318–322, 324–327
Complex ocean wakes, idealized laboratory flows and, 274–275
Computational fluid dynamics (CFD), 1, 65–66
  wave interactions in, 70–71
Computer simulations, 1–5
Conceptual models, 9
Conductive heat transport, 27
Conjugate pairs, of eigenvalues, 49
Continental shelf waves, 146
Continuation methods/techniques, 15, 56–57
Continuation parameter (α), 57
Continuity equations, 120, 141, 155
Contour plots, 246–248, 332
Contravariant velocity, 182
Control parameter values, 303–305
Convection
  sloping, 11, 12, 14
  thermal, 10
Convergence, under grid refinement, 243

Coriolis acceleration, 13, 26, 140
Coriolis force, 2, 85, 86
Coriolis parameter (f), 102, 105, 106, 108
Coriolis platform, 144, 145, 147
Coriolis terms, 63, 66, 104, 298
Couette flow, 85, 86, 87. *See also* Plane Couette flow (PCF); Rotating plane Couette flow entries
Couette flow apparatus, 91
Critical layer, 180–181
Cross-flow instability, 89–90
Cross-stream Reynolds stress, 97
Curl-free field, 331, 332
Curl-free matrix-valued kernels, 331
Currents, buoyant gravity, 251
Curved channel flow, 86
Cyclogenesis, anticyclonic and cyclonic, 18
Cyclogeostrophically balanced states, 127
Cyclogeostrophic equilibrium, 120
Cyclone–anticyclone asymmetry, 274
Cyclones, 269, 271
Cyclonic cyclogenesis, 18
Cyclonic eddies, 271, 272
Cyclonic rotation, 89
Cyclonic vortices, 145, 147–148
  over topographic ridges, 149
Cyclonic vorticity, 108, 111

Data management issues, in particle tracking, 281
Data matrices, 320
Dead water phenomenon, 1
Dean flow, 86
Decaying dipolar vortices, evolution of, 155–156
Decaying grid turbulence experiments, 160
Deep-water wakes, 265–266
Demodulation techniques, 285–286
Density change, 258–259
Density gradients
  inclined, 213
  measuring, 205
Density interface thickness, 219–220, 221, 225
Density–shear interfaces, frontal instabilities at, 213–228
Deterministic model predictions, 37–38
Differentially heated rotating annulus, 316, 333–334
  of fluid, 315
  experiments, 45
Diffusive boundary layers, 37
Diffusivity, 267
Digital imaging, 279
Dimensional analysis, of island wakes, 266–267
Dimensionless drift frequency, 306
Dimensionless equations, 161
Dimensionless parameters, 45
Dipolar vortices, 153
  evolution of decaying, 155–156
Dipoles, 114
Direct numerical simulations (DNSs), 3, 32, 34, 87, 89, 90, 160, 163–171, 179, 187, 297, 310–311
  of baroclinic cavity instability/transition, 297–313
  of geophysical laboratory experiments, 179–191
  of laboratory analogue QBO, 182–183
  of laboratory-scale stratified turbulence, 159–175
  results of, 173
Discretization, 56, 57, 58
Dispersion curves, 129
Dispersion diagrams, 121, 123–125
Dispersion relations, 106, 115, 147
Dissipation, 16
Dissipation number (d), 216
Dissipation parameter, 67
Dissipation terms, 167–168
Distortion prediction, 205
Disturbance field, 201, 202
Divergence-free field, 331, 332
Divergence-free matrix-valued kernels, 331
DNS code, 217
DNS simulation, 183
Dominant wave flow, 17
Doppler shift, 287–289, 291, 293
  in acoustic tracking, 286, 287–290
Doppler shift extraction, 288–289
Double-diffusion processes, 262
Double Fourier transform, 115
Double-Hopf bifurcation, 75
Downmixing, 288–289
Drift frequency, dimensionless, 306
Drift periods, 301
Dufort–Frankel scheme, 298
Dye lines, 207, 210, 235, 236, 245
Dynamical parameters, of island wakes, 266–267
Dynamical systems, linear, 330–331
Dynamical systems approach, 46, 56
Dynamic fields, global measurements of, 101

Eady model, 14
Eady-type models, 67
Earth rotation, impact on oceanic flow, 266
Eastward flows, 109–111
Eddies, 18
  anticyclonic, 272
  cyclonic, 271, 272
  stability of, 269
Eddy diffusion coefficient, 28
Eddy diffusivity, 33–34
Eddy heat flux, horizontal, 33
Eddy heat transfer, in oceans and atmospheres, 27–28
Eddy-induced transport, 31–32
Eddy-induced velocity, 28
Eddy kinetic energy, 64, 65
Eddy parameterizations, 32–33
  in an annulus model, 34–36
Eddy potential energy, 65
Eddy transport, quantifying baroclinic, 31–32
Eddy variances/fluxes, 28, 33, 329
Effective eddy diffusivity, 33–34
Effective thermal forcing, 74, 75
Eigenfunctions, 48, 49
Eigenmodes, 127–130

Eigenvalues, 48
  conjugate pairs of, 49
Eigenvalue spectrum, 321
Ekman
  boundary layer, 151
  circulation, 220–221
  condition, 141–142
  corrections, nonlinear, 152
  friction theories, 152
  layers, 26, 29–30, 72, 87, 143, 221
    internal, 225
  number (E), 29, 62, 140, 267, 273
    inverse squared, 39
  spiral, 86–87
  study, 1
  time scale, 142, 144, 152
  transport, 26
Electromagnetic forcing, 153
Empirical orthogonal function (EOF) analysis, 316, 318–331. *See also* EOF filtering technique
  of wave interactions with PIV and LDV, 320–321
Empirical singular vectors, 329–331
Energy budget, 166–169
Energy cascades, 24–25
  inverse, 150
Energy exchanges, 25
Energy flow, 65
Energy spectra, 165–166, 169
  mesoscale, 159
Energy transfer, 68–69
Enstrophy spectrum, 293
EOF filtering technique, 330. *See also* Empirical orthogonal function (EOF) analysis
Equations of motion, 125–127, 161
  for stratified fluids, 161
Equatorial stratosphere, variability in, 180
Equatorial waves, 179
Equilibration, 15–16. *See also* Radiative–convective equilibrium
Essential balance, in rotating systems, 140
Eulerian techniques, 279–280
Evectors, 18
Evolution equation, 141, 250
Experimental
  flow comparison, 240–241, 243–245
  flows, real flows vs., 329
  fluid mechanics, Lagrangian methods in, 279–296
  modeling, of fronts, 217
  techniques, in fluid mechanics, 295
  topography variations, 233
Experiments. *See also* Laboratory experiments; Turbulence experiments
  with large-amplitude shelf waves, 234–236
  large-scale, 144–148
  medium-scale, 148–152
  as physical system models, 9–10
  shallow-layer flow, 153–155
  small-scale, 152–156
Extended laser Doppler velocimetry (ELDV), 280, 283–286, 288, 295

Fast demodulation algorithm, 285–286
Field programmable gate array (FPGA) technology, 295
Finite-amplitude waves, 73
Fixed points, 51
Flows. *See also* Annulus flows; Baroclinic flows; Boundary layer flow; Laboratory flows; Mean flow entries; Oceanic flow entries; Rotating plane Couette flow entries; Rotating disk flows; Rotating flows; Shallow-water flows; Shallow flows; 3D flow measurements; Turbulent flows; Two-layer flows; Vortex flow; Wall-bounded flows; Wave
  axially rotating pipe, 87, 90–91, 94, 95
  axisymmetric, 13, 26, 75, 152
  azimuthal, 26, 325
  balanced, 112
  barotropically unstable, 39
  behavior, time-dependent, 316
  BEK, 96
  over bottom topography, 139–158
  buoyancy-driven, 1
  channel, 87
  complexity of, 279
  Couette, 85, 86, 87, 91
  curved channel, 86
  Dean, 86
  dynamics, governing equations for, 298–299
  eastward, 109–111
  evolution, of instabilities, 218–224
  experimental *vs.* real, 329
  fields, measuring, 205
  geophysical, 1–2, 139–140, 143–144, 279
  Hagen–Poiseuille, 90
  interactions, wave–zonal, 19–22
  laminar boundary layer, 89–90
  long-term state of, 150–151
  measurement approaches, 279–280
  measurements, nonintrusive, 3
  mixed-mode, 54
  observing with AIV, 107
  around obstacles, 109–111
  oceanic, 265–276, 325
  parameters, of rotating flows, 88–91
  phenomena dynamics, 187
  Poiseuille, 85
  on polar β-plane, 101, 102, 108, 111
  problems, numerical methods for, 1
  radial, 26
  regimes, 13–14, 38–40, 315
    classes of, 301–302, 303, 306–307, 307–309
    irregular, 316, 321–322
  rotation-induced, 87
  shallow, 152
  S-mode, 257, 258
  Taylor–Couette, 86
  3D simulated, 35
  T-mode, 257, 258, 259
  turbulent, 279
  turbulent pipe, 94
  westward, 108, 109–110
  zonal, 15–16, 39–40, 68, 75, 107, 108, 324–325

Flow–topography phenomena, 139
Fluctuations, small-scale, 306–308, 309
Fluid density (ρ), 46
Fluid dynamics, 184
    computational, 65–66
Fluid dynamics processes, 279
Fluid interfaces, choice of, 214
Fluid mechanics
    experimental techniques in, 295
    Lagrangian methods in, 279–296
Fluid mechanics experiments, history of, 1
Fluid motion, 140
Fluid phenomena modeling, 144
Fluid properties, dependence on, 4
Fluo-line technique, 206–207
Fluorescence, laser-induced, 206–207
Flux form formulations, 188
F_m extended matrix, 320
Forcing, 164–165, 232
    boundary, 187
    effective thermal, 74, 75
    electromagnetic, 153
    harmonic, 69–70
    rotational, 164–165
    small-scale, 162
Forcing methods, 150
Forecast simulations, 185
Fourier analysis, 285
Fourier–Chebyshev model, 66
Fourier-convolution approach, 203
Fourier modes, 68
Fourier transform images, 198
Fourier transforms, 115, 290, 291
f-plane approximation, 142
Frames of reference
    nonrotating, 85
    rotating, 85
Free surface outcropping, 125–127
Free-surface water-filled cavity, open, 310
Frequency, 303
Frequency vacillation, 62
Frictional effects, reducing, 2
Friction coefficient, 144
Friction velocity, 94, 96
Frontal instabilities, at density–shear interfaces in rotating two-layer stratified fluids, 213–228
Frontal modes, 128, 130
Frontal stability results, 226
Fronts
    baroclinic, 217
    experimental modeling of, 217
    oceanic and atmospheric, 297
Front stability, 119–125
Froude number (F), 15, 31, 63, 67, 71, 122–124, 159–160

Gal–Chen coordinate transformation, 182–183
γ-plane approximation, 102
Gaussian hill, 209
Gent–McWilliams parameterization, 34–35

Geophysical flows, 1–2, 139–140, 279
    experiments on, 143–144
Geophysical fluid dynamics (GFD), 101–102. *See also* GFD laboratory
Geophysical laboratory experiments, numerical simulations of, 179–191
Geostrophic balance, 266
Geostrophic turbulence, 316, 331
    transition to, 22–25
Geostrophic velocity ($\mathbf{V}_g$), 104
GFD laboratory, altimetry in, 101, 102–105. *See also* Geophysical fluid dynamics (GFD)
Ginzburg–Landau equation, 96
Global circulation, of atmospheres and oceans, 10–12
Global circulation models (GCMs), 39, 181
Global instability, 90, 96
Global measurements, of dynamic fields, 101
Global scattered wave, 291
Global wake pattern, 267
GMRES iteration, 57–58
Gradient threshold, 245
Grassberger–Procaccia dimension, 77
Gravity current, 232, 233
Gravity wave emission, 331
    interfacial, 112
Gravity wave processes, 187
Gravity waves, 127, 179
    shorter-scale, 180
Grid refinement, convergence under, 243
Grid resolution, 50
Grid spacing, 164
Growth rates, instabilities and, 121–125
Gulf Stream, 231

Hagen–Poiseuille flow, 90
Harmonic forcing, 69–70
Hawaiian wake, 275
Heat convection, 74
Heated rotating annulus, 315, 316, 333–334
Heat transfer. *See also* Heat transport
    advective, 27
    eddy, 27–28
Heat transfer behavior, 72
Heat transfer dimension, 77
Heat transfer parameterization, 32
Heat transport, 25–36, 28–31
    advective, 25–26
    conductive, 27
    total, 28
Heaviside step function, 238
Heisenberg principle, 285
Helmholtz–Hodge decomposition, of annulus flows, 331–333
Heterodyne downmixing, 288–289
Hide criterion, 316
High-capacity computing, 3
Higher-order wave interactions, 70
Highly turbulent flows, 24
High-Prandtl-number fluids, 64
High-resolution numerical models, 271

High-resolution optical tracking, 295
High-resolution spectral technique, 298
Hilbert transform (HT), 286
Hilbert transform method, 198, 200, 321
Hölmböe instability, 214, 215, 222, 223–224
Hölmböe-type mode, 4
Homogeneous fluid scenario, 72
Hopf bifurcation, 49, 64, 75, 303. *See also* Double-Hopf bifurcation
  subcritical, 54, 56
  supercritical, 53, 54
Horizontal eddy heat flux, 33
Horizontal length scale, 236
Horizontal motions, 2
Horizontal velocity, 320
Horizontal wave number spectra, 165, 168
Hot plumes, 310
Hot-wire measurements, 92–93, 94, 280
Hovmöller diagram, 113
Hyperbolic tangent profile, 223
Hyperbolic tangent transformation, 298
Hyperviscosity, 173
Hysteresis, 53, 54, 255, 257–258, 262, 303
Hysteresis cycle, 306

Idealized laboratory experiments, 274–275
Idealized laboratory flows, complex ocean wakes and, 274–275
Idealized laboratory setups, for oceanic island configuration, 267–269
Image displacements, 196–198, 200–204
Imaging, digital, 279. *See also* Schlieren technology
Implicit large-eddy simulations (ILESs), 186, 188
Inclined density gradient, 213
Inertia–gravity waves (IGWs), 127, 222, 224, 307, 308, 311
Inertial–centrifugal instability, of shallow stratified anticyclones, 273
Inertial range, 279
Inertial wave emission, 112–113
Inertial waves, 105–107
Influence matrix technique, 299
Instabilities. *See also* Baroclinic cavity instability/transition; Frontal instabilities; Linear instability entries; Parametric subharmonic instability (PSI); Shear instabilities
  absolute, 90
  baroclinic, 11, 14, 16, 18, 27, 39, 45, 65, 121, 124, 129, 213–214, 222, 310, 315
  Benjamin–Feir, 70
  boundary layer, 95–96
  cross-flow, 89–90
  flow evolution of, 218–224
  global, 90, 96
  growth rates and, 121–125
  Hölmböe, 214, 215, 222, 223–224
  impact of bathymetry on, 134–136, 137
  Kelvin–Helmholtz, 161, 162, 173, 214, 215
  local absolute, 96
  nonmodal, 329
  resonances and, 127–131
  RF, 132–134
  Rossby–Kelvin, 121, 125, 136, 214–215, 219–220, 224, 225–226
  of shallow stratified anticyclones, 273
  of shallow-water flows, 119–138
  sideband, 70
  zigzag, 162
Instability experiments, 131–134
Instability theory, 14–15, 34, 329
Instability wavelength, 213
Instrumented particles, 293–294
Interface conditions, 220–221
Interface thickness, 221, 222
Interfacial gravity wave emission, 112
Interfacial waves, 214
  small, 222–224
Interference fringes, traveling, 284
Interference vacillation, 62, 75
Interior dynamics, 236–237
Internal Ekman layers, 225
Internal rotation, solar, 180
Internal waves, 199, 200
  breaking, 210
  generation, 207–209
  generation model, 203
  in laboratory experiments, 193–212
  low-mode, 208–209
  measurements of, 203, 205–206
  study of, 209–210
  tidally generated, 193
  upward *vs.* downward propagating, 198
Intransitivity phenomena, 308
Intraseasonal atmospheric fluctuation, 184
Inverse
  cascade theory, 162
  cascading, 24
  energy cascade, 150
  squared Ekman number, 39
  tomography, 203–205, 209
Inviscid quasi-geostrophic equations, numerical solution of, 241–245
Inviscid topography effects, 141
Irregular flow regimes, 316, 321–322
Irregular topographies, 150
Irregular wave regime, 309–310, 318, 321
Island wake flows, in laboratory, 265–276
Island wakes, dynamical parameters and dimensional analysis of, 266–267
Isopycnal slopes, 134–135

Jacobian operator, 241, 242
Jupiter, stratosphere of, 180

Kelvin gravity waves, 180
Kelvin–Helmholtz instability, 121, 125, 161, 162, 173, 214, 215
Kelvin–Helmholtz-type mode, 4
Kelvin waves, 128, 219
Kernel replacement technique, 332
Kernels, matrix-valued, 331, 332

Kinetic energy, 64–65, 168, 170, 171
    eddy, 64, 65
    zonal, 64, 65
Kinetic energy spectra, 162, 165–166, 171, 172
Kolmogorov–Kraichnan theory, 24
Kolmogorov scales, 165, 217
Kolmogorov theory, 162

Laboratory altimetry, 102–103
Laboratory analogue QBO, direct numerical simulations of, 182–183. *See also* Quasi-biennial oscillation (QBO)
Laboratory experiments, 3, 101. *See also* Annulus experiments; Experiments; Geophysical laboratory experiments; Open cylinder experiments; Rotating experiments; Stratified experiments; Thermal annulus experiments; Thermally driven rotating annulus laboratory experiment
    with abrupt thermohaline transitions and oscillations, 255–263
    advancement in, 193–194
    on flows over bottom topography, 139–158
    geophysical applicability of, 173
    idealized, 274–275
    internal waves in, 193–212
    Plumb–McEwan, 181–182
    results of, 193
    role of, 10
    rotating annulus, 36–38
Laboratory flows, complex ocean wakes and, 274–275
Laboratory measurements, 37
Laboratory results, scaling of, 258
Laboratory-scale
    climate simulations, 186
    experiments, for MJO-like tropical dynamics, 185–187
    stratified turbulence, numerical simulations of, 159–175
Laboratory setups
    for oceanic island configuration, 267–269
    virtual, 187
Laboratory simulations, 1–5
Laboratory studies, 36–38
    of oceanic island wake flows, 265–276
    of shelf waves and coastal currents, 231–234
Lagrangian
    acoustic tracking, 286, 287–290
    approach, 280, 294
    dynamics, 295
    measurements, 294
    methods, in experimental fluid mechanics, 279–296
    systems, 281
    temperature measurement, 293–294
    tracking, 282, 283
    tracking technique, 288
Laminar boundary layer, 89
Laminar boundary layer flow, 89–90
Landau equation, 16, 17
Laplacian operator, 242
Large-amplitude coastal shelf waves, 231–253
Large-amplitude shelf waves, 233
    experiments with, 234–236
Large-buoyancy Reynolds number (Re_b), 160

Large-eddy simulation (LES), 34, 179, 187
    of geophysical laboratory experiments, 179–191
    of MJO-like tropical dynamics, 185–187
Large rotating facilities, 274
Large-scale
    baroclinic waves, 308
    coastal current flow, 250
    experiments, 144–148
    geostrophic wake, 269–270
    numerical simulations, 181
    wakes, 269–270
Laser Doppler velocimetry (LDV), 280, 283–286, 316. *See also* LDV studies
Laser Doppler velocimetry measurements, 94, 320–321, 334
    analysis of, 321–324
Laser-induced fluorescence, 206–207
Lateral boundary meander simulation, mapping, 187
Layered apparatus, 260–261
Layered experiment, 256, 257
Layer
    thicknesses, 72–74
    thickness ratio (δ), 266
    thickness scales, 171
LDV studies, 92. *See also* Laser Doppler velocimetry (LDV)
Lee waves, 107, 108
Length scales, 169–171
    azimuthal, 239
Life cycles, baroclinic, 325
Lindborg cascade theory, 163
Linear
    dynamical systems, 330–331
    equatorial wave motions, 179
    instabilities, of rotating disk flows, 90
    instability theory, 14–15, 34
    multivariate statistical techniques, 316
    shallow-water theory, 179, 184
    shelf wave theory, 231
    stability analyses, 53, 54, 55, 87, 90
        numerical, 47–49
    stability theory, 96, 219
Linearization, 330, 334
Linearized boundary conditions, 127
Liquid-filled cavity, 303–310
Local absolute instability, 96
Local closures, testing for baroclinic eddy transport, 32–34
Local wind stress, 275
Long-wave theory, 250, 251
Lorenz annulus model, 331
Lorenz equations, 68
Lorentz force, 153
Lower symmetric regime, 39
Low-mode internal waves, 208–209
Low-order models, 67
    of boundary layer feedback, 78–79
    of wave–boundary layer interaction, 72–75
    wave interaction scenarios in, 71
Luv color space, 103
Lyapunov exponent, 303

Madden–Julian oscillation (MJO), 4, 179, 184–189. *See also* MJO entries
  defined, 186
Madeira wake, 275
Magnetic resonance imaging (MRI), 203
Marsigli's experiment, 2
Mathematical box models, 255
Matrices, 320
Matrix inversion method, 203–204
Matrix-valued kernel approximation, 332
Matrix-valued kernels, 331, 332
Meanders, 112–113
Meander simulation, mapping, 187
Mean-flow correction, 67, 68, 69
Mean-flow oscillation
  stratospheric, 181
  zonal, 181
Mean-flow reversal, zonal, 182
Mean flows
  retrograde azimuthal, 250
  wave-driven, 232
Mean-flow structure, 68
Mean flow–wave interactions, 71, 180
Mean zonal flow, 107, 108, 180, 181
  reversal, 183, 187
Measurement problems, 101
Mechanical integrators, 11
Medium-scale experiments, 148–152. *See also* Mesoscale entries
Mesh-free reconstruction method, 331
Mesoscale
  cascade theories and atmospheric, 162–163. *See also* Medium-scale experiments
  energy spectrum, 159
  motions, 2, 3
  vortex wake, 269–270
Meteorological analysis methods, 38
Metrology, advances in, 279
Mixed-mode
  flow, 54
  solution, 55, 56
MJO case studies, 186. *See also* Madden–Julian oscillation (MJO)
MJO-like tropical dynamics, LES of laboratory analogue for, 185–187
Modal amplitude equations, 17
Mode interaction points, 51–53
Model equations, quasi-geostrophic, 236–238
Model predictions, deterministic, 37–38
Models, 4–5. *See also* Numerical modeling
  of physical systems, 9–10
Moderate rotation regime, 30
Modulated amplitude vacillation (MAV), 62
  2MAV flow regime, 303
  3MAV flow regime, 307
  chaotic, 77
Modulated amplitude vacillation flow regime, 301, 303
Modulation theory, 183
Moist physics, 184–185
Molecular viscosity, 269

Momentum equations, 120
Momentum transport, 28–31
Monopolar vortices, β-drift of, 144–145
MORALS (Met Office/Oxford Rotating Annulus Laboratory Simulation) code setup, 78
MORALS model, 65, 72, 298
MSSA software package, 320. *See also* Multivariate singular system analysis (MSSA)
Multivariate singular system analysis (MSSA), 316, 319–321, 324
  of wave interactions with PIV and LDV, 320–321
Multivariate statistical techniques, 318
  linear, 316

Natural-flow simulations, in laboratory and on computer, 1–5
Navier–Stokes equations, 1, 3, 15, 18, 35, 38, 46–47, 68, 89, 90, 95, 140, 215
Navier–Stokes model, 65
Nearest neighbor algorithm, 283
Neimark–Sacker bifurcation, 56
Neutral stability curves, 48
Newton–Krylov continuation, 58
Newton–Raphson iteration, 57
Nondimensional equations, 120, 126, 155
Nondimensionalization, 47
Nondimensional parameters, 62–63, 215, 216
Nondimensional rotation rate, 88
Nonintrusive flow measurements, 3
Nonlinear
  analysis, 49–56
  Ekman corrections, 152
  interactions, 19
  shelf wave equation, 239, 240, 245, 246–247, 250, 251
  shelf wave theory, 235, 238–241
  steady states, three-dimensional, 88–89
  theory, 15–16
  time series analysis, 316
  wave equation, 233
  wave interactions, 68–69
  wave theory, 246, 248, 250
Nonmodal instability, 329
Nonoscillatory forward-in-time (NFT) approach, 183
Nonrotating frame of reference, 85
Nonrotating problem, 29
Non-spanwise-uniform disturbances, 200–205
Normal form coefficients, 51, 52
Normal form equations, 50–51
No rotation regime, 30
Novel wave generator, 207–209
Null point, 104
Numerical
  approaches, 217–218
  continuation methods/techniques, 15, 56
  experiments, 164
  integration, 242–243
  investigations, of baroclinic waves, 297–301
  linear stability analysis, 47–49
  methods, for flow problems, 1
  modeling, 37

Numerical (*Cont'd*)
  models, 182–183, 186–187, 331
    high-resolution, 271
  ocean models, 231
  procedures, confidence in, 187
  simulations, 224, 297
    of baroclinic cavity instability/transition, 297–313
    of geophysical laboratory experiments, 179–191
    of laboratory-scale stratified turbulence, 159–175
    large-scale, 181
    3D, 155
  solutions, of inviscid quasi-geostrophic equations, 241–245
  studies, of stratified turbulence, 163
  viscosity, 241, 242
Nusselt number, 26–27, 29, 31

Obstacles, flows around, 109–111
Oceanic
  flows, 325
    impact of earth rotation on, 266
  flow system, 86
  fronts, 297
  island configuration, idealized laboratory setups for, 267–269
  island wake flows, in laboratory, 265–276
Oceans
  circulation models, 255
  dynamics, atmospheric dynamics and, 297
  eddy heat transfer in, 27–28
  global circulation of, 10–12
  temperatures, 255
  wakes, idealized laboratory flows and, 274–275
One-parameter bifurcation diagram, 53
Open cylinder experiments, 11–12
Open free-surface water-filled cavity, 310
Optical altimetry, 103, 116
Optical arrangement, 284
Optical density method, 101, 105
Optical techniques, 280–286
Optical thickness method, 3
Optical tracking
  example of 3D, 283
  high-resolution, 295
Optics, developments in, 3
Optimal perturbations, 329
Ordinary differential equations (ODEs), 16, 21–22, 49, 50, 132
Orthogonal decomposition methods, for thermally driven rotating annulus laboratory experiment, 315–336
Oscillating
  bodies, 193
  cylinder, 198
Oscillations, 4. *See also* Amplitude oscillation; Thermohaline transitions/oscillations; Zonal flow oscillations
  laboratory experiments with, 259–261
  cycle, 260–261
  patterns, 329
  periodic, 307
  periods, 182, 183
  system, 256
  wave, 326
Outcropping, 125–127
Overturning circulation, buoyancy-driven, 25
Ozmidov scale, 160, 163

Parabolic
  bottom plate, 144
  profile, 94
Parameterization(s), 27–28. *See also* Continuation parameter (α); Control parameter values; Coriolis parameter (f); Dimensionless parameters; Dissipation parameter; Dynamical parameters; Eddy parameterizations; Flow parameters; Nondimensional parameters; One-parameter bifurcation diagram; Regularization parameter; Smoothness parameters; Topography parameter (To); Two-parameter bifurcation diagram; **u*** parameterization
  for annulus geometry, 51
  of bottom friction, 240
  Gent–McWilliams, 34–35
  heat transfer, 32
Parameter space sampling, 39, 40
Parametric subharmonic instability (PSI), 210
Partial differential equations (PDEs), 49–50
Partial transmission, of waves, 198–200
Particle detection, 282, 284
Particle image velocimetry (PIV), 101, 193–194, 205–206, 210, 280, 290, 316. *See also* PIV studies
Particle image velocimetry measurements, 320, 332, 334
  analysis of, 321
Particle seeding issues, 286
Particle tracking, 280–283
  implementation challenges in, 281
Particle tracking velocimetry (PTV), 280–283, 294, 295
Péclet number (Pe), 26, 27, 29, 31–32, 216
Period doubling, 19, 21, 67
Periodic
  amplitude oscillations, 16
  amplitude vacillation, 62
  orbits, 49, 51, 53, 55, 57
  oscillations, 307
  solutions, 58
Perturbation density gradient, 198, 199
Perturbation equations, 48
Perturbations
  measuring, 196, 197
  optimal, 329
  small-scale, 222–223, 224
Perturbed surfaces, reflection from, 103, 104
Phase-locking function, 70–71
Phase velocities, 122
Phillips model, 120, 124
Physical systems, experiments as models of, 9–10
Pitot tube, 92
PIV studies, 92. *See also* Particle image velocimetry (PIV)
Plane acoustic wave, scattered amplitude of, 290–291
Plane Couette flow (PCF), 89. *See also* Rotating plane Couette flow (RPCF)

Planetary
  atmospheres, circulation of, 9–44
  β–planes, 144
  circulation regimes, 38–40
  waves, 105–107
Plumb–McEwan laboratory experiment, 181–182
Poincaré modes, 129, 130
Poiseuille flow, 85
Poisson equations, 142, 300–301
Polar β-plane, flows on, 101, 102, 108, 111
Potential energy
  eddy, 65
  zonal, 65
Potential vorticity (PV), 102. *See also* PV entries
Potential vorticity diffusion hypothesis, 34
Power law range, 166
Power spectrum, 292–293
Prandtl number (Pr), 13, 27, 46, 71, 75, 78–79, 298, 299, 301, 302, 305, 311, 318
  decrease in, 53, 54, 56
  dependent coupling, 73–74
  effects, 45–59, 71–75
  fluids, 64
Pressure variations, 129–130
Primary flow transition, in baroclinic annulus, 45–59
Primary transition curve, 46
Princeton Ocean Model (POM), 272
Principal component analysis, 318
Principal oscillation patterns (POPs), 316–317, 319, 329
  analysis, 319
Projection
  models, 282
  scheme, 300
Propagating wave structures, 327
Proper orthogonal decomposition, 318
Pseudo-arc length continuation method, 51, 56–57
PV conservation, 108. *See also* Potential vorticity (PV)
PV gradient inversion, 134

QBO analogue simulations, 188. *See also* Quasi-biennial oscillation (QBO)
QBO reversing winds, 180
Qualitative synthetic schlieren methods, 195
Quantitative synthetic schlieren methods, 195–198, 199, 200
Quasi-biennial oscillation (QBO), 4, 179–183. *See also* QBO entries
Quasi-geostrophic (QG)
  approximation model, 142
  approximation, 66, 104–105, 237
  dynamics, over a step, 238
  equations, 236, 237
    numerical solution of inviscid, 241–245
  model equations, 236–238, 241–242
  models, 23, 247
    two-layer, 67–68, 216
  shallow-water model, 250
  theory, 32, 224, 233, 234, 236, 237
  turbulence, 331
  wake, 269

Quasi-periodic solutions, 58, 303
Quasi-quadrennial oscillation (QQO), 180
Quasi-regular wind regimes, 180
Quasi-steady state, 150
Quasi-two-dimensional models, 142, 147–148
Quasi-two-dimensional turbulence, 161

Radial basis functions (RBFs), 317
Radial
  flow, 26
  spectra, 76
  temperature difference, 317
  velocity, 221, 320
  velocity profiles, 121
  wave number, 79
Radiative–convective equilibrium, 27
Radio frequency (RF) emitter, 293. *See also* RF entries
Rare earth magnets, 113–114
Rayleigh–Bénard convection, measurement in, 294
Rayleigh
  criterion, 273
  friction coefficient, 144
  number (Ra), 29, 300
Ray path calculations, 103–104
Ray theory, 198–199
Real flows, experimental flows *vs.*, 329
Reallocation technique, 285
Reflection, of waves, 198–200
Reflection law, 103, 104
Regime diagrams, 63, 64, 315–316
Regional Oceanic Model System (ROMS), 271
Regularization parameter (μ), 204
Regular wave regimes, 301–302, 305, 306
Relative vorticity, anticyclonic, 108
Resolution issues, in particle tracking, 281
Resonances, 214
  instabilities and, 127–131
Resonant
  triads, 69
  wave triads, 316
Retrograde
  azimuthal mean flow, 250
  coastal current, 233, 234, 235, 251
Reverse vortex street, 111
Reynolds number (Re), 2, 3, 63, 67, 88, 89, 90, 159–160, 163–164, 217, 218, 221, 265, 266, 267, 271–272, 279. *See also* Buoyancy Reynolds number (Re_b)
  viscous effects and, 163
Reynolds stress, cross-stream, 97
RF
  instability, 132–134. *See also* Radio frequency (RF) emitter
  mode, 129, 132
Rhines scale, 28, 111, 115, 116
Richardson number (Ri, J), 216, 222, 223
Rigid lid
  approximation, 141, 142
  configurations, 119
  front stability under, 119–125

Rossby
  deformation radius (R_d), 2, 28. See also Thermal Rossby number
  gravity waves, 180
  modes, 125, 128–129
    dispersion curves of, 121–122
  normal modes, 232
  number (Ro), 88, 140, 216–217, 266, 268, 270, 271, 273, 299, 317. See also Thermal Rossby number
  radius (R_d), 265
  shelf wave dynamics, 231, 232
  wave-driven mean flows, 232
  waves, 105–107, 108, 115, 116, 125, 145–146, 219, 232, 234, 251
    vortex streets and, 111–112
Rossby–Haurwitz (RH) mode, 17
Rossby–Kelvin (RK) instability, 121, 125, 136, 214–215, 219–220, 224, 225–226
Rossby–Rossby (RR) interaction, 129
Rotating annulus
  annuli, 31
    with barrier, 324–329
    differentially heated, 316, 333–334
    vertical shear in, 119–138
  laboratory experiments, 36–38, 315. See also Rotating fluid annulus experiments
    orthogonal decomposition methods for, 315–336
  of fluid, differentially heated, 315
  system, 12
Rotating disk
  boundary layer flow, 89
    simulations of, 97
    turbulent, 96–97
  experimental studies, 92–93
  flows, linear instabilities of, 90
  flow transition, 96
Rotating
  experiments, global circulation of atmospheres and oceans and, 10–12
  flows, 139
    flow parameters and features of, 88–91
    results in, 95–97
  flow systems, categorization of, 85–87
  fluid annulus experiments, 45. See also Rotating annulus laboratory experiments
  frame of reference, 85
  plane Couette flow (RPCF), 88–89. See also RPCF entries
  plane Couette flow experiments, 91
    results in, 95
  systems, 116
    essential balance in, 140
    topography effects in, 139–158
  turbulence, 113
    over topography, 150
  two-layer stratified fluid experiments, equations and scales for, 215–218
  two-layer stratified fluids, frontal instabilities in, 213–228
  wakes, 268

Rotation
  anticyclonic, 88, 95
  background, 155, 217
  bottom friction with, 151–152
  cyclonic, 89
  destabilization, 90
  effect on laboratory experiments, 91–95
  effects on wall-bounded flows, 85–100
  effects, 29–31
  examining effects of, 2
  role of, 97
  solar internal, 180
  solid-body, 97, 143
Rotational forcing, 164–165
Rotation-induced flows, 87
Rotation rate, 120, 121, 308. See also Angular velocity
  annulus, 317
  increasing, 303
  nondimensional, 88
  of vortices, 111
Rotation regimes, 30–31
RPCF
  apparatus, 91, 92. See also Rotating plane Couette flow (RPCF)
  simulations, 97
Runge–Kutta method, 132

Salinity mode (S-mode) flow, 257, 258
Salt stratification, 268–269
Satellite altimetry, 208
Saturn, 180
Scalar quantity tracking, 293
Scale analysis, 9, 161–162
Scaled normal form equations, 50
Scaling
  approach, 28–29
  arguments, 155–156
  of laboratory results, 258
Scattering angles, 291, 292
Scattering techniques, 284
Schlieren technology, 3, 194–198
Schmidt number, 215, 216, 217, 218, 219–220, 221, 225
Self-organizing turbulent flows, 151
Semi-Lagrangian schemes, 188
Shadowgraphs, 194, 197, 257
Shallow flows, 144, 152
Shallow-layer flow experiments, 153–155
Shallow stratified anticyclones, inertial–centrifugal instability of, 273
Shallow vortex flow, axisymmetric, 155
Shallow-water
  approximations, 150, 216
    two-layer, 119
  equations, 104, 106, 251
    two-layer, 126
  flows, instabilities of, 119–138
  models, quasi-geostrophic, 250
  parameter, 266, 267, 269
  QG equations, 236. See also Quasi-geostrophic entries

QG theory, 236–237
   theory, 184
      linear, 179
      wakes, 265
Shape vacillation, 22
Shear flow, 310
Shear instabilities, 163, 223
   Kelvin–Helmholtz, 121, 125
Shelf slopes, 134–135
Shelf waves, 232, 250–251. *See also* Coastal shelf waves
   breaking, 245–250
   dynamics, 231, 232
   equations, nonlinear, 239, 240, 245, 246–247, 250, 251
   in the laboratory, 231–234
   large-amplitude, 233
   theory, 231
      nonlinear, 235, 238–241
Shorter-scale gravity waves, 180
Sideband
   instability, 70
   phase-locking function, 70–71
Signal
   acquisition, 285
   processing, 285–286, 288–289
Simulated temperature fields, 35
Simulations, 97. *See also* Direct numerical simulations (DNSs)
   natural-flow, 1–5
Single-time-series (SSA) method, 319
Singular vectors (SVs), 316–317, 329–331, 334
Sink vortices, 145
   axisymmetric, 152
Skew-symmetric form, 299
Skin friction, 92, 94, 96
Slope tapering method, 35
Sloping convection, 11, 12, 14
Slot experiment, 256–257, 258
Small interfacial waves, 222–224
Small-scale
   disturbances, 272
   experiments, 152–156
   fluctuations, 306–308, 309
   forcing, 162
   invariance, 169, 170
   perturbations, 222–223, 224
   waves, 214, 225
smartCENTER, 294
Smart particles, 293, 294
Smoothness parameters, 332
Snap-through bifurcation, 22
Snell's law, 195, 200
Solar internal rotation, 180
Solenoidal velocity, 182
Solid-body rotation, 97, 143, 234, 237
Solid boundary, rotation of, 85
Solitary wave studies, 205, 210
Solution branches, 57

Space-time EOFs (ST-EOFs), 321. *See also* Empirical orthogonal function (EOF) analysis
Spanwise-uniform disturbances, 195–200
Spatial dimensions, choice of, 156
Spatial spectra
   azimuthal, 76
   time-averaged, 305
Spectral
   bumps, 166, 168, 170, 173
   collocation method, 121, 127
   energy balance, 168
   energy transfers, 24
   ranges, 168
   techniques, 298
Spin-up process, 143
Spontaneous gravity wave emission, 331
Spontaneous imbalance process, 222
Squared buoyancy frequency, 197
Square wave, 293, 294
Stability. *See also* Inertial–centrifugal instability; Three-dimensional destabilization
   of buoyancy-driven boundary currents, 125–134
   of eddies, 269
   analysis, 96, 119, 132, 136
   curves, neutral, 48
   diagrams, 121, 136, 137
Stationary solutions, 58
Stationary waves, 108
Statistical techniques
   linear multivariate, 316
   multivariate, 318
Steady axisymmetric solution, 47–48, 51
Steady baroclinic waves, 298
Steady states, three-dimensional nonlinear, 88–89
Steady (S) waves, 15–16, 45, 64, 65, 69, 301, 302–303
Stellar physics, 180
Step function, 238
Stewartson layers, 72, 73, 74, 76, 78, 79
Stirring vortices, 145
Stratification, 2
   salt, 268–269
   strong, 161
   vertical, 266
Stratification levels, thermal, 305
Stratification parameter (S), 266
Stratified
   anticyclones, inertial–centrifugal instability of, 273
   experiments, global circulation of atmospheres and oceans and, 10–12
   fluids, 194
      equations of motion for, 161
   scenarios, 72
   shear instability, 223
   turbulence, 159
      laboratory-scale, 159–175
      numerical studies of, 163
   turbulence hypothesis, 162
   turbulence theory, 160
Stratosphere, variability in equatorial, 180

Stratospheric
  mean-flow oscillation, 181
  motions, 179
  vacillation, 61–62
Stream
  function, 239, 242, 243
  function approach, 48
Strongly stratified scenario, 72
Strongly supercritical baroclinic flows, 34
Strong rotation regime, 30
Strong stratification, 161
Strong vortex deformation, 149
Structural vacillation (SV), 22–25, 65, 66, 75–76, 77, 302, 316, 319. *See also* SV analysis
  3SV flow regime, 308
Structural vacillation flow regime, 301, 307–309, 311
Subcritical Hopf bifurcation, 54, 56
Subdivided annulus, 328–329
Submesoscale vortex wake, 270–273
Successive wave numbers, transition between, 302–303
Supercritical
  baroclinic flows, 34
  Hopf bifurcation, 53, 54
Supercriticality, 31, 32
Surface outcropping, 125–127
Surface temperature
  data, 329
  temperature images, 325
SV analysis, 330–331. *See also* Structural vacillation (SV)
Symmetric regimes, lower, 39
Synthetic schlieren methods, 3, 195, 209
  axisymmetric, 201–203
  qualitative use of, 195
  quantitative use of, 195–198, 199, 200
System oscillations, 256
System rotation categories, 85–87

Taylor–Couette flow, 86
Taylor–Goldstein equation, 200, 222
Taylor number, 13, 14, 45, 53, 62, 63, 78–79, 299, 300, 317, 318
Taylor–Proudman theorem, 72, 140
Technological advances, 295
Temperature anomalies, 327–328
Temperature fields, simulated, 35
Temperature measurement, Lagrangian, 293–294
Temperature mode (T-mode) flow, 257, 258, 259
Temperature sensors, 317
Temperature variance spectra, 24
Temporal spectra, 75–76
Theoretical models, 9
Thermal annulus experiments, 12
Thermal annulus problem, 14–15
Thermal boundary layers, 72, 73–74
Thermal convection, 10
Thermal forcing, effective, 74, 75
Thermally driven annulus, 61

Thermally driven rotating annulus laboratory experiment, orthogonal decomposition methods for, 315–336
  results of, 318–333
  setup, parameters, and flow regime for, 317–318
Thermal Rossby number, 13, 39, 45, 53, 62–63, 64, 66, 78–79
Thermal stratification levels, 305
Thermohaline circulation regime, 255
Thermohaline transitions/oscillations, laboratory experiments with, 255–263
Thickness ratio, 223
Tidally generated internal waves, 193
Tilted-through vacillation, 22
Time-averaged
  spatial spectra, 305
  spectrum, 23–24
Time-dependent
  coefficients, 321, 322
  flow behavior, 316
Time series analysis, nonlinear, 316
TOPEX/Poseidon satellite, 101
Topographic
  β–effect, 143–144, 145, 147, 149
  β–plane, 145, 149
  experiments on flows over bottom, 139–158
  features, vortices over, 148–150
  inversion, 203
  inverse, 203–205, 209
  irregular, 150
  obstacles, vortex attraction toward, 150
  ridges, cyclonic vortices over, 149
  rotating turbulence over, 150
  Rossby waves, 232, 234
  turbulence over variable, 150–151
  waves, 105–107, 135, 145–147
Topography parameter (To), 134–136, 137
Topography/topographic effects, 141–142, 143
  inviscid, 141
  in rotating systems, 139–158
  viscous, 141–142
Topography variations, experimental, 233
Total heat transport, 28
Tracking, of scalar quantities, 293
Trajectory reconnection, 283
Transformed Eulerian Mean (TEM), 35
Transients, 301
Transition curves, 47–49
Transitions. *See also* Baroclinic cavity instability/transition; Thermohaline transitions/oscillations
  abrupt, 255–256
  abrupt thermohaline, 256–259, 261–262
  for rotating disk flow, 96
  between successive wave numbers, 302–303
Transition to turbulence, 89, 90, 91
Transmission, of waves, 198–200
Transparent belt mechanism, 91
Transport, Ekman, 26
Transport function, 142
Traveling baroclinic wave, 67

Traveling interference fringes, 284
Triads, 68–69
   resonant, 69
Tropical dynamics, LES of laboratory analogue for, 185–187
Truncated normal form equations, 50
Turbulence, 329. *See also* Stratified turbulence entries
   amplitude vacillation toward, 76–77
   experiments, decaying grid, 160
   geostrophic, 316, 331
   intensity, 94
   models, 34
   quasi-geostrophic, 331
   quasi-two-dimensional, 161
   rotating, 113
   stabilization, 95
   studies, 92–93
   transition to, 89, 90, 91
   transition to geostrophic, 22–25
   two-dimensional, 113, 161
   over variable topography, 150–151
Turbulent
   boundary layers, 87, 97
   energy equation, 95
   flows, 24, 279
      self-organizing, 151
   pipe flow, 94
   rotating disk boundary layer flow, 96–97
   spots/stripes, 89
   statistics, 171–172
   vortical motion, 114–115
Turntables, 2
Two-layer
   flows, in annuli, 120
   fluid case, 105
   models, 67–68
      baroclinic, 78–79
   quasi-geostrophic model, 216
   shallow-water approximations, 119
   shallow-water equations, 126
   shallow-water models, 216
   stratified fluids, frontal instabilities in rotating, 213–228
   studies, 15
Two-parameter bifurcation diagram, 53, 54, 55, 56
Two-wave model, 180

Ultrasonic probes, 210
Ultrasonic scattering, 291
Unstable anticyclones, 272
Unstable modes, 121, 122, 124–125, 130, 132, 135, 136, 137
Unstably stratified body force, 85
u^* parameterization, 35
Upstream current, 275

Vacillating waves, 19–22
Vacillation, 15–16, 22–25, 61. *See also* Amplitude vacillation (AV); Frequency vacillation; Interference vacillation; Structural vacillation (SV)
   forms of, 75–76
   interference, 75
   stratospheric, 61–62
   wave number, 62, 75
Variable topography, turbulence over, 150–151
Variance spectrum, 321–323
Velocimetry, 293
Velocity
   angular, 320
   azimuthal, 240
   boundary layer, 73
   data, 320
   horizontal, 320
   measurements, 3
   profiles, 222
      coastal current, 113
   radial, 320
   scale, 236
Venus, atmospheric superrotation phenomena on, 180
Vertical
   advection, 139
   aspect ratio (δ), 268
   displacements, 201, 202
   motions, 2, 147–148
   scale, 163, 169–170
      of zigzag instability, 162
   shear, 213, 222
      in a rotating annulus, 119–138
   stratification, 266
   velocity, 142, 148
   viscosity, 163, 237
   vorticity, 153, 237
Vettin experiments, 10–11
Virtual laboratory setups, 187
Viscosity
   molecular, 269
   numerical, 241, 242
   vertical, 163, 237
Viscous
   boundary layers, 15, 193
   coupling, 163
   effects, 163
   scale, 163
   topographic/topography effects, 141–142, 143
von Kármán boundary layer flow, 89–90
von Kármán boundary layer flow experiments, 91–94
   results in, 95–97
von Kármán layer, 87
Vortex attraction, toward topographic obstacles, 150
Vortex decay, 152
Vortex deformation, 149
Vortex flow, axisymmetric shallow, 155
Vortex-ring model, 153
Vortex shedding, 265
Vortex street regime, 109–111
Vortex streets, Rossby waves and, 111–112
Vortex wakes
   mesoscale, 269–270
   submesoscale, 270–273

Vortices, 273, 327–328
    dipolar, 153, 155–156
    rotation rates of, 111
    over topographic features, 148–150
Vorticity
    acoustic measurement of, 292
    anticyclonic, 111, 147
    anticyclonic relative, 108
    bottom-induced, 153
    cyclonic, 108, 111
    vertical, 153, 237
Vorticity diffusion hypothesis, 34
Vorticity equation, 29, 140, 142, 151
Vorticity fluxes, 33
Vorticity measurements, 290–293
    acoustical, 286
Vorticity stream function, 298–299
Vorticity vector, 144

Wake flows, in laboratory, 265–276
Wake patterns, global, 267
Wake region, 96
Wakes
    deep-water, 265–266
    Hawaiian, 275
    idealized laboratory flows and ocean, 274–275
    large-scale, 269–270
    Madeira, 275
    mesoscale vortex, 269–270
    observing, 4
    quasi-geostrophic, 269
    rotating, 268
    shallow-water, 265
    submesoscale vortex, 270–273
Wall-bounded flows, rotation effects on, 85–100
Wall-following coordinates, 242
Water-filled cavity, open free-surface, 310
Wave amplitude, 249–250
Wave–boundary layer interaction, low-order model of, 72–75
Wave breaking, 240–241, 251. *See also* Shelf wave breaking
    onset of, 246–247
Wave capture phenomenon, 222
Wave dispersion, 316
Wave drift, 301, 307
Wave-driven mean flows, 232
Wave dynamics, 193
Wave emission, inertial, 112–113
Wave equations, nonlinear, 233
Wave flow, 45
    dominant, 17
Wave generator, 207–209
Wave interactions, 68–71
    EOF and MSSA analysis of, 320–321
    in experiments and computational fluid dynamics, 70–71
Wave interaction scenarios, in low-order models, 71
Wavelike disturbances, 27
Wave–mean flow interactions, 71, 180
Wave modes, 69, 71
    zonal, 71

Wave motions, linear equatorial, 179
Wave number, 316
    azimuthal, 19, 23, 24
    transition between successive, 302–303
Wave number
    mode, 17
    selection, 17–18
    spectra, horizontal, 165, 168
    transitions, 49
    vacillation, 62, 75
Wave oscillations, 326
Wave radiation, 222
Wave regimes
    baroclinic, 39
    irregular, 309–310, 318, 321
    regular, 301–302, 305, 306
Wave resonance, 214
Waves. *See also* Coastal shelf waves; Gravity wave entries; Internal wave entries
    adjustment, 222
    baroclinic, 25–36, 45, 73, 74, 297–301, 308, 324–329, 334
    bifurcating, 49
    continental shelf, 146
    emission of, 116
    finite-amplitude, 73
    global scattered, 291
    inertia–gravity, 127, 222, 224, 307, 308, 311
    inertial, 105–107
    interfacial, 214
    Kelvin, 128, 219
    large-scale baroclinic, 308
    lee, 107, 108
    numerical investigations of baroclinic, 297–301
    partial transmission and reflection of, 198–200
    planetary, 105–107
    Rossby, 105–107, 108, 111–112, 115, 116, 125, 145–146, 219
    scattered amplitude of plane acoustic, 290–291
    simulations, baroclinic, 33
    small interfacial, 222–224
    small-scale, 214, 225
    square, 293, 294
    stationary, 108
    steady, 45, 64, 65, 301, 302–303
    structures, propagating, 327
    topographic, 105–107, 135, 145–147
    triad interactions, 68–69
    triads, resonant, 316
    vacillating, 19–22
    weak, 303, 305–306, 311
Wave–wave interactions, 17
Wave–zonal flow interactions, 19–22
Weak dissipation, 16
    weak rotation regime, 30
    weak waves, 303, 305–306, 311
Weakly
    nonlinear analysis, 50
    nonlinear baroclinic instability theory, 19, 21
    nonlinear theory, 15–16, 34, 56
    stratified scenario, 72

Wedemeyer model, 152
Westward flows, 108, 109–110

Zigzag instability, vertical scale of, 162
Zonal flow, 15–16, 39–40, 324–325
    correction terms, 68
    mean, 107, 108
    oscillations, 75
Zonally axisymmetric noise, 183
Zonal mean-flow
    correction, 69–70
    oscillation, 181
    reversal, 182
Zonal mean zonal flow, 180, 181
    reversal, 183, 187
Zonal
    kinetic energy, 64, 65
    potential energy, 65
    wave modes, 71
    wave number, 78–79